Small Satellites:
Past, Present, and Future

Small Satellites:
Past, Present, and Future

Henry Helvajian and Siegfried W. Janson
Editors

The Aerospace Press • El Segundo, California

American Institute of Aeronautics and Astronautics, Inc. • Reston, Virginia

The Aerospace Press
2350 E. El Segundo Boulevard
El Segundo, California 90245-4691

American Institute of Aeronautics and Astronautics, Inc.
1801 Alexander Bell Drive
Reston, Virginia 20191-4344

Library of Congress Cataloging-in-Publication Data

Small satellites : past, present, and future / Henry Helvajian and Siegfried Janson, editors.
p. cm.
Includes bibliographical references and index.
ISBN-13: 978-1-884989-22-3 (alk. paper)
ISBN-10: 1-884989-22-5 (alk. paper)
1. Microspacecraft. I. Helvajian, Henry. II. Janson, Siegfried W.

TL795.4.S64 2008
629.43'4--dc22

2008039221

Cover design by John Griffith and Joseph Hidalgo

Contents

Contributors

William Ailor is Principal Director of the Center for Orbital and Reentry Debris Studies (CORDS) at The Aerospace Corporation in El Segundo, California.
Chapter 22: Small Satellites: Space Debris and Reentry Hazards

David J. Barnhart (Maj USAF) is Assistant Professor at the Space Systems Research Center in the Department of Astronautics, United States Air Force Academy, in Colorado.
Chapter 7: The U.S. Air Force Academy FalconSAT Small Satellite Program
Chapter 19: Toward Space-based Wireless Sensor Networks

Dirk Bernaerts is the Proba-2 Project Manager at the European Space Technology Centre, Noordwijk, The Netherlands.
Chapter 13: Proba: ESA's Series of Small Satellite Testbeds

Bob Bruninga (Cdr USN [ret.]), teaches at the Satellite Lab at the United States Naval Academy in Annapolis, Maryland (WB4APR).
Chapter 8: The U.S. Naval Academy Small Low-Cost APRS Satellites

Matthias Buhl is a Science Associate at the Institut für Luft- und Raumfahrttechnik at the Technische Universität Berlin, Germany.
Chapter 12: TUBSAT: The Technical University of Berlin Satellite Program

Vladimir Chobotov is a member of the Center for Orbital and Reentry Debris Studies (CORDS) at The Aerospace Corporation in El Segundo, California.
Chapter 22: Small Satellites: Space Debris and Reentry Hazards

Srimal W. Choi is Senior Principal Systems Engineer at Orbital Sciences Corporation in Dulles, Virginia.
Chapter 14: A Revolutionary Approach: Orbital's MicroStar Satellite

Thomas Doyne (Col USAF) is Deputy of Space Programs and Policy Directorate, United States Air Force, Arlington, Virginia.
Chapter 6: Coalition Operationally Responsive Space

Steven Fredrickson is Chief of the Intelligent Systems Branch, Automation, Robotics, and Simulation Division at NASA Johnson Space Center in Houston, Texas.
Chapter 16: Mini AERCam

William W. Hansen is Senior Research Associate in the Physical Sciences Laboratories at The Aerospace Corporation in El Segundo, California.
Chapter 18: Mass-Producible Glass/Ceramic Micropropulsion Unit for COSA

David Hardy is Proba Systems Engineer at the European Space Technology Centre, Noordwijk, The Netherlands.
Chapter 13: Proba: ESA's Series of Small Satellite Testbeds

Steven L. Hast is Senior Project Leader in the Astrodynamics Department at The Aerospace Corporation in El Segundo, California.

Chapter 21: Small Satellite Mission Design and Operations

Dean Hawes is the MicroStar Product Line Lead at Orbital Sciences Corporation in Dulles, Virginia.

Chapter 14: A Revolutionary Approach: Orbital's MicroStar Satellite

Henry Helvajian is a Senior Scientist in the Physical Sciences Laboratories and runs the Laboratory for Laser Microengineering Science at The Aerospace Corporation in El Segundo, California. He is the editor of *Microengineering Aerospace Systems* (AIAA/The Aerospace Press, 1999).

Chapter 18: Mass-Producible Glass/Ceramic Micropropulsion Unit for COSA
Chapter 24: Satellites as Mass-Producible Functionalized Modules

David Hinkley is Project Leader in the Physical Sciences Laboratories at The Aerospace Corporation in El Segundo, California.

Chapter 20: Picosatellites at The Aerospace Corporation

Masafumi Hirahara is Professor of Space and Planetary Science, Department of Earth and Planetary Science, Graduate School of Science, The University of Tokyo. He served as Principal Investigator of Science Instruments for the Reimei mission.

Chapter 15: Japan's Small Scientific Satellite, "INDEX" (Reimei)

Jim Horning is a Research Associate with the Space Systems Academic Group, Graduate School of Engineering and Applied Sciences, Naval Postgraduate School, Monterey, California.

Chapter 9: The Naval Postgraduate School Small Satellites Program

Adam Huang is an Assistant Professor in the Department of Mechanical Engineering at the University of Arkansas in Fayetteville, Arkansas.

Chapter 18: Mass-Producible Glass/Ceramic Micropropulsion Unit for COSA

Siegfried W. Janson is a Senior Scientist in the Physical Sciences Laboratories at The Aerospace Corporation in El Segundo, California.

Chapter 2: The History of Small Satellites
Chapter 18: Mass-Producible Glass/Ceramic Micropropulsion Unit for COSA
Chapter 23: The Future of Small Satellites

Frederick T. Kiley is Professor at the Space Systems Research Center in the Department of Astronautics at the United States Air Force Academy in Colorado.

Chapter 7: The U.S. Air Force Academy FalconSAT Small Satellite Program

Timothy J. Lawrence (LtCol USAF) is Director of the Space Systems Research Center and Assistant Professor for the Department of Astronautics at the United States Air Force Academy in Colorado.

Chapter 7: The U.S. Air Force Academy FalconSAT Small Satellite Program

Jaymie M. Matthews is MOST Mission Scientist and Professor of Physics and Astronomy at the University of British Columbia in Vancouver, Canada.
Chapter 11: Microvariability and Oscillations of STars (MOST) Mission

David W. Miller is Professor of Aeronautics and Astronautics at the Massachusetts Institute of Technology, and Director of the MIT Space Systems Laboratory in Cambridge, Massachusetts.
Chapter 17: Microsatellite Formation Flight Technology

R. Gilbert Moore is Director of Project Starshine and Adjunct Professor of Physics at Utah State University in Logan, Utah. He was formerly Director of the Rocket Section, Physical Science Laboratory, New Mexico State University and Schriever Professor of Space Systems Engineering at the U.S. Air Force Academy.
Chapter 1: The First Small Satellites: Sputnik, Explorer, and Vanguard

Simon Nolet is a Post-Doctoral Associate in the Space Systems Laboratory at the Massachusetts Institute of Technology in Cambridge, Massachusetts.
Chapter 17: Microsatellite Formation Flight Technology

Daniel L. Oltrogge is Senior Analyst and CEO of 1Earth Research, LLC, in Colorado Springs, Colorado.
Chapter 21: Small Satellite Mission Design and Operations

Rudy Panholzer is Chair of the Space Systems Academic Group, Graduate School of Engineering and Applied Sciences, Naval Postgraduate School, Monterey, California.
Chapter 9: The Naval Postgraduate School Small Satellites Program

Russell Patera is Senior Engineering Specialist in the Center for Orbital and Reentry Debris Studies (CORDS) at The Aerospace Corporation in El Segundo, California.
Chapter 22: Small Satellites: Space Debris and Reentry Hazards

Glenn Peterson is Manager of the Orbit Performance and Space Debris Section at The Aerospace Corporation in El Segundo, California. He is the author of *Dynamics of Meteor Outbursts and Satellite Mitigation Strategies* (AIAA/The Aerospace Press, 1999).
Chapter 22: Small Satellites: Space Debris and Reentry Hazards

Pete Rustan is Director, Ground Enterprise Directorate, National Reconnaissance Office, Chantilly, Virginia. He was the Clementine mission manager at the Ballistic Missile Defense Organization and later served as Director of Small Satellite Systems in the Office of the Secretary of the Air Force for Space Technologies.
Chapter 3: Clementine: The Rebirth of Smaller, High-Value Missions

Hirobumi Saito is a member of the Institute of Space and Astronautical Science, Japan Aerospace Exploration Agency (ISAS/JAXA), Sagamihara-shi, Japan.
Chapter 15: Japan's Small Scientific Satellite, "INDEX" (Reimei)

Contributors

Alvar Saenz-Otero is a Research Scientist with the Space Systems Laboratory at the Massachusetts Institute of Technology (MIT) in Cambridge, Massachusetts.
Chapter 17: Microsatellite Formation Flight Technology

Dan Sakoda is a Research Associate with the Space Systems Academic Group, Graduate School of Engineering and Applied Sciences, Naval Postgraduate School, Monterey, California.
Chapter 9: The Naval Postgraduate School Small Satellites Program

Luke M. Sauter (Capt USAF) is an Instructor in the Systems Division of the Department of Astronautics at the United States Air Force Academy in Colorado.
Chapter 7: The U.S. Air Force Academy FalconSAT Small Satellite Program

Raymond Sedwick is Assistant Professor of Aerospace Engineering and Director of the Space Power and Propulsion Laboratory at the University of Maryland.
Chapter 17: Microsatellite Formation Flight Technology

Kenneth E. Siegenthaler is Professor of Astronautics at the Space Systems Research Center at the United States Air Force Academy in Colorado.
Chapter 7: The U.S. Air Force Academy FalconSAT Small Satellite Program

G. Gould Smith (WA4SXM) is AMSAT Vice President of User Services, AO-51 Command Station, Knoxville, Tennessee.
Chapter 4: The Role of AMSAT in the Evolution of Small Satellites

Lee F. Steffeney is a Research Associate in the Physical Sciences Laboratories at The Aerospace Corporation in El Segundo, California.
Chapter 18: Mass-Producible Glass/Ceramic Micropropulsion Unit for COSA

Frederic Teston is the Proba-series Program Manager at the European Space Technology Centre, Noordwijk, The Netherlands.
Chapter 13: Proba: ESA's Series of Small Satellite Testbeds

Bob Twiggs is Professor Emeritus in the Department of Aeronautics and Astronautics at Stanford University in Sunnyvale, California.
Chapter 5: Origin of CubeSat

Craig I. Underwood is Deputy Director of Surrey Space Center and is head of the Planetary Environments Group at the University of Surrey, Guildford, UK.
Chapter 10: The UK's First Nanosatellite—SNAP-1

Tanya Vladimirova leads the VLSI Design and Embedded Systems Research Group at the Surrey Space Centre, University of Surrey, Guildford, UK.
Chapter 19: Toward Space-based Wireless Sensor Networks

Robert E. Zee is Director of the Space Flight Laboratory at the University of Toronto Institute for Aerospace Studies in Toronto, Canada.
Chapter 11: Microvariability and Oscillations of STars (MOST) Mission

C'est que pour accomplir de grandes choses il ne suffit pas d'agir, it faut rêver, il ne suffit pas de calculer, il faut croire.

To accomplish great things we must not only act but also dream, not only plan but also believe.

—Anatole France (1844–1924)

Preface

The Space Age started with the successful launch of Sputnik on 4 October 1957 by the former Union of Soviet Socialist Republics (USSR). Fifty years have past since this small satellite initiated a major event in human history, and we are still flying similar-sized satellites for a number of military, civil, commercial, and educational missions. It is therefore fitting that on this semi centennial anniversary, we compile a brief record of the past, provide a snapshot of the present, and propose what may be possible in the future with respect to small satellites. We have arbitrarily chosen to define a small satellite as a spacecraft with an initial on-orbit mass of less than 150 kg. Small satellites therefore include microsatellites (10–100 kg), nanosatellites (1–10 kg), picosatellites (0.1–1 kg), and even lighter femtosatellites and attosatellites. Over the last 50 years, more than 860 microsatellites, 680 nanosatellites, and 38 picosatellites have been launched worldwide. Small satellites have recorded data on the terrestrial and space environment near the Earth and moon, helped in the search for planets on other star systems, and demonstrated various telecommunications systems that we enjoy today. Finally, these satellites have served as test beds for the development of new space technologies, and as hands-on educational tools for countless students, scientists, and engineers.

Small satellites played a key role at the dawn of the Space Age, but since then, the evolution of ever-larger launch vehicles have produced "typical" spacecraft with masses in the 1000–5000 kg range that cost hundreds of millions of dollars or more. Interestingly, the evolution of microelectronics over the same time has produced ever-smaller, faster, and cheaper transistors and the incorporation of these components as integrated circuits and systems. The ever-increasing functionality brought about by the microelectronics revolution (increase of component density per unit time, Moore's Law, and a commensurate reduction in unit power consumed) triggered a rebirth of microsatellite activity during the 1980s. Today, the continuation of Moore's Law, the development of microelectromechanical systems (MEMS) and the application of microengineering systems concepts have enabled the development of highly-capable microsatellites. Those same drivers will enable highly-capable nanosatellites and picosatellites in the near future. Small satellites serve as low-mass platforms that can be sent into orbit for well under a few million dollars, thus allowing nonspacefaring nations, corporations, educational institutions, and even individuals low-cost access to space. A nanosatellite or picosatellite can be developed by any nation and launched as a secondary payload on a wide range of launch vehicles. As individual spacecraft, micro-, nano-, and picosatellites are limited by power and aperture constraints; however, some space missions could be better served by mass-produced small satellites in large constellations or local clusters. Constellations or clusters with a hundred or more spacecraft will typically require microsatellites or smaller spacecraft to keep total mission cost practical.

Preface

The contributing authors and others in the small satellite community firmly believe that the proliferation of small satellites will increase our understanding of the near-Earth environment and provide a nearly real-time assessment of that changing environment. For space exploration, small satellites can provide an efficient and economical means to identify regions of interest before sending larger space systems with more instrumentation. In the far term, mass production of small satellites could provide autonomous space assembly of large systems comprised of thousands of "smart" nano- and picosatellite "Lego" blocks.

This book records seminal ideas, concepts, experiments, and space-test results relevant for small satellites in order to provide an overview of the past, present, and future. It is an international compendium of articles by leading experts in the field and is intended to be a technical journal that presents interesting engineering aspects in the development of small satellites and small satellite missions. However, it also contains chapters on space policy and alternative modes of space systems development. Entrenched policies and practices in the space industry sometimes hinder or even halt the exploitation of new concepts and technologies. This issue is independently addressed by several of the chapter authors, but two specific chapters address this head on; one chapter describes a successful mission in which the typical operating approaches and practices were culled in favor of streamlining, and in another chapter the procurement and operations strategies currently used in other industries are presented with a view toward applying them to the space systems development/procurement process.

The book is organized in rough chronological order. Four initial chapters give a broad historical overview of small satellites. The first focuses on the historical and political environment which led to the development and launch of the Russian Sputnik, U.S. Explorer, and U.S. Vanguard satellites. A historical perspective with interviews and personal insights of the accidental discovery of the Van Allen belts by the U.S. satellites is a feature. The second chapter provides a broad overview and historical trends for all small satellites launched up to the present. Chapter 3 presents a historical accounting of the re-emergence of small satellites in the United States in the 1980s, with a focus on the successful Clementine mission. The history of amateur radio satellites from its humble beginnings in the United States in the late 1950s to a successful international body is described in the fourth chapter.

The next eleven chapters focus on current missions and concepts. Chapter 5 presents the history of the highly-successful CubeSat program which started in 1998 and has become an international educational tool. Chapter 6 addresses an issue of space policy. Recognizing that space has strategic military value, and that many nations can now afford individual spacecraft, this chapter addresses how best to establish a cooperative satellite constellation that serves the international community while increasing general global security. The next three chapters present three academic approaches to the education of students in satellite development and operations. In all three approaches, the goal is to provide hands-on experience of all facets of satellite development and operations (i.e. design, manufacturing, assembly and test, and launch deploy and operations). These academic approaches are described through examples of a series of successful student-led projects; the FalconSAT program (U.S. Air Force Academy); the PCSAT, ANDE, and RAFT missions (U.S. Naval Academy); and the PANSAT and NPSAT-1 (U.S. Naval Post Graduate School). The next three chapter contributions are from educational institutions that have both the desire to train members of the trade and the capability to develop very advanced small satellites. Chapter 10 describes the SNAP-1 mission, which raised worldwide awareness in 2000 after it proved that a 6.5 kg satellite could perform a sophisticated task (remote inspection and orbital maneuvering) and could be built for a mere million dollars or so. Chapter 11 presents the MOST mission, an ambitious attempt to measure minute oscillations in the brightness of distant stars to deduce their ages, core compositions, and sizes. MOST continues to provide data on 30 distant stars by achieving an extraordinary pointing stability of ±1 arcsec. Chapter 12 presents the TUBSAT missions, which mark the development of seven successful microsatellite launches starting in 1991. These missions include store-and-foreword communications and Earth observation. The next three contributions are from larger organizations that are capable of deploying a much wider range of interdisciplinary talent onto a problem. Chapter 13 describes the European Space Agency Proba missions, which sought to develop an affordable development environment in which to prove new ideas and technologies and to develop the technologies for onboard autonomy. The Proba-1 mission carried a suite of novel technologies (e.g., a high-resolution optical spectrometer) but also demonstrated the transfer of autonomy to the onboard systems. Chapter 14 provides details of the MicroStar spacecraft bus, which was used for the ORBCOMM satellite-based communications network. Particular issues addressed are the development of an economical spacecraft with a production process that can manufacture them in 24 months. Chapter 15 is a contribution from the Japanese Aerospace Exploration Agency (JAXA) and Institute of Space and Astronautical Science (ISAS) that presents the design and development of the novel INDEX or REIMEI scientific satellite. INDEX has been making auroral observations for more than 35 months and is also serving as a bus to test advanced space systems technologies.

The final nine chapters cover technologies and missions that will affect future small satellite applications. Chapter 16 presents the development of the NASA/Johnson Space Center (NASA/JSC) Mini AERCam space inspector vehicle, which is intended to meet the remote viewing and inspection tasks of future human space flight missions. The spherical Mini AERCam is a 5 kg reusable vehicle intended to provide remote viewing that cannot be undertaken with a fixed camera. Mini AERCam is designed to be relatively autonomous. Chapter 17 expands on the concept of remote sensing or operations by describing the control systems necessary to enable satellite formation flying. Formation flying enables distributed spacecraft composed of a number of free-flying "nodes." Chapter 18 deals with a specific manufacturing technology that could enable mass production of small satellites. It describes the development of the Co-Orbital Satellite Assistant (COSA) prototype propulsion module using photostructurable glass ceramic materials in lieu of metal structures. Chapter 19 presents the case for distributed wireless space sensor networks; it provides an introduction to current distributed space missions and goes on to describe the development of a system-on-a-chip, which leads to a "satellite-on-a-chip." The "chip" has the functionality of all satellite systems integrated on a common platform. Chapter 20 gives a historical and technical overview of the picosatellite development effort at The Aerospace Corporation. The objectives of this picosatellite project are to serve as a test bed for space applications of micro/nanotechnology and to develop a manufacturing process for design, assembly, and testing of small space systems; 11 miniature spacecraft have now been launched (the 12th is awaiting launch) by this group during the past decade.

Chapters 21 and 22 present two fundamental issues concerning all small spacecraft operators; tracking, communications, and space debris. Chapter 21 presents the strategies that small satellite operators can implement to optimize capabilities and minimize mission risk, given the difficulties of tracking small objects. This chapter addresses aspects of satellite design to improve communications, and proper design and operation of ground station hardware and software. Chapter 22 on the space debris issue should be a wake-up call to all satellite operators and small satellite manufacturers. Many governmental organizations and environmental purists view small satellites as near-term orbital debris. Some of the accusations have been unfair, but more could and should be done to ensure that small satellites in low Earth orbit limit release of debris on orbit and re-enter in a timely fashion.

The final two chapters look into the far future. Chapter 23 describes missions for future nanosatellites and smaller spacecraft that are not possible today. It assumes that microelectronics and microengineering technologies will continue to evolve, allowing for systematic integration of components, followed by subsystems and finally, full systems. This chapter also assumes that nanosatellite manufacturing technology will evolve, such that it will be possible to manufacture hundreds if not thousands of small spacecraft over a reasonable time period. Chapter 24 discusses what maybe required for "the generation after next"; it presents the case for developing satellites as an assembly of mass producible functionalized modules: a smart Lego block. The goal is to realize economies of scale in mass production and permit the adoption of manufacturing procedures/practices that can increase overall reliability. The chapter also explores the criteria required in the types of materials and manufacturing methodologies that need to be chosen to acquire the benefits that the microengineeering and microelectronics industries have enjoyed.

The editors are aware that this book is at best a snapshot outlook, and consequently, other meaningful perspectives are lacking. We do not have direct contributions from the Russians and their pioneering work in space systems. We do not have chapters from Johns Hopkins Applied Physics Laboratories (APL) and Naval Research Laboratories (NRL), both of which played pivotal roles in the early days of the U.S. space program and continue to be involved in advanced space systems technologies. In addition, the readership would have benefited with a contribution from NASA/Goddard on the Multi-mission Modular Spacecraft (MMS, ca 1975), the Small Explorer Program Spacecraft (SMEX, ca 1988) and the Modular, Adaptive Reconfigurable Systems (MARS, ca 2002) program. Missing also from these pages is information on other small satellite programs; for example, Xi4, (sai-four) and Xi5 (sai-five), Munin, AAUSAT1, CUTE, COMPASS-1 and Delphi-C^3.

This book does not include in-depth discussions on specific technologies that could enable systems miniaturization, allow for better integration via advanced packaging, or explore the utility of "self-correcting" software and the application of advanced materials. We chose to exclude these discussions in order to focus more on the spacecraft. The applications of miniaturization technologies to space systems have been covered in prior work.[a] However, in hindsight, this book would have benefited from a contribution on modern robotics to illustrate what could be possible using mobile autonomous systems.

Naturally, the world of small spacecraft continued to evolve while this book was being assembled. The Boeing Corporation and NASA flew CubeSats as inexpensive space flight testbeds. NASA-Ames orchestrated the launch of GeneSat-1, Nanosail-D, and the PRESat triple CubeSats during the last two years, and is currently working on PharmaSat. Even though Nanosail-D and PRESat did not reach orbit due to a launcher anomaly, they demonstrated that CubeSats enable the "smaller, faster, cheaper" approach to space exploration. Nanosail-D involved two NASA centers (Ames and Marshall), plus the University of Alabama. During 2008, the Chinese BX-1 nanosatellite, released by the Chenzhou-7 manned spacecraft, passed within 25 km of the International Space Station. The 34 kg Indian Moon Impact Probe, ejected by the 1380 kg Chandrayaan-1 lunar orbiter, successfully impacted the southern lunar polar region on 14 November 2008. Had this book been initiated in 2008, we would certainly have solicited chapters on these efforts.

The success of the two Experimental Satellite System (XSS-10 [2003], and XSS-11 [2005]) microsatellites developed by the U.S. Air Force Research Laboratory has encouraged the U.S. National Science Foundation (NSF) to sponsor a space science research program using a CubeSat bus. The NSF program comes

---

[a] *Micro- and Nanotechnology for Space Systems: An initial Evaluation,* H. Helvajian and E. Y. Robinson, Eds. (1993); *Microengineering Technology for Space Systems,* H. Helvajian, Ed. (1995); and *Microengineering Aerospace Systems,* H. Helvajian, Ed. (1999); Janson, S.W., Helvajian, H. and Robinson, E.Y., "The Concept of Nanosatellite for Revolutionary Low-Cost Space Systems," Paper IAF-93-U.5.573, *44th Congress of the International Astronautics Federation* (Graz, Austria, October 1993).

at the heels of a report released by U.S. Air Force Scientific Advisory Board that gives a favorable nod to the use of small satellites for space investigation. Finally, it appears that launch opportunities for small satellites (e.g. CubeSats, nanosatellites, microsatellites) will be possible not only on the Russian Dneper, the European Ariane, or the Indian PSLV launch vehicles, but also on U.S. launchers (the Evolved Expendable Launch Vehicles, or EELVs, as well as the Minotaur-1, Minotaur-IV, and Falcon-1).

The editors hope that this work will spawn similar compendia to allow a broader and deeper perspective on small satellite contributions to space exploration. Furthermore, the authors hope that on the centennial celebration of the dawn of the Space Age, 50 years hence, the ideas, concepts, and approaches presented in these pages will have found their way into the general lexicon of space technology. The editors, the chapter authors, and countless other engineers and scientists believe that small satellites will continue to shape the future of space exploration. Ultimately, those small satellites may be become mass-produced commodities complete with their own bar codes as shown on the back cover.

Henry Helvajian
Siegfried W. Janson

Acknowledgments

The editors would like to take the opportunity to thank not only the few who appear in these pages, but also the many that have contributed to the development of small satellites. Many of these champions are no longer with us, and we honor them and their perseverance, which has allowed the remaining few to make a more coherent case to the world community regarding small satellites.

This book would not exist without the generous contribution of the authors, their supporting organizations, and management. The editors would like to extend their gratitude and thanks. Furthermore, this book could not have even been born without the commitment and support of The Aerospace Press and the creative talents of its Director Gabriel Spera and his team. We thank Mr. Spera and the Press for their continued support and guidance. In a book where historical matter is documented, it is inconceivable that a quality product can be assembled without tapping into historical archives worldwide and in this regard, the editors would like to thank Ms. Bonita Smith, The Aerospace Corporation's archivist, for her sleuthing talents.

Finally and on a personal note, the editors would like to thank their families and friends for their patience and support during this two year venture.

Chapter 1

The First Small Satellites: Sputnik, Explorer, and Vanguard

R. Gilbert Moore*

1.1 Introduction

Three small satellites were the first to be launched into low Earth orbit during the International Geophysical Year (IGY) of 1957–1958 to begin the Space Age: Sputnik I, Explorer I, and Vanguard 1 (Figs. 1.1 and 1.2). At the time, the United States and the Soviet Union were engaged in a "Cold War." Both countries were developing thermonuclear warheads and intercontinental ballistic missiles (ICBMs) capable of delivering their horrific weapons of mass destruction onto the other's territory.

The fear and suspicion that pervaded the relations between these countries during the Cold War was so intense that the early Soviet satellite program was conducted in complete secrecy. In fact, the identity of Sergei Pavlovich Korolev, the leader of the Soviet space program, was hidden under the title "Chief Designer," apparently in fear that he would be kidnapped or assassinated. In the United States, the Explorer program was also conducted partially in secrecy. Only the U.S. Navy's Vanguard program was conducted completely in the open. As a result, much of the early history of the world's first three small satellites became badly misreported in the public press at the time, and many misconceptions about them still linger today. It is the purpose of the opening chapter of this book to describe this history in an objective manner, so the reader can judge the programs' true accomplishments, from the perspective of 50 years of intervening history.

1.2 The Historical Background

1.2.1 Sounding Rockets

The size of the launch vehicles and satellites that were flown during IGY were heavily influenced by the predecessor sounding rocket and missile programs that had been conducted by the United States and the Soviet Union before, during, and after World War II. The world's first sounding rockets were built and launched by Robert H. Goddard of Worcester, Massachusetts. His short-term objective was to develop a family of liquid-propelled rockets that could extend measurements of the atmosphere's characteristics to altitudes above those reachable by research balloons (~30 km). In the long term, he hoped to launch a rocket to the moon and

*Dept. of Physics, Utah State University
Logan, Utah

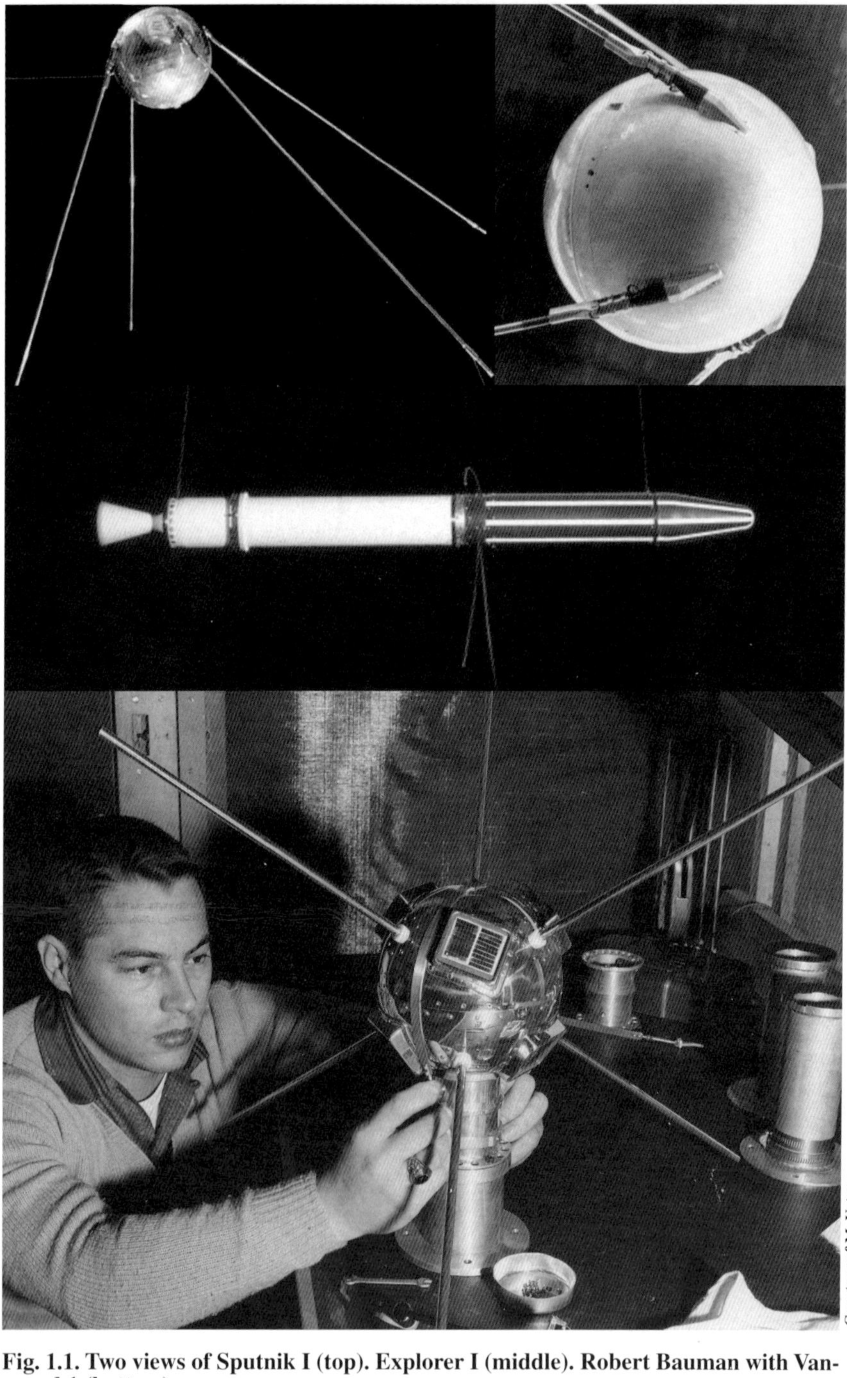

Courtesy of NASA

Courtesy of NASA

Courtesy of M. Votaw

Fig. 1.1. Two views of Sputnik I (top). Explorer I (middle). Robert Bauman with Vanguard 1 (bottom).

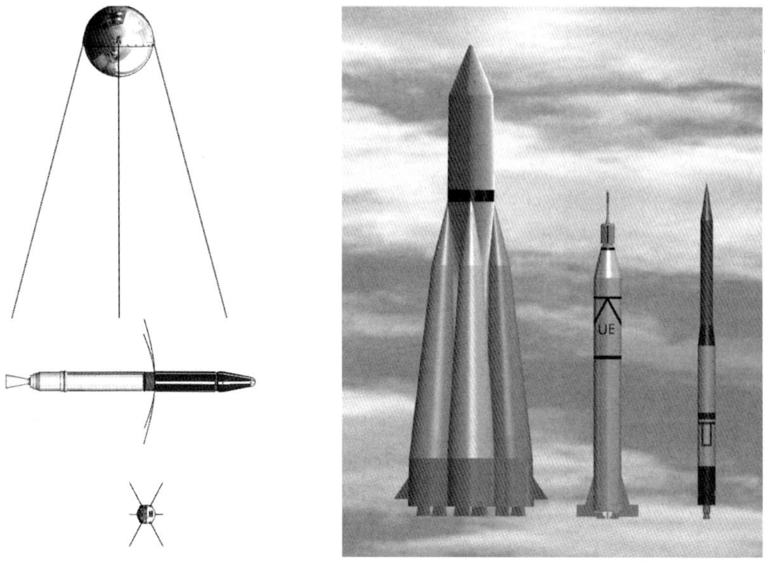

Courtesy of B. Fraze

Fig. 1.2. Comparative sizes of the Sputnik I, Explorer I, and Vanguard 1 satellites (left, top to bottom) and their respective launch vehicles, R-7, Juno I, and Vanguard (right, left to right).

set off a magnesium powder flash that would be visible to telescopes on Earth. He published his first formal paper on rocketry and space flight in a report to the Smithsonian Institution in 1919. He launched his first liquid-propelled rocket on 16 March 1926 (Fig. 1.3) from a farm near Auburn, Massachusetts, and followed up with a series of additional flights from that site; these flights were made possible by a small grant from the Smithsonian Institution in Washington, DC. However, under pressure from the local fire marshal and ridicule from local newspapers (including the *New York Times*), he moved his small team of technicians in 1930 to a ranch near Roswell, New Mexico, where he continued a series of launches out of the public eye, until 1941. During that time, he refused to meet or share data with other groups.

During his Roswell period, he demonstrated the use of turbine pumps to deliver liquid oxygen and alcohol to a regeneratively cooled rocket combustion chamber, gyro-stabilized control of coupled nozzle-mounted jet vanes and external fins, gimbaled nozzles, photography of Earth from a rocket, onboard recording of readings from scientific instruments, and parachute recovery of vehicle and payload—in short, most of the elements required to conduct rocket investigation of Earth's upper atmosphere.

Throughout this phase of his work, Goddard was sponsored by intermittent, small grants from the Guggenheim Foundation, to whom he also made periodic progress reports that eventually made their way into the public literature. In 1941, the U.S. Navy Bureau of Aeronautics and the U.S. Army Air Corps prevailed

Courtesy of NASA

Fig. 1.3. Robert H. Goddard with his first successful liquid-fueled rocket.

upon Goddard to interrupt his private work and move his team to the Naval Research Station at Annapolis, Maryland, to develop a controllable, liquid-propelled jet-assisted-take-off rocket motor for use by search and rescue seaplanes and for short-field takeoffs by fighter aircraft. Goddard intended to move back to Roswell after the war to pursue his dream of launching a rocket to the moon; however, he died in July of 1945. If it were not for the subsequent efforts of his wife, Esther, to publicize his work through a series of public lectures and to publish his voluminous notes and patents, Goddard would never have received credit for his pioneering efforts in the propulsion field.

In the early1930s, after learning of Goddard's work through his progress reports to the Smithsonian and Guggenheim organizations, members of the

American Interplanetary Society (later renamed the American Rocket Society, or ARS) attempted to visit him in New Mexico, but were refused access to his laboratory and launch facilities.

The ARS members developed and tested their own liquid rocket engines in New Jersey and launched complete rocket vehicles from Staten Island in New York harbor. They also visited members of amateur rocket societies in Germany, such as the Verein fur Raumschiffart (VfR), with whom they exchanged ideas on rocket engine development and flight testing. In the Soviet Union, similar groups started work under the leadership of Sergei Korolev, based on the early writings of a schoolteacher named Konstantin Tsiolkovsky. For reasons of military security, the work of the Soviet group was not divulged to the outside world until after the successful launch of Sputnik I.

The work of the German amateur rocket societies was inspired by the fictional writings of Jules Verne, the theoretical writings of Hermann Oberth (to whom Goddard sent his Smithsonian report, "A Method of Reaching Extreme Altitudes," in 1922, at Oberth's request, when Oberth was a student at the University of Heidelberg), and the writings and demonstrations of Austrian rocket enthusiast Max Valier. In 1934, the work of these groups and, in particular, the doctoral thesis of one of Oberth's graduate students, Wernher von Braun, came to the attention of high-level German army officers. They decided to develop rocket weapons to circumvent the restrictions that had been placed on Germany regarding the development of artillery weapons after World War I. They placed the work of the VfR group under military security and assigned a captain named Walter Dornberger to establish a rocket research center on Usedom Island near the mouth of the Peene river (Peenemunde) on the Baltic coast.

Dornberger selected young von Braun as his technical director to develop a family of large ground-to-ground ballistic missiles. The most successful of these was the V-2 (Fig. 1.4). It had a maximum diameter of 1.5 m and a length of

Fig. 1.4. Tactical V-2 rocket in Germany with mobile ground support vehicles (left). V-2 launch at White Sands Proving Ground on 22 August 1951 (right).

14 m and weighed 12,800 kg at launch. Its fuels were liquid oxygen and alcohol, pumped by a hydrogen peroxide steam turbine into an injector with 18 burner cups at the head end of a steel combustion chamber that produced a sea-level thrust of 245 kN. It was equipped with an amatol explosive warhead weighing 1000 kg; more than 3000 were launched at military and civilian targets in England, Belgium, France, and Holland in 1944 and 1945.

Near the conclusion of the war in 1945, crates of parts for approximately 100 V-2 rockets and complete sets of drawings were smuggled out of the Russian Zone of defeated Germany by the U.S. Army and transported to the White Sands Proving Ground in New Mexico. A contingent of 118 German rocket engineers and scientists, led by von Braun, were also taken to Ft. Bliss, Texas, to transfer their V-2 knowledge and ideas for future rocket weapon systems to American engineers. Additional missile parts and engineers were also transferred from Germany to the Soviet Union for similar purposes.

While the V-2 parts and engineers were being assembled in the United States, the California Institute of Technology's Jet Propulsion Laboratory (JPL) and Douglas Aircraft Company launched Private A and Tiny Tim boosters and Wac Corporal sounding rockets from White Sands (Fig. 1.5).

The U.S. Naval Research Laboratory (NRL) built new instrument sections to replace the warheads of the captured V-2 rockets and equipped them with radiotelemetry subsystems to support scientific studies of Earth's upper atmosphere.

Courtesy of U.S. Army

Fig. 1.5. Frank Malina of JPL with Wac Corporal sounding rocket.

They invited research scientists from across the country to a meeting in Washington, DC, on 16 January 1946, to discuss participation in these endeavors. The V-2 Rocket Panel was formed at this meeting, under the chairmanship of Ernst Krause of NRL, to set the agenda for upper atmospheric and space research in the United States. This panel underwent a series of name changes and member institutions during the following two decades, while its new chairman, James Van Allen of the Johns Hopkins University's Applied Physics Laboratory, continued Krause's efforts to assign rockets to its participating scientists. Between 1946 and 1955, 67 V-2 rockets were launched from White Sands by the U.S. Army and General Electric Company (GE) to altitudes as high as 210 km, carrying 1000 kg of instruments and electronic support systems (Fig. 1.6); one was also launched at sea. In the higher flights, investigators were able to make brief measurements of the spectrum of radiation from the sun and other stars in wavelengths that are normally absorbed by Earth's atmosphere. They were also able to measure the way in which Earth's atmospheric properties changed in response to dynamic processes on the sun, throughout an 11-year sun-spot cycle.

Still, there was one major drawback to the V-2 as a research vehicle: it depended on carbon vanes in its nozzle, plus linked tabs at the tips of a set of four fixed aerodynamic fins, for attitude control. When its engine shut down, and it rose above the stabilizing influence of the atmosphere, the rocket tended to roll uncontrollably about its longitudinal axis and then tumble end over end. This

Courtesy of APL

Fig. 1.6. Applied Physics Laboratory cosmic ray telescope in V-2 instrument section.

meant that the instruments in the payload section swept rapidly past discrete targets such as the sun and stars and allowed only a limited quantity of data to be collected from them. Attempts were made to compensate for this deficiency by developing sun-follower and star-tracker payloads to stabilize individual instruments at the tip of the V-2 nose cone (Fig. 1.7). Other problems were created by the substantial cost of refurbishing and launching these large rockets, coupled with the fact that there was a finite quantity of them available.

Anticipating the end of the V-2 program, Milton W. Rosen and Homer E. Newell of NRL and their contractor, Glenn L. Martin Company, developed two versions of a much smaller sounding rocket, which they named the Viking. They launched 11 of them from White Sands and one at sea between 1949 and 1957. The first version was 0.8 m in diameter and 12 m tall (Fig. 1.8). Its monocoque structure was made from aluminum, rather than the sheet steel of the V-2, which resulted in a launch mass of 4877 kg (as opposed to the V-2's 12,300 kg). The sea-level thrust of its turbine-fed, alcohol-liquid oxygen engine from Reaction Motors, Inc., was 92.7 kN, and its burning time was 72 sec. It carried 250 kg of instruments to a maximum altitude of 218 km.

The second version of the Viking was 1.2 m in diameter and 13 m tall, and it weighed 6820 kg. The sea-level thrust of its turbine-fed alcohol-liquid oxygen engine from GE was also 92.7 kN but its burning time was extended to 103 sec.

Courtesy of NRL

Fig. 1.7. Richard Tousey with NRL V-2 sun-follower payload.

It carried 250 kg of instruments to a maximum altitude of 253 km. Both versions were equipped with pitch and roll hydrogen-peroxide jets and gyroscopes for stability outside the atmosphere, thereby increasing the quantity and quality of data recovered from discrete sources such as the sun and stars without the weight penalty of a sun follower or star tracker.

During the same period, at the instigation of Van Allen, JPL and Aerojet General Corporation initiated development of an even smaller and more economical rocket, the Aerobee, which was powered by pressure-fed unsymmetric dimethyl hydrazine and white fuming nitric acid hypergolic fuels. This rocket was basically an evolution of the Wac Corporal. When mounted on top of an Aerojet solid propellant booster, it carried payloads of 70 kg to an average altitude greater than 160 km (Fig. 1.9a). More than 150 of them were launched between 1948 and 1956.

This trend toward smaller launch vehicles was made possible by continuous miniaturization of electronic components and improvements in batteries and detectors. The trend continued, as multistage solid-propellant rockets, boosted by surplus motors from the Nike air defense program and topped by Deacon and Cajun upper stages, were developed specifically for upper atmospheric and astronomical research by the National Advisory Committee on Aeronautics and the University of Michigan (Fig. 1.9b). These rockets carried 15 cm diameter payloads weighing 25 kg to altitudes of 160 km from portable launchers in the United States and Canada and ships at sea.

Courtesy of U.S. Army

Fig. 1.8. Viking 2 (left) and Viking 9 (right) launches at White Sands Proving Ground.

Fig. 1.9. Dual Aerobee launches at White Sands Missile Range (a), Richard Custer with Nike-Cajun rocket and HUGO payload at Wallops Island (b), Van Allen's Rockoon being launched from the icebreaker USS Eastwind (c), and Commander Gus Ebel, Carl McIwain, and Frank McDonald with a Loki rocket at sea (d).

When Van Allen moved to the State University of Iowa to become the head of the physics department in 1951, he lowered the cost of launches even further and greatly extended the lateral extent of the atmosphere being studied by developing the Rockoon (rocket-balloon). In this novel approach, a high-altitude research balloon carried a single-stage, 15 cm diameter, 2.5 m long, solid propellant Deacon rocket to an altitude of 30 km and ignited it to propel a 25 kg payload to a peak altitude of 70 km, without the need for a formal launch range (Fig. 1.9c). A second version of the Rockoon, in which a solid propellant Loki antiaircraft rocket was substituted for the Deacon to carry a 7.5 cm. diameter, 3.1 kg payload to an altitude of 91 km, represented the final stage of the miniaturization process (Fig. 1.9d).

During these years, the Soviet Union also adapted captured German V-2 ballistic weapons for upper atmospheric and space research and launched them from the Baikonur launch range in Kazakhstan (Fig. 1.10). However, the USSR did not

Fig. 1.10. Russian sounding rocket.

follow the U.S. lead in developing smaller launch vehicles, because they were not yet capable of building small payloads—probably owing to a lack of access to miniaturized electronic components. In fact, as late as 21 February 1958, the Soviet Union launched a 1500 kg atmospheric research payload on a single-stage sounding rocket to an altitude of 470 km.

1.2.2 IGY and the Birth of the Space Race

All during the decade described above, the engineers and scientists in the nascent space program dreamed of the day that they could place their instruments aboard artificial satellites and launch them into long-duration orbits. That way, they could make continuous measurements all around Earth, instead of performing them for only a few minutes at a time and at only a few discrete locations. They finally found the impetus for achieving that goal in the International Geophysical Year (IGY). This was an 18-month period from July 1957 to December 1958 that was designated by an international consortium of nations to be devoted to a coordinated study of Earth's physical characteristics and its relationship to the sun, centering on the most active part of a solar sunspot cycle.

The impetus for IGY took place years earlier, on the evening of 5 April 1950, at a dinner party in Van Allen's suburban Maryland home. One of the guests at that dinner, Lloyd V. Berkner, suggested to Sydney Chapman, a world-renowned geophysicist, that it was time to organize another International Polar Year. The usual interval between polar years had been 50 years, but Berkner felt that recent advances in geophysical measurements and sounding rocket launches into the upper atmosphere would justify shortening the interval to 25 years and centering it on an upcoming peak in solar activity in 1957–1958. A comprehensive program of rocket launchings was included in the planning for IGY, to be directed by the Upper Atmosphere Rocket Research Panel, headed by Van Allen.

Discussions about launching an artificial satellite into orbit were held in subcommittees of the IGY, and on 24 November 1954, the Space Flight Technical Committee of the ARS published a position paper entitled, "On the Utility of an Artificial Unmanned Earth Satellite," directed to the attention of the U.S. National Science Foundation. The committee, chaired by Rosen of NRL, recommended that the United States fund an effort to design and build a series of 12 scientific spacecraft and attempt to launch them into orbit during IGY. These spacecraft would carry instruments designed and built by the various scientific groups that were active in the U.S. sounding rocket program. The committee members were well aware of the enormous technical difficulties inherent in such an ambitious undertaking, but they felt that if one of these launches were successful, it would justify the expense.

On 28 July 1955, President Dwight Eisenhower's press secretary announced that the United States would conduct a program to launch an artificial satellite around Earth. Specifically, the objectives were to place a satellite in orbit during IGY, accomplish a scientific experiment in orbit, and track the satellite in orbit. The Soviet Union followed up with a similar announcement four days later. In the United States, a Technical Panel on the Earth Satellite Program was then established, with Richard Porter of GE as its chairman, to review and fund selected proposals for satellite experiments.

On 26–27 January 1956, the Upper Atmosphere Rocket Research Panel held a symposium in Ann Arbor, Michigan, to discuss the scientific aspects of the IGY satellite program; 33 satellite experiment proposals were presented at that meeting. Van Allen proposed flying a Geiger-Muller counter to study the cosmic-ray intensity above the appreciable atmosphere on a comprehensive geographical and temporal basis. He felt that this study would be especially valuable in helping to interpret the extensive but widely spaced ground, balloon, and rocket observations being planned for IGY.

Van Allen's proposal was selected for full funding, as were proposals by Herbert Friedman and James Heppner of NRL, William Stroud of the U.S. Army Signal Engineering Laboratory, Maurice Dubin and Edward Manring of the Air

Force Cambridge Research Center, and Werner Suomi of the University of Wisconsin. Other proposals were funded at a lower level for possible launch in the event that the primary experiments could not be completed in time.

In advance of the president's announcement about launching a satellite during IGY, two teams had vigorously competed to develop the launch vehicles to accomplish that goal. One team, led by Rosen of NRL, based its design on the experience gained from flying V-2 and Viking sounding rockets for upper atmospheric research. This team proposed using a modified Viking for their first stage, a modified Aerobee as their second stage, and a solid-propellant rocket for their third stage. This program would later be named Vanguard.

The second team was led by Maj. Gen. Bruce Medaris, von Braun, and Ernst Stuhlinger of the U.S. Army Ballistic Missile Agency (ABMA) and William Pickering of JPL. This team initially proposed to use a Redstone SRBM as their first stage and a total of 31 Loki solid rocket motors divided up into clusters to power their second, third, and fourth stages. These upper stages would be contained in an electrically powered "spin bucket," and their scientific instrument section would remain attached to the final stage. This program would later be named Explorer. During the evolution of the design, the team substituted 15 JPL-built Scale Sergeant rocket motors for the 31 Lokis, which greatly improved the performance of their launch vehicle. Since they had no payload of their own, Stuhlinger, the principal scientist, asked Van Allen to modify his Geiger-Mueller counter experiment to fit on the end of their fourth-stage Scale Sergeant. After some initial reluctance, Van Allen decided to hedge his bets by having his Iowa graduate student, George H. Ludwig, design an experiment to fit both the Vanguard and Explorer spacecraft.

Actually, a third team within the U.S. Air Force also wanted to offer a proposal to use the Atlas missile to launch a satellite into orbit, but Air Force top leadership, including Gen. Bernard Schriever, squelched the idea because of the negative impact that the work would have on the crucial schedule for developing an ICBM, which was the military's top priority at the time.

To settle the competition between the ABMA/JPL and NRL teams, U.S. Assistant Secretary of Defense Donald Quarles formed an Ad Hoc Committee on Special Capabilities, composed of eight distinguished scientists recommended by the Army, Navy, and Air Force and chaired by Homer Joe Stewart of JPL. After considerable deliberation, the committee recommended on 4 August 1956 that the NRL proposal be selected. Medaris and von Braun vigorously objected and demanded a second vote. After giving the matter due consideration, the committee voted in a special meeting on 23 August to confirm its selection of the NRL proposal.

The committee emphasized that their selection was partially based on the scientific breadth and quality of the experiments from various scientific groups around the country that the NRL team proposed to host in their satellite. Another important factor was that the NRL proposal contained a comprehensive "Minitrack" tracking system that would be installed in several countries to receive telemetry data from the orbiting satellites and to measure their orbits precisely by a phase comparison technique. Eisenhower concurred with the committee's selection—largely because the Army's proposal employed military weapons for all of its propulsion units, whereas the NRL proposal employed stages that had been used only for peaceful research. He felt that a civilian program would enable him to establish the legitimacy of peaceful overflight of all nations, without arousing the objections of those nations, especially the Soviet Union.

Following the President's decision, NRL formed Project Vanguard (named by Rosen's wife, Sally) to build a scientific satellite and its three-stage launch vehicle, under the overall leadership of John P. Hagen and the technical direction of Rosen, with Glenn L. Martin Company as the integrating contractor. In spite of serious funding delays from a patchwork of sponsoring agencies, together with a low national priority (because the program by definition had no military significance), Vanguard finally got off to a good start, with the successful launches of suborbital single-stage and two-stage test vehicles, designated TV-0 and TV-1 (TV for "Test Vehicle"), from Cape Canaveral in Florida during 1956 and early 1957. The first stages of these vehicles were made up of Vikings 13 (Fig. 1.11) and 14, originally intended for launch from White Sands. A test transmitter was ejected from TV-1 at the peak of its suborbital flight and was successfully tracked by a Minitrack system installed at the Cape.

However, the Soviet Union, which most people had not taken seriously when it announced its intention to launch a satellite during IGY, shocked the world with the launches of Sputnik I on 4 October 1957 and Sputnik II on 3 November 1957. To compound matters, an attempt to launch a Vanguard engineering test satellite ended in a spectacular explosion on the launchpad on 6 December 1957. This launch was not expected to attract significant press attention, since the TV-0 and TV-1 launches had been rather routine; however, the Sputnik launches changed all that, and a large media contingent was present to relay images of the explosion to the nation and the world.

Fig. 1.11. Vanguard TV-0 (Viking 13).

Eisenhower was initially unconcerned by the Sputnik launches; in fact, he was privately pleased that those launches had automatically resolved the right of international overflight, which he had been so determined to establish. However, the enormous international press coverage generated by Sputnik I led him to authorize the ABMA/JPL team to undertake a backup effort to the Navy Vanguard program. An all-out effort was mounted to take a Jupiter C three-stage test vehicle out from under canvas in a hangar (where it had been set aside earlier by an ever-optimistic and determined von Braun) and add to it a fourth stage to create the Juno I orbital launch vehicle. At the same time, a 200-person team at JPL raced to complete the design and fabrication of a satellite.

Unfortunately, Van Allen, whose cosmic-ray experiment they hoped to feature in this satellite, was on a Navy ship in the South Pacific during that time, launching a series of Rockoons to define the geographic distribution of cosmic rays above the atmosphere. Since the Army program was conducted under strict military secrecy, meaningful conversations between ABMA, JPL, and Van Allen by radio were nearly impossible. Pickering therefore decided to hire George H. Ludwig, Van Allen's graduate student who had built the original Vanguard cosmic ray experiment as a master's thesis project, and move him to JPL in Pasadena to modify his Vanguard experiment as necessary to fit into the new satellite, which was called "Deal I" up until the time of its launch and "Explorer I" thereafter (Fig. 1.12). Pickering also got permission from Newell at NRL to transfer a Minitrack spacecraft transmitter and receiver to the Explorer program—otherwise, there would have been no way to measure the Explorer I satellite's orbit

Courtesy of G. H. Ludwig

Fig. 1.12. George H. Ludwig with the Deal I (Explorer I) instrument section at the University of Iowa.

precisely, since the JPL Microlock telemetry system was not designed for orbit determination.

Ludwig moved to Pasadena and worked as a junior engineer at JPL to complete fabrication, integration, and testing of his cosmic-ray experiment—just in time to ship the package to Cape Canaveral for launch. He stayed behind in Pasadena until the last possible minute to complete some critical calibrations and arrived at the Cape on the day of the launch. To his surprise, he was not permitted to enter the Army blockhouse or any of the JPL telemetry receiving areas to monitor his instrument's performance during launch. Fortunately, his old Vanguard friends, Roger Easton and Marty Votaw, invited him into their NRL telemetry trailer to monitor his signals. They did so in spite of the fact that the ABMA Florida launch crew (not including George) had stood on the roof of their hangar and cheered loudly when the Vanguard TV-3 launch perished in a ball of fire seven weeks earlier.

In any event, the aggressive spirit of the ABMA/JPL/Iowa team was finally rewarded on the night of 31 January 1958, when the Juno I launch vehicle ascended into the Florida night sky with its striped payload spinning and winking in the launchpad's floodlights.

With this historical context in mind, we turn now to the satellites themselves.

1.3 Sputnik I

Sputnik I was launched from the Baikonur launch complex on 4 October 1957. It was propelled by a modified R-7 ICBM into an elliptical orbit, inclined 65.1 deg to Earth's equator, with a perigee of 215 km, an apogee of 939 km, and an orbital period of 96 min.

The mass of the assembled Sputnik I was 83.6 kg. Its exterior shell was a polished aluminum-magnesium-titanium sphere, 2 mm thick and 58.5 cm in diameter (Fig. 1.13). It was pressurized to 1.3 atm with dry nitrogen. It contained a 1 W, 3.5 kg vacuum-tube radio transmitter, operating alternately on two frequencies, 20.005 MHz and 40.002 MHz, through two external pairs of spring-loaded whip antennas 2.9 m and 2.0 m long. The antennas were initially restrained inside the nose cone of the launch vehicle and were then released in orbit into a configuration swept aft 65 deg from the longitudinal axis of the satellite.

The power supply for the spacecraft weighed 51 kg and contained two sets of silver-zinc batteries that powered its transmitter and a third set that powered an electric fan to help control the spacecraft's internal temperature. The fan was commanded on by a thermostat when the temperature rose above 36°C and turned off when it dropped below 20°C. The modulation scheme for the spacecraft's telemetry transmitter consisted of sending out an alternating series of pulses of 0.3 sec duration on its two radio frequencies, so long as the spacecraft's internal temperature was between 0°C and 50°C. When the temperature exceeded that range, the pulse duration changed on both frequencies. The duration was also

Courtesy of NASA

Fig. 1.13. Internal view of Sputnik I components.

designed to change if the pressure inside the spacecraft dropped below one third of an atmosphere, to indicate that its shell had been punctured by a meteorite. A view of the satellite's interior is given in Fig. 1.13. The octagonal objects are the battery packs; the radio transmitter was mounted in a cavity in their center.

Sputnik I's launch vehicle, designated 8K71PS, was a two-stage, parallel-burn vehicle with a mass of 267 metric tons, and it was propelled by four sets of liquid propellant rocket boosters wrapped around a liquid propellant core stage (Fig. 1.14). Each booster was powered by an RD-107 engine, comprising four main thrust chambers and two vernier chambers, all fed from a common turbopump assembly. The core stage was powered by an RD-108 engine, comprising the same four main thrust chambers operating at a lower pressure and four vernier chambers, again fed from a common turbopump assembly. All 20 main thrust chambers and 12 verniers were ignited at launch, producing an aggregate thrust of 3,900 kN. Its strap-on boosters fell away when their fuel was depleted 116 sec after launch. Its core stage continued to burn for an additional 178 sec, after which the nose cone was shed. At 314 sec after liftoff, Sputnik I separated from the spent core stage, and its radio transmissions were immediately received by one of a string of government ground stations located downrange from the launch site within the Soviet Union.

Signals from the transmitter were again received 96 min after launch by the official tracking stations, and a formal announcement was made that Sputnik I

Courtesy of NASA

Fig. 1.14. R-7 launch vehicle for Sputnik satellite.

was successfully in orbit. The frequencies on which the transmitter was operating were publicly announced, and amateur radio operators all over the world started tuning into the signals and measuring their Doppler shifts to help estimate the satellite's orbit. Operators of a newly installed NRL "fence" of Minitrack ground stations, which had been installed to track upcoming Vanguard satellites on 108 MHz, also reconfigured their equipment to receive Sputnik's frequencies, so they could measure the satellite's orbit with much greater precision.

Sputnik I, although quite simple, was a complete technical success, proving that an orbiting spacecraft could survive in the hostile environment of space, without being punctured by meteorites or having its vacuum-tube electronics damaged by solar or cosmic radiation above the shielding effects of Earth's atmosphere. It also proved that the internal temperature of an orbiting object could be controlled within limits acceptable for standard electronic equipment. It radioed useful internal and skin temperature data to the ground for three weeks, before its batteries became depleted, while international teams of physicists measured alterations in the paths of its radio transmissions to determine the density of ions in Earth's upper atmosphere.

It was also observed optically at twilight by "Moonwatch" volunteers, who had been organized into teams all over the world earlier in 1957 by Fred Whipple of the Smithsonian Astronomical Observatory at Harvard University. These teams used arrays of small, homemade telescopes, built to Whipple's specifications, to determine the meridian crossing times of the satellite and the core stage of its rocket. They reported their sightings to Observatory staff members, who worked with the Massachusetts Institute of Technology (MIT) to calculate their orbits and publish forecasts of when they would pass over various places around the world.

In addition to its engineering and scientific accomplishments, Sputnik I was an even greater propaganda success. People around the world were astounded that, armed with the Moonwatch forecasts, they could go outside at morning or evening twilight and watch bright flashes of sunlight reflecting from the tumbling core stage of Sputnik's launch vehicle, which was 29 m long by 3 m in diameter and covered with special reflectors to increase its visibility, as it swept across the sky at an amazing rate. Most of them erroneously thought they were seeing the satellite itself; however, at only 58.5 cm in diameter, it couldn't reflect enough sunlight to be seen without a telescope. Scientists around the world heaped praise on the Soviet Union for achieving an engineering and scientific triumph. The man in the street and the leaders of the Western nations, however, were chilled by the thought that if the Soviets could orbit a spacecraft around Earth, they must surely be able to launch thermonuclear weapons anywhere in the world.

Even after Sputnik's batteries died, Moonwatch teams continued tracking the satellite and its rocket body. The Harvard/MIT team used the amateurs' sightings to calculate the gradual shrinkage of the orbits caused by aerodynamic drag. In this manner, they were able to estimate the density of the upper atmosphere at the various altitudes through which the objects passed. Sputnik I orbited Earth 1440 times until 4 January 1958, a total of 90 days after its launch, before descending deep enough into the upper atmosphere to become vaporized by aerodynamic heating. After Sputnik, no additional small satellites were flown by the Soviets, who instead increased the size of their later spacecraft enormously.

1.4 Explorer I

Explorer I (Fig. 1.15) was launched from Cape Canaveral on 31 January 1958 on a four-stage Juno I launch vehicle into an elliptical orbit, inclined 33.3 deg to Earth's equator, with a perigee of 360 km, an apogee of 2535 km, and an orbital period of 114 min.

The launch mass of the spacecraft was 13.9 kg, which included 8.3 kg of payload that remained attached to the 5.6 kg cylindrical steel casing of its expended fourth-stage Scale Sergeant rocket motor. Three scientific instruments were incorporated into the payload. The first, as mentioned previously, was the Anton 314 omnidirectional Geiger-Mueller counter built by Ludwig; its purpose was to measure the intensity of cosmic radiation above the atmosphere and define the extent of Earth's magnetic field. Ludwig made extensive use of transistors in his circuits. The other two scientific instruments were detectors built by Maurice Dubin and Edward Manring of the Air Force Cambridge Research Center to

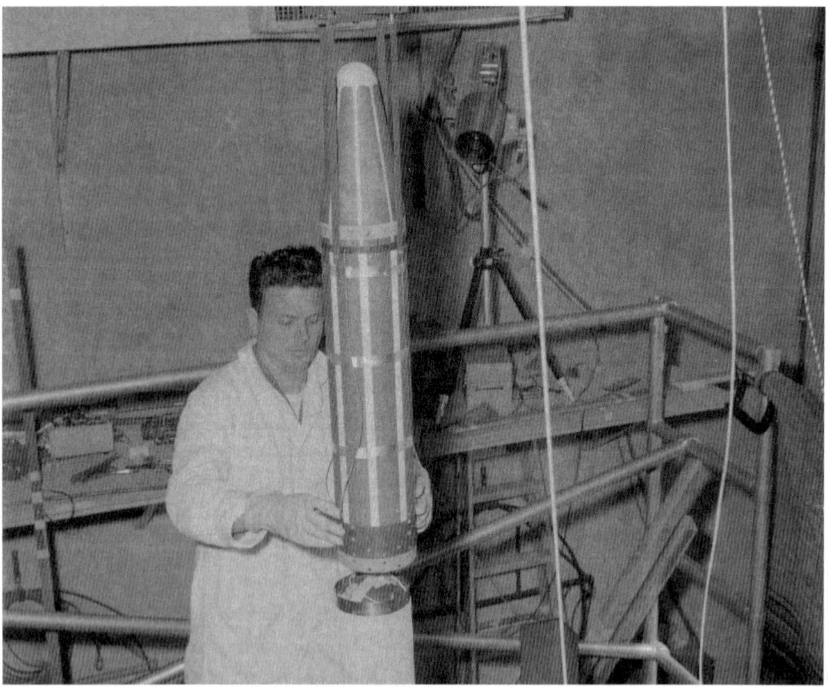

Courtesy of NASA

Fig. 1.15. Explorer I instrument section being installed on Scale Sergeant fourth-stage motor.

measure the frequency and energy distribution of meteorites above Earth's atmosphere. The first detector employed a crystal transducer and solid-state amplifier to acoustically record the energy of meteor impacts. The second was a wire-grid detector that recorded micrometeorite impacts by the fracture of wires in its grid.

The Explorer I instrument section contained two telemetry transmitters. The first was a 10 mW JPL Microlock operating on 108.0 MHz and feeding a turnstile antenna that consisted of four flexible steel wires, each 56 cm long. The second was the 60 mW Minitrack, originally built by NRL for Vanguard, operating on 108.030 MHz and connected to two fiberglass slot antennas in the body of the satellite. A package of mercury batteries and packages of control electronics were also contained in the payload. The satellite's exterior shell was an aluminum cylinder, 16.5 cm in diameter and 64 cm long, tipped by a blunt cone 32 cm long. It was painted in a pattern of alternating white and green longitudinal stripes (Fig. 1.16) to control the temperature of the satellite.

The low-powered JPL telemetry transmitter operated for 105 days, during which time, Microlock ground stations received housekeeping and scientific data from the satellite. NRL Minitrack ground stations also received signals from its high-powered transmitter for 31 days and relayed them to the Vanguard Computing Center, which was operated on a volunteer basis by IBM Corporation in Washington, DC, to compute the satellite's orbit with high precision. After its batteries were depleted, Explorer I remained in orbit for 12 years, and Moonwatch volunteers continued to track it through small telescopes. In addition, specially designed professional Baker-Nunn tracking cameras were installed around the

Courtesy of NASA

Fig. 1.16. Explorer I components and external shell. The cosmic ray experiment is mounted on the upper portion of the compartment, and the telemetry section, flexible antennas, and micrometeor detectors are mounted at the bottom. The white stripes are designed for thermal control of the spinning satellite.

Courtesy of U.S. Army/ Redstone Arsenal

Fig. 1.17. Night launch of Explorer I aboard the Juno I launch vehicle.

world not long after Explorer I was launched. Their observations allowed the Harvard/MIT team to determine the satellite's orbit much more precisely than they had been able to do using the values calculated from the amateur sightings.

Explorer I's launch vehicle, Juno I, was 21.2 m tall and 1.78 m in diameter and weighed 29,060 kg (Fig. 1.17). Its first stage was a modified Redstone containing one North American Rocketdyne A-7 motor that used liquid oxygen as its oxidizer and hydyne (60% unsymmetrical dimethylhydrazine and 40% diethylenetryamine) as its fuel to generate 369.4 kN of thrust for 155 sec. Its three upper stages consisted of clusters of Scale Sergeant solid-propellant (polysulfide-aluminum/ammonium perchlorate) rockets mounted in a "spin bucket" (Fig. 1.18) that was attached by a bearing to the top of the Redstone booster and rotated at a rate of 750 rpm by an electric motor to provide gyroscopic stability to the upper stages and average out their nozzle misalignments during thrusting.

Its second stage comprised a cluster of eleven 15.2 cm diameter Scale Sergeants that were ignited simultaneously to generate a 78.3 kN total thrust for 6.5 sec. Its third stage contained a cluster of three similar Scale Sergeants that generated

Fig. 1.18. Upper-stage spin bucket on top of a modified Redstone first stage.

a total of 21.4 kN for 6.5 sec. Its fourth stage was a single Scale Sergeant that generated 8 kN for 6.5 sec and stayed attached to the Explorer I satellite in orbit.

Shortly after achieving orbit, Explorer I and its attached fourth stage stopped spinning about its longitudinal axis and, instead, tumbled end over end, as a result of a phenomenon well known to sounding rocket engineers of that era as "roll-pitch coupling." A long, slender, cylindrical body, such as the combined satellite and rocket motor, is most stable while rotating about its transverse axis. If an attempt is made to spin it about its longitudinal axis instead, any asymmetry in its transverse mass distribution will cause its spin energy to become coupled into pitch energy, and it will precess into larger and larger coning angles until all the energy is transferred into its pitch mode. The flexible antennas on Explorer I served as the coupling mechanism, in this case. The tumbling behavior caused intermittent dropouts in the satellite's data transmissions to receiving stations on the ground and complicated the analysis of measurements made by its instruments.

Further complicating the analysis was the fact that telemetered data were received from its instruments for only the few minutes during each orbit when the satellite was passing over any of the Microlock or Minitrack receivers on the ground. In spite of these limitations, the Air Force Cambridge Research Center scientists deduced that their micrometeorite acoustic detector reported 145 hits of cosmic dust in a 12-day period, which equates to an impact rate of $8 \times 10^{-3}/(\text{m}^2 - \text{sec})$.

The satellite's eccentric orbit, coupled with Earth's rotation, caused its brief periods of data transmission to the Microlock stations to occur when the satellite was at varying altitudes, ranging from 358 km to 2550 km. When the satellite was passing over a receiving station at altitudes near perigee or up to 1000 km or so, the Geiger-Mueller counting rates remained in the expected range, increasing steadily with altitude—but when the satellite was at higher altitudes, the counting rates dropped to zero. This unexpected behavior caused Van Allen and Ludwig to suspect that the instrument might be failing intermittently. However, Carl McIlwain, another of Van Allen's graduate students, had recently returned from launching an instrumented cosmic-ray payload (Fig. 1.19) on a Nike-Cajun sounding rocket

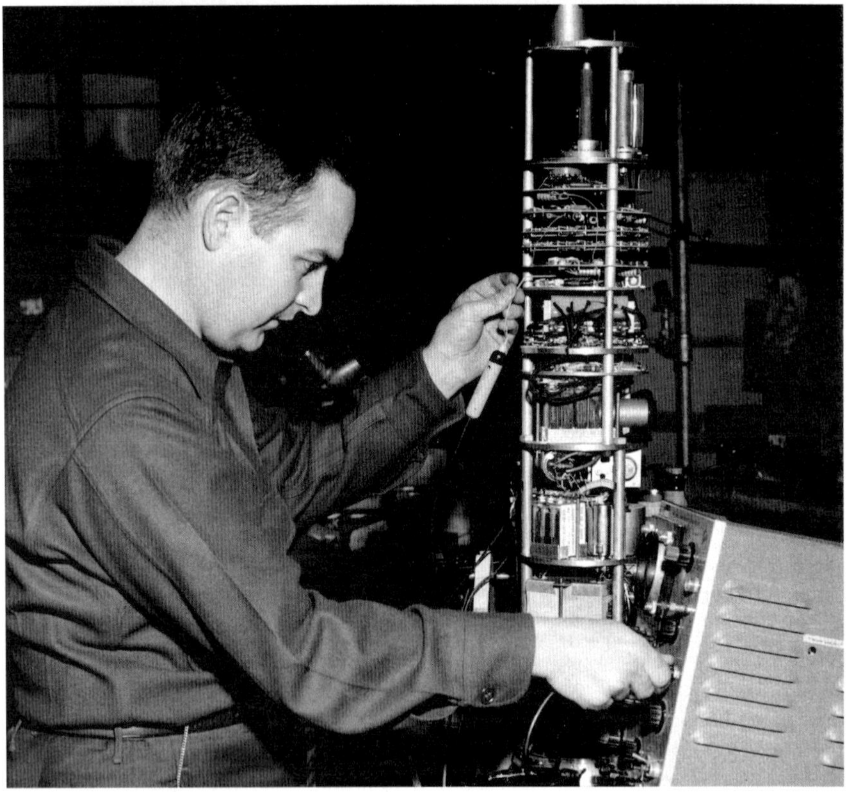

Courtesy of C. McIlwain

Fig. 1.19. Iowa graduate student Carl E. McIlwain with the cosmic ray experiment that he flew on a Nike-Cajun rocket from Ft. Churchill, Canada, into an auroral substorm at about the same time that Explorer I flew. His analysis of the flux of energetic particles detected on his Nike-Cajun flight allowed him to explain to Van Allen why the Explorer I Geiger counter saturated.

into an active aurora at Canada's Ft. Churchill Launch Range on the shore of Hudson's Bay, and he reported that his instrument had encountered unusually large fluxes of energetic particles above the atmosphere at high geomagnetic latitudes. His interpretation of the confusing results from the Explorer I instrument was that it was being overwhelmed by a huge flux of energetic particles at the satellite's higher altitudes that was driving its Geiger tube into such hard saturation that it did not count at all. Van Allen and Ludwig agreed with that interpretation and changed the counting rate of their next instrument.

1.5 Vanguard 1

In preparation for the first Vanguard orbital launch attempt, three successful suborbital test vehicles, TV-0, TV-1, and TV-2, were launched to flight certify the individual stages and the overall configuration of the launch vehicle.

Then, two unsuccessful attempts to launch a Vanguard satellite preceded its first success. The first attempt, designated TV-3, briefly rose from its launchpad on 6 December 1957, fell back, and exploded, due to a first-stage motor turbine underpressurization problem. The second attempt, TV-3 Backup, took place on 5 February 1958, but yawed and broke up 57 sec after launch, due to a control system failure.

On 17 March 1958, six weeks after the successful launch of Explorer I, the Vanguard 1 satellite (Fig. 1.20) was successfully launched from the Cape Canaveral Air Force Station on a three-stage Vanguard launch vehicle (Fig. 1.21) into

Fig. 1.20. Vanguard 1 during final installation and checkout on its TV-4 launch vehicle.

an elliptical orbit, inclined 34.25 deg to Earth's equator, with an initial perigee of 654 km, an apogee of 3969 km, and an orbital period of 134 min. The launch vehicle was designated TV-4.

The launch of the Vanguard 1 satellite was used to verify the performance of the launch vehicle and to define the environment in which future siblings would be expected to operate. Vanguard 1 was spherical in shape, 16.5 cm in diameter, and weighed 1.47 kg. Its exterior shell was made from two spun aluminum hemispheres, coated with alternating layers of silicon dioxide and aluminum for thermal control, and it housed a cylindrical aluminum instrument compartment, 10 cm in diameter, that was suspended within the craft by Teflon rods that served as electrical and thermal insulators (Fig. 1.22). The instrument compartment contained two radio telemetry transmitters, one of which was powered by seven

Fig. 1.21. Successful launch of Vanguard 1 on 17 March 1958.

the effects of extreme UV radiation from solar flares on the density of the upper atmosphere and the way that this density variability influences satellite orbits.

1.6 Additional Sputnik Launches

A few days after Sputnik I's launch, Soviet Premier Nikita Krushchev, greatly pleased by the overwhelming international acclaim for his nation's pioneering achievement in launching the world's first artificial satellite, directed Korolev to put into orbit another, even more impressive satellite, designated Sputnik II, to celebrate the 40th anniversary of the Soviet revolution on November 7. In the astonishingly short timespan of a few weeks, Korolev's team took the backup satellite for the Sputnik I launch and installed it in a three-part payload (Fig. 1.26). The second part was an aluminum, self-contained, hermetically sealed life-support capsule that contained a small dog named Laika ("Barker" in English). The temperature of the capsule was regulated by a combination of reflectivity of the spacecraft's external surfaces, thermal insulating blankets, and a forced-air circulating system. The team taped instruments to the dog to measure her pulse, respiration, and blood pressure and take electrocardiograms. They also mounted other instruments inside the capsule to monitor its pressure and temperature. The

Fig. 1.26. Sputnik II three-part payload, with science instrument package on top, spherical Sputnik I type shell below that, and Laika the dog's capsule on the bottom.

batteries for the entire payload were designed to last seven days in orbit, so the designers provided a seven-day supply of food and water for the dog.

The third section of the payload consisted of a separate scientific instrument package that would measure cosmic radiation and solar UV and x-ray emissions and was mounted on top of the spherical Sputnik package. The entire payload was designed to be deployed as a unit from the final stage of a modified R-7, which would also accompany it in a separate orbit. The total mass of the payload was 508 kg. It was launched on 3 November 1957 from Baikonur into a 212 × 1660 km orbit with a 65.3 deg inclination. Its orbital period was 103.7 min. The spherical capsule that had been a backup satellite for the Sputnik I mission served as the data transmission system for the combined Sputnik II payload. Its 15 MHz transmitter provided the same series of beeps that its predecessor had, but its 7.5 MHz transmitter generated a continuous signal that was commutated to relay the much larger data set of the Sputnik II mission to the ground stations.

Following a successful launch, the launch vehicle's nose cone, which protected the triple stack of capsules during ascent, separated on schedule, but a layer of thermal insulation was torn loose in the process. Postflight telemetry analysis revealed that Laika survived the effects of launch but died a few hours afterward from the loss of temperature control in her capsule—a fact that was not announced to the public until decades later. Sputnik II's scientific instruments and telemetry system continued to operate for seven days, until their batteries were exhausted. The Moonwatch network and Baker-Nunn cameras provided optical measurements of Sputnik II's orbit until it burned up in the atmosphere on 13 April 1958, a little more than five months after launch.

The Soviets had been preparing an enormous payload, which Korolev referred to as the "Object D" spacecraft, for many months, but had not been able to complete it until after Sputnik I and II had been flown. It was finally launched from Baikonur on 27 April 1958. However, it failed to reach orbit, due to a breakup of its R-7 launch vehicle 88 sec after ignition. Undeterred by this failure, the Soviet team installed a backup Object D spacecraft on another R-7 and successfully sent it into a 215 × 1863 km orbit inclined 65.2 deg to the equator on 15 May 1958 in a mission designated as Sputnik III.

This spacecraft was conical in shape, 3.57 m long and 1.73 m wide at its base. It weighed 1327 kg, including 968 kg of scientific instruments and power and radio telemetry subsystems. It used a combination of silver-zinc batteries and silicon solar cells to power its electrical subsystems. Its multichannel telemetry system transmitted pulses ranging from 150 to 300 msec in duration on a frequency of 20.005 MHz. Its internal temperature was controlled by a system of adjustable louvers on the surface of the spacecraft that opened and closed over a series of thermal radiators. The spacecraft contained 12 scientific instruments that measured pressure and composition of the upper atmosphere, concentration of charged particles, photons in cosmic rays, heavy nuclei in cosmic rays, magnetic and electrostatic fields, and meteorites. Its tape recorder failed prior to launch, so although it detected enhanced radiation at high altitudes, it was unable to map the radiation belt that had been encountered earlier by Explorer I. Sputnik III remained in orbit, with its radio transmitter continuing to operate, until 6 April 1960. However, useful measurements were not transmitted from its instruments after the first few weeks of its mission, apparently due to a failure of its modulation system.

1.7 Additional Explorer Launches

Buoyed by the success of the Explorer I launch, the JPL/Iowa team moved forward with their original plans to incorporate a tape recorder that could be commanded to play back complete orbits of data every time the satellite passed over the Vanguard fence (Fig. 1.27). This satellite, designated Explorer II, was launched on a second Juno I on 5 March 1958. It did not achieve orbit, due to a failure to ignite the fourth-stage Scale Sergeant motor. (Note: The Explorer program office originally assigned numbers to all the Explorer launch attempts, regardless of whether they were successful, until Explorer VI, at which time they numbered only the successful launches. They also used Roman numerals in their titles, while the Vanguard program used Arabic.)

Courtesy of George H. Ludwig

Fig. 1.27. Henry Richter of JPL (left) and Iowa graduate student George H. Ludwig with prototype equipment for the Explorer II cosmic ray experiment, which added a tape recorder to the basic Explorer I package. The satellite failed to reach orbit, but an identical payload was flown successfully on the Explorer III mission on 26 March 1958.

A duplicate payload was prepared and launched successfully as Explorer III on 26 March 1958, one week after the start of the first successful Vanguard mission. Explorer III's initial perigee was 190 km, its apogee was 2784 km, and its orbital inclination was 34.4 deg. Ludwig's tape recorder and the Minitrack tracking system worked flawlessly, and Van Allen was able to perceive a pattern in the puzzling "zero count" behavior of his cosmic-ray instrument. His graduate student, McIlwain, exposed a Geiger counter, identical to the one flying in Explorer III, to a 250 kV x-ray machine in his laboratory and produced a data set that matched the slope of counting rates from the Geiger tube in orbit. He proved that fluxes that would ideally produce more than 35,000 counts per second instead drove the count rate to zero. The Van Allen team therefore concluded that telemetered data from the Explorer III mission indicated an enormous band of trapped energetic particles circled Earth at altitudes higher than 550 km, and that the shape of the band was a function of Earth's magnetic field. Van Allen announced this discovery to the world at a joint session of the American Physical Society and the

Fig. 1.28. A Juno II rocket.

National Academy of Sciences in Washington, DC, on 1 May 1958. This region subsequently became known as the inner of two "Van Allen radiation belts."

The next satellite in the series, Explorer IV, was propelled into a much more highly inclined orbit than usual by a Juno I launch vehicle on 26 July 1958. Its objectives were to extend its measurements over a much broader latitude range of the trapped particle radiation belt that had been discovered on the previous Explorer flights and to measure the effects on that natural radiation belt of the three 1.7 kt W-25 "Argus" nuclear devices that were detonated above the atmosphere in an experiment conducted by the U.S. Advanced Research Projects Agency (ARPA). Explorer IV's initial apogee altitude was 2180 km, its perigee was 260 km, and its inclination was 50.3 deg. Its 11.5 kg payload contained two Geiger-Mueller counters, two scintillation counters, and several temperature sensors. It proved that the effects of the nuclear detonations were transient and not the source of the original belts of trapped particles, as had been alleged by the Soviets.

Explorer V was a twin spacecraft to Explorer IV, but it failed to reach orbit on 24 August 1958 because its Juno first stage collided with its upper-stage cluster at separation.

The next spacecraft in the Explorer series contained an inflatable aluminized Mylar balloon, 3.7 m in diameter, which was intended to be tracked by radar to provide atmospheric density data. It was launched on a modified Juno I that contained a fifth stage; however, the spacecraft failed to reach orbit on 22 October 1958 because it broke free from its upper-stage motor 112 sec after liftoff, due to spin resonance effects.

By the end of 1958, the International Geophysical Year had produced such comprehensive results and had raised so many new questions about geophysics that an agreement was made to extend the coordinated worldwide research for an additional year.

Fig. 1.29. The Explorer VI satellite.

At this point, a new Juno II launch vehicle (Fig. 1.28) was developed, consisting of the substitution of a much larger Jupiter IRBM for the modified Redstone, or Jupiter C, first stage that was used in the Juno I. A double-truncated conical Explorer spacecraft, with a mass of 42 kg, was launched on the first Juno II on 16 July 1959, but a failure of the guidance inverter caused the vehicle to yaw sharply and break up some 5 sec after liftoff.

Explorer VI (Fig. 1.29) was a small, spheroidal satellite designed to study trapped radiation of various energies, galactic cosmic rays, geomagnetism, radio propagation in the upper atmosphere, and the flux of micrometeorites. It also tested a scanning device for photographing Earth's cloud cover, and it transmitted the first televised images of Earth from orbit. The satellite was launched on 7 August 1959 aboard a Thor-DM-18 Able III launcher that consisted of a military Thor IRBM as a first stage and the top two stages of the Vanguard launch vehicle as its second and third stages (Fig.1.30). Its initial orbit had a perigee of 237 km, an apogee of 41,900 km, and an inclination of 47 deg.

Fig. 1.30. The Thor Able III launch vehicle.

Four solar-cell paddles mounted near its equator recharged the storage batteries in orbit. Each experiment except the television scanner had two outputs, digital and analog. A UHF transmitter was used for the digital telemetry and the television signal. Two VHF transmitters were used to transmit the analog signal. The VHF transmitters were operated continuously, while the UHF transmitter was operated for only a few hours each day. Only three of the solar-cell paddles fully erected, and this occurred during spin up, rather than before spin up as planned. Consequently, initial operation of the payload power supply was 63% nominal, and this decreased with time. The decreased power caused a lower signal-to-noise ratio, which affected most of the data, especially near apogee. One VHF transmitter failed on 11 September 1959, and the last contact with the payload was made on 6 October 1959, at which time the solar-cell charging current had fallen below that required to maintain the satellite equipment. The satellite's orbit decayed on 1 July 1961.

Fig. 1.31. The Explorer VII satellite.

Explorer VII, a double-truncated conical spacecraft with a mass of 42 kg (Fig. 1.31), was successfully launched on 13 October 1959 on a Juno II into an orbit with an inclination of 50.3 deg, a perigee of 552 km, and an apogee of 1090 km. It contained instruments that measured micrometeoroid frequency, cosmic-ray distribution, heavy-nuclei spectrum, Earth energy balance, solar x rays, and solar Lyman alpha radiation. It was powered by an array of solar cells and 15 nickel-cadmium batteries. It transmitted useful scientific data for 15 months, and its transmitter continued to broadcast a carrier frequency for an additional five months. Explorer VII is still in orbit today.

1.8 Additional Vanguard Launches

In addition to the Vanguard 1 engineering test satellite, several full-scale, fully instrumented, 51 cm diameter scientific spacecraft were prepared by NRL (Fig. 1.32). In addition to the normal Minitrack telemetry and tracking transmitter and receiver, the first one of them contained photon counters designed by Herbert Friedman and his team in the NRL Electron Optics Branch to measure the x-ray spectrum of the sun in the 1-8 Å band, while a second and third satellite contained similar photon counters to measure the solar spectrum in the 1100–1300 Å band. All three payloads also contained devices to monitor the space environment. Each spacecraft weighed 10 kg.

The first of these spacecraft was launched on Vanguard TV-5 on 28 April 1958. The flight was normal through second stage burning, but the second-stage shutdown sequence was not completed electrically, so the third stage was not properly separated and fired. The spacecraft did not achieve orbit.

The second of the scientific spacecraft was launched on 27 May 1958, on the first production Vanguard launch vehicle, designated Satellite Launch Vehicle 1, or SLV-1, which was basically the same as one of the test vehicles, minus some of the vehicle performance instrumentation. Once again, the flight was normal through second stage burn, but at that point, an attitude disturbance caused the third stage to be fired at an incorrect angle. The spacecraft did not achieve orbit.

The third spacecraft in the series was launched on SLV-2 on 26 June 1958, but failed to reach orbit because of a shut down of the second-stage engine after only 8 sec of thrusting.

The fourth and fifth spacecraft in this scientific series contained photoelectric instruments designed by William Stroud and his team at the U.S. Army Signal Corps Engineering Laboratory to measure the global distribution and movement of cloud cover and to contribute to the basic knowledge of Earth's energy budget. Each of these satellites weighed 10.7 kg. The first was launched on SLV-3 on 26 June 1958, but failed to achieve orbit because of a slight underperformance of the launch vehicle's second-stage engine, which resulted in the satellite being 76.2 m/sec short of its required orbital velocity of 7622 m/sec. The second (Fig. 1.33) achieved orbit on SLV-4 on 17 February 1959 and was designated Vanguard 2. Its initial orbit had a perigee of 554 km and an apogee of 3301 km. It was the

Courtesy of NASA

Fig. 1.32. Transparent view of 51 cm Vanguard satellite interior.

pathfinder for the highly successful TIROS weather satellites that followed in the next decade. Vanguard 2 and its 21.4 kg third-stage motor case are still in orbit today. The National Aeronautics and Space Administration (NASA) had been created on 29 July 1958, so although the satellite was built by NRL, the mission is recorded as a NASA success.

The next in the series of Vanguard science spacecraft contained a proton precessional magnetometer experiment built by James Heppner, originally of NRL and later of the NASA Goddard Space Flight Center. It was designed to survey Earth's magnetic field and to determine whether the predicted Stormer-Chapman ring current existed. Its external configuration consisted of a magnesium hemisphere, 51 cm in diameter, and a matching fiberglass hemisphere with a fiberglass cone that contained the magnetometer protruding out its top. An inflatable aluminized Mylar sphere designed by William J. O'Sullivan Jr. of the National Advisory Committee on Aeronautics and, later, of the NASA Langley Research

Courtesy of NASA

Fig. 1.33. Vanguard 2 satellite on launch vehicle.

Center, was also carried on this mission in a deflated condition, placed between the magnetometer satellite and its ejection system. Its objective was to provide a smooth, lightweight, spherical reflective target, 72 cm in diameter, that would be tracked optically and by radar to measure atmospheric density at very high altitudes. These two satellites were launched on the Vanguard SLV-5 on 13 April 1959, but failed to achieve orbit as a result of a loss of the launch vehicle's second-stage pitch attitude control.

At this point, an upgraded solid-propellant motor, the Allegheny Ballistics Laboratory X-248 Altair, was substituted for the Grand Central Rocket Company motor in the Vanguard SLV third stage. The X-248 used a fiberglass case and a nitroglycerine, nitrocellulose "double-base" propellant to provide a significant improvement over the former third-stage motor.

The penultimate spacecraft in the Vanguard program was a fully instrumented, 51cm diameter, 11 kg sphere that contained a radiation balance experiment

Fig. 1.34. Vanguard 3 satellite with magnetometer boom protruding from top.

devised by Suomi of the University of Wisconsin. It was intended to be launched into an orbit inclined 48 deg to the equator, so as to cover a wide latitudinal band of Earth and its atmosphere. Its instruments were designed to measure direct solar radiation as well as radiation reflected and reemitted from Earth and its atmosphere. This satellite was launched on the Vanguard SLV-6 on 22 June 1959, but failed to achieve orbit due to a rapid decay of pressure in the second-stage propellant tanks that occurred shortly after second-stage ignition. The radiation balance experiment was later flown successfully on the Explorer VII satellite.

The final satellite in the Vanguard program was a fully instrumented, 51 cm diameter, 24 kg magnesium sphere, with a fiberglass boom (Fig. 1.34) that contained a second magnetometer built by Heppner. An additional solar x-ray payload built by Friedman and his NRL associates, basically identical to a payload flown earlier on an unsuccessful Vanguard mission, was integrated with the magnetometer payload. This satellite, designated Vanguard 3, was successfully propelled into a 507 × 3721 km orbit with a 33.3 deg inclination on 18 September 1959. Measurements it made helped to define the lower boundary of the inner Van Allen radiation belt. Again, this satellite was built by NRL, but the mission was recorded as a NASA success.

The total mass placed in orbit by the launch, including the 24 kg spacecraft and its 19 kg third-stage motor, was 43 kg—a record for the Vanguard program. All three of the Vanguard satellites and Explorer VII are still in orbit today, because their launch vehicles placed them in orbits with higher perigees than did all the other launches of that era. This approach significantly reduced atmospheric drag and extended the orbital lifetimes of these four satellites.

1.9 Conclusions

The Sputnik, Explorer, and Vanguard programs were all highly successful in opening the door to spaceflight in the late 1950s (see Table 1.1). Collectively, they pioneered in-space thermal control, radio telemetry transmission from orbit, Earth-based telemetry reception, precision interferometer ground-based tracking, solar cells for onboard power collection, transistors to minimize weight of spacecraft electronics, and an array of scientific instruments that could measure the characteristics of Earth and space while surviving the rigors of launch and orbital flight.

The Sputnik and Explorer programs have received their fair share of accolades over the past 50 years; however, the Vanguard program, which launched three times as many satellites into long-lived orbits as it was originally charged to do, all of which contributed materially to our knowledge of the geophysics of Earth and its surroundings, has been basically written out of the history books, with one or two exceptions. In fact, the Executive Committee of the U.S. National Academy's Space Studies Board refused repeated requests in 2007 by myself and others to acknowledge the existence of the Vanguard program in a brochure commemorating the 50th anniversary of IGY. The stated rationale for their refusal was: "Because we don't want to insult the Russians by claiming we had two satellite programs during IGY when they had only one." Hopefully, this chapter will have confirmed that we definitely did have two successful satellite programs during IGY.

The Explorer program continued to flourish and make major discoveries for many years following the end of the extended IGY, under the auspices of the NASA Goddard Space Flight Center. The Vanguard program came to an end in late 1959, but many of the technologies it demonstrated were adopted by other programs. Interestingly, the two upper stages of the Vanguard launch vehicle were adopted by the Thor Able program to launch Explorer VI to an apogee of 41,900 km and provide additional details about the outer Van Allen radiation belt. These stages were also used in the early versions of the popular Delta series of launchers. In addition, the Minitrack system evolved into the Navy Space Surveillance System "fence" that added powerful transmitters to track noncooperative satellites for the next 50 years. This, in turn, led to the Timation satellite navigation program, which laid the technological foundation for the Global Positioning System.

Table 1.1 Summary of Launches from July 1957 to December 1959

	Sputnik	Explorer	Vanguard
Orbital launch attempts	4	9	11
Satellites placed in orbit	3	5	3
Satellites still in orbit	0	1	3
Launch vehicle failures	1	4	8

1.10 Reflections

With the luxury of 50 years of hindsight, we can now say that the Sputnik launches were probably the best thing that ever happened to the United States after World War II. Military successes during the war, coupled with a burgeoning economy and growing national pride in the postwar years, made the country extremely complacent and even arrogant about its place in the world. The U.S. general public had absolutely no inkling that the "backward" Soviet Union had advanced so far in developing large launch vehicles and satellites until Sputnik came beeping overhead. The resulting fear and outrage on the part of that public sparked a revolution in science and engineering and allowed the U.S. government to increase the budget for civilian and military space to levels unthinkable before Sputnik.

Shaking off its lethargy, the United States mounted a furious effort in space program development. For a time, the Russians continued to pull off a series of space "firsts"—first man in space, first woman in space, first man to perform an extravehicular activity (EVA), first woman to perform an EVA, first spacecraft to photograph the backside of the moon, first automated vehicle on the moon, first samples of lunar regolith returned to Earth, etc. However, the momentum generated in response to Sputnik I, II, and III resulted in the United States' finally plucking off the big prize—landing the first men on the moon and returning them safely to Earth. That momentum also resulted in this country's building up over the past 50 years the finest civilian and military space programs in the world, and in working together with most of the world's spacefaring nations on a staggering variety of manned and unmanned ventures in near-Earth and deep space. America's recognition of its debt to Sputnik I is best epitomized by the photograph (Fig. 1.35) of Konrad Dannenberg, Homer Hickam, William Lucas, Ernst Stuhlinger, and Julian Davidson offering a champagne 50th birthday "thank you" toast to a replica of Sputnik I at NASA's Marshall Space Flight Center on 4 October 2007.

Courtesy of NASA

Fig. 1.35. NASA Marshall Space Flight Center team toasting Sputnik I.

1.11 Acknowledgments

I am grateful to many individuals for providing information contained in this chapter. In particular, conversations and meetings over the past 50 years with members of the original NRL Vanguard team, such as Milton Rosen, Homer Newell, Roger Easton, Martin Votaw, John W. Townsend Jr., Alton Jones, Edmund Habib, John Neilon, Albert Nagy, and Ronald Muller, plus Paul Karpiscak of Glenn L. Martin Company and Bruce Bollerman of Atlantic Research, have greatly filled in the gaps in my memory. Other conversations many years ago with Wernher von Braun, Ernst Stuhlinger, and Konrad Dannenberg of ABMA, James Van Allen of the University of Iowa, and William Pickering of JPL, plus meetings and extensive discussions during the past year with George H. Ludwig and Carl E. McIlwain have done the same for gaps in memory about the Explorer program. I am especially grateful to George H. Ludwig for permission to extract material from his forthcoming book, *Opening Space Research—Dreams, Technology, and Scientific Discovery*. Robert L. Ammons, Alton E. Jones, Roger and Richard Easton, Ronald Muller, George H. Ludwig, and Carl McIlwain proofread early drafts of this chapter and provided much constructive criticism.

Conversations with U.S. Air Force Gen. Bernard Schriever and with Yvonne Brill, one of the participants in the original 1946 Project RAND study, shed light on the U.S. Air Force's early plans for Earth orbiting satellites and the difficult decisions that had to be made concerning missile and satellite program priorities. I particularly treasure discussions with Esther Goddard, Edward Francisco, Lowell Randall, Charles Mansur, and Fred Durant about Robert Goddard's Roswell years. Peter Wilhelm, Leo Slater, Gayle Fullerton, James Tugman, and Marybeth Dowdell of the current staff of NRL have been most helpful in providing documentation and many of the NRL photos reproduced here. A special thanks to Raymond (Bud) Fraze of Design Consulting, Tomas Svitek of Stellar Exploration Inc., Al Whitaker of the U.S. Space and Rocket Center, and Doyle Piland of the White Sands Missile Range archives for reference materials and photos. Thanks also to George H. Ludwig and Carl McIlwain for the use of their photos.

1.12 Bibliography

R. H. Goddard, *A Method of Reaching Extreme Altitudes* (Smithsonian Institution, 1919).

R. H. Goddard, *Liquid Propellant Rocket Development,* (Smithsonian Institution, 1936).

Douglas Aircraft Company, Preliminary Design of an Experimental World-Circling Spaceship (Project RAND Report No. SM-11827, U.S. Army Air Force Contract W33-038, May 2, 1946).

W. von Braun, "A Proposal for Project Slug, A Minimum Satellite Vehicle, Based on Components Available from Missile Developments of the Army Ordnance Corps" (Guided Missile Development Division, Ordnance Missile Laboratories, Redstone Arsenal, Huntsville, AL, 15 Sept. 1954).

American Rocket Society Space Flight Technical Committee, chaired by M. W. Rosen, "On the Utility of an Artificial Unmanned Earth Satellite," *Jet Propulsion,* pp. 71–78 (The American Rocket Society, Feb. 1955).

Naval Research Laboratory, A Scientific Satellite Program—Memorandum Report No. 487 (Rocket Development Branch, Rocket Sonde Branch and Electron Optics Branch, Optics Division, 5 July 1955).

M. W. Rosen, *The Viking Rocket Story* (Harper & Brothers, New York, 1955).

R. L. Easton and M. J. Votaw, "Vanguard I IGY Satellite 1958 Beta," *The Review of Scientific Instruments,* **30** (2) (February 1959).

B. Klawans and J. Baughardt, "The Vanguard Satellite Launching Vehicle, an Engineering Summary," Engineering Report No.11022, Contract Nonr-1817 (The Martin Company, Baltimore, MD, April 1960).

J. T. Shea and R. T. Bauman, "Vanguard I Satellite Structure and Separation Mechanism" (NASA Technical Note D-495, Goddard Space Flight Center, March 1961).

J. T. Wilson, *IGY, The Year of the New Moons* (Longmans, Green and Co., Toronto, 1961).

M. Lehman, *Robert H. Goddard, Pioneer of Space Research* (Da Capo Press, Inc., New York, 1963).

C. McLaughlin Green and M. Lomask, *Vanguard—A History* (NASA Special Publication 4202, Washington, DC, 1970).

F. C. Durant III, *Robert H. Goddard, Accomplishments of the Roswell Years, 1930–1941* (National Air and Space Museum, Smithsonian Institution, 1973).

E. Bergaust, *Wernher von Braun* (National Space Institute, Washington, DC, 1976).

H. E. Newell, *Beyond the Atmosphere, Early Years of Space Science—The NASA History Series* (Scientific and Technical Information Branch, National Aeronautics and Space Administration, Washington, DC, 1980).

J. A. Van Allen, *Origins of Magnetospheric Physics* (Smithsonian Institution Press, Washington, DC, 1983).

D. H. Devorkin, *Science With a Vengeance, How the Military Created the U.S. Space Sciences After World War II* (Springer Verlag, New York, Berlin, Heidelberg, 1992).

Russian Scientific Research Center, *Roads to Space, An Oral History of the Soviet Space Program* (Aviation Week Group, A Division of the McGraw-Hill Companies, 1995).

J. Harford, *Korolev—How One Man Masterminded the Soviet Drive to Beat America to the Moon* (John Wiley and Sons, Inc., 1997).

C. E. McIlwain, *Music and the Magnetosphere* ("Discovery of the Magnetosphere" in *History of Geophysics*, Vol. 7, American Geophysical Union, 1997).

A. A. Siddiqi, *Sputnik and the Soviet Space Challenge* (University Press of Florida, Gainesville, FL, 2000).

P. Dickson, *Sputnik: The Shock of the Century* (Walker Publishing Company, New York, 2001).

D. A. Clary, *Rocket Man, Robert H. Goddard and the Birth of the Space Age* (Hyperion, New York, 2003).

M. Bille and E. Lishock, *The First Space Race, Launching the World's First Satellites* (Texas A&M University Press, College Station, TX, 2004).

G. H. Ludwig, *The First Explorer Satellites* (prepared for James Van Allen's 90th birthday celebration at the University of Iowa, 9 Oct. 2004).

R. Easton and M. Votaw, *Vanguard I, Proposal, Installation, Launch and Selected Results* (privately published, 2006).

Richard Easton, "TIMATION and the GPS: 1964–1973," *Quest, the History of Spaceflight Quarterly*, **14** (3), pp 12–18 (2007).

A. Foerstner, *James Van Allen, the First Eight Billion Miles* (University of Iowa Press, Iowa City, 2007).

M. J. Neufeld, *Von Braun, Dreamer of Space, Engineer of War* (Smithsonian Institution, Alfred A. Knopf, New York, in association with the National Air and Space Museum, Smithsonian Institution, 2007).

W. P. McCray, *Keep Watching the Skies! The Story of Operation Moonwatch and the Dawn of the Space Age* (Princeton University Press 2008).

Chapter 2

The History of Small Satellites

Siegfried W. Janson*

2.1. Introduction

Small satellites were first built more than 50 years ago but were not classified according to size until 1992, when spacecraft with mass between 10 to 100 kg were identified as "microsatellites" and those with mass less than 10 kg as "nanosatellites." These small satellites may be active or passive, but are intentionally designed and fabricated to orbit as individual entities. The terms microsatellites and nanosatellites originated at the Centre for Satellite Engineering Research at the University of Surrey in England.[1] In 1995, I proposed changing the lexicon to redefine the microsatellite, nanosatellite, and "picosatellite" as a satellite with a mass of 1–1000 kg, 1–1000 g, and 1–1000 mg, respectively, to conform with the International System of Units (SI) prefixes.[2] Although these definitions were never fully adopted, they did stimulate thinking on subkilogram-mass spacecraft.

Current small-satellite definitions are listed in Table 2.1, where each SI prefix such as "micro," "nano," and "pico" spans only a single order of magnitude in mass. This classification scheme can readily be extended to include attosatellites (1–10 g) through xennosatellites (1–10 mg). So far, only passive spacecraft have populated these ultrasmall satellite classes.

At the start of the Space Age (circa 1957), computers occupied an entire room, and personal wireless communication systems such as "walkie-talkies" could fit in automobile glove boxes, not shirt or pants pockets. Nevertheless, microsatellites and even nanosatellites performed useful functions. With advancements in solar cells, batteries, micro/nanoelectronics, and miniaturized sensors, picosatellites today can outperform the microsatellites of the 1960s.

Table 2.1. Small Satellite Classification by Mass

Spacecraft Class	Mass Range
Microsatellite	10–100 kg
Nanosatellite	1–10 kg
Picosatellite	0.1–1 kg
Femtosatellite	0.01–0.1 kg

*The Aerospace Corporation
El Segundo, California

2.2 Micro/Nano/Picosatellites

2.2.1 Launch Record

More than 860 microsatellites, 680 nanosatellites, and 38 picosatellites have been launched over the last 50 years. Launch rates for microsatellites have been significantly higher than those of nanosatellites or picosatellites and provide statistically significant data. Figure 2.1 shows the annual worldwide launch rates since 1955 of microsatellites, nanosatellites, and picosatellites, with and without the prolific Soviet Strela series or the passive Soviet radar calibration Romb satellites. The 384 Strela-1 and Strela-1M microsatellites, to be discussed later, were part of a sizable Soviet communications satellite constellation that skew the microsatellite launch history. Removing Strela microsatellites from the yearly totals yields a better indication of general, worldwide microsatellite growth and decline over the last 50 years. Microsatellites experienced a well-defined peak of about 30 launches per year in the mid-1960s, a decline to very low launch rates between 1977 and 1987, and a resurgence to about 9 per year starting in the late 1980s.

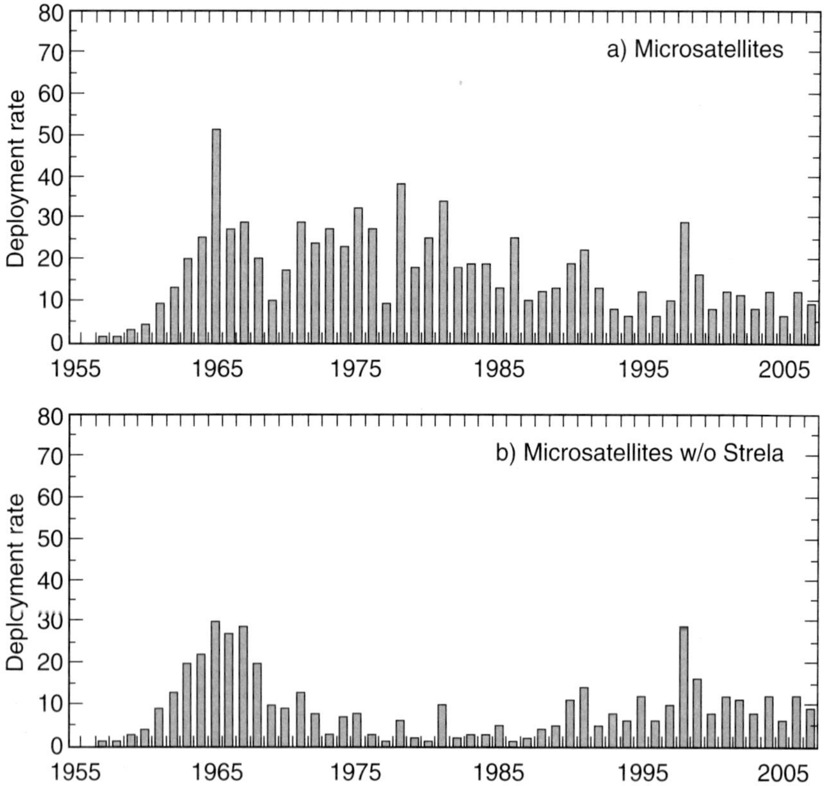

Fig. 2.1. Yearly launch rates of microsatellites (a) and microsatellites minus the Strela series (b) from 1955 to 2007.[3–5]

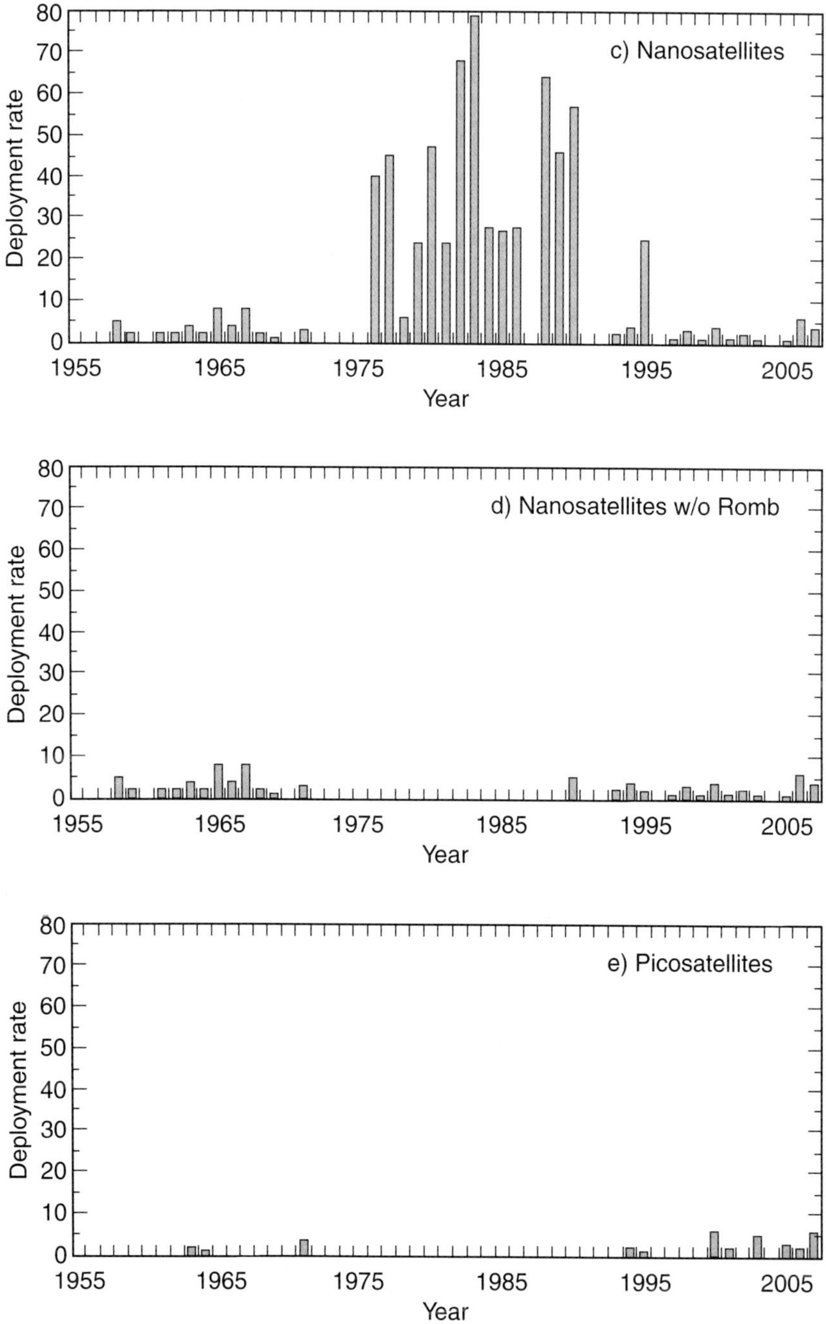

Fig. 2.1 (cont.). Yearly launch rates of nanosatellites (c), nanosatellites minus the Romb series (d), and picosatellites (e) from 1955 to 2007.[3–5]

The first spacecraft launched into orbit was a microsatellite. As expertise in space technologies increased, satellites would generally increase in mass and leave the microsatellite class. Figure 2.2 shows the mass of NASA's Explorer series spacecraft from 1958 through 2005. The first Explorers were microsatellites, but their mass gradually increased until 1975, when they stayed almost exclusively above the microsatellite mass range. Explorers 19, 24, and 39—the only nanosatellites in this series—were launched during the 1960s and were 3.6 m diameter balloons with a solar-cell-powered 136 MHz tracking beacon used to determine atmospheric density as a function of altitude. Explorer 81, also known as "CHIPSat" (Cosmic Hot Interstellar Plasma Spectrometer Satellite) has been the only microsatellite of this series for the past three decades.[6]

As shown in Fig. 2.1c, nanosatellite launch rates were high from 1976 to 1995, but the majority of these spacecraft (601) were Soviet Romb series passive radar calibration satellites carried onboard larger spacecraft that deployed them over time. Subtracting these from the worldwide nanosatellite launch rates—Fig. 2.1d—reveals a pattern similar to that shown in Fig. 2.1b, but with about 5 times fewer launches per year.

A few picosatellites were launched between 1963 and 1971 and none between 1972 and 1994, but a significant number have been launched since 2000—Fig. 2.1e. The general trend of increasing satellite mass with time appears to be reversing in the 21st century. Over the last five years, the average yearly launch rates of microsatellites, nanosatellites, and picosatellites were 9.4, 2.4, and 3.2, respectively. Note that the picosatellite launch rate currently exceeds the nanosatellite launch rate. The rise, fall, and recovery of small satellite launch rates were caused by political, technological, and economic factors. These factors will be covered in the next sections.

2.2.2 The First Two Decades: 1957–1977

Figure 2.3 shows mass and launch date for the first 30 successfully launched spacecraft. Soviet spacecraft are identified by triangles, U.S. spacecraft by circles. The first satellites launched by the USSR and the United States were microsatellites: Sputnik I had a mass of 83.6 kg; Explorer I had a mass of 14.0 kg. Had the Vanguard launch vehicle carrying Test Vehicle 3 (TV-3) been successful on 6 December 1957, the United States would have entered the space race with a nanosatellite. The 152 mm diameter, 1.7 kg spherical Vanguard 1 was the first nanosatellite from any country to reach orbit, the first satellite with solar cells, and the oldest human-made object still in space. The other two nanospacecraft from the first 2½ years of space exploration were Pioneer-3 and Pioneer-4. Both were designed for lunar flyby missions, but Pioneer-3 had an upper stage malfunction and failed to reach the moon.

Note that the USSR started with a higher mass-to-orbit capability. This is a result of the Eisenhower administration's desire to have civilian management of America's first satellite program to establish an "open skies" policy for space. The relatively small Vanguard launch vehicle (10,250 kg launch weight, 9 kg

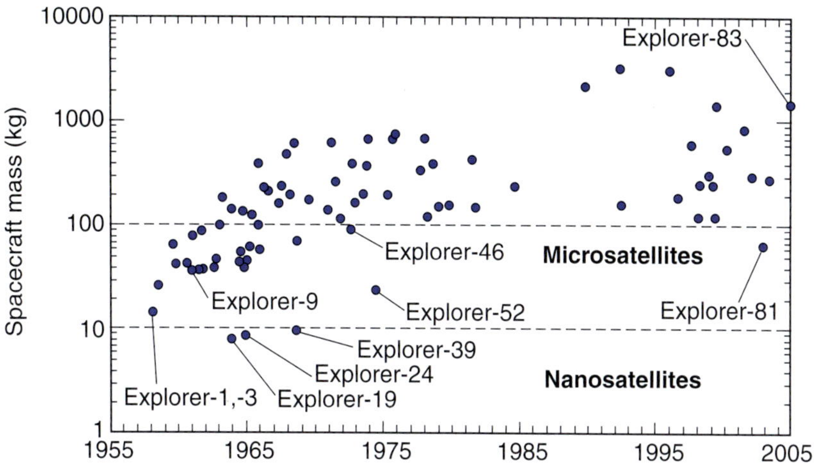

Fig. 2.2. Launch mass vs. launch year for the Explorer series of satellites.[7]

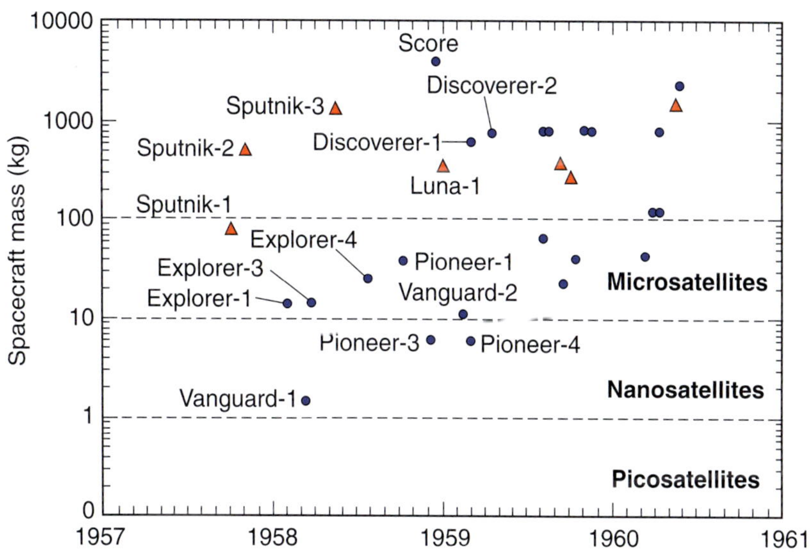

Fig. 2.3. Spacecraft mass vs. launch year for the first 30 spacecraft with successful launches.[3]

to orbit)[8], used the Viking research rocket as the first stage rather than one of the larger intercontinental ballistic missiles (ICBMs) that were developed during the Cold War (see chapter 1). The Vanguard decision resulted in a longer vehicle development schedule that vaporized when the Soviets launched their first satellite much earlier than anyone in the West expected. Vanguard, and the hastily-assembled Jupiter C/Juno I (which launched Explorer I), were no match for the larger

Soviet launch vehicles, although the United States caught up to Soviet launch capability within two years.

As the race for space progressed, larger launch vehicles with larger payload capabilities were developed, and ever-larger spacecraft were designed to ride on those vehicles. While U.S. spacecraft were initially low-mass microsatellites, average spacecraft mass continued to grow over time.

A large number of microsatellites and nanosatellites, and a few picosatellites, were launched throughout the 1960s to obtain space environment data, flight test various technologies, and provide operational communications. The U.S. Naval Research Laboratory, which had developed the Vanguard series of spacecraft, developed and launched 10 SOLRAD spacecraft between 1960 and 1971. These spacecraft were officially designed to monitor ultraviolet and x-ray radiation from the sun with operational lifetimes ranging from 2 to 84 months. Figure 2.4 plots the mass of each successful SOLRAD spacecraft as a function of launch date. Once again, there is a clear trend toward higher mass as launch-vehicle throw weight improved and the spacecraft designers became more familiar with the space environment and space systems technology.

Multiple satellite and "piggyback" launches, the current method of getting microsatellites and smaller satellites into orbit, can be traced to the early years of space exploration. SOLRAD-1 was part of the first tandem satellite launch with the Transit-2A navigational satellite in 1960. In 1963, SOLRAD-6 was launched with five other spacecraft as part of the first six-satellite launch.

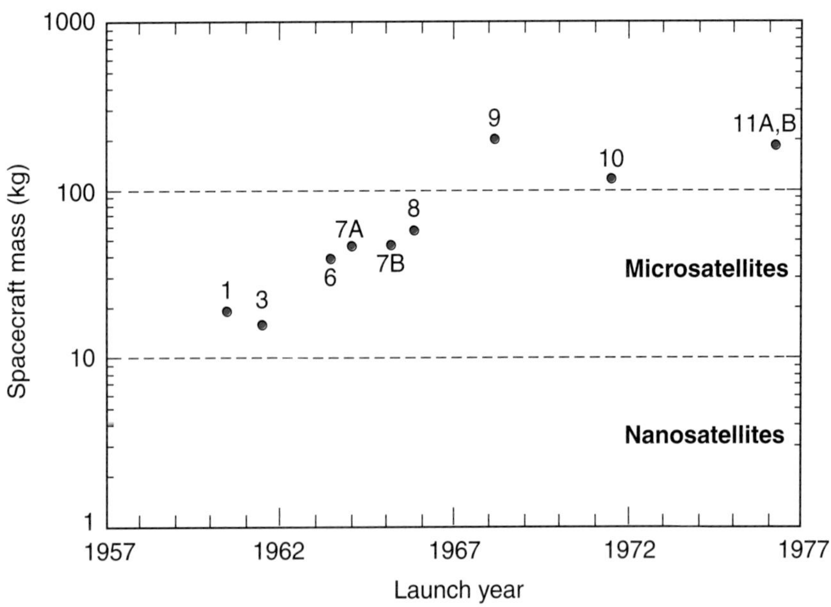

Fig. 2.4. The SOLRAD series spacecraft. Numbers indicate designation of successfully launched spacecraft.[9]

Communication satellites also started as experimental microsatellites. The first active communication satellite was the experimental 45 kg SCORE (Signal Communications by Orbiting Relay Experiment) microsatellite developed by DARPA (Defense Advanced Research Projects Agency) and launched on 18 December 1958. Interestingly, the first passive communication satellites followed after the first active communication satellite, not before. Passive satellites could have extremely high bandwidths and could be much more reliable than active satellites due to their electronic simplicity. ECHO-1A, launched on 12 August 1960, was a 100 ft (30.5 m) diameter spherical aluminized Mylar balloon with a total mass of 75.5 kg.[10] The Mylar was 13 μm thick, and the balloon was inflated using 13.6 kg of subliming solid powder. The ECHO-1A tests and demonstrations used 960 MHz and 2390 MHz frequencies; ECHO-1A provided the first two-way telephone conversation via satellite and the first satellite relay of a signal across the Atlantic Ocean. ECHO-1A fit inside a 26 in. diameter canister for launch, yet proved that a microsatellite could be physically large. Figure 2.5 (left) shows an inflated ECHO-1 balloon inside a hanger in Weeksville, North Carolina.

Passive satellites did not have to be large spherical reflectors. Project West Ford, launched on 9 May 1963, used 480 million copper dipoles, each 1.78 cm long by 17.8 μm in diameter, as a distributed reflector for communications at approximately 8 GHz.[11] The individual dipoles, shown in Fig. 2.5 (right), had a mass of 41 μg, which is too small to be classified using our current small-satellite naming convention (we run out of SI prefixes just below 1 mg). After several months on orbit, the constellation formed a toroidal-like volume with a circumference of 63,000 km (3650 km altitude), a mean width of 15 km, and a mean radial depth of ~30 km.[12] Between 16% and 39% of the dipoles were orbiting as individual, unconnected units.[13] West Ford used the smallest passive satellites ever made to create the first distributed spacecraft. Due to the high ratio of surface area to

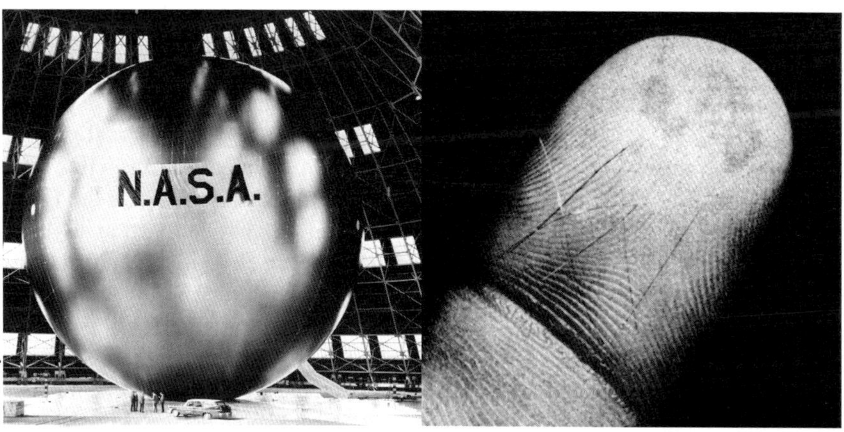

Courtesy of NASA

Fig. 2.5. Big and small passive satellites from the early 1960s: ECHO-1 (left) and West Ford (right).

mass for these thin-wire dipoles, the solar pressure-induced accelerations of up to 6×10^{-5} m/sec^2 were orders of magnitude larger than what typical spacecraft experienced. Solar pressure forces typically increase orbit eccentricity, thus raising orbit apogee and lowering perigee. The orbital parameters for the West Ford dipoles were carefully chosen to provide a resonance condition where the perigee would drop almost linearly with time, until the dipoles encountered enough atmospheric drag to deorbit. The expected orbital lifetime was initially estimated at seven years, but subsequently corrected to three years.[14,15] All individual needles decayed within five years, but approximately 60% of the original approximate 100 clumps (clusters of dipoles containing between 71–84% of the original 480 million) had not reentered by the year 2000.[16]

Active communication satellites proved superior to passive satellites because of vastly increased communications link budgets; active satellites can boost returned signal power levels by more than 12 orders of magnitude compared with a passive satellite of equal size. Both the United States and USSR recognized the advantages of active satellites and embraced their use for operational communications starting in the mid-1960s. The U.S. Department of Defense (DOD) launched 26 Initial Defense Communication Satellites between June 1966 and June 1968. These 45 kg, spin-stabilized (at 2.7 rps!), approximately 1 m diameter polyhedral satellites were put into slightly subsynchronous orbits with an altitude range of 32,400–34,000 km.[17] They provided five voice channels, 11 "tactical" voice channels, or 1550 "TTY" (teletype) channels between large (60 ft diameter) ground stations.[18]

The USSR used microsatellite-based communications in a different way. Satellites in low Earth orbit (LEO: 200–1500 km altitude) are much closer to ground users than satellites in geosynchronous Earth orbit (GEO). LEO communication satellites have shorter time delays when communicating with ground users and require less antenna gain and/or transmitter power (the communications link budgets are at least 100 times better for LEO satellites than GEO satellites). Orbcomm and Iridium are current examples of LEO communication satellite systems that use the very favorable link budget to enable portable, mobile satellite communication systems that can use simple, low-power ground terminals. While three GEO satellites can cover the entire globe for continuous coverage of all ground users (except those at the North and South Poles), individual LEO satellites are visible to less than 10% of Earth at any given time. Figure 2.6 shows the Earth coverage for a satellite as a function of altitude and minimum elevation angle as seen by a ground-based observer. While 0 deg minimum elevation has the highest percent coverage and would require the fewest satellites for any given altitude, ground topography limits practical minimum elevations to the 10–30 deg range for land-based users. Maritime users, however, can approach the 0 deg elevation limit. Thirty to several hundred LEO satellites are typically required for complete, continuous global coverage, depending on altitude and application. LEO communication systems with continuous or near-continuous coverage requirements are

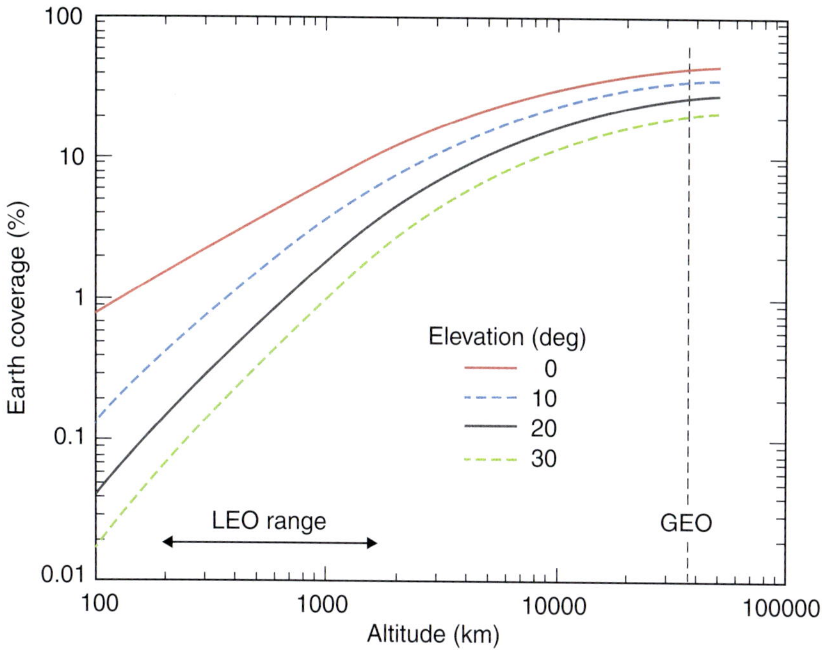

Fig. 2.6. Percent Earth coverage for a satellite as a function of altitude and minimum elevation angle.

therefore typically driven to smaller spacecraft to keep total system costs reasonable. GEO communication satellites are typically large spacecraft because they require larger, high-gain antennas, associated attitude and orbit control systems, and a high degree of system and subsystem redundancy to enhance reliability. If one GEO satellite fails, one third to one half of Earth loses coverage. If a LEO satellite fails, communications for any particular user is lost for perhaps 15 minutes, two to three times per day. LEO communications systems have graceful degradation modes.

Many decades ago, the USSR chose to design, build, launch, and operate a LEO constellation for mobile communications. Strela (Russian for "arrow") spacecraft were designed to provide medium-range real-time communications and store-and-forward communications for the entire planet.[19] Between August 1964 and September 1965, the USSR launched 26 of the 50 kg experimental Strela-1 satellites, placed two, three, and five at a time into various orbits at inclinations of 48.9 or 61.1 deg. Starting in 1970, the 61 kg mass operational Strela-1M satellites were launched on Kosmos-3M boosters, eight at a time, into approximately 1500 km altitude orbits at 74 deg inclination about twice a year. They had an expected operational lifetime of about two years (in comparison, GEO communication satellites now have 7–15 year expected lifetimes) and had to be replaced on that same time scale; 360 Strela-1M spacecraft were launched between 1970 and

1993. At an approximate 1500 km altitude, each satellite had a ground footprint covering about 9.5% of Earth's surface. About 30 operational satellites, at any given time, gave fairly complete Earth coverage using randomly distributed satellites within individual orbital planes. The Strela-1M satellites were later replaced by heavier Strela-2M and Strela-3M satellites. Today, the Strela constellation is an orbital debris hazard due to the large number of spacecraft and boosters within a small altitude range.

Like their military cousins, civil and commercial communication satellites started out as small satellites, as shown in Table 2.2. They were spin-stabilized and had power levels ranging from 15 W (Telstar) to 85 W (Intelsat II). AT&T Inc. designed and fabricated the Telstar-1 microsatellite at Bell Laboratories.[20], Telstar-1 began transmitting live television across the Atlantic Ocean on 10 July 1962 and became the most well-known satellite of this group. Note that Telstar-1 was not geosynchronous; the first geosynchronous satellite was Syncom-2, and the first geostationary (geosynchronous with a 0 deg inclination equatorial orbit) communication satellite was Syncom-3 in 1964. Even large geosynchronous communication satellites can trace their heritage to microsatellites.

Table 2.2. The First Civil and Commercial Communications Microsatellites[3,20]

Name	Launch Date	Mass (kg)	Orbit	Organization
Telstar-1	10 July 1962	77.3	935 × 5,553 km, 45 deg inc.	AT&T
Relay-1	13 Dec 1962	78.2	1,566 × 7,302 km, 47.5 deg inc.	NASA
Telstar-2	7 May 1963	79.5	956 × 10,611 km, 45 deg inc.	AT&T
Syncom-2	26 July 1963	35.5	Geosynchronous, 33 deg inc.	NASA
Relay-2	21 Jan 1964	78.2	2,057 × 7,280 km, 46 deg inc.	NASA
Syncom-3	19 Aug 1964	35.5	Geosynchronous, 0.1 deg inc.	NASA
Intelsat-I	6 April 1965	38.6	Geosynchronous, 0.1 deg inc.	ITSO
Intelsat-IIA	26 Oct 1966	87.3	3,161 × 37,244 km, 17.9 deg inc.	ITSO
Intelsat-IIB	11 Jan 1967	87.3	Geosynchronous, 6.6 deg inc.	ITSO
Intelsat-IIC	7 April 1967	87.3	Geosynchronous, 7.3 deg inc.	ITSO
Intelsat-IID	27 Sep 1967	87.3	Geosynchronous, 0.9 deg inc.	ITSO

Between 1965 and 1972, space research became multinational as France, Australia, Japan, China, and the United Kingdom developed their own space launch capabilities. Unable to match the economic and technological resources expended by the two superpowers in their quest for space dominance, the rest of the world followed a few steps behind. Canada, the UK, Italy, and West Germany took advantage of their close political ties with the United States to get their first spacecraft launched on U.S. boosters. Microsatellites provided a relatively inexpensive way to test emerging technologies and national engineering capabilities. Table 2.3 lists the initial spacecraft built by the UK, Canada, Italy, France, Australia, West Germany, Japan, and China. Launches provided by the United States were on Delta, Thor-Agena B, or Scout rockets. Note that most of the spacecraft in Table 2.3 are microsatellites.

By 1977, the Netherlands (Astronomical Netherlands Satellite, 130 kg), Spain (Intasat-1, 20 kg), India (Aryabhata, 360 kg), and Indonesia (Palapa-1, 300 kg) joined the growing community of nations with satellites in orbit.[22] Aryabhata was launched by the USSR, while the others were launched by the United States. Historically, microsatellites were the first payloads for most nations testing their first domestically developed launch systems. The first Indian satellite launched by an Indian-built booster, for example, was the 35 kg Rohini-1 microsatellite launched on 18 July 1980 by the Satellite Launch Vehicle.

While microsatellites performed a wide variety of scientific, civilian, and military missions during the 1960s and 1970s, nanosatellites and a few orbiting picosatellites performed simpler scientific and technology demonstration missions such as radar calibration, measuring atmospheric density by monitoring the orbital decay of various spheres, characterizing on-orbit radiation, and measuring solar cell degradation as a function of time and orbit. The U.S. Air Force Environmental Research Satellite (ERS) series contained a significant portion of the nanosatellites and picosatellites from this era. Table 2.4 lists the successfully launched ERS spacecraft, which were used as technology test beds for spacecraft systems and components at medium Earth altitudes (greater than 1500 km apogee). Due to their high apogees, ERS spacecraft were in sunlight most of the time, thus eliminating the need for batteries and battery charge regulators. The lightest spacecraft in this series (ERS-5 and -6) were Tetrahedral Research Satellites (TRS). These picosatellites were four-sided tetrahedrons with 13 cm side lengths.

Figure 2.7 shows the larger (15 cm side length), 2 kg ERS-11 satellite on display at the Smithsonian's Udvar-Hazy Air and Space Museum in Chantilly, Virginia. It is identical to the ERS-12 satellite flown in 1963. Note the large number of solar cells on each face, the coiled "carpenter tape" antennas (bottom right and left) that form a dipole when released, and the internal wiring harness.

Table 2.4 shows yet another example of ubiquitous satellite mass growth during the first decade of the Space Age. The lightweight TRS series evolved into the heavier Octahedral Research Satellite (ORS) series with eight triangular sides to provide more internal volume for electronics and experiments. ERS-27,

Table 2.3. Initial Non-U.S. and Non-Soviet National Satellites[3,21]

Name	Country	Launch Date	Mass (kg)	Orbit	Launcher
Ariel-1	UK	26-Apr-1962	60	389 × 1,214 km, 53.9 deg inc.	Delta
Alouette-1	Canada	29-Sep-1962	145	991 × 1,025 km, 80.5 deg inc.	Thor-Agena B
Ariel-2	UK	27-Mar-1964	68	180 × 843 km, 51.6 deg inc.	Scout
San Marco 1	Italy	15-Dec-1964	115	198 × 846 km, 37.8 deg inc.	Scout
Asterix-1	France	26-Nov-1965	42	530 × 1,736 km, 34.3 deg inc.	Diamant A
Alouette-2	Canada	29-Nov-1965	145	503 × 2,800 km, 79.8 deg inc.	Thor-Agena B
FR 1	France	6-Dec-1965	60	717 × 728 km, 75.9 deg inc.	Scout
D1A	France	17-Feb-1966	18.5	503 × 2,576 km, 34.1 deg inc.	Diamant A
Diademe-1	France	8-Feb-1967	23	557 × 1,198 km, 34.1 deg inc.	Diamant A
Diademe-2	France	15-Feb-1967	23	589 × 1,790 km, 39.4inc.	Diamant A
San Marco 2	Italy	26-Apr-1967	129	135 × 498 km, 2.9 deg inc.	Scout
Ariel-3	UK	5-May-1967	90	497 × 608 km, 80.2 deg inc.	Scout
Wresat-1	Australia	29-Nov-1967	45	193 × 1,259 km, 83.2 deg inc.	Sparta
ISIS-1	Canada	30-Jan-1969	241	575 × 3,494 km, 88.4 deg inc.	TAID
Azur-1	W. Germany	8-Nov-1969	71	379 × 2,550 km, 102.8 deg inc.	Scout
Skynet-1	UK	22-Nov-1969	243	Geosynchronous 2.4 deg inc.	Delta
OSCAR-5	Australia	23-Jan-1970	18	1,432 × 1,476 km, 101.8 deg inc.	TAT-Delta M
Osumi	Japan	11-Feb-1970	12	333 × 3,589 km, 31.1 deg inc.	Lambda 4S-5
Dial-1-WIKA	W. Germany	10-Mar-1970	63	301 × 1,631 km, 5.4 deg inc.	Diamant B
MIKA	France	10-Mar-1970	52	313 × 1,665 km, 5.4 deg inc.	Diamant B
DFH-1	PRC	24-Apr-1970	173	436 × 2,256 km, 68.4 deg inc.	CZ-1
Skynet-1B	UK	19-Aug-1970	243	270 × 36,041 km, 28 deg inc.	TAT-Delta

Table 2.3 (cont.). Initial Non-U.S. and Non-Soviet National Satellites[3,21]

Name	Country	Launch Date	Mass (kg)	Orbit	Launcher
PEOLE-1	France	12-Dec-1970	70	517 × 747 km, 15.0 deg inc.	Diamant B
Tansei	Japan	16-Feb-1971	62	987 × 1,108 km, 29.7 deg inc.	Mu-4S
SJ-1	PRC	3-Mar-1971	221	268 × 1,830 km, 69.9 deg inc.	CZ-1
ISIS-2	Canada	1-Apr-1971	264	1,354 × 1,423 km, 88.2 deg inc.	Thor-Delta
Tournesol-1	France	15-Apr-1971	96	457 × 697 km, 46.3 deg inc.	Diamant B
San Marco 3	Italy	24-Apr-1971	164	222 × 718 km, 3.2 deg inc.	Scout
Eole-1	France	16-Aug-1971	84	664 × 870 km, 50.1 deg inc.	Scout
Shinsei	Japan	28-Sep-1971	65	872 × 1,870 km, 32.1 deg inc.	Mu 4S
Prospero	U.K.	28-Oct-1971	66	537 × 1,482 km, 82.1 deg inc.	Black Arrow

Table 2.4. Successfully Launched Nano/Picosatellites from the ERS Series[3,23,24]

Name	Launch Date	Mass (kg)	Orbit	Other Names
ERS-5	9 May 1963	0.7	2,927 × 4,317 km, 87.6 deg inc.	TRS-2
ERS-6	9 May 1963	0.7	3,607 × 3,675 km, 87.3 deg inc.	TRS-3
ERS-9	19 July 1963	1.5	3,680 × 3,717 km, 88.4 deg inc.	TRS-4
ERS-12	17 Oct 1963	2.0	220 × 103,500 km, 36.8 deg inc.	TRS-5
ERS-15	19 Aug 1966	5.0	3,681 × 3,699 km, 89.8 deg inc.	ORS-1
ERS-17	20 Jul 1965	5.4	153 × 112,694 km, 34.4 deg inc.	ORS-3
ERS-18	28 Apr 1967	9	8,400 × 111,100 km, 32.8 deg inc.	ORS-4
ERS-20	28 Apr 1967	9.1	8,600 × 111,200 km, 32.8 deg inc.	OV5-3
ERS-27	28 Apr 1967	6.2	8,400 × 111,100 km, 32.8 deg inc.	OV5-1
ERS-28	26 Sep 1968	10	184 × 35,787 km, 26.4 deg inc.	OV5-2

for example, had a regular octahedral shape with a side length of 29 cm, plus two antennas that formed a half-wave dipole. Figure 2.8 shows this satellite with the triangular solar panels removed. Each panel contained 102 solar cells of 1 × 2 cm area with an efficiency of 9%; each panel could provide 2.55 W at normal incidence, resulting in an available power of 4.1–5.9 W, depending on orientation. This spacecraft was initially spin-stabilized at 12 rpm, but no active or passive attitude control was included in the design.

ERS-27 was designed to monitor solar x rays from 2–14 Å, solar bremsstrahlung from 33 to 90 keV, solar flare and trapped protons from 1 to 80 MeV, and trapped electrons from 0.04 to 3.3 MeV. Eleven detectors, ranging from Geiger-Mueller tubes to photomultipliers coupled to scintillators, produced pulse outputs that were conditioned and fed into logarithmic rate meters. The resulting analog outputs from the rate meters, which ranged from 5.05 to 5.8 V, were sequentially selected for a period of 4.4 sec each using a 16-channel solid-state commutator. One of these channels was further commutated into 8 channels, resulting in a total of 23 channels of telemetry. The selected output voltage controlled the frequency of a subcarrier oscillator between 1252 and 1400 Hz. Detector outputs, reference voltages, solar array voltages, spacecraft temperature, and solar aspect sensor outputs for sun angle determination were all output as a series of tones, which were transmitted at 136.77 MHz at 1 W. Ground stations recorded the tones on 2400 ft reels of audiotape for later analysis. No receiver was included in the spacecraft; it had an electrochemical suicide timer to cease transmissions after 14 months of operation. Since the electronics were too crude to reprogram (e.g., no microprocessors existed), ground control of the satellite was unnecessary. These mostly analog satellites were crude by today's standards, but they provided a wealth of information.

By the start of the 1970s, small satellite launch rates were decreasing. Average satellite mass had increased well beyond the micro/nanosatellite range. While some microsatellites were still being launched during the 1970s, the only U.S. nanosatellites to reach orbit were three radar calibration balloons launched in August 1971 (Rigid Sphere-1 and Grid Sphere-1 and -2). Picosatellites were limited to three 26 cm diameter, 730 g hollow aluminum spheres (Calsphere-3, -4, and -5) launched in February 1971 to study the effect of surface composition on atmospheric drag, and a 2.13 m diameter, 0.8 kg Mylar balloon that was deployed in August 1971.[25,26] These passive spacecraft would be the last U.S. kilogram-class spacecraft for the next 18 years.

The USSR, however, started using large numbers of small passive satellites for radar calibration starting in 1976. Between 1976 and 1995, more than 600 Romb subsatellites were deployed by the larger Taifun-2 and -3 spacecraft.[27] Cosmos 965, launched on 8 December 1977, was a Taifun-2 satellite with a total mass of 550 kg that deployed 25 radar calibration satellites. Although specific data on the deployed Romb subsatellites are hard to obtain, they were most likely nanosatellites, based on the initial Taifun-2 spacecraft mass and number deployed. The Romb satellites

Courtesy of the Smithsonian Institution

Fig. 2.7. Photograph of a prototype ERS-11 nanosatellite with one side panel removed.

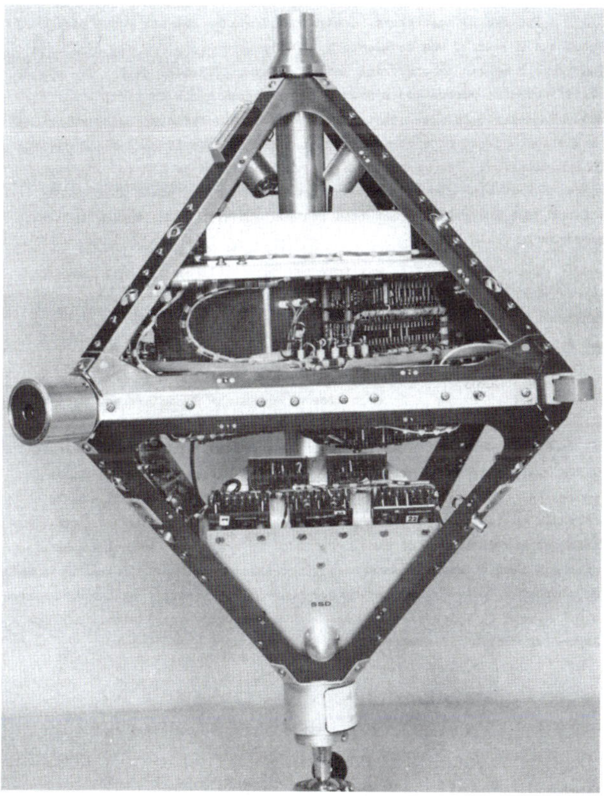

Courtesy of U.S. Air Force

Fig. 2.8. Photograph of ERS-27 with the body-mounted solar panels removed.

are not included in Fig. 2.1d because of the uncertainty in mass and the high average launch rate of these spacecraft (about 32 a year over 19 years).

2.2.3 The Small Satellite Doldrums: 1978–1987

While a wide variety of microsatellites and nanosatellites were launched in the 1960s and early 1970s, launch rates dropped off markedly starting in the mid-1970s as most nations focused on heavier spacecraft. The USSR continued launching hundreds of Strela-1M spacecraft, but the West focused on fewer, larger, heavier satellites. Another factor was the drop in civil space activity by the United States once the Apollo moon landing program was finished in 1972. NASA's funding, in inflation-adjusted 1996 U.S. dollars, hit a peak of approximately $24 billion in 1966, followed by declining funding until it bottomed out at approximately $8 billion between 1974 and 1980, with a very gradual rise to approximately $11 billion in 1987.[28] A plot of NASA's annual budget authority[28] shows many of the ups and downs in the microsatellite launch rate data presented in Fig. 2.1b.

Expensive, multiyear projects became harder to start in the 1980s; interest rates were high, and a larger fraction of the U.S. federal budget was allocated to servicing the national debt.[29] Large interplanetary or military spacecraft, with their 5–10 year development times and billion-dollar price tags, were more difficult to fund. Existing large satellite programs, however, were still consuming a large fraction of the available resources, thus stifling the smaller programs. Note the dearth of Explorer missions shown in Fig. 2.2 between 1980 and 1995.

Fortunately, a confluence of complementary economic, political, and technological factors came together in the 1980s favoring small satellite development in the United States. DARPA was now investigating small satellites or "lightsats" as low-cost units that could complement the capabilities of much larger military spacecraft. Space was becoming more crowded, and the United States needed a tactical, cost-effective, rapid-response satellite capability in the future. DARPA developed and launched the 68 kg Global Low Earth Orbit Message Relay (GLOMR) satellite in 1985.[30] With the successful launch of GLOMR, DARPA proved that sub-100 kg spacecraft could perform useful functions for the DOD. These "microsats" were much more advanced than their 1960s predecessors; they had microprocessor-controlled digital communications payloads.

Interestingly, the "small satellite doldrums" did not seem to affect experimental and educational spacecraft launched by the Amateur Satellite Corporation (AMSAT; see chapter 4). By 1978, AMSAT had launched seven educational and amateur radio satellites (OSCAR 1 to OSCAR 7) with masses ranging from 4.5 to 29 kg. By 1987, they launched five more, with masses ranging from 50 to 91 kg. Two of these spacecraft, OSCAR-9 and –11 (also called UoSAT-1 and -2), were built by the Centre for Satellite Engineering Research at the University of Surrey. These box-shaped spacecraft used several different microprocessors to perform, among other functions, command and control, digital communications, and power-system control. The term microsatellite seemed appropriate for these microelectronics-enabled small spacecraft, and this terminology finally entered the lexicon years later in 1990 to describe sub-100 kg spacecraft.

During the doldrum years of 1977–1987, no nanosatellites (excluding the Romb series) or picosatellites were launched. Strela microsatellites were launched at an average rate of 17.5 per year over this time period, but other microsatellites from around the world had an average launch rate of only 3.3 per year. Strela satellites were the primary payload for their launch vehicles, but almost all other microsatellites were now being launched as secondary payloads. Getting a launch for a microsatellite as a secondary payload was difficult enough, but getting a launch for a secondary nano/picosatellite with limited capabilities and a limited budget was even harder. Fortunately, that was about to change.

2.2.4 The Second Wave: 1988–1999

In July 1988, Ross M. Jones of NASA's Jet Propulsion Laboratory (JPL) hosted the Microspacecraft for Space Science workshop at JPL. The main conclusions of this workshop were that microspacecraft (we would call these nanosatellites today) of 1–10 kg were technically feasible. They would enable missions that require spatially distributed simultaneous measurements or very high velocity changes, but they were not applicable to all space missions.[31] Their small size would make it possible to launch them with cannons or electromagnetic launchers. Jones[32] offers a conceptual design for a 1 kg interplanetary spacecraft with a 3 in. (75 mm) diameter imaging optic that would be "launched" at speeds approaching 10 km/sec from an orbiting railgun (an electromagnetic accelerator). This injection velocity is more than enough to reach Pluto within 30 years; tens of thousands of target bodies like asteroids, planetary moons, and the nearer Kuiper belt objects (like Pluto) would be available for flybys or impacts. Instead of launching large interplanetary spacecraft at rates less than once every few years, tens to hundreds of identical, small, and relatively inexpensive spacecraft could be launched each year. The concept of using micro/nanosatellites for space exploration was rediscovered, but with a bolder vision because of the significant advancements in microtechnologies.

On 21 January 1990, an Ariane-4 put the French SPOT-2 satellite into orbit, along with three ~10 kg AMSAT MicroSats (Pacsat, Dove, and Lusat), a 12 kg modified AMSAT MicroSat called WeberSat, and two University of Surrey microsatellites (UoSAT-OSCAR 14 and UoSAT-OSCAR 15). This was the first flight of the Ariane Structure for Auxiliary Payloads (ASAP) ring that could hold up to six small satellites with a maximum mass of 50 kg each and maximum dimensions of 35 × 35 × 60 cm. The maximum mass that could be put onto the ASAP was 200 kg, and the cost was about $1 million.[33] Microsatellites and nanosatellites now had a standard low-cost launch service that could put six satellites into orbit at a time.

The AMSAT MicroSats were 23 cm cubes covered with solar cells with a maximum output power of 15.7 W.[34] The energy storage system used eight commercial aviation-grade NiCd batteries rated at 6 Ahr. AMSAT MicroSats used

NEC V-40 microprocessors, had up to 10 Mb of solid-state memory, and used a 15 cm long local area network (LAN) to link the five electronics trays together. They were basically personal computers with some "unusual" peripherals (radio modems, digital cameras, magnetometers, optical spectrometers, etc.). Several digital satellites with a mass of less than 10 kg were now flight proven.

DARPA continued with its Lightsat program, launching two 66 kg Multiple Access Communications satellites (MACSATs) in 1990 and seven 22 kg microsats in 1991.[35,36] The microsatellites had microprocessor-controlled digital communications payloads. The U.S. Navy launched the 68 kg Small Experimental Communications Satellite (SECS) in April 1990, and the U.S. Air Force launched the 95 kg Radar Calibration (RADCAL) microsatellite in June 1993. U.S. military microsatellites were finally making a comeback.

While NASA had enjoyed an increasing budget during the late 1980s (it rose 50% in inflation-adjusted dollars between 1987 and 1991), the budget flattened out in 1991 and declined slightly during the rest of that decade. In parallel, space science and interplanetary exploration budgets were being squeezed during the 1990s to support construction of the International Space Station (ISS). In 1992, NASA administrator Daniel S. Goldin introduced the "smaller, faster, cheaper" approach that used advanced technologies, reduced mission objectives, lower-cost launch vehicles, and simpler management techniques. While this concept was somewhat controversial—a number of larger spacecraft suffered spectacular failures—the concept was sound for microsatellites and their smaller siblings. The Explorer-class missions, shown in Fig. 2.2, began to include more 100–300 kg spacecraft.

The 1990s also saw the establishment of large LEO commercial communications constellations such as Iridium and ORBCOMM. The ORBCOMM system required the launch of 34 microsatellites between 1995 and 2000 into different orbital planes. DARPA had Orbital Sciences Corp. develop the revolutionary air-launched Pegasus booster, which was used to launch the seven 22 kg DARPA MicroSats in 1991; ORBCOMM also used this booster to place their spacecraft, six at a time, into orbit. The ORBCOMM system is operational to this day (see chapter 14).

Collapse of the Soviet Union in 1991 had a major impact on small satellite builders around the world. Russia adapted to free-market economics—converted ICBMs became available to the worldwide community as low-cost launch vehicles. TUBSAT N (8.5 kg) and TUBSAT-N1 (3 kg), launched in July 1998 from a Russian submarine, were early beneficiaries of the post–Cold War meltdown.

In the United States, a series of passive nano-, pico-, atto-, and zeptosatellites were ejected from space shuttles in 1994 and 1995 as part of the Orbital Debris Radar Calibration Sphere (ODERACS) experiment. ODERACS 1, flown on STS-60 in February 1994, deployed six spheres: two 15 cm diameter, 5 kg aluminum spheres; two 10 cm diameter, 1.49 kg aluminum spheres; and two 5 cm diameter, 0.53 kg stainless-steel spheres.[37,38] ODERACS 2, flown on STS-63 in February 1995, deployed a 15 cm, 5.00 kg aluminum sphere; a 10 cm, 1.49 kg aluminum sphere; a 5 cm diameter, 0.53 kg stainless-steel sphere; two 13.3 cm long by 0.102 cm diameter dipoles with approximately 1.5 g mass; and one 4.42 cm long by 0.102 cm diameter dipole with approximately 0.5 g mass. While these were inert metal objects, it showed that a number of nanosatellites through zeptosatellites could be deployed from the U.S. space shuttle. It also showed that at least 10 cm diameter objects could be tracked and that even an electronically "dead" nanosatellite could be used to provide atmospheric density measurements by monitoring its altitude as a function of time.[39] The ODERACS satellites included the only picosatellites flown between 1972 and 2000.

2.2.5 Mighty Midgets for the New Millennium: 2000–Present

An Orbital Sciences Minotaur rocket lifted JAWSAT into orbit on 27 January 2000. JAWSAT released the 22 kg Optical Calibration Sphere Experiment (OCSE, a 3.5 m diameter balloon), the 52 kg FalconSAT-1 from the U.S. Air Force Academy, the 5 kg ASUsat-1, and the 25 kg OPAL. OPAL subsequently ejected three picosatellites from Santa Clara University (the 0.2 kg JAK, the 0.5 kg Thelma, and the 0.5 kg Louise), a 0.23 kg amateur radio picosatellite called Stensat, and two 0.3 kg DARPA/The Aerospace Corporation PicoSats (see chapter 5).[40] JAK, Stensat, and the two PicoSats were 1 × 3 × 4 in. (25 × 75 × 100 mm) in size, while Thelma and Louise were 1 × 3 × 8 in. (25 × 75 × 200 mm). Two more DARPA/The Aerospace Corporation PicoSats rode into orbit on the second Minotaur launch on 19 July 2000 inside the 124 kg MightySat II.1 spacecraft built by the Air Force Research Laboratory and ejected in September 2001. More picosatellites were orbited in 2000 than in any previous year (see chapter 20).

The success of OPAL eventually led to the establishment of the CubeSat program (see chapter 5) by Stanford and the California Polytechnic State University that will probably put 100 nano/picosatellites into orbit over the next decade. All picosatellites launched since 2003 (see Fig. 2.1e) have been CubeSats. A CubeSat is basically a 10 cm cubic spacecraft less than 1.0 kg in mass. Establishment of specific CubeSat dimensional, mass, and electrical standards allowed fabrication and flight qualification of picosatellite containment and ejection systems called the P-POD (Poly Picosatellite Orbital Deployer).[41] A P-POD holds three CubeSats, or one double-length CubeSat and a standard CubeSat, or one triple-length CubeSat. An example of a triple-CubeSat ejected by a P-POD was the NASA Genesat-1 in December 2006.[42]

The P-POD minimizes potential interactions with the primary payload(s) on a launch vehicle by physically enclosing the CubeSats and requiring that they be launched in a dormant "off" state. The Aerospace Corporation developed the A-POD (The Aerospace Corporation Picosatellite Orbital Deployer) 4410 with a 10 × 10 × 25 cm internal volume and the A-POD 5510 with a 12.5 × 12.5 × 25 cm internal volume that are flight qualified for use on the space shuttle. Other POD variants include the University of Tokyo T-POD (Tokyo Picosatellite

Orbital Deployer), which ejects a single CubeSat, and the University of Toronto X-POD for a single CubeSat and X-POD-II for three CubeSats.

An important nanosatellite milestone was achieved in June 2000 when the 6.5 kg SNAP-1 was launched along with the 50 kg Tsinghua-1 microsatellite, which was to serve as a target platform for a satellite inspection mission (see chapter 10). SNAP-1 was the first nanosatellite to demonstrate three-axis attitude control, orbital maneuvering, nearby spacecraft imaging, and on-orbit GPS position and velocity determination.

Small microsatellites of approximately 12 kg, or "almost nanosatellites," are now beginning to populate LEO. A recent store-and-forward communication system called AprizeSat, formerly called LatinSat, has four 11.4 kg spacecraft in LEO.[43] These passively stabilized, inexpensive, 20 cm cubic spacecraft are the first elements of a planned 48-satellite system. A similar system from Saudi Arabia, called SaudiComSat, now has seven 12 kg store-and-forward spacecraft on orbit. Five of these were launched on 17 April 2007 on a Russian Dnepr rocket (a converted ICBM) that placed 16 small satellites into LEO.

2.2.6 From Microsatellites to Femtosatellites

Other nations and many international organizations followed the USSR and the United States by using microsatellites for their first orbiting spacecraft. Larger, more capable, and more costly spacecraft typically evolved from these humble beginnings. Microsatellites offered low-cost access to space and enabled cost-effective LEO constellations with 30 or more satellites.

A large number of passive nanosatellites were launched by the USSR as radar calibration targets, but active nanosatellites were flown far less often. A resurgence in active nanosatellite launch activity started about 15 years ago as continuing advancements in the functional density of microelectronics made fairly intelligent and capable nanosatellites such as SNAP-1 possible. Many previous microsatellite missions, such as LEO communications satellites, could now be performed by nanosatellites.

Active picosatellites (ERS-5 and -6) first entered orbit in 1963. All picosatellites launched between 1964 and 1999 were passive, but that changed in the year 2000. Further increases in the functional density of microelectronics plus improvements in solar-cell efficiency, the energy storage density of secondary batteries, and miniaturized optical and inertial sensors now made intelligent and capable picosatellites possible. More important, the cost of space access using picosatellites or CubeSats is low enough where universities and even individuals can fly their own spacecraft. This should accelerate technology development for small satellites. With continuing advancements in technology, full three-axis control will become available for picosatellites, thus opening up many more mission possibilities using significantly cheaper spacecraft. Based on historical trends, the next evolutionary steps should result in active femtosatellites.

2.3 Nano/Picosatellite Development at The Aerospace Corporation

2.3.1 The Silicon Satellite, a Personal Story

The first two University of Surrey UoSATs enabled reception of real-time spacecraft telemetry using an inexpensive FM scanner (under $100) and a small antenna. I started listening to these satellites in 1989. There is no experience quite like working hands-on with satellite tracking programs, building electronic circuits such as modems, assembling a ground station, hearing your first satellite, and decoding telemetry to learn spacecraft design and space systems engineering. I saw firsthand what the current-generation microsatellites could do and wondered about the potential capabilities of even smaller satellites. Advancements in digital microelectronics, radio frequency (RF) circuits, solar cell technology, and microelectromechanical systems (MEMS) could give kilogram-class satellites a fair degree of functionality. Even though Jones's microspacecraft[32] (a picospacecraft by today's standards) was just a conceptual design, it illustrated what was possible.

Figure 2.9 shows a prototype Relay-1 communications microsatellite from the early 1960s on display at the Smithsonian's Udvar-Hazy Air and Space

Fig. 2.9. Photograph of Relay-1 prototype communications microsatellite at the Smithsonian Air and Space Museum.

Museum in Chantilly, Virginia. Spacecraft were typically assembled by hand using hundreds to thousands of individual components soldered to circuit boards, assembled into systems inside individual boxes, attached onto the basic spacecraft structure, and interconnected using a wiring harness. If a propulsion system was needed, propellant tanks, valves, thrusters, feed manifolds, and other components, were also welded together inside the structure. Figure 2.9 shows the tubular aluminum skeleton, a large number of individual black boxes that contain electronics (command and control system, command decoders, transmitters, receivers, etc.), solar cells mounted on aluminum honeycomb panels, and the internal wiring harness. More details on the systems and subsystems can be found in Metzger and Pickard.[44] Note that most of the internal volume is empty.

Spacecraft fabrication techniques have ranged from hand crafting unique spacecraft such as Relay-1 to assembly-line production for larger constellation spacecraft such as Iridium and most likely Strela-1M. I wanted to move basic spacecraft fabrication technologies from the early 1900s-era assembly line to 1990s batch-fabrication, as illustrated in Fig. 2.10. If a person were to solder transistors at a rate of one per second, it would take more than 3 years to fabricate a current-generation microprocessor, and multiple human errors would probably prevent its operation. Batch fabrication is a photolithography-based process used to produce the highly complex, multibillion-transistor integrated circuits that we enjoy today. Integrated circuits are designed as a stack of thin layers of varying thickness and material composition, where the two-dimensional geometry of each layer is defined by a photolithographic mask. A photocopier is a good example of single-layer batch fabrication; the machine does not care whether you are duplicating the Declaration of Independence or graph paper as long as all of the features are larger than the machine's optical resolution. If spacecraft could be designed as multilayer stacks, batch-fabricated copies would be cheap and reliable.

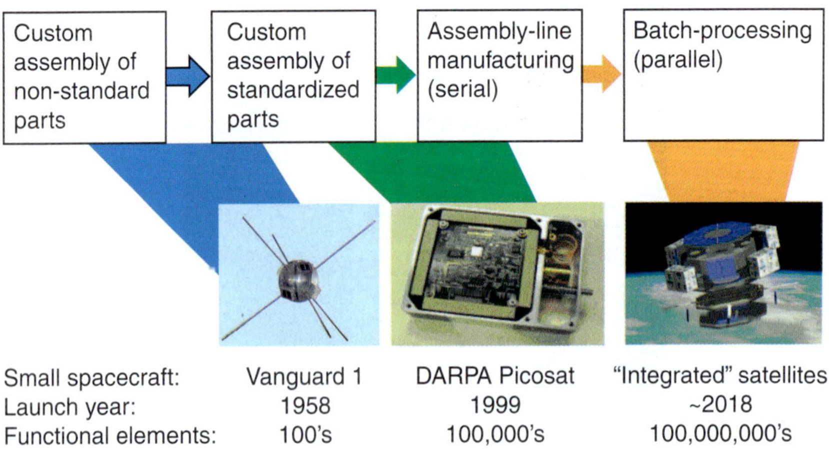

Fig. 2.10. The evolution of satellite manufacturing.

The "functional elements" listed in Fig. 2.10 are the basic building blocks of systems and subsystems and include both electronic and structural elements such as bolts, nuts, transistors, capacitors, and others. Fabricating individual functional elements using batch processing would interconnect them as a single piece part, thus reducing the number of components and interfaces required to produce a spacecraft. Modified semiconductor batch fabrication techniques had been used to produce MEMS actuators and sensors such as inkjet printer heads, pressure sensors, accelerometers, and mechanical switches. Many of these were integrated with sensing or control electronics on the same silicon die. Madou,[45] in the IEEE Special Issue on Integrated Sensors, Microactuators, and Microsystems;[46] and Gad-el-Hak[47] provide good overviews of MEMS fabrication processes, devices, and applications, while Helvajian[48] and Osiander et al.[49] focus on aerospace applications of MEMS. Rather than produce a single integrated circuit with transistors in the tens of millions, I wanted to produce larger, 100 mm diameter substrates with transistors in the hundreds of millions plus supporting circuitry and MEMS that integrated primary spacecraft systems. Silicon was an ideal choice since it was already used as a microelectronics substrate for complementary metal-oxide semiconductor (CMOS) circuits and MEMS.[50–52]

The silicon nanosatellite, which I introduced with my colleagues Henry Helvajian and Ernie Robinson,[53,54] presented a radical new technology for space system design, construction, testing, architecture, and deployment. I conceived the all-silicon, subkilogram satellite in 1989 based on wafer-scale integration over an entire 4 in. (10 cm) or 6 in. (15 cm) diameter silicon wafer. Silicon would serve as spacecraft structure, radiation shield, and thermal control system, plus optical substrate for mirrors or infrared lenses. Integrated MEMS sensors and actuators would provide attitude determination (microbolometer arrays for Earth horizon sensors, accelerometers and angular rate gyros for short-term inertial guidance) and propulsion (microvalves and micromachined thrusters) for highly capable miniature spacecraft. Some conventional macroscopic components such as batteries and individual solar cells would still be required, but the total number of parts and assembly time would be drastically reduced. The silicon nanosatellite is shown schematically in Fig. 2.11.

Low cost per electronic function was a direct result of the CMOS fabrication process; semiconductor batch fabrication techniques had evolved within the constraints of consumer-driven market economics. Unfortunately, these fabrication techniques have evolved to mass produce components (dice) at the 10-million-unit level and higher. Multimillion dollar nonrecurring costs for a single component are acceptable only when amortized over these large production runs.

Significant reductions in nonrecurring costs could be enabled by using commercial microelectronics and MEMS dice integrated onto silicon circuit boards, but this would increase nonrecurring costs. The real question was, "How many identical nano/picosatellites would be required for various missions?" This would establish the design and fabrication technique that made the most sense. This

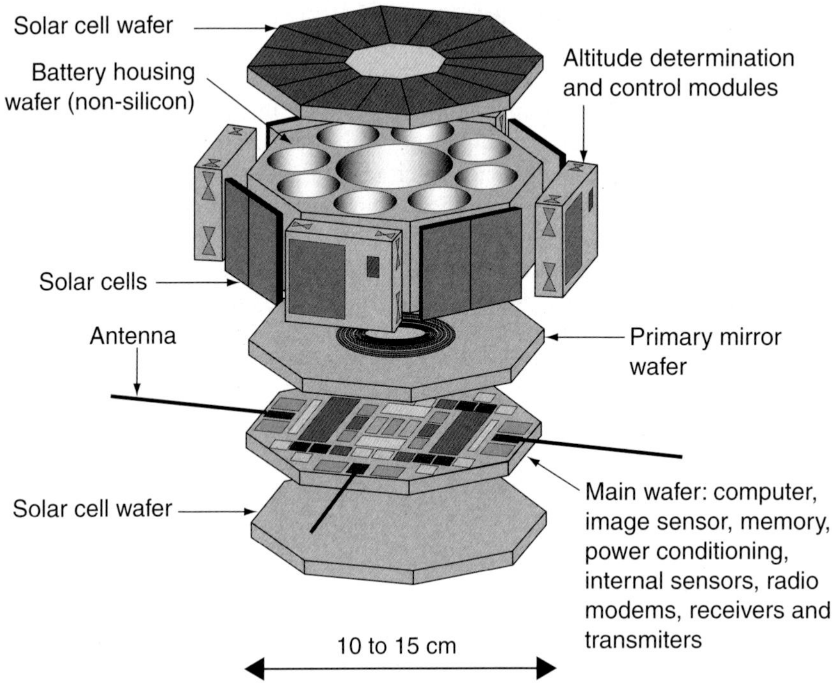

Fig. 2.11. The original silicon nanosatellite concept.

question could only be answered when individual spacecraft capabilities could be reliably predicted. The major uncertainty at the time was what role miniaturization and MEMS could play in small satellites. Since most spacecraft components such as horizon sensors, inertial measurement units, star cameras, momentum wheels, and thrusters, had masses exceeding 1 kg, significant miniaturization was necessary. Basic physics provides some physical limits such as minimum aperture size, but many of these limits are not directly related to mass. It was a challenge, but not an impossible challenge.

One of the benefits of having geographical centers of excellence (similar to the Los Angeles area for U.S. Air Force and NASA spacecraft) is that cross-pollination of ideas occurs more rapidly when people at different institutions can interact on a daily basis. Henry Helvajian had attended a workshop on Microtechnologies for Space Applications at JPL in May 1992.[55] He was enamored with the idea of using MEMS for space applications and proceeded to infect others at The Aerospace Corporation, including Ernie Robinson, with this radical concept. Our insightful management (Bob Fillers) decided that the three of us would work as a team to foster MEMS for space applications with silicon nanosatellites as a long-term application. We evolved into the eight-person Center for Microtechnology, headed by Seymour Feuerstein, to promote microtechnology for space applications and the silicon nanosatellite. This ultimately resulted in a significant amount of internal research and development funding for MEMS and nanosatellites, several Aerospace technical reports that reviewed microtechnology for space applications, two Integrated Microtechnology for Space Applications conferences co-hosted with NASA-Houston and NASA-JPL, and a number of conference papers on MEMS, microthrusters, and silicon nanosatellites.[2,56–67]

My expertise was in CMOS-based MEMS fabrication, microthrusters, orbital mechanics, and mission design. In the early 1990s, I wrote that miniaturized spacecraft systems as application-specific integrated microinstruments (ASIMs) could be batch fabricated.[59] The term ASIM was coined at The Aerospace Corporation to provide a MEMS equivalent to the industry standard ASIC (application-specific integrated circuit). Figure 2.12 shows an early concept of a multilayer thruster system,[59] suitable for batch fabrication, while Fig. 2.13 shows an example of a digital thruster array mounted on a 24-pin chip carrier.[68] The digital thruster was composed of three microfabricated layers bonded together, where the bottom silicon layer contained individually addressable igniters, the middle glass layer contained cavities filled with solid propellant, and the top silicon layer contained burst disks and nozzles. This work was funded by DARPA and executed by TRW, The Aerospace Corporation, and the California Institute of Technology.[69] Each microcavity provided a 0.1 mN-sec impulse when the contained propellant was ignited. Micromachining would enable fabrication of thousands of similar microthrusters on a single 10 cm^2 tile.

The development of silicon satellites would require the design, testing, and verification of millimeter-scale components such as the digital thrusters and centimeter-scale spacecraft systems. In 1994, we envisioned the initial integration

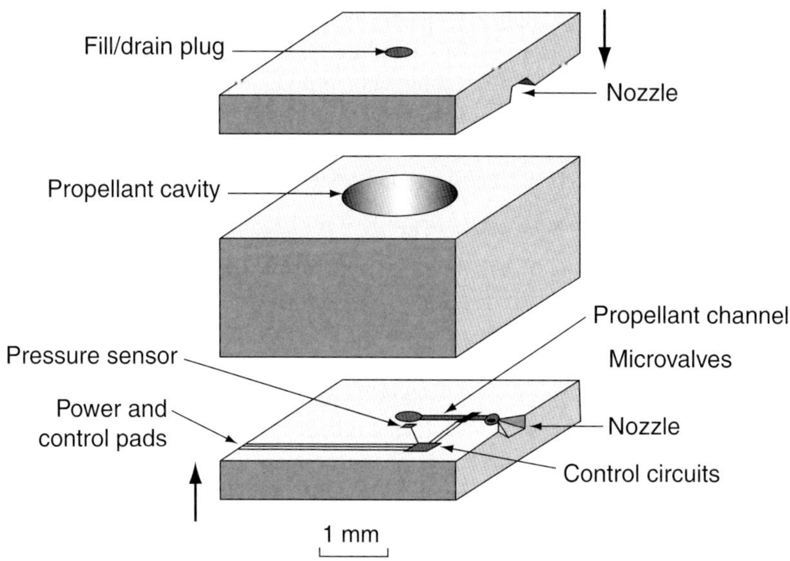

Fig. 2.12. Early concept for a batch-fabricated thruster based on etched silicon dice.

Fig. 2.13. Photograph of a 15-thruster "digital propulsion" chip.

of ASIMs into wireless sensor modules that would be mounted on launch vehicles and spacecraft to provide additional environmental information.[59] The host vehicle would supply a wireless LAN controller to communicate with one or more wireless sensors, but it would not supply power or other physical electrical connections. This allowed inexpensive integration of new sensors onto existing platforms by minimizing physical interfaces.

A free-flying sensor platform ejected from a satellite was the next step in ASIM-based satellite evolution. This would require development of attitude determination and control ASIMs, along with propulsion ASIMs. The free-flyer would have a limited on-orbit lifetime and would operate using simple cold-gas thrusters. A 1 kg free-flying sensor platform appeared to be feasible based on a simple preliminary analysis,[59] but it needed review by other aerospace engineers.

2.3.2 The Untethered Flying Observer

At Helvajian's suggestion, the first conference in Houston in late 1995 included a workshop on the feasibility of developing a 1 kg class satellite inspector, initially called the Untethered Flying Observer (UFO), that would be deployed from a larger spacecraft on command.[70] The idea was to have a deployable, maneuverable, nano/picosatellite that would provide additional spacecraft health and status information after an on-orbit anomaly such as a failed solar-array deployment. NASA was already developing a 46 cm diameter spacecraft called AERCam (see chapter 16) that would fly around the space shuttle cargo bay under astronaut control, so we knew that at least microsatellite-class inspectors were possible. If three or four of these smaller UFOs were attached to every large spacecraft to improve overall mission reliability, hundreds of them would have to be produced every year.

Initially, only a few workshop participants thought that the UFO was possible. Unlike AERCam, the UFO would potentially maneuver about a dead or unstable host vehicle and be controlled by a ground-based operator. It would have to communicate video, images, and telemetry over a communications link several thousand kilometers long. The UFO mission lifetime was expected to be at least two days and require insertion into specific formation-keeping orbits about the host vehicle. This required accurate position (10 cm accuracy) and velocity (better than 1 mm/sec) data for both the host vehicle and the UFO, and autonomous collision avoidance. After a few days of calculations and deliberations, everyone agreed that a small, three-axis stabilized nano/picospacecraft carrying a digital imager could maneuver about a cooperative host spacecraft and send images directly to appropriate ground stations.[70] A kilogram-class satellite inspector was possible.

2.3.3 MEMS Experiments on the Space Shuttle

While the silicon satellite looked like a good, long-term goal, we realized that microtechnology and MEMS would have to fly in space before it would be adopted by traditional spacecraft programs. One approach we followed was integrating low-impact, stand-alone MEMS experiments onto spacecraft already manifested for flight. We obtained funding from NASA-Houston to build and fly a MEMS test bed that would fit within a middeck locker on the space shuttle. We flew MEMS accelerometers, rate gyros, chemical sensors, and the digital propulsion thruster system on space shuttle flight STS-93 in July 1999.[71] Figure 2.14 shows measured

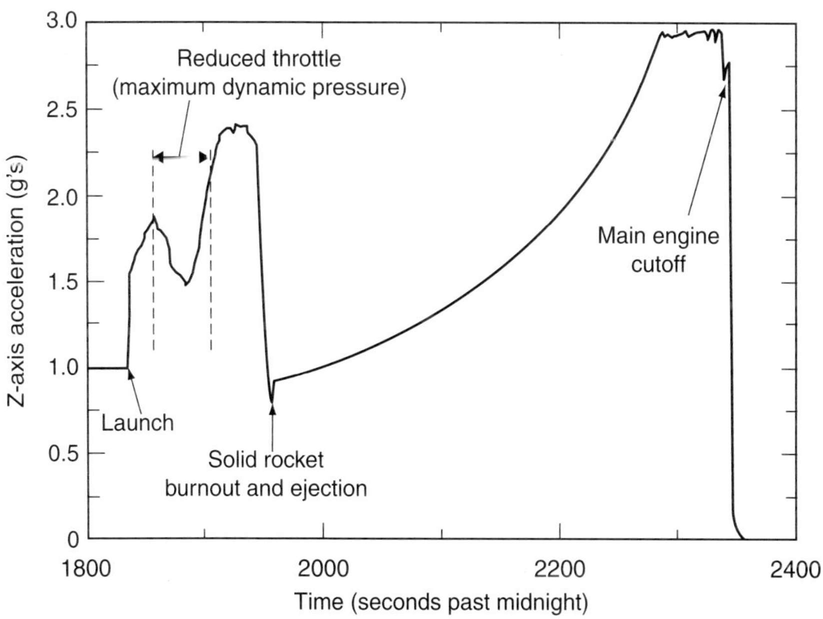

Fig. 2.14. Accelerations measured in the "z" direction by a MEMS accelerometer during the launch of STS-93.

z-axis accelerations for the shuttle as a function of time using a commercial-off-the-shelf (COTS) Silicon Designs model 1010-10J MEMS accelerometer.[72] The z-axis was in the flight direction, and since the shuttle was sitting on the launchpad with the launch direction vertical, the z-axis acceleration was 1.0 g until launch. The measured acceleration profile shows launch occurring at 1840 seconds past midnight, EDT. Flight direction acceleration increases to about 1.8 g, followed by a reduction in main engine thrust during the period of maximum aerodynamic pressure. Flight direction acceleration then increases until the solid rocket engines burn out, followed by a gradual acceleration to 2.9 g as the propellant is used up. The accelerometer sensor, about the size of Lincoln's nose on a U.S. penny in this device, had a sensitivity of 1 milli-g over a 1 Hz bandwidth. This MEMS sensor could be used in a picosatellite to monitor thrust levels as low as 1–10 mN.

2.3.4 The PicoSat

Interest in microtechnologies for small satellites was rising within DOD during the late 1990s, partially as a result of our numerous briefings to the DOD community. Helvajian and Maj. John Comtois of the Air Force Research Laboratory helped organize an Air Force/DARPA workshop—Micro/Nanotechnology for Micro/Nanosatellites—held in Albuquerque, New Mexico, in April 1998.[73] Key microtechnology and small satellite researchers, and various U.S. government program managers such as Drs. Howard Schlossberg and Joseph Janni of the Air Force Office of Scientific Research (AFOSR) and Prof. Al Pisano (UC Berkeley) of DARPA Microsystems Technology Office were invited. Al Pisano was managing our digital thruster effort and was looking for inexpensive ways to flight test MEMS technologies. In his presentation, Prof. Bob Twiggs of Stanford University challenged the government program managers to fund his Orbiting Picosatellite Automated Launcher (OPAL) microsatellite and get it into orbit. Within minutes (at 9:20 a.m. on April 21, to be exact), he had funding from AFOSR and DARPA. The right people at the right time and location made the right decision. OPAL would get funding and a launch. In addition, DARPA would fund a new effort to develop PicoSats that could be used as flight test beds for MEMS technologies in space using OPAL as a low-cost delivery system. This workshop initiated our PicoSat program, resurrected Stanford's OPAL program, and stimulated the AFOSR University Nanosatellite Program to train future space professionals. Since OPAL and the PicoSats were the precursor to the CubeSat program, this workshop turned out to be a milestone event in the worldwide development of picosatellites (see chapter 5).

Aerospace designed and built the initial PicoSats with Rockwell Science Center. Rockwell was developing MEMS RF switches under contract to DARPA, so these would be the primary payload. In addition, Rockwell Science Center designed and built the communications boards for the spacecraft. These transceivers demonstrated a low-power communications-hopping protocol, originally developed for unattended ground sensors, between a pair of spacecraft. Two PicoSats were therefore attached by a 30 m tether and ejected as a pair. Figure 2.15 shows

Fig. 2.15. Photograph of The Aerospace Corporation 1 × 3 × 4 in. (2.5 × 7.5 × 10 cm) PicoSat held by the author.

a PicoSat with one side panel replaced by a partially transparent sheet to show the interior. A pair of these were launched on OPAL in January 2000, and another pair were launched on MightySat II.1 in July 2000.

The initial PicoSat effort by The Aerospace Corporation evolved into the DARPA-funded MEMS-Enabled PicoSatellite Inspector (MEPSI) program that launched a pair of 4 × 4 × 5 in. (10 × 10 × 12.5 cm) nanosatellites in 2002 and another pair in 2006. This effort was prompted by the UFO workshop held at the First International Conference on Integrated Micro/Nanotechnology for Space Applications.[70] In addition, The Aerospace Corporation fabricated a nanosatellite called the PicoSatellite Solar Cell (PSSC) Testbed, launched in November 2008, plus two CubeSats called AeroCube-1 and AeroCube-2 that were launched in July 2006 and April 2007. AeroCube-1 failed to reach orbit. In all, we've had five picosatellites and five nanosatellites reach orbit since 2000. These represent significant fractions of the total nano- and picosatellites launched since 2000 shown in Figs. 2.1c and 2.1d. We are building another CubeSat (AeroCube-3) for launch in 2009 that will include sun and Earth sensors with passive magnetic attitude control. For more engineering details on Aerospace's small satellite programs, see chapter 20, "Picosatellites at The Aerospace Corporation."

2.3.5 The Photostructurable Glass/Ceramic Satellite

The silicon nanosatellite stimulated thinking on mass-producible small satellites, but the projected nonrecurring costs were too high to justify any significant bus development without a requirement for 10,000 or more spacecraft. We therefore switched to a photostructurable glass/ceramic (PSGC) material that we could process in our laboratory. Helvajian and coworkers had developed novel laser-based patterning techniques that eliminated the need for expensive photolithographic masks and enabled fabrication of MEMS-like mechanical and optical structures directly within the spacecraft structure. We could build individual satellites at reasonable cost, yet still provide mass-production options using photolithographic masks where possible and high-throughput laser systems.

Internal research and development funding allowed us to develop a PSGC cold-gas propulsion module with a 5 cm^2 footprint, which we flew on an air table. With DARPA seed money, we were able to build and test a more advanced 10 cm diameter propulsion module suitable for a kilogram-class satellite inspector, which we called the Co-Orbiting Satellite Inspector (COSA) shown in Fig. 2.16. Details on the design, fabrication, and testing of the COSA are given in chapter 18.

2.3.6 Formation Flying

After discussing proximity operations with NASA-Houston AERCam developers at the 1995 Houston conference, I created an electronic spreadsheet plotting trajectories of subsatellites about a host spacecraft as a function of initial conditions,

Fig. 2.16. PSGC propulsion module. Lee Steffeney (left) is holding a completed COSA that shows the top-mounted electronics board. Bill Hansen (right) is holding a partially assembled COSA with integrated battery and valves.

based on the linearized equations of co-orbital motion. Exploring the parameter space, I discovered certain initial conditions that allowed a daughter satellite to maintain a constant range with respect to a host satellite while revolving around the host satellite once per orbit. The host would see the daughter satellite orbiting about it in a circular orbit even though there was essentially no gravitational attraction between the two. This was ideal for a satellite inspector. The co-orbital rotation period was equal to the satellite orbital period about Earth, so co-orbital period was independent of the daughter satellite-to-host range. This enabled arbitrary placement of daughter satellites within a specific plane around a host vehicle, and the physical distribution of multiple daughter satellites would revolve about the host as if it were a rigid body. Figure 2.17 shows the geometry.

The Aerospace Astrodynamics department worked with me to include real-world effects such as the nonspherical mass distribution of Earth, and we concluded that kilometer-diameter formations could be maintained in LEO using modest propulsion requirements.[74–76] Formation flying was enabled. This would lead to distributed satellite concepts that used hundreds to thousands of nano/picosatellites to provide new mission opportunities.

2.4 Nanosatellite Technologies

A number of technologies have already given nano/picosatellites significantly enhanced capabilities during the past decade. These include digital microelectronics, image sensors, magnetic field sensors, MEMS inertial sensors, triple-junction solar cells, and lithium-ion batteries. Emerging technologies that will further enhance nano/picosatellite capabilities include microthrusters, miniature optical attitude sensors, and improved MEMS accelerometers and rate gyros. This section will review these available or emerging technologies.

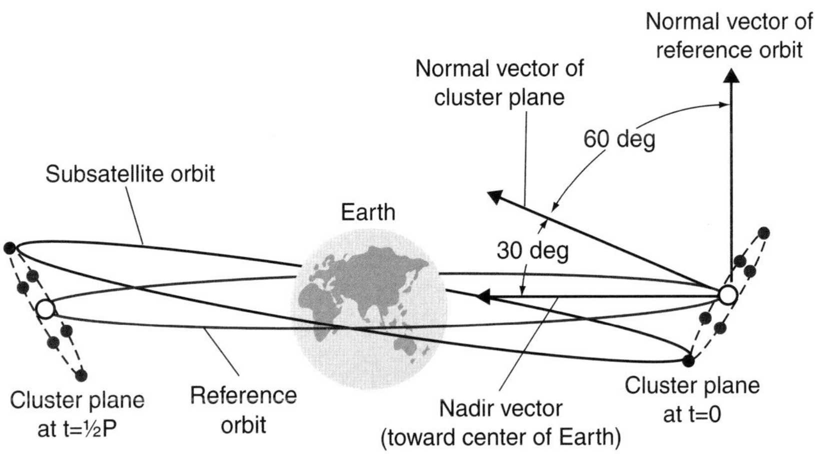

Fig. 2.17. Orbital configuration for formation flying. The normal to the cluster (subsatellite) plane must point 30 deg above or below the nadir vector.

2.4.1 Microelectronics

Gordon Moore, cofounder of Intel Corporation, noticed that the complexity of integrated circuits, for lowest cost per component, doubled roughly every year because of advancements in technology.[77] That general trend, now known as Moore's Law, has held for central processing units (CPU) and dynamic random-access memory (DRAM) chips for the last four decades. We normally equate Moore's Law with the doubling of microprocessor performance every 24–30 months, but this is a side effect of producing ever-smaller and faster transistors.

The world's first microprocessor, the Intel 4004, was introduced in 1971 with a mere 3200 transistors based on a 10 μm minimum feature size (MFS). It could process 70,000 instructions per second or 0.07 MIPS (million instructions per second). Table 2.5, based on Hansen's data,[78] shows that this processor could perform many basic small satellite functions. All satellites designed before 1971 did not use microprocessors; they used multiple functional logic circuits for command and control. After 1971, an entire board full of 2–4 cm long ceramic integrated circuit packages could be replaced by a single package roughly 2 cm long.

The Intel 80286 microprocessor became available about a decade later in 1982. It was capable of 1 MIPS using 134,000 transistors that were fabricated using 1.5-μm MFS on a 69 mm^2 silicon die. This processor was used in the IBM AT series of personal computers. In 2000, Intel introduced the Pentium IV, with 42 million transistors fabricated using a 0.18 μm MFS, resulting in 3000 MIPS performance and a transistor area less than a square micron.[79] In 2008, Intel will release the Tukwilla microprocessor with 2 billion transistors, fabricated using a 65 nm MFS process.[80] If historical performance trends continue, this processor should provide about 50,000 MIPS. The International Technology Roadmap for Semiconductors (ITRS), updated every two years, predicts 45 nm MFS by 2010, 25 nm MFS by 2015, and 14 nm MFS by 2020.[81] Significant challenges exist

Table 2.5. Throughput Requirements for Basic Small Satellite Bus Functions

Function	Throughput (MIPS)
Command Processing	0.007
Telemetry Processing	0.003
Attitude Sensor Processing	0.025
Attitude Determination and Control	0.105
Autonomy	0.020
Fault Detection	0.020
Power Management	0.005
Thermal Control	0.003
Kalman Filter	0.080

in meeting these goals, but Moore's Law should continue to operate for at least another decade.

How much silicon area is required for a 1 MIPS microprocessor suitable for basic small satellite command and control functions? Figure 2.18 shows the past, present, and future size of an 80286-class processor based on historical and ITRS predictions of minimum feature size. Space-qualified processors will be somewhat larger because of extra integrated structures to improve radiation tolerance (e.g., transistor guard bands and fault detection and correction circuitry), but 80286-class processors can still be extremely small. By 2016, this class of processor could be fabricated on a silicon die with a side length about twice that of a human hair. Figure 2.18 also shows the DRAM die size required to hold 1 million words (8 bits of data plus 4 extra bits for error detection and correction) of memory. Today, the processor and memory could fit on a 1 mm^2 die with room to spare. The Microchip PIC10F222 is an example of a current generation ultrasmall microprocessor available to the masses. It comes in an 8-lead DFN package (2 × 3 × 0.9 mm) and includes two 8-bit analog input channels.[82]

Nanosatellites and smaller spacecraft have limited power, so processor efficiency can become important when multiple processors are used in a single spacecraft and are active most of the time. The continuing reduction in MFS has also enabled an increase in processor efficiency based on smaller transistors with reduced operating voltage. Low-power microprocessors and microcontrollers are now available with processing performance on the order of 3000 MIPS/W. The

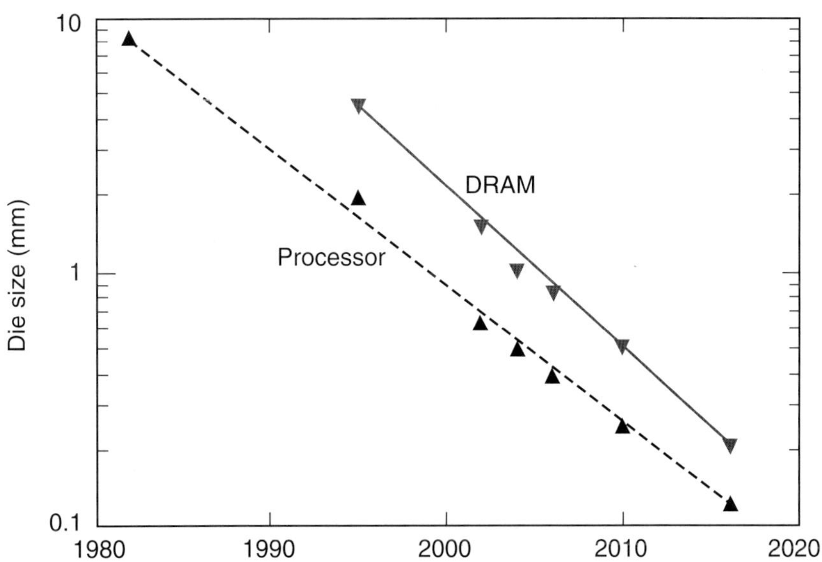

Fig. 2.18. Die size for an Intel 80286-class microprocessor and a 1 MB DRAM as a function of time.

Microchip PIC10F222 requires less than 350 μW (175 μA at 2 V) when operating at 4 MHz (1 MIPS). While the computational power efficiency is high at 2800 MIPS/W, this microcontroller has only 768 bytes of program memory. It's suitable for simple tasks, such as timing functions and converting analog sensor outputs into digital outputs. The more capable Microchip PIC18F1320 microcontroller, with 7 analog inputs, 16 input/output lines, and 16 kb of program memory, requires about 300 μW (150 mA at 2 V) when operating at 1 MHz (0.25 MIPS).[83] This results in a performance rating of 833 MIPS/W; a 1 MIPS command and control computer would consume about 1.2 mW. Other examples of ultralow-power processors include NEC's VR4131 microprocessor (340 MIPS at 220 mW, 1545 MIPS/W) and the Atmel AT91R40807 processor used on the CanX-1 (University of Toronto) CubeSat (~1.4 mW/MHz and 36 MIPS at 40 MHz, 643 MIPS/W).[84–87]

2.4.2 CMOS Image Sensors

One of the benefits of an aggressive semiconductor fabrication industry is the development of related products that leverage existing fabrication infrastructure. Active pixel sensors were invented in the early 1990s at NASA/JPL to circumvent multiple problems with charge coupled device (CCD) image sensors.[88,89] CMOS-fabricated active pixel sensors use transistors collocated with 2–8 μm^2 p-n photodiode light sensors to amplify the signal and route it to an appropriate addressing line for read-out. Inexpensive mass production of active pixel image sensors for digital cameras, web cameras, wireless telephones, and others, is a by-product of CMOS compatibility. Larger pixels (~7 μm^2) are typically offered on VGA-resolution cameras (640 × 480 pixels), while ~2 μm^2 pixels are currently offered on 5-megapixel imagers.

Inexpensive, low-power image sensors can be used in sun sensors, star cameras, visible Earth imagers, and near-infrared Earth imagers in the 700 to 1100 nm wavelength band.[90] Figure 2.19 shows a photograph of a COTS camera board that includes JPEG compression.[91] This 2.0 × 2.8 cm board has a mass of 10 g and a maximum power usage of 200 mW and uses a 0.25 in. (6.4 mm) 640 × 480 color CMOS imager with a lens. We have successfully used this imager on our MEPSI and AeroCubes.[92]

2.4.3 Magnetic Field Sensors

In LEO, crude (about a degree) orientation about two axes can be determined by measuring the local magnetic field vector. An example of a microfluxgate magnetic field sensor based on a modified CMOS process is given in Schneider.[93] This 0.8 × 1.5 mm sensor uses a ferromagnetic layer deposited on top of the die passivation layers and has a linear response to magnetic fields below 50 μT with a maximum sensitivity of 2.7 V/T. MEMS-based magnetic field sensors have also been fabricated.[94,95] Magnetic field sensors using the giant magnetoresistive field effect (resistance is a function of applied magnetic field) are commercially available from Honeywell.[96] A number of CubeSats have used the Honeywell

Fig. 2.19. Photograph of the C328 camera board with lens, held by the author.

HMC2003 sensor, which provides analog output voltages that can be read by analog-to-digital converters embedded in many microprocessors.[97] This 2.7 × 2.00 × 1.2 cm hybrid module contains sensors and signal amplifiers. It uses 120 mW of power at 6 V and can resolve 40 μG; this corresponds to 0.02% of the ambient magnetic field at the equator at an altitude of 700 km.

2.4.4 MEMS for Inertial Sensing

Accelerometers can be used to measure instantaneous accelerations due to thruster operation, air drag, solar pressure, and like causes. MEMS accelerometers are particularly convenient for small satellites because of their small size and low power requirements. Many manufacturers exist, and the maximum sensing range varies from ±1.5 to ±100 g.

Inexpensive commercial accelerometers such as the Analog Devices ADXL103 are small chip-size devices (5 × 5 × 2 mm) that can measure spacecraft accelerations down to the 0.1 milli-g level using less than 5 mW of power. This would provide 1 mN resolution on a 1 kg CubeSat, adequate for cold-gas and chemical thrusters in the 10 mN thrust range and higher. More expensive MEMS devices such as the Colibrys Si-Flex SF-1500S, a large chip-sized device (24.4 × 24.4 × 16.6 mm), can measure accelerations down to the 0.3 μg level using approximately

70 mW of power. This would provide 3 μN resolution for monitoring high specific impulse thrusters on a CubeSat. A 1-W, 3000 sec specific impulse ion engine, for example, would produce 34 μN of thrust at 50% efficiency.

Rate gyros are used on small spacecraft to monitor angular rate changes due to magnetorquing, momentum wheel operation, or thruster operation. Commercial MEMS gyros are not suitable for inertial navigation beyond a few minutes, but they are suitable for monitoring spacecraft rotation rates significantly faster than orbit rates in LEO (360 deg in ~90 min; 0.067 deg/sec). For example, the Analog Devices ADIS16250 comes in an 11.1 × 11.0 × 5.3 mm package, consumes only 40 mW of power, and has a noise density of 0.05 deg/sec/Hz$^{1/2}$.[98] It can be used to provide a 1 deg pointing accuracy over a 2 min period.

2.4.5 Solar Cells

Solar cells used during the mid-1960s were small single crystal silicon cells with sunlight-to-dc conversion efficiencies of approximately 10% on orbit. The 1960s era ERS-27 nanosatellite used 1 × 2 cm rectangular cells because substrate diameters were limited to 1 in. Microelectronics-grade silicon substrates were used for silicon solar-cell production, and the standard substrates grew over the years to diameters of 2 in. (5 cm), 4 in. (10 cm), 6 in. (15 cm), 8 in. (20 cm), and 12 in. (30 cm) as techniques for growing larger, ultrapure single crystals of silicon evolved. Space-grade silicon solar cells produce approximately 0.6 V at maximum output power and now have beginning-of-life sunlight-to-dc conversion efficiencies of about 15% (20 mW/cm^2 for in-space solar irradiation at normal incidence).

GaAs solar cells produce approximately 0.85 V at maximum output power and have beginning-of-life sunlight-to-dc efficiencies of about 18%. The solar spectrum has most of its energy at wavelengths between 300 and 1500 nm, while GaAs can only absorb and convert photons with energies higher than its bandgap energy of 1.43 eV (e.g., wavelengths shorter than 870 nm). Each absorbed photon with energy E higher than 1.43 eV produces a single electron/hole pair with the excess energy (E – 1.43 eV) ending up as heat.

Triple-junction cells get higher efficiency by essentially stacking two other thin solar cells, optimized for different wavelengths, on top of a base solar cell. The top layer absorbs the shortest wavelengths, while the bottom cell, typically germanium with a bandgap of 0.67 eV (1850 nm), absorbs the longest wavelengths. The different cell layers are connected in series, so the output voltage is much higher than what silicon or GaAs cells generate. Triple-junction solar cells were first produced in 1999, so they represent a relatively new technology. Today's commercially available triple-junction cells produce about 2.5 V and have beginning-of-life sunlight-to-dc conversion efficiencies of approximately 28% (38 mW/cm^2 for in-space exposure at normal incidence). A single cell can now be used as an entire solar panel on a picosatellite or smaller spacecraft; the output voltage is high enough to directly drive microprocessors and memory circuits, recharge pairs of NiCd batteries, or power up-converter circuits to generate higher voltages.

2.4.6 Rechargeable Batteries

Mercury batteries were the first primary batteries used on spacecraft, while NiCd batteries were the first secondary (rechargeable) batteries. NiCd was supplanted by nickel-metal-hydride batteries for small spacecraft and NiH batteries for larger spacecraft. Starting in the early 1990s, lithium-ion secondary batteries became available for consumer products and are now used in many small satellites. The energy density of lithium-ion cells is more than twice that of NiCd. The lithium-ion 18650 battery, used in many laptop computers, has a storage capacity of approximately 2200 mAhr, a 3.7 V output voltage, and a mass of 47 g; the energy storage density is 160 Whr/kg. A low voltage (3 or 3.3 V) bus can be driven by a single cell, thus eliminating the need to select matched cells for use in series; for example, three NiCds in series would be required to produce a 3.6 V battery pack. Finally, due to the proliferation of battery-powered cell phones, prismatic lithium-ion batteries are readily available in a wide variety of dimensions. This facilitates integration of batteries into the cramped interiors of CubeSats and other picosatellites.

2.4.7 Microthruster Systems

The basics of traditional rocket propulsion include velocity increment and specific impulse (I_{sp}). Every propulsion maneuver requires changing the spacecraft velocity, and the sum of the absolute magnitudes of all maneuvers for a given mission is called the velocity increment (ΔV). Table 2.6 shows some ΔV requirements for representative maneuvers. Spacecraft ΔV requirements are strongly mission

Table 2.6. Velocity Increment for Some Representative On-orbit Satellite Maneuvers

Mission	ΔV (m/sec)
Increase circular orbit altitude by 1 km at GEO	0.04
Move 10 km ahead in a 700 km orbit over 2 days	0.04
Increase altitude from 1000 to 1001 km	0.50
Increase altitude from 700 to 701 km	0.53
Increase altitude from 400 to 401 km	0.57
Move 10 km ahead in a 700 km orbit over 3.3 hr	1.1
Boost altitude by 100 km at GEO	3.65
Change inclination by 1 deg at GEO	54
Change inclination by 1 deg at 700 km altitude	131
Deorbit from 700 km using an impulsive transfer	160
Increase altitude from 400 to 800 km	217
Deorbit from 1000 km using an impulsive transfer	235
Deorbit from 700 km using a low-thrust spiral	330
Change inclination by 10 deg at 700 km altitude	1308

dependent and can range from fractions of a meter per second to well over a thousand meters per second.

Velocity increments can be converted into required propellant masses once specific thruster system parameters are selected and the postmission spacecraft mass has been established. The propellant mass required for a given mission is related to the ΔV and I_{sp} of the thrusters by the Rocket Equation:

$$\Delta V = g_0 I_{sp} ln\left(m_i / m_f\right) \tag{2.1}$$

where g_o is the gravitational acceleration at Earth's surface (9.8 m/sec^2), m_i is the initial spacecraft mass, and m_f is the final spacecraft mass. I_{sp} is defined as the thrust divided by the mass flowrate of propellant into the thruster, and higher values are better (more time integrated thrust per unit mass loss of propellant). Cold-gas systems have the lowest I_{sp} with a range of 20–150 sec, depending on the propellant species, while existing chemical propulsion systems range from about 220 sec (solid rocket engines) to 440 sec (liquid oxygen/hydrogen engines). Chemical thrusters basically create a hot gas and expand that gas through a converging/diverging nozzle to convert the thermal energy into directed kinetic energy and hence thrust. They have a maximum theoretical I_{sp} of about 500 sec due to the limited amount of energy stored in the chemical bonds of the reactants. Electric thrusters use electric power from solar arrays or nuclear generators to accelerate ions and/or atoms to much higher velocities that yield a specific impulse between 600 and 10,000 sec. Figure 2.20 shows the propellant mass fraction (required propellant mass per initial spacecraft mass) as a function of I_{sp} and ΔV.

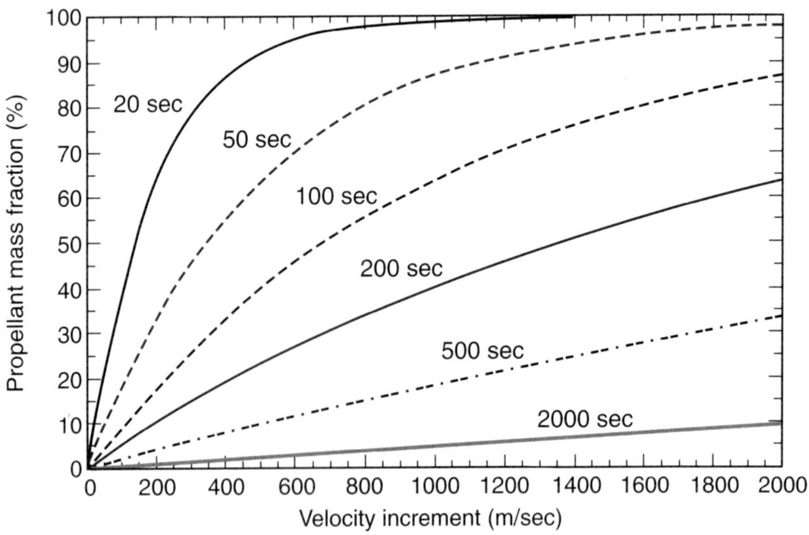

Fig. 2.20. Propellant mass fraction as a function of mission ΔV requirement for different I_{sp}.

Cold-gas systems are the easiest to make and can be purchased from a number of vendors. Needed are a tank containing pressurized gas or a vaporizable liquid, a particle filter and/or liquid gas separator, one or more valves, and a converging/diverging nozzle to convert gas pressure into thrust. Figure 2.21 shows a photograph of a cold-gas propulsion system fabricated by VACCO Industries for the MEPSI program. This 0.5 kg, 9.1 cm^2 integrated propulsion system included five thrusters, a propellant tank, a propellant feed manifold, valves, and sensors. It had a specific impulse of 65 sec using isobutane and could generate a total impulse of 34 N-sec. We have also used laser-based processing to create fully integrated cold-gas propulsion systems in photostructurable glass/ceramic materials and plastic.[92,99] If low pressure (under 7 bar) propellants are chosen, the latter option is relatively inexpensive using commercial stereolithography fabrication services.

Small solid-rocket engines with I_{sp} of approximately 200 sec are available from AeroTech Consumer Aerospace.[100] Unlike the smaller, carbon/potassium nitrate/sulfur based engines typically sold in the United States to model rocketry enthusiasts, these are ammonium perchlorate based engines similar to those used in larger rockets. Individual propellant grains and thruster casings are available for total impulses ranging from 20 to 14,000 N-sec. A pair of "F" class engines with a total impulse of 160 N-sec, for example, could be used to actively deorbit a 1 kg CubeSat from an initial 700 km circular orbit, boost it into a higher 1000 km altitude circular orbit, or change the orbit inclination by 1.2 deg.

Solid rockets are simple, but the number of maneuvers is severely limited by the single-shot nature of these devices. A number of researchers in France, the United States, and Japan have now fabricated solid-rocket thruster arrays, such as the one shown in Fig. 2.13, that could be used to provide a large number of smaller impulses.[101,102] None of these digital thrusters have yet been fired on-orbit, but that should change over the next five years.

Fig. 2.21. A five-thruster cold gas micropropulsion system fabricated by VACCO Industries for the MEPSI program.

Conventional hydrazine monopropellant thruster systems have been fabricated and flown with thrust levels down to 0.2 N. Aerojet offers the MR-103 thruster with a thrust of 0.2–1.1 N, specific impulse of 210–220 sec, and mass of 330 g. EADS Astrium offers the model CHT 0.5 thruster with a thrust of 0.5 N, specific impulse of 227 sec, and mass of 195 g.[103] Both of these are too large to fit in a single CubeSat, but would be appropriate for larger nano/microsatellites. An excellent review of thruster options for small satellites can be found in Mueller.[104]

For even higher specific impulse, the best choice appears to be electric propulsion. Arcjet and ion engine systems are typically too large and massive for many small satellites, and operate best at continuous power levels exceeding several hundred watts. Pulsed plasma thrusters (PPTs) have similar instantaneous power requirements, but they require much lower average power levels by operating at low duty cycles. A PPT uses a high-voltage arc to break down exposed surface material into a plasma that contains both ionized and neutral species. A combination of gas-dynamic and electromagnetic acceleration typically provides specific impulses in the 1000–3000 sec range. The Micro Propulsion Attitude Control System (MPACS), built by AFRL and flown on FalconSAT-3, uses these thrusters (see chapter 7). A PPT variant, called the vacuum arc thruster (VAT), was integrated into the 2 kg double CubeSat called ION (Illinois Observing Nanosatellite) from the University of Illinois.[105] Four thrusters were integrated into the structure to provide two-axis attitude control plus translation. The fuel was aluminum from the spacecraft structure. Unfortunately, the Dnepr launch vehicle didn't reach orbit, so flight data are not available. Another vacuum arc thruster system, shown in Fig. 2.22a, was designed by Alameda Applied Sciences Corporation (AASC) and had a mass of 150 g, an input power of 4 W, and a thrust of 54 μN. Eight of these thrusters will be integrated onto the 30 kg HuskySat spacecraft being fabricated by the Michigan Technological University.[106] A magnetically enhanced vacuum arc thruster (MVAT), designed and fabricated by AASC and shown in Figure 23b, has a 1 kg mass, an average thrust of 100 μN, and a mass-to-

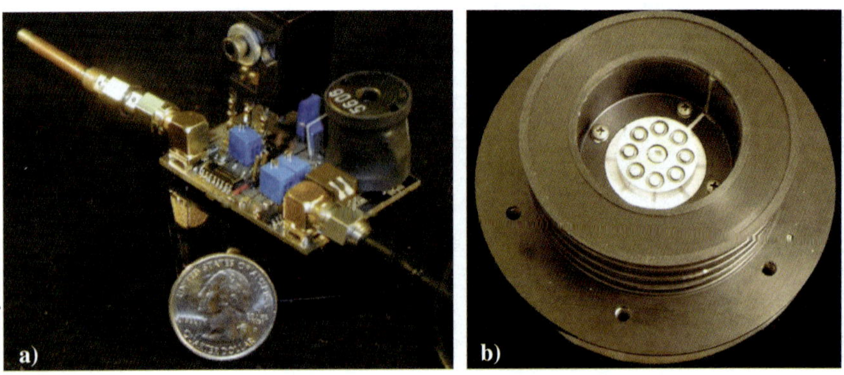

Courtesy of AIAA

Courtesy of Alameda Applied Sciences Corp.

Fig. 2.22. Photograph of a vacuum arc thruster (a) and a magnetically enhanced vacuum arc thruster (b). Left image from Carpenter et al.[106]

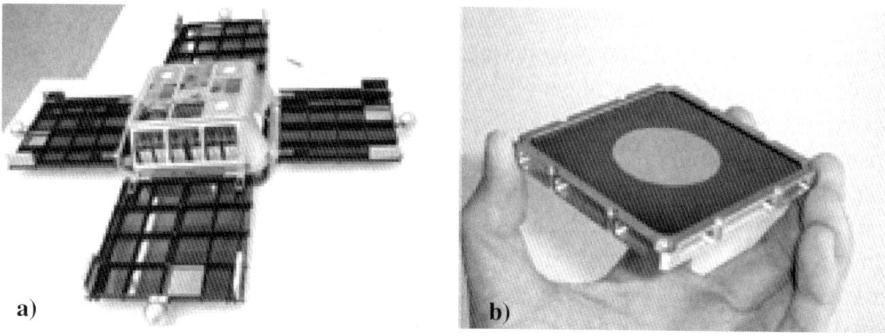

Fig. 2.23. The NanoSpace-1 (now MicroLink-1) spacecraft body with some silicon modules removed to show the interior (a) and a silicon module (b). Image from Nguyen et al.[111]

power ratio of 50 g/W. This thruster integrates the power processing electronics close to the arc so that the axial magnetic field generated by the main power inductor collimates the plasma plume for enhanced propulsion efficiency. This thruster gives ~5 million, 1 μNsec impulse bits at 2000 sec specific impulse, with no propellant feed or any moving parts.

2.4.8 Integrated Silicon Spacecraft

Researchers at Sandia National Laboratories, the Shanghai Institute of Metallurgy, and the Ångström Space Technology Centre at Uppsala University in Sweden have worked on integrated silicon modules for silicon spacecraft.[107–109] The Ångström Laboratory research was sufficiently well developed to create a separate company called NanoSpace to provide MEMS-based products for the space community.[110] One of those products is a 60 g, golfball-size, four-thruster cold/warm gas propulsion module fabricated on silicon wafers.

The Ångström Space Technology Centre originally developed an 8 kg nanosatellite called NanoSpace-1 (now called MicroLink-1) that uses the MEMS silicon thruster pods and 8.3 cm^2 silicon panels (MEMS modules) as part of the basic spacecraft structure. Up to 30 MEMS modules can fit on the outer surface of the spacecraft. Figure 2.23a shows an aluminum body with some silicon modules installed.[111] The base of the spacecraft is 33 cm^2, the body is 10.5 cm high, and four 30 × 32 cm solar panels are attached to the bottom. Of the total dry mass of 7.6 kg, 1.7 kg are in the form of silicon MEMS modules, while only 1.0 kg is aluminum structure.[112] The silicon module shown in Fig. 2.23b is an S-band patch antenna that uses micromachined silicon wafers to reduce the effective dielectric constant of the silicon substrate.[113] MicroLink-1 could fly within a few years, if sufficient funding were available.

2.5 Summary

Over the last 50 years, more than 860 microsatellites, 680 nanosatellites, and 38 picosatellites have been launched. Most of the microsatellites and nanosatellites were launched by the former Soviet Union. Microsatellites were the first spacecraft launched by the USSR, the United States, most countries, and many international, civil, and educational organizations. As expertise in space systems engineering improved, most of these organizations started launching heavier, more capable spacecraft that could no longer be considered microsatellites. Most of the nanosatellites and picosatellites launched before 2000 were passive spacecraft used for radar calibration and atmospheric density monitoring. The evolution of micro/nanoelectronics, solar cell technology, battery technology, and MEMS during the last decade has enabled capable nanosatellites and picosatellites such as CubeSats. Mission applications for nano- and picosatellites now range from flight testing components and subsystems, monitoring the space environment, to store-and-forward communications. Launch rates of active picosatellites currently exceed the launch rates of nanosatellites. More technology development is still needed to provide star trackers, GPS receivers, and microthruster systems that are required for more demanding missions such as satellite inspection and formation flying.

2.6 Acknowledgments

I thank The Aerospace Institute for sponsoring this chapter, and I gratefully acknowledge The Aerospace Corporation's Independent Research and Development program and its former Corporate Research Initiative in MEMS and Microtechnology for funding research in silicon nanosatellites, micropropulsion, formation flying, CMOS-based spacecraft sensors, and parts of The Aerospace Corporation's PicoSat program. Finally, I also want to thank Henry Helvajian, Ernest Y. Robinson, the late Seymour Feuerstein, and David Hinkley for their hard work and steadfast belief in the bright future for nano/picosatellites. When it comes to fortitude, these are men of steel.

2.7 References

[1] M. N. Sweeting, "UoSAT Microsatellite Missions," *IEE Electronics & Communication Engineering Journal*, **4** (3), 141–150 (June 1992).

[2] S. W. Janson, "Silicon Satellites: Picosats, Nanosats, and Microsats," *International Conference on Integrated Micro/Nanotechnology for Space Applications* (Houston, TX, Nov. 1995).

[3] T. D. Thompson, (ed.), *Space Log 1996* (TRW Space and Electronics Group, Redondo Beach, CA, 1997).

[4] Surrey Satellite Technology Laboratories, "Small Satellites Home Page" (2004), http://www.SmallSatellites.org (23 Oct. 2008).

[5] Encyclopedia Astronautica, Quarterly Chronology web pages, first quarter 2003 to last quarter (2007), http://www.astronautix.com/chrono/20031.htm, http://www.astronautix.com/chrono/20032.htm, http://www.astronautix.com/chrono/20074.htm (23 Oct. 2008).

[6] J. L. Janicik, "CHIPSAT: An Affordable & Capable Microsatellite Solution," *Space 2003* (Long Beach, California, 23–25 Sept. 2003), AIAA Paper 2003–6290.

[7] National Space Science Data Center, "NASA's Explorer Missions," (NASA Goddard Space Flight Center, Greenbelt, MD, 2007), http://nssdc.gsfc.nasa.gov/multi/explorer.html (23 Oct. 2008).

[8] S. J. Isakowitz, *International Reference Guide to Space Launch Systems, 2nd Edition*. (American Institute of Aeronautics and Astronautics, Washington, DC, 1995), p. 341.

[9] Naval Research Laboratory, SolRAD Solar Radiation web page, http://code8200.nrl.navy.mil/solrad.html (23 Oct. 28).

[10] L. Jaffe, "Project Echo Results," *Astronautics*, **6** (5), 32–33, 80 (May 1961).

[11] D. C. MacLellan, W. E. Morrow, and I. I. Shapiro, I.I., "Effects of the West Ford Belt on Astronomical Observations," *Proceedings of the IEEE*, **52** (5), 564–570 (May 1964).

[12] C. F. J. Overhage and W. H. Radford, "The Lincoln Laboratory West Ford program—An Historical Perspective," *Proceedings of the IEEE*, **52** (5), 452–454 (May 1964).

[13] F. E. Heart, D. Karp, A. A. Mathiasen, F. Nagy, W. R. Crowther, and W. B. Smith, "Measured Physical Characteristics of the West Ford belt," *Proceedings of the IEEE*, **52** (5), 519–533 (May 1964).

[14] I. I. Shapiro and H. M. Jones, "Lifetimes of Orbiting Dipoles," *Science*, **134**, 973–979 (6 Oct. 1961).

[15] I. I. Shapiro, "Last of the West Ford Dipoles," *Science*, **154**, 1445–1448 (16 Dec. 1966).

[16] B. R. Bowman, W. N. Barker, and W. G. Schick, "Orbital Perturbation Analysis of West Ford Needles Clusters," *AAS/AIAA Space Flight Mechanics Meeting* (2000), AIAA paper 2000–4236.

[17] D. H. Martin, P. R. Anderson, and L. Bartamian, *Communication Satellites, Fifth Edition* (The Aerospace Press, El Segundo, CA, 2007), pp. 187–189.

[18] W. L. Pritchard, "The Initial Defense Communication Satellite," *The Microwave Journal*, **9** (11), 43–47 (Nov. 1966).

[19] B. Harvey, *Russia in Space: The Failed Frontier?* (Springer, New York, 2001) p. 126.

[20] D. H. Martin, P. R. Anderson, and L. Bartamian, "*Communication Satellites, Fifth Edition*" (The Aerospace Press, El Segundo, CA, 2007), pp. 7–11 and 57–62.

[21] J. A. Bleeker, J. Geiss, and M. Huber, editors, *The Century of Space Science*, Vol. 1 (Springer, 2002), pp. 51–53.

[22] NASA National Space Science Data Center, ANS (NSSDC ID: 1974–070A), (2007), http://nssdc.gsfc.nasa.gov/nmc/masterCatalog.do?sc=1974–070A (23 Oct. 2008).

[23] M. D. Kenny, "Environmental Research Satellite ERS-20," (13 Oct. 2006), http://md-kenny.customer.netspace.net.au/ERS-20.pdf (23 Oct. 2008).

[24] M. C. Chapman, J. B. Gardner, and R. J. Martin, "Octahedral Research Satellite Mark III OV5-1 (ERS 27)," Air Force Cambridge Research Laboratories report AFCRL-68-0265 (Bedford, MA, April 1968).

[25] R. R. Zirm, "The Triple Calibration Sphere (CALSPHERE) Experiment," *Proceedings of the IEEE*, **59** (12) 1725–1726 (Dec. 1971).

[26] National Aeronautics and Space Administration, "Mylar Balloon" National Space Science Data Center Catalog (2007), http://nssdc.gsfc.nasa.gov/nmc/masterCatalog.do?sc=1971–067F (23 Oct. 2008).

[27] N. L. Johnson, D. O. Whitlock, P. Anz-Meador, M. E. Cizek, and S. A. Portman,

"History of On–Orbit Satellite Fragmentations, 13th Edition," NASA Johnson Space Center Report JSC–62530 (May 2004).

[28] NASA History Division, "NASA's Budget Authority in 1996 Dollars," (2007), http://history.nasa.gov/pocketstats/sect%20D/NASA%20Budget.pdf (23 Oct. 2008).

[29] U.S. Federal Reserve Statistical Release, Selected Interest Rates, Annual Federal Funds, http://www.federalreserve.gov/releases/h15/data/Annual/H15_FF_O.txt (23 Oct. 2008).

[30] J. L. Slack and N. R. Helm, "Global Low Orbit Message Relay (GLOMR) Satellite Applications Development Program," Final Technical Report (26 March–30 Sept. 1986, Ardak Corp, McLean, VA).

[31] R. M. Jones, "Microspacecraft Missions and Systems," *Journal of the British Interplanetary Society*, **42**, 448–454 (1989).

[32] R. M. Jones, "Electromagnetically Launched Microspacecraft for Space Science Missions," *Journal of Spacecraft and Rockets*, **26** (5), 338–342 (Sept.–Oct. 1989).

[33] B. J. Horais, "The Ariane ASAP Ring: Workhorse Launcher for Small Satellites," *SPIE, Platforms and Systems*, **2317**, 124–130 (Jan. 1995).

[34] J. A. King, R. McGwier, H. Price, and J. White, "The In–orbit Performance of Four Microsat Spacecraft," *Proceedings 4th Annual AIAA/USU Conference on Small Satellites*, Vol. 1 (Logan, UT, 27–30 Aug. 1990).

[35] B. Steele, C. McCormick, K. Brandt, W. Fornwalt, and R. Bonometti, "Utilization of the Multiple Access Communications Satellite (MACSAT) in Support of Tactical Communications," *Proceedings 5th Annual AIAA/USU Conference on Small Satellites* (Utah State University, Logan, UT, 26–29 Aug. 1991).

[36] R. J. Bonometti, "Tactical Applications of Defense Microspace Communications," *Proceedings of the Tactical Communications Conference 1992, Vol. 1: Tactical Communications: Technology in Transition* (28–30 April 1992, Fort Wayne, IA).

[37] NASA Orbital Debris Program Office, Orbital Debris Radar Calibrations Spheres, 2007, http://www.orbitaldebris.jsc.nasa.gov/measure/oderacs.html (23 Oct. 2008).

[38] G. H. Cress, A. E. Potter, T. J. Setticeri, G. P. Sherrill, E. G. Stansbery, and D. L. Talent, "Radar and Optical Ground Measurement Final Report, Orbital Debris Calibration Spheres." NASA Report JSC–27241 (NASA Johnson Space Center, Houston, 1996).

[39] A. Tan and G. D. Badhwar, "Detecting the Dynamical State of the Atmosphere from the Orbital Decay of the ODERACS Spheres," *Journal of Atmospheric and Solar–Terrestrial Physics*, **59** (4), 431–437 (March 1997).

[40] J. Cutler and G. Hutchins, "OPAL: Smaller, Simpler, and Just Plain Luckier," *Proceedings of the 14th annual AIAA/USU Conference on Small Satellites* (Ogden, UT, Aug. 2000).

[41] W. Lan, "Poly Picosatellite Orbital Deployer MK III ICD," California Polytechnic State University—San Luis Obispo (Feb. 2008), http://CubeSat.atl.calpoly.edu/media/P–POD%20Mk%20III%20ICD.pdf (23 Oct. 2008)

[42] NASA, GeneSat–1 Technology Demonstration Mission home page, 2007, http://genesat.arc.nasa.gov (23 Oct. 2008).

[43] Aprize Satellite, Space Segment Technology web page, 2007, http://www.aprizesat.com/space.php (23 Oct. 2008).

[44] S. Metzger, S. and R. H. Pickard, "Relay," *Astronautics and Aerospace Engineering*, **1** (8), 64–67 (Sept. 1963).

[45] M. Madou, *Fundamentals of Microfabrication* (CRC Press, New York, 1997).

[46] Special Issue on Integrated Sensors, Microactuators, & Microsystems (MEMS), *Proceedings of the IEEE, 86* (Aug. 1998).

[47] M. Gad–el–Hak (ed.), *The MEMS Handbook* (CRC Press, New York, 2002).

[48] H. Helvajian (ed.), *Microengineering for Aerospace Systems* (Aerospace Press, El Segundo, CA, and AIAA, Reston, VA, 1999), ISBN 1–884989–03–9.

[49] R. Osiander, M. A. Garrison Darrin, and J. L. Champion, *MEMS and Microstructures in Aerospace Applications* (CRC Press, Taylor & Francis Group, New York, 2006).

[50] R. J. Baker, H. W. Li, D. E. Boyce, *CMOS: Circuit Design, Layout, and Simulation* (IEEE Press, New York, 1998).

[51] P. Van Zant, *Microchip Fabrication, 3rd Edition* (McGraw–Hill, New York, 1997).

[52] L. Ristic, *Sensor Technology and Devices* (Artech House, Norwood, MA, 1994).

[53] S. W. Janson, "Mini–, Micro–, and Nanosatellite Concepts," in *Micro– and Nanotechnology for Space Systems: An Initial Evaluation*, edited by H. Helvajian and E. Robinson, Aerospace Technical Report ATR–93(8349)–1 (The Aerospace Corporation, El Segundo, CA, March 1993), pp. 67–75.

[54] S. W. Janson, H. Helvajian, and E. Y. Robinson, "The Concept of Nanosatellite for Revolutionary Low–Cost Space Systems," *44th Congress of the International Astronautics Federation* (Graz Austria, Oct. 1993), Paper IAF–93–U.5.573.

[55] B. A. Wilson, *Proceedings of the Workshop on Microtechnologies and Applications to Space Systems* (Pasadena, CA, 27–28 May 1992).

[56] The Aerospace Corporation, *Proceedings of the International Conference on Integrated Micro/Nanotechnology for Space Applications* (Houston, TX, 30 Oct.–3 Nov. 1995).

[57] The Aerospace Corporation, *Proceedings of the 2nd International Conference on Integrated Micro/Nanotechnology for Space Applications* (Pasadena, CA, 11–15 April 1999).

[58] S. W. Janson, "Chemical and Electric Micropropulsion Concepts for Nanosatellites," *Micro/Nano–Technologies for Space Roundtable* (ESTEC, Nordwijk, The Netherlands, 27 March 1995), *European Space Agency Proceedings* ESA WPP–091.

[59] S. W. Janson, "Spacecraft as Assembly of ASIMs," in *Microengineering Technology for Space Systems*, edited by H. Helvajian (The Aerospace Press and AIAA, El Segundo, CA, Oct. 1995), pp. 181–258.

[60] S. W. Janson and H. Helvajian, "Batch–Fabricated Microthrusters: Initial Results," *32nd AIAA/ASME/SAE/ASEE Joint Propulsion Conference* (Orlando, FL, July 1996).

[61] E. Y. Robinson, H. Helvajian, and S. W. Janson, "Small and Smaller: The World of MNT," *Aerospace America* (Sept. 1996) pp. 26–32.

[62] E. Y. Robinson, H. Helvajian, and S. W. Janson, "Big Benefits from Tiny Technologies," *Aerospace America* (Oct. 1996), pp. 38–43.

[63] H. Helvajian, S. W. Janson, and E. Y. Robinson, "Big Benefits from Tiny Technologies: Micro–Nanotechnology Applications in Future Space Systems," *Advancement of Photonics for Space: SPIE Critical Reviews of Optical Science and Technology, SPIE CR66*, Edward Taylor (ed) (San Diego, CA, 28–29 July 1997), pp. 3–34

[64] S. W. Janson and H. Helvajian, "Batch–Fabricated Microthrusters for Kilogram–Class Spacecraft," *Government Microcircuit Applications Conference (GOMAC '98),* (Arlington, VA, March 1998).

[65] S. W. Janson, "Silicon Satellites for 21st Century Missions," *2nd International Conference on Integrated Micro/Nanotechnology for Space Applications* (Pasadena, CA, 11–15 April 1999).

[66] H. Helvajian and S. Janson, "Microengineering Space Systems," in *Microengineering for Aerospace Systems,* edited by H. Helvajian (Aerospace Press, El Segundo, CA, and AIAA, Reston, VA, 1999), chapter 2.

[67] S. Janson, H. Helvajian, and K. Breuer, "Micropropulsion Systems for Aircraft and Spacecraft," in *Microengineering for Aerospace Systems*, edited by H. Helvajian (Aerospace Press, El Segundo, CA, and AIAA, Reston, VA, 1999), chapter 17.

[68] S. W. Janson, "Mass–Producible Silicon Spacecraft for 21st Century Missions," *AIAA Space Technology Conference & Expo* (Albuquerque, NM, 28–30 Sept. 1999), AIAA paper 99–4458.

[69] D. H. Lewis Jr., S. W. Janson, R. B. Cohen, and E. K. Antonsson, "Digital Micropropulsion," *Sensors and Actuators*, **80**, 143–154 (2000).

[70] D. Sutton and R. Stroud, "Workshop 9: Untethered Flying Observer," *Proceedings of the International Conference on Integrated Micro/Nanotechnology for Space Applications* (Houston, TX, 30 Oct.–3 Nov. 1995).

[71] D. G. Sutton, R. H. Smith, L. Wiedeman, B. H. Weiller, N. Marquez, J. V. Osborn, S. W. Janson, L. Kumar, and P. D. Chaffee, "MEMS Spaceflight Testbed," *AIAA Space Technology Conference & Expo* (Albuquerque, NM, 28–30 Sept. 1999), AIAA Paper 99–4604.

[72] Silicon Designs Model 1010 Datasheet, Silicon Designs Inc. (Issaquah, WA, 2007), http://www.silicondesigns.com/Pdf/1010.pdf (28 Oct. 2008).

[73] A. Das (ed.), *Proceedings of the AFOSR/DARPA/AFRL/VSD Micro/Nanotechnology for Micro/Nanosatellites Workshop* (Albuquerque, NM, 21–22 April 1998).

[74] J. E. Pollard, C. C. Chao, and S. W. Janson, "Populating and Maintaining Cluster Constellations in Low Earth Orbit," *35th AIAA/ASME/SAE/ASEE Joint Propulsion Conference* (Los Angeles, CA, June 1999), AIAA Paper 99–2871.

[75] C. C. Chao, J. E. Pollard, and S. W. Janson, "Dynamics and Control of Cluster Orbits for Distributed Space Missions," *AAS/AIAA Space Flight Mechanics Meeting* (Breckenridge, CO, 7–10 Feb. 1999), Paper AAS 99–126.

[76] S. W. Janson, J. E. Pollard, and C. C. Chao, "Method for Deploying an Orbiting Sparse Array Antenna," U.S. patent #6,725,012 (20 April 2004).

[77] G. E. Moore, "Cramming More Components Onto Integrated Circuits," *Electronics*, **38** (8), 114–117 (April 1965).

[78] L. J. Hansen, "Spacecraft Computer Systems," in *Space Mission Analysis and Design, 2nd Edition*, edited by J. Wertz and W. J. Larson (Kluwer Academic Publishers, Boston, 1992), pp. 616–629.

[79] Intel Corp., "Moore's Law" 2002, http://www.intel.com/technology/mooreslaw (28 Oct. 2008).

[80] Intel Corp., "World's First 2 Billion Transistor Microprocessor" (2008), http://www.intel.com/technology/architecture–silicon/2billion.htm?iid=tech_arch+body_2b (28 Oct. 2008).

[81] International Technology Roadmap for Semiconductors, Executive Summary (2007), http://www.itrs.net/Links/2007ITRS/ExecSum2007.pdf (28 Oct. 2008).

[82] Microchip, PIC10F220/222 datasheet, Microchip Technology Inc. (Chandler, AZ, 2007), http://ww1.microchip.com/downloads/en/DeviceDoc/PIC10F220222%20Data%20 Sheet_41270E.pdf (23 Oct. 2008).

[83] Microchip, PIC18F1220 datasheet, Microchip Technology Inc. (Chandler, AZ, 2007), http://www.microchip.com/1010/pline/picmicro/category/embctrl/8kbytes/devices/18f1220.

[84] Motorola Inc., "M*CORE Review," Motorola Semiconductor Products Sector (2007), http://e–www.motorola.com/brdata/PDFDB/docs/MCORECUST.pdf.

[85] NEC Corp., VR 4131 product brochure, (2007), http://www.necelam.com/docs/files/ U15396EU3V0PB00.pdf (28 Oct. 2008).

[86] Atmel Corp., "AT91R40807 Electrical Characteristics" (2007), http://www.atmel.com/dyn/resources/prod_documents/DOC1367.PDF.

[87] G. J. Wells, L. Stras, and T. Jeans, "Canada's Smallest Satellite: The Canadian Advanced Nanospace eXperiment (CanX–1)," *Proceedings 16th AIAA/USU Conference on Small Satellites* (2002).

[88] E. R. Fossum, "Active Pixel Sensors: Are CCDs Dinosaurs?" *Proceedings of the SPIE, Charge–Coupled Devices and Solid State Optical Sensors III*, Morley M. Blouke, ed., vol. 1900 (July 1993), pp. 1–14.

[89] E. R. Fossum, "CMOS Image Sensors: Electronic Camera–on–a–Chip," *IEEE IEDM Tech. Dig.*, 17–25 (Dec. 1995).

[90] S. W. Janson, "Micro/Nanotechnology for Picosatellites," *22nd AIAA/USU Conference on Small Satellites* (Logan, UT, Aug. 2008), paper SSC08–VII–6.

[91] Joint Photographic Expert Group (2007), http://www.jpeg.org.

[92] D. Hinkley, "A Novel Cold Gas Propulsion System for Nanosatellites and Picosatellites," *22nd Annual AIAA/USU Conference On Small Satellites* (2008), paper SSC08–VII–7.

[93] M. Schneider, S. Kawahito, Y. Tadokoro, and H. Baltes, "High Sensitivity CMOS MicroFluxgate Sensor," *International Electron Devices Meeting Technical Digest* (Washington DC, Dec. 1997), pp. 907–910.

[94] D. K. Wickenden, R. B. Givens, R. Osiander, J. L. Champion, D. A. Oursler, and T. J. Kistenmacher, "MEMS–based Resonating Xylophone Bar Magnetometer," *Proceedings SPIE Micromachined Devices and Components IV*, Vol. 3514 (Santa Clara, CA, 20–24 Sept. 1998), paper 3514–34.

[95] V. Beroulle, Y. Bertrand, L. Latorre, and P. Nouet, "Micromachined CMOS Magnetic Field Sensors with Low–Noise Signal Conditioning," *Proceedings of the 15th IEEE International Conference on Micro Electro Mechanical Systems* (Las Vegas, NV, 20–24 2002), pp. 256–259.

[96] Honeywell, "1, 2 and 3 Axis Magnetic Sensors, HMC1051/HMC1052/HMC1053," Honeywell Solid State Electronics Center (Plymouth, MN, 2007), http://www.ssec.honeywell.com/magnetic/datasheets/HMC105X.pdf (28 Oct. 2008).

[97] Honeywell, "Three Axis magnetic Sensor Hybrid," Honeywell International Solid State Electronics Center (Plymouth, MN, 2007), http://www.ssec.honeywell.com/magnetic/datasheets/hmc2003.pdf (28 Oct. 2008).

[98] Analog Devices, ADIS16250/16255 Programmable Low Power Gyroscope data sheet, Analog Devices (Norwood, MA, 2007), http://www.analog.com/UploadedFiles/Data_ Sheets/ADIS16250_16255.pdf (28 Oct. 2008).

[99] S. Janson, A. Huang, W. Hansen, L. Steffeney, and H. Helvajian, "Development of an Inspector Satellite Using Photostructurable Glass/Ceramic Materials," *AIAA Space 2005 Conference* (Long Beach, CA, 30 Aug. 2005), AIAA paper 2005–6802.

[100] AeroTech Rocketry, "2007–2008 Product Catalog," AeroTech Consumer Aerospace Division, RCS Rocket Motor Components Inc. (Cedar City, UT, 2007), http://www.aero-tech–rocketry.com (28 Oct. 2008).

[101] C. Rossi, B. Larangot, P.-Q. Pham, D. Briand, N. F. de Rooij, M. Puig-Vidal, and J. Samitier, "Solid Propellant Microthrusters on Silicon: Design, Modeling, Fabrication and Testing," *Journal of Microelectromechanical Systems*, **15** (6), 1805–1815 (Dec. 2006).

[102] K. Kondo, S. Tanaka, H. Habu, S. Tokudome, K. Hori, H. Saito, A. Itoh, M. Watanabe, and M. Esashi, "Vacuum Test of a Micro–Solid Propellant Rocket Array Thruster" *IEICE Electronics Express,* **1** (8), 222–227 (2004).

[103] Astrium Propulsion Division, Monoprellant Thruster Data Page, EADS Astrium (2008), http://cs.astrium.eads.net/sp/SpacecraftPropulsion/MonopropellantThrusters.html #CHT05 (28 Oct. 2008).

[104] J. Mueller, "Thruster Options for Microspacecraft: A Review and Evaluation of the State-of-the-Art and Emerging Technologies," in *Micropropulsion for Small Spacecraft*, Progress in Astronautics and Aeronautics, Vol. 187 (AIAA, Reston, VA, 2000), pp. 45–137.

[105] F. Rysanek, J. W. Hartmann, J. Schein, and R. Binder, "MicroVacuum Arc Thruster Design for a CubeSat Class Satellite," *16th Annual USU Conference on Small Satellites* (Logan, UT, 2002), paper SSC02–I–2.

[106] J. J. Carpenter, T. R. Haas, and L. B. King, "Michigan Technological University Aerospace Enterprise," *AIAA Space 2004 Conference* (San Diego, CA, 28–30 Sept. 2004), AIAA paper 2004–6119.

[107] T. E. Zipperian and L. R. Dawson, "The Second Revolution—Mixed–Technology Integrated Microsystems," *Digest of the 60th Device Research Conference* (June 2002), pp.129–131.

[108] L. Xuwen, L. Li, L. Huawang, C. Yinjian, H. Dexin, and Y. Genqing, "Silicon Solid–State Small Satellite Design Based on IC and MEMS," *Proc. of the 5th International Conference on Solid State and Integrated Circuit Technology* (Oct. 1998), pp. 392–395.

[109] L. Stenmark and J. Köhler, "NANOSPACE–1, The MNT–Based Nanosatellite," *Proceedings of the 4S Symposium: Small Satellites, Systems, and Services* (ESA SP–571) (La Rochelle, France, 20–24 Sept. 2004).

[110] T. Grönland, P. Rangsten, H. Johansson, and J. Bejhed, "Development of MEMS–Based Spacecraft Propulsion" *42nd AIAA/ASME/SAE/ASEE Joint Propulsion Conference & Exhibit Agenda* (2006), AIAA paper 2006–4873.

[111] H. Nguyen, R. Thorslund, G. Thornell, J. Kohler, and L. Stenmark, "Structural Integrity of Flat Silicon Panels for Nanosatellites: Modeling and Testing," *Journal of Spacecraft and Rockets*, **43** (6), 1319–1327 (Nov.–Dec. 2006).

[112] H. Nguyen, "Microsystem Interfaces for Space," Ph.D. thesis, Uppsala University, Uppsala, Sweden, 2006.

[113] H. Kratz, E. Öjefors, and L. Stenmark, "Micromachined S–band Patch Antenna with Reduced Dielectric Constant," *Sensors and Actuators A*, 130–131, 478–484 (2006).

Chapter 3

Clementine: The Rebirth of Smaller, High-Value Missions

Pete Rustan*

3.1 Introduction

As the space industry continued to mature throughout the 1970s and 1980s, spacecraft users demanded significantly increased capabilities. This led the U.S. spacecraft industry to focus on large, multiple-mission systems with increased capacity; however, these spacecraft were taking longer to build and more money to deploy. By 1991, many of those involved with the day-to-day management of government-funded spacecraft began actively working to reverse this trend, using innovative technologies and streamlined management procedures.

The Clementine mission was a result of these efforts. At the time, researchers at the Ballistic Missile Defense Organization (BMDO, formerly the Strategic Defense Initiative Organization) were developing the technology base for a constellation of space-based interceptor satellites (known as Brilliant Pebbles) and a satellite-based surveillance system (Brilliant Eyes). They were therefore continually looking for opportunities to fly these new technologies, which would provide new and improved capabilities for missile defense and other applications. The Clementine mission was conceived from a desire to flight-qualify the latest technology being developed by BMDO and to help to change the trend in satellite design, production, and operation. As part of my duties at BMDO, I managed the development of advanced technologies for a new generation of smaller and more capable satellites, and ultimately became program manager for Clementine.

3.2 The Mission

The Clementine spacecraft is shown in Fig 3.1. NASA supported BMDO to demonstrate recently developed advanced technologies while enhancing the scientific community's body of knowledge through imaging and mapping the entire lunar surface. The NASA science team, led by Gene Shoemaker and Paul Spudis, gave a strongly positive report about the scientific utility of such a mission,[1] and recommended flying the newly developed miniature cameras and laser altimeter. The Clementine spacecraft would also demonstrate autonomous navigation and potential interceptor capabilities of interest to BMDO by flying to a near-Earth asteroid (NEA). As recommended by The Johns Hopkins University's Applied

*Ground Enterprise Directorate, National Reconnaissance Office
Chantilly, Virginia

Fig. 3.1. The Clementine spacecraft. A full-size engineering model is on display at the National Air and Space Museum in Washington, DC.

Physics Laboratory following a detailed trajectory study, a flyby of the NEA 1620 Geographos was planned into the mission.

Other important BMDO programs in existence prior to 1991 had an enormous impact on the Clementine program. First, the Lawrence Livermore National Laboratory (LLNL) was building sensors for the Brilliant Pebbles program.[2] Second, the Naval Research Laboratory (NRL) had built the Relay Mirror Experiment and the Low-power Atmospheric Compensation Experiment, achieving desired performance on budget and on schedule.[3] BMDO wanted to take advantage of these successful programs, and therefore asked LLNL to build and test the payload sensors; NRL was designated as the overall system integrator, bus builder, and mission operator. In March 1992, BMDO signed a contract with NRL for the Clementine mission at a cost of $55 million, including the LLNL payload sensors. Since BMDO had been managing the LLNL effort directly, the management philosophy with LLNL was not altered. BMDO, LLNL, and NRL mutually developed the interface control documents. Further, BMDO negotiated a contract with the Air Force to use a refurbished Titan 2G for launch at a price of $20 million. Finally, NRL built a ground station in Alexandria, Virginia, for command and control and initial mission data processing. The building was an unused National Guard facility near the Braddock metro station; it came to be known as

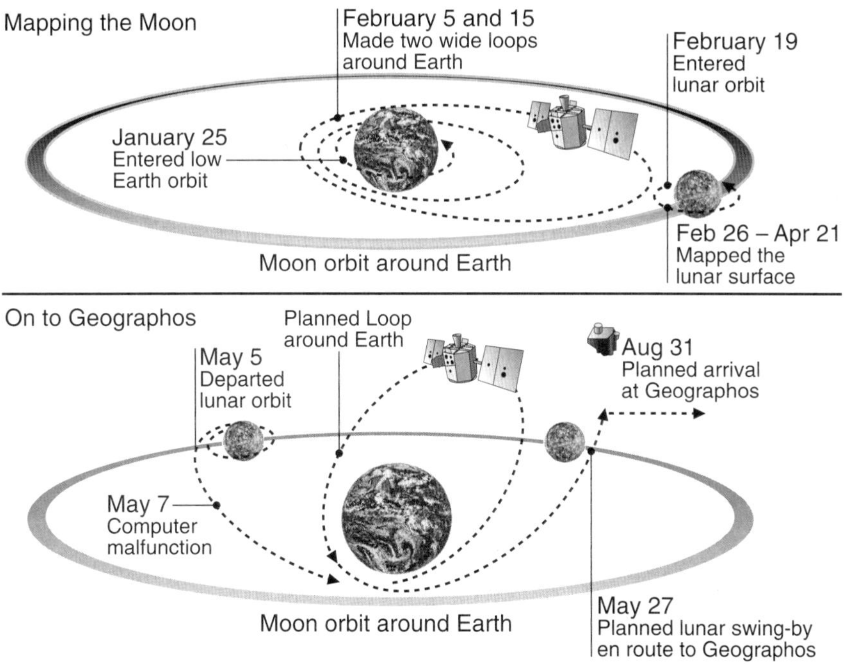

Fig. 3.2. Clementine flight profiles.

"the Batcave" because bats were found in it during reconstruction and remodeling. The total cost of the Batcave and the initial six months of satellite operations was $5 million. Therefore, the direct cost to BMDO for the Clementine program was $80 million. Of course, there were many nonrecurring costs paid by previous BMDO programs, especially sensor development costs for Brilliant Pebbles and bus development costs for the Relay Mirror Experiment and Low-power Atmospheric Compensation Experiment.

The actual Clementine flight profile was divided into five segments:

- Launch the satellite into low Earth orbit and perform initial satellite checkout.
- Burn the solid rocket motor, sending the satellite into a lunar transfer orbit, and thence into an elliptical orbit around the moon.
- Perform the lunar imaging and mapping mission for two to three months.
- Exit the lunar orbit, swing by Earth to get a "kick," and fly by NEA 1620 Geographos.
- Image Geographos and transmit the navigation/interception data back to Earth.

3.3 Mission Outcome

Clementine was launched 25 January 1994, from Vandenberg Air Force Base in California. The mass of the spacecraft at launch was 1690 kg, including the launch adapter, the kick motor, and the interstage. The actual spacecraft dry weight was 228 kg (424 kg wet) after ejection of the solid rocket motor.

Figure 3.2 shows the Clementine flight profiles. After a period of outgassing and calibration in low Earth orbit, the solid rocket motor was burned and separated from the main craft, sending the satellite to lunar transfer orbit. Subsequently, main-engine burns were used to insert the satellite into a 5-hour lunar orbit on 19 February 2004. This orbit was selected to provide complete imaging and mapping of the lunar surface during 40% of the orbit and to transmit command and control and imaging data to one of three NASA Deep Space Network terminals (at Goldstone, CA; Madrid, Spain; and Canberra, Australia) during the remaining time. The lunar mission was an unqualified success, not only flight-qualifying many new technologies and sensors, but also providing a trove of new scientific information about the moon. After less than three months in lunar orbit, on 5 May, Clementine commenced transit to NEA 1620 Geographos. The satellite successfully exited lunar orbit and swung around Earth for the desired boost. Unfortunately, during its initial flight to the asteroid, a bad command sent from the Batcave on 7 May caused multiple firings of the attitude-control thrusters, expending most of the fuel needed for orbit adjustment and making it impossible to perform the planned flyby; however, mission operators continued to monitor the satellite for several months to evaluate the various technologies.

3.4 New Payload Technologies

The payload sensors consisted of two lightweight star trackers (which served as wide-field cameras when not used for attitude control), an ultraviolet-visible (UV-VIS) camera, a near-infrared/shortwave-infrared (NIR/SWIR) camera, a long-wave infrared (LWIR) camera, and a combination laser ranger (LIDAR) and high-resolution imager.[4] All the collected data, along with spacecraft health and status information, were stored in a solid-state data recorder with 2 Gb of random access memory (RAM). It was then transferred at 128 kbps to the Batcave using the spacecraft's S-band transponder and via NASA's Deep Space Network.

The sensors contained 156 unique part types. Of these, 125 were high-reliability components (883B, JANTX, R reliability level), and the other 31 were commercial. Of the commercial parts, 7 were ceramic, and 24 were plastic-encapsulated devices. The parts were screened by LLNL and NRL to ensure compliance with SSD-D-AS214. Table 3.1 describes the characteristics of the payload sensors. There are many significant features that should be noted:

- These cameras were very light—about an order of magnitude lighter than prior state-of-the-art units, ranging from 0.29 kg for the star tracker to 2.1 kg for the LWIR camera.

Table 3.1. Characteristics of Clementine's Instruments

	UV-VIS Imager	Star Tracker	NIR Imager	LWIR Imager[a]	High-Resolution Imager[a]	LIDAR Receiver[b]	LIDAR Transmitter[c]
Power (W)	4.5	4.5	11.0	13.0	9.5	N/A, housed in high-resolution imager	6.8 at 1 Hz; 2.6 quiescent
Mass (kg)	0.41	0.29	1.92	2.10	1.12	N/A, housed in high-resolution imager	1.25
Focal-plane array	CCD	CCD	InSb	HgCdTe	CCD	Si APD	
Pixel format	384 × 288	384 × 576	256 × 256	128 × 128	384 × 288	Single cell	
Pixel size (μm)	23 × 23	23 × 23	38 × 38	50 × 50	23 × 23	0.5 mm^2	
Clear aperture (mm)	46	14	29	131	131	Shared with high-resolution imager	38
Focal length (mm)	90	17.5	96	350	350	Shared with high-resolution imager	99
Field of view (deg)	5.6 × 4.2	28 × 43	5.6 × 5.6	1.0 × 1.0	0.4 × 0.3	0.057	
Resolution per pixel (arcsec)	52.6 × 52.6	262.4 × 268.9	78.8 × 78.8	28.1 × 28.1	3.6 × 3.6	205.2	
Filter bandpass (nm)	415 ± 20 750 ± 5 900 ± 15 950 ± 15 1000 ± 15 400 to 1000	400 to 1100	1102 ± 30 1248 ± 30 1499 ± 30 1996 ± 31 2620 ± 30 2792 ± 146	8.0 to 9.5	415 ± 20 560 ± 5 650 ± 5 750 ± 10 400 to 800 opaque	0.4 to 1.1	1.064 and 0.532

[a] Amber Engineering provided the focal planes and the dewars for these cameras.

[b] The A/D resolution of the LIDAR receiver was 14 bits (40 m/bit), whereas all of the cameras had a resolution of 8 bits.

[c] McDonnell Douglas provided the laser transmitter.

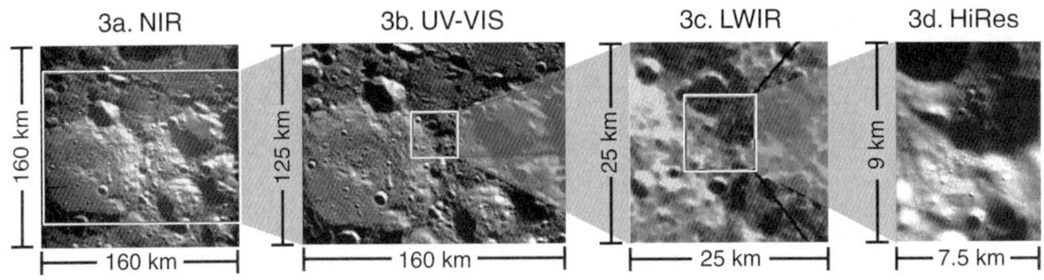

Fig. 3.3. Example of images collected by the four Clementine cameras while looking at the same lunar background.

- These cameras consumed about one-third the power of traditional cameras, ranging from 4.5 W for the star tracker and UV-VIS camera to 11 W for the LWIR camera.
- The UV-VIS, LWIR, and high-resolution cameras were boresighted to the NIR/SWIR camera. Figure 3.3 illustrates the field of view for each camera. The NIR/SWIR provided the widest field-of-view at 5.6 × 5.6 deg, or roughly 160 × 160 km. The UV-VIS camera covered slightly less area, at 5.6 × 4.2 deg or about 160 × 125 km. The LWIR camera, aligned with the center of the UV-VIS camera, had a field of view of 1 × 1 deg, or about 25 × 25 km. The high-resolution camera captured the 0.4 × 0.3 deg field of view in the center of the LWIR camera, or 9 × 7.5 km. Because the high-resolution camera covered the visible spectrum from 400 to 800 nm, the complete sensor suite allowed zooming from 0.4 to 9.5 μm over an area from 160 km to a few kilometers.
- The UV-VIS, NIR/SWIR, and LWIR cameras had filter wheels providing 11 different spectral bands, plus broadband imaging. Tens of thousands of images were collected over these 11 bands, providing for a comprehensive 1.8-million-image atlas of the lunar surface.
- The spacecraft orbit was selected to provide two passes over every square kilometer of the moon during two months of mapping operations. These cameras imaged the entire lunar surface at a resolution between 10 m (for the high-resolution camera) and 200 m (for the LWIR camera). The total imaged area is comparable to the surface area of Africa.

Figure 3.4 shows the compact LWIR camera with the required cryocooler. The unit measured 15 × 15 × 40 cm, and in the early 1990s, this was considerably smaller than most units in use. The only recurring problem with camera operations was related to the Sterling-cycle cryocoolers, used to maintain the temperatures of the NIR/SWIR and LWIR focal-plane arrays at 77K and 65K, respectively. A normal cycle of continuous lunar mapping lasted nearly an hour, but it was difficult to maintain the desired LWIR focal-plane array temperatures for that entire period. In practice, the NIR/SWIR and LWIR sensors were turned on about

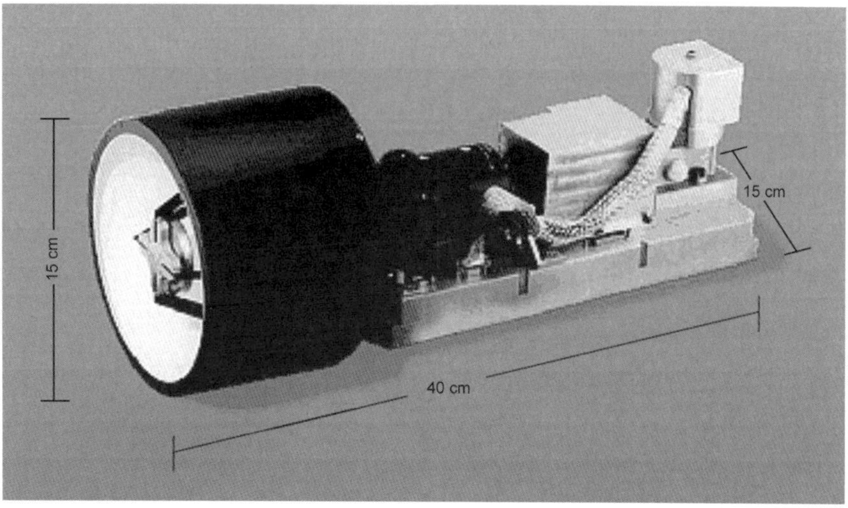

Fig. 3.4. Actual dimensions of the LWIR camera.

30 min and 20 min prior to collection, respectively, to reduce the effects of dark currents on imaging. Dark currents often continued to change after the initialization time—probably because of slow warming of the cold filter located at the end of the cold shield and/or radiative thermal loading from the moon. In particular, the LWIR cryogenic cooler began to drift after about 10 min of continuous mapping, and if the dark currents did not allow good imaging, the camera was turned off. In retrospect, both cryocoolers should have been designed to use more power and to operate at lower temperatures.

Figure 3.5 shows the actual dimensions of the LIDAR units. The LIDAR receiver was housed with the high-resolution camera. The LIDAR used an Nd-YAG pulsed laser that could transmit in two modes. When operating at 1.06 μm, it had less than 10 msec pulsewidth, energy of 171 mJ, and a divergence less than 500 microradians. When operated at 0.532 μm, the pulse had energy of 9 mJ and a divergence of 4 milliradians. The laser transmitter could be operated continuously at a rate of 1 Hz or in bursts at a much higher rate. The resolution of the LIDAR receiver's analog-to-digital converter was 14 bits, or about 20 m/bit. As will be discussed, the lunar-surface ranging data was one of the most important scientific contributions of the Clementine spacecraft.

3.5 New Bus Technologies

The spacecraft bus introduced a number of new technologies. Of particular note was the debut of parallel processing architecture via a MIL-STD-1750A and a new R3081 processor. The attitude control system featured a newly developed ring laser gyro and interferometric fiber gyro. For the first time, a NiH_2 common-pressure-vessel battery and a frangibolt release mechanism were used in space.

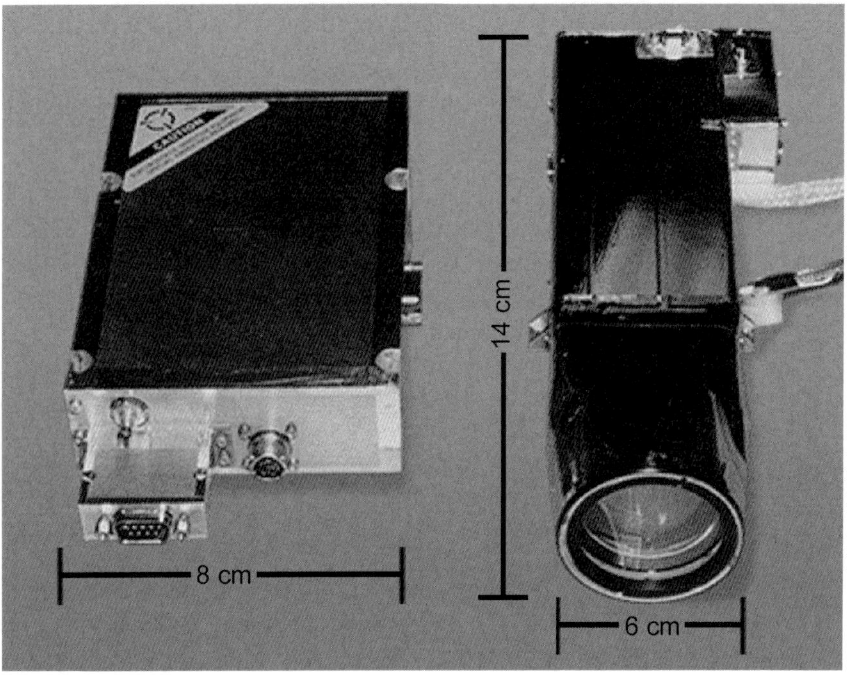

Fig. 3.5. The LIDAR units: the laser transmitter (left) and the laser receiver (right), which also housed the high-resolution camera.

3.5.1 Processors

It is difficult in 2008 to speak about significant innovation in command and data handling systems and space microprocessors for a spacecraft built in the early 1990s. Given that processor clock speed has followed Moore's Law (doubling every 18 months), innovations back then seem archaic by today's standard. It may be useful to evaluate the spacecraft computers by looking at their contemporaries. At the time, some of the more notable space-qualified microprocessors included the Inmos T800 transputer, which operated at 20 MHz and was used on the 1991 UoSAT-5 remote sensing mission, as well as the Intel 80C86 and the Motorola 68302 chips, both with 8-bit registers, clock speeds of 7 MHz, and processing capability of 2 MIPS (millions of instructions per second). For comparison, the commercial terrestrial market was using the 32-bit 486 DX microprocessor introduced in 1989 with a clock speed of 50 MHz, and the 32-bit Pentium microprocessor introduced in 1993 with a clock speed of 66 MHz. While the space industry normally lags behind the terrestrial processor market by about six to eight years, the Clementine program wanted to reduce this gap and take advantage of the latest advances in computer processing.

The program also sought an error-tolerant processor that could handle anticipated total ionized radiation and single events upsets in the Clementine orbit. In

Table 3.2. Clementine Spacecraft Controllers

	RH 1750 Chipset	UTMC69 R6000 Processor	R3081 32-bit RISC Processor
Command and telemetry	Primary		Backup
Housekeeping	Primary		Backup
Data formatting and sequencing	Backup	Primary	Backup
Image compression		Controls JPEG chipset	Backup
Optical sensor controller		Primary	Backup
Image processing	Star tracker only		Primary
MIPS	1.7	8.0	18
MFLOPS	0.7	0	3.5
Radiation immunity (rad Si)	> 100 k	> 100 k	> 15 k
Mass (kg)	0.5	< 1.0	< 1.0
Volume (cm^3)	575	800	800
Comments	Advanced MIL-STD-1750A architecture	16-bit RISC rad-hard microcontroller	Commercial 32-bit RISC processor

addition to the standard attitude control system command and telemetry functions, the processor had to perform signal processing, data handling and formatting, on-board processing, and data reduction. A MIL-STD-1750A 16-bit processor with 1.7 MIPS was used for safe mode, attitude control, and housekeeping operations, while a newly developed R3081 32-bit processor with RISC (reduced instruction set computer) architecture capable of 18 MIPS was used for image processing (it also served as a backup for the UTMC69 16-bit RISC radiation-hardened micro-controller). This was the first space application of a 32-bit RISC processor, and it provided critical data for the future development of the RH3000 (20 MIPS) and RH6000 (40 MIPS) processors, which have been used extensively in space during the last 15 years. The Clementine processors worked well during the short mission, surviving a total dose greater than 20 krad at a cost of only $400,000. Table 3.2 describes the characteristics of the selected processors.

3.5.2 Solid-State Data Recorder

A 2-Gb solid-state data recorder (SSDR) composed of 500 commercially available 4-Mb DRAM was selected for the mission. This memory was four times larger than any previous flight-qualified SSDR, and it provided storage at a reception rate greater than 20 Mbps and with a bit-error rate of 10^{-10}. The SSDR

Table 3.3. Clementine Data Recorder

Capacity	2.0 Gb
Size	2130 cm^3
Mass	3.4 kg
Reliability	0.99 (1 year)
Power – maximum	15 W
Power – standby	< 1 W
Radiation tolerance	30 krad (Si)
Memory	4 Mb DRAM

provided high reliability via redundant error detection and correction, wide active fault management, and built-in test. The goal was to validate the use of high-density-interconnect multichip module packaging and radiation-hardened ASICs, which could provide approximately five times better power per unit area and cost savings over programmable logic arrays. Table 3.3 provides some of the SSDR metrics. The SSDR, built by SEAKR Engineering, worked extremely well during the mission.

3.5.3 Spacecraft Command Language Software

The software to control the 1750A processing functions was written in C, but a new spacecraft command language was developed to provide "English-like" programming for autonomous command and control. It enabled command-and-control applicable syntax and semantic language via scripts and rule-based scenarios; embedded kernel software was available to support real-time execution of directed operations. Because it was "English-like," it promoted rapid prototyping and implementation and could be used daily during the lunar mapping mission. This approach has wide applicability for autonomous systems and unmanned command and control applications. Figure 3.6 shows the three spacecraft command language elements and the three interfaces: the language, the development environment, and the kernel.

3.5.4 Solar cells

Clementine was powered by 2.3 m^2 of gallium arsenide on germanium (GaAs/Ge) solar-cell arrays, providing 460 W at 30 V dc. Clementine flew a thin version (5.5 mil) of this substrate, and although it did not beat the 19% efficiency of previously flown 8-mil-thick GaAs/Ge, it was significantly lighter. It also weighed less and generated more energy than conventional silicon cells. The GaAs/Ge solar cells had been developed by the Air Force Wright Aeronautical Lab (now the Air Force Research Laboratory) MANTECH program. They performed well throughout the mission.

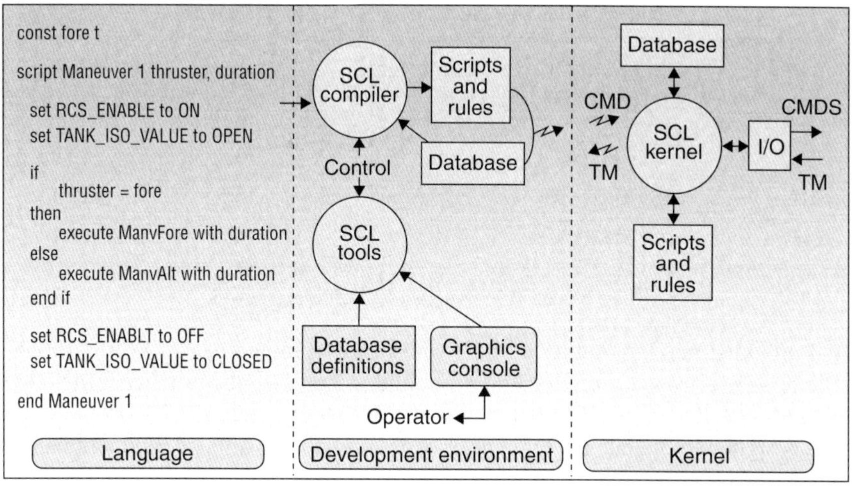

Fig. 3.6. Example of the spacecraft command language (SCL) instructions set, development environment, and kernel.

3.5.5 NiH_2 CPV Battery

The Clementine spacecraft flight-qualified a 15 Ahr NiH_2 battery with a unique containment and fin system design. At that time, about 2500 NiH_2 independent-pressure-vessel (IPV) battery cells had been flown by the Department of Defense and the commercial industry. Common-pressure-vessel (CPV) batteries offered twice the power and half the mass of comparable IPV batteries. As such, they were the preferred option for Clementine, providing high energy density, low volume-to-mass ratio, and high cycle abilities. At 47.1 Whr/kg, the Clementine CPV battery represented a significant advance in the state of the art. Table 3.4 summarizes the various batteries being used in the 1980s and the early 1990s. The Clementine battery worked far better than expected. Even though there were concerns about recharging it after total discharge, in practice, the battery exceeded the manufacturer's specifications concerning depth of discharge without any degradation after recharge. The Johnson Controls CPV battery, first tested on Clementine, has become today's industry standard. While the advertised energy density has risen to only about 60 Whr/kg, the batteries have been built to much larger sizes (including a 100 Ahr version).

3.5.6 Attitude Control System and Reaction Wheels

The spacecraft bus was three-axis stabilized. The attitude control system consisted of two lightweight inertial measurement units (a ring laser gyro and an interferometric fiber-optic gyro), two star trackers, twelve thrusters, and four reaction wheels. With the exception of the thrusters, all the attitude control system components were newly developed and flight-qualified for the first time. The reaction wheels had 0.05 deg control and 0.03 deg attitude knowledge. They weighed 2 kg

Table 3.4. Clementine Candidate Battery Systems

	Configuration	Volume (cm^3)	Mass (kg)	Whr/kg
NiCd	22 cells	5112	17.0	26.5
NiH_2	22 cells IPV	49,685	19.7	22.8
NiH_2	11 cells (2-cell IPV)	19,271	16.5	27.4
NiH_2	1 cell (CPV)	10,160	9.5	47.1

[a] CPV: common pressure vessel; IPV: independent pressure vessel.

and provided maximum torque of 2 N-m at 2500 rpm and peak power of 21 W. The drive electronics were located inside the wheel assembly and were included in the 2 kg allocation. Each wheel measured 22.9 cm wide by 5.9 cm high and was designed to operate for three years with a reliability of 0.989 and a radiation tolerance of 20 krad. The Ball Aerospace reaction wheels were used to slew the spacecraft during the lunar mapping mission; they performed well.

3.5.7 Interstage Adapter Radiation Experiment

A STAR 37FM solid rocket motor was used for the translunar transfer injection maneuver, placing the spacecraft in a lunar phasing loop orbit. The combination of the solid rocket motor and the mating equipment was referred to as the interstage adapter. After separation, it remained in a highly elliptical orbit and performed specific radiation experiments. Figure 3.7 shows the interstage adapter apart from the Clementine spacecraft. Figure 3.8 shows the interstage adapter's orbital profile.

The interstage adapter had instruments to measure the total ionized dose and proton level in its orbit. The experiment was named the Radiation and Reliability Assurance Experiment (RRELAX). Henry Garrett of the Jet Propulsion Laboratory directed RRELAX and analyzed the results. RRELAX included the characterization of SRAM and CMOS field-effect transistor chipsets. These SRAMs were monitored to determine the incidence of high-energy protons, alpha particles, and heavy cosmic ray single-event upset between 0.9 and 20 MeV. RRELAX command and control was performed by the U.S. Navy's Blossom Point satellite tracking facility.

3.5.8 Lightweight Release Systems

For the first time, a frangibolt release mechanism was used to eliminate pyrotechnic ordnance. The frangibolt reduced weight about 40% over conventional pyrotechnic ordnance and harness. A frangibolt operates by heating a shape-memory alloy actuator, which elongates and breaks a notched bolt held in tension. The advantages of frangibolt are straightforward: they provide a high-strength release mechanism, eliminate the safety concerns of explosives, improve reliability by eliminating sliding friction and explosive charge, reduce ground flow

Fig. 3.7. The Clementine interstage adapter.

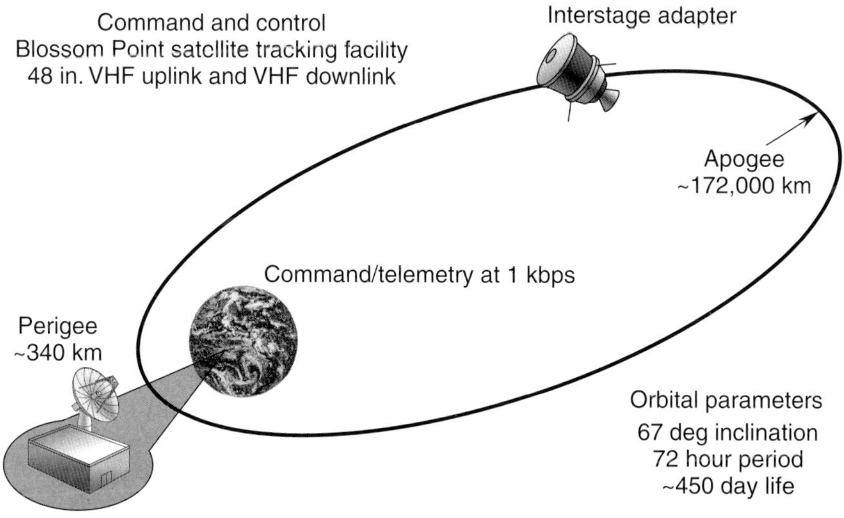

Fig. 3.8. The Clementine interstage adapter flight profile.

testing and handling, and employ simplified control electronics. The frangibolt mechanism worked perfectly on Clementine. Frangibolts have since been adopted throughout the spacecraft industry.

3.5.9 Composite Structures and Materials

Table 3.5 describes the satellite structural components. Aluminum alloy, aluminum honeycomb, composite isogrid, and carbon-fiber/epoxy were used in various places to minimize structural weight while providing the stiffness required for both the spin-balanced injection burn and the three-axis stabilization for the remaining part of the mission. The spacecraft center of gravity was controlled along the ΔV axis to minimize the torques induced during large thruster burns.

In all, the Clementine program integrated 23 advanced, lightweight technologies in a small spacecraft, both to flight-qualify them and to determine their utility for scientific and defense missions. Clementine ultimately mapped 100% of the lunar surface using 11 different spectral bands, allowing scientists to determine lunar topography, mineralogy, and gravity. While many of these technologies had been developed for other missions, the Clementine spacecraft was used for their flight-qualification. These technologies were critical to the advancement of a new generation of spacecraft, allowing for reduced size and weight, increased power,

Table 3.5. Clementine Structural Components

Structure	Materials of Construction
Spacecraft Primary Structure	
Primary structural frame	Aluminum alloy
Mid-deck and 8 enclosure panels	Aluminum honeycomb panel with aluminum alloy face sheets
Solar-panel support structure	6061-T6 extrusion and sheet with titanium hardware
Solar-array assembly substrates	Composite isogrid
Sensor Bench Assembly	Aluminum honeycomb panel with beryllium stand-off
Interstage Adapter	
Forward and aft adapter ring	Milled aluminum alloy
Skin	Carbon (graphite) fiber/epoxy composite
Longerons	Carbon fiber/epoxy
Launch Adapter	
Forward and aft payload adapter ring	Milled aluminum alloy
Skin and stringers	Aluminum alloy

and enhanced capabilities. With the few exceptions already noted, these technologies performed admirably; however, the mission was not just about technology, but also about streamlined management practices.

3.6 Mission Management

The Clementine mission implemented many innovative streamlined management techniques. There was a clear intent to challenge the management practices being applied in the 1980s and early 1990s and to resurrect the 1960s philosophy of building spacecraft faster and cheaper. Two major tenets concerning satellite construction in the 1960s were that satellites did not possess a marked degree of complexity and that failure was acceptable (because the space industry was in its nascent stage). In the early 1990s and even today, government managers and industry leaders have become extremely reluctant to accept the risk of failure.

Many groups today tout the advantages of small spacecraft, and some even define the acceptable mass, power, time to market, cost, etc. Some groups—especially those advocating "operationally responsive space"—not only subscribe to these specific parameters, but also describe specific procedures that must be followed. The Clementine philosophy does not advocate this approach, but simply asserts that satellites should be built to achieve the highest possible value (greatest performance at the lowest end-to-end cost), regardless of size, weight, cost, time to market, etc. Decision makers and program managers should push for the integration of advanced technologies while streamlining management practices to achieve better capabilities at acceptable cost. The operationally responsive space charter focuses on demonstrating the ability to build, integrate, and launch satellites on a short cycle, instead of concentrating on demonstrating new capabilities with a high return on investment. Hopefully, the confluence of the various schools of thought in this area will drive the space community toward value and streamlined management principles. The Clementine team implemented some important management practices that can be used for the development of any satellite. Perhaps the five most important are:

- **Use the most advanced, readily available technologies.**

Several significant technology development programs within BMDO preceded the actual Clementine program. The Brilliant Pebbles office developed the lightweight cameras and the laser altimeter, while many of other 23 technologies were built as part of other space programs (such as the Relay Mirror Experiment and the Theater High Altitude Area Defense program). At that time, BMDO was a new organization, and everyone was concentrating on developing new enabling technologies for a very challenging mission. This provided the stage for integrating technologies into specific missions. The major challenge was to adapt newly developed technologies that were already at NASA Technology Readiness Level 6 (that is, a system/subsystem model or prototype had been demonstrated in a relevant environment) into a single integrated mission that was not identical to the initial application. Hardware and software changes and extensive testing were required to integrate the various technologies into Clementine. To significantly increase space capabilities, Clementine planners also strove to take full advantage of advances in information technology by implementing the best and latest commercial computational tools.

The Clementine process of technology development and integration is generally not followed by government organizations today. The usual approach for a new space program is to develop most of the new technologies by the program office after authorization to proceed (ATP). Thus, challenges or setbacks in technology development can significantly increase mission cost and schedule and can necessitate substantial staffing additions.

Figure 3.9 illustrates the difference between this approach and the Clementine model (referred to as the technology maturity approach). In the Clementine model, all the development and qualification tests to attain Technology Readiness Level 6 occur separately, including any potential changes to adapt commercial-off-the-shelf applications, before authorization to proceed. This is not to say that a "technology office" or "R&D organization" should be responsible for maturing all the technologies before authorization to proceed, but all major technology challenges must be addressed by having technology maturity programs prior to authorization to proceed.

- **Integrate only complementary payloads in the same spacecraft.**

As the spacecraft industry matured through the 1970s and 1980s, there was a natural tendency to add new payloads to previous spacecraft to minimize launch

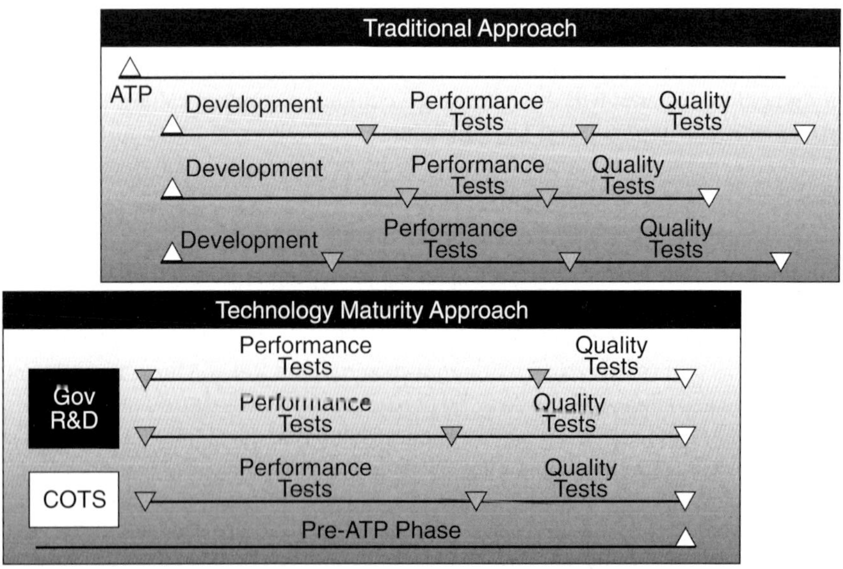

Fig. 3.9. Comparison of the traditional approach to government satellite procurements versus the Clementine technology maturity approach.

vehicle costs and to provide more capabilities through a system-of-systems configuration. This trend drove spacecraft size, power, and weight to always match the launch vehicle's maximum lift capacity. It sounds like a good thing to do, but it turned out to increase spacecraft complexity, time to launch, and system cost. By many measures, the advantages were not commensurate with the penalties.

Figure 3.10 shows how spacecraft cost increases more rapidly with the addition of unrelated payloads, as opposed to complementary payloads. This argues for smaller, focused spacecraft designs to deliver limited capabilities. Additional capabilities can be obtained by increasing the number of satellites. The satellite constellation can consist of identical spacecraft or spacecraft subsystems with plug-in modules to provide other capabilities. Users should consider constellations of smaller satellites to deliver those needs that are often integrated on medium or large single satellites. Other important advantages of satellite constellations include shorter revisits and increased architectural redundancy and survivability.

- **Balance risk with mission assurance.**

To deliver the highest return on investment, spacecraft customers and producers must not necessarily adopt existing management practices, but must continue to innovate, seeking a balance between risk and mission assurance. The tremendous advances in spacecraft and terrestrial technologies during the first 50 years of the space age have created numerous challenges and opportunities.

Perhaps the greatest challenge is to understand the potential risk introduced by new technologies and to manage that risk by using streamlined processes to provide an acceptable risk/reward ratio. Mission planners do not get partial credit for being innovative and industrious if the spacecraft fails to deliver. To be successful, they must manage and limit the risk on all aspects of spacecraft design, manufacturing, assembly, integration, and testing.

Perhaps the greatest opportunity is to transform the spacecraft industry so that it functions more like a mass market than a collection of unique development projects. Continuous or "hot" production lines could significantly reduce

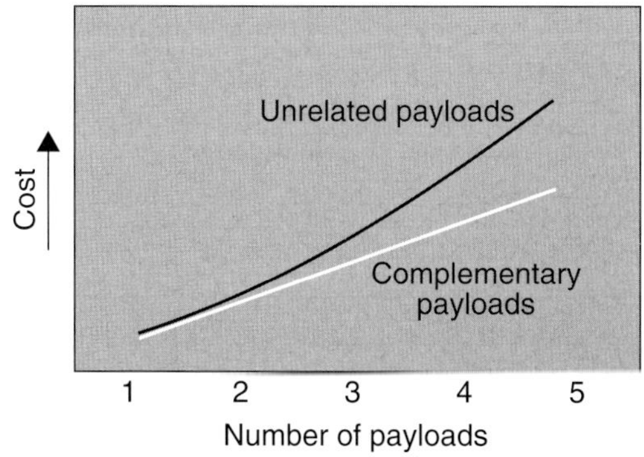

Fig. 3.10. Estimate of spacecraft cost as a function of payloads.

component failures, which account for roughly one-third of today's spacecraft problems. Many of these component failures occur due to a new application in a novel configuration, without sufficient testing to understand potential failure modes. Production lines have been successfully demonstrated for Iridium, Globalstar, GPS, ORBCOMM, and other satellite constellations. However, their successful management practices have not been transferred to other satellite constellations. Another opportunity when building multiple systems is to develop standard buses for various payload weights, powers, and desired capabilities. Even though these buses will not be optimized to the specific set of requirements of any given configuration, they will provide a good balance between risk and mission assurance if used with a hot production line concept.

- **Develop an engineering model and build spare flight units for the critical subsystems.**

Before building actual flight units, it is prudent to build brassboard units and full engineering models of the actual subsystems. The time and money spent building these units will significantly reduce overall program cost and schedule because they allow early identification and mitigation of the mechanical, electrical, and thermal problems likely to occur during assembly and integration. Today, there seems to be a strong reliance on modeling and simulation instead of actual qualification and testing. There is no substitute for a thorough qualification program. No testing should ever be compromised or shortened to support a "faster, better, cheaper" mentality. The results of full functional testing of the engineering units should serve as prerequisites before design finalization and construction of the flight units. The engineering units should also go through environmental testing with the same rigor as the final flight units. When this is done, the final flight unit will often go through environmental testing without any new problems being identified.

It is crucial to build two spare engineering models for each of the critical spacecraft subsystems. If problems are encountered with the flight unit, the engineering units can be swapped in. One engineering unit must be kept intact to evaluate test results during operations. If malfunctions develop, or if new software needs to be built to enhance capabilities or correct a problem, then an engineering model and a full testbed are required to validate and verify the new work. To increase mission assurance, ground functional and environmental tests must reflect the spacecraft internal and external environment. That is, the spacecraft should be tested the way it will be flown.

- **Improve subcontracting management.**

When a commercial company is awarded a government satellite contract, it is because this company has provided the best proposal. This proposal includes subcontracting information, with sufficient detail to instill confidence in the government evaluators. However, the initial agreements between the prime contractors and the subcontractors do not cover all the required specifications, and often, months or even years pass between the time that the prime contract is awarded and the time that the subcontracts are finalized. During that uncertain period, neither the government nor the prime contractor can develop an integrated master plan or an integrated master schedule. Without the final subcontracts, one cannot estimate cost, schedule, or expected performance. Government and commercial program managers working on a "cost-plus" contract must mandate a detailed requirement flowdown to all the subsystem and major component subcontractors shortly after contract award. This information must contain performance at the various subsystem and component levels and expected cost and schedule, and must be integrated into the master plan and schedule. When the various subcontractors' work schedules are integrated with the prime contractor, the integrated master schedule can be the customer's most important management tool. It should include performance, cost, and schedule for the various aspects of development, manufacturing, assembly, integration, and testing. This information must be generated within two or three months after contract award. When it is available, an astute program manager can study it carefully to determine potential problems and develop required contingency plans.

Because it is much easier to build an integrated master plan and schedule for smaller satellites and for a constellation of satellites using a hot production line, this process is commonly used. But for complex multifunction satellites, it is either lacking or takes many months or years to develop. The early availability of an integrated master plan or schedule does not necessarily bode well for program success unless the program manager and the contractors continuously investigate potential problems and develop effective workaround techniques.

3.7 Scientific Results

The NASA science team had a profound impact on the Clementine mission. Within a few weeks after the team members were selected, they recommended the orbit to be flown around the moon, the specific wavelengths for the filter wheels of the various cameras, the required calibrations for all the cameras, and the data compression and storage required for the data.

Clementine collected 1.8 million images of the lunar surface with an average resolution of 100 m using 11 different wavelengths. These images have been analyzed extensively and have been the subject of many graduate research programs, publications, and books. These are by far the best data available of the lunar surface.[5] Although much higher resolution images exist for some specific areas related to the Apollo landings, a complete mapping of the moon had not been previously accomplished. Additionally, the extensive use of the star-tracker cameras to provide wide-angle perspectives of the moon, Earth, and other planets was unprecedented. Figure 3.11 illustrates one of the many images that captured the imagination of many. Clementine obtained this image after completing lunar orbit 102 on 13 March 1994.

Fig. 3.11. View of Earth over the lunar north pole.

Clementine determined the mineral type for about 99% of all rocks on the lunar surface. The careful selection of 11 spectral bands assisted the lunar scientists in determining the soil surface material and distinguishing specific lava patterns from defunct volcanoes[6] and the lithologic diversity of various craters.[7] The evaluation of the composition and distribution of the materials in and around the craters assisted the community in understanding critical information about the impact process, which can be used for any other crater impact within the solar system.

Clementine mapped the global topography of the lunar surface. The laser altimeter (LIDAR) provided a 20 m depth resolution, illustrating a marked difference between the near and far side of the moon.[8] It expanded and corroborated a previous finding by showing that the near side stands an average of 3.2 km below the far side.[9] The images, when compared with the topography, clearly indicate that the intensely cratered far side lacks the basaltic plains found in the near side.[8] A comprehensive analysis of the altimeter data showed the actual topography of Aiken basin at the south pole of the far side. It was measured as 2500 km in diameter and as much as 12 km deep, making it the largest known impact basin in the solar system—much larger than the famous 1400 km diameter Caloris Basin on Mercury.

Clementine discovered an impact basin more than 300 km in diameter and about 30,000 km^2 near the lunar south pole, which remains in perpetual darkness during a full lunar rotation. This area had been blank on previous global maps because it remained in shadow.

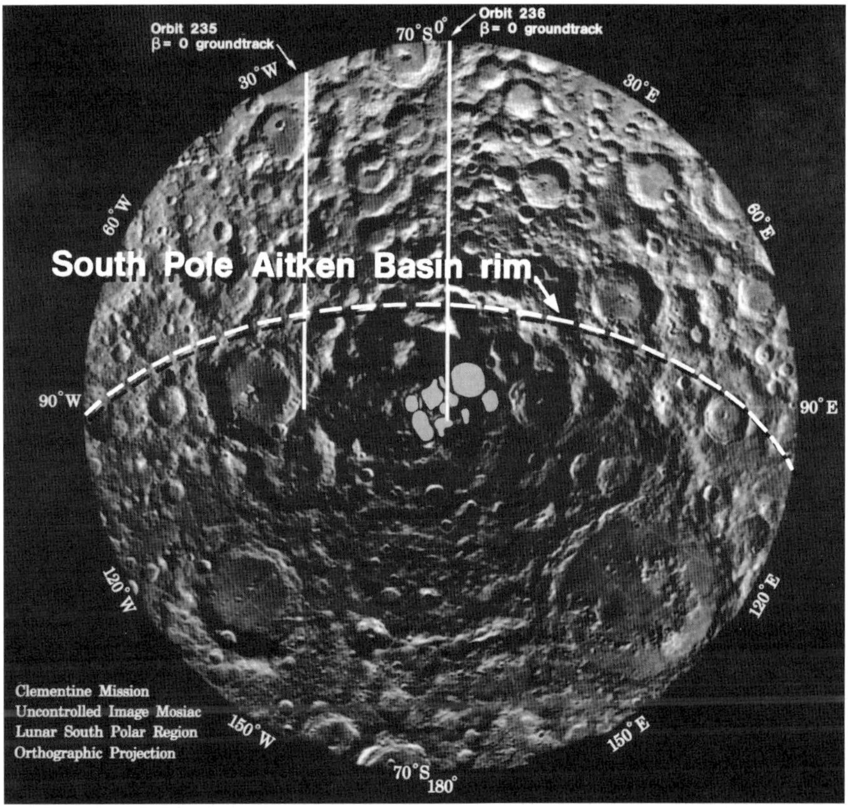

Fig. 3.12. Clementine mosaic of the lunar south pole noting areas (in light grey) of possible ice formations. This famous image was actually constructed manually on the floor of the Batcave. The photographer then climbed near the ceiling and took the picture. The projection is centered on the south pole.

Figure 3.12 shows a mosaic of 1500 UV-visible images (750 nm band) of the south polar region of the moon.

After researchers realized the size and depth of the shadowed regions, they decided to investigate them further using bistatic radar measurements. Figure 3.13 illustrates the bistatic radar experiment. The spacecraft was aligned to transmit with the antenna pointing to the shadowed regions, and the reflected signals were collected both at the spacecraft and at the respective NASA Deep Space Network antenna closely aligned with the selected configuration. The experiment was repeated many times, and the data were analyzed for about a year. The science team concluded that the signal return suggested the presence of ice in the permanently shadowed regions. As there are other potential chemical combinations that could produce similar results, the team was careful to indicate that other lunar missions had to confirm the suggested results. Eventually, the NASA Lunar Prospector mission verified the Clementine results.[10,11]

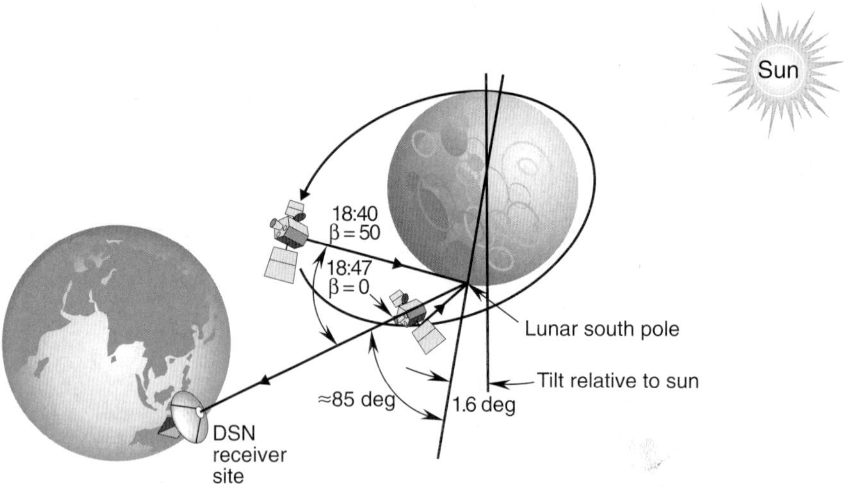

Fig. 3.13. Clementine bistatic radar experiment.

3.8 Conclusions

In evaluating the Clementine mission, it is vital to emphasize the circumstances that surrounded it. BMDO was a relatively new organization, with strong, dynamic, and flexible leadership; it was therefore possible to incorporate new business practices and a leadership style without being bogged down by traditional management procedures or existing independent cost estimates (which tend to stifle innovation). Basically, as long as they did not violate the law or misuse taxpayer resources, innovative business practices could be implemented. The BMDO senior managers on the various phases of the program (Michael Griffin, Pete Worden, and Henry Cooper) supported the mission and empowered the team to do what was needed. Therefore, the team formulated and implemented a vision without being burdened by some of the traditional no-value-added bureaucratic processes. This vision included all performance objectives, including the overarching strategy required to achieve them.

The Clementine team members did not want to repeat what had recently been done (that is, taking longer periods to build spacecraft), but instead, wanted to do something that had not been done for many years. As such, they threw away their fear of failure and became astute students and implementers of the streamlined business practices of the 1960s. BMDO represented a change of direction for missile defense, and it was essential to demonstrate that the space portion could rise to the challenge. But these changes were introduced not just to make changes; they were a calculated response to a changing world that needed to be reawakened by bringing back the old paradigm.

The small dynamic team managing the Clementine mission stands in sharp contrast to the many government large program offices today, managing both small and large space programs. On Clementine, there was minimal system engineering analysis because the spacecraft was specifically built to integrate 23 readily available technologies. The challenge was to integrate all the newly developed subsystems and components to answer specific scientific problems. The NASA-selected science team helped tune the various cameras and design the entire payload and bus functions to maximize the scientific results.

Unfortunately, the rebirth of smaller high-value missions, as implemented on the Clementine Program and many other subsequent NASA missions (via Dan Goldin's well established agenda of "cheaper, faster, and better" missions) have not affected the entire commercial and government space industry. Many NASA and some DOD missions followed the management practices established in the Clementine program—but not always successfully. There was a lot of misunderstanding about what it really meant to build faster and cheaper missions. Some program managers failed when they thought they could cut or reduce functional or environmental testing. Today, the industry is continuing to regroup without a clear direction. Some leaders are trying to reduce time to market through strict adherence to a given weight, power, cost, and schedule without understanding that what really matters is to increase the overall return on investment. It's hard to go wrong when you focus on value.

3.9 Acknowledgments

The BMDO Clementine program office consisted of two government people and a support contractor. My deputy, Steward Nozette, addressed all the NASA-related issues, while my support contractor, Brian Willard, responded to all the administrative and routine organizational tasks.[12] The program managers who were building and integrating the various subsystems at the Naval Research Lab and Lawrence Livermore National Lab were the best in the industry. Lyn Pleasance managed the design, manufacturing, assembly, integration, and testing of the payload sensors. He was fully committed to mission success and developed not only the two engineering-model units, but also oversaw the integration and testing of the cameras. Paul Regeon managed the design of the bus and the integration of all the LLNL cameras on the spacecraft. He worked tirelessly to lead an outstanding NRL team to accomplish the job.

3.10 References

[1] L. Lanzerotti, "Space Studies Board letter," National Academy of Sciences (August 1992).

[2] "Lawrence Livermore Timeline," http://www.llnl.gov/timeline/80s.html (July 2007).

[3] "Strategic Defense Initiative," Wikipedia, http://en.wikipedia.org/wiki/Strategic_Defense_Initiative (July 2007).

[4] National Research Council Space Studies Board, "Lessons Learned from the Clementine Mission," (National Academy Press, Washington, DC, 1997).

[5] S. Nozette, P. Rustan, L. P. Pleasance, J. F. Kordas, I. T. Lewis, H. S. Park, R E. Priest, D. M. Horan, P. Regeon, C. L. Lichtenberg, E. M. Shoemaker, E. M. Eliason, A. S. McEwen, M. S. Robinson, P. D. Spudis, C. H. Acton, B. J. Buratti, T. C. Duxbury, D. N. Baker, B. M. Jakosky, J. E. Blamont, M. P. Corson, J. H. Resnick, C. J. Rollins, M. E. Davies, P. G. Lucy, E. Malaret, M. A. Massie, C. M. Pieters, R. A. Reisse, R. A. Simpson, D. E. Smith, T. C. Sorenson, R. W. Vorder Breugge, and M. T. Zuber, "The Clementine Mission to the Moon: Scientific Overview," *Science,* Vol. 266 (December 1994).

[6] A. S. McEwen, M. S. Robinson, E. M. Eliason, P. G. Lucey, T. C. Duxbury, and P. D. Spudis, "Clementine Observations of the Aristarchus Region of the Moon," *Science,* Vol. 266 (December 1994).

[7] C. M. Pieters, M. I. Staid, E. M. Fischer, S. Tompkins, and G. He, "A Sharper View of Impact Craters From Clementine Data," *Science,* Vol. 266 (December 1994).

[8] P. G. Lucey, P. D. Spudis, M. Zuber, D. Smith, and E. Malaret, "Topographic-Compositional Units on the Moon and the Early Evolution of the Lunar Crust," *Science,* Vol. 266 (December 1994).

[9] W. M. Kaula, G. Schubert, R. E. Lingenfelter, W. L. Sjogren, and W. R. Wollenhaup, "Apollo Laser Altimetry and Inferences as to Lunar Structure," *Proceedings of the Lunar and Planetary Scientific Conference,* Vol. 5, p. 3049 (1974).

[10] P. D. Spudis, S. Nozette, C. Lichtenberg, R. Bonner, W. Ort, E. Malaret, M. Robinson, and E. Shoemaker, "The Clementine Bistatic Radar Experiment: Evidence for Ice on the Moon," *Solar System Research,* Vol. 32 (1998).

[11] "Ice on the Moon," NASA, http://nssdc.gsfc.nasa.gov/planetary/ice/ice_moon.html (July 2007).

[12] P. L. Rustan, "Spacecraft Project Management," *Technology Management,* **1** (1), (1994).

Chapter 4

The Role of AMSAT in the Evolution of Small Satellites

G. Gould Smith* (WA4SXM)

I invite all nations to participate in a communications satellite system, in the interest of world peace and closer brotherhood among peoples of the world.

—John F. Kennedy, 24 July 1961 in a report to Congress

4.1 Introduction

The first manmade satellite, Sputnik, orbited Earth on 4 October 1957. The simple beeps from Sputnik fired the imagination of engineers and amateur radio operators around the world and sparked a resolve to enter and expand this new field. It is widely held that the Soviets chose the 20 MHz band for the Sputnik transmissions to exploit the fact that most amateur radio operators would be able to receive the signals and report on them. In any case, a short four years after Sputnik, the first amateur satellite, OSCAR-1, transmitted a beacon from space on 12 December 1961. Since that time, the global community of amateur radio satellite enthusiasts has amassed an impressive record of more than 80 spacecraft in orbit.

4.1.1 What are Amateur Satellites?

AMSAT (the Radio Amateur Satellite Corporation) is a worldwide group of amateur radio enthusiasts interested in designing, building, operating, and maintaining satellites using amateur radio frequencies for communication, control, and data transfer. These satellites were first championed in the United States by Project OSCAR and AMSAT. Other groups followed, and amateur satellites and components were developed in Australia (Melbourne University), the United Kingdom (the University of Surrey), Germany, the Soviet Union, and Japan. Today, there is no global AMSAT organization, but groups from different countries do cooperate and share information, technical expertise, and hardware.

Some amateur radio packages have been hosted aboard larger commercial or governmental satellites. Some were developed by universities or military academies to train their students in the design, development, and operation of specific scientific experiments. Amateur satellites developed by the major AMSAT organizations worldwide continue to have amateur radio communications packages

*AO-51 Command Station
Knoxville, Tennessee

as the primary payload. These groups also experiment with new technologies in space and digital communications techniques.

One needn't be a licensed amateur radio operator to work on an amateur satellite or receive transmissions from it, but one does need a license to transmit through one. As such, most of the members of AMSAT are licensed. They are typically motivated by the challenge and educational experience of designing, building, commanding, and operating these amateur satellites. The task of getting them into space is left to the professionals. A good deal of the effort in getting a satellite operational for space involves long hours of discussions, paperwork, fund raising, and political wrangling to obtain a ride.

4.2 Early Amateur Satellites

Don Stoner (W6TNS)[a] wrote an article in the April 1959 issue of *CQ* (an amateur radio magazine) suggesting that amateurs undertake the construction of a radio relay satellite capable of supporting two-way communication between amateur ground stations. This seemed somewhat fanciful at the time, since work on Telstar I (the first commercial relay satellite) had not yet begun. Nonetheless, this article did strike home to a group in Sunnyvale, California, who organized the OSCAR (Orbiting Satellite Carrying Amateur Radio) Association, later to become Project OSCAR.[b] The group decided to build a continuous-wave beacon to transmit Morse code, hook it up to batteries, place it into metal box with an antenna coming out of the top, and have it launched into space (Fig. 4.1). Working with the U.S. Air Force, they arranged for a free ride in December 1961 for the 4.5 kg (10 lb) satellite in place of inert ballast. OSCAR-1 broadcast a 145 MHz beacon, sending a "HI" greeting in Morse code at a rate proportional to the internal temperature of the spacecraft. (The "HI" greeting has been a part of nearly every amateur satellite telemetry stream since then.) The internal temperature was higher than expected, and thus the broadcast rate was faster than expected and somewhat difficult to copy. OSCAR-2, launched in June 1962, used the same superstructure, improved the exterior coating to reduce the internal temperature, and employed a lower-powered beacon to increase battery life.

---

[a] Amateur Radio callsigns (such as W6TNS) are issued by each country after passing a radio theory and operating practice examination. They include a country prefix (e.g., W for United States, DL for Germany), a number to determine a subregion of the country (e.g., 6 for California), and additional letters to uniquely identify that operator in the region.

[b] The OSCAR appellation has become a part of nearly every amateur satellite. The other part of the name generally applies to the group building the satellite. The following requirements must be satisfied to qualify for an OSCAR designation: First, the spacecraft's use of frequencies in the amateur bands must have been coordinated before launch through the International Amateur Radio Union. Second, the spacecraft must have successfully achieved orbit or deployment. Third, once in orbit, one or more transmitters must have been successfully activated in the Amateur Satellite Service.

Fig. 4.1. Lance Ginner holding the OSCAR-1 satellite in 1961.

OSCAR 1 and 2 were simply beacon transmitters in a metal box, but they transmitted that beacon from space. OSCAR-3 was much more ambitious: It was designed to answer the challenge of Don Stoner. It would carry a 50 kHz wide transponder transmitting with a power of 1 W. It would fly in a higher orbit that would allow amateurs to communicate within a 5000 km footprint. A few solar cells were added to help power the two beacons. The uplink was on 146 MHz, and the downlink was on 144 MHz. Launched in March 1965, OSCAR-3 operated as a transponder for only 18 days, but the beacons lasted months longer. Radio amateurs had successful communications between the United States and Europe and the eastern United States and Alaska, demonstrating the potential for reliable long-distance communication.

In late 1964, an opportunity arose to ride a rocket headed for a 33,800 km (21,000 mile) equatorial orbit that would keep half the planet in the communications footprint at any time—but builders would have less than a year to design and build a satellite. The Project OSCAR team was consumed with getting OSCAR-3 ready and didn't have the resources to work on a new satellite. So, a group from the TRW Radio Club in Redondo Beach, California, took on the task of designing and building OSCAR-4. The TRW group kept things simple and was able to complete the satellite in time for launch in December 1965. Unfortunately, the final stage of the rocket failed, and the orbit turned out to be highly elliptical—very different than what the satellite was designed for. Nevertheless, a number of amateurs managed to complete international communications—even one between the United States and Russia.

That same year, a group of students at the University of Melbourne in Australia began developing a satellite. They dubbed it A-O-5 (Australis-OSCAR-5). It would explore new territory by including a 29.45 MHz downlink (in the amateur radio 10 m band) to test high-frequency propagation from space and a passive magnetic stabilization system. It would also accept commands from Earth. It would incorporate a 144.05 MHz (2 m band) beacon sending satellite telemetry. The students produced the satellite and arranged with Project OSCAR to complete the testing and acquire a launch. Unfortunately, the launch that Project OSCAR had arranged was postponed indefinitely, leaving A-O-5 in limbo.

George Jacobs (W3ASK) addressed the COMSAT Amateur Radio Club of Washington, DC, in early 1969 and discussed the plight of two satellites on the west coast without a launch. This excited Perry Klein (W3PK), who assembled a group of radio amateurs at his apartment to discuss the formation of an east coast amateur space group. At the meeting were Klein, Jan King (W3GEY), Harry Helfrich (W3ZM), Bill Tynan (W3XO), Jim Puglise (W3CBJ), Cap Petry (W3AWN), and George Kinal (K2MBU). After a few organizational meetings, the group incorporated on 3 March 1969 as a nonprofit organization named the Radio Amateur Satellite Corporation, or AMSAT. Their first mission was to get A-O-5 approved and launched.

The first OSCARs had been launched on military rockets, with the launch agency handling many of the details. AMSAT had to convince NASA that including A-O-5 on a launch (for free!) would be in the public interest, would not jeopardize the main mission, and would not affect the cost of the launch. A tall order, but it was completed.

A-O-5 had spider webs in it when it arrived in Washington, and new batteries were needed. Additional modifications were also needed to get A-O-5 to pass the rigorous NASA test procedures. Although this satellite would have a short lifetime, it was deemed important to establish a relationship with NASA. The process would prove that a piggyback launch from a third party such as AMSAT was feasible; it would also serve to define thermal-vacuum and vibration procedures, develop methods for integrating the payload to the launch vehicle, refine telecommand and data collection methods, and highlight the myriad of details to get a satellite launched. These details included construction, testing, documentation, integration, launch procedures, licensing, and operations. A-O-5 was finally launched in January 1970.

4.3 The Phase 2 Satellites

In 1969, the AMSAT group met with Karl Meinzer (DJ4ZC) of Germany to discuss flying his 2M-2M repeater (uplink and downlink in the 2 m band), but Meinzer had a new repeater design with an improved 432 MHz uplink and 146 MHz downlink (2 m, 70 cm). Plans were made to fly this on a future spacecraft. This became AO-6, the first of the so-called "Phase 2" satellites (Fig. 4.2). It marked a giant step from the first five OSCARs, retroactively dubbed "Phase 1"

Fig. 4.2. The AO-6 spacecraft ready to travel in 1972.

satellites. Designed to last several years in a circular low Earth orbit (LEO), it contained several new experiments, such as a Morse code telemetry encoder by John Goode (W5CAY) and a 2 m to 10 m transponder built by Dick Daniels (W4PUJ). The repeater/transponder would receive a 100 kHz bandwidth centered on 145.95 MHz and retransmit the amplified signal on 29.5 MHz. Since amateur radio voice transmissions using SSB (single side band) modulation have a bandwidth of ~2.4 kHz, this would allow multiple conversations simultaneously. Noticeable in Fig. 4.2 are the coiled carpenter tape antennas. This is one of many instances of a clever solution by AMSAT designers to solve a problem simply. Here, the problem was to restrain the antennas during transport and launch and extend them in orbit. The designers came up with a common, inexpensive, flexible material that performed well. Another solution was to use fishing line to bind the antennas; the nylon would quickly degrade from sunlight in space, thus freeing the antennas. AO-6 was a 16 kg satellite, with dimensions of 43 × 30 × 15 cm. It was launched on 15 October 1972 as a piggyback to ITOS-D, a NOAA satellite, from Vandenberg Air Force Base in California. AO-6 became the second AMSAT satellite to remain in service longer than four years. Because the transponder on AO-6 was the first one used on an amateur radio satellite, its 2 m uplink/10 m downlink frequency pairing came to be known as "mode A."

AO-7, launched in 1974, was an evolutionary step in satellite technology but it was a revolutionary step with regards to international collaboration (Fig. 4.3). It contained components from Germany, United States, Canada, and Australia.

Fig. 4.3. AO-7 ready to undergo thermal-vacuum testing in 1974. *From left*: Dick Daniels, Jan King, Marie Marr, and Perry Klein.

AO-7 had two transponder packages, redundant command systems, two telemetry systems and beacons, and a store-and-forward message system. The first transponder operated in mode A, like the one on AO-6; the second, from AMSAT Germany, had an uplink on 432 MHz and a downlink on 145 MHz (this pairing was subsequently known as "mode B"). In addition, a 2304.1 MHz transmitter was onboard, but it was not turned on because of international treaty constraints. The international amateur satellite frequency allocations were later changed to allow use of these higher frequencies.

The now famous HELAPS (High Efficiency Linear Amplification by Parametric Synthesis) amplifier made its debut aboard AO-7. Meinzer developed this as part of his Ph.D. work.[1] This amplifier design has been used on all the Phase 3 amateur satellites so far and is being specified for new Phase 3 and Phase 4 units.

Some of the very first ground-satellite/satellite-ground communication links were made using AO-6 and AO-7. Battery failure rendered AO-7 inoperable in 1981, but in 2002, one of the shorted batteries opened, and the satellite sprang to life. As of 2009, each time AO-7 enters sunlight, it becomes operational randomly in either mode A or mode B. In constant sunlight, the onboard timer will still switch modes at the set 24 hour interval.

AO-8 was designed to serve as a bridge between AO-6/AO-7 and the new Phase 3 satellites. It was built using a $50,000 donation from the ARRL (American Radio Relay League). AMSAT Japan offered to build a second transponder for AO-8 with an uplink on 146 MHz and downlink on 435 MHz;

Fig. 4.4. FujiSat A (FO-12) in 1986.

this frequency pair became known as "mode J." Both transponders could be operated at the same time.

Japan contributed components to many of the early satellites. Harry Yoneda (JA1ANG) championed the development of the JAS series of Japanese amateur satellites. These three satellites—FO (Fuji OSCAR)-12, FO-20, and FO-29—were used by thousands of amateur satellite enthusiasts over the course of two decades (Fig. 4.4). The Fuji satellites were launched on Japanese launch vehicles from Japan and had a distinctive 26-sided polyhedron shape. The use of a digitized voice to convey the onboard telemetry values to 2 m FM amateur radios introduced many to amateur satellites and proved to be a great teaching device. These satellites had both state-of-the-art mode-J analog transponders and digital data communications.

Russia also launched a number of amateur satellites (generally known as RS satellites, for Russian Sputnik/Radio Sport). The most popular were the RS-10/11 and RS-12/13 analog transponders housed in the Cosmos navigation/rescue satellites. The amateur payloads had the advantage of a large power budget because the primary satellites were large and able to provide a good deal of power to these piggyback payloads. The RS satellites were popular because they used HF (1–30 MHz) and VHF (30–300 MHz) frequencies, which opened up the world of amateur satellites to thousands of amateur radio operators worldwide (because they already had the necessary ground station equipment). A total of 22 Russian amateur satellites have achieved orbit.

Fig. 4.5. UoSAT-1 (UO-9) with Martin Sweeting (lower right) and team.

In the mid 1970s, graduate student Martin Sweeting (G3YJO) came to the United States from England to participate in design meetings for the Phase 3 satellites and acquired the information to set up a command station for AO-6 and AO-7. In 1979, when he had completed his Ph.D. at the University of Surrey, he proposed that the university embark on a program to develop cost-efficient satellite engineering techniques and build operational microsatellites. His passion convinced the administration, and a satellite curriculum and lab were funded, with Sweeting at the head. The Centre for Satellite Engineering was the academic and research department for graduate and doctoral students. The first satellite built at Surrey was OSCAR-9/UoSAT-1, launched in 1981 (Fig. 4.5). SSTL (Surrey Satellite Technology Limited) was formed in 1985 as a university-owned company to build satellites for commercial interests. Not all of the satellites they produced have been amateur, but they have certainly retained the spirit of amateur satellites. Most have used amateur radio frequencies and have done some valuable scientific investigations. Many have also been converted to amateur satellite use after their primary missions were completed. The UoSATs have pioneered many small satellite technologies and have encouraged many future amateur satellite enthusiasts.

4.3.1 Microsats

In the 1980s, commercial interests began to purchase most of the available launches for small satellites. This presented a problem for AMSAT.

Jan King identified a way to place small secondary payloads in unused space in the Ariane launchers. This space severely constrained the maximum size of

Fig. 4.6. ASAP for the Ariane 4 with four microsats and two UoSATs in 1990.

these new satellites; nonetheless, King determined that a 23 cm (9 in.) satellite cube could house the necessary components, and that solar cells covering the outside of the cube could supply enough power to recharge batteries and subsidize power requirements. He presented his findings at the 1987 AMSAT annual meeting, calling these satellites "microsats." The concept was well received, and work began. The next job was to convince the European Space Agency (ESA) that selling excess launch vehicle capacity for secondary payloads would be worthwhile. Thus, the ASAP (Ariane Structure for Auxiliary Payloads) was born. AMSAT worked with ESA to design this structure for auxiliary payloads, and in return received a discount on the launch cost. The ASAP has a total loading capacity of 200 kg, with a limit of 50 kg per spacecraft. A typical cost for the ASAP launch is about $1.2 million, which is shared amongst the secondary payloads. The primary payload is deployed first, followed by the secondary payloads, using their own deployment mechanisms.[2]

Six new microsats (AO 14–19) carrying diverse payloads and experiments were mounted to the ASAP structure and launched in 1990 (Fig. 4.6). Five of these worked well for a decade. New technologies were proven, and the new microsat structure would be used in more than 10 future amateur and commercial satellites. Tom Clark (W3IWI) designed the microsat architecture, receiver, AART (internal data bus), and the infrared sensor. Bob McGwier (N4HY) was instrumental in helping get four microsats operational and ready for launch. He also spent a great deal of time writing and loading the onboard software to enhance their operation.

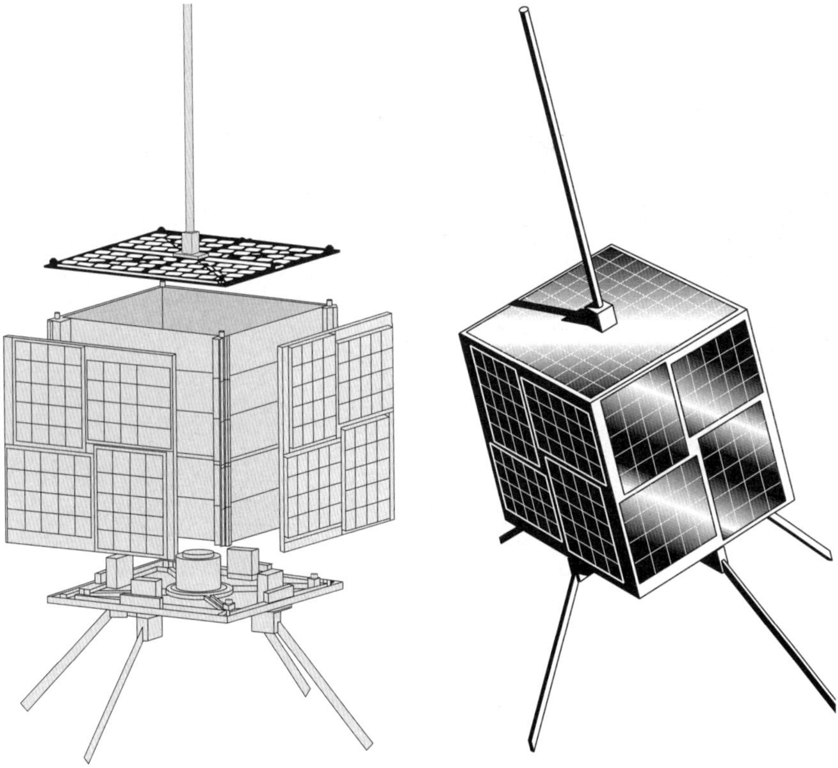

Fig. 4.7. Microsat exploded and standard view.

The microsats were an innovative concept. Each contained five stacked modules, or trays: a receiver tray, a transmitter tray, a power tray, an onboard computer tray, and a payload tray (Fig. 4.7). The first four microsats shared the standard four trays, but each carried a different payload tray. The VHF antenna was on the top, and four canted turnstile UHF antennas were on the bottom. The cost of a microsat in 1990 was approximately $65,000.[3] The Pacsat (AO-16) microsat introduced the new Pacsat protocol for digital communications via satellite. It was based upon the AX.25 protocol[4] made popular by TAPR (Tucson Amateur Packet Radio), which was an amateur radio version of the ISO 7498 seven-layer communications protocol.[5] The satellite stored a number of digital files, uploaded by various users, that could be broadcast by request from a ground user. Some of these files contained AMSAT spacecraft orbital element data that most stations would want to record for future use. Each ground station within view of the satellite simultaneously received the broadcast digital information, but only recorded it if the ground station operator had previously selected that file for download. In many cases, a ground station could receive a desired file based on a previous request by another ground station. Since many people would be interested in many of the same files, this was a much more efficient method of transferring data than having each individual station connect to the satellite and download duplicate files. Since communications links are dynamic, and since different ground stations would be in view at different times as the satellite continued on its orbital trajectory, a single ground station could miss parts of a data file. Ground stations would identify any missing packets and request a retransmission of those particular packets. A ground station control program called WiSP (Windows Satellite Program, available from AMSAT) was typically used to automate the file transfer process, point antennas, and autotune amateur radio transceivers to correct for Doppler shift. The data transmissions used 1200 bps phase-shift keying (PSK) modulation because it has a much better data recovery rate[6] than frequency-shift keying (FSK). Later digital microsats would use FM at 9600 and 38,400 bps FSK.

A number of additional microsats were developed by many countries and organizations during the 1990s. Most used standard VHF/UHF FM voice repeaters and digital modes. This would allow ground operators to use dual-band, handheld FM radios and a small handheld antenna to talk through an amateur satellite. New experiments and modes were also included in these satellites.

Notable microsatellites, somewhat larger than the AMSAT microsats, were GO-32, MO-36, and AO-51:

GO-32 (Gurwin-Oscar 32) from the Technion Institute of Technology in Israel was launched on a Russian Start rocket in 1998. It carried experiments measuring cosmic particles, a superconductivity experiment, an Earth imaging system, an ozone meter, and an x-ray detector. After the experiments were concluded, the satellite was converted to a digital store-and-forward bulletin-board for radio amateurs in 2004; it was still in service in 2007.

MO-36 (TiungSat-1, MySat [Malaysian Satellite]), built by SSTL, had an exceptional Earth imaging system, GPS navigation and attitude control system, 384 kbps data-transfer experiment, digital signal processing, cosmic ray detection experiment, and a momentum wheel for stabilization. The April 2000 issue of *QST* magazine featured some of the incredible images from MO-36 on the cover and inside.

AO-51 is the latest Phase 2 microsat. It is a 24 cm cube, composed of six trays and covered with solar cells. Most of the satellite was contracted to SpaceQuest, LLC, in Virginia. AMSAT provided the L-band (1.2 GHz) receiver, S-band transmitter, and the operational software. It was launched in 2004 to fill in while the new Phase 3 satellites (Phase 3E and Eagle) were being designed and built. The AO-51 microsat has two UHF (435 MHz) transmitters that can transmit simultaneously (see Fig. 4.8). They can be configured as FM analog repeaters or for 9.6 or 38.4 kbps digital data. There is also a wideband, multimode receiver that can operate between 20 MHz and 1.3 GHz and can be matrixed to any of the transmitter outputs. This is a truly versatile 9.5 in. cube. AO-51, as all the other amateur satellites, is controlled by volunteer command stations who capture telemetry, analyze the telemetry, and change satellite configurations and modes.

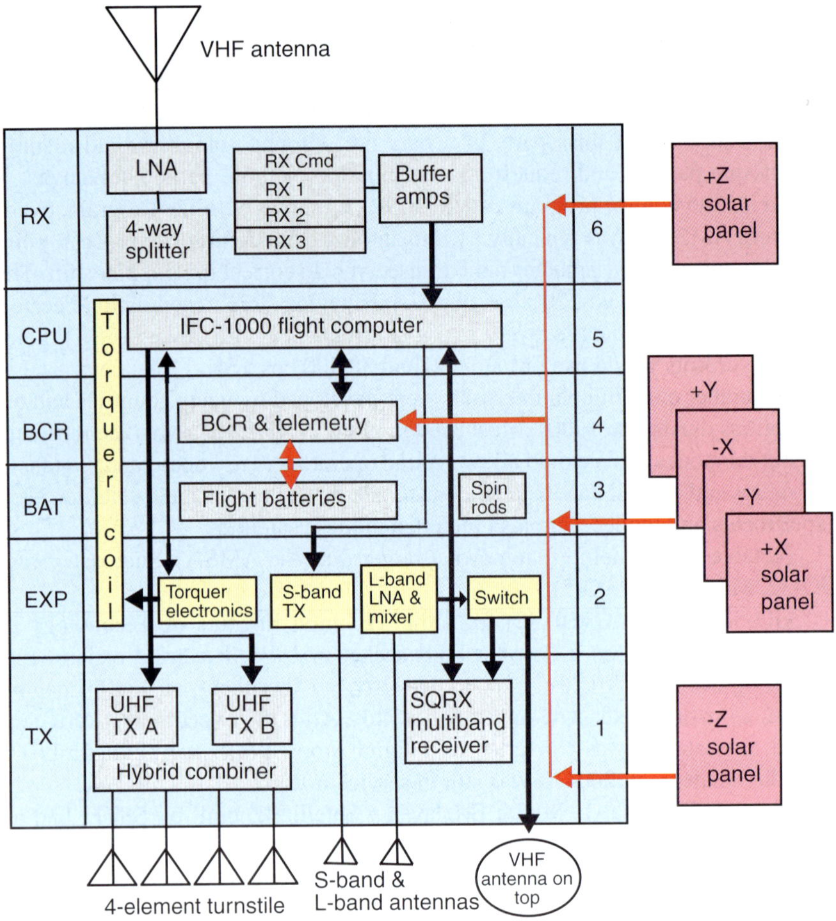

Fig. 4.8. AO-51 microsat configuration.

4.4 The Phase 3 Satellites

The next leap in amateur satellites came with the Phase 3 satellites. These operational satellites would feature a long operational lifetime (8–10 years) and fly in elliptical high Earth orbits (HEO; 1000–60,000 km).

The HEO satellites are the most popular amateur satellites. The main reasons are the coverage area (nearly half the planet) and the long operational window (most stations can operate 10–13 hours a day). Also, the elliptical orbit reduces satellite complexity and launch costs compared to a geosynchronous satellite. Stationkeeping is not required, and Earth's magnetic field at perigee is strong enough to enable magnetic torque rod control of satellite attitude; thrusters and momentum wheels are not required on these HEO satellites.

Fig. 4.9. P3A checkout complete in the lab at Goddard Space Flight Center lab.

The design, construction and launch arrangements for the first Phase 3 satellite (P3A) consumed many hours and most of the spare time of many AMSAT volunteers between 1971 and 1980. This complex and expensive project was led jointly by King and Meinzer. P3A featured a distinctive tristar shape, measured 60 × 40 × 20 cm, and had a mass of about 92 kg (Fig. 4.9). It had a kick motor to take it into a new orbit plane; it had a 50 W mode-B transponder and a power system to support it; it had a computer that allowed control of the spacecraft and multiple telemetry channels. A ground-based command network was established to control and monitor the satellite. Flight hardware came from the United States, Germany, Canada, Japan, and Hungary.

In May of 1980, a new Ariane 4 rocket that was to take AMSAT to a new operational level had a malfunction and ended up at the bottom of the Atlantic Ocean along with P3A. Tom Clark, president of AMSAT, relaying information from the launch site in Kourou, French Guiana, felt the despair of losing years of work, but understood his mission to lead the organization and continue. Later that evening he issued a statement to the worldwide launch network:

> What we lost on Black Friday was sheet metal, solar cells, batteries, transistors, a lot of sleep and a major portion of our lives the last few years. What we gained over those same years was knowledge; knowledge that we could make a complicated spacecraft. Knowledge in areas of aerospace technology that none of us had before. Knowledge that we could work as a team, despite national boundaries, differences in our cultures, lifestyles and personalities.

> Knowledge that, from within the ranks of Amateur Radio, we could draw upon enough resources to attempt a project with a complexity rivaling commercial satellite endeavors funded at levels of tens of millions of dollars. The knowledge is still intact. We even have a second set of solar panels, batteries and sensors. We have on hand the documentation and art work necessary to replicate all the printed-circuit boards. We have in place and ready to go a network of ground telecommand stations.

Three years later, P3B launched successfully from Kourou on an Ariane rocket and attained orbit. The elation was quickly subdued when the first telemetry from the satellite showed some anomalies. It turned out that ESA was testing a new separation procedure, and the third stage of the rocket actually bumped the satellite twice. As a result, one of the VHF antennas was damaged, the satellite was spinning faster than anticipated, it was in an unplanned attitude that generated less solar power, and the liquid propellant had been cooled more than desired. The command team worked to assess the situation and recover as best as possible. The burn to higher orbit didn't go as expected, and the newly named AO-10 satellite had an inclination of 26 deg, as opposed to the planned 57 deg. AO-10 also carried a new transponder (1269 MHz uplink, 435 MHz downlink) called mode L. It was known before launch that the components used in the computer might not handle the space environment well, giving AO-10 a projected lifetime of 5–7 years. About three years into service, radiation damage began to show up, and by the end of 1986, the command part of the computer was declared dead. Nonetheless, AO-10 continued to operate into the late 1990s.

In 1988, the Phase 3C satellite was launched by an Ariane rocket. It had the same tristar shape as its two predecessors, but had a radiation-hardened computer, new experiments, and transponders aboard. The burn to a new orbit went as planned, and the satellite was performing well in the desired orbit. A new mode-S transponder (435 MHz uplink, 2.4 GHz downlink) and a German digital transponder (called RUDAK) were tested. The mode-B transponder aboard AO-13 proved to be exceptionally popular, and AMSAT ranks swelled worldwide. Because of its initial orbit with a low perigee, and the effects of solar and lunar perturbations, AO-13 prematurely reentered the atmosphere in December of 1996 after a very successful stint in space.

These Phase 3 satellites were spin stabilized, they used electromagnets in the spacecraft frame to adjust the attitude during perigee by interacting with Earth's magnetic field. The spacecraft attitude was adjusted to give it a good solar angle to power the satellite and recharge the batteries and to optimize the antennas pointing toward Earth. Often, there were compromises made between these two functions.

While thousands of amateurs were enjoying AO-10 and AO-13, the AMSAT engineering teams were working on a new Phase 3 satellite, P3D. This would be a bigger and better satellite—maybe too big. P3D was hexagon shaped, 230 × 230 × 70 cm, with a mass of 244 kg (Fig. 4.9). It had two different thruster systems and multiple transmitters and receivers that could be used in many combinations and numerous experiments.

The development of P3D took more than 10 years and millions of dollars. It was an extremely large and complex satellite, being built by different groups on different continents—a real challenge for integration. In 1996, the computer (known as the integrated housekeeping unit) for P3D came alive, but it would not be until November of 2000 that P3D would roar into space. AO-40 was its new name, and it was plagued with problems early on. Once the spin stabilized satellite had become stable, it was time to move it into a better orbit. A plugged valve in the 400 N bipropellant thruster (used to move the satellite into the new orbit) precipitated an explosion, causing the satellite to go silent. It took weeks to get the satellite to respond. The popular mode-J transponder no longer worked. A positive result of this was that thousands of satellite operators were introduced to mode S (2.4 GHz downlink, 425 MHz uplink), and found that it was not as difficult to use as they had assumed. Mode S was used for the rest of the short lifetime of AO-40. In January 2004, AO-40 suffered a catastrophic failure of its main battery system. It is thought that the batteries shorted, preventing the system power level from reaching a high enough voltage to activate the onboard computer. If the computer were working, operators could switch to a secondary set of batteries, and presumably the satellite would function—another lesson learned in satellite design. AO-40 may one day "return" as AO-7 did.

An interesting experiment aboard AO-40 (funded by NASA) was to test whether the GPS satellites could be used for navigation and location. There were four GPS antennas placed on the surface of AO-40 (the four small disks at the edge of the satellite in Fig. 4.10 at positions 1, 3, 7, and 11 o'clock). The AO-40

Fig. 4.10. P3D (to be AO-40) in Kourou with Peter Guzlow in November 2000.

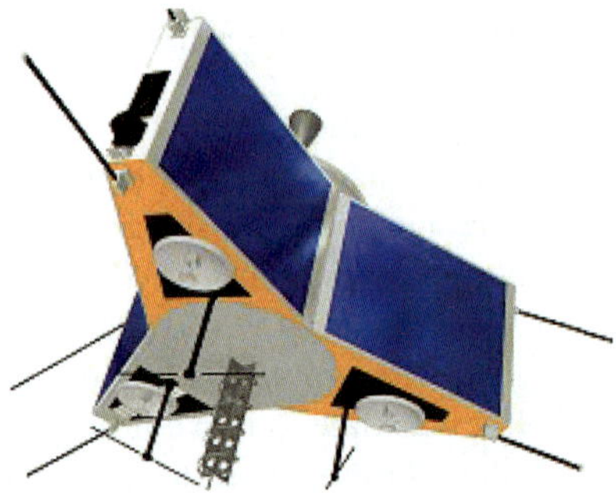

Fig. 4.11. P3E, the next Phase 3 satellite built by AMSAT-DL.

satellite captured GPS data and sent it to Earth. NASA was pleased with the results and determined that there was enough signal to use the GPS satellites on the far side of Earth.[7]

AMSAT conducted a membership survey in 2004 and found that a large majority of the members were interested only in the HEO type satellites. The Phase 3 satellites give long operational windows with reduced Doppler shift, and operation time is determined by the operator rather than by the satellite pass (as it was with Phase 2 satellites). In response, AMSAT changed its mission statement to concentrate on HEO satellites while continuing to support those building Phase 2 satellites.

The period between 1983 and 2004 was a great time for Phase 3 HEO satellite users. The launch of the German P3E (Fig. 4.11) and U.S. Eagle Phase 3 beginning in 2009 might bring a return to those exciting times.

4.5 The Phase 4 Satellites

Although Jan King proposed a Phase 4 satellite in 1986, it seemed a daunting and very expensive undertaking. The satellites would be operational and designed for a long lifetime in high-altitude geosynchronous orbits (GEO), which would provide a constant platform for communications over 40% of the world. However, the slots in this orbit are highly prized by commercial satellites organizations. The prospect of acquiring one seemed a dream until 2007, when Intelsat introduced a "rideshare" program in which they would host additional equipment aboard their GEO communications satellites. At the end of 2007, AMSAT and Intelsat were negotiating the transport of an AMSAT payload into GEO, with Intelsat supplying the launch, power, and stationkeeping. AMSAT can concentrate on doing what they do well: designing communication systems and antennas. AMSAT expects to fund their part of the launch by grants from government agencies. AMSAT would provide an emergency communications system for disasters; this would consist of small, portable ground stations doing high-speed digital communication through the AMSAT transponders onboard the GEO satellite available 24 hours a day.

4.6 The Phase 5 Satellites

Phase 5 satellites are defined as lunar or planetary. This, too, seems like a dream, but Meinzer is pursuing that dream. The P5A project is a Mars mission. The difficult part of this mission is not the engine burn to Mars—this can be done with a Phase 3 type motor—but hearing and controlling the spacecraft. The German AMSAT arranged to have priority access to a decommissioned 20 m dish at Bochum, Germany, in exchange for making the dish controllable.

A European team was assembled (Fig. 4.12). Much work was done to redesign and update the control mechanism, feeds, and preamps. In 2003, the dish moved into position to hear noise from the moon. Since then, all of the NASA/ESA spacecraft have been tracked, and their signals have been heard. Then, in 2006, the team was successful in actually receiving the Voyager spacecraft at a distance of 14.7 billion km![8] They now have a receiving and control antenna for P5A.

The group has obtained permission from France Telecomm to use their highly efficient turbo codes (forward error correction technology) for communications. These and other Phase 5 technologies will be tested on the Phase 3E satellite.

4.7 Amateur Satellite Launches and Launch Sites

Amateur satellite launches have increased dramatically during the last decade—mainly because of the increase in microsats and CubeSats. There are still a surprisingly small number of sites that have launched amateur satellites. These launch agencies are chosen because they can reliably launch a small satellite for the least amount of money (see Figs. 4.13 and 4.14 and Table 4.1).

Fig. 4.12. Reception of the Voyager 1 signal at the Bochum facility March 2006. ***From left:*** **Freddy de Guchteneire from Belgium (ON6UG), Hartmut Paesler (DL1YDD) and Achim Vollhardt (DH2VA/HB9DUN) from Germany, and James Miller from England (G3RUH).**

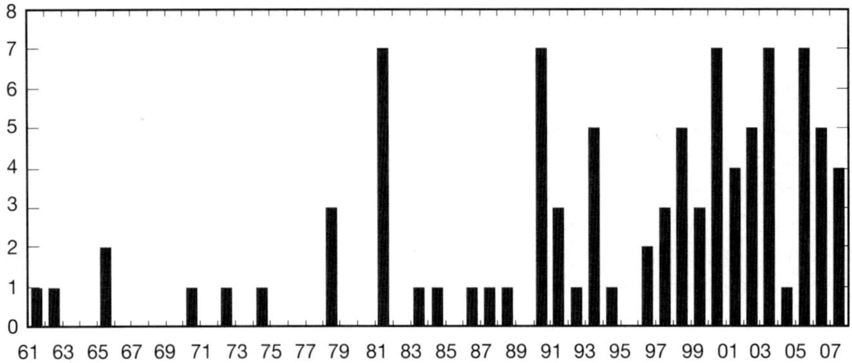

Fig. 4.13. Operational amateur satellites launched by year.

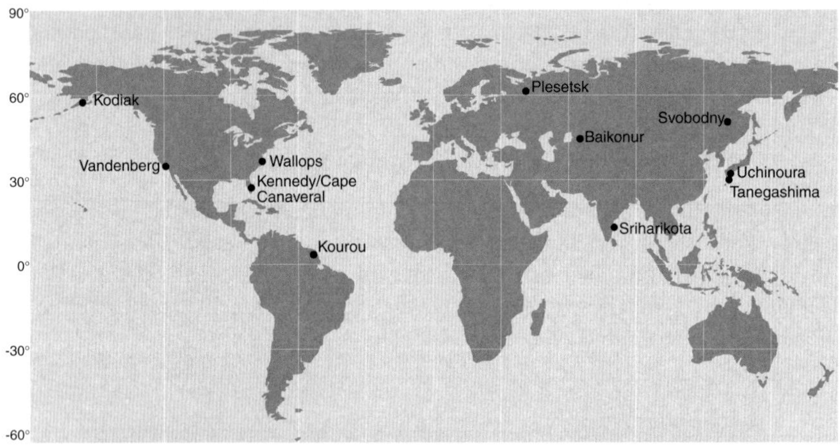

Fig. 4.14. Launch sites for amateur satellites.

Table 4.1. Amateur Satellite Launch Record by Launch Site

Vandenberg, California, USA	12 launches	14 satellites
Baikonur/Tyuratam, Russia	9 launches	22 satellites
Kourou, French Guiana	8 launches	18 satellites
Plesetsk, Russia	7 launches	20 satellites
Cape Canaveral, Florida, USA	4 launches	6 satellites
Tanegashima, Japan	3 launches	3 satellites
Kodiak, Alaska, USA	1 launch	3 satellites
Uchinoura (Kagoshima), Japan	1 launch	2 satellites
Sriharkota, India	1 launch	1 satellite
Wallops Island, Virginia, USA	1 launch	1 satellite

4.7.1 Funding

The amateur satellites are basically funded by donations from interested users and organizations, and from companies wanting to test their products in space. Some components have also been donated by companies and governmental organizations. By far, the largest contribution is the time and expertise donated by the team designing, building, and testing the satellite. AMSAT has become a victim of its own success. The early launches were free. These small, less-expensive amateur satellites proved that small satellites could perform useful scientific experiments and provide reliable communication, store-and-forward messaging, and file transfer. Commercial groups became interested in developing constellations of these smaller, inexpensive satellites, and were willing to pay for launches. There are no more "free" launches.

It is often quite difficult to determine actual launch costs. The launch agencies negotiate fees for each payload independently, and the fees are seldom published. AMSAT has had to be very creative in finding and funding new launch vehicles. With the large increase in the number of small satellites seeking launch opportunities and the ever increasing launch fees, launch vehicles are at a premium. The larger Phase 3 satellites at 140–240 kg are much more expensive than the microsats. In the past, there was a four to five year cycle time to schedule a launch. This gave the AMSAT group time to identify a prospective launch opportunity, court the launch agency, and build to specifications that matched the launch vehicle. Since 2002, the contract time to launch has decreased rapidly—they are now on the order of one to two years. This means a satellite must be almost ready to fly before a launch can be identified and negotiated. This puts a great deal of pressure on fund-raising and satellite design.

There have been extensive fund raising campaigns. Often, AMSAT has arranged for bargain launches on new rockets during their test launches. This proved costly in 1990, when the P3A satellite was lost during an early Ariane test flight. Lately, Russia has been using reconfigured ICBMs to launch educational and commercial satellites at a reduced cost. These Dnepr rockets are proving quite popular at about $10,000 per kilogram.

Another creative approach to getting a launch was to develop a component that the launch organization could market. AMSAT worked with Weber State University to have students produce a launch adapter for the Ariane launch vehicle for AO-40. This became the SBS adapter (Fig. 4.15). The university students got experience building a real-world product, the ESA got a new adapter unit, and AMSAT got a reduced launch fee.

Some satellites have sponsored a "fly your callsign in space" fund raiser; others have looked to commercial interests. The ARRL has been of great assistance and participated in fund-raising. The German AMSAT is sponsoring a "purchase a pixel" on a graphic of the P3E satellite to support its development.

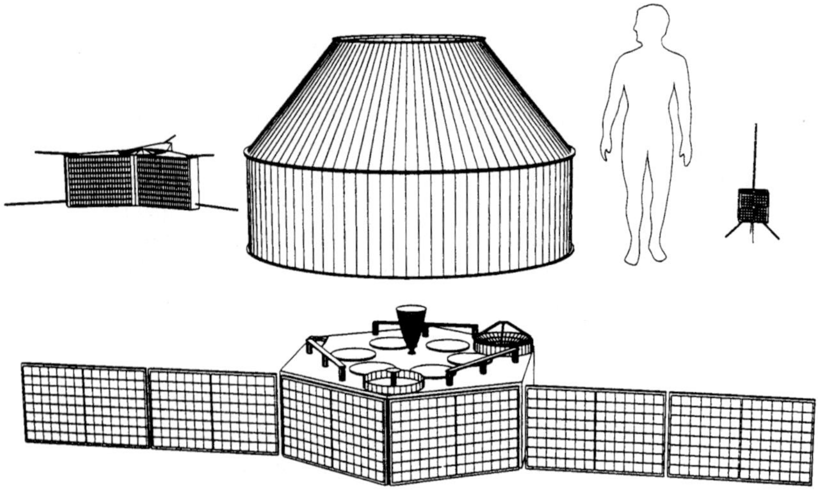

Fig. 4.15. SBS adapter (center) for P3D (bottom) and the relative sizes of other amateur satellites AO-13 (left) and AO-16 (right), along with a typical 5 ft 10 in. person.

4.7.2 Qualifying Amateur Satellites for Space

All of the amateur satellites must undergo thorough testing before launch so as not to endanger the primary payload. These tests include thermal and vacuum tests to make sure the satellite will perform in a vacuum and be able to handle extremes of heat and cold in space (Fig. 4.16). Often, these will be on the order of +40°C to –40°C on opposite sides of the spacecraft. A shake table is used to make sure that all of the parts of the satellite are tight and nothing will shake lose during the extreme vibrations experienced during liftoff.

These tests are generally done on weekends by volunteers from commercial space organizations. These donated services save many thousands of dollars.

4.8 Operations

4.8.1 Amateur Satellite Orbits

The Phase 1 satellites were in low orbits that would decay rapidly. The Phase 2 satellites are in somewhat higher orbits, giving them a wider footprint and longer lifetime. To get maximum help from the rotation of Earth, it is best to launch due east—the closer to the equator, the better. So, many of these early satellites were in orbits at an angle about 30 to 60 deg to the equator because of the longitude of the launch site. The alternative choice for the Phase 2 satellites is a near polar, sun-synchronous orbit. These orbits permit the satellite to come over each location on Earth the same time each day. All of the LEO satellites allow only a 12–14 minute maximum period of visibility per pass, but there are four to six passes per day. The polar orbits are very popular for commercial mapping and

Fig. 4.16. Echo (AO-51) undergoing vacuum testing in 2004.

Earth photography. AMSAT is looking at ways to mitigate orbital debris in these prime orbits. The Phase 3 satellites were placed into a highly elliptical Molniya-like orbit. These have the advantage of providing a continual 10–13 hours of visibility per day and a footprint covering almost half of Earth. The Phase 4 satellites will be in a geosynchronous orbit and be visible in the same location 24 hours a day.

4.8.2 Tracking Amateur Satellites

Tracking amateur satellites in the early years was difficult. The *AMSAT Newsletter* would publish pass times for the upcoming four months for major U.S. locations. People could subscribe to a monthly mailing of satellite passes for their location.

The mechanical ARRL Oscarlocator and Satellabe[10] made their appearance in the mid 1970s to help determine when a satellite pass would occur based upon satellite equator crossing times and their orbit. In 1979, HP offered the HP67 and HP97 programmable calculators, and Roy Welch (W0SL) wrote a tracking program for these calculators based upon a Tom Clark algorithm.[11]

The late 1970s to early 1980s saw the appearance of personal computers like the TRS-80, Commodore, Apple, and eventually the IBM PC. Clark published his IWI tracking program (written in BASIC) in the 1981 issue of the *AMSAT Orbit* magazine. James Miller (G3RUH) published a tracking program in AMSAT-UK's *OSCAR News* in 1985. Much of Miller's algorithm is still used in tracking programs today. These programs allow satellite users to easily calculate orbits at home. The Keplerian orbital elements that were used by these programs

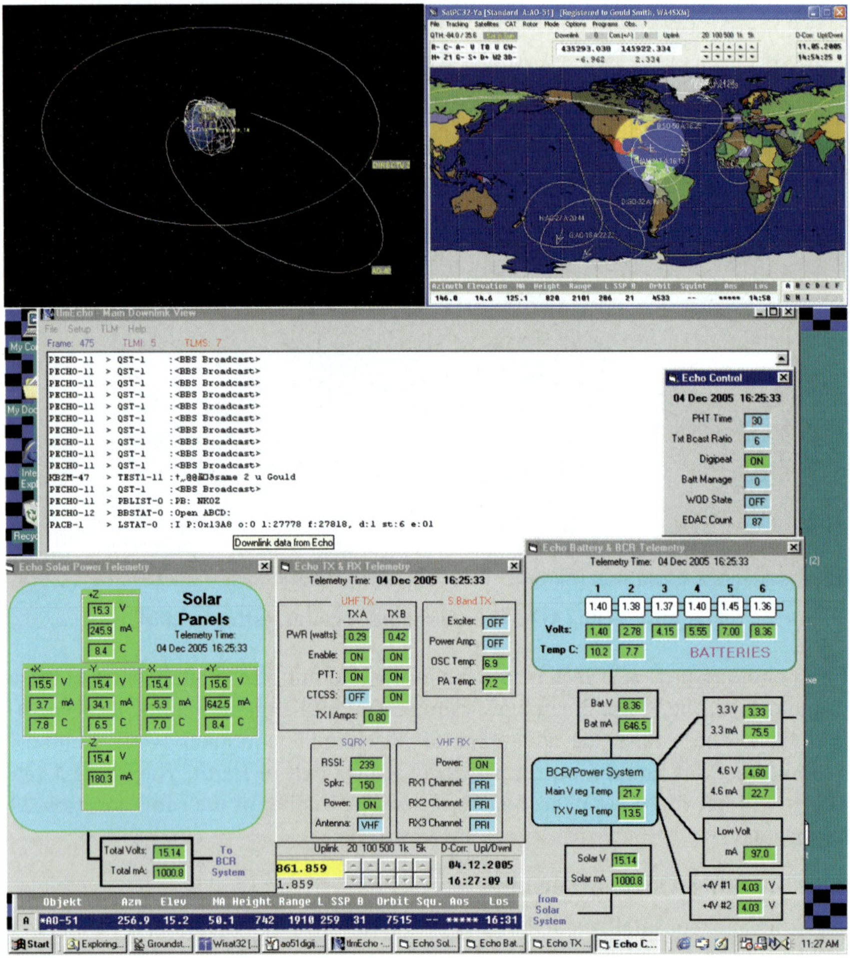

Fig. 4.17. Nova for Windows software showing LEO, GEO, and HEO. NOVA is an inexpensive satellite tracking program available through AMSAT (top left). The main display window of the current amateur satellite tracking software, SatPC32 (top right). Typical AO-51 9600 bps telemetry display (bottom).

were mainly distributed by reading 10-digit values over HF amateur radio nets: Users would manually enter them into their tracking program. It wasn't until the late 1980s that programs would accept text files of Keplerian elements, removing the error-prone human entry method. Welch wrote a tracking program with a graphical display called Orbits. There were a series of these Orbits programs that mirrored the evolution of computer monitors, from monochrome to CGA to EGA and finally VGA. Around 1989, Bob McGwier (N4HY) released his well-received QuikTrak program. This was followed by the very popular Instant*Track* program by Franklin Antonio and Paul Williamson. These tracking programs were all donated to AMSAT as a fund-raiser. The next leap in tracking programs came with Nova for Windows and SatPC32 (Fig. 4.17). These full-featured programs now provide great visuals, as well as rotor control and satellite radio automatic Doppler correction.

4.8.3 Satellite Telemetry

Amateur satellite telemetry has taken many forms over the last 46 years, from simple Morse code to teletype, spoken voice, and specialized digital mode. Many of the digital modes required uncommon, specialized modems. This reduced a potentially large worldwide data-capture system. With the increased computing power available since 2000, many of the satellite telemetry data formats can be demodulated using the computer sound card. To allow the command stations to better monitor and control the satellite, most of the satellites for the last 20 years store telemetry data in files for a day at a time. The data are sampled at specific intervals (generally 30 to 60 seconds) and then stored in a file for download and analysis. These files, known as "whole orbit data," are available for many satellites through the AMSAT website.

The freely distributed EchoTlm program (available from the AMSAT website) takes the real-time digital data broadcast by the Echo (AO-51) microsatellite from the ground-station receiver and demodulator, then displays the current state of most of the onboard systems as well as the latest temperature, current, and voltage values. Viewers can see the current solar-cell voltage, current, and temperature; how much power each transmitter is producing; the state of each of the six batteries; and the receive signal strength of the SQRX multimode receiver.

4.8.4 Amateur Satellite Ground Stations

To make amateur satellites attractive to amateur radio operators, the satellites have to be usable without extravagant ground stations. The LEO FM satellites are entry-level satellites that radio amateurs can communicate through using only a dual-band, handheld FM radio and a small handheld dual-band Yagi-Uda antenna (Fig 4.18). The HEO satellites require a more substantial ground receiving station, but not a large transmitting station.

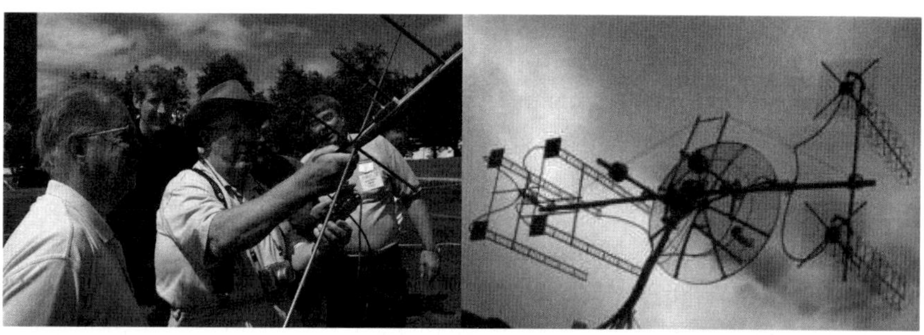

Fig. 4.18. Operators deploy an amateur satellite portable with a handheld radio and antenna (left). UHF/SHF amateur satellite antenna array from VE3NPC (right).

As with any hobby, a great deal of pleasure is achieved by many people from the challenge of learning new methods and experimenting by putting together the best combination of system components possible. Another important aspect is sharing this information with other enthusiasts. Often, many of these innovators don't spend much time communicating via the satellites, just getting to the point that they can communicate reliably.

4.9 AMSAT Educational Projects

AMSAT has a long history of including educational activities in conjunction with their satellites. In 1970, AMSAT worked with the Talcott Mt. Science Center in Avon, Connecticut, where more than 300 students were exposed to A-O-5. AMSAT has actively presented educational training sessions on all of their satellites and has produced and distributed lesson plans to many schools throughout the nation. Many articles have been written over the last 40 years describing satellite operation and theory to educate the general amateur radio public.

During the 1990s, Rich Ensign (N8IWJ) was the AMSAT science education advisor and encouraged science education using amateur satellites. The physics students at Chaminade High School in West Hills, California, produced an educational newsletter that was shared with interested educators throughout the world.

AMSAT has been a strong partner of the Amateur Radio on the International Space Station (ARISS) program. The astronauts and cosmonauts aboard the space station have talked live with school groups around the world. More than 15,000 students per year have benefited from this experience.[12] These school contacts are also broadcast over the Internet via VoIP links such as EchoLink and IRLP to foster as wide an audience as possible.

One of the most famous projects was SuitSat-1 in 2006, when a retired Russian Orlan space suit became an amateur satellite (Fig. 4.19). The spacesuit was equipped with an amateur radio transmitter, an old set of batteries, and an antenna attached to the helmet. The suit satellite was manually launched from the International Space Station during a spacewalk. This event captured worldwide attention. Voice messages from students in five languages were transmitted. The battery voltage and internal temperature of the suit were also transmitted as spoken values, so students could record and plot the changes. SuitSat-2 will have many additional modes and should be able to communicate for a month to six weeks as well as send slow-scan TV pictures to Earth.[13] AMSAT and ARISS are sponsoring this project. The most probable "launch" date is mid 2009.

4.10 Conclusion

A number of technological achievements have been derived from amateur satellites. For example, the AO-6 and AO-7 satellites tested and validated the use of Doppler shift analysis to locate ground-based beacons. These investigations directly led to the international COSPAS-SARSAT satellite-based emergency-

Courtesy of NASA

Fig. 4.19. SuitSat shortly after its release from the International Space Station on 3 February 2006.

beacon location system.[8,9] The AO-7 spacecraft also demonstrated the first use of a "photon propeller" to spin a spacecraft and the first use of CMOS integrated circuits in space.

The term "battery charge regulator" came from amateur satellites. The HELAPS amplifier has provided very efficient transmissions of satellite signals for 30 years. The P3A satellite was the first to use a microprocessor (RCA 1802 COSMAC) with a real-time operating system as the sole means of mission control.

At the first Small Satellite Conference in 1980, AMSAT and Project OSCAR had already designed, developed, launched, and operated nine spacecraft. The ASAP structure envisioned by Jan King revolutionized the ability of commercial launch vehicles to carry multiple small payloads along with the primary payload. The microsat concept changed the way that small satellites could be built and easily replicated. Amateur programs demonstrated that small satellites could do real science, and constellations of these small satellites could provide overlapping coverage. A summary of amateur satellite launches is provided in Table 4.2.

The future of AMSAT is bright. New Phase 3 satellites are being built, and talks progress in getting AMSAT a position on a GEO satellite. Phase 5 to Mars is a great challenge. AMSAT is always looking for talented engineers and scientists to contribute. New technologies are constantly being explored to enhance the performance of the amateur satellites for users worldwide.

Table 4.2. Amateur Satellites (Dec. 2007)

Name	Mass	Origin	Type	Launched	Mode(s)	Aliases/Notes
OSCAR-1	5 kg	Project OSCAR	Analog	12 Dec '61	Telemetry	
OSCAR-2	10 kg?	Project OSCAR	Analog	2 Jun '62	Telemetry	
OSCAR-3	14 kg	Project OSCAR	Analog	9 Mar '65	V band transponder	
OSCAR-4	13 kg	Project OSCAR	Analog	21 Dec '65	Mode J	
OSCAR-5	18 kg	Melbourne Univ./ AMSAT-NA	Analog	23 Jan '70	Telemetry	A-O-5, Australis-OSCAR 5
OSCAR-6	16 kg	AMSAT	Analog	15 Oct '72	Mode A	A-O-6
OSCAR-7	29 kg	AMSAT-NA & AMSAT-DL	Analog	15 Nov '74	Mode A & B, telemetry	A-O-7
OSCAR-8	27 kg	AMSAT-NA	Analog	5 Mar '78	Mode B	A-O-8
RS-1		USSR	Analog	26 Oct '78	Mode A	
RS-2		USSR	Analog	26 Oct '78	Mode A	
P3A	92 kg	AMSAT-DL & AMSAT-NA	Analog	23 May '80	(Did not reach orbit)	Phase 3 A
Iskra-1		USSR	Analog	10 Jul '81	Mode K	
UO-9	52 kg	SSTL	Beacons	6 Oct '81	S & X band transmitters	UoSAT OSCAR 9, UoSAT 1
RS-3 to RS-8		USSR	Analog/ digital	17 Dec '81	Mode A, CW robot	Radio Sputnik, RadioSport
Iskra-2		USSR	Analog	18 Apr '82	Mode K, telemetry	
Iskra-3		USSR	Analog	18 Nov '82	Mode K, telemetry	
AO-10	70 kg	AMSAT-NA & AMSAT-DL	Analog	16 Jun '83	Mode B, L	P3B, Phase 3-B AMSAT-OSCAR 10
UO-11	60 kg	SSTL,	Digital	2 Mar '84	Store & forward data, telemetry	UoSAT-2, UoSAT OSCAR-11
FO-12	50 kg	AMSAT Japan	Analog/ digital	12 Aug '86	BBS, repeater	JAS-1a, Fuji-OSCAR 12
RS-9		(never launched)				
RS-10/ RS-11		USSR	Analog	23 Jun '87	Mode A, K, Robot	Radio Sputnik, RadioSport
AO-13	150 kg	AMSAT-DL & AMSAT-NA	Analog	15 Jun '88	Mode B, J, L, S	Phase-3C, AMSAT-OSCAR 13
UO-14	48 kg	SSTL	Digital/ analog	22 Jan '90	Store & for-ward, repeater	UoSAT-3, Vitasat, HealthSat
UO-15	47 kg	SSTL	Digital	22 Jan '90	Experiments	UoSAT-15, UoSAT 4
AO-16	12 kg	AMSAT-NA	Digital	22 Jan '90	Store & forward mailbox, S band transmitter	PACSAT, Microsat-A, AMSAT OSCAR 16
DO-17	12 kg	AMSAT-NA & Brazil	Analog/ digital	22 Jan '90	Digitalker, V band downlink	DOVE, Microsat-B, OSCAR 17

Table 4.2 (cont.). Amateur Satellites (Dec. 2007)

Name	Mass	Origin	Type	Launched	Mode(s)	Aliases/Notes
WO-18	12 kg	AMSAT-NA & Weber State Univ.	Digital	22 Jan '90	Camera, digital data, Mode J	WeberSat, Microsat-C, OSCAR 18
LO-19	12 kg	AMSAT-NA & AMSAT-LU (Argentina)	Digital	22 Jan '90	Mode J digital, CW telemetry	LUSAT, Microsat-D, OSCAR 19,
FO-20	50 kg	AMSAT Japan	Analog	7 Feb '90	Mode J analog & digital transponder	Fuji OSCAR 20, JAS-1b
AO-21/ RS-14		USSR & Germany	Analog/ digital	29 Jan '91	Many digital modes	RS-14, OSCAR 21, RUDAK II
RS-12/13		USSR	Analog	5 Feb '91	Mode A, K, T transponder, robot	RS-12, RS-13
UO-22	48 kg	SSTL	Digital	17 Jul '91	Earth imaging, digital data, Mode J; VITA	UoSAT-F, OSCAR 22, UoSAT 5
KO-23	48 kg	SSTL & Korea	Digital	10 Aug '92	Mode J digital	KITSAT A, KITSAT OSCAR 23
ARSENE	150 kg	AMSAT-F (France)	Digital	13 May '93	Mode S transponder	ARSENE-OSCAR 24
KO-25	48 kg	SSTL & Korea	Digital	26 Sep '93	Mode J digital, CCD camera	KITSAT B, KITSAT OSCAR 25
IO-26	12 kg	AMSAT Italy	Digital	26 Sep '93	Mode J digital	ITAMSAT, IO-26, Italy OSCAR 26
AO-27	12 kg	AMRAD (US)	Analog	26 Sep '93	Mode J, 2007 D-Star	EYESAT-1 AMRAD OSCAR 27
PO-28	49 kg	Portugal	Digital	26 Sep '93	CCD imaging, GPS unit	PoSAT-1
RS-15	70 kg	USSR	Analog	26 Dec '94	Mode A	Radio Rosto, RadioSkaf-15
FO-29	154 kg	AMSAT Japan	Analog/ digital	17 Aug '96	Mode J	Fuji OSCAR 29, JAS-2
MO-30	10 kg	AMSAT-Mexico	Digital	5 Sep '96	Mode J	Mexico OSCAR 30
RS-16		USSR	Analog	5 Mar '97	Mode A	Radio Sport 16, RadioSkaf-16
RS-17		Russian, French students	Analog	3 Nov '97	Sputnik mode, V band beacon	Radio Sputnik-17, RadioSkaf-17
TO-31	55 kg	Thailand	Digital	10 Jul '98	Imaging, store & forward msg.	TMSAT-1
GO-32	70 kg	Israel	Digital	10 Jul '98	Mode J BB, experiments	TechSat-1B, Gurwin OSCAR 32
SO-33	36 kg	Univ. of Alabama Huntsville	Digital	24 Oct '98	Mode A, L; experiments	SEDSAT, SEDSAT OSCAR 33
PO-34	65 kg	Naval Post-graduate School	Digital	30 Oct '98	U band spread spectrum	PANSAT
SO-35	60 kg	Univ. of Stellenbosch, SA	Analog/ digital,	23 Feb '99	Mode J, B; S & L band	SUNSAT
UO-36	300 kg	SSTL	Digital	21 Apr '99	High-resolution camera, GPS experiment	UoSAT-12, UoSAT-OSCAR 36
AO-37	6 kg	Arizona State Univ.	Digital/ analog	26 Jan '00	Mode J	ASUSat-1

Table 4.2 (cont.). Amateur Satellites (Dec. 2007)

Name	Mass	Origin	Type	Launched	Mode(s)	Aliases/Notes
OO-38		Stanford Univ.	Digital	26 Jan '00	Mode J	OPAL, OPAL-OSCAR 38
WO-39		Weber State Univ.	Digital	26 Jan '00	Mode J, S band beacon	JAWSAT, WEBER-OSCAR 39
AO-40	244 kg	AMSAT International	Analog/ digital	16 Nov '00	Mode S transponder, GPS, new technology	P3D, Phase 3D, AMSAT-OSCAR 40
SO-41	~11 kg	SSTL & Saudi Arabia,	Analog/ digital,	26 Sep '00	Mode J repeater	OSCAR 41
SO-42	~11 kg	Saudi Arabia & SSTL	Analog/ digital	26 Sep '00	Not operated in amateur service	SAUDISAT-1B
SO-43	91 kg	Naval Reseearch Lab & US/Canadian students	Analog/ digital,	30 Sep '01	Passive mirror, 2 m beacon	Starshine
NO-44	67 kg	Naval Postgraduate School	Digital	30 Sep '01	Mode B	PCSAT, Nav OSCAR-44
NO-45	18 kg	Stanford & Washington Univs.	Digital	30 Sep '01	digipeater, BB	Sapphire, Nav OSCAR-45
MO-46	50 kg	Malaysia & SSTL	Digital	26 Sep '00	High-resolution imaging	TIUNGSAT-1
RS-20	64 kg	USSR	CW telemetry,	28 Nov '02	V & U band beacons	Mozhayets
RS-21	18 kg	Russian, Australian students	Analog	20 Mar '02	U band CW data	Radio Sport 21, Kolibri-2000
BO-47		AMSAT-F (France)	Analog/ digital,	3 May '02	U band beacon	BreizhSat-OSCAR 47
BO-48		AMSAT-F (France)	Analog/ digital	3 May '02	U band beacon	BreizhSat-OSCAR 48
AO-49		OBH Bremen, Germany	Digital	20 Dec '02	Mode B	AATiS
SO-50	10 kg	Saudi Arabia & SSTL	Analog	20 Dec '02	Mode J repeater	SAUDISAT-1C, Saudi OSCAR 50
AO-51	11 kg	AMSAT-NA	Analog/ digital	29 Jun '04	Mode J, V/S, L/U, L/S	ECHO, AMSAT OSCAR-51
VO-52	42 kg	AMSAT India	Analog	5 May '05	Mode B transponder	USAT, HAMSAT
XO-53	62 kg	European students & ESA	Analog/ digital,	27 Oct '05	Mode U/S, V band data/ telemetry	SSETI Express, eXpress OSCAR
AO-54		ARISS	Analog/ digital	3 Feb '06	V band beacon	SuitSat-1, AMSAT OSCAR-54
CO-55	1 kg	Tokyo Inst. of Technology	Digital	30 Jun '03	CS, U band telemetry/data	Cute-1, CubeSat OSCAR-55
CO-56	10 kg	Tokyo Inst. of Technology	Digital	21 Feb '06	U band telemetry, L/U packet	Cutie-1.7 (2x CubeSat)
CANX-1	1 kg	Univ. of Toronto, SFL	Digital	30 Jun '03	CS, U band telemetry	
DTUSat	1 kg	Technical Univ. of Denmark	Digital	30 Jun '03	CS, U band telemetry/data	
AAU-CubeSat	1 kg	Univ. of Aalborg, Denmark	Digital	30 Jun '03	CS, U band telemetry/data	

Table 4.2 (cont.). Amateur Satellites cont. (Dec. 2007)

Name	Mass	Origin	Type	Launched	Mode(s)	Aliases/Notes
RS-22		Mozhaisky Military Space Univ., USSR	Telemetry	27 Sep '03	U band CW beacon	Mozhayets 4
CO-57	1 kg	Univ. of Tokyo	Digital	30 Jun '03	CS, U band telemetry/data	CubeSat XI-IV, CubeSat OSCAR-57
PCSAT-2		US Naval Academy	Digital	3 Aug '05	Mode J	attached to ISS
CO-58	1 kg	Univ. of Tokyo	Digital	27 Oct '05	CS, U band telemetry/data	CubeSat XI-V, CubeSat OSCAR-57 (SSETI-Express)
UWE-1	1 kg	Univ. of Wurzburg	Digital	27 Oct '05	CS, U band telemetry/data	(Ejected from SSETI-Express)
NCube-2,	1 kg	Univ. of Oslo	Digital	27 Oct '05	CS, U band telemetry/data	(Ejected from SSETI-Express)
HO-59	2.7 kg	Hokkaido Inst. of Technology	Digital	22 Sep '06	U band telemetry/data	HITSat
GeneSat-1	4.5 kg	NASA Ames & Sant Clara Univ.	Digital	16 Dec '06	U band telemetry/data, S band beacon, biology exp.	
NO-60		US Naval Academy	Digital	21 Dec '06	V band data downlink	RAFT-1, attached to ISS & returned
NO-61	50 kg	US Naval Academy	Digital	21 Dec '06	V band digipeater	ANDE, Navy OSCAR-61
RAFT		US Naval Aacademy	Digital	21 Dec '06	V band digipeater	MAA, deployed from STS-116
NO-62	62 kg	Naval Research Lab	Digital	21 Dec '06	Radar Fence calibration	FCAL, NRL OSCAR-Sat, depl. from STS-116
PO-63	6 kg	Argentina	Analog/ digital	10 Jan '07	V band digital data	Pehuensat-1
CAPE-1	1 kg	Univ. of Louisana	Digital	17 Apr '07	U band telemetry/data	
Libertad-1	1 kg	Univ. Sergio Arboleda, Columbia	Digtial	17 Apr '07	Mode J digipeater	
Mir		USSR	Analog/ digital	Nov '88	Voice, SSTV	
SAREX		US	Analog/ digital	Nov '83	Voice & digital data	Space shuttle, SAFEX
ISS		ISS	Analog/ digital	Sep '00	Voice & digital data	ARISS

APRS: Automatic Position Reporting System
BB: Bulletin Board, store & forward mailbox
CS: CubeSat
CW: Continuous Wave, used to send Morse code
Digipeater: device that retransmits APRS data packet
ESA: European Space Agency
ISS: International Space Station
L band: 1.2 GHz band
Mode A: uplink 145 MHx, downlink 29 MHz
Mode B: uplink 435 MHx, downlink 145 MHz
Mode J: uplink 145 MHx, downlink 435 MHz
Mode L/U: uplink 1.2 GHz, downlink 435 MHz
Mode L/S: uplink 1.2 GHz, downlink 2.4 GHz
Mode K: uplink 21 MHz, downlink 29 MHz
Mode V/S: uplink 145 MHz, downlink 2.4 GHz
Mode T: uplink 21 MHz, downlink 145 MHz
Packet: AX.25 digital protocol
Robot: automatic CW reply to CW request
S band: 2.4 GHz band
SSTL: Surrey Satellite Technology Ltd.
SSTV: slow scan television
U band: 435–438 MHz
V band: 145–146 MHz
VITA: Volunteers in Technical Assistance

4.11 References

[1] K. Meinzer, "A Linear Transponder for Amateur Radio Satellites," *VHF Communications* (1973); available at http://www.amsat-dl.org/dl/HELAPS_en.pdf.

[2] http://centaur.sstl.co.uk/SSHP/launcher/launch_asap.html#ASAP.

[3] T. Clark, C. Duncan, J. King, B. McGwier, "The First Flock of MicroSats," *AMSAT Journal* (May 1989).

[4] AX.25 Protocol, http://www.tapr.org/pub_ax25.html.

[5] http://www.iso.org.

[6] G. Smithson, *Introduction to Digital Modulation Schemes,* http://www.plextek.com/papers/schmsv6.pdf.

[7] G. Davis, M. Moreau, R. Carpenter, F. Bauer, *GPS-Based Navigation and Orbit Determination for the AMSAT AO-40 Satellite*, http://www.emergentspace.com/pubs/AIAA_GNC_2002_AMSAT_A040.pdf.

[8] L. Kayser, "Emergency Location Experiments Via OSCAR," *AMSAT Newsletter*, p. 14 (September 1976).

[9] J. V. King, "Overview of the Cospas-Sarsat Satellite System for Search and Rescue,"

http://satjournal.tcom.ohiou.edu/pdf/king.pdf.

[10] K. Deskur, "Shoot Oscar with a Satellabe," *AMSAT Newsletter* (March 1976).

[11] D. Ingram, *OSCAR: The Ham Radio Satellites* (Tab Books, 1979).

[12] F. Bauer, http://www.amsat.org/amsat/ariss/Presentations/AMSAT Symposiums/2007 Symposium/AMSAT 2007.pdf.

[13] L. McFadin, "Suitsat-2/RadioSkaf-2: The Second Amateur Radio Space Suit Project and Stepping Stone to Future Small Amateur Satellites," *Proceedings of the 2007 AMSAT-North America Space Symposium* (Pittsburgh, PA).

4.11.1 Additional Works Consulted

AMSAT Newsletter, 1971–1979.

AMSAT Orbit magazine, 1983–1984.

AMSAT Journal 1985.

AMSAT Journal 1995–2007.

AMSAT Reports 1989–1994.

K. Baker (KB1SF) and D. Jansson (WD4FAB), "Space Satellites from the World's Garage—The Story of AMSAT," *National Aerospace and Electronics Conference* (Dayton, Ohio, 23-27 May 1994).

T. Clark (K3IO [W3IWI]; AMSAT president emeritus), correspondence.

D. Daniels (W4PUJ), interviews and correspondence.

M. R. Davidoff, *The Satellite Experimenter's Handbook* (ARRL, 1984).

M. R. Davidoff, *The Radio Amateur's Satellite Handbook* (ARRL, 1998).

P. Güzlow (DB2OS; AMSAT-DL president), presentation on P3E and P5A (Dayton, 2007).

http://ariss.gsfc.nasa.gov.

http://www.amsat.org.

http://www.astronautix.com.

http://www.ee.surrey.ac.uk/postgraduate/msc/pathways/satcomms.

http://www.IARU.org.

http://www.jamsat.or.jp/index_e.html.

http://www.projectoscar.net.

http://www.spacetoday.org/Satellites/Hamsats.

http://www.sstl.co.uk.

J. King (W3GEY; former AMSAT VP engineering), interviews and correspondence.

J. King et al., "OSCAR at 25: The Amateur Space Program Comes of Age," *QST* (Dec. 1986).

J. A. King, R. McGwier, H. Price, and J. White, "The In-Orbit Performance of Four, Microsat Spacecraft," *Proceedings of the 4th Annual AIAA/USU Conference on Small Satellites* (Houston, TX, 1990).

P. Klein (W3PK; former AMSAT president/founder), interviews and correspondence.

P. Klein and R. Soifer, "Intersatellite Communication Using the AMSAT-OSCAR 6 and AMSAT-OSCAR 7 Radio Amateur Satellites," *Proceedings of the IEEE Letters* (Oct. 1975), pp. 1526–1527.

R. Ludwig, "JPL Acknowledges Voyager Reception at Bochum," *AMSAT Journal* (July/August 2007), p. 31.

H. Paesler (DL1YDD; AMSAT-DL vice president), interview.

G. Smith (WA4SXM), "Decoding Telemetry from the Amateur Satellites," (AMSAT-UK, 1991).

G. Smith (WA4SXM), "Amateur Satellite Launch Sites, Launches, and Orbits," *Proceedings of the AMSAT-NA 21st Space Symposium* (Toronto, 2003).

D. Stoner, "Semiconductors," *CQ* (April 1959), p. 84.

B. Tynan (W3XO; former AMSAT president), interviews and correspondence.

R. Welch (W0SL), correspondence.

Chapter 5

Origin of CubeSat

Bob Twiggs

5.1 Introduction and Concept History

The CubeSat is a standardized miniature satellite measuring 10 × 10 × 10 cm and weighing up to 1 kg. It was developed primarily for use as an educational tool. The general concept for such a satellite arose in 1998 as a result of work by students at Stanford University's Space Systems Development Laboratory (SSDL) on the OPAL (Orbiting Picosatellite Automated Launcher) microsatellite. OPAL, first conceived by students in 1995, was designed to demonstrate mother-ship and daughter-ship technologies: A storage, deployment, and communication scheme was designed and implemented on the OPAL microsat mother ship, which ultimately deployed six picosat daughter ships.

The OPAL project coincided with a special graduate research class that I directed with Prof. Tom Kenny of the mechanical engineering department. This class explored the possibility of developing picosats and launching them from a mother craft. Kenny had received a grant from the Jet Propulsion Laboratory (JPL) for his MEMS work, and this grant supported the class research. Kenny had been developing a MEMS magnetometer that used electron tunneling to measure the deflection of a current-carrying beam. JPL wanted to use the magnetometer in a picosat such as that shown in Fig. 5.1 to measure Earth's magnetic field. This "hockey puck" shaped picosat needed to spin to detect the magnetic field vectors.

The launcher developed by the students, however, was hopelessly complicated (see Fig. 5.2). While students could, with a great deal of effort, make it work "most of the time," it was seldom successful in a practice launch of a picosat. The launcher could hold three "hockey puck" picosatellites stacked in a canister. Each picosatellite had Velcro on one face. The load arm had the mating Velcro on a flat face oriented toward the picosats. When a picosat needed to be set up for launching, the drive screw on the canister moved the picosatellites forward until the first one pressed against the Velcro on the load arm. When the drive screw stopped, the load arm pivoted 45 deg to align the attached picosat with the spinner. The spinner was then moved forward with its carriage screw until its clamps engaged over the picosatellite. The spinner carriage then reversed, pulling the picosat off the load-arm Velcro. At launch time, the spinner's rotation motor started the picosat

Space Systems Development Laboratory, Stanford University
Palo Alto, California

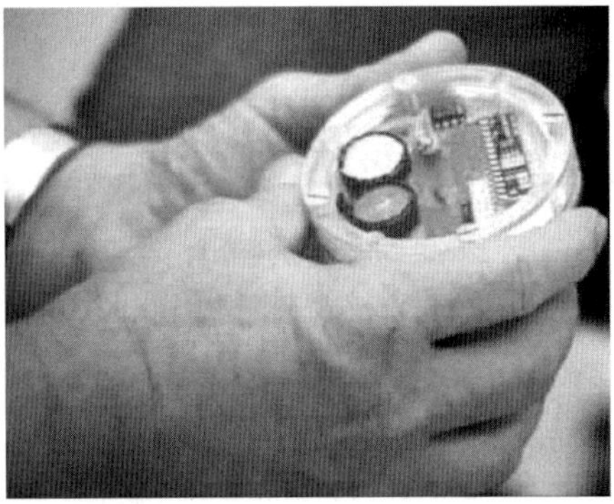

Fig. 5.1. Early picosatellite developed by Michael Hicks, Jason Suchman, and Paul Graven while graduate students at Stanford.

rotating at about 10 rpm. The linear acceleration screw on the spinner carriage would drive the spinning picosatellite forward. At the end of the carriage travel, the spinner clamps would open, the carriage would stop, and the picosatellite would fly out into space through a hole in the side of the microsatellite.

While we were working on this program for JPL, Henry Helvajian at The Aerospace Corporation in El Segundo, California, asked us to participate in a nanotechnology conference in Pasadena; however, Kenny had concerns about attending this conference and advised against it. We continued work on the round picosatellite until it could be demonstrated at JPL early in 1998. Shortly after our demonstration, JPL told Kenny to cease work on the project because it had received NASA funding to do the picosat work in-house and to fly it on a sounding rocket. However, this was an ongoing student project, so canceling it completely was not a good educational option. I called Helvajian to see if Aerospace would be interested in collaborating on a picosat project. He put me in touch with his colleague, Ernie Robinson.

Helvajian, Robinson, and Siegfried Janson had been pioneers in the concept of the picosat, with papers published as early as 1993, and were the inspiration for our work. Robinson came to Stanford shortly after our initial discussions to look at our work on OPAL and the picosat launcher. He was not impressed with the launcher, and proposed that we collaborate on a new design.

Shortly thereafter, at a conference at the Air Force Research Laboratory (AFRL) at Kirtland Air Force Base in Albuquerque, New Mexico, in April 1998, Robinson arranged a meeting with Al Pisano, director of the MEMS program for DARPA. At the meeting—in a hallway, between sessions—Pisano agreed that DARPA would fund a picosat mission.

It was also at this conference that the Air Force and DARPA nanosat program got its start. At the end of a presentation about the Stanford microsatellite program, in which I stated that Stanford was developing microsatellites for material costs of about $50,000, I challenged government agencies in the audience to

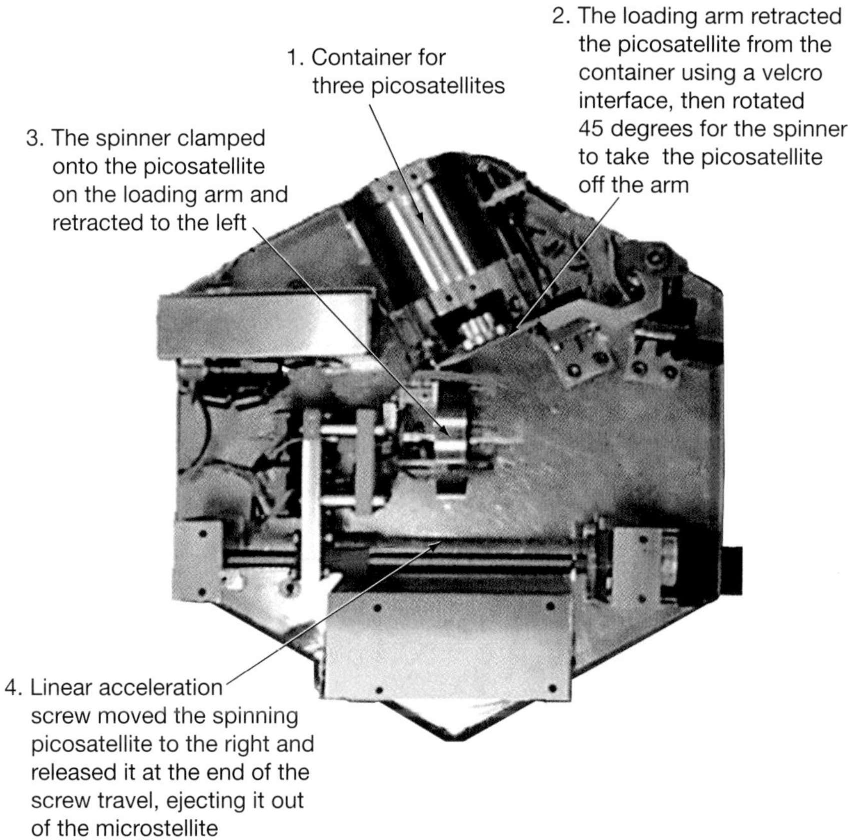

Fig. 5.2. "Hockey puck" picosat launcher for OPAL.

provide funding and launches for university satellite programs. Immediately, Joe Janni from the Air Force Office of Scientific Research stood up and announced to the audience that he and his deputy, Jack Agee, would fund five university programs for one year with $50,000 grants each and provide launches.

After the morning break, Janni again stood up to announce that he and Agee decided that one year and $50,000 was not sufficient, so they would increase funding to two years, bringing the total to $500,000. Then, Joe Mitola from the DARPA Strategic Technology Office announced that DARPA would match Janni's offer. So, in less than an hour, decisions were made to fund 10 universities for two years with $1 million. And the "University Nanosat" program was born.

With new funding from DARPA, Aerospace started work on a picosat design that could be launched from a spacecraft, and a Stanford SSDL team of students began exploring new launcher concepts for OPAL. Unlike JPL, Aerospace did not call for a round picosat that was spun up at launch. Rather, its concept was

Fig. 5.3. First picosatellite launcher made from Corian.

to use two tethered picosats to test communications and MEMS. The new design was a flat picosat about 4 × 3 × 1 in. After considering several designs for holding the picosats in the launcher, the launcher prototype shown in Fig. 5.3 was built. The launcher was made from Corian countertop material in the shop of Advanced Counter Technology in Santa Clara, California. The shop was owned by my son, Mark Twiggs, which gave me access to the tools and material in the workshop for building models late at night.

In this design, the picosats were held on a chamfer on the corners to minimize surface contact. A pusher foot propelled by a spring ejected the picosats when the launcher door opened. To keep the picosats from rattling in the launcher, an angle plate was made on the pusher foot and the door to push the picosats to the side and lock them in place when the door was closed. A diagram of the launcher is shown in Fig. 5.4. The final launcher and the Corian prototype are shown in Fig. 5.5. A diagram of the final OPAL launcher is shown in Fig. 5.6; it had two rows of two launcher tubes, each with a volume of 8 × 3 × 1 in. Figure 5.7 shows OPAL during the final integration before launch. After the OPAL mission, the launcher was tested for a future mission—but it was discovered that repeated testing made score marks on the launcher rails, which could cause an ejection failure.

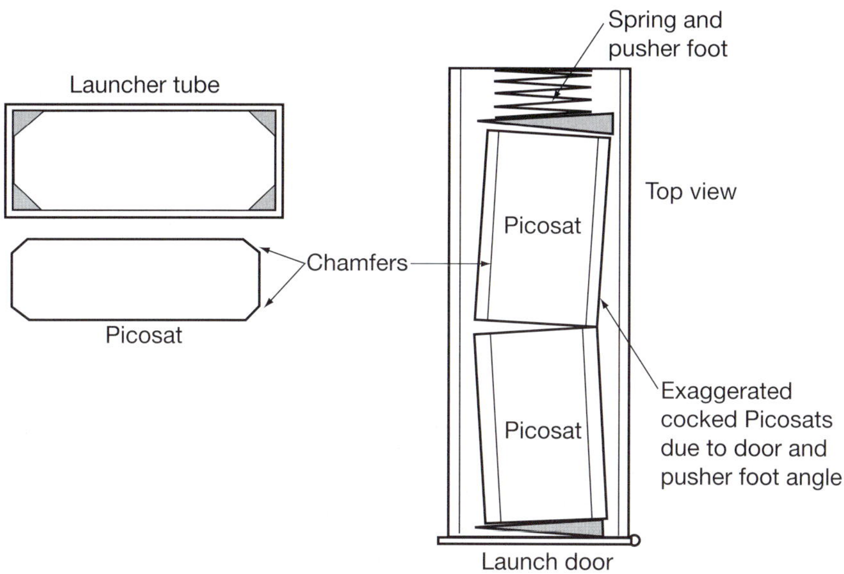

Fig. 5.4. OPAL launcher diagram.

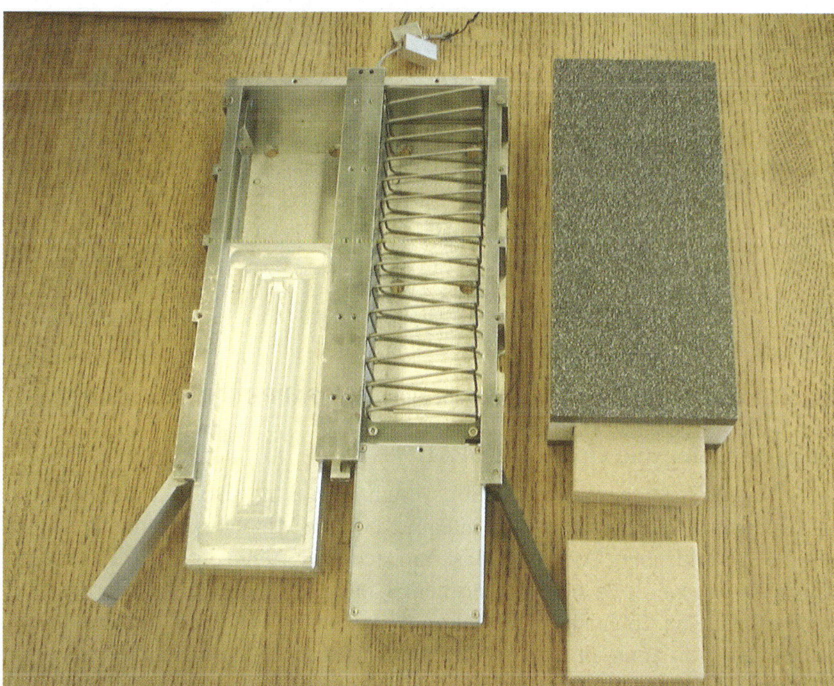

Fig. 5.5. OPAL picosatellite launcher (left and center) shown with the prototype Corian model (right).

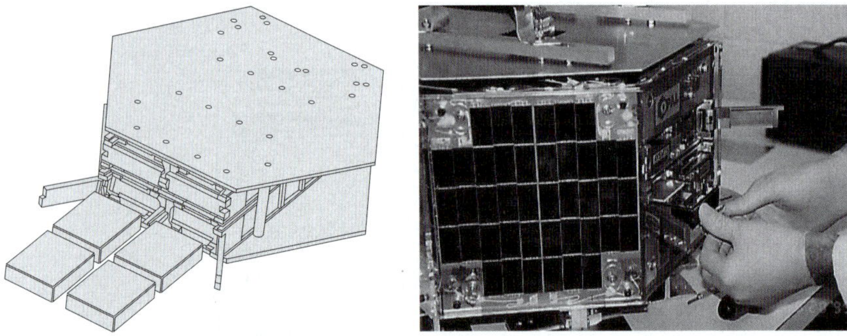

Fig. 5.6. Illustration of the final version of the OPAL picosat launcher (left) and photo of a picosat being loaded in OPAL for the satellite integration (right).

Fig. 5.7. OPAL before launch integration, along with SAPPHIRE, a sister satellite.

5.1.1 Launching and Operating OPAL

OPAL was a hexagonal prism 23.5 cm high with a mass of 23.1 kg and a usable volume of 21,300 cm^3. It was integrated onto a Minotaur expendable launch vehicle at Vandenberg Air Force Base in December 1999 (Fig. 5.8). The rocket, which also carried two other experimental microsatellites plus a nanosatellite, was launched on 26 January 2000 into a sun synchronous, sun terminator, circular polar orbit at an altitude of 773 km. This marked the maiden flight of the Minotaur. After some initial difficulty with the ground operations at Stanford, OPAL released all picosats on 6 February 2000. There were the two tethered $4 \times 3 \times 1$ in. picosats developed by The Aerospace Corporation; one picosat called

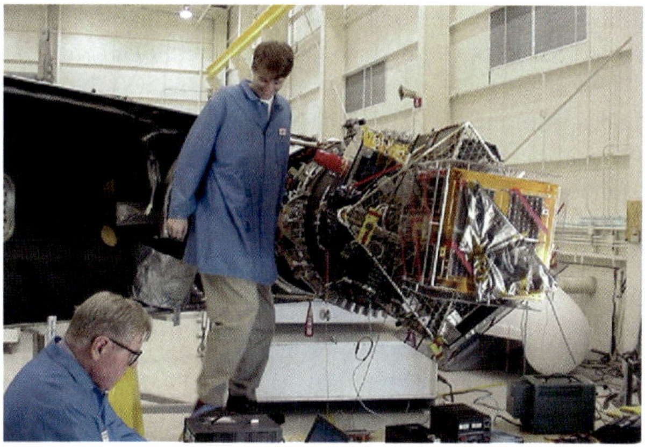

Fig. 5.8. Integration of OPAL onto the Minotaur launch vehicle.

Stensat built by a group of Ham radio enthusiasts in the Washington, DC, area; and three picosats (Thelma, Louise, and JAK) built by an engineering team of women students at Santa Clara University in California.

The Aerospace picosats, which were powered by primary batteries, proved to be very successful. OPAL was the only microsatellite launched on this mission that continued to operate after a few weeks. All others seemed to experience battery-charging problems after launch. OPAL operated for two years and was used to train students in satellite operations.

5.2 A New Direction

After the success of the Aerospace picosat mission, I realized that to make the student satellite program more successful, two changes in the satellite designs were necessary. First, development time would have to be shortened—five years had passed between the start of OPAL in 1995 to its launch in 2000—and second, launch costs would have to be reduced. I figured that if we could reduce the size of the satellite, that would limit the number of experiments that students could fly. That would reduce the satellite mass, which would in turn help lower the launch cost.

5.2.1 Design Considerations for a Picosat

The design question became: "How much could you reduce the size and still have a practical satellite?" The picosats for OPAL were 4 × 3 × 1 in., but charging batteries was difficult with these flat unstabilized spacecraft. If solar cells were put on all surfaces, some power could be generated—but it was not practical to put solar cells on the edges of a satellite measuring only 1 in. thick.

A picosat that could successfully charge batteries would need solar cells on all sides. Eight cells at 1.2 V per cell would provide nominal panel voltage of 9.6 V, sufficient to charge two lithium-ion batteries in series. I laid a number of

2 × 4 cm GaAs cells out on the table and measured what area eight cells would require—it turned out to be 3.5 inch square. I wanted a launcher that did not require the picosats to be locked down during launch, as was the case with OPAL. I estimated that a clearance of 0.010 in. between the picosats and the launcher should be sufficient to prevent damage from the accelerations or vibration during launch. A distance of 0.250 in. was allowed for each rail to capture the corners of the satellites (the chamfer design from the OPAL picosat launcher was not used). Two rails and surface of 0.250 in. made the width of the picosat 4 in. We looked for a box to use as a model, and found a 4 in. cubic plastic box normally used to display "Beanie Baby" dolls in toy stores. This box, covered with photos of solar cells, is shown in Fig. 5.9. Note the spacers on the corners: These were added to prevent the solar-cell surfaces from touching when several satellites were stacked serially in a launch tube, as the OPAL picosats were.

At the time CubeSat was being developed, a Mars Lander suffered a failure, which was found to be caused by a difference in the measuring units used by JPL and Lockheed Martin—one used English units, and the other, metric. In poking some minor fun at this error, we decided to make the cube in metric units. Since 10 cm was very close to 4 in., we designed it to be a 10 × 10 × 10 cm cube, which

Fig. 5.9. First plastic model of CubeSat.

would provide a volume of 1000 cm^3 (which would weigh 1 kg if filled with water—the designated upper limit for picosat mass).

5.2.2 Developing the CubeSat Launcher

With the new CubeSat dimensions defined, a new launcher could be developed. A design similar to the one used in OPAL was evaluated, but for a 10 cm cube (with attached spacers to protect the solar panels, as previously mentioned). A diagram of the CubeSat launcher concept is shown in Fig. 5.10.

A plastic model of the launcher was built for three cubes with a spring-activated pusher foot and a door. The cube model and the launcher were taken to California Polytechnic State University (Cal Poly) at San Luis Obispo in the summer of 1999 and shown to Prof. Jordi Puig-Suari. We agreed that a standard for a picosat and launcher to be used by many universities could lead to the launch of a large number of picosats at one time from several launcher tubes. If Cal Poly could handle the integration of the CubeSat launcher with the launch vehicle, then it seemed possible to get affordable launches (less than $50,000) for university programs. The launcher developed by Cal Poly and flown on four missions so far is shown in Fig. 5.11. Cal Poly students named the launcher the P-POD for Poly Picosat Orbiting Deployer.

This concept was shown at the Japan-U.S. Science, Technology, and Space Applications Program (JUSTSAP) conference in Hawaii in November 1999. At that time, the cube was given the name S^3Sat for Student Space Science Satellite. Students developed a webpage for S^3Sat after the conference, but in December 1999, they decided to simplify the tedious typing of the name by renaming it CubeSat.

From a Beanie Baby box, a new picosat was developed that has become an international standard—a serendipitous outcome from a concept that was introduced to solve time and cost barriers for using satellites as a project-based learning tool for engineering students.

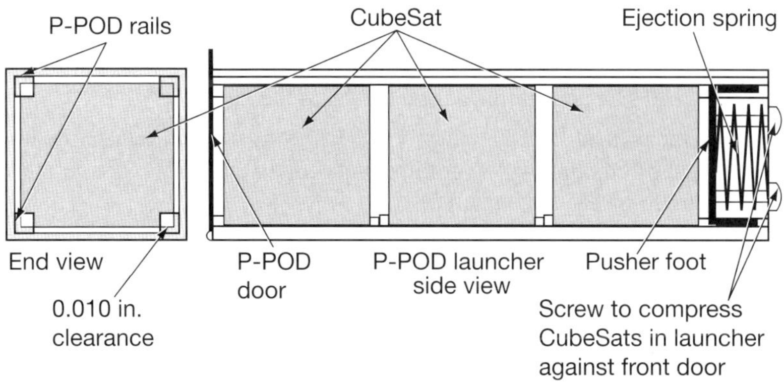

Fig. 5.10. CubeSat launcher concepts.

Fig. 5.11. Concepts of the P-POD launcher (left) and standard CubeSat (right).

5.2.3 The Unique Attributes of a CubeSat

Some issues related to size in the design and construction make CubeSat unique. Positive issues are:

- The 10 cm, 1 kg cube is a reasonable size for one person to carry.
- The frame is of a simple shape and construction.
- A CubeSat can be displayed at full size.
- Limited area for solar cells reduces manufacturing costs (the solar panels are the most expensive components for a small satellite).
- A CubeSat's low weight, which allows it to be combined with other CubeSats in a single launch, helps defray costs.
- CubeSats can take advantage of new technologies for consumer electronics such as cell phones and other portable devices.
- Low power dissipation minimizes internal thermal issues.
- A CubeSat can take advantage of the latest wireless devices and technologies.
- The small size and simple design appeal to the public.
- With the present P-POD launcher, the CubeSat size can actually be increased by combining two or three units end to end. (Sizes are defined by lengths of 1U, 2U, and 3U.)

Some negative issues are:

- The size is always pushing the designer to want more volume and power than can be provided.
- Currently, the CubeSat has limited capability because no proven attitude control systems are available. (Expect to see this change within the next two years.)

- It is difficult to get a good packing factor for components in the allowed dimensions.
- Surface area for body-mounted solar panels is limited.
- Subsystem requirements limit payload volume.

Despite these limitations, one of the major advantages of CubeSats is the structural standardization and cost of launch. If a designer builds to the CubeSat standard, several launch integration services will provide launches. These include Cal Poly in San Luis Obispo, California; the UTIAS Space Lab at the University of Toronto in Canada; and Space Access Technologies LLC at Lakewood, Colorado. Costs are reduced through the collaboration of the collective community to build quantities to a standard.

5.3 History of CubeSat Launches

Table 5.1 shows seven launch missions for CubeSats, starting in June 2003. As of this writing, 33 CubeSats have been delivered for launch, including double (2U) and triple (3U) units. In one launch failure in July 2006, 14 CubeSats were lost. Of the 19 CubeSats that have achieved orbit, 5 have had some contact and 10 have had full contact; only 2 were not sponsored by universities, and both of those were successful.

Table 5.1. CubeSat Launches and Contact Statistics

Batch #	LV Failure	No Contact	Some Contact	Full Contact	Total
1 (June 2003)	0	2-1U	1-1U	2-1U + 1- 3U	5-1U + 1-3U
2 (Oct. 2005)	0	1-1U	0	2-1U	3-1U
Solo (Feb. 2006)	0	0	0	1-2U	1-2U
3 (July 2006)	13-1U + 1-3U	0	0	0	13-1U + 1-2U
Solo (Dec. 2006)	0	0	0	1-3U	1-3U
4 (Apr. 2006)	0	1-1U	3-1U + 1-3U	2-1U	6+1-3U
5 (Dec. 2006)	0	0	0	1-1U	1-1U
6 (Dec. 2007)	5 awaiting launch				
Total	13-1U + 1-2U	4-1U	4-1U + 1-3U	6-1U+1-2U +3-3U	27-1U + 2-2U + 4-3U

Courtesy Quake Finder LLC and Stellar Solutions

Fig. 5.12. The disassembled QuakeSat shown with the boom, antennas, and solar panels. On the right is the P-POD launcher.

5.3.1 First CubeSats Launched

The first CubeSat mission, launched on 26 June 2003, included six individual CubeSats. Five were single-unit CubeSats, and one was a three-unit CubeSat (which stretched the full length of the P-POD) called QuakeSat.

QuakeSat, shown in Fig. 5.12, was the first CubeSat-derived spacecraft launched for a scientific mission. The primary payload was a magnetometer with the sensor mounted in the extended boom. This passive, magnetically stabilized satellite mission was to look for ultralow-frequency (ULF) signals. These were considered potential "earthquake predictors" because data from terrestrial sensors near the 6.9 magnitude Loma Prieta earthquake near Santa Cruz, California, in October 1989 showed a large rise in ULF signals over a period of several weeks before the quake. QuakeSat did not generate enough solar power to run the sensor all of the time, so the spacecraft processor was programmed to turn on the sensor only over known earthquake regions.

Over its six-month operation life, QuakeSat did not detect anomalous signals in advance of an earthquake, but it did detect signals immediately after one: a magnitude 6.5 earthquake occurred near San Simeon, California, on 22 December 2003. The QuakeSat instrument had been activated to examine this area and did provide evidence of ULF signals produced at the faults, but it did not detect precursor or warning signals at sensing times before the quake. The area of the sensing passes and dates are shown in Fig. 5.13.

5.4 Your Own Personal CubeSat

5.4.1 Who Builds CubeSats and Why?

At this time, those who are building and launching CubeSats are primarily students at universities that have aerospace programs. A few aerospace companies

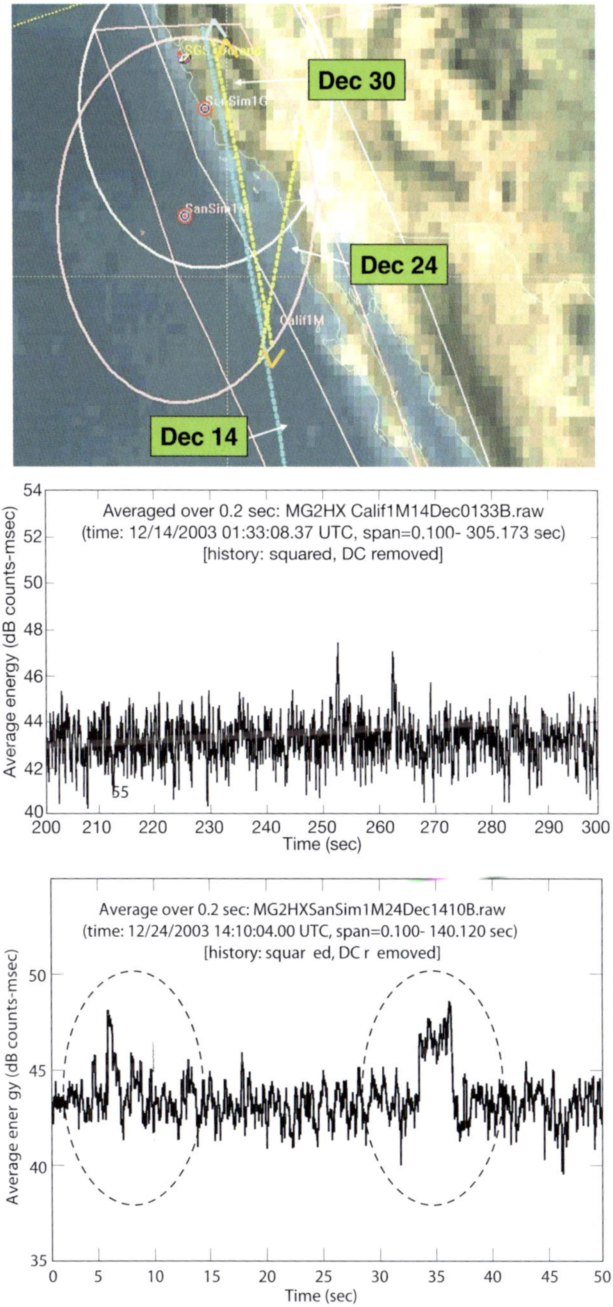

Fig. 5.13. QuakeSat passes over the San Simeon area before and after the magnitude 6.5 earthquake on 22 December 2003 (top); ULF signal levels on 14 December (middle) and 24 December 2003 (bottom).

also have small teams building and demonstrating CubeSat space operations, such as The Boeing Company and The Aerospace Corporation.

University students are eager to build CubeSats because, in general, such projects are within their technical and financial means. The cost for materials to build a CubeSat can be as low as multiple \$10,000 amounts and a launch of \$40,000. The most expensive part of the CubeSat is the cost to have a commercial or aerospace company build the solar panels. Building something that goes into space and works is also a frontier that not many engineers get the opportunity to pursue. It is "cool" to have your fingerprints in space.

So, for students, the "why" is the excitement and the learning experience. The payoff is space experience across a broad range of technologies that takes many years of engineering at a major aerospace company. For aerospace companies, the CubeSat program provides a cheaper means of placing satellites into space and an opportunity to complete some space qualification of small parts at low cost in a matter of a year, in contrast to expensive programs that take several years.

5.4.2 Is it Hard to Build and Get a Launch for a CubeSat?

Building a CubeSat without prior experience can be a daunting task. Not that building something to go into space is difficult, but the advice you may get from those without experience in building "inexpensive" satellites in a nontraditional manner may make it seem impossible.

New satellite builders, for example, are often confounded by space radiation and the thermal environment in space. With regard to space radiation, it's quite likely that as a CubeSat builder, you cannot afford space-qualified parts—nor do you want them. The newer parts that you may want to use have probably not been space qualified. The market for these space-qualified, electronic components is so small that manufacturers usually don't bother. So, on the one hand, you can forget about space-qualified parts, but on the other hand, you must consider designing some restart capability in the processor, with a watchdog timer in case the processor hangs due to a bit flip in the RAM code or other detectable conditions.

For the thermal design, some consideration must be given to connecting internal high-heat sources—such as the transmitter final amplifier stage—to the frame of the satellite for radiative heat dissipation. In our experience, the temperature ranges in space were not anything near the –150°C to +200°C that "experts" predicted. So, the major guideline in the design of a CubeSat is just to use good common-sense engineering.

If you want to build a CubeSat, first build one that will function on the desktop, without worrying about all the space requirements. Having done this, you will understand where the effort must be placed to build a space-qualified version. Many university student programs start out with grand ideas for satellites to impress others. It is more important to get the first satellite built and launched than to have an innovative satellite that never gets finished. Build a second one to innovate. If you must, make the first one a "Sputnik" to get that complete system experience.

When you decide to build a CubeSat, check the Cal Poly CubeSat website (http://CubeSat.atl.calpoly.edu/). Look at all cited references and review websites of universities that have documented their CubeSat projects. Not being unique is not an engineering sin. Everything you learn building a CubeSat is unique to you, even if others have done it before. An open-source philosophy was one of the guiding principles in starting the CubeSat program at Stanford and Cal Poly. No fees are charged for information to start a CubeSat program, and advice is freely given to all participants—who are encouraged, in turn, to advise new students.

At the beginning of your project, count costs not only in funds required to build and launch a CubeSat, but also in terms of the timeline and students' learning needs. If you make the first one as simple as possible to complete in the shortest time, you will find the task much easier than if you end up changing students several times throughout the project. What's more, you do a disservice to those students who start the project and leave before it is launched. The complete systems experience is very important for students.

Waiting for a launch can be frustrating for students who have built a launch-ready satellite. That may change in the future, with the greater demand for full utilization of launcher weight capacity and the mandate for better educated students. In the meantime, the best way to get a launch is to work with an integration provider such as Cal Poly. The integrator takes the hassle out of launching a CubeSat. You wouldn't want to make arrangements yourself to ship a package when a carrier such as FedEx does that for you—you don't get engineering experience from that. Start the CubeSat program with a definite date for completing a flight-ready satellite. Waiting for a firm launch date can actually discourage completion of a flight-ready satellite. The benefit of the program is education and experience—most of which can be accomplished by completing that flight-ready satellite. The launch and operations will be the final "icing on the cake."

Build a CubeSat while a student in a university, as part of a private group, or alone. It is an experience of a lifetime.

5.4.3 What are the Integration Requirements for a CubeSat?

Integration for a satellite can refer to several tasks. One major integration task is assembling all of the parts of a satellite into its final form and testing it for flight readiness. Integration is also the process of mounting the satellite on the launch vehicle.

The Cal Poly CubeSat website provides a list of suggested testing requirements to qualify the CubeSat before integration with the P-POD. These requirements are meant to make sure that if your CubeSat is loaded into a P-POD with other CubeSats, it will pose no harm to them in terms of outgasing during launch or having a mechanical failure that would preclude them from being successfully deployed. A typical vibration profile and thermal processing for a CubeSat in testing for integration are shown in Figs. 5.14 and 5.15.

Integration onto the launch vehicle will vary depending upon the selection of launch vehicle. Most CubeSats to date have been launched on the Russian Dnepr

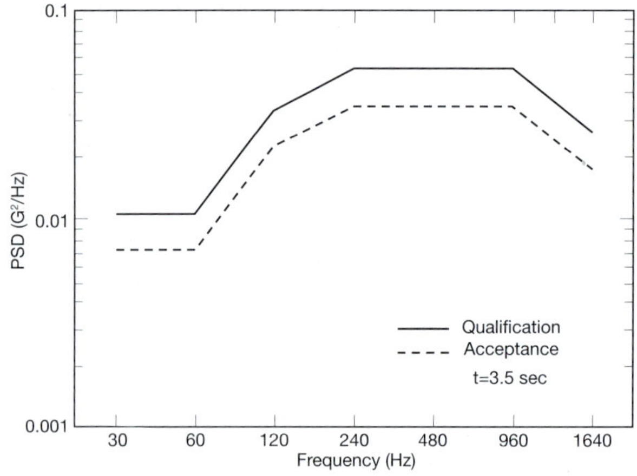

Figure 5.14. Vibration testing for a CubeSat.

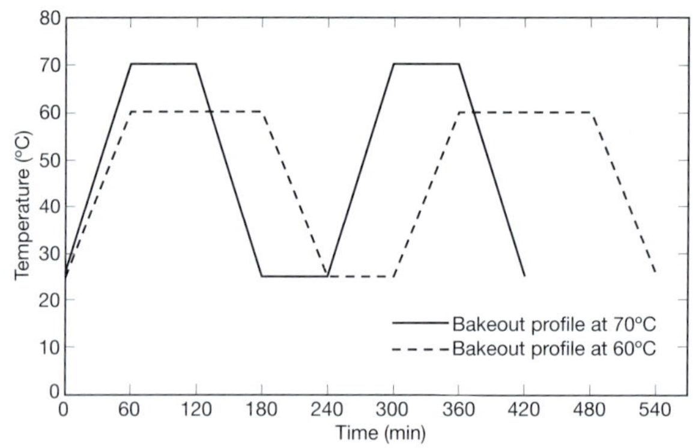

Fig. 5.15. Thermal outgasing of a CubeSat before integration.

Fig. 5.16. Artist's conception of a P-POD on the Dnepr launch vehicle.

from the Baikonur launch site. A typical launch for the Dnepr is from a silo. The P-PODs are loaded into the Space Head Module, then transported to the launch silo and dropped onto the Dnepr launch vehicle. An artist's conception of the P-POD attached in the payload section of the Dnepr is shown in Fig. 5.16.

5.4.4 What Equipment is Needed to Communicate with a CubeSat?

By the nature of its small size and limited surface area for solar panels, the CubeSat has limited power—generally 1 to 2 W (comparable to one of the small lights on a Christmas tree). Communication is difficult for CubeSats in low Earth orbits. The power for communications is dependent upon two factors: how fast you want the data to transfer from your CubeSat to the ground, and at what distances you can communicate with your CubeSat. The CubeSat is farthest away when it is just coming over the horizon into view and closest when it passes directly overhead.

The most common equipment used for communication with the CubeSats are directional antennas that are mounted on rotors that allow pointing at the satellite. These antennas, which work at lower frequencies, are called Yagi antennas and look like long TV antennas that you see mounted on some homes for local TV reception. They work at the VHF frequencies (146 MHz range) and UHF frequencies (in the 440 MHz range). These antennas and "Ham" radios are readily available from local amateur radio stores. To broadcast using an amateur radio transmitter, however, you must obtain an amateur radio license. The components of a typical CubeSat ground station are shown in Fig. 5.17.

You also must coordinate your use of these radio frequencies with the IARU (International Amateur Radio Union) and the FCC (U.S. Federal Communications Commission). This information is available on the Cal Poly CubeSat website.

With your antennas and radios set up, you can communicate with most CubeSat satellites. Data rates are as low as 1200 and 9600 baud—slow, even compared with the old 56K telephone modems for personal computers. Cost for a complete satellite ground station can run as much as $7500. The radios required will allow you to operate not only your own CubeSat, but a number of in-orbit amateur radio satellites as well.

5.4.5 How are CubeSats tracked?

The 10 cm CubeSat is a rather small object to track when it is several hundred miles above Earth. Government services track all objects larger than a certain size orbiting Earth and publish ephemeris data that can be used in tracking formulas that will give your CubeSat's location at any time. Fortunately, the CubeSat is within the lower limit for tracking size. Tracking information on university satellites can be found at the www.amsat.org website. Engineering design choices can be implemented to enhance tracking (see chapter 21).

CubeSats are generally launched in a group—which presents an interesting problem for the tracking services to identify individual satellites. The tracking service can determine that many objects are in orbit from your launch, but cannot distinguish yours from one of its fellow travelers. How do you identify your

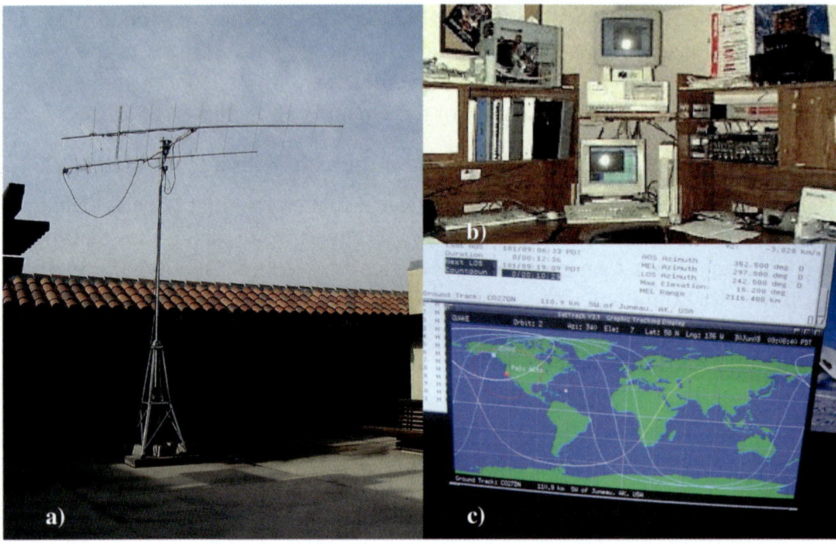

Fig. 5.17. Typical CubeSat antenna (a), ground control station (b), and tracking display (c).

CubeSat? Based on their experiences, students and amateurs encourage developers of small satellites to use a beacon transmitter that periodically transmits a unique code. This beacon enables individual CubeSats to be tracked as they spread apart after launch. The beam cone of the tracking antenna, through physical pointing, can identify each transmitting beacon, thus locating your CubeSat.

When you identify your CubeSat with an object number provided by the tracking service, you then need to use the ephemeris data they provide to track it. There are many programs available that use this ephemeris data to provide graphic displays of where your satellite is located (see chapter 4). These programs can also provide output that is used to control your tracking antenna to follow the satellite as it passes over your ground station.

5.4.6 Your Own Personal Satellite is a Reality

Because of the number of universities, aerospace companies, and government laboratories that are taking interest in the CubeSat for space activities, a number of vendors are now making spacecraft components to build a CubeSat. Some of these components now have now been flight-tested on orbit.

These commercial vendors make it practical and affordable for an individual or hobbyist club to have its own satellite in orbit. An example of a commercial kit from Pumpkin, Inc., San Francisco, with a space-qualified flight structure is shown in Fig. 5.18. An example of a single-unit CubeSat structure is shown in Fig. 5.19. This structure meets all of the mechanical requirements of the Cal Poly CubeSat documentation. The structure can also be purchased in 0.5U, 2U, and 3U lengths. The 2U would be the double CubeSat; the 3U is referred to as the triple CubeSat.

Courtesy Pumpkin Inc.

Fig. 5.18. Commercially available CubeSat kit.

Courtesy Pumpkin Inc.

Fig. 5.19. Commercially available CubeSat structure.

There are also electronic power system (EPS) boards available from Clyde Space in Scotland, and UHF and VHF transceivers from Stensat LLP in Virginia. Kits start at $6,000, and nearly all standard satellite bus components for a CubeSat cost near $25,000. Then, with a launch for a single CubeSat from Cal Poly now costing $40,000, your personal satellite is a reality. Why would you want a personal satellite and what would you do with it? Only the future will reveal. Before the advent of the Apple computer, no one foresaw the many possible applications for the personal computer. Will there be such a revolution in space with the CubeSat?

5.5 Developing New Technologies Using CubeSats

Many opportunities exist for CubeSats to significantly increase technological development within the space community. The reduced cost for a space test mission, the short time through development to launch, and the innovation of unbiased students will bring new technologies to benefit space exploration.

For example, students are now developing new technologies to increase the utility of the CubeSat with attitude determination and control, radio modulation and efficiency, more efficient power systems, and systems to mitigate space debris. The low cost of CubeSats have started program developments that consider satellite clusters and collaborative satellite swarms for space science, which have not been affordable or practical with large satellites.

The requirement for reduced size and power for CubeSat components will produce devices that will ultimately improve performance for larger missions and foster development of science instruments that allow low-cost missions. These factors are being increasingly recognized by many science organizations, such as the National Science Foundation, which is interested in using CubeSats for space weather measurements.

Consider the rapid development cycle of new commercial electronic components for the consumer industry in personal music and communication devices. The use of these new components for space missions can only be validated by using them in the real space environment. This can be done with a program that would regularly place CubeSats in highly elliptical orbits that traverse the high radiation conditions of the Van Allen belts. For this mission, some special electronics and techniques would be required.

One of the main concerns about placing clusters of CubeSats in low Earth orbits is their contribution to the debris population once they cease operation. Most secondary launch opportunities are on launch vehicles that place satellites in orbit more than 500 km in altitude, which can leave these satellites in orbit for more than 25 years. The CubeSat university community is fully aware of this problem. Probably more than one-third of the universities that now have CubeSat experience are researching or designing deorbiting technologies to be tested in future missions.

5.6 CubeSat as a Teaching Tool

When the CubeSat concept was conceived, the major objective was to provide a space program that would shorten the development time to launch readiness, reduce the launch integration difficulties for inexperienced university students, and significantly reduce the launch cost. If this objective could be achieved, it would also reduce the time for students to experience the full system cycle from conception to operations. One year was established as the cycle time at Stanford. Because the Stanford Aeronautics and Astronautics program is only graduate level, the spacecraft design program was intended for master's students. Stanford does not require a thesis at the master's degree level, and most candidates stay only one year.

The limited power from the small surface area of the CubeSat, the uniqueness of this small satellite, and a known launch support service gives students the confidence that the CubeSat's complexity is within their technical capability. Interestingly, the major difficulty students have in a CubeSat project is program management and team dynamics. The CubeSat program is unlike any other "project based" university interdisciplinary program in that satellite development within an engineering curriculum requires a broad spectrum of technologies and is constrained by the powerful requirement that "once it's launched, you can't get it back to fix it." This is not a normal lab exercise—you must produce a complete, tested satellite before you'll be willing to pay the cost for launch. The effectiveness in education of the CubeSat program concept is evidenced by the number of universities internationally that now have satellite programs as part of their curriculum at both the undergraduate and graduate level. The CubeSat gives engineering students a chance to participate in real space programs, which would have been impractical a few years ago.

5.7 Acceptance of the CubeSat Standard

The purpose of a CubeSat standard was to establish an external physical structure that would work in a standard launcher. Other universities were not consulted because their satellite programs concentrated mostly on larger rather than smaller satellites. We were not concerned whether the university community would accept it as a standard; the design worked for us—Stanford and Cal Poly. Our only concern was that universities would be interested enough to produce multiple picosats to fill the launchers and lower the costs. What would have been the outcome if we had set out to develop a standard that would benefit everyone? Maybe it would have resulted in something much more complicated, or maybe we would not yet have any standard at all.

In any case, the CubeSat concept has met with astonishingly strong acceptance by universities internationally and has fostered much excitement within the educational community eager to involve engineering students actively in space technology. The educational benefits of the university CubeSat program are of far greater significance than the contributions to technical knowledge from the picosat design, fabrication, launch, and operation. Students work in interdisciplinary teams, develop leadership skills, and learn systems engineering, as specifically emphasized in satellite projects. Simply put, there is a special excitement for students to work on space projects.

The acceptance of a CubeSat standard by aerospace organizations has been much faster outside the United States. NASA and other government agencies virtually ignored this concept until about 2006. Since then, there has been a significant change, with conferences being promoted by government agencies such as the National Science Foundation.

The launch of the first CubeSats in 2003 sparked some very adverse reactions within the amateur satellite community. For universities using amateur communication frequencies, this was as much a concern as the utility of the satellites. The AMSAT community is now very supportive in establishing a committee in the International Amateur Radio Union (IARU) to help with frequency coordination in the amateur frequency bands for universities.

5.8 Future Plans for Larger or Smaller CubeSats

As with all standards, modifications are often sought by those who require something different to satisfy their mission requirements. Plans are now being assessed for 20 cm and possibly larger cubes. This increased size would move this cube into the nanosat class between 1 kg and 10 kg. I'm not aware of any current proposals for a smaller satellite than the CubeSat, although it may be practical with the rapid development of commercial parts that can be used in short-life satellites. Maybe we could make a miniature CubeSat—1 cm—that would broadcast Morse code after being ejected from a mother spacecraft and placed on a path to do a Lunar fly-by and return past Earth. It may even be a femtosat called Moon Beam.

5.9 Acknowledgments

The many students who worked at the Stanford Space Systems Development Laboratory since 1994 all have made major contributions to the development of OPAL, SAPPHIRE, and subsequent microsatellites and picosats built at the university. Most are still in space careers. Two students in particular who have been major contributors and who are still working with the academic community are Prof. Mike Swartwout at Washington University in St. Louis, Missouri, and Prof. James Cutler, who now directs the Stanford Space Systems Development Laboratory.

At the time of this writing, the recognition and success of the CubeSat has been far more than was expected at the time of its inception. Its major purpose was to provide a better way to educate engineering students with real project-based space engineering.

5.10 Bibliography

Nanosatellites

S. W. Janson, H. Helvajian, and E. Robinson, "The Concepts of 'Nanosatellite' for Revolutionary Low-Cost Space Systems," *44th Congress of the International Astronautics Federation*, (Graz, Austria, 16–22 Oct. 1993), IAF-93-U.5.5573.

OPAL

D. S. Clarke, M. T. Hicks, A. M. Fitzgerald, J. J. Suchman, R. Twiggs, T. W. Kenny, and J. Randolf, "Picosat Free Flying Magnetometer Experiment," *Proceedings of the 10th Annual AIAA/USU Small Satellite Conference* (Logan, UT, 16–19 Sept. 1996).

B. Engberg, J. Ota, and J. Suchman, "The OPAL Satellite Project: Continuing the Next Generation Small Satellite Development," *Proceedings of the 9th Annual AIAA/USU Conference on Small Satellites* (Logan, UT, 19–22 Sept. 1995).

R. Twiggs, J. Cutler, G. Hutchins, and J. Williams, "OPAL: A First Generation Microsatellite Providing Picosat Communications For The Amateur Radio Community," *Proceedings of the 1999 AMSAT-NA Symposium* (San Diego, CA, Oct. 1999).

J. Cutler and G. Hutchins, "Opal: Smaller, Simpler, Luckier," In *Proceedings of the AIAA Small Satellite Conference* (Logan, UT, Sept. 2000).

M. Swartwout, C. Kitts, and J. Cutler, "Sapphire: Case Study for Student-Built Spacecraft," *Journal of Spacecraft and Rockets,* **3** (3) 1136–1139 (2006).

CubeSat

F. Ince, "A Role for CubeSats in Responsive Space," *Proceedings of 2nd International Conference on Recent Advances in Space Technologies, 2005. RAST 2005* (Istanbul, Turkey, 9–11 June 2005), pp. 106–108.

M. Swartwout, C. Kitts, and J. Cutler, "Sapphire: Case Study for Student-Built Spacecraft," *Journal of Spacecraft and Rockets,* **3** (3) 1136–1139 (2006).

G. Chandler, D. McClure, S. Hishmeh, J. Lump Jr., J. Carter, B. Malphrus, D. Erb, W. Hutchinson, G. Strickler, J. Cutler, and R. Twiggs, "Development of an Off-the-Shelf Bus for Small Satellites," *2007 IEEE Aerospace Conference* (Big Sky, MT, 10 March 2007), DOI: 10.1109/AERO.2007.352752.

QuakeSat

J. Bortnik, J. W. Cutler, C. Dunson, T. E. Bleier, and R. L. McPheron, "Characteristics of Low Latitude Pc1 Pulsations During Geomagnetic Storms," *Journal of Geophysical Research,* **13** (April 2008), A04201, doi:10.1029/2007JA012867, 2008.

Chapter 6

Coalition Operationally Responsive Space: A "100 Satellite" Solution

Tom "Dingo" Doyne*

6.1 Introduction

Operationally responsive space (ORS) is about providing guaranteed space power fast enough to meet warfighter needs. Small spacecraft known as tactical satellites (TacSats) are one capability now under development that could be quickly deployed to support joint military operations. TacSats could provide space power at a cost of only tens of millions of dollars each. On the other hand, orbital mechanics impose limitations on use of the low Earth orbits preferred for small satellites. For example, a constellation of TacSats in low Earth orbit requires more satellites than a constellation in a higher orbit because a satellite's residence time over a particular area is shorter for lower orbits. Thus, the use of small satellites presents a problem: how can the United States afford TacSat constellations while continuing the acquisition of major space programs such as EELV, SBIRS, NPOESS, and GPS?

One answer is to enlist close allies as partners in the operational use of TacSat-derived constellations. The Air Force has often partnered with allies in the acquisition and sales of aircraft systems; examples include later blocks of the F-16, F-15, E-3A, and F-35 Joint Strike Fighter. Expanding these partnerships to the space arena would allow the United States and its allies to harness the potential of small satellites in support of theater military operations.

6.2 TacSat History

The U.S. Dept. of Defense (DOD) Office of Force Transformation began its ORS initiative in 2003 with TacSat-1. Since then, additional TacSat missions have begun. Recent accomplishments include the launch of TacSat-2, commencement of the ORS Payload Technologies Initiative, and completion of the Congressionally mandated "120-Day Study," which led to the establishment of a Joint ORS office in May of 2007 at Kirtland Air Force Base, New Mexico. Additionally, acquisition funding for ORS was added to the existing Science & Technology/ Research & Development budget, and the implementation plan for the Joint ORS office is being drafted.

*Office of the Director of Defense Research and Engineering, Office of the Secretary of Defense, Washington, DC

6.2.1 The TacSat Experiments[1]

TacSat experiments are an important part of ORS. Each one combines key ORS elements, often for the first time, allowing the ORS community to interact with new technologies and explore new possibilities. This experimentation fosters a natural chain reaction, as people think of more and better ways to use such a system. This creativity generates momentum and quickly moves related ideas, in DOD and industry, out of the viewgraph stage and into implementation.

Five TacSat missions are currently in operation or development. The emphasis of each one varies: All provide some level of operational utility, but some have a stronger emphasis on technology development, while others focus more on system development. This emphasis results from a logical pairing of mission objectives to meet a desired operational capability and to advance enabling technologies and systems. Figure 6.1 illustrates the emphasis for the first four TacSat experiments.

TacSats are rooted in the principle of operational experimentation, bringing the research community together with the operational community to effectively and rapidly channel promising concepts and technologies into useful capabilities. This experimentation allows the operational community to exercise new technologies with regard to current problems and current capabilities—often resulting in applications, concepts of operations, tactics, techniques, and procedures that the research community would have never have envisioned. This experimentation also helps the research community understand and consider the realities and constraints of the operational environment while conceptualizing and designing new systems. This direct community contact, in the context of tangible experiments, allows a valuable exchange of ideas not achievable over a set of viewgraphs, and it must be maintained and expanded to ensure that steady and operationally relevant spirals of capability are introduced into the ORS acquisition pipeline.

The following sections summarize each TacSat mission.

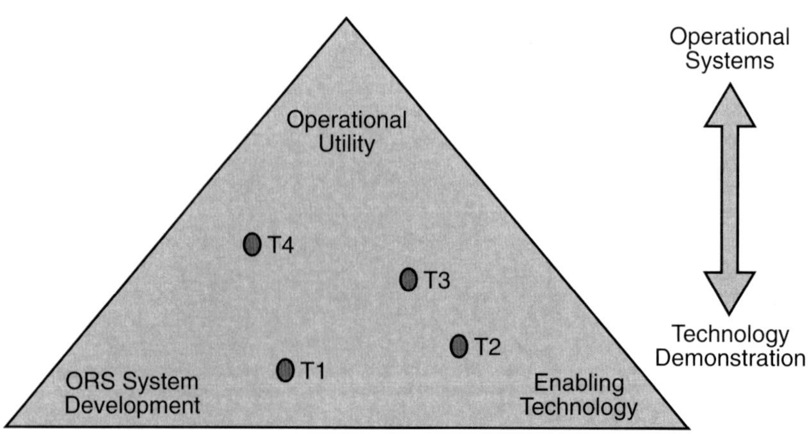

Fig. 6.1. Each TacSat has a different emphasis balance.

6.2.1.2 TacSat-1

The TacSat-1 experiment is shown in Fig. 6.2. The overall goal of TacSat-1 is to demonstrate the utility of a broader complementary business model and provide a catalyst for energizing defense and industry in the operational space area. TacSat-1 integrates three key elements of an ORS system: a capable microsatellite, a low-cost launch vehicle, and a link to the Secret Internet Protocol Routing Network (SIPRNet) via the Virtual Mission Operation Center (VMOC) for tasking, data access, and collaboration. The 125 kg spacecraft carries a radio-frequency (RF) payload, a low-resolution visible (70 m) imager, an infrared (850 m) imager, and a direct tactical link to reconnaissance aircraft. The Air Force Space and Missile Systems Center (SMC) used the TacSat-1 mission to tailor the formal launch vehicle mission assurance processes and will continue to provide technical assistance as appropriate. The U.S. Navy's Blossom Point ground station will provide spacecraft command and control and will host the VMOC server. Additional antenna coverage is planned from the U.S. Air Force Satellite Control Network (AFSCN). U.S. Pacific Command is the primary combatant command for operational experimentation and warfighter feedback.

TacSat-1 was supposed to launch from the U.S. Army's Kwajalein test site on a SpaceX Falcon-1 launch vehicle in 2004. However, Falcon-1 developmental delays, coupled with the successful launch and operation of TacSat-2 (containing a newer, more capable version of the TacSat-1 payload), prompted a decision to consider upgrading TacSat-1 based on TacSat-2 lessons learned. Therefore, TacSat-1 remains, for now, on the ground.

6.2.1.3 TacSat-2

The TacSat-2 experiment (Fig. 6.3) grew out of the Air Force Research Laboratory's Roadrunner project, which sought to demonstrate the feasibility of developing and launching a militarily significant satellite in just one year. TacSat-2 represents a spiral development in several ways. First, the payload provides approximately 1-meter imaging and significantly better RF capability than TacSat-1. Second, TacSat-2 provided the first tactical common data link from space, serving as a

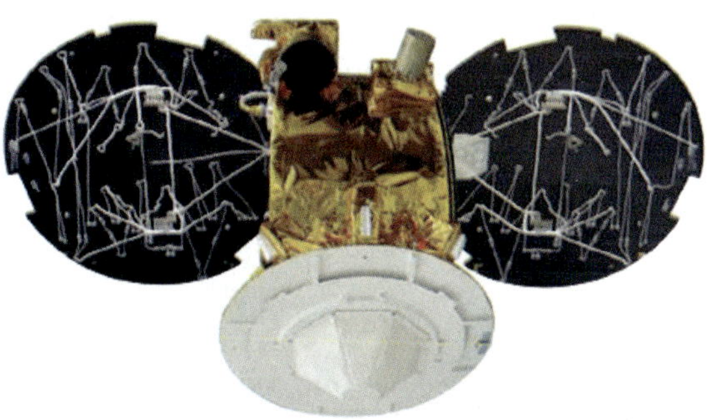

Fig. 6.2. TacSat-1.

pathfinder for a high-rate, tactical data link for ORS. It also carries several science payloads. The spacecraft was successfully launched on a Minotaur-1 rocket on 16 December 2006, from the Wallops range in Virginia. The rocket was pulled from a delayed program and prepared for the TacSat-2 launch in only six months; it used a new launchpad and facility. TacSat-2 completed its on-orbit checkouts and has participated in military exercises. Thus, it is helping to force new implementations of policies and authorities regarding DOD spacecraft collecting data historically gathered only by intelligence spacecraft. These new policies will set the stage for ORS utility and operations in the future.

6.2.1.4 TacSat-3

The TacSat-3 experiment (Fig. 6.4) was the first mission selected jointly by the various combatant commands. The process was led by Air Force Space Command and consisted of white paper submissions by combatant commands and services for current needs, payload concept development, mission rankings, and steering group selection. The selected payload is a land-focus hyperspectral imager with 1-meter resolution funded by the Air Force Research Laboratory and the Army. The Office of Naval Research is providing a small data-exfiltration (Data-X) payload as well. The Office of Force Transformation funded the spacecraft bus, which will provide the first implementation of bus standards with priority on avionics standards. TacSat-3 received go-ahead selection in October 2004, and launch is planned for summer 2009. The Air Force Research Laboratory's Space Vehicle Directorate is the lead organization for TacSat-3.

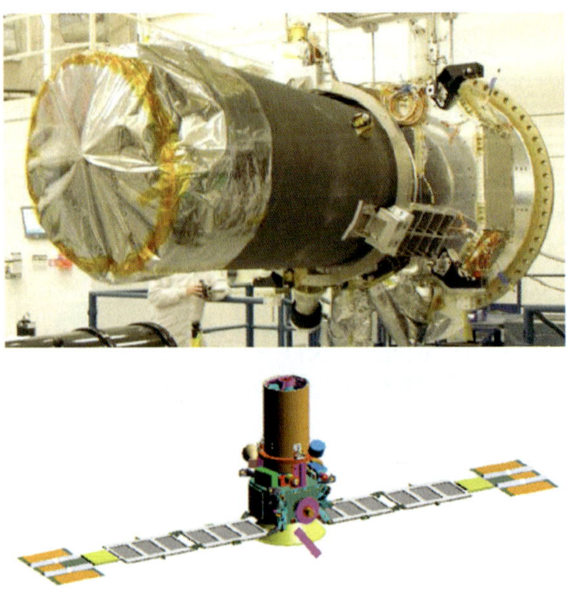

Fig. 6.3. TacSat-2.

6.2.1.5 TacSat-4

The TacSat-4 experiment (Fig. 6.5) was also selected by the joint community in October 2005. This second iteration refined the process created during TacSat-3 selection, and the process was formally documented. The payload capability includes Communications on the Move, Blue Force Tracking, and Data-X. The Communications on the Move capability provides UHF legacy radio support and capability similar to the Navy's Mobile User Objective System (successor to the UFO narrowband communications system). The Blue Force Tracking capability, designed to prevent "friendly fire" incidents, collects information from devices carried by troops on the ground, with tasking priority given to underserved areas. The Data-X capability focuses on data collection from Navy buoys, which are typically remotely located on the seas and coastal areas. TacSat-4 will provide ten UHF channels to simultaneously support all three primary payload packages.

TacSat-4 will be placed in a 4-hour highly elliptical orbit, allowing 1- to 2-hour dwells per pass. Ultimately, this orbit will enable persistent theater coverage with a minimum number of satellites (typically 4) and a minimum number of launches. These orbits also provide excellent coverage at high and low theater inclinations, as well as in mountainous terrain, which geosynchronous satellites do not cover well. TacSat-4 will use the bus resulting from the Phase 3 bus standardization effort (see below), allowing the bus standards to be refined throughout the design and testing process. Launch is expected in late 2009, with the Air Force Space Command and the Space Development and Test Wing providing a Minotaur-IV launch with mission assurance and range support. A Minotaur-IV and TacSat-4 success will be a significant advancement supporting ORS system

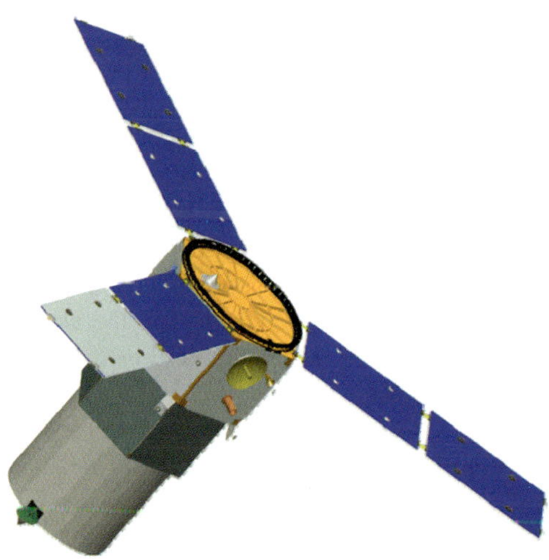

Fig. 6.4. TacSat-3.

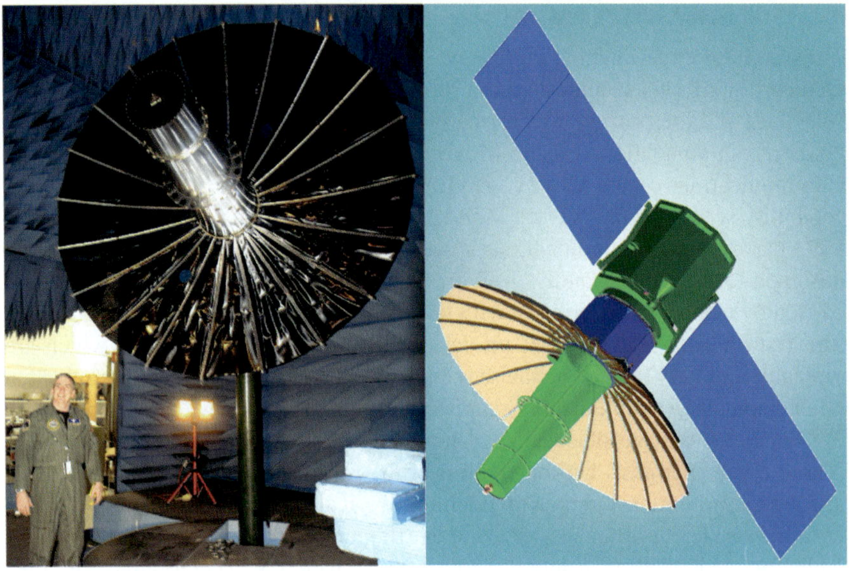

Fig. 6.5. TacSat-4 payload UHF antenna in RF test chamber (left) and satellite (right).

development. The Navy is leading the TacSat-4 experiment for the joint community; the Office of Naval Research is funding the payload and flight operations, and the Naval Research Lab is serving as program manager.

6.2.1.6 TacSat-5

TacSat-5 is in the mission-development and selection process. The selection team has continued to enhance the selection process to ensure commitment from all military services and U.S. government organizations, including the intelligence community. The newly created Joint ORS office has assumed responsibilities for mission selection. A key change made during this selection process was to publish the concept/white papers along with a request for ideas to further streamline the payload selection. A total of 47 TacSat-5 proposals were received from both large and small companies seeking to assist with the payload determination. TacSat-5 selection will satisfy as many ORS objectives as possible, including responsive launch, "plug-and-play" capability, and modularization; it will also satisfy a tactical warfighter need such as signals and electronics intelligence gathering, electro-optical imaging, tactical surveillance, space situational awareness, or space defensive countermeasures.

In the spirit of international collaboration, the TacSat-5 team is investigating potential cooperation between friendly countries in exchange for space products and capabilities. This effort is also related to a Trilateral Technology Research & Development Program agreement between the United States, Canada, and the UK.

6.2.1.7 ORS Bus Standards Development

Standardization of the spacecraft bus interface is necessary for an ORS system. The Office of Force Transformation and the Space and Missile Systems Center have therefore undertaken a four-phase initiative to develop and test bus standards and then transition them for acquisition. This effort involves multiple government laboratories and contractors, which is notably different from most past efforts. Almost by definition, for a standard to be accepted, a relatively large portion of the affected community must have participated in its development. Therefore, the Office of Force Transition has used its position to align the efforts of MIT's Lincoln Laboratory, the Air Force Research Laboratory, the Naval Research Laboratory, The Johns Hopkins University's Applied Physics Laboratory, the Space and Missile Systems Center, and many industry and university participants in this area.

Lincoln Laboratory led Phase 1, which provided the analytical foundation to determine the proper class of spacecraft needed to be militarily useful. Phase 1 began in October 2004 and finished with a report in June 2005.[2]

The Air Force Research Laboratory is leading Phase 2 of this bus standardization initiative. This phase has two critical objectives: to provide avionics standards between the bus and the payload and to provide a spacecraft bus for the TacSat-3 hyperspectral payload. In addition, many other approaches to internal bus modularity and plug-and-play bus standards are being explored in this phase as well as through the Air Force Research Laboratory's plug-and-play technology initiatives.

The Naval Research Laboratory and the Applied Physics Laboratory are leading Phase 3 of the bus standardization effort to help bridge the gap between a scientific and an operational bus capability.[3] Phase 3 uses a broad consortium of industry and government representatives to develop bus standards along with the associated costs and business considerations for acquisition. More than 10 companies are involved in both the integrated system engineering team and the business team. The results of Phase 2 and Phase 3 will form the basis for standards and provide a credible baseline for block procurements of TacSat-4, which is to be led by the Space and Missile Systems Center. Spiral block buys of five are anticipated, but this quantity is notional at this early stage. The importance of stable volume is realized by all and emphasized by industry.

6.3 Coalition ORS: The Logic for Cooperation

Warfare in the 21st century continues to be a coalitional effort. Admiral Mike Mullen, former Chief of Naval Operations (and as of summer 2007, Chairman of the Joint Chiefs of Staff), articulated the logic for cooperation with his "1000-ship navy" construct:

> We talk about a "thousand-ship navy." That's not just our ships. It's an international fleet of like-minded nations participating in security operations around the world.[4]

> No matter how large or small your navy or coast guard may be, we all face similar internal constraints like shrinking budgets, aging equipment, and populations that may not be attracted to military service. Our level of cooperation and coordination must intensify in order to adapt to our shared challenges and constraints. We have no choice in this matter, because I am convinced that nobody—no nation today—can go it alone, especially in the maritime domain.[5]

Mullen bases his case for cooperation with allies on shared challenges and common goals. When satellites are substituted for ships and space for the maritime domain, the rationale for an international 100-satellite constellation is as strong as the rationale for an international 1000-ship navy:

> We talk about a "hundred-satellite ORS." That's not just our satellites. It's an international constellation of like-minded nations participating in security operations around the world.... No matter how large or small your nation may be, we all face similar internal constraints like shrinking budgets, aging equipment, and transforming our militaries. Our level of cooperation and coordination must intensify in order to adapt to our shared challenges and constraints. We have no choice in this matter, because I am convinced that nobody—no nation today—can go it alone, especially in space.

An examination of the most recent U.S. National Space Policy reveals that international cooperation for national security purposes is allowed:

> The United States Government will pursue, as appropriate, and consistent with U.S. national security interests, international cooperation with foreign nations and/or consortia on space activities that are of mutual benefit and that further the peaceful exploration and use of space, as well as to advance national security, homeland security, and foreign policy objectives.
>
> —*U.S. National Space Policy,* 31 August 2006

While current space policy clearly allows for international cooperation, policy alone is not sufficient for implementing coalition ORS because policy is not binding. Policy, however, can establish the conditions for implementation. Moreover, many precedents can be found for international cooperation in space systems acquisition and operation. For example, since 1967, NATO has used communications satellites as part of its command and control architecture.[6] The initial NATO I satellites evolved into the NATO II, III, IV-A, and IV-B satellites. These were procured and operated as part of the U.S. Defense Satellite Communications System. NATO's latest program for its geostationary communications satellites, known as NATO Satcom Post 2000, is being provided by the UK, France, and Italy.

6.4 NATO: A Hypothetical Case Study for Coalition ORS

NATO provides an illuminating example of how coalition ORS might be implemented on a large scale (although the logic applies equally well to other historically close U.S. allies such as Australia, Japan, and South Korea). Today, as NATO's 26 members are engaged in operations from Bosnia to Afghanistan, NATO's need for space support to theater operations is as strong as ever. In addition, NATO members are actively transforming their militaries and developing their own space capabilities.

NATO members have demonstrated the technical skill and knowledge to design, build, and launch imaging systems with resolution better than 1 m. For example, Germany has successfully launched two synthetic-aperture radar satellites as part of a satellite reconnaissance system known as SAR-Lupe. These satellites each weigh 720 kg and provide day or night all-weather imagery at approximately 0.5 m resolution; five satellites are planned for the constellation.[7,8] SAR-Lupe is Germany's contribution to the emerging European Reconnaissance System, in which France will be allowed to use the German SAR-Lupe radar system and in return, Germany will be able to access the French optical HELIOS II satellite system.

A NATO version of coalition ORS could take advantage of international business models much in the same manner that Surrey Satellite Technology did for its Disaster Monitoring Constellation.[9] Surrey's business plan was based on the premise that many countries can afford a single small satellite, but most cannot afford an entire constellation. Therefore, Surrey established an international consortium for the constellation: Each partner (Algeria, China, Nigeria, Spain, Turkey, and the UK) agreed to buy and operate a small multispectral imaging satellite and share the data with the other partners during and after natural disasters. This arrangement would provide each partner the benefits of a constellation at the price of only a single small satellite.

The Surrey business model provides opportunities for larger block purchases of small satellites than would be the case of a U.S.-only TacSat constellation, for example. These block purchases would stimulate the satellite and launch industries while providing economies of scale and reducing satellite cost. Applying the Surrey business model, if each of NATO's 26 members agreed to buy and operate a single TacSat, then 26 small satellites could be available for use in NATO operations, such as in Afghanistan.

6.4.1 Potential Military Benefits

Potentially, these 26 TacSats could be used to populate two different types of constellations—one for communications, and one for intelligence. The two constellations, each composed of 10–12 TacSats, could provide significant coverage and support to NATO forces. Simultaneously, NATO members would gain invaluable experience with space support to their fielded forces. These constellations would provide additional communications and intelligence, resulting in enhanced situational awareness. Hence, NATO-based ORS would also provide a venue for improving interoperability, continuing transformation of allied militaries, and further cementing geopolitical ties. Furthermore, allied ORS could provide member nations an added means of making meaningful contributions to NATO operations, without the domestic political challenges of troop contributions.

The relatively large constellation in this hypothetical example allows graceful degradation. The more satellites in a constellation, the less impact there is on revisit times if a satellite is lost in launch or on orbit. An individual nation might, at some point, have valid political constraints that prohibit the use of its satellite to support an allied operation. In such a case, that nation would lose access to the constellation, while the coalition would retain significant capability.

An allied ORS also has the potential to maintain the free use of space. A coalition constellation presents challenges to a would-be attacker in ways that a national constellation does not. Quite simply, an attack on a coalition constellation is an attack on the coalition, not an attack on a single nation. A coalition can apply diplomatic, economic, and military responses to an attack in space on a scale greater than any single nation. Consequently, coalition ORS offers the potential to maintain the freedom of space simply via the collective deterrence that it engenders.

6.4.2 Potential Economic Benefits

This hypothetical NATO example demonstrates the opportunity for block purchases of small, yet capable, satellites. The potential economies of scale created from such block purchases could result in a price tag per satellite in the low tens of millions of dollars or euros. This level of cost could make purchasing individual small satellites more attractive than purchasing a constellation or purchasing a single larger, more expensive satellite, especially for smaller countries. As demonstrated by the Surrey business model, NATO could get the benefits of a constellation, with the costs of individual satellites paid by individual members. Fundamentally, NATO ORS could allow for cost sharing, which would lower the cost to individual member nations.

An alternative acquisition model would be for NATO to procure a constellation and create organic NATO space capabilities, as with NATO satcom. Furthermore, one constellation could be produced for NATO members by the U.S. industry, and the other by European industry. This type of acquisition strategy would create the mutual benefit articulated in current U.S. space policy.

6.4.3 Potential Implementation Challenges

While NATO ORS could provide significant military and economic benefits, the implementation must overcome significant challenges, both political and technical. One of the primary challenges is in acquiring ORS systems without violating the national acquisition processes of individual NATO members. For the United States, the International Traffic in Arms Regulations (ITAR) pose the greatest challenge. However, precedence for coalition ORS has been set by NATO communications satellites and the international nature of other programs such as the F-35. A second significant challenge would be the time required to reach a consensus among the 26 members for the establishment of any type of new NATO space capability.

Lastly, implementation of a NATO version of coalition ORS would require the development of new command relationships as well as tactics, techniques, and procedures to take full advantage of the capabilities provided. To establish these new models, NATO could build upon the efforts of member nations to create a space-based reconnaissance program for the European Union.

6.5 Summary

In spite of the challenges, coalition ORS is likely to provide a myriad of economic and military benefits, as illustrated by the hypothetical NATO case study. Perhaps of more importance, especially to currently nonspacefaring nations, is the fact that coalition ORS offers the potential to maintain the freedom of space simply via engendered deterrence of the coalition. Coalition ORS will enable the United States to constructively engage its allies, take advantage of world-class capabilities in space, and bring international cooperation to new levels. Moreover, coalition ORS can help cement partnerships between like-minded nations while solving the problem of funding new, small satellite systems along with existing space programs.

6.6 References

[1] T. Doyne, P. Wegner, C. Olmedo, G. Moretti, M. Hurley, M. Johnson, T. Duffy, and C. Huffine, "ORS and TacSat Activities, Including the Emerging ORS Enterprise," *5th Responsive Space Conference*, paper AIAA-RS5 2007-4001 (April 2007); available at http://www.responsivespace.com.

[2] "Phase 1 Bus Utility Final Report" and "Phase 1 Bus Utility Final Briefing," Phase 3 Bus Standards Initiative, https://projects.nrl.navy.mil/standardbus (June 2007).

[3] T. Doyne, E. Finnegan, J. Chris Garner, M. Hurley, M. Johnson, G. Moretti, W. Raynor, G. Sandhoo, C. Schein, P. Schwartz, P. Stadter, K. Turner, S. Vernon, "ORS/JWS Phase 3 Bus Standards Status," *5th Responsive Space Conference,* paper RS5-35-2007 (April 2007); available at www.responsivespace.com.

[4] Remarks made at a town hall meeting, Naval Support Activity Mid-South, in Millington, TN, 7 Sept. 2006.

[5] Remarks made at the 17th International Seapower Symposium of the Naval War College (21 Sept. 2005, Newport, RI).

[6] "Comparison of MILSATCOM Systems," Federation of American Scientists, http://www.fas.org/spp/military/docops/army/ref_text/chap07b.htm (11 October 2007).

[7] "Surrey Missions: DMC: Disaster Monitoring Constellation," Surrey Satellite Technology Ltd., http://centaur.sstl.co.uk/datasheets/Mission_DMC.pdf (11 October 2007).

[8] "SAR-Lupe," Wikipedia, http://en.wikipedia.org/wiki/SAR-Lupe (10 Oct. 2007).

[9] "German SAR-Lupe Satellite Safely in Orbit," OHB System AG, http://www.ohb-system.de/gb/News/presse/1912_06.html (20 Oct. 2007).

Chapter 7

The United States Air Force Academy FalconSAT Small Satellite Program

Timothy J. Lawrence,* David J. Barnhart,* Luke M. Sauter,* Frederick T. Kiley,* Kenneth E. Siegenthaler*

7.1 Introduction

The Space Systems Research Center at the United States Air Force Academy is building a cadre of satellite space professionals. Its aim is to provide a program for cadets to "learn space by doing space." Each year, approximately 45 cadets majoring in astronautical engineering and related disciplines participate in a year-long course that entails designing, fabricating, testing, launching, and operating a satellite in space—FalconSAT.

7.2 Program Methodology

The FalconSAT program imitates a real Air Force program in which a contractor works under an Air Force program manager—except in this case, the cadets act as the contractor, and the faculty act as the Air Force manager. With nearly 100% cadet turnover every year, a standardized program is essential. Figure 7.1 illustrates a typical FalconSAT team in size and composition. The team creates a flexible spacecraft platform that can be readily adapted and enhanced to meet developing payload requirements and take advantage of secondary launch opportunities within budget and time constraints. Part of this approach is to use commercial off-the-shelf (COTS) hardware. Although such purchases obviously ease basic design problems, considerable effort remains in the areas of payload design and development, structures, attitude control, thermal control, solar panels, testing, and operations—more than enough to challenge even the most ambitious students.

7.2.1 Practical Learning Approach

The curriculum of the Department of Astronautics ensures that by the time students enter the capstone course, they have enjoyed considerable hands-on education in every aspect of satellite development and operations. For example, a focus of the introductory Space Systems Engineering course, required of all astronautical engineering and space operations majors, is a fully functional 2.5 kg nanosatellite

*HQ USAFA/DFAS
USAF Academy, Colorado

Fig. 7.1. A typical FalconSAT team.

known as EyasSAT (Fig. 7.2). This tabletop satellite is a revolutionary development in space systems education, as it enables large numbers of students to build, test, and "fly" a satellite in the classroom with very little risk. EyasSAT demonstrates six traditional satellite subsystems: structural, electrical power, data handling, communications, attitude determination and control, and thermal. It can also integrate student payloads (such as a camera) and other subsystems (such as propulsion or GPS).

After EyasSAT, the FalconSAT capstone course is where academic knowledge is put to the ultimate practical test, as cadets apply their acquired skills to the entire satellite development process—from identifying a payload need, to developing requirements and flowing them into a design. The program accomplishes its overall objective to have cadets learn space by doing space by providing real-world, hands-on experience with a complete space mission design. Additionally, the FalconSAT program supports Department of Defense (DOD) space science and technology objectives through real Air Force and DOD experiments. Through involvement with space agencies and oversight panels such as the

Fig. 7.2. External and internal views of EyasSAT (18 × 18 × 25 cm).

Table 7.1. Total Budgets of Academy Spaceflight Missions

Mission	Total Budget in Thousands of Dollars
FalconGOLD	$250
FalconSAT -1	$750
FalconSAT-2	$1500
FalconSAT-3	$2100
FalconSAT-4	unfunded mission
FalconSAT-5	$2400

Space Experiments Review Board, cadets come to understand the larger picture of the acquisition and development process. By the time they graduate, they not only possess the practical knowledge needed for satellite development and operations in the Air Force, they have also experienced the invaluable opportunity of hands-on application throughout every step of their academic journey. The entire program runs within a typical academic calendar and is further constrained by a relatively small annual budget (shown in Table 7.1).

7.2.2 Design Process

In addition to using COTS items as a foundation for the design of the satellite, another improvement is the requirement for cadets to build three models of the satellite: First, an engineering model is constructed to ensure that all components fit and are compatible with the mission. Next, a qualification model is constructed and tested to limits in excess of what the flight model will encounter; any weaknesses in the design are corrected until the model meets MIL-STD acceptance levels. Finally, the actual flight model is produced. Each model must complete the entire review process—including the preliminary design review, critical design review, and flight readiness review—before the fabrication and testing of the next model can begin.

By nature, any design class is open-ended and difficult to program lesson-by-lesson, compared with a traditional lecture-based course. However, by requiring students to follow prescribed, industry-standard systems-engineering processes, some formal structure can be imposed—and the design reviews serve as major assignments for grading purposes, as well.

7.2.3 Mission Operations

After orbit insertion, the satellite is operated by cadets working in the FalconSAT ground station at the Academy. FalconSAT does not make use of the Air Force Satellite Control Network. Instead, dedicated or amateur frequencies are used. The operations team is responsible for creating and perfecting the processes and procedures required to commission and maintain the space vehicle. They develop quality documentation, checklists, and training plans that will be used by their team or the team of the following year to conduct satellite operations.

Satellite operations are based on the model used by the Air Force. Cadets are organized into a squadron composed of a satellite operations officer, a ground control officer, a satellite operations engineer, and a crew commander. The satellite operations officer is the spacecraft expert; this person understands the telemetry and knows how the space vehicle is supposed to be configured. If the satellite is in an anomalous situation, the operations officer should be the first to identify it and offer a solution. The control officer understands the ground equipment and is responsible for contacting the satellite and maintaining the radio link. If the link is broken, the control officer will respond and reestablish contact. The satellite operations engineer is responsible for the technical quality of the experimental data collection, storage, and distribution. The crew commander runs the shift, establishes a working pace during a satellite pass, and ensures that all mission requirements are met.

The entire program thus requires cadets to organize and manage themselves. Because the principal objective is to turn college thinkers into space leaders, cadets are given a "commander's intent" and then turned free. Faculty are involved as mentors rather than instructors. Initially, the cadets struggle to understand their new roles, but after a month, they typically begin to take ownership of the program.

At a minimum, five complete crews (20 operators) with one instructor each are required for successful operation of a FalconSAT. Crew members may serve the dual role of operator and instructor. The operations crew training is accomplished in a similar fashion to space operations training in the Air Force. An instructor is assigned to each trainee, who is required to complete all objectives stated in the training manual. There are several levels of training: initial qualification training, unit qualification training (which includes positional qualification training), instructor training, corrective training, and monthly recurring training.

Initial qualification training imparts comprehensive knowledge of FalconSAT operations and is required for every crew member. Upon completion, students begin the unit qualification training, which involves extensive hands-on training with both the simulator and the actual equipment. Trainees thus gain in-depth knowledge of equipment pertinent to their positions and a working knowledge of all checklists. To complete positional qualification training, students must pass several progress checks, a comprehensive positional test, and an evaluation check based on three scenarios: one standard pass, and two that include abnormal conditions. Instructor training is accomplished by the training flight. To pass, instructors must train someone to handle three scenarios, one normal and two adverse. Corrective training is mandated for errors committed during operations. This training, which involves knowledge and performance training as necessary, is also accomplished by the training flight. Exams and scenarios are generated by the training flight as needed. The monthly recurring training can occur more

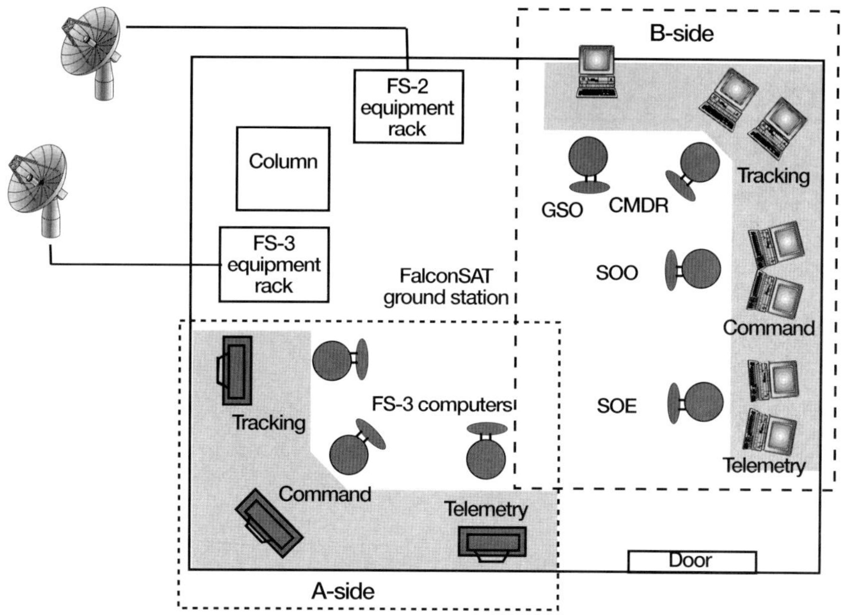

Fig. 7.3. The FalconSAT ground station at the USAF Academy.

often as necessary to maintain crew skills while awaiting launch; it is also used to review skills that crew members may not use often. It may consist of either written tests, scenarios, or both.

FalconSAT satellites are flown from the ground station at the Academy. Communication is only with line-of-sight passes. The layout of the ground station is illustrated in Fig. 7.3. There are two sides to the station: The "A" side is for training new crews using the qualification model in the cleanroom; the "B" side is for the crews to fly the satellite in orbit.

7.3 History of the FalconSAT Program

The Academy started experimenting more than a decade ago with small satellites via cadet-built prototypes "launched" on high-altitude balloons. These projects gave students immediate hands-on experience and spurred the Department of Astronautics to seek more ambitious space projects. A major milestone was the launching of FalconGOLD, a 15 kg bolt-on secondary payload on an Atlas-Centaur launch vehicle in 1997. FalconSAT-1 (FS-1) was the program's first free-flying satellite, with a mass of 52 kg, launched on a Minotaur in 2000. The 19.5 kg FalconSAT-2 (FS-2), ready for launch in 2002, was delayed several times until it launched in 2006. FalconSAT-3 (FS-3), with a mass of 50 kg, was launched in 2007. Table 7.2 gives a brief overview of the program history.

Table 7.2. History of USAF Academy Spaceflight Activities

Date	Launch Vehicle	Satellite	Research Mission	Mass (kg)
May 1995	Balloon	USAFASAT-B	Attitude control	
March 1996	Balloon	Glacier	GPS & magnetometers	
September 1996	Balloon	PHOENIX	Laser communication	
1997	Balloon	Blue Moon	Test satellite subsystems	
April 1997	Balloon	FalconGOLD	GPS signal	15
October 1997	Atlas-Centaur	FalconGOLD	GPS signal	15
January 2000	Minotaur	FalconSAT-1	Spacecraft charging hazards	52
March 2006	SpaceX Falcon-1	FalconSAT-2	Ionospheric plasma depletions	19.5
March 2007	Atlas V	FalconSAT-3	Ionosphere plasma, attitude control, and propulsion	50

7.4 Balloons

USAFASAT-B was launched on 21 May 1995 on a helium-filled balloon. While the balloon floated at 35,000 m, the cadet team sent radio commands to a receiver in the spacecraft, and an onboard computer responded by firing small nitrogen gas thrusters to orient a solar panel toward the sun. Performance data from the solar panel, the computer, a sun sensor, a battery pack, and a passive thermal control subsystem were radioed to the Academy ground station. The mission was successful in testing an attitude control system designed and fabricated by cadets.

The first spacecraft designed by Academy students was Glacier, originally scheduled to launch on a converted Minuteman II rocket. It carried four payloads: the Glacier Imager to take pictures of Earth; the Space Modular Materials Experiments to gain knowledge of low Earth orbit (LEO) environmental effects on instruments; a GPS attitude determination subsystem; and the Charging Hazards and Wake Studies—Long Duration (CHAWS-LD) experiment to assess the hazards for spacecraft operations in the wake of larger bodies. The Minuteman launch was canceled, but Glacier was flown and tested on a high-altitude balloon in 1996.

Blue Moon was a project to design and test the world's first hybrid-rocket-powered satellite. It was designed and built entirely by cadets. The goal was to use a hybrid-rocket motor to place the Blue Moon spacecraft into a lunar orbit or lunar flyby, then take pictures of the moon and send them back to Earth. It was scheduled to launch on an Ariane 4 from French Guiana in the spring of 1997. When the launch did not materialize, Blue Moon was also flown and tested on a high-altitude balloon.

7.5 FalconGOLD

FalconGOLD (Fig. 7.4) marked the transition from balloon launches to actual rocket launches. The mission was to investigate the feasibility of GPS-aided navigation by satellites higher than the GPS constellation. After a test flight on a high-altitude balloon in April 1997, FalconGOLD was launched as a fixed secondary payload on an Atlas-Centaur on 24 October 1997. A GPS receiver was the payload. Figure 7.5 details the bus architecture. FalconGOLD sampled the GPS spectrum once every 5 min. Data were transmitted on a 400.475 MHz UHF frequency at 9.6 kbps to the 18 m dish antenna at the Boulder, Colorado, ground station. Orbit features are illustrated in Fig. 7.6.

Fig. 7.4. The FalconGOLD satellite.

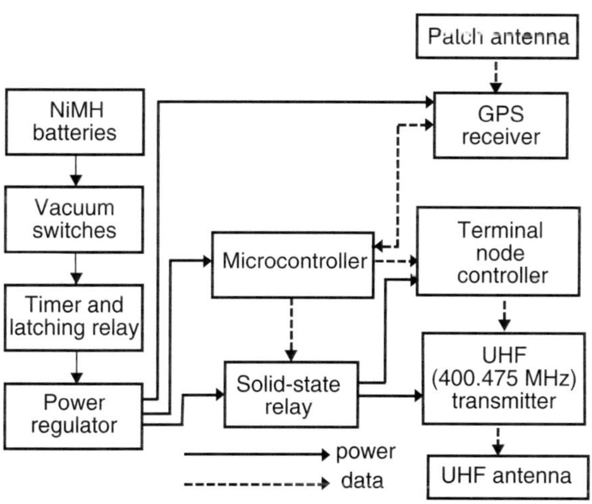

Fig. 7.5. FalconGOLD bus architecture.

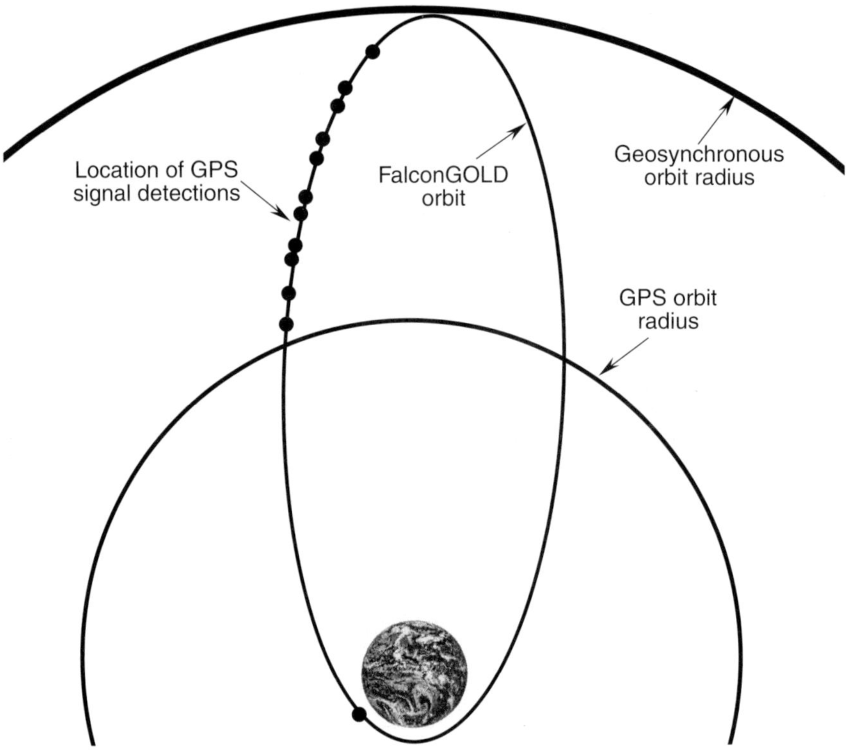

Fig. 7.6. FalconGOLD orbit. 100 data frames were obtained, with 357 "good" fragments. GPS signals were detected in 12 of the "good" fragments. Semimajor axis: 24,070 km; eccentricity: 0.7275; inclination: 26.296 deg; apogee: 35,200 km; perigee: 181 km.

NORAD aided in determining the true position of FalconGOLD. With an apogee altitude of 35,200 km, numerous measurements were obtained beyond the GPS constellation's 20,200 km altitude. All systems operated nominally for 15 days until primary battery power was depleted. The successful operations and data recovery from FalconGOLD concluded that GPS signals could be used for high-altitude spacecraft position determination, even beyond the altitude of the GPS constellation. Cadets participated in all phases of the mission, including systems design, fabrication, launch vehicle integration, qualification testing, launch site operations, and mission operations.

7.6 FalconSAT-1

The FalconSAT program's first "free-flyer," FS-1 was launched on 14 January 2000 aboard the first Minotaur launch vehicle (a modified Minuteman II ICBM) along with several other university-built microsatellites (Fig. 7.7). FS-1 flew the DOD-supported CHAWS-LD experiment in a continuation from the earlier balloon work.

Fig. 7.7. The FalconSAT-1 satellite.

FS-1 was deployed into orbit without problem. During the following weeks, however, cadets struggled to bring the satellite totally on-line. Initial contacts went well, but it soon became apparent that the spacecraft's power system was not functioning correctly to charge the batteries during daylight. After only a month, the mission was terminated. While FS-1 was a technical failure, it was a resounding academic success. Cadets participated in all phases of the mission from conceptual design through assembly, integration, testing, launching, and on-orbit operations. The disappointing technical failure carried its own real-world lessons for all involved. The schedule for building FS-1 was less than one year—not enough time for proper design, building, testing, and correction of problems. The satellite was built from scratch, including the electronics, despite limited facilities and experience for doing so. With insufficient funds for buying backup or replacement parts, weak improvisations were used that may have contributed to the satellite failure. There was not enough time to make some of the necessary modifications indicated in the testing mode.

7.7 FalconSAT-2

The lessons learned from FS-1 motivated significant change to the Academy's long-term program, prompting a decision to build a program first and a satellite second. This shift in focus required a reduction in the overall mission risk and time needed for design development. So, rather than have students design the entire spacecraft from scratch, the design problem was constrained, and COTS hardware was used in place of difficult core subsystems.

Another hard lesson from the FS-1 failure was that proceeding directly from concept to flight vehicle in one step—the so-called protoflight approach—was not a sustainable option in an undergraduate environment. Not only does this approach necessitate a large body of corporate knowledge (something very difficult to sustain at a military institution prone to high faculty turnover), but it means that once the mission is launched, there is no hardware left on the ground to use for teaching and future engineering development. For these reasons, the idea of a modular satellite seemed more appropriate.

FS-2 benefited from these lessons and developed an approach that started with a straightforward mission statement and laid out a robust multiyear development process consisting of an engineering model, qualification unit, and flight unit. A longer-term objective became validation of key technologies, design concepts, and processes that could be used for subsequent FalconSAT missions. To bound the problem faced by the students, the program used COTS hardware to supplement the cadet-built systems. By buying SNAP (Surrey Nanosatellite Applications Program) hardware from Surrey Satellite Technology Ltd. in England, the Academy achieved an "out-of-the-box" solution for several critical subsystems, including power, communications, and, most important, data handling. This development allowed cadets to focus on integrating a working system rather than fixating on a few complex subsystems. With this approach and objective, the FS-2 program started in 2000, and the satellite was ready to fly in summer 2002.

FS-2 was a 19.5 kg cubical satellite, 32 cm on each side, as illustrated in Fig. 7.8. The scientific research payload was a miniature electrostatic analyzer (MESA) to investigate low-latitude ionospheric plasma bubble depletions and their effects on radio waves such as GPS signals. FS-2 would establish flight heritage for the MESA instrumentation, which was produced by cadets and faculty in the physics department.

The COTS core components established a hardware-in-the-loop integrated modeling and simulation tool for the integration of payloads and other spacecraft subsystems. These core components consisted of an onboard computer, the SNAP S-band transmitter, the SNAP VHF receiver, a spacecraft integration module, and a power system, together with their associated assembly hardware. The electrical power system provided a switched, regulated, and protected power supply for the

Fig. 7.8. Photo and solid CAD rendering of FS-2 flight model.

spacecraft bus, as well as battery charge regulators, a power conditioning module, power switches, and a COTS battery. The onboard computer was based on a RISC (reduced instruction set computer) processor, which gave increased computational power with relatively low power consumption; it came with a bootloader and could be programmed in C using a low-cost commercial compiler. Finally, the integration module board provided a number of input/output sources for connectivity to payload and other subsystems built upon the CAN (control area network) framework with both RS-422 and RS-232 standards for connectivity. These core COTS systems from SNAP provided a backbone, quite literally, for several critical subsystems remaining, including power generation, communications, and data handling. This initial set of the core COTS hardware was known after delivery as the FalconSAT Avionics and Simulation Testbed (FAST). It became the basis of a "flatsat" in which subsystems could be built and tested throughout the development process (see Fig. 7.9).

The engineering-model phase of the program was complete in April 2001. This phase included the design, fabrication, and testing of an initial design for form, fit, and function. This was the first cut at combining customer requirements with payload requirements and flowing them into a rudimentary design. The culmination of the initial design work was a campaign to test the structure, including mass simulators; the purpose was to reduce overall program risk by verifying key design parameters and developing operational and testing procedures. This test campaign required about 20 cadets to travel for a week to Albuquerque, New Mexico, to conduct solar array testing and vibration testing at the Air Force Research Laboratory (AFRL). The vibration tests, levels, and criteria for the engineering model, qualification, and flight testing are presented in Table 7.3. Through testing, the engineering model was able to validate 80% of mechanical

Fig. 7.9. FalconSAT Avionics and Simulation Testbed (FAST) COTS hardware, without battery.

Table 7.3. Planned Tests, Levels, and Criteria for FS-2

Test Phase	Test Article	Planned Tests	Test Levels and Criteria
Development (April 2001)	Engineering model	Sine sweep Random vibration Sine burst	Low level to determine natural frequencies 3 dB above acceptance levels in each axis 1.4× limit acceleration in each axis
Qualification (Feb. 2002)	Qualification model	Sine sweep Random vibration Sine burst Shock	Low level to determine natural frequencies 3 dB above acceptance levels in each axis 1.4× limit acceleration in each axis 3 nominal firings of separation system
Acceptance (July 2002)	Flight model	Sine sweep Random vibration Sine burst Shock	Low level to determine natural frequencies Acceptance levels for 1 min. per axis Limit acceleration in each axis 1 nominal firing of separation system

design, 80% of thermal design, 85% of flight software, 60% of payload functionality, and 85% of key procedures and processes. This work was completed by the cadet class of 2001. With the completion of this phase, the program had nearly arrived at a preliminary design review level.

This official review took place in the fall of 2001 with a new class and paved the way to detailed design analysis and specification development. With an initial design and extensive Academy facilities, the new class of 26 cadets and a team of faculty were faced with the challenges of completing the system-engineering process and producing and testing both a qualification model and a flight spacecraft.

The critical design review was completed in late fall 2001, which paved the way for fabrication of the qualification model. The resulting qualification unit block diagram illustrates the extent to which the satellite was developed around the core COTS components (Fig. 7.10).

The qualification model, identical to the flight unit, would be vibration tested to higher-than-expected flight conditions. This testing was designed to reduce overall program risk and fulfill NASA's safety requirements; FS-2 was originally scheduled for a shuttle launch. The testing campaign was completed in February 2002 at AFRL. Initial test data are also shown in Fig. 7.11. The fundamental frequency—a very large 154 Hz—was greater than expected and allowed the spacecraft structure to gain mass, which helped ease the manufacturing of components. The pretesting spacecraft weight was 19.7 kg; the moments of inertia and center of gravity are identified in Table 7.4.

The testing included a random vibration test exceeding space shuttle loads by 6 dB with sine bursts in all three axes, as illustrated above in Table 7.3. These tests, conducted by about 20 cadets over a two-week period at AFRL, were at higher loads to verify the margins of safety. Results showed that the structure withstood all tests without evidence of failure; that there were no visible cracking,

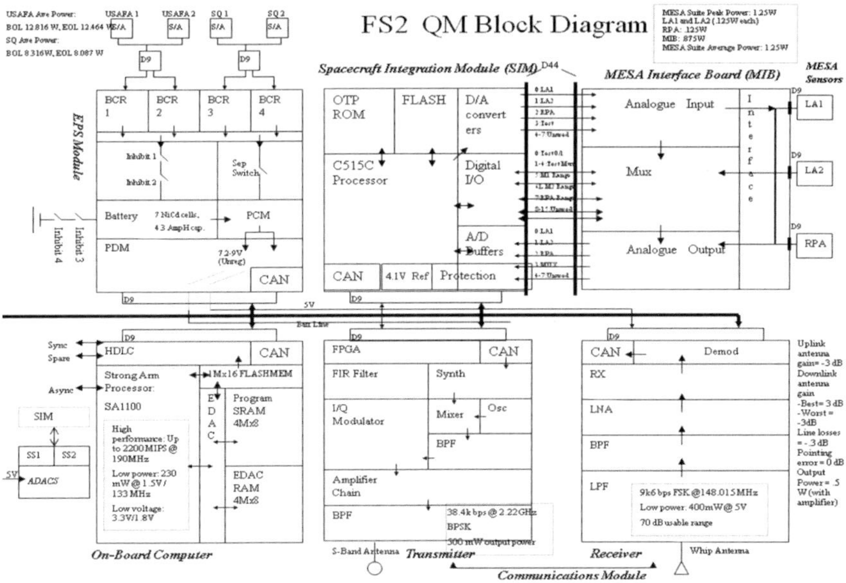

Fig. 7.10. FS-2 Qualification model block diagram.

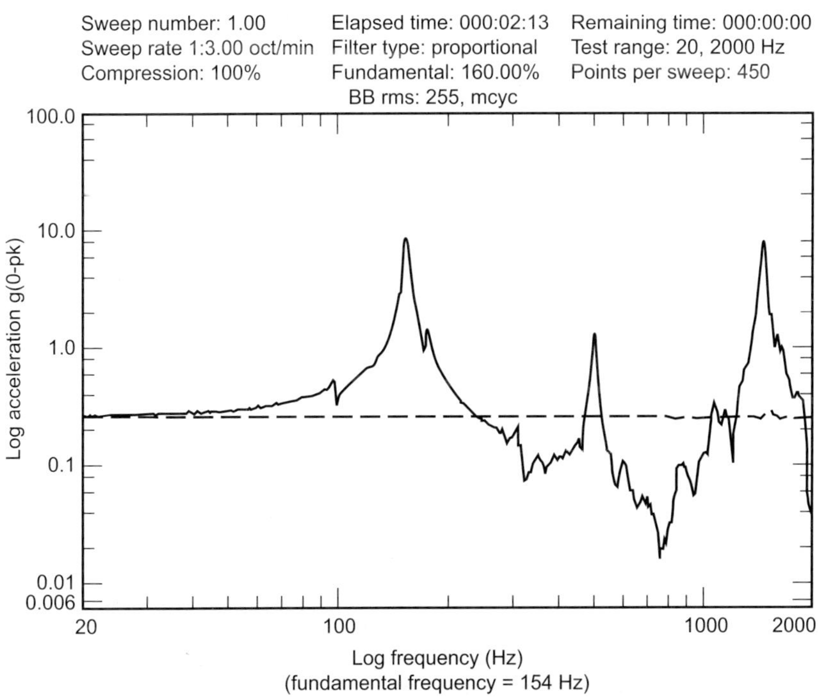

Fig. 7.11. Initial low-level sine sweep in the x-axis fundamental frequency at 154 Hz.

Table 7.4. FS-2 Center of Mass Location and Mass Moment of Inertia

Center of mass measurement	1	2
X (inches from ring center line)	0.1030	0.0962
Y (inches from ring center line)	–0.0196	–0.0202
Z (inches from baseplate bottom)	4.9633	4.9649
Moment of inertia measurement		
I_{xx}	1337.89 lb-sq in. (from center of mass)	
I_{yy}	1403.42 lb-sq in. (from center of mass)	
I_{zz}	1171.10 lb-sq in. (from center of mass)	

no fasteners that visibly backed out, and no significant reduction in natural frequencies; that random-vibration tests appeared to stress the critical areas of the structure more severely than the sine-burst tests; and most importantly, that all shuttle safety requirements were met. These results paved the way for fabrication of the flight model. The only required redesign involved replacing a cracked S-band antenna. In addition, the satellite was subjected to thermal vacuum testing with five thermal cycles from –20°C to +50°C. The flight-model phase was completed in July 2002, including design, fabrication, and testing. The testing included thermal vacuum and vibration. A final center-of-gravity/moment-of-inertia test was also completed. The last test was a solar-panel functional test under one-sun conditions (Fig. 7.12). All tests showed that the system met or exceeded flight requirements. The satellite was complete and ready for delivery to NASA with a projected launch date of 16 January 2003.

Fig. 7.12. FS-2 flight model undergoing solar-array testing.

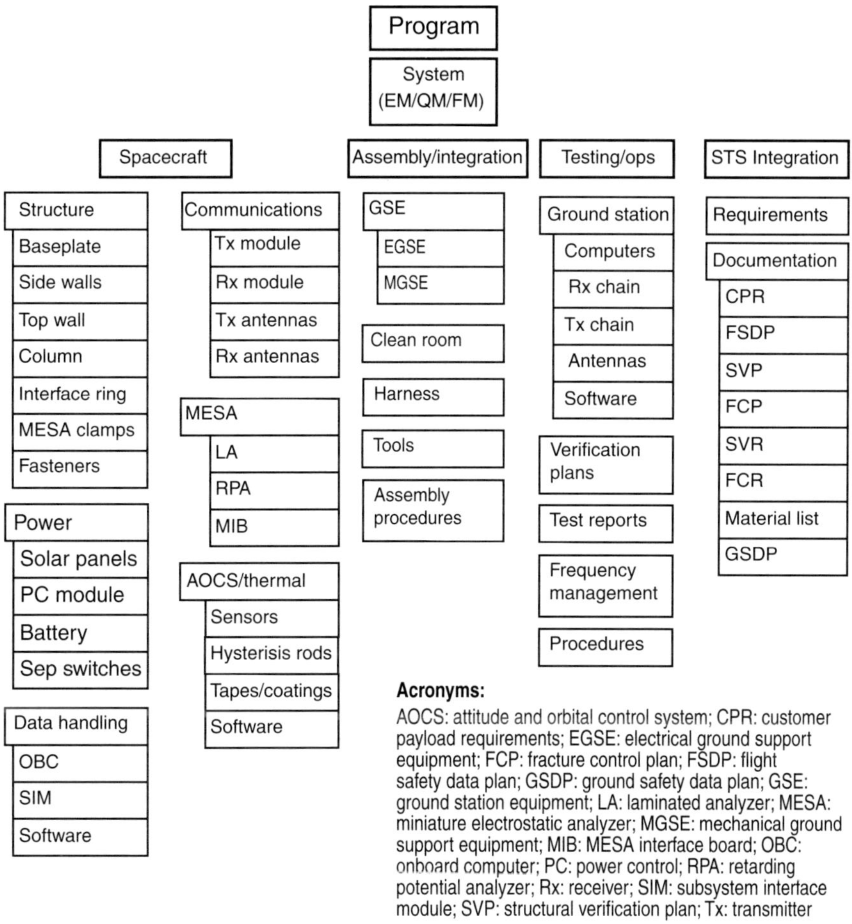

Fig. 7.13. FS-2 work breakdown structure.

The system was developed under the work breakdown structure illustrated in Fig. 7.13. The following sections provide a more detailed description of FS-2 subsystems according to the work breakdown structure.

7.7.1 FalconSAT-2 Structure

The primary requirement of the structural subsystem was to serve as the backbone to the other spacecraft subsystems. This structure had to support the MESA payload sensors, the solar arrays, and the SNAP modules as well as the antennas, separation switches, sun sensor, battery pack, and interface ring. It also had to protect the bus components inside the shuttle's Pallet Ejection System. The structural subsystem requirements are derived chiefly from those of the other subsystems in the spacecraft; in fact, only the attitude-control, thermal-control, and data-handling subsystems do not require anything from the structure itself. In addition, the structure needed to survive many extreme environments.

Shown in Table 7.5 are the margins of safety for each of the structural subsystems. "Large" is defined as a margin of safety greater than 500%. As shown, the spacecraft was sufficiently engineered to handle the expected environments.

The entire structure weighed 19.5 kg, which classified FS-2 as a small microsatellite. The initially overdesigned shape had to be reworked and reduced to decrease some of the rigidity. This conclusion was determined only after testing was completed on the finished model.

7.7.2 FalconSAT-2 Power

The electrical power subsystem was responsible for providing power to all modules and payload sensors and for ensuring that adequate power would be available to recharge the batteries during eclipse. Power system failures are the number-one problem with satellites in orbit. For this reason, ongoing analyses were conducted on the electrical subsystem as designs were refined.

Table 7.5. FS-2 Structural Margins of Safety

Structural Subsystem	Margin of Safety
1. Bolt circle attaching baseplate to adapter ring	
1a. Bolt tension, ultimate	3.20
1b. End-pad shear, ultimate	4.16
1c. End-pad bending, ultimate	0.80
1d. End-pad bending, yield	0.51
1e. Bolt shear, ultimate	Large
2. Bolts attaching column to baseplate	
2a. Bolt tension, ultimate	Large
2b. Failure of insert, ultimate	Large
2c. End-pad shear, ultimate	4.34
2d. End-pad bending, ultimate	3.48
2e. Bolt shear, ultimate	Large
3. Bolts attaching battery to baseplate	
3a. Bolt tension, ultimate	Large
3b. Failure of insert, ultimate	Large
3c. End-pad shear, ultimate	3.02
3d. End-pad bending, ultimate	2.37
3e. Bolt shear, ultimate	Large

The electrical power subsystem consisted of four body-mounted solar panels, a seven-cell battery, one battery charge regulator attached to each solar panel (for a total of four), and a power-distribution module. In calculating the average power needed by the spacecraft, one needs to consider both the power needed on average and at peak times and the power that can be provided by the power system. These calculated requirements for FS-2 are given in Table 7.6. The system was sized based on a needed average power of 3.396 W with a peak power of 8.1 W.

7.7.3 FalconSAT-2 Attitude/Thermal Control

The attitude performance requirements were limited—the only attitude that had to be avoided was the "death mode," where a side of the satellite without solar panels is constantly pointing toward the sun. This orientation occurs when the *z* axis of the satellite points to the sun or away from the sun in the opposite direction. MESA payload sensors did not have pointing requirements.

Thermal control requirements were based on the need to maintain satellite operability in the extreme temperatures of space. Basic thermal requirements imposed a temperature range of 5°C to 30°C for operation and –30°C to 50°C for

Table 7.6. FS-2 Power System Component Breakout

Subsystem	Component	Description
Solar panels	SpaceQuest (two panels)	20 4.1 × 4.1cm GaAs cells Bonded to honeycomb panels with fiber-glass face sheets Power—beginning of life: 8.316 W Power—end of life: 8.087 W
	USAF Academy-built (two panels)	16 5.5 × 6.5 cm GaAs Cells (TecStar) PCB substrate Power—beginning of life: 12.816 W Power—end of life: 12.464 W
Battery	Battery (seven Sanyo NiCd cells-N4000DRL)	Available capacity: 28 Ahr Needed capacity: 3.754 Ahr Max discharge rate: 0.314 A/hr Supplied by SSTL Built to NASA specs
	Power conditioning/ distribution	5 V regulated / 8.4 –10.5 V unregulated COTS hardware supplied by SSTL Provides peak power tracking Assumed depth of discharge of 11.9 %
Switches	Honeywell separation switch (one)	
	Honeywell inhibit switches (four)	

Table 7.7. FS-2 Subsystem Temperature Limitations

	MESA	Electrical	Data Handling	Comm.	Structure
Minimum	−40°C	−30°C	−40°C	−40°C	N/A
Maximum	+85°C	+50°C	+85°C	+85°C	N/A

nonoperation. The limiting restraint on these temperatures was the power subsystem. Full temperature ranges for each subsystem are noted in Table 7.7.

FS-2 was designed without an active attitude control mechanism; therefore, only passive means of control were used, in conjunction with the thermal control system. The first simulation used for the thermal analysis was a finite differential analysis model. Created in MATLAB, it explored flux history (calculating the heat flux caused by direct solar radiation and solar radiation reflected from Earth) and nodal temperature analysis (calculating the change in temperature at each node at each time step and iterating through one orbit 20 times to get a true equilibrium temperature profile). The analysis led to the use of two thermal coatings: Kapton tape, and aluminum tape. Tape was placed on half of each external facet on areas not taken up by other components as well as on solar panel facets (to induce spin around z axis) and on top and bottom facets (to induce spin around x axis). The tape configuration is shown in Fig. 7.14.

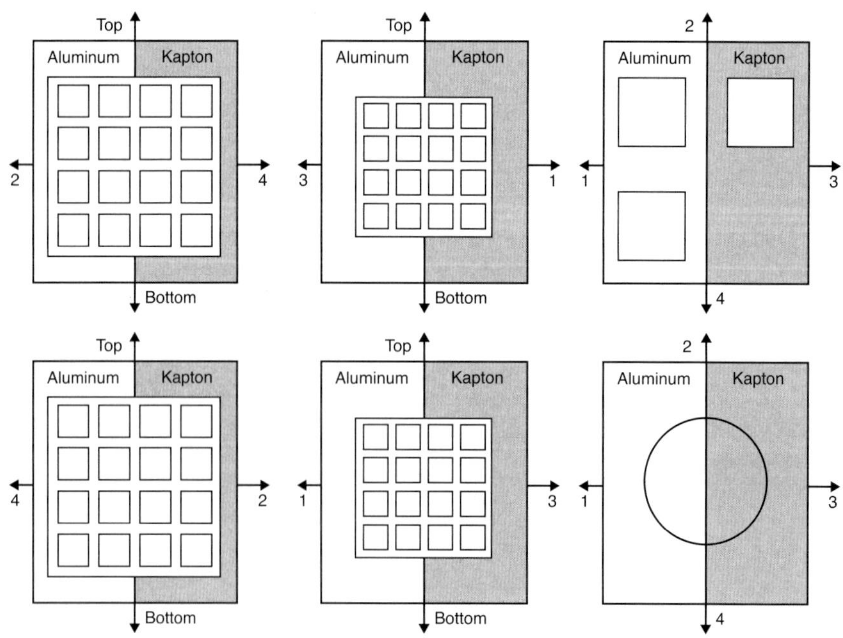

Fig. 7.14. Thermal tape placement.

The predicted nodal temperatures met the required temperature range of –30°C to 50°C for nonoperation. The operating temperature range was slightly less at 5°C to 30°C. Actual results from the engineering model thermal vacuum tests and the simulated MATLAB results can also be seen in Fig. 7.15. The predicted values with the use of thermal tapes were:

- Solar panels cycle between –10°C and +73°C.
- Modules cycle between +17°C and +30°C.
- Batteries cycle between +16°C and +28°C.

Various perturbation effects tend to cause torques in the orbital motion. Several perturbations were analyzed:

- Drag torque = a min. of 4.119×10^{-6} and a max. of 4.535×10^{-6} N-m.
- Gravity gradient torque = a maximum of 1.933×10^{-7} N-m.
- Magnetic torque = a maximum of 5.22×10^{-6} N-m.
- Solar pressure torque = a maximum of 1.152×10^{-6} N-m.
- Hysteresis-rod torque = 1.056×10^{-6} N-m.

7.7.4 FalconSAT-2 Communications

The communications subsystem took a complex approach to simple spacecraft communication. The downlink antenna on the satellite had an approximate 120 deg angular range in which the gain would be 2–3 dB, and a 20 deg range, where it would be 1 dB; there was also a 20 deg range in which the satellite body

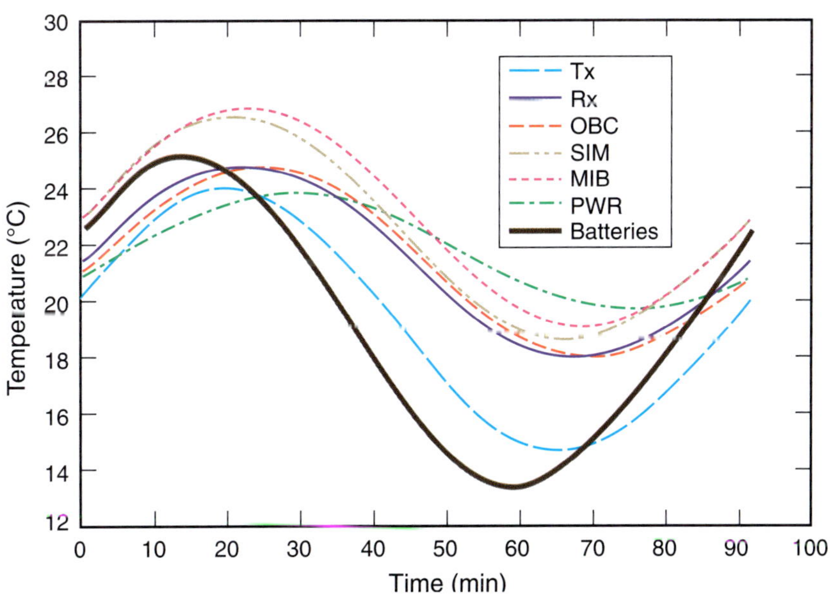

Fig. 7.15. Internal subsystem thermal analysis results.

would come between the ground station and the antenna, in which case the gain would be about –3 dB.

For the uplink, worst case is defined as 5 deg elevation (maximum viewable range), best case is defined as 90 deg elevation (directly overhead and shortest range), and reasonable case is defined as 30 deg elevation (conditions representative of what we expect 80% of the time).

Downlink elevations for each case were the same, with the addition of –3 dB spacecraft antenna gain for worst case, 3 dB for best case, and 1 dB for a reasonable case. The uplink would work at any time the spacecraft is viewable. The downlink, however, would not work in the worst-case conditions due to its limited antenna gain. At the 5 deg elevation for worst case, the required signal-to-noise ratio of 13.3 dB was not quite met. That was the absolute worst case, however, and only 2 deg higher elevation would again bring a positive margin. The spacecraft would rarely see the worst-case condition, and if so it will not be for long.

The FS-2 communication subsystems consisted of an S-band transmitter and a VHF receiver. Both were purchased from SSTL as COTS PCBs housed in SSTL-standard 6.5 × 4.8 × 0.82 in. aluminum trays.

The transmitter was designed to operate near a frequency of 2.2 GHz at an output power of 500 mW. Power consumption would be 4.9 W when the two RF2126 amplifiers were turned on and 1.5 W when they were powered down. A single S-band ground-plane antenna was mounted outside the top panel of the spacecraft structure and had a height of 6.6 in. and a maximum diameter of 2.2 in. The transmitter employed a CAN controller and interface, and data could be sent to the transmitter via the CAN bus or directly from the onboard computer via the D44 connector. The downlink modulation scheme used binary phase shift keying at 38.4 kbps at the transmitter frequency. The transmitter also used an FPGA to implement the encoders and scrambler, and the "I" and "Q" generation.

A regulated 5 V bus power line powered the VHF receiver module, and the receiver section itself ran off an internally regulated 4 V line. The power consumption was approximately 400 mW. A VHF monopole whip antenna was used, mounted to the outside top panel of the spacecraft extending 6.9 in. above the spacecraft surface. The front end of the receiver consisted of a low-pass filter, a 20 MHz bandpass filter, and a low-noise amplifier circuit. The receiver employed a CAN controller and interface, and recovered data could be distributed via the CAN bus or sent directly to the onboard computer. The uplink modulation used 9.6 kbps frequency shift keying at an uplink frequency of approximately 140 MHz. The performance of the uplink and downlink modules are summarized in Table 7.8.

7.7.5 FalconSAT-2 Payload

The MESA payload directly measures the microelectrostatic plasma depletions in the ionosphere over the equatorial bands just before, during, and after eclipse. The primary functional requirement of the MESA suite of instruments was to make in situ measurements of the electron density. The overall requirement was

Table 7.8. FS-2 Link Analysis

Link	Case	Signal-to-Noise Analysis and Margin
Uplink	Best case (90 deg elevation)	Eb/No = 51.955 dB Margin = 35.655 dB
	Worst case (5 deg elevation)	Eb/No = 38.874 dB Margin = 22.574 dB
	Reasonable case (30 deg elevation)	Eb/No = 46.624 dB Margin = 30.324 dB
Downlink	Best case (3 dB spacecraft antenna gain, 90 deg elevation)	Eb/No = 24.802 dB Margin = 18.202 dB
	Worst case (−3 dB spacecraft antenna gain, 5 deg elevation)	Eb/No = 11.215 dB Margin = -1.385 dB
	Reasonable case (1 dB spacecraft antenna gain, 30 deg elevation)	Eb/No = 23.417 dB Margin = 10.817 dB

to provide voltages (within specified limits and representative of local electron densities) to the spacecraft computer throughout the operational life of the spacecraft mission.

The functional requirement of the MESA interface board was to provide power to the suite of three sensors—two laminated analyzers, and a retarding potential analyzer—and to provide analog voltages representative of the local electron density to the onboard computer via the spacecraft integration module. The laminated analyzers would accept setpoint voltages from the interface board and return voltages that represent energy-resolved electron densities. These sensors and interface boards would yield an approximation of the plasma field density. Through further analysis on the ground, researchers could achieve a better understanding of plasma behavior in the atmosphere.

The analytical approximations of the sensitivity and power consumption were key performance parameters that affected the rest of the mission. The biggest concern with the MESA suite, from the standpoint of spacecraft integrations, was the power requirement. For this reason, additional tests were conducted on the engineering models of the circuits to provide an accurate model of the actual power requirements. The MESA suite peak power level was determined to be 1.65 W, which reflected the laminated analyzers at 0.125 W each, the retarding potential analyzer at 0.125W, and MESA integration board at 0.915W.

7.7.6 FalconSAT-2 Data Handling and Software

The FS-2 software was written in-house by one of the faculty engineers with extensive software experience. It resided on the SNAP onboard computer, which was based on the Intel StrongARM 1100 processor. The BZL real-time, cooperatively multitasking operating system was developed by SSTL specifically for the

SNAP bus. The ARM developer suite with the CodeWarrior integrated development environment was used for software development. More than 14,000 lines of code programmed in C were required.

One of the unique features of FS-2 was its power-up sequence. When the spacecraft separates from the launch vehicle, the electrical subsystem and receiver are the only two components that activate (so that the batteries can begin charging while the receiver waits for ground contact). These modules communicate internally via the CAN bus as shown in Fig. 7.16. During the early commissioning steps, the ground station sends to the receiver commands that are relayed on the CAN bus to the appropriate modules. Default telemetry and command firmware reside on each module, each of which has a smart interface on the CAN bus, so the onboard computer is not required for basic spacecraft operations. First, the transmitter is activated and deactivated manually to verify operation. When basic operation is established, the onboard computer is powered up and remains on to assist with the remaining tasks.

Once the onboard computer is activated, the software defaults to commissioning mode, performing tasks 1–4 in Table 7.9. In this mode, the satellite must be commanded manually from the ground to perform all required tasks. Once all commissioning actions are complete, the satellite can be commanded to enter a semiautonomous operational mode, looping through tasks 1–8.

Once the satellite is in operational mode, these satellite software tasks manage the operations. The typical configuration would be for the orbit task to determine when the satellite enters eclipse based on uploaded orbital elements, then set the payload task to collect data from the MESA sensor at the rate specified by a commanded value from the ground. (Note that the sun sensors are not used as a second check to determine the eclipse state because this data is available only in the telemetry and used for orbit task verification on the ground). When the satellite exits eclipse, the payload task is terminated. Next, the orbit task determines the next ground station access, at which time the transmitter is activated and the satellite waits for commands from the ground. This is where autonomy ends, as

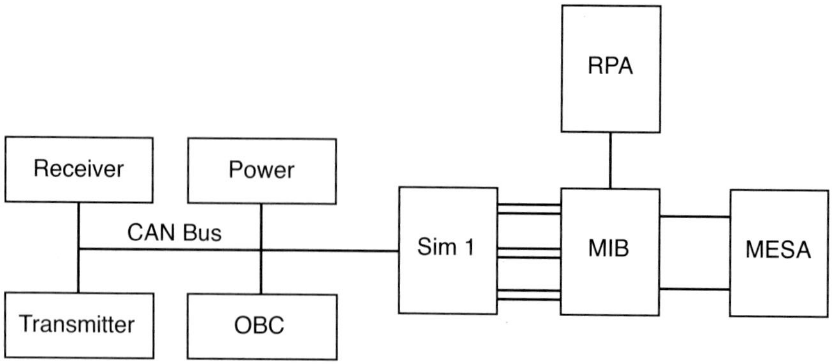

Fig. 7.16. Block diagram of FS-2 avionics connected via the CAN bus.

Table 7.9. FalconSAT-2 Software Tasks

Order	Task	Lines of Code	Description
1	Watchdog	267	Resets hardware watchdog
2	Memory wash	348	Manages memory
3	Asynchronous boot loader	1,067	Handles ground-initiated CPU reset, upload, and dumps
4	Control	2,013	Sets commissioning or operational mode
5	Payload	5,333	Collects payload data based on configuration
6	Telemetry	2,259	Collects telemetry data based on configuration
7	Orbit	2,211	Determines orbit eclipse and ground station access times based on uploaded two-line element sets and clock synchronization
8	Download	922	Manages download of payload and telemetry data
	Total:	14,420	

commands from the ground station must be sent by an operator. The short telemetry file is downloaded, followed by the longer payload data.

When a ground station access occurs during eclipse, payload data collection ceases so that download activities can take place. Similarly, multiple eclipse cycles of payload data collection can take place in the absence of a ground station contact. Several modes exist to handle contingencies such as running out of memory or detecting an anomalous temperature, voltage, or current condition. For further protection, the electrical power subsystem module had battery voltage setpoints where the system could be shut down gracefully in the event that not enough solar power was available to maintain charge on the battery.

7.7.7 FalconSAT-2 Launch Vehicle Integration and Operations

FS-2 was originally scheduled to fly on shuttle flight STS-108, but as the spacecraft was being loaded for delivery to Cape Canaveral in July 2002, a control moment gyro failed on the International Space Station. STS-108 was reconfigured for an emergency repair mission, leaving no room for FS-2. The spacecraft was remanifested on STS-114, but the Columbia tragedy and other budget cuts resulted in the termination of the shuttle's secondary payload program for some time.

One lesson that had not yet been learned from FalconGOLD and FS-1 was the need for a sustainable operations program. While the search continued for a flight for FS-2, much of the team's efforts were redirected toward FS-3; however, a small team of cadets was selected each year for the next three years to maintain FS-2's flight readiness and develop the operations concept. FS-2 was designed with operations in mind, as the original vision of the satellite was to use it to grow a cadet operations program. However, without the active input of an experienced operational program, a few operational aspects were overlooked.

Both the satellite software and supporting ground station hardware were intentionally not designed for autonomy. Each satellite pass would require the presence of operators in the ground station. At first glance, it appeared that this approach would provide plentiful opportunities to expose cadets to space operations.

The FS-2 qualification model was primarily used as the "flight simulator" during the development of the operational training program. Operational checklists were developed from the full functional test documentation developed by the FS-2 engineers. These preliminary checklists did not account for many of the anomalies that could be encountered during real-world satellite communications during a pass. By the end of the 2003 academic year, a small team of five graduating cadets and five faculty received in-house training in the program, with only two additional cadets who would graduate the following year receiving training to provide program continuity.

During the following two academic years, the training program was revised with the help of experienced U.S. Air Force space operators working in concert with the faculty engineers. All basic commissioning and operational phase objectives were broken down into more realistic groupings of tasks that could be accomplished in a satellite pass, accounting for communication anomalies. Because the FS-2 qualification model was exercised so heavily over these three years, a list of expected technical anomalies with workaround solutions was also developed. For example, the boot loader software in the flight computer became corrupted, so procedures were developed in case such an event were to happen on orbit. An anomaly was also discovered in the analog-to-digital converters of the payload. The payload software had to be modified to account for their previously unnoticed nonlinear behavior.

The FS-2 engineering and operations teams from 2003 to 2006 also had the responsibility of periodically servicing the FS-2 flight model. Each month, the battery was charged, and a full functional test was run by the most experienced operators. Only after the full operations program was developed did it become clear that administrative overhead of faculty and cadets was nearly unmanageable on a long-term basis. Because of this, all future FalconSAT missions were required to have an autonomous operations capability when enough cadet crews are not available, such as during the summer or normal academic holidays.

It was not until January 2005 that FS-2 was finally manifested on the inaugural SpaceX Falcon-1 launch vehicle. Because of the change in launch vehicles, FS-2 had to be recertified for flight. This required a vibration test, fit check, creation of an interface control document, and a separation test. The engineering team also upgraded the VHF receive antenna to a higher-gain version because of the larger encapsulation area of the Falcon-1.

The class of 2006 was to have the unique experience of supporting spacecraft integration activities at the SpaceX launch facilities on Kwajalein, in the Marshall

Fig. 7.17. The FS-2 launch campaign at Kwajalein (left and center). The recovered FS-2 satellite following the Falcon-1 launch failure (right).

Islands. Unfortunately, the inaugural Falcon-1 launch in March 2006 was not successful. Figure 7.17 shows photograhs of this campaign.

7.8 FalconSAT-3

FS-3 was the fourth satellite built by the Academy. The primary scientific motivation was to provide greater knowledge and understanding of the plasma environment in the ionosphere through two payloads, PLANE (Plasma Localized Anomalous Noise Environment) and FLAPS (Flat Plasma Spectrometer). The PLANE objectives were to characterize the plasma turbulence in the environment surrounding the space vehicle, distinguish turbulence in the ambient environment (global effects) from variations in the plasma population moving with the space vehicle (local effects), and quantify plasma perturbations caused by active systems. Only one PLANE sensor—developed by AFRL—was on the satellite, located on the side corresponding to the ram direction.

The FLAPS payload was built by NASA Goddard Space Flight Center and AFRL. It had only one sensor, also located on the panel facing the ram direction. The primary objectives for FLAPS were to support the following efforts.

- Establish flight heritage of a MEMS plasma spectrometer based on microfabricated smart-skin technology.
- Characterize the effects of non-Maxwellian charged-particle distributions on formation, propagation, and decay of ionospheric plasma bubbles.
- Demonstrate the capability of the FLAPS experiment to make fine-resolution (10%) measurements of ionospheric plasma spectra differential in energy and angle.
- Contribute to the validation of the plasma bubble and radio wave scintillation "nowcasting" and forecasting system associated with the DOD Communication/Navigation Outage Forecasting System (C/NOFS).

- Support simultaneous multipoint in situ measurements of a single plasma bubble structure with another space vehicle, such as the C/NOFS spacecraft, if the opportunity arises.

Additionally, the satellite contained the small thruster payload, MPACS (Micro Propulsion Attitude Control System), which was being tested as a possible attitude actuator on small satellites. MPACS was built by AFRL and rated by the Space Experiments Review Board in 2002. The objectives were to establish flight heritage for MPACS technology, quantify the change in momentum caused by MPACS on a microsatellite in space, quantify beginning-of-life versus end-of-life performance characteristics in space, detect and quantify any adverse operational interactions between MPACS and a microsatellite in space, and to demonstrate that MPACS could be used as an effective actuator by operating it as part of the attitude determination and control system for 100 cumulative hours (objective) and 40 cumulative hours (threshold). There were four MPACS clusters on the top of FS-3, each with three different tubes. MPACS worked by sparking across solid Teflon, creating a small plasma discharge and giving an average thrust of about 150 μN at a frequency of approximately 2 Hz.

To fulfill the requirements of all payloads, FS-3 needed a functional bus with important mechanical, avionics, electrical, data handling, communications, and attitude control subsystems, as well as proper operations and software in place to manage and handle the data from the satellite.

The mechanical subsystem consists of the satellite's structure and physical design. The satellite design and major components are shown in Fig. 7.18. The cube structure is made of 6061-T651 aluminum orthogrid panels. The lightband

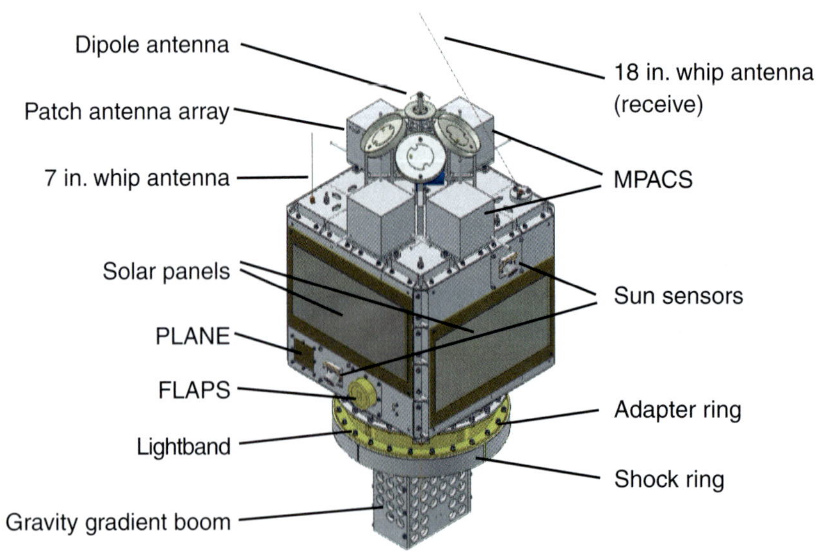

Fig. 7.18. FS-3 satellite design and major components.

and gravity-gradient boom are located in the bottom ($-z$) side of the satellite, with the MPACS clusters and primary antennas on the top ($+z$). The plasma payloads PLANE and FLAPS are located on the ram ($+x$) side. Sun sensors and solar panels are located on each of the four 46 cm sides.

The avionics subsystem is composed of a single stack that has a base of 23.8 cm and a height of 3.8 cm. In this stack are all necessary components and processors to ensure the functionality and operations of the satellite. Figure 7.19 gives the general diagram for how this stack is layered. As shown, the trays making up the stack (from the bottom) are the battery tray, the battery charge regulator, the in-flight computer tray, the transmitter tray, a torque tray, a receiver tray, and a tray at the top for the PLANE interface module.

The electrical power subsystem must provide sufficient power to the satellite to allow multiple payloads and other subsystems to run concurrently. The power is provided to the satellite by seven Sanyo NiCd secondary batteries as well as four 20 W SpaceQuest GaAs solar arrays. Both the batteries and the solar arrays feed into the battery charge regulator, which regulates the charge and discharge of the batteries and distributes power to different boards and trays in the avionics stack.

The data-processing subsystem's primary function is to collect telemetry, process commands, and perform routine housekeeping tasks via stored or uplinked software. This is done primarily through the in-flight computer and modem. The in-flight computer is composed of a NEC V-53A CMOS 16-bit microprocessor

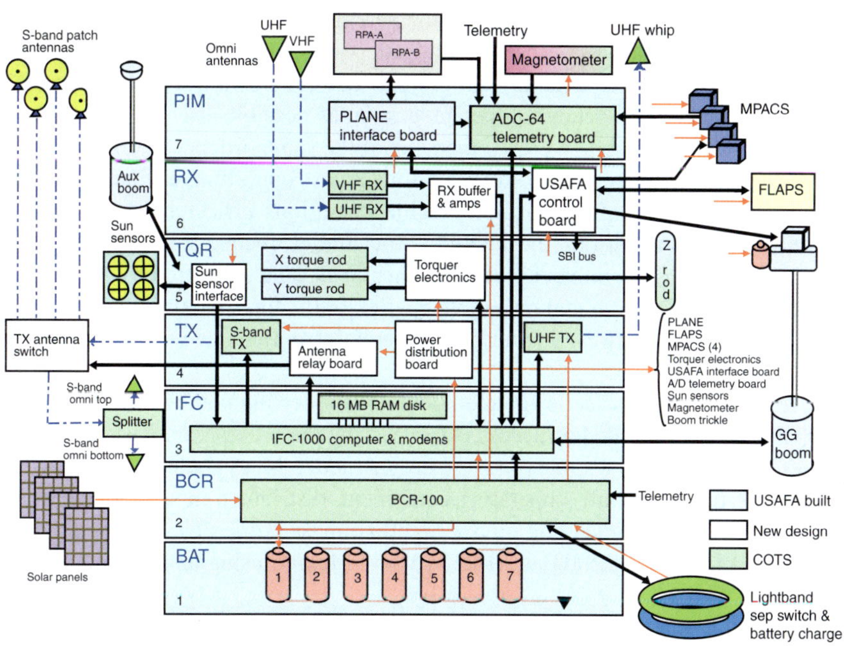

Fig. 7.19. FS-3 avionics block diagram.

with a 29.5 MHz crystal oscillator. It has 16 Mb of RAM and 16 Mb of flash available, along with 3 Mb of physical SRAM for error detection.

The communications subsystem is composed primarily of two transmitters and two receivers. The primary communications architecture is a UHF uplink and an S-band downlink. The S-band signal can be transmitted through the omnidirectional canted turnstile, giving 0 dB gain, or one of four patch antennas, giving 8.5 dB average gain. The UHF receiver is a 7 in. whip antenna located on the top of the satellite. The secondary communications structure includes communicating through a UHF transmitted signal and a VHF received signal through an 18 in. whip antenna. The S-band downlink can be sent at speeds of 38.4 or 76.8 kbps and the UHF downlink at a speed of 9.6 kbps. Both the receiver and transmitter trays were supplied by SpaceQuest.

The attitude determination and control subsystem is responsible for maintaining accurate attitude knowledge as well as effectively controlling the satellite within the specifications required for the payloads. The satellite needs to be able to be ram pointed within 5 deg, and must have a clock accurate to at least within 1 sec. To achieve these results, the satellite has one Billingsley TFM100G2 fluxgate magnetometer and four SQ Model SS-256 sun sensors for attitude sensing. Two main actuators are used to control the satellite, a passive stabilization device in a 175 in. Northrop Grumman gravity-gradient boom (two other booms by different vendors failed in the engineering model and qualification model campaign) and three SpaceQuest torque rods.

The primary goal was to operate the satellite for at least a year using the Academy ground station. Before launch, cadets wrote operational plans to commission and operate the satellite, and ground station hardware was prepared and operators trained to be ready to operate as soon as possible after launch. Operations are run by four-person crews, made up of the same officers as FS-2.

Each day, there are three to four significant pass opportunities that last about 10 min each. The commander and satellite operations officer have only these 10 min windows to turn on the satellite's transmitter, record critical health, and then accomplish all command plans and checklists. The software on FS-3 is uploaded from the ground station, allowing flexibility for changing the software if necessary. The software included commercial programs as well as programs written by the cadets.

As with FS-2, a strict design methodology and testing plan was followed for FS-3 to ensure flight model success. After passing the preliminary design review, cadets built and tested a structural engineering model. Mass simulators were used in place of the spacecraft's operating equipment, distributed in such a way as to give the structural engineering model flightlike mass properties. The model was tested on 14–15 April 2003 at AFRL in New Mexico using an electrodynamic shaker.

After the structural model tests, the design was reevaluated, and changes were documented into a critical design review. This paved the way for a qualification model to build confidence in the final design. The qualification model testing took place at the Academy; however, all primary testing of the satellite took place at AFRL between 28 January and 6 February 2005. The model was tested in thermal vacuum beyond the expected bounds imposed by the space environment. It was then tested for low-level sine sweep to determine the fundamental and other natural frequencies, and random vibration and sine burst to confirm strength and structural integrity.

At the conclusion of these tests, the design was modified as necessary for the final flight model. The qualification model has all of the internal characteristics of the flight model and can be used to test any procedures or recreate anomalies while the flight model is in orbit.

The flight model was assembled at USAFA, then tested 17–28 April 2006 at AFRL in Kirtland. The mass properties of the flight model were first measured to verify the design data. The spacecraft was then tested for low-level sine sweep to determine the fundamental and other natural frequencies, and random vibration and sine burst to confirm strength and structural integrity. Additionally, the flight model was put through thermal-vacuum testing to verify the avionics and software in a spacelike environment.

The different tests revealed component problems, which were fixed; however, after testing on each model, the boom had to be replaced. Because of this stringent design and testing process, no hardware problems have been encountered since launch.

FS-3 was launched aboard an Atlas V rocket from Cape Canaveral Air Force Station on 9 March 2007. The satellite was one of five deployed from the first ever EELV Secondary Payload Adapter Ring, which held the small satellites as well as the main payload, Orbital Express. The launch put FS-3 into a nearly circular orbit with an altitude of 560 km and an elevation of 35.4 deg. Operations from the Academy ground station were planned to start the next day.

Immediately after launch, the satellite could not be contacted through the ground station. Multiple explanations were considered, such as ground problems, a mislabeled two-line element, or satellite problems. Upon further investigation, a problem was discovered in the FS-3 operating system that interfered with proper communication. A software patch was created by SpaceQuest. After nearly a month without being able to operate the satellite, the satellite was contacted from the Academy ground station on 4 April 2007, and the commissioning phase began.

The commissioning phase involves testing the primary and secondary communications, uploading the software, testing the file system and file storage, downloading data, testing torque rods, testing each payload, testing the onboard controller, deploying the boom, and then finally orienting the satellite for normal operations. Cadets learn much during this phase. The initial communications structure of UHF uplink and S-band downlink failed because of the low power of the S-band omnidirectional antenna and the tumbling state. The backup communications of VHF uplink and UHF downlink have since been used without any problems. Each payload's hardware was tested and shown to be functional, and PLANE has even given useful data for determining rotational rates, which was not an expected capability.

Software problems caused a few setbacks during the commissioning process. When the satellite had to be reset after the memory was filled, many problems occurred while the software was being loaded. Once the software was loaded correctly, the primary focus turned to developing enough attitude knowledge to deploy the gravity gradient boom with no sun sensors because of the software problem. A few possibilities were analyzed, while more PLANE, attitude, and status-of-health data were collected and evaluated. The conclusion was that without the gravity-gradient boom deployed to slow the tumble rate, accurate attitude knowledge would not be possible using only magnetometers due to the small angle approximation required by the Kalman filter state estimator. However, it would be possible from the magnetometer reading to determine whether the satellite is in a favorable position to deploy the boom.

With the end of the academic year rapidly approaching, it was decided that the Academy class of 2008 would have the responsibility for deploying the boom and stabilizing FalconSAT-3. The class of 2007 had established communications and verified that all hardware on the satellite was in good working order.

During the 2007–2008 academic year, the cadet operations team developed a method for modeling the angle off of nadir using only magnetometer data and the IGRF model of Earth's magnetic field. The gravity-gradient boom was deployed in November 2007 using magnetometer data to predict favorable geometry for deployment. The boom deployment was validated by telemetry, estimated tumble rates, and radar cross section information. Unfortunately, the attitude model still showed the spacecraft was tumbling after boom deployment—but it was tumbling at a much slower rate. From this exercise, students applied the knowledge from several courses on control theory and cemented understanding of the concepts through rigorous testing and through then implementing the operational procedures on the actual orbital vehicle. While the spacecraft was tumbling, it caused cadets to reexamine everything they thought they knew about gravity-gradient stabilization.

Magnetorquers were employed to slow the satellite spin rate, with little to no effect. The cadet operations team discovered the binary commands that controlled the relative strength of the torque rods were inverted before they were processed. This meant that the torque rods were on their weakest setting while trying to torque the satellite. Once the commands were understood, it was found that commanding the torque rods did have an effect on tumble rates. Several experiments were accomplished with various b-dot controllers and torque-rod polarity settings. Unfortunately, while the torque rods are active, they corrupt the magnetometer measurements. This problem was not unforeseen during development, but it has become apparent that software and processor limitations prevent the original implementation scheme.

Despite the attitude control challenges, the cadet operations team kept the mission in focus. The spacecraft tumble rate is about 1 revolution every 20 min. Under this assumption, cadets turned on the PLANE payload to collect data. The PLANE sensor only collects useful information when the sensor is in the ram direction. During the course of several orbits, cadets have collected useful science information for the principal investigators. The MPACS thrusters have also been operated to prove that they are working. Attitude limitations do prevent measuring the impact of the MPACS, but the very valuable data showing power consumption can be collected as well as establishing lifetime and spaceflight heritage.

The class of 2008 made great strides in the operation of FalconSAT-3, both from the perspective of their understanding as astronautical engineers and from the perspective of FalconSAT-3 operations. The task of operating an on-orbit satellite was regarded as the most challenging, the most rewarding, and the best learning experience by the cadets. The class of 2008 provided a greater understanding of the satellite workings than has ever been achieved. It is estimated that FalconSAT-3 will maintain full operational capability for approximately two more academic years, provided the electrical power subsystem performance is closely tracked, and may provide limited capabilities long after that.

The satellite attitude model still shows that the satellite is tumbling, but cadets do collect useful PLANE and MPACS mission data. The FLAPS instrument should eventually be able to provide some useful data in a similar manner to the PLANE sensor, but is currently hampered by software control issues. In theory, attitude stabilization is still possible, so it will be a class of 2009 project to filter the magnetometer data and work around the torque-rod issues.

Despite the operational issues of FalconSAT-3, it is impossible to view the mission as anything but a smashing success. The learning opportunities it provides and the excitement it generates at the undergraduate level make it an incredibly valuable asset. The fact that it is also contributing to the understanding of the ionosphere reflects great credit on the classes that designed, tested, built, and operated the satellite as well as their faculty mentors.

7.9 FalconSAT-4

The design of FS-4 began in the summer of 2006, with a preliminary design review in December 2006. Unfortunately, because of funding constraints, no sponsors were identified for the mission. Therefore, the Academy found other payloads and started work on FS-5. This was a learning experience for cadets, who had a good technical design but no sponsor. FS-4 will remain a satellite in the program as a lesson learned.

7.10 FalconSAT-5

Conceptual design of the next FalconSAT satellite began in January 2005 with a symposium at which a variety of organizations from industry, government, defense, and academia presented 25 different possible payloads. Cadets completed the critical design review in December 2006 and started building the structural engineering model. It will have four payloads:

- WISPERS, an experiment to measure ions emanating from a source on the satellite to validate Academy and AFRL plume models.
- iMESA, an experiment to detect the temperature and density of ambient ions to validate ionospheric data assimilation models.

Table 7.10. FS-5 Top-Level Mission Requirements

Altitude	650 km	
Inclination	72 deg	
Satellite OAP	38 W	
Satellite peak power	32.86–91.03 W	
	Threshold	**Objective**
Attitude knowledge (all directions)	±0.5 deg (3σ) (WISPERS) ±1 deg (3σ) (iMESA) 1 observation/10 sec	±0.5 deg (3σ) (WISPERS) ±1 deg (3σ) (iMESA) 1 observation/sec
Attitude control (all directions)	±2.5 deg (3σ) (WISPERS) ±4 deg (3σ) (iMESA)	±1.25 deg (3σ) (WISPERS) ±2 deg (3σ) (iMESA)
Data rate (acquisition mode)	1.3 Mb storage/orbit (iMESA) 4.9 Mb storage/orbit (WISPERS)	12.6 Mb storage/orbit (iMESA) 49.4 Mb storage/orbit (WISPERS)
Orbit	Any	> 45 deg inclination (polar orbits preferred)
	Threshold	**Objective**
	30 survey orbits over 30 days	360 survey orbits over 360 days
	Survey Mode	**Acquisition Mode**
Payload duty cycle	iMESA: survey full orbit at low fidelity	iMESA: survey full orbit at high fidelity
	WISPERS: survey full orbit at low fidelity	WISPERS: survey full orbit at high fidelity. Thruster firing ±quarter orbit
	SPCS: does not run in survey mode	SPCS: 2–10 min/orbit
	RUSS: 10 min/orbit	RUSS: 100%

- RUSS (RF uplink signal strength meter), an experiment to characterize VHF signal distortion to improve ionospheric models.
- SPCS (Space Plasma Characterization Source), an experiment using a Hall-effect thruster and cold-gas thruster to demonstrate orbital maneuvers.

FS-5 is a small satellite that will be launched 1 December 2009 as a secondary payload on the STP-S26 Minotaur IV. The satellite will be launched to a 650 km by 72 deg circular orbit. Using three-axis stabilization, FS-5 will help characterize the ambient plasma atmosphere and detect satellite thrusters in space. The iMESA and WISPERS payloads will support the FS-5 science objectives. For this reason, the payload requirements drive the top-level mission requirements, which are shown in Table 7.10.

7.10.1 Satellite Subsystems

FS-5 is composed of four main subsystems, each serving a distinct purpose with different requirements.

The payload subsystem is the driving force of the FS-5 design and mission. It will be made up of three iMESA units, three WISPERS units, and their interfaces to the onboard computer. These units are responsible for all data collection during operations. The payloads will run in two modes, survey and acquisition, which characterize the ambient atmosphere at varying levels of fidelity.

The avionics subsystem includes the electrical power assembly, responsible for generating and supplying power to each electrically operated component. FS-5 requires thermal control to ensure that the payloads and electronics are operating at appropriate temperatures. The SPCS will be integrated as a black box on the typically nadir-pointing end of FS-5, and the satellite will provide power, data, structural, and thermal interfaces to support it. Three-axis stability is needed to ensure a certain threshold of attitude knowledge and control.

The operations subsystem is responsible for all software needed to operate the various components of the satellite as well as the data storage and processing and the communication to and from the Academy ground station.

The mechanical subsystem is composed of the satellite's structure, hardware, launch vehicle interface, separation system, and connectors and fasteners. The launch vehicle adapter constraints are listed in Table 7.11.

Table 7.11. FS-5 Launch Vehicle Payload Adapter Constraints

Envelope	60.96 × 71.12 × 96.52 cm
Loading	12 g lateral and axial
Mount	31.8 cm mounting ring
Mass	<181 kg
Frequency	>35 Hz
Center of gravity	within 50.8 cm of interface plane

7.10.2 Satellite Configuration

Figure 7.20 shows the complete external and internal configuration of the satellite. The mission characteristics of the FS-5 space segment stem directly from the concept of operations. Figures 7.21 and 7.22 show the four phases of the FS-5 mission as well as a diagram of the satellite's ground trace.

The physical characteristics of FS-5 are summarized in a specification tree, created and maintained by cadets to provide detailed pictures and layouts of each component on the satellite. The analysis to support the design and integration of each unit is summarized in a system engineering report and linked to the specification tree. All testing procedures and requirements, software, and interface control documents are also accounted for within the specification tree.

FS-5 is designed to be easily maintainable and reproducible. The six-panel design allows the satellite to be taken apart or worked on from any angle in any configuration; it enables easy access in case a component needs to be added, substituted, or removed.

7.10.3 Design and Construction

The satellite has been designed to meet customer requirements and comply with launch vehicle specifications. Standard engineering procedures dictate a design, build, and test sequence to ensure that all requirements have been properly met. Assembly, integration, and testing plans will dictate the manufacturing and assembly procedures.

A complete list of materials and parts is maintained, containing every component of each subsystem, including the COTS hardware and all assembly equipment. The list also contains all support hardware, such as assembly tools, harnesses, storage devices, and transportation devices. This document can be

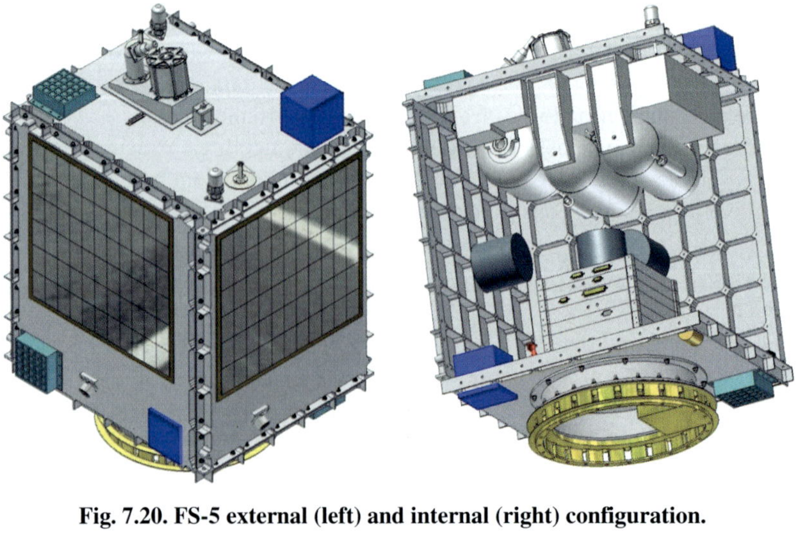

Fig. 7.20. FS-5 external (left) and internal (right) configuration.

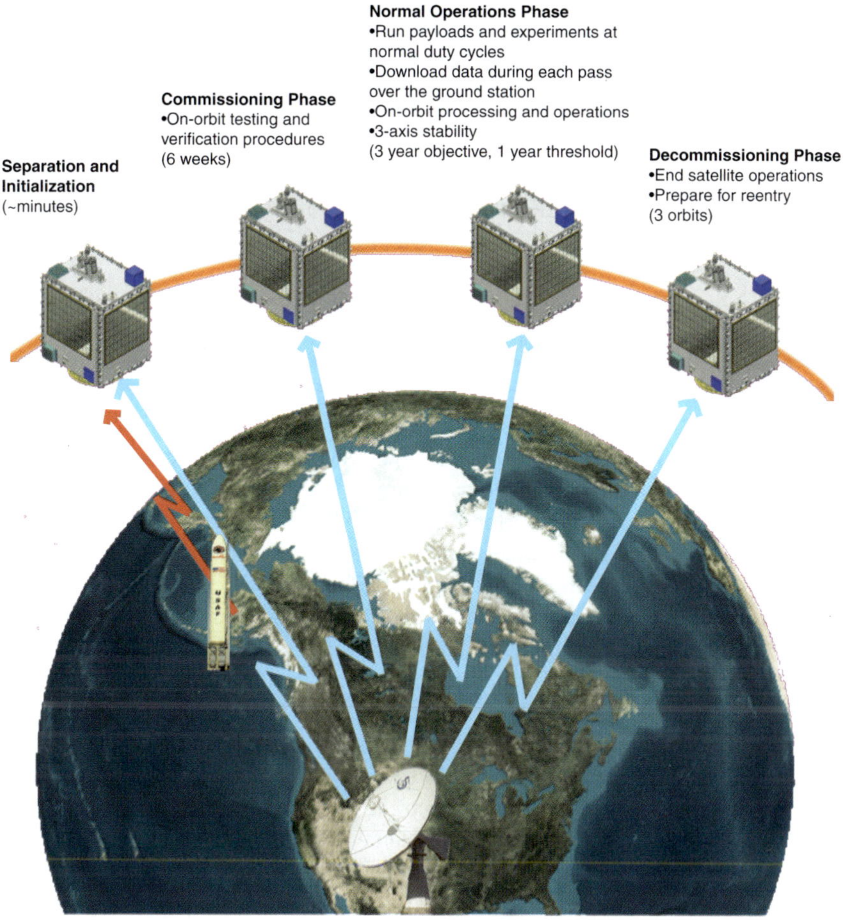

Fig. 7.21. FS-5 mission lifeline.

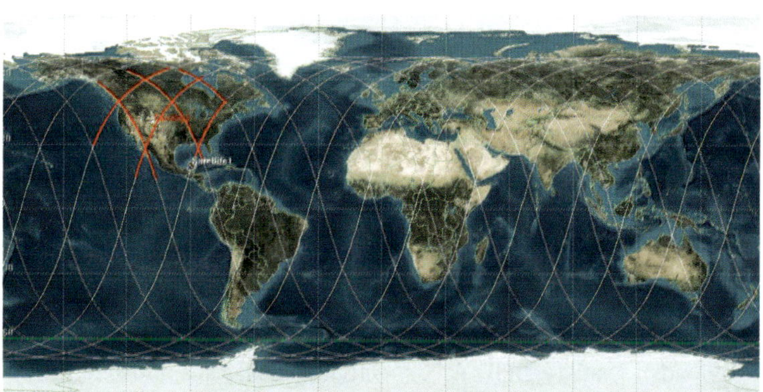

Fig. 7.22. FS-5 ground pass view.

updated by anyone in the program at any time. The manufacturing and assembly process documents will be the formal guidance for construction of the qualification and flight models.

Electromagnetic radiation is a concern for FS-5. The current design incorporates enough tolerance (200 V/m) to resist any detrimental RF energy. Additionally, the use of COTS hardware with flight heritage further reduces the likelihood that FS-5 will be affected by harmful radiation.

7.10.4 Quality Assurance

The COTS hardware will undergo acceptance testing upon procurement to validate adherence to stated system specs. All testing requirements will be summarized in the requirements verification matrix. An acceptance test report will summarize the findings of each test and document the functionality of each unit.

The qualification model test campaign will ensure the functionality of in-house units (software, structure, and payload) and the interface between all COTS hardware and internally developed hardware. Preliminary testing will be done in flatsat mode in the cleanroom prior to assembling a qualification or flight model. The qualification model will undergo a rigorous testing scheme to include thermal vacuum and vibration tests. A qualification test report will capture the results of these tests and document the functionality of the qualification model.

For environmental testing, the qualification model will be subjected to a temperature range of –20°C to 50°C. The flight model will be tested from –10°C to 30°C. All vibration loads, thermal requirements, electromagnetic specifications, and vacuum conditions will be tested and accounted for prior to delivery.

The flight model test campaign will be the final series of tests used to ensure satellite functionality and operability with the ground station. The validation and verification of the flight model will be documented in both the requirements verification matrix and the flight test report. Successful completion of all tests will mitigate potential launch and flight hazards and will demonstrate that FS-5 is fully operational and space qualified. The FS-5 schedule is shown in Fig. 7.23.

7.11 FalconSAT-6

The approach for FS-6, which is currently unsponsored, is to fly new ion-propulsion systems, smaller plasma sensors, and several deployable picosats to demonstrate distributed networks and formation flying, and take varying plasma readings over small distances.

7.12 Lessons Learned

With only a shoestring budget and a tight three-year schedule, the Academy's small-satellite program has become one of the premier undergraduate programs flying payloads to satisfy DOD needs. One factor that has allowed this program to thrive is its streamlined, small-quantity acquisition process greatly supported

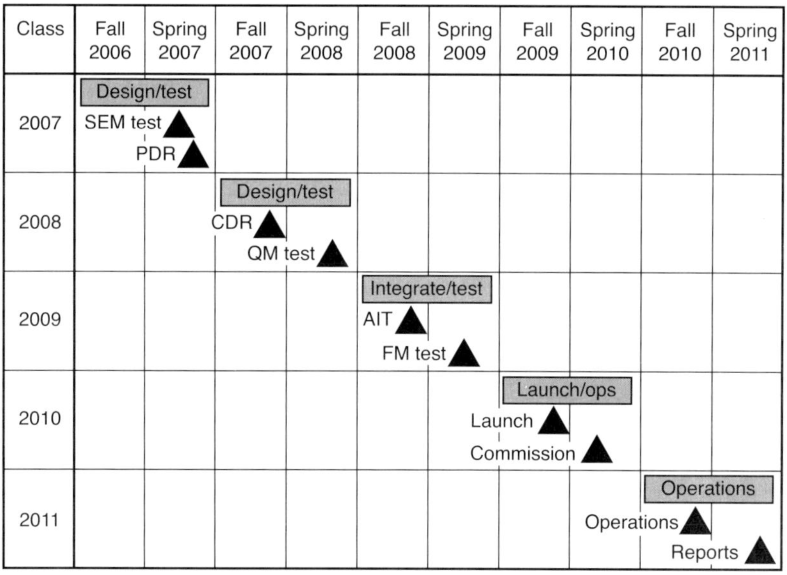

Fig. 7.23. FS-5 schedule.

by its ability to rapidly identify and implement lessons learned. The FalconSAT program uses an evolutionary design approach in which cadets employ or refine technologies and procedures developed by their predecessors. Lessons learned are captured by year, integrated into the culture, and built into catalogs of technical procedures for future missions.

The Academy's first experiments with prototype satellites flown on high-altitude balloons gave students immediate, hands-on experience and prompted the astronautics department to take on increasingly more ambitious space projects. While these programs suffered from the harsh realities of securing a ride into space, they showed that cadets could build and operate the rudimentary subsystems of a satellite. One of the largest lessons learned was the need to establish a relationship with the launch or rideshare provider and justify the importance of the mission to DOD needs.

The success of FalconGOLD demonstrated that the Academy could build a small satellite on a tight schedule to support a militarily useful mission. It established a precedent and laid the ground for a continuing program to develop the Falcon satellites.

Although the hardware may not have functioned as planned, the FS-1 mission was by no means a failure. From an academic perspective, it was a success because cadets participated from cradle to grave in a real-world mission with an all-too-real-world outcome. Cadets designed and built the payload and subsystems, devising operations plans and participating in the launch campaign. Cadets also staffed the Academy's ground station when the satellite was not operating under normal conditions. Through troubleshooting the anomalies, they gained insight into system functions and operations. The cadets and faculty learned valuable lessons to carry forward concerning mission relevancy, launch opportunity, and fundamental requirements for future small-satellite development, acquisition, and operations. Some of the many lessons captured from the FS-1 experience include:

- Resist unrealistic schedule constraints. If the rocket people say "Be here by x date or you don't fly," and you know you aren't going to make that date, look for another ride.
- Purchase at least one spare of everything for use in design, development, and testing. Use flight components only for flight.
- It takes more than three people working 12-hour days for 7 months to build a viable satellite.
- When a vendor falters, don't delay in finding a replacement.
- Resist politically driven design decisions.
- Use the absolutely lowest-power everything. You never have as much power as you think you do.
- Don't cut corners in testing because of unrealistic schedules.
- Don't assume all the parts will come together at the end. Test serially unless you are absolutely certain the interfaces will work.
- Analyze your orbit very carefully, and plan for the worst case.
- Allow time for a fully integrated test during environmental testing and simulated solar illumination conditions.
- Test every single function of the ground station, then put it under configuration control.
- Complete the design decisions and purchase hardware early. Integration, calibration, and testing always take longer than anticipated.
- Obtain computer hardware early and give programmers a set to develop and test on.

The lessons learned from FS-1 motivated a significant structural change to the program. Thus, the new approach starting with FS-2 was to focus on building up infrastructure, including design and development tools. The FS-2 design effort was aimed at developing a flexible platform that could be readily adapted and enhanced to meet future payload requirements and secondary launch opportunities.

As part of this, FAST provided both a long-term facility for cadets to gain hands-on experience with spacecraft hardware and software, as well as a facility for reducing overall program risk through development and testing of subsystems, software, and operational procedures.

A concerted effort was also made to involve cadets from a variety of departments, not just astronautics, to expand the knowledge base of participants and give all cadets, regardless of their majors, an opportunity to contribute. Teaming scientists with a keen interest in designing space experiments with engineers who want to build missions provided great synergy to the program. The CHAWS-LD experiment flown on FS-1 was conceived and built by faculty and students from the Academy's physics department. Since then, select computer science, electrical engineering, mechanical engineering, systems engineering, space operations, and management majors have joined the program. This change has not only given an interesting scientific focus to the missions, but it has also helped to assert real-world credibility.

These improvements in the process resulted in a satellite that met its targeted delivery date, met all performance specifications as demonstrated through the numerous test campaigns, and (after its launch vehicle shifted) served as a simulator to train several years' worth of cadets in satellite operations.

The big lesson from FS-4 was that although a technically sound design is developed, it may not necessarily obtain funding from a sponsoring agency. The primary mission was to fly a low-cost propulsion system to demonstrate tactical rapid maneuvering in low Earth orbit—ΔV on the order of 350 m/sec via 35 firings of 10 m/sec, mainly to demonstrate revisit times. Cadets thought this would fit in well with the newly formed Operationally Responsive Space Program Office. Unfortunately, it was too early for the program office.

7.13 Conclusions

The most important lessons learned on the FalconSAT program overall can be summarized as follows:

- Use COTS technology if applicable (it lowers cost, and allows undergrads to focus on integration).
- Make sure you have sponsor commitment early in the program.
- Testing with an engineering model, qualification model, and flight model is an excellent way to reduce risk and maximize student involvement, because each class gets its own deliverable instead of building just one satellite every three years.
- Design your satellite to the harshest environments to open up as many launch opportunities as possible.
- Complete your safety plan before the critical design review.
- Schedule and funding should be the drivers in small satellites; if a 50 kg satellite costs as much as a 350 kg satellite, you've missed the boat.
- Make your satellite inert during the launch phase (no power, no communication) and have minimal functionality testing at integration (just charge batteries). Again, this opens up secondary payload opportunities and convinces primaries that you can integrate with them safely.

- Treat software development as any other satellite design process. Software has been the toughest obstacle at the Academy; working it early in the process will save your program.
- Having a complete qualification model is important in training future students in operations and allows you to troubleshoot problems with the flight model and update your software.

7.14 References

D. J. Barnhart, J. J. Sellers, C. A. Bishop, J. R. Gossner, J. J. White, and J. B. Clark, "EyasSAT: A Revolution in Teaching and Learning Space Systems Engineering," *Proceeding of the AIAA Space Systems Engineering Conference* (Atlanta, GA, Nov. 2005).

D. J. Richie and Jerry J. Sellers. "FalconSAT-3: A Military University Satellite for Innovative Department of Defense Experiments," *Proceedings of the 2003 Core Technologies for Space Systems Conference* (Colorado Springs, CO, 4–6 Nov. 2003).

W. Saylor and M. France, "Design of an Undergraduate 3-Axis Space Science Satellite," *Proceedings of the Small Satellites Systems and Services Symposium* (Rhodes, Greece, May 2008).

W. Saylor and M. France, "Test and On-Orbit Experiences of FalconSAT-3," *Proceedings of the Small Satellites Systems and Services Symposium* (Rhodes, Greece, May 2008).

K. E. Siegenthaler, D. J. Barnhart, J. J. Sellers, J. J. White, and T. White, "Educational Methods/Aids Used in the Astronautics Core Course at the USAF Academy," *Proceedings of the 2004 American Society for Engineering Education Annual Conference & Exposition* (Salt Lake City, UT, 20–23 June 2004).

K. E. Siegenthaler, J. J. Sellers, D. A. Miller, T. J. Lawrence, D. J. Richie, and D. J. Barnhart, "The Undergraduate Satellite and Rocket Design, Fabrication and Launch Program at the USAF Academy," *Proceedings of the 33rd International Symposium IGIP/IEEE/ASEE* (Fribourg, Switzerland, Sept. 2004).

K. E. Siegenthaler, J. J. Sellers, D. J. Richie, and T. J. Lawrence, "Student Design, Development and Operations of Small Satellites at the USAF Academy," *Proceedings of the 2004 American Society for Engineering Education Annual Conference & Exposition* (Salt Lake City, UT, 20–23 June 2004).

K. E. Siegenthaler, T. J. Lawrence, D. A. Miller II, D. E. Swanson, M. J. Meerman, D. J. Barnhart, M. G. McHarg, and J. White, "Nurturing Our Satellite Space Workforce at the USAF Academy," *Proceedings of the AIAA Space 2005 Conference* (Long Beach, CA, 30 Aug.–1 Sept. 2005). AIAA-2005-6779.

D. Swanson, K. E. Siegenthaler, D. J. Barnhart, J. J. Sellers, D. J. Richie, and E. Bruno. "Spacemanship at the USAF Academy: Developing a Satellite Ground Station Crew Training Program for Non-Technical Students," *Proceedings of the 2004 American Society for Engineering Education Annual Conference & Exposition* (Salt Lake City, UT, 20–23 June 2004).

J. White, "Analysis of the FalconSAT-1 Program," *Colorado Satellite Services* (June 2000).

Chapter 8

The United States Naval Academy Small Low-Cost APRS Satellites

Bob Bruninga*

8.1 Introduction

This chapter describes a series of inexpensive student satellite projects from the United States Naval Academy that were designed to maintain a continuous digital transponder in space. These projects gave students hands-on experience in the design, development, construction, integration, launch, and operation of small satellites. The five satellites— PCSAT-1, PCSAT-2, ANDE, RAFT, and MARScom—had somewhat different primary and secondary missions, but all carried a transponder in support of the Automatic Packet Reporting System (APRS; also known as the Automatic Position Reporting System).[1]

The other common thread was the emphasis on inexpensive commercial off-the-shelf (COTS) components—including the use of low-cost terrestrial solar cells. The basic communications transponder hardware cost about $2000, not counting all the test items and spares or the cost of integration. This low-cost, high-risk approach was appropriate because the primary mission was education. Failure, if it occurred, was an option, and would be considered a significant learning experience. In addition, all but one of the satellites were in short-lived orbits (less than a year), so long-term reliability was not a significant design requirement. All achieved their primary mission objectives with little or no problems.

Three of these projects—PCSAT-1, RAFT, and MARScom—were developed completely by students, from the spaceframe all the way through the power system, electronics, payloads, and communications. The other two—PCSAT-2 and ANDE—were primarily Naval Research Laboratory (NRL) spacecraft, with the Naval Academy students responsible for most of the electronics and all of the command, control, and communications.

8.2 The APRS Mission

APRS is a single-channel multiuser communications system analogous to the Defense Department's Blue Force Tracking system, but operating in the Amateur Satellite Service.[2] APRS exchanges GPS position data among all participants in the network so they can map their position and maintain situational awareness

*U.S. Naval Academy Satellite Lab
Annapolis, Maryland

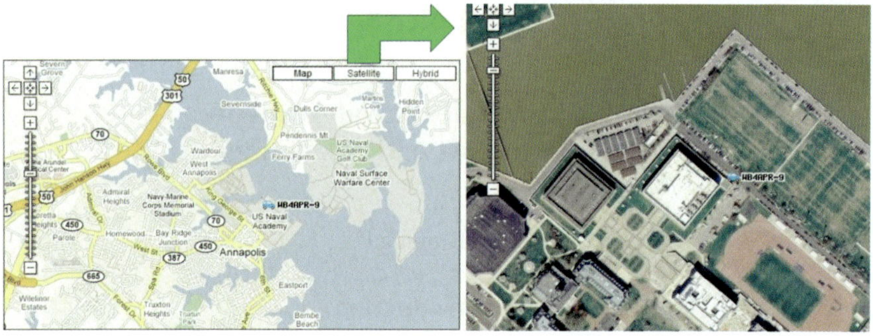

Fig. 8.1. The live APRS global data stream can be mined by any application. Here, findu.com presents the real-time position of any station on either maps or satellite images. This page shows the author's car in the Naval Academy parking lot.

through brief messages and status packets. Nearly 10,000 or so APRS users are on the air on any particular day in the United States. As with cellular phones, very few of these actually need to report by satellite because most are in range of the terrestrial network, but those that are out of range or in the wilderness do need a satellite constellation to maintain connection. Any of these users can switch to the satellite network with a simple twist of a knob because the terrestrial network and the satellite uplink frequency are in the same band and use the same waveforms.

All of the APRS data are combined into a single global feed available to all users via a telnet connection. Numerous applications have been written to display this data. For example, one application (findu.com) merges the data with Google maps and allows a user to zoom all the way into the precise position and even select a satellite image of the station's reported position (Fig. 8.1).

8.2.1 The APRS Satellite Tracking and Reporting System

The satellite component of APRS, known as the APRS Satellite Tracking and Reporting System, or ASTARS, evolved through a number of previous satellite experiments.[3] First were the 1200 baud phase-shift keying (PSK) experiments using the amateur AO-16, LO-19, and IO-26 satellites (known as TRAKNET).[4] These satellites required specialized PSK modems and linear receivers, which were not generally found on amateur radio vehicles. The next experiment used the 1200 baud audio-frequency-shift keying (AFSK) packet system on the space station Mir. This configuration was compatible with the most common digital system and FM transceivers in the Amateur Radio Service, and this brought ASTARS to the common user.[5] The Naval Academy conducted a weeklong experiment via Mir using the Kenwood TH-D7 walkie-talkie, which resulted in more than 55 stations with two-way handheld satellite message communications.[6] Similar experiments were conducted using the Space Shuttle Amateur Radio Experiment (SAREX) on several missions.[7]

In 2000, the SUNSAT and UO-22 satellites supported APRS relay at 1200 and 9600 baud, and the Kenwood 1200/9600 baud TM-D700 radio with integrated

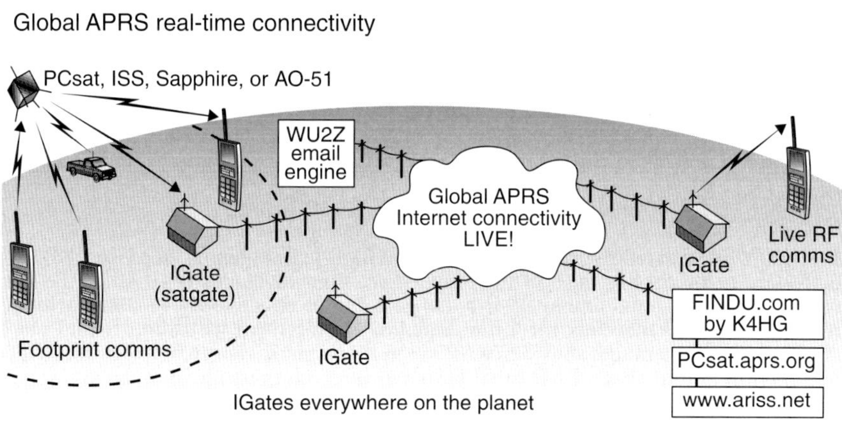

Fig. 8.2. IGates (Internet Gates) and SatGates and the volunteer APRS Internet system give APRS users global connectivity from handheld-to-handheld anywhere on the planet, as well as connectivity to ground stations and satellite telemetry.

modem finally allowed mobile satellite digital communications. With this radio, text messaging via satellite was now possible, front panel to front panel. Further, the radio could switch between the terrestrial network and the satellite network with a one-knob frequency change within the same band. In 2001, PCSAT-1 was launched, and then PCSAT-2 in 2005. Subsequent missions on ANDE and RAFT in 2006 continued these experiments, and about the same time, the AMSAT AO-51 was also enabled for APRS digital relay. By 2007, the GO-32 Israeli amateur satellite was enabled for APRS experiments, and success has been demonstrated using a TH-D7 walkie-talkie with only a 19 in. whip antenna.[8]

8.2.2 The Volunteer Ground Station Network

ASTARS is implemented through a global network of volunteer ground stations using inexpensive equipment linked together via the Internet (Fig. 8.2). These stations, called SatGates, combine all downlink telemetry packets into the existing worldwide APRS-Internet System for delivery to any APRS operator anywhere in real time.

Although the classical amateur satellite ground station consists of high-gain tracking antennas and linear receivers integrated by a PC, the goal of ASTARS is to have enough link margin so that these ground stations can use fixed omnidirectional antennas with no moving parts and inexpensive FM receivers.

The range-gain difference of a typical low Earth orbiting satellite is on the order of 6 to 10 dB above 20 deg from the horizon, as shown in Fig. 8.3. A tracking gain antenna can work the satellite down to the horizon, but this requires an excellent site with high altitude, moving antennas in azimuth and elevation, and a tracking computer. On the other hand, a fixed omni antenna with no moving parts and no requirement for tracking has 6 dB less gain, but can still have an effective link budget to satellites above 20 deg. The reduced range of each such omni

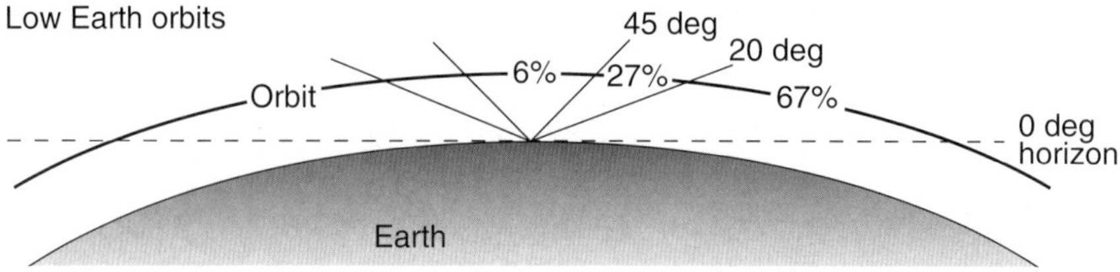

Fig. 8.3. Typical satellite ground stations need high-gain antennas because LEO satellites are below 20 deg most of the time, where the range adds 6–10 dB to the link requirement. But omni antennas with less gain can be effective above 20 deg with no moving parts and therefore meet the same link requirement with simpler and more stations. The percentages shown here between the angular ranges represent the average amount of time that a LEO satellite spends between those angular ranges for any particular ground station.

station can be mitigated by simply having more stations. The combined stream from numerous stations can result in a better than 99.96% chance of capturing every packet over the United States. Because the ground stations are all volunteers, it's more reasonable to seek adequate coverage through larger numbers of simple stations than through fewer but more complex tracking stations.

8.2.3 The Mission Link Budget

The objective of ASTARS is to close the link to mobile and handheld radios using simple omnidirectional antennas. The AX.25 packet protocol is used to advantage here because the short, bursty packets can provide the needed peak transmit power but with a much lower average power. This allows a 10 dB or so improvement in link budget for each packet, compared with the average power consumed by comparable transmitters with 100% duty-cycle downlinks.

For the Naval Academy missions, students determined the average duty cycle of the APRS transponders by studying the geographic distribution of prospective users. About 10% of Earth has any such populations. In these areas, the maximum possible throughput when the uplink and downlink share the same channel was assumed to be about 40% because the uplinks are blocked while the downlink is active. The product of these suggests an overall 4% transmitter duty-cycle average to support the user mission. Additional packets for command and control and telemetry bring the average up to about 10%. This yields a 10 dB advantage in link budget over average power, and was a fundamental design feature of all of the Naval Academy's APRS transponders.

A second factor in the APRS mission link budget is the use of VHF. The VHF downlink provides a 9 dB advantage in the link equation over UHF links (between omni antennas). For uplinks, the VHF link budget is similarly advantageous for low-power devices and other remote experiments.†

†There are several student projects using stand-alone tracking devices or data-collection buoys or remote WX stations, such as the one built by Ronald Ross (KE6JAB) in Antarctica,[9] or the WX data from the Planetary Societies habitat in the Arctic.

Fig. 8.4. PCSAT-1 was the first Naval Academy satellite to support the APRS communications mission. The white/black antennas around the equator provided about a 0.5 rpm spin due to radiometric solar pressure. This spin, plus a z-axis magnet which also assured two tumbles per orbit, provided thermal balance.

8.3 PCSAT-1

PCSAT-1 (Prototype Communications Satellite), shown in Fig. 8.4, was built at the Naval Academy with student participation throughout—from concept, through design, construction, integration, and launch.[10] The ultimate goal of this mission was to demonstrate a low-cost approach to satellite design and to encourage other universities to consider similar and compatible AX.25 satellites.[11] PCSAT-1 pioneered the use of COTS AX.25 packet radio terminal node controllers (TNCs) to not only provide the primary APRS relay capability, but also to provide all necessary telemetry, command, and control.[12] The use of stock TNCs for all flight functions minimized risk because these units were well known and understood in daily operations and had been shown to be extremely reliable in remote terrestrial applications. As a transponder, the TNCs provided real-time messaging, position, and status relay via satellite to a worldwide Internet-linked amateur radio tracking system. The two TNCs also provided 20 channels of telemetry and 10 on/off control functions each.

PCSAT-1 carried a GPS unit designed by DLR in Germany to measure the effectiveness of GPS on small satellites.[13] Not only did the GPS generate detailed position and processing information in the MITEL format for downlink at 9600 baud, it also generated standard National Marine Electronics Association $GPGGA data sentences on the 1200 baud APRS downlink channel. This format would display the satellite's real-time position on all APRS user terminals.

8.3.1 PCSAT-1 Operational Status

PCSAT-1 hitched a ride on an Athena launch vehicle in the first (and last, as of 2007) launch-to-orbit from Kodiak, Alaska, in 2001. The satellite's bent-pipe AX.25 transponder has supported thousands of users and continues to operate after 7 years as of 2008, with no on-orbit failures except for the launch failure of the $-z$ solar panel. The failure of the panel was anticipated (after two failures during manufacture by an outside vendor), but it was flown anyway because it was a custom unit, with no spares. The solar-panel failure reduced the average power budget by 16%, which caused the battery system to deep cycle during most eclipses. PCSAT-1 was fully operational for the first 18 months, and then battery cycling reduced its operations in eclipse. Through the time of this writing, PCSAT-1 is still usable on every orbit if it is in midday sun and is recoverable to normal operations three times a year during winter full sun periods.[†]

8.3.2 PCSAT-1 Design Validation

The following elements of the PCSAT-1 design were validated and performed flawlessly:

- Dual redundant payloads/systems.
- Commands and hardware redundancy.
- Commercial low-cost plastic Tedlar-coated solar panels.
- Orbital temperatures within 10°C variance.
- Thermal balance within 5°C.
- Radiometric spin between 0.5 and 1 rpm.
- Fixed magnetic stabilization.
- Good link budgets.
- Ground station network.
- Failsafe circuits and single-event upset recovery.
- Automatic reboot on loss of communications for 72 hr.
- Discipline of user service agreement.

8.3.3 PCSAT-1 System Design

An advantage of the AX.25 protocol is that any node in the system can be used for relaying data between any other nodes. Thus, the satellite's TNCs could not only provide the dedicated uplinks and downlinks and command and control channels, but also serve as a generic relay for other applications on a secondary basis. Although the AX.25 protocol has flown on many missions,[‡] PCSAT-1 was the first

---

[†]Live downlink activity for PCSAT-1 can be monitored on the http://pcsat.aprs.org webpage.

[‡]The AX.25 protocol was flown, for example, on SAREX, SPRE, Mir, the International Space Station, SUNSAT, OPAL, SAPPHIRE, STARSHINE-3, ANDE, and RAFT.

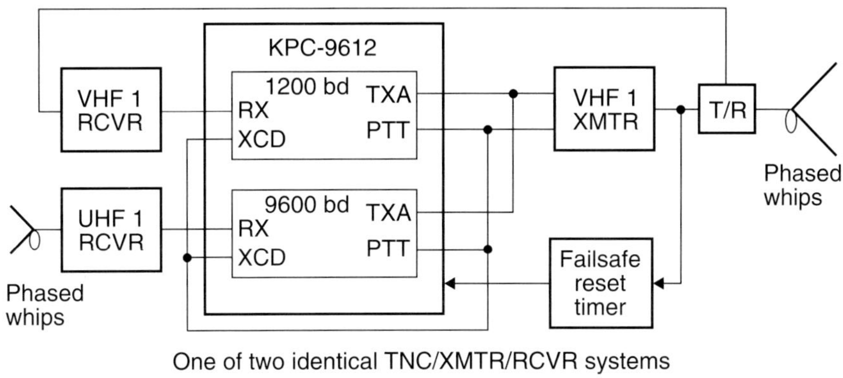

Fig. 8.5. One of the PCSAT-1 dual redundant systems consisting of a KPC-9612 TNC with 1200 and 9600 baud uplink receivers and a VHF downlink transmitter. For redundancy, the transmitters of the two systems were cross-connected as well.

to use the commercial TNCs as the complete spacecraft system controller, with no other CPUs onboard.

The PCSAT-1 design was based on two completely independent TNC transmitter, receiver, battery, and solar-panel systems to provide dual redundancy throughout the spacecraft. The two identical systems also had an auxiliary UHF receiver to serve as both a 9600 baud high-speed input as well as a backup receiver in case of primary receiver failure (Fig. 8.5).

Each system had its own battery pack, which was isolated to protect against cross-system failures but which also could connect to the other system to share the load during nominal operations, thereby extending battery life. The design also allowed for cross connecting the transmitters in case of transmitter failure. This design allowed for continued operation of either system for any single receiver, transmitter, or battery failure. It would take dual failures of two or more main elements to lose the mission.

8.3.3.1 PCSAT-1 Solar Panels

One of the most dramatic cost savings was the use of conventional "plastic" covered solar panels. These 1.5 W solar panels are mass produced for low-power terrestrial applications. The cells are single-crystal silicon and do not have the best packing factor; however, they are very inexpensive and very robust because they are designed to withstand the worst weather. Their materials are not space-rated, and the final coating is a very thin (0.001 in.) layer of Tedlar, which is not on the approved list of NASA materials because it is flammable. However, the volume of this material is actually quite small, and it was approved on that basis. The Naval Academy has now flown the solar panels on three missions, and the data suggest they may have lost about 30% of their power in five years—about twice the typical rate. Also, their power density is only about half that of commercial space-rated panels, but at only 2% of the cost. At a cost of $20 per watt—compared to about $1000 per watt for a fully integrated space panel—they can be a good match for short low-cost missions.

8.3.4 PCSAT-1 Failsafe Reset Circuits

To allow recovery from a single-event upset or other lockup of the commercial TNCs, PCSAT-1 used three methods to reset the hardware to launch defaults. First, there was a 72-hour timer that would reset the TNC if it was not contacted at least once every three days. (This timer was increased to 120 hours on PCSAT-2 based on experiences with timed-out resets over occasional three-day weekends.) Second, a hardware command via one TNC could reset the other. Third, a 2 min 555 timer chip served as a failsafe mechanism to perform a cold boot of each TNC if it failed to transmit its once-a-minute telemetry beacon. This circuit was also used on PCSAT-2, RAFT, and MARScom (Fig. 8.6).

Finally, all reset signals were capacitively coupled, so any failure in any reset mechanism would not permanently block the others. It appears that the reliability of the COTS hardware was good enough, so that these reset capabilities were rarely used—maybe two or three times in five or six years. Of course, the prevalence of single-event upsets and lockups in general would have been masked by the orbit-by-orbit resets caused by the low power budget during the satellite's later years. In effect, the satellite was cold booted on every orbit when sun angles were not optimal. When full sun returned, continuous operations for three to four weeks were common.

8.3.5 PCSAT-1 Lessons Learned

8.3.5.1 Material Construction

As the first Naval Academy spacecraft, PCSAT-1 was built using mostly student workbench and hand-tool construction. A welded frame and internal bent-aluminum electronics chassis were used. The night before weigh-in and vibration testing, the spacecraft was almost 1 kg over budget. A midnight session drilling holes in the frame and cutting excess metal from the chassis got within budget—but the result wasn't pretty, as can be seen in Fig. 8.7. Although these passed all flight tests and no problems were noted on orbit, later missions shared with aerospace professionals would prohibit these hand-tool techniques.

8.3.5.2 Space Paint

For thermal balance, the equatorial antennas (made from tape measures) were painted front and back with white and black coatings, respectively, to provide a slow radiometric spin. Researching "space" paint revealed a very detailed, costly, and time-consuming process, performed only by approved personnel using special facilities. The team was told that the yellow tape-measure paint would first have to be removed so that the proper space-rated paint could be applied. The team negotiated with an outside agency to do so, and spent weeks coordinating and scheduling—but in the end, the agency had extreme difficulty removing the

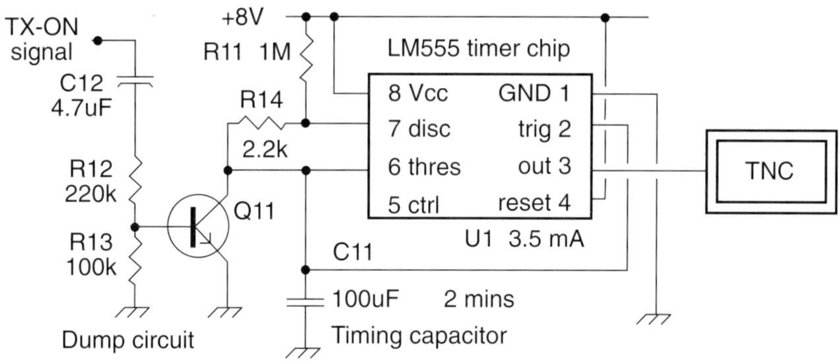

Fig. 8.6. The TNC was powered by the output of a simple 555 timer chip. The Q11 transistor dumps the timing capacitor each time a 1 min telemetry packet is transmitted. This keeps the 555 from ever timing out as long as the TNC has not locked up. If the TNC locks up, the loss of beacons would reset the 555 and cold boot the TNC.

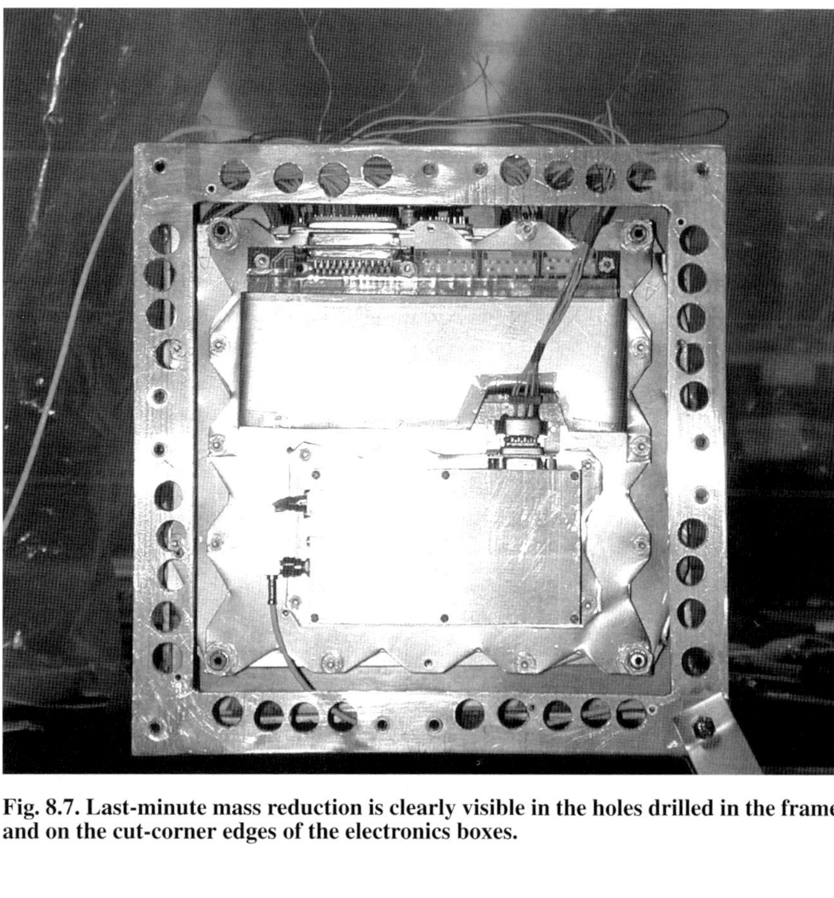

Fig. 8.7. Last-minute mass reduction is clearly visible in the holes drilled in the frame and on the cut-corner edges of the electronics boxes.

yellow paint, and just painted over it for the equatorial antennas and left the other antennas untouched, as shown in Fig. 8.4. Evidently, the paint for a good-quality tape measure designed to last for years with constant flexing, bending, stress, temperature changes, and abrasion was as good or better than the space paint—but there was just no formal way to prove it.

8.3.5.3 Backup Receivers

The design element that contributed most to the low-power recovery issues on PCSAT-1 was the provision of two extra backup receivers and transmitters that would be enabled in the event of a reset. It made sense to enable these backups after an unknown reset so that command and control could be reestablished if the primary receivers or transmitters failed. It was assumed that if these backups were not needed, they would be commanded off during the next pass over a command station to save power. Good plan—except of course, if the reason for the reset was low-power in eclipse. In this case, it takes the first 15–20 min in sun just to reestablish communications. Then, a command to turn off the extra loads leaves not enough time in the sun (45 min) to accumulate enough charge in the battery to make it through the next 35 min eclipse. The spacecraft remains in this reboot-every-orbit mode until the orbit precesses around to full sun every month or so. Then it can be recovered for several weeks, until the next accidental overload or single-event upset forces a reboot. Future backup receivers will operate in a 90% sleep/wake cycle so that they can still provide backup access, but at only a 10% cost in power.

8.3.5.4 Facilities

The team's tabletop "clean tent" was constructed from a wood frame, lots of plastic, and a filter from a hypoallergenic vacuum cleaner. As assembly proceeded, all components and surfaces were cleaned and then finally assembled in the tabletop tent until shipped. Surface inspection in accordance with VCL-1 (Visual Cleanliness Level One) was satisfactory in this homemade tent.

8.3.6 PCSAT-1 Successes

The spacecraft continues to operate in the proper sun conditions. The low-cost plastic-coated solar panels have only lost 30% of original power in the first five years. The link budget does allow communications with the users. The command and control function and telemetry is fully operational, and the global Internet-linked system of ground stations continues to provide data access at no cost to system operators. The live downlink from PCSAT-1 and subsequent spacecraft is available at http://pcsat.aprs.org.

8.4 PCSAT-2

PCSAT-2 was an effort to follow up on the success of PCSAT-1 as an inexpensive telemetry, command, and control system. The launch opportunity came from

NRL, which was working on the fifth Materials International Space Station Experiment (MISSE5). The PCSAT-2 payload was essentially a suitcase measuring 2 sq ft by 6 in. thick that astronauts would attach to the outside of the International Space Station (ISS) and return a year later. The primary mission was the Forward Technology Solar Cell Experiment (FTSCE), designed to make precise measurements of the latest solar-cell technology during one year of exposure to the space environment. Figure 8.8 presents a bottom view of the PCSAT-2/MISSE5 payload, which shows the antennas and the static samples that were attached. The FTSCE solar cells are on the other side.

The original NRL concept was to have no communications; the data would only be retrievable from internal storage after return to Earth. The project was initiated in 2002 and was on a very tight nine-month schedule from concept to launch. NRL was willing to accept the slight schedule risk to add the PCSAT-2 transponder because it would not only help in the education of midshipmen at the Naval Academy, but would also give NRL a back channel for real-time telemetry and command and control. Thus, MISSE5 managers could monitor the progress of the experiment and possibly adjust parameters via the PCSAT-2 link if needed. The Naval Academy was then able to operate its PCSAT-style transponder as long as there was adequate power available. The hardware was complete and in integration when almost three years of delays were introduced by the Columbia accident.

PCSAT-2 was attached to the outside of the ISS and activated on 3 August 2005; it was deactivated on 2 September 2006 and returned to Earth after about a year in orbit.

Fig. 8.8. PCSAT-2 on outside of the International Space Station showing the deployed configuration and location of the antennas after being unfolded.

8.4.1 PCSAT-2 Operations in the Amateur Satellite Service

Because PCSAT-2 (and ANDE later) had additional primary missions sponsored by the Department of Defense, the issue of operations in the Amateur Satellite Service had to be carefully addressed.[14] Generally, the frequency allocations in the Amateur Satellite Service are fiercely defended against any exploitation by commercial or government use. The service exists only to provide experimental and public-service communications, including education for individuals solely interested in the art of radio communications and without pecuniary interest. But since individuals have a hard time financing rides into space for experimentation, the missions of PCSAT and ANDE were carefully crafted to fit cleanly within the rules for the Amateur Satellite Service.

The defense missions for PCSAT-2 and ANDE were to study the space environment and its impact on materials in space. These objectives could fit within the missions authorized by the ITU for amateur satellites, and so the synergy between the APRS transponder for command and control and the space environment experiments combined for an acceptable mission. The justification and analysis for meeting these rules should be required reading for any satellite project that is considering the use of these frequencies and licenses.[15]

In addition to coordination via the ITU, the frequencies to be used by PCSAT-2 had to be coordinated and ensured compatible with existing amateur radio equipment on the ISS. This required collaboration with ARISS, the international team of mostly volunteers who coordinate the hardware and operations of amateur radio on the ISS. The most visible aspect of ARISS is the weekly live

Photos courtesy of NASA

Fig. 8.9. PCSAT-2 was installed on the outside of the ISS by the crew during an EVA for exposure to the space environment and retrieved one year later for return to Earth. Inset: The PCSAT-2 "suitcase" in the left hand of the astronaut during installation.

contact made with schools around the world, where kids use amateur radio to talk directly with the ISS crew. The PCSAT-2 operations had to be coordinated so as not to cause any interference with the frequencies used by the crew during their routine school contacts. Further, the frequencies and bands had to be compatible with future systems and experiments planned for ARISS.

8.4.3 PCSAT-2 and the ISS

PCSAT-2 was attached to the outside of the ISS by an astronaut during an extravehicular activity (EVA). This presented many design challenges in the areas of power, safety, and thermal control.

The PCSAT-2 communications payload is in the back half of a suitcase-like box that is opened on orbit to expose the solar cells to the space environment, as shown in Fig. 8.9. The solar panel for PCSAT-2 and the solar cells in the FTSCE payload are pointed straight "up," as shown in Fig. 8.9, and only get whatever sun is available depending on ISS attitude. Fortunately, NASA likes to fly the ISS "up" most of the time, so this gives good average power.

Because PCSAT-2 was installed by an astronaut during an EVA, the human safety requirements were significant. The 2 W transmitters were considered catastrophic hazards to an astronaut in an EVA suit. To ensure safety, one switch and three more redundant power-inhibit contacts in a safe/arm EVA plug were required while PCSAT-2 was in the payload bay and while being handled by an astronaut. Further, once PCSAT-2 was installed, there had to be four additional transmit inhibits to prevent any inadvertent activation from the ground until the astronaut was clear of the device. This was accomplished via an additional 8-hour timer and three more ground command inhibits, as shown in Fig. 8.10.

Once the EVA switch and EVA plug were armed, the PCSAT-2 receivers and special command decoders were activated, and an 8-hour count-up timer was initiated. At the end of those 8 hours, the two separate commands could be transmitted

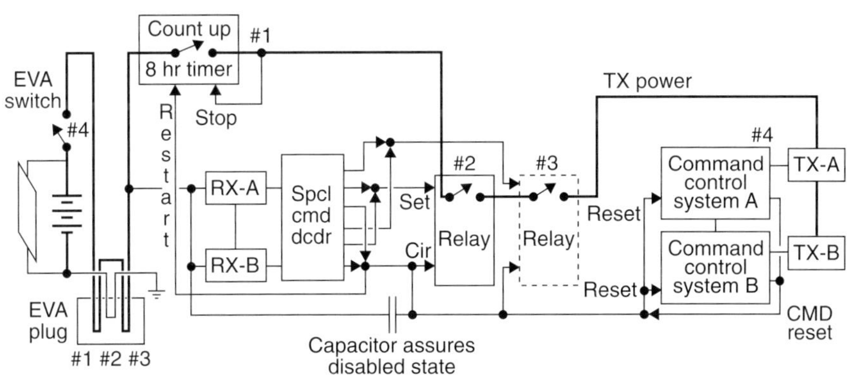

Fig. 8.10. PCSAT-2 required multiple transmit inhibits for astronaut safety during EVA. There were four inhibits consisting of the EVA safe/arm plug and the EVA switch. After the system was installed and the EVA switch activated, an additional four inhibits were required before activation of the transmitters.

in the blind to activate inhibit relays #2 and #3. At this point, the system was enabled, and a proper access code on the uplink would cause the command and control system to respond via the transmitters with the password challenge. Once the passwords were matched, the ground station could command the spacecraft.

8.4.2 PCSAT-2 Hardware Requirements

The PCSAT-2 communication system developed by the students was designed around two KPC-9612+ dual-port TNCs.[16] These controllers support 1200 and 9600 baud and have all the latest APRS generic communications relay capabilities as well as telemetry, command, and control. They can even route packets between ports and baud rates. The KPC-9612+ also offers five analog telemetry channels, eight configurable command or I/O bits, plus four on/off command bits and one input bit. These features were sufficient to handle all of the telemetry, command, and control for PCSAT-1 and PCSAT-2. By using stock TNC hardware and firmware in use by tens of thousands of existing users, on-orbit risk was minimized because of the track record of these systems in existing service across the country.

The two baud rates of the TNCs were used for different purposes. The verbose data from the FTSCE solar panel experiment was transmitted at 9600 baud to maximize throughput, but was broken up into 21 one-second bursts of seven lines of data each, separated by one second of off time. This bursty style allowed the channel to be shared even during these periodic three-minute FTSCE data dumps. Ground stations collected this data and distributed it via the global APRS network. The 1200 baud was more popular for user data because the link budget was better by 7 dB. The dual receivers and transmitters allowed for other experimental communications modes to be supported, such as PSK-31 multiuser narrowband transponder and a voice FM repeater, as shown in Fig. 8.11.

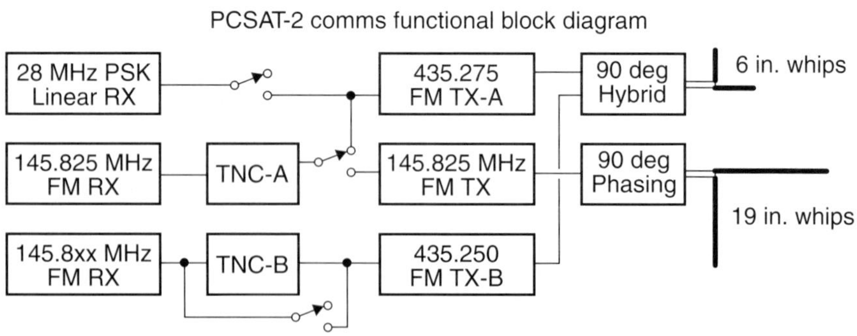

Fig. 8.11. PCSAT-2's dual AX.25 command and control system transponders can also be switched into modes to support PSK-31 and FM voice operation. Simply bypassing the receiver audio around the TNC-B to the transmitter permitted FM voice relay. The antenna systems are also connected to the three receivers on the left, but these lines were omitted from this drawing for clarity.

PCSAT-2 used the APRS five-channel telemetry format used in the TNC with an added 24-to-5 hardware multiplexer to allow telemetry to read as many as 20 analog values and four status bits transmitted in four consecutive telemetry packets. Two of the output lines of the KPC-9612+ were configured to count through the four frames as 00, 01, 10, 11 to drive the 74LV4052 multiplexer chips, as shown in Fig. 8.12.

8.4.5 PCSAT-2 Failsafe Reset System

To recover from a single-event upset or other lockup condition in these commercial TNCs, PCSAT-2 used three methods to reset hardware to launch defaults—similar to PCSAT-1. First, a 120-hour timer would reset the TNCs if the satellite had not been contacted at least once every 5 days. Second, a hardware command via one TNC could reset the other. Third, an independent dual-tone-multifrequency command tone decoder gave a backdoor reset capability for each TNC. The

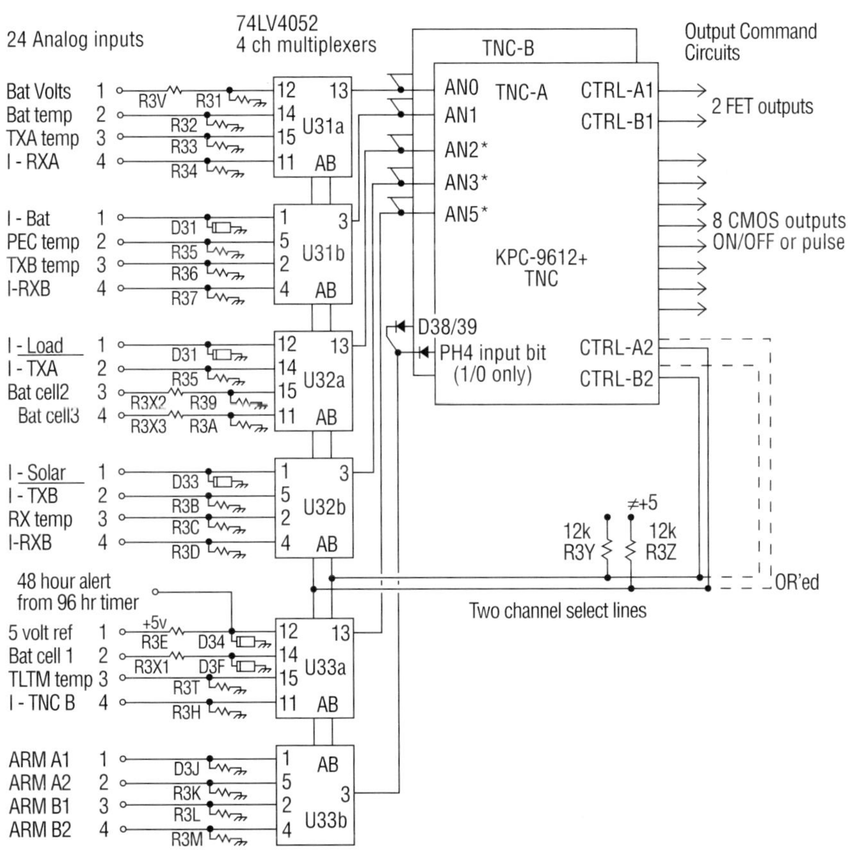

Fig. 8.12. The 24-channel multiplexer used to condense the PCSAT-2 telemetry into the five available analog channels of the KPC-9612+ TNC.

reliability of the COTS hardware was so good, that these reset capabilities were never required, although they were exercised during each remote shut-down for each EVA and docking evolution.

8.4.6 PCSAT-2 PSK-31 Multiuser Transponder

To fulfill one of the requirements for operations in the Amateur Satellite Service, there needed to be a communications transponder for general access. In the past, narrowband single-channel FM transponders were very popular, but also very restrictive, because only one station at a time could transmit. The design objective of the user communications experiment on PCSAT-2 was to retain the single FM channel downlink, but expand user throughput by using the PSK-31 waveform, which could support as many as 30 individuals within the single 3 kHz bandwidth of the FM transmitter. Thus, PCSAT-2 could support multiple users while still having the simplicity of a single FM downlink channel.

Although PCSAT-2 was the first satellite designed to support a dedicated PSK-31 digital transponder, the three-year delay in getting to orbit meant that the experiment was briefly carried out earlier using AO-51.[17] But the AO-51 experiment was only partially successful because it was not designed for this task and suffered a penalty of 20 dB or so. The PCSAT-2 experiment seemed to work initially, but then degraded after the first week of operation, such that the uplink power requirement on 29.4 MHz went from 4 to 100 W, and after about a month, no reports of any additional signal relays were received. This was disheartening because the PSK-31-to-FM mode offered so many advantages to users:

- Free downloadable software (available to everyone).
- No unique hardware requirements (used PC sound card).
- FM downlink, so discrete 5 kHz Doppler was easy.
- 10 m uplink with less than 1200 Hz total Doppler.
- Simultaneous operations by dozens of users.
- Lack of FM capture and uplink congestion.

Postflight inspection revealed that the failure of the PSK-31 uplink was caused by a crack in the receive antenna. This was discovered as the astronaut went to retrieve the MISSE5/PCSAT-2 system; while she was preparing to fold the antennas into the "suitcase" for retrieval, the PSK-31 uplink antenna fell off into her hand. Apparently, the only thing that had been holding it in place was the layer of Kapton tape used to coat the tape-measure antenna.

The reason for the break was obvious. A last-minute pair of hard-right-angle bends in that antenna blade were made by the integration team as a means of minimizing any wear of the antennas onto the FTSCE solar cells while the "suitcase" was folded up. These bends should have been made with a smooth radius, but apparently were made with no radius. In the brittle beryllium-copper antenna, this stress point and subsequent bending during deployment broke the antenna.

8.4.7 PCSAT-2 EMI/EMC Issues

Because PCSAT-2 was to fly on the ISS, not only did it have to meet all of the human-safety issues as a radiation hazard, it also had to be fully compliant with NASA requirements for electromagnetic interference and compatibility (EMI/EMC). To meet those requirements, NRL conducted EMI/EMC tests to verify that MISSE5/PCSAT-2 did not exceed the radiated emissions requirement and was not susceptible to RF emissions from the ISS.

The EMI/EMC testing was performed at NRL on 3–6 November 2003, but there was a question about which test had to be performed. The EMI test RE02 applies to radiated emissions from the device, enclosure, wiring, and permanent antennas, but with transmitters off. These limits increased from 44 to 59 dBμV/m from 326 to 2176 MHz. PCSAT-2 passed this in all respects.

NASA specified RE02, but there were no written requirements to test PCSAT-2 while radiating and connected to an antenna. Therefore, PCSAT-2 was subjected to the CE106 test detailed in MIL-STD-461, which measures the transmitter spectrum on its antenna cable. Since there was no access to the transmitter cable in flight configuration, the project team conducted the tests to the levels "relative to the carrier" indicated in CE106 as being the best-faith test but applied to the RE02 levels. The NASA requirements are 40 dB more stringent than the FCC requirement on these COTS radios. There were ten spurs above the −80 dB limits, as shown in Table 8.1, and two were of concern to the Russian partners because they had the potential to interfere with the Soyuz docking TV system.

Table 8.1. Transmitter A—20 kHz Bandwidth Carrier 435.275 at 124.6 dBμV

Spur Frequency	dBμV	CE06 Allowed	CE06 Over	RE02 Allowed	RE02 Over	Comments
326.4 ± 0.02	56.6	44.6	12.0	44	+12.6	Military UHF band
399.002 ± 0.03	66.1	44.6	+21.4	45	+21.1	Military UHF band
423.184 ± 0.01	60.8	44.6	+16.2	46	+14.8	Amateur radio band
447.366 ± 0.01	58.7	44.6	+14.1	47	+11.7	Amateur radio band
471.548 ± 0.03	64.6	44.6	+20.0	47	+17.6	Land mobile band
544.100 ± 0.02	56.6	44.6	+12.0	49	+7.6	TV broadcast band
870.550 ± 0.02	65.5	71.6	−6.1	NA	NA	OK; 2nd harmonic
1305.825 ± 0.03	79.3	71.6	+7.7	NA	NA	OK; 3rd harmonic
1741.100 ± 0.04	74.7	44.6	+30.1	58	+16.4	
2176.375 ± 0.05	72.8	44.6	+28.2	59	+13.8	"S" uplink band

Apparently, this system was also derived from COTS components, and was never tested for compatibility with these amateur radio frequencies.

These results were presented to the NASA electromagnetic engineering panel on 21 January 2004. With the exception of a couple of wording changes, the panel approved of the MISSE5/PCSAT-2 plan with the understanding that the PCSAT-2 transmitters would be turned off for all EVAs and for all Soyuz and Progress docking maneuvers and with the coordination of the ARISS community.

8.4.8 PCSAT-2 Lessons Learned

8.4.8.1 Human Safety

The human safety requirements for flying on the space shuttle and ISS were an order of magnitude more complex and stringent than the previous requirements for PCSAT-1. Planning, testing, and meeting those requirements took the majority of the effort by Naval Academy and NRL personnel. The requirement to shut down PCSAT-2 for human safety during every EVA and docking maneuver increased the necessary operations effort by an order of magnitude over PCSAT-1. The fact that PCSAT-2 was on the critical path for every ISS evolution was an inconvenience for both NASA and DOD coordinators as well as Naval Academy operators. NASA concluded that this kind of coordination would not be approved on future missions. Either the crew would have direct control over the transmitters, or the transmitters would not fly again.

8.4.8.2 Coordination

Although coordination between the Naval Academy operations team and the ARISS team on the ground was good, coordination of actual results on ISS were mixed. This is because the ARISS team could only effect changes on station via the crew, which took a significant administrative path. Similarly, PCSAT-2 commanding was not instantaneous either, as it was dependent on the timing of orbit passes over command stations in different parts of the world. In the future, modes of operations should be designed to be inherently independent of ARISS and crew operations.

8.4.8.3 Design Redundancy

PCSAT-2 consisted of dual independent redundant systems to ensure reliability. A few months into the mission, occasional unexplained resets were observed on the side B system, but not on side A. For this reason, side A was used exclusively for the rest of the mission. Side B remained on line and operational, but it experienced random resets approximately once a week. It's possible that a chip in the side B system was more vulnerable to radiation than the chip in side A and was experiencing more frequent single-event upsets. Long-term conclusions about the frequency of single-event upsets causing resets on side A could not be made because the system was typically shut down and rebooted every few weeks for EVAs and dockings.

8.4.9 Successes

The PCSAT-2 was an outstanding success. It delivered more than 3 Mb of data per day to the FTSCE team, provided needed uplink commanding of FTSCE, and supported APRS communications among users worldwide. It was fully commandable at all times.

8.5 ANDE

The Atmospheric Neutral Density Experiment (ANDE) was the third payload in the Naval Academy's APRS satellite series.[18] The ANDE launch opportunity again came by way of NRL, which was preparing to fly a series of 19 in. spherical satellites to precisely measure the orbital decay by means of laser tracking stations. These hollow spheres were to be passive, with no electronics onboard. When the students at the Naval Academy requested a ride, the response was a qualified "yes," provided that the experiment did not have any external antennas or solar panels that would impact the nearly perfect spherical shape.

The students responded with a clever design using an insulator between the two halves of the sphere so that the spacecraft shell itself could be used as the antenna (Fig. 8.13). The power system then was designed around a package of 112 size D lithium primary cells in four battery modules in the internal structure shown in Fig. 8.14 so that no external solar panels were required. The capacity of the battery system would provide a battery life to match the approximate 1.5 year

Fig. 8.13. The ANDE spacecraft during final tests at the Naval Research Labs in Washington, DC. This photo shows the final tweaks of the antenna matching network to match the sphere to the 145.825 MHz operating frequency.

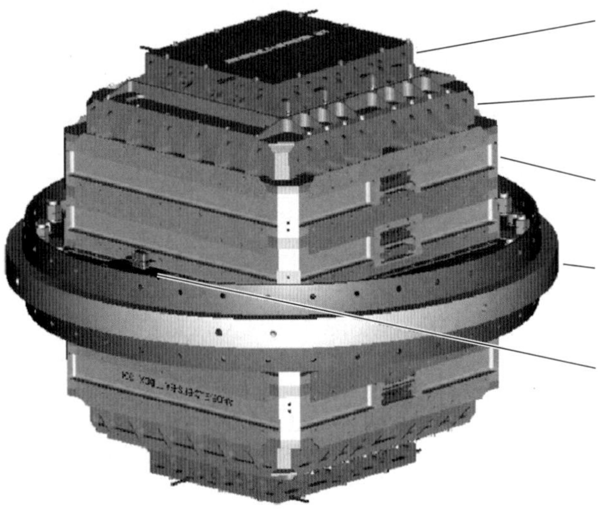

Fig. 8.14. The internal electronics and battery boxes were all structurally mounted to the center insulated antenna disk. This provided excellent whole orbit thermal isolation within 2 deg or less.

mission life. The combination of shell-antenna and primary batteries allowed the Naval Academy communications payload to fly on this spacecraft with no impact on the primary atmospheric drag mission. As a benefit, NRL gained some nominal telemetry on spacecraft temperature and attitude.

Similar to PCSAT-2, the primary science mission of ANDE was the space environment, so operation of the telemetry, command, and control system in the Amateur Satellite Service was possible while also providing the communications system to fulfill the APRS mission.

8.5.1 ANDE Science Requirements

Mechanical and mission design and construction was conducted at NRL in Washington, DC. The communication/electronics system was the responsibility of the Naval Academy. The overall ANDE project consisted of four satellites: the risk reduction sphere (ANDE-RR with Naval Academy electronics), the Fence calibration sphere (ANDE-FCL), the passive sphere (ANDE-PS), and the Active Sphere (ANDE-AS).[19] The ANDE-RR and ANDE-FCL spheres were flown on STS-116 in December 2006. The other two spheres would follow when a flight was available.

The surface of each spacecraft was processed to 99.8% of a perfect sphere. With the exception of ANDE-FCL, the exterior of each sphere was configured with a beach-ball pattern consisting of Aeroglaze A276 paint and black anodize Type 2 Class 2 coating. This pattern differs for each sphere, and forms one of the experiments of the mission known as the Polarization Return Experiment, designed to see if the attitude of the spheres could be determined optically.

For precise laser tracking, NRL designed corner-cube retroreflectors with a beam deviation of less than 5 arcsec evenly spaced around the sphere, taking into account received power, coherence effects caused by overlap of returns from multiple retroreflectors, and atmospheric loss. However, to reduce machining costs, NRL changed to a simpler pattern in three latitudes at 15, 55, and 75 deg with little compromise.

Cost was a severe constraint for the ANDE mission. The proposed parts cost of each microsatellite (with the exception of the ANDE-AS) was on the order of \$50,000 to \$100,000. Modularity of design allowed for commonality of parts and assemblies for the ANDE spheres. With this commonality, the main structures and instruments could be built early, thus leaving more time and resources for the custom components of the individual microsatellites.

For launch on the space shuttle, the new Canister for All Payload Ejections (CAPE) was developed by Muniz Engineering of Houston, Texas, under the auspices of the U.S. Air Force Space Test Program. This system was designed to hold two of the dense spheres without any hard mechanical mounting other than compression of two annular rings. The rings detach from the CAPE structure using the Lightband system by Planetary Systems Corp. of Silver Spring, Maryland. The CAPE system also launches the spheres and, if necessary, imparts a spin.

8.5.2 ANDE-RR Polarization Return Experiment

The Polarization Return Experiment was to be conducted using the High-Performance CO_2 Ladar Surveillance Sensor (HI-CLASS), which was integrated with the Advanced Electro-Optical System 3.67 m telescope at the Air Force Maui Optical and Supercomputing facility.[20] The HI-CLASS system would illuminate the ANDE-RR sphere with linearly polarized light at 11.5 μm. The reflected signal from the transmitted 12 J pulses would be picked up using a heterodyne receiver with two orthogonally oriented channels. The strengths between these two channels and their rate of change would be used to determine the orientation and rotation rates of the ANDE-RR sphere.[21] A plot of one set of laser returns is shown in Fig. 8.15.

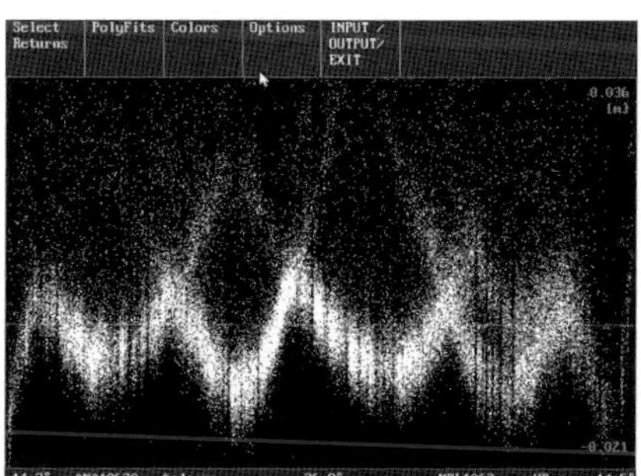

Fig. 8.15. Laser returns from ANDE-RR from the Gras, Austria, laser ranging site, suggesting a spin rate of about 0.2 rpm.

8.5.3 ANDE-RR Communications Experiment

Similar to PCSAT-2, to meet the spirit of the rules of the Amateur Satellite Service, an additional communications experiment was included as a secondary payload. This consisted of the APRS packet data transponder and a text-to-speech module for packet-to-speech communications. This mode not only provided a simple voice downlink for reception by students around the world using simple radio receivers, but it also minimized uplink congestion by the inherent shortness of the uplink packet. A single 1 sec packet on the uplink could contain a complete sentence that would take 10 sec or more on the downlink.

8.5.4 ANDE-RR Hardware Requirements

To minimize power drain, ANDE-RR was designed with two redundant single-port KPC-3+ TNCs instead of the dual-port KPC-9612+ TNCs used on the PCSATs. The KPC3+ TNCs have all the latest APRS generic digital repeating advantages as well as telemetry, command, and control, but they operate at half the power of the PCSAT units because they have only a single port. Again, by using stock TNC hardware and firmware, developers minimized on-orbit risk because of the track record of tens of thousands of identical hardware devices in use all across the country. Also, like the PCSATs, the ANDE-RR mission used COTS transmitters and receivers from Hamtronics. Although these RF boards appear to be 1980s vintage discrete components, they are the only COTS transmitters and receivers still available that use crystal control without any microprocessors in their frequency synthesizers. It was crucial to avoid any state-dependent circuits that could change the frequency of operation of the command and control system as a result of a single-event upset.

The block diagram of ANDE-RR is shown in Fig. 8.16. To save power, the entire communications payload was operated in a 90% sleep/wake duty cycle. The sleep/wake timers save power by keeping everything, including the receivers, essentially off when not in view of ground stations.

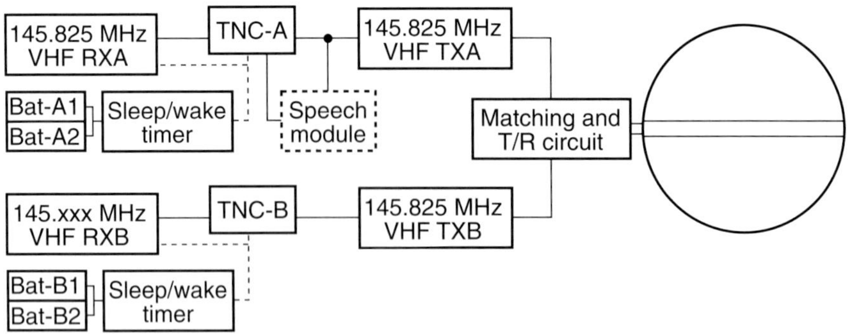

Fig. 8.16. ANDE's dual AX.25 command and control system transponders based on KPC-3+ TNCs can also be switched into modes to support a packet-to-voice experiment.

Both transmitters feed the same spacecraft shell antenna system via a system of half-wave and quarter-wave matching lines and PIN diode T/R arrangements, as shown in Fig. 8.17. The shell antenna system is actually more of a slot antenna formed by taking the 36 bolts that connect the two conducting hemispheres and making half of them conductors and half of them insulators and feeding the resulting insulated slot in the middle.

8.5.5 ANDE-RR Communications and Telemetry

The digital communications mission for ANDE-RR was nearly identical to that implemented in the PCSATs, except that ANDE-RR would not have solar power

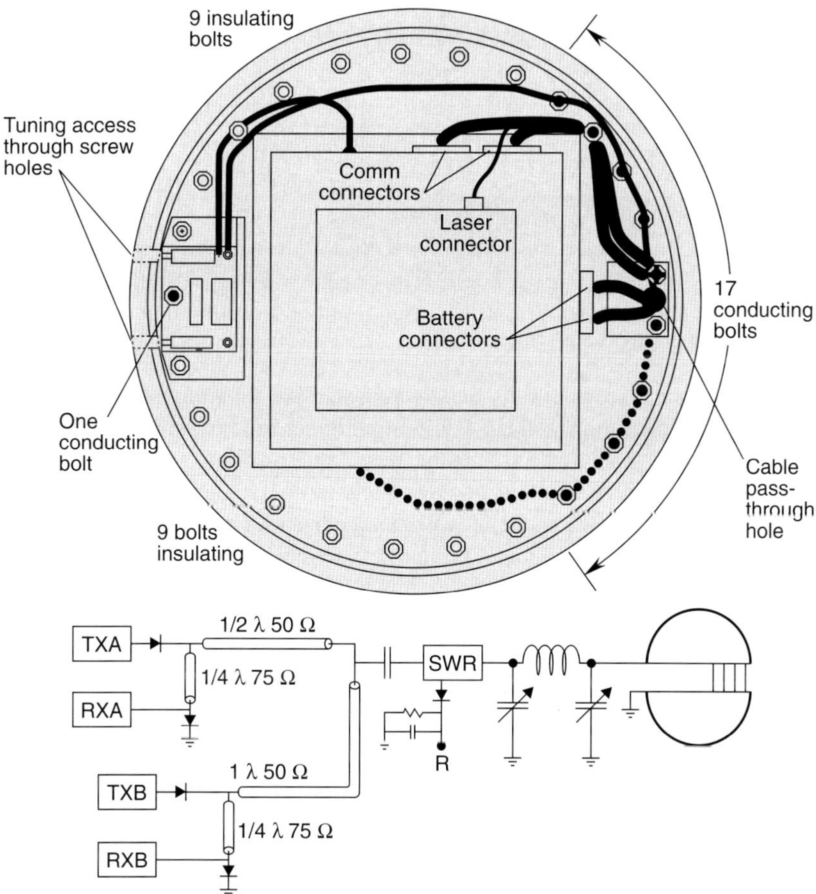

Fig. 8.17. The ANDE-RR system antenna was the spacecraft itself, fed as a half-wave slot in the insulator disk between the two hemispheres. The slot was formed by selecting the number of conducting and nonconducting bolts between the disks. Quarter-wave and halfwave lines switched with PIN diodes shared the antenna between the A and B systems.

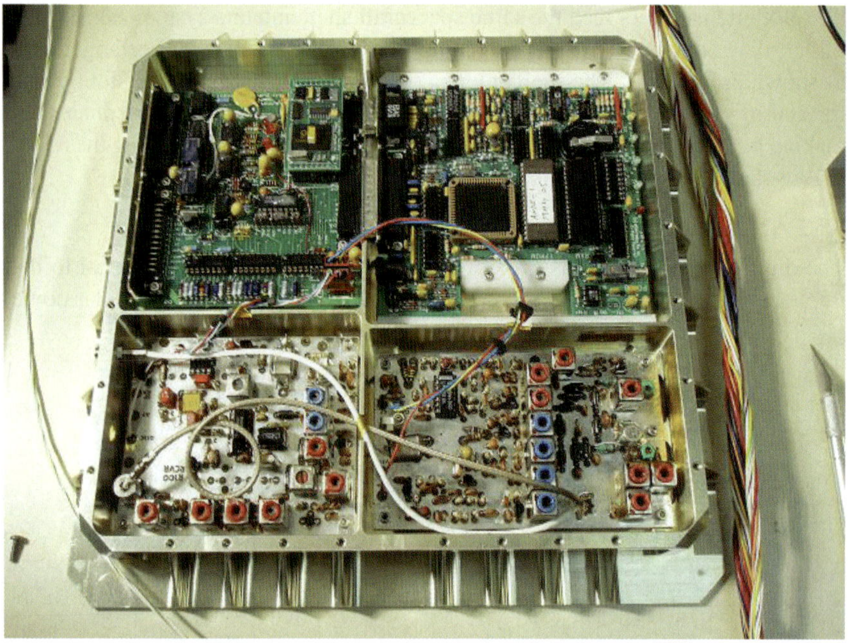

Fig. 8.18. The KPC-3+ TNC system with Hamtronics transmitter and receiver inside the ANDE communications tray.

and would have to conserve its battery power. Like previous Naval Academy payloads, the digital transponder would provide not only mission telemetry and command and control but also real-time messaging, position, and status relay via satellite to a worldwide Internet-linked amateur radio tracking system. The objective was to have all AX.25 satellites work together as a constellation to provide global connectivity in the Amateur Satellite Service. The complete side A communications system for ANDE-RR is shown in Fig. 8.18. The side B system is identical.

The stock Kantronics KPC-3+ TNCs had a serial communications port at 1200 baud as well as five analog telemetry channels and three on/off command bits. These features were sufficient to handle all of the telemetry, command, and control. The five analog channels were multiplexed in four groups for a total of 20 telemetry channels, similar to that shown for PCSAT-2 in Fig. 8.12. This multiplexing scheme requires receipt of four sequential telemetry packets to assemble a complete snapshot of spacecraft health and condition.

8.5.6 ANDE-RR Electrical Power System

The electrical power system for ANDE-RR used specially designed battery boxes that output a nominal 12 to 16 V dc. The 112 batteries were Tadiran-type TL5930 lithium D cells, which operate at a nominal 3.6 V and have a 19 Ahr lifetime.

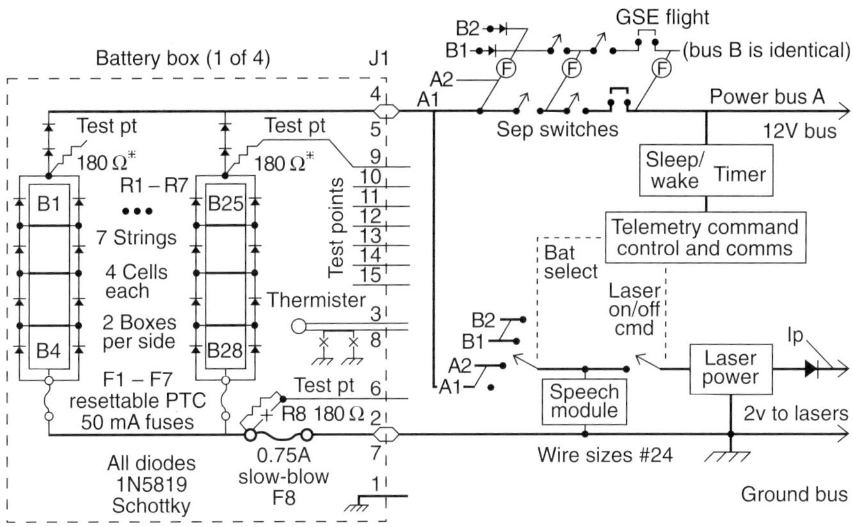

Fig. 8.19. ANDE battery system schematic showing one of the four parallel battery packs and the battery switching arrangement for the laser power system.

Each battery box was designed with an outer 6061-T6 aluminum housing that surrounded and protected the 28 individual cells mounted onto a printed circuit board. As shown in Fig. 8.19, each circuit board consisted of seven parallel strings of four cells in series (only two strings are shown). This arrangement provided the needed 12–16 V system voltage, and the seven parallel strings gave the needed power capacity and redundancy. The voltage is specified as 12–16 V because these high-capacity, long-life lithium cells have a relatively high internal resistance. Under no load, the series string is nominally about 16 V, but under system load peaks, this drops to around 12 V.

When assembled, all the cells were held in a slight uniform compression to transfer loads to the housing walls (Fig. 8.20). The modular design ensured that each battery module was interchangeable and could be stacked or mounted with L brackets. The square shape of the modules allowed them to be mounted in any direction, independent of an adjacent module. This robust mounting form also enabled the battery modules to act as mounting surfaces for the other electronics payloads.

To meet the safety requirements of the NASA battery group, two Schottky diodes were placed in parallel with each cell to ensure that the cell could not be driven more negative than the conduction threshold of the diodes. A reverse-driven cell could be an explosion hazard. Also, in series with each string were two diodes and a 50 mA resettable fuse designed to prevent reverse bias and overload. A single 0.75 A slow-blow fuse was mounted in series with the power return of the module to protect against catastrophic shorting during integration and test.

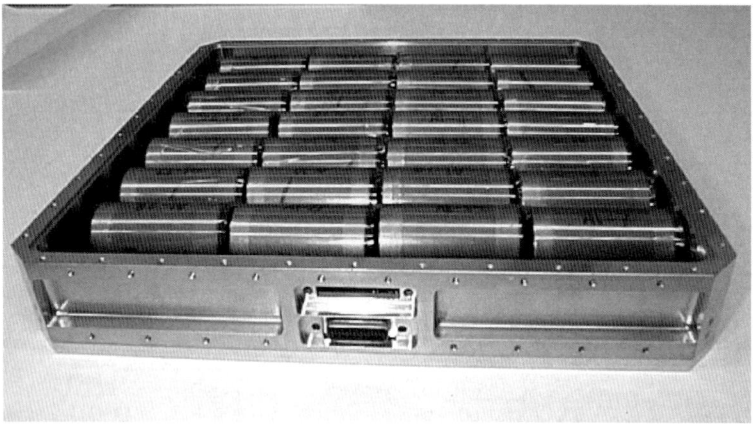

Fig. 8.20. ANDE lithium-thionyl-chloride battery module with top coverplate and compression sheet removed.

To space-qualify these cells, the midshipmen at the Naval Academy tested them against overload, reverse charging, and single-cell failure. The single-cell test involved inserting a spent cell in a four-cell string to determine if the current-limiting diodes would prevent current or at least limit the reverse current in the spent cell to safe levels. Environmental testing for leakage or damage was also performed in vacuum and at high temperature (80°C). Additional safeguards mandated by NASA were the addition of two vents with a Gore-Tex screen to allow outgassing while containing electrolyte leakage and the use of a Durafelt-gold electrolyte absorbent material to fill the space between the cells.

8.5.7 ANDE-RR Battery Life Management

Because the mission was designed to last about a year, and the load on the batteries could vary significantly depending on user load, there needed to be a means for monitoring the remaining capacity of the battery system during the mission. This is impossible for these lithium cells because their discharge curve is absolutely flat up to the 97% point, as shown in Fig. 8.21. So, depending on load, the voltage curves are too flat to make any meaningful indication of mission life remaining.

To overcome this problem, a power-management scheme was devised to force an additional 0.5 V drop from the second, third, and fourth battery packs. This would provide a 1.5 V taper from the initial life voltage, where mostly the first battery pack was carrying the load. As the mission progressed, the system battery voltage would indicate the additional 0.5 V long-term voltage drop as each battery pack was used up. This telemetry would allow the operations team to anticipate the end of life and thereby plan the user operating load to match.

Unfortunately, during integration testing, it was discovered that the relatively high internal resistance of these lithium primary cells resulted in so much internal voltage drop during laser operation, that this additional 1.5 V drop would not

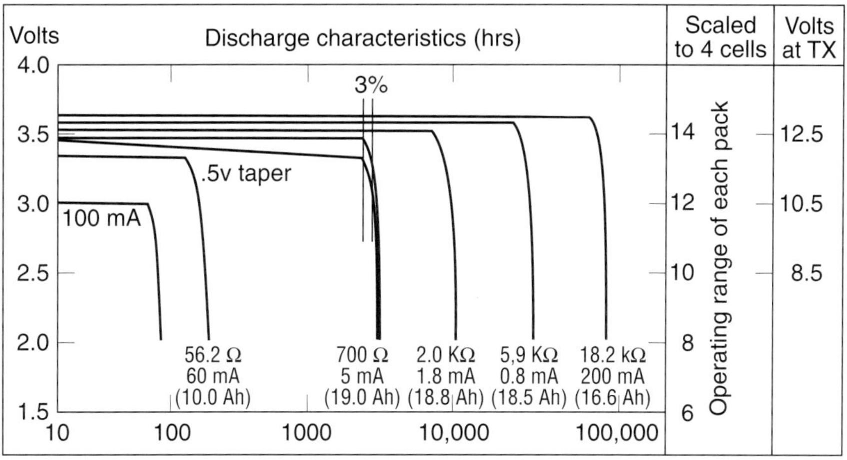

Fig. 8.21. Lithium primary cells' discharge curve is perfectly flat until 97% of life. A 0.5 V taper was accomplished by adding a Schottky diode drop for the B battery system. This provides some degree of battery life monitoring through the mission.

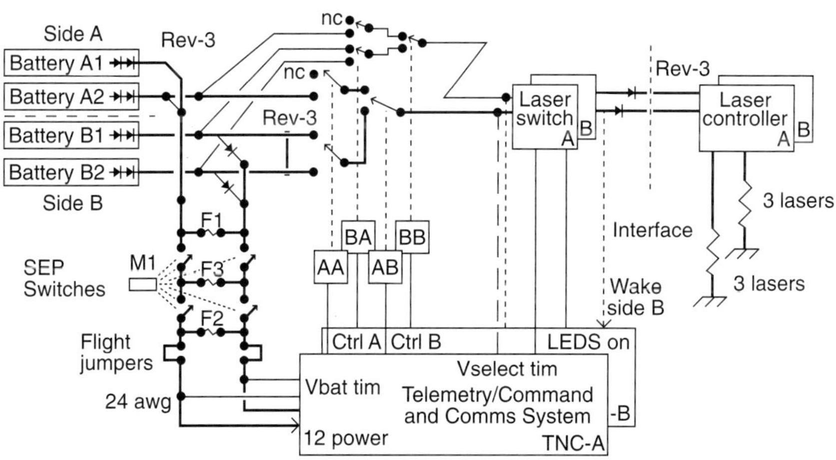

Fig. 8.22. The ANDE battery system consists of four battery packs of 28 cells each. These redundantly power the two spacecraft systems and transmitters, but they can be individually switched to power the lasers so that their performance can be separately evaluated through the life of the mission as individual battery packs fade.

leave ample system voltage to power the lasers. To solve this problem, a wiring harness modification was made to bypass three of these battery management diodes and to permanently connect battery packs 1 and 2 and battery packs 3 and 4 in parallel. The resulting two parallel systems retained only one 0.5 V drop for battery assessment and management. Figure 8.22 shows the final arrangement of the switching between the battery packs.

This figure also shows how the separation switches were redundantly connected in series-parallel and cross-connected to ensure redundant inhibit and/or activation in case of a switch failure. The relay system is also shown that allows independent selection of each of the four battery packs for laser operation. This was intended to also allow for battery assessment and management by controlling which battery was consumed first by the lasers. Most of this flexibility was lost because of the late integration wiring harness modification noted above that placed in parallel the four battery packs into only the A and B side systems to make sure there was enough current to run the lasers.

One additional revision was made in the wiring harness during those late integration tests. Because peak currents were required when both the A and B systems of lasers were activated, there was the potential near end of life for these high currents to drive the bus voltage so low that the laser driver modules would go into a kind of laser "fibrillation," rapidly flashing in synch but at extremely reduced power. To prevent this, the A and B side lasers were isolated so that they could be turned on one side at a time.

8.5.8 ANDE-RR Attitude Determination System

The basic ANDE sphere was required to have an attitude determination system in anticipation of the follow-on requirements of the MODRAS (modulating retroreflectors) experiment that would be flown on the ANDE-AS mission. Because the MODRAS experiment requires the sphere to spin, the CAPE system was designed to impart a 5 rpm spin upon ejection.

The ANDE-RR sphere used three methods for determining attitude and spin: the Polarization Return Experiment, the laser driver beacons, and the CIGS Photovoltaic Cells Experiment. Both the laser-driver experiment and the photovoltaic cells used identical Universal Mounting System form factors. The locations for

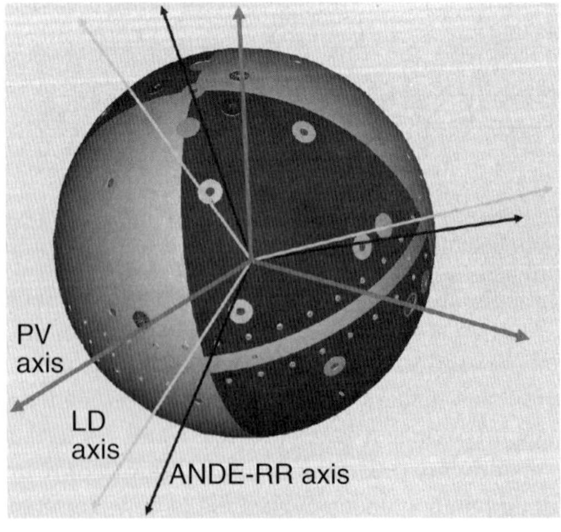

Fig. 8.23. ANDE-RR attitude determination system axis.

these plugs were on two sets of near orthogonal axes 90 deg ± 5 deg that are rotated with respect to each other and with respect to the sphere's geometrical axis, which is aligned most closely to the Polarization Return Experiment paint pattern as shown in Fig. 8.23.

8.5.9 ANDE-RR Thermal Considerations

The black and anodized paint scheme for ANDE-RR was designed to make the spacecraft visible to high-power telescopes, and the small laser reflectors were for tracking by the Maui laser tracking station. Originally, this coating scheme was black and white to maximize the contrast, but it was determined that the higher emission and lower absorption of the white surfaces would lead to a colder nominal spacecraft temperature. Late in mission design, the white areas were changed to an anodized finish that had a higher solar absorption and lower emission to provide a warmer spacecraft. These changes were prudent, as the resulting spacecraft on-orbit temperatures were excellent, as shown in Fig. 8.24.

The 90-min orbital variation in temperature was on the order of 15°C on the shell, but only a degree or so on the internal electronics stack. The average temperature of the stack then varied according to the overall sun beta angle as the orbit precessed through the mission. The minimum and maximum shell temperatures (after deployment) were between about 4°C and 42°C. The core temperatures (after deployment) were between about 15°C and 37°C.

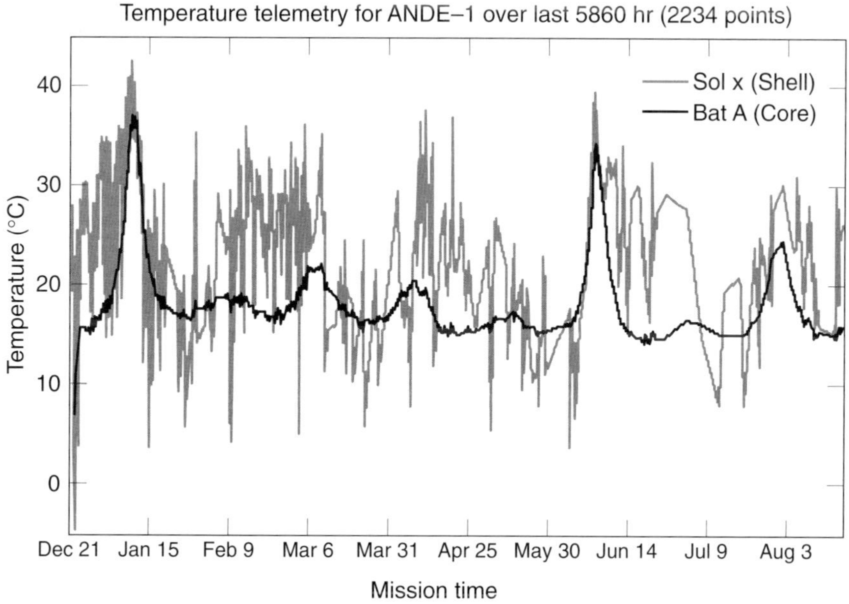

Fig. 8.24. Ten months of ANDE-RR temperatures show the orbit-by-orbit temperature fluctuation of the shell and the relatively stable core temperature that only drifted with long-term orbit sun beta angle.

8.5.10 ANDE-RR Failsafe Reset Mechanism

To save power and provide immunity from a single-event upset or other lockup condition in space, ANDE-RR was designed to operate with a simple 555 integrated circuit timer to power-up and cold-boot the TNC and receiver every 16 sec to check for channel activity. If no ground station activity was detected, it would return to sleep in 1.5 sec. If activity was detected, it would remain awake for the next minute to support communications. At the next loss of signal and after 1 min of no activity, it would go back into the power-saving on/off cycle. Not only did this save power, but it also ensured a cold boot on every cycle, keeping the system immune from latchup and software lockup.

8.5.11 ANDE-RR Power and Link Budget

Similar to the PCSATs, ANDE-RR took advantage of the 9 dB link budget advantage of VHF and several dB of transmitter power advantage stemming from the average 10% transmit duty-cycle requirements of the APRS mission. Although each user packet consumed as much as 8 W peak for one second, the average transmitter power when in use was less than a watt. Also, the receiver benefited from a 90% power saving sleep cycle, yielding a combined overall power system average power of only 0.8 W. This low average power allowed the yearlong mission life on primary batteries, including the secondary communications mission in the Amateur Satellite Service. The design power budget for ANDE-RR is summarized in Table 8.2.

8.5.12 ANDE-RR Lessons Learned

8.5.12.1 Design

The use of the spaceframe as an antenna worked well, and was within about 6 dB of expectations for what a full-size external antenna would have provided. This 6 dB, however, did reduce access times, typically requiring elevation angles of 10 deg or so for effective commanding.

8.5.12.2 Batteries

The primary lithium–thionyl-chloride batteries had a surprisingly high internal resistance. It took the parallel combination of 28 strings of these high-capacity D-cell batteries to meet the worst-case current demands (all communications and lasers on). One string of NiCd batteries would have easily handled the 3 A load. Batteries with such high internal impedance were beyond the experience of everyone—from designers to testers and inspectors. Despite the hazards of lithium batteries, there were no testing accidents, integration anomalies, or flight issues. The rigorous testing confirmed the cells were flightworthy; however, two incidents bear mention.

First, a student soldered the first row of cells in the battery pack backwards. Noticing that they never seemed to cool down after soldering, the student soon realized the mistake. Being backwards, the reverse-protection diodes provided a

Table 8.2. ANDE Power System Energy Budget

Battery system:	Lithium-primary
Battery mass/volume:	11.2 kg/7562 cm^3
Battery construction:	112 "D" size cells—Tadiran TL-5930
Battery configuration:	28 4-cell strings in parallel
Battery volts:	12 V (14.4 V no load)
Battery capacity:	532 Ahr (7448 Whr at 14 V)

System	Nominal Current	Duty Cycle Standby	10% On Time w/ users	Result Usage	Average Current	Percent
Receiver A	35 mA	10% (1.5 sec on)	10%	20%	7 mA	14%
TNC A	30 mA	10% (15 sec off)	10%	20%	6 mA	12%
XMTR A	600 mA		20%	2%	12 mA	25%
Lasers	350 mA	3% over Maui	10%	0.3%	1 mA	2%
Timers	3 mA	100%	100%	100%	3 mA	6%
Volt TLM	0.2 mA	100% ea (2)	100%	100%	0.4 mA	1%
Temp TLM	0.7 mA	10% ea (6)	100%	20%	0.4 mA	1%
5 V ref	0.9 mA	10%	100%	10%	0.1 mA	0%
Counter	1.2 mA	100%	100%	100%	1.2 mA	2%
Totals from Side B:					18 mA	
Total Spacecraft:					50 mA	

Mission life: 532 Ahr/50 mA = 10,000 hr = about 420 days

direct short across the cells. But, the inherently high internal resistance and low current capability of these cells prevented either burn-out of the diodes or any overheating of the cells or any components. This served as a good demonstration of the inherent safety of these cells.

Second, because the formal testing of cells had induced no casualties, the students took a set of cells and heated them well beyond specs, without catastrophic failures. After the test, the batteries were placed in a box and forgotten. A month later, the cabinet was opened to find all metallic parts throughout the cabinet were corroded by the slow escape of gasses from these cells. Yes, these cells are toxic and dangerous!

8.5.12.3 Testing

Never leave flight hardware in a thermal chamber—even a dual redundant thermal chamber—without an observer present. The morning of initial delivery of flight hardware began with all flight units burned beyond recognition during a final overnight bake-out at 60°C. Turns out the dual thermostats failed and temperatures went well beyond several hundreds of degrees. Fortunately, spare parts and COTS hardware allowed a completely new build in about one month.

8.5.13 Successes

PCSAT-style communications, command, and control were again completely successful. The sleep/wake system conserved battery power to the end of orbit life (about 12 months). Data collection by the worldwide system of ground stations was also successful. The primary lithium batteries performed to design expectations.

8.6 RAFT

The advent of CubeSats, microsats, picosats, and other small satellites launched in clusters has brought a new problem: The large number and small size of these satellites makes it difficult to identify the specific units in orbit. The RAFT (Radar Fence Transponder) mission was conceived as a Naval Academy project to test the ability of the Space Surveillance System radar to track small CubeSats.[24] The plan was to launch a pair of small CubeSats, one heavier than the other, so that one would drift to the front of a pack of CubeSats and the other would drift behind. One would have an active transponder, and the other would have an augmented passive radar cross section. Comparing the tracking and identification of the two objects would provide useful data on the ability of the Space Surveillance System radar to track such small objects.

8.6.1 The Space Surveillance Radar Fence

The Space Surveillance System is a network of radar transmitting and receiving stations across the southern United States operating around 216.98 MHz. The transmitting sites generate a continuous-wave fan beam, called "the Fence," that is very narrow in the north/south direction but extends straight up and stretches from the east to west horizons.[25] Any object that crosses the Fence will generate a brief echo that will be detected by the receiving stations.[†]

The Space Surveillance System includes three transmitting stations and six receiving stations, as shown in Fig. 8.25. The main transmitting station on 216.980

[†]Students and experimenters across the southern United States can hear these radar pings using simple antennas and a linear single-sideband receiver tuned to 216.98 MHz. The Naval Academy in Maryland assembled a simple 4 sq ft fence wire reflector and dipole and was able to hear the tone of the moon passing through the Fence. Unlike the 1 sec or so duration of a satellite passage, the moon takes about 2 min, giving ample time to tune in the signal.

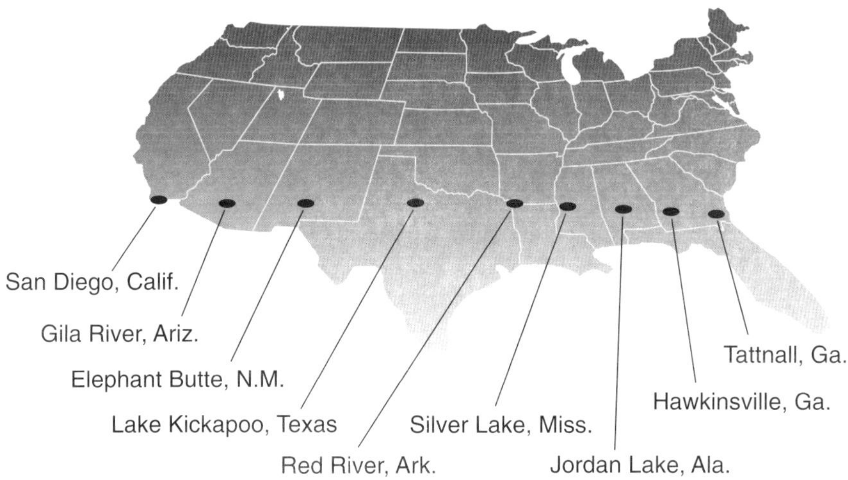

Fig. 8.25. Space Surveillance System transmitting and receiving stations.

MHz in Lake Kickapoo, Texas, consists of a linear array of 2556 inverted-V antenna elements, each with its own 300 W power amplifier, over a reflector screen. The total effective isotropic radiated power is greater than 6000 MW. The other two transmitting sites in Arizona and Alabama operate at lower power and at 216.970 MHz and 216.990 MHz, respectively, to fill in the east and west horizon of the Fence. The six receiving stations in California, New Mexico, Arkansas, Mississippi, and Georgia each have an interferometer antenna array and use a variety of signal processing techniques, including measuring the Doppler shift, to extract the orbital vectors from the received echo.[26,27] This data is fed to NORAD, which generates the Keplerian elements needed to track satellites orbiting Earth.

8.6.2 RAFT Spacecraft Design

The two small test satellites, named RAFT and MARScom, were deployed from the space shuttle mission STS-116 in December 2006. RAFT had the active radar transponder, and MARScom was passive. Their ground control terminals consisted simply of a laptop and TH-D7 walkie-talkie.

RAFT included a 217 MHz radar transponder, a VHF downlink, and HF and VHF uplinks. The Space Surveillance System radar experiment was the primary mission. RAFT also carried an amateur radio APRS transponder payload with a text-to-voice synthesizer experiment as secondary missions. MARScom also had a secondary communications payload for the Military Affiliate Radio System, but it was unsuccessful because the HF wire antenna failed to deploy properly.

The RAFT block diagram is shown in Fig. 8.26. The radar experiment included a direct-conversion receiver and a beacon oscillator operating at a frequency of 216.98 MHz. Output from the 216.98 MHz receiver was fed into the VHF FM downlink transmitter operating at 145.825 MHz so that ground stations could hear

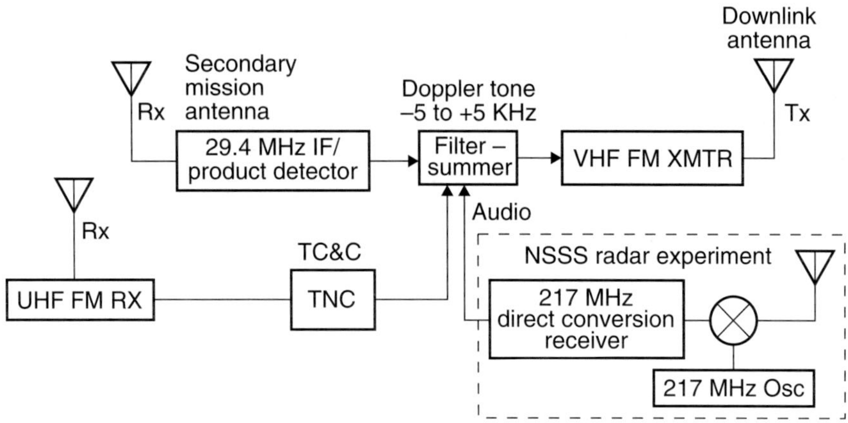

Fig. 8.26. RAFT satellite block diagram.

the actual radar Doppler signal as received on the satellite. The beacon oscillator provided a signal to help the radar ground receivers positively identify the satellite.

8.6.3 RAFT Command and Telemetry

The communications, telemetry, command, and control system of the RAFT CubeSats was very similar to the previous Naval Academy designs of PCSAT-1, PCSAT-2, and ANDE—and like ANDE, it used a single KPC-3+ TNC for all command, control, and communications. Surrounding the central TNC and modem/RF modules were several simple interface circuits to control the spacecraft functions and provide telemetry. On RAFT, each payload on/off command was designed with a backup 555 timer chip as a failsafe-off one-shot flip-flop mechanism. The simplicity of these circuits is shown in Fig. 8.27.

Unlike PCSAT-2 and ANDE, which had 20 telemetry channels, RAFT used only the five channels for battery voltage, solar current, battery current, load current, and temperature. The temperature channel was shared among three different thermisters, which were switched into the circuit depending on the mode of the spacecraft.

8.6.4 RAFT Antennas

The design of the spring-rod antennas for RAFT enabled a deployed antenna length of 10.5 in. from a 5 in. satellite by embedding the tip of one whip in the body of the other satellite. This conserved space and also provided the force for separating the spacecraft and giving them enough momentum to unspool the two 4 ft HF wire antennas. The pair of antennas on each CubeSat provided equal separation forces on the four corners of the spacecraft. Further, the inductance of the long springs on the tips of the 10.5 in. whips served to make them electrically longer for better matching at 145.825 MHz. The lower springs were insulated from

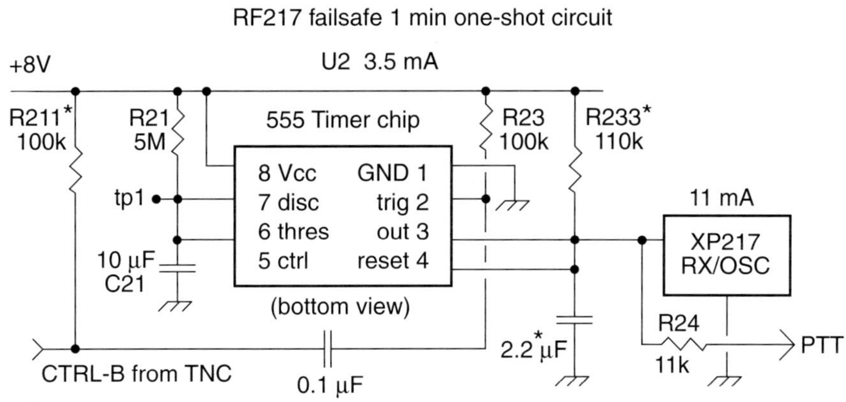

Fig. 8.27. Several simple circuits interfaced the standard TNC I/O lines to perform the RAFT mission. Here the CTRL-B command bit is used to trigger the XP-217 radar transponder on for one minute. A 555 timer assures that it will fail-safe off, because while it is on, the command link is blocked by the transmitter.

the antenna rods, but a sliding contact point ensured contact of the rod assembly to the antenna coax connection when fully deployed.

8.6.5 RAFT Structure

The photo in Fig. 8.28 shows how the KPC-3+ TNC is wedged on the diagonal along with the PSK-10 transceiver board, both sandwiching the HF long-wire deployment spool. The tiny HF whip tip magnet is visible in the exact center of the assembly. The two diagonal VHF whip antennas protrude upwards in this photo, and the opposite corners contain the brass tubes to receive the antennas from the other inverted CubeSat when the two are stacked. The antennas of each CubeSat nest into the brass tubes of the other. The right and left panels support the transmitter and receiver circuit boards.

The trapezoidal battery boxes are visible in this photo on the other two sides. On two faces of these battery boxes, several pieces of lead were added to increase the mass of the RAFT satellite as much as possible to differentiate its orbital position from that of the MARScom satellite. This added mass also helped extend the orbit lifetime by about 15%.

8.6.6 RAFT Transponder

RAFT has the same type of VHF FM receiver and transmitter on the space APRS frequency of 145.825 MHz and an AX.25 TNC similar to the other Naval Academy APRS satellites. RAFT also includes a secondary experimental PSK-31 receiver with a multiuser uplink on 28.121 MHz, similar to that flown on PCSAT-2. The PSK-31 signals are combined with the audio output of the TNC and fed into the VHF FM transmitter.

Fig. 8.28. RAFT carried the transmitter and receiver boards on two side panels, the two battery boxes on the other side panels, and the PSK-31 and TNC on a diagonal panel. The side panels dovetail each other, and only the top and bottom were assembled with #4-40 fasteners. The antennas and mating sleeves mount in the four corners.

While most of the electronic subsystems for RAFT were assembled from stock components, the XP-217 radar transponder was designed and constructed by a team of AMSAT volunteers with the hope of providing a design suitable for future spacecraft.† The complete transponder uses only four transistors, an off-the-shelf oscillator module, and a single IC voltage regulator, keeping the circuit simple and inexpensive. The audio output of the receiver is the Doppler tone of the passage of the spacecraft through the Space Surveillance System radar beam.

Although the primary circuit of the XP-217 is a receiver, it is called a transponder because it was also designed to have a 4 mW simultaneous transmit emission on the same frequency. With the high power of the Space Surveillance System radar, the receiver did not need to be particularly sensitive, but it did need to tolerate a high dynamic range of RF. The XP-217 receiver used a direct-conversion approach, with the local-oscillator operating at the same 216.980 MHz frequency as the Space Surveillance System radar transmitter. The recovered audio output of the XP-217 receiver, which was offset from the local oscillator by the Doppler

†Three AMSAT members in the Boston area, David Goncalves (W1EUJ), Joe Fitzgerald (KM1P), and Anthony Monteiro (AA2TX), designed and built the 217 MHz radar transponder unit. Additionally, Tom Kneisel (K4GFG) helped the team understand the operation and requirements of the Space Surveillance System radar.

shift of the satellite's crossing geometry, was then transmitted to the ground by RAFT's 2 m FM transmitter with a bandwidth of about 3 kHz.

Because the XP-217's transmit power only needed to be around 4 mW, the circuit was designed to allow its local oscillator to have enough leakage to be radiated. To account for antenna gain and internal losses, the XP-217 transmitter was designed to provide +7 dBm_0 output.

8.6.7 RAFT Transponder Link Budget

The antenna element for the XP-217 was a dedicated 10 in. whip protruding from one side of the satellite. An EZNEC model of this antenna showed an impedance of around 10.5+ j293 Ω at 217 MHz and a gain of about –1 dB.[28]

To calculate the link budget for the XP-217 transponder, the following assumptions were made: The RAFT circular orbit had an altitude of about 360 km. The Space Surveillance System radar transmits a continuous-wave signal, with a power density of $-7.5 dBm_0/m^2$ at this altitude. An isotropic antenna at this frequency has an effective aperture of 0.15 m^2, so the maximum received signal power level at the satellite is –15.7 dBm_0 for an overhead pass. At the horizon, the path loss would be an additional –15.6 dB, and the transmitting antenna pattern would be down by –13 dB. Adding in a polarization mismatch loss of –3 dB would yield signal levels in the range of about –17 dBm_0 to –50 dBm_0.

Depending on the satellite pass geometry, the Doppler shift on the received signal could be ±5 kHz. Therefore, the XP-217 local oscillator had to be less than ±5 kHz from the 216.980 MHz center frequency. This represents a tolerance requirement of 25 parts per million. Additionally, the XP-217 local oscillator frequency needed to be predictable to allow accurate Doppler-shift measurements. A temperature sensor was provided on the XP-217 crystal so that the frequency-vs-temperature curve could be used to accurately determine the oscillator's frequency on orbit.

8.6.8 RAFT Signal Rejection

The XP-217 receiver needed sufficient isolation and filtering of the 1 W 145.825 MHz downlink transmitter. The worst-case power transfer between the downlink antenna and the receive antenna was on the order of –30 dB, leaving about 0 dBm of 145.825 MHz energy into the receiver. The XP-217 receiver was coupled to its antenna through a diplexer with a 20 dB rejection of the 145 MHz transmit signal. Combining these loses required the XP-217 receiver itself to tolerate about –20 dBm residual of the 145 MHz transmitter.

8.6.9 RAFT Transponder Architecture

The resulting XP-217 circuit consists of six major subsystems: a voltage regulator, a local oscillator, a power amplifier, an RF filter, an infinite-Z mixer, and an audio amplifier, as shown in Fig. 8.29. Because the XP-217 needed to both transmit and receive at the same time, an infinite-Z mixer was connected in parallel with the power amplifier so as not to load down either the transmit or receive signals.

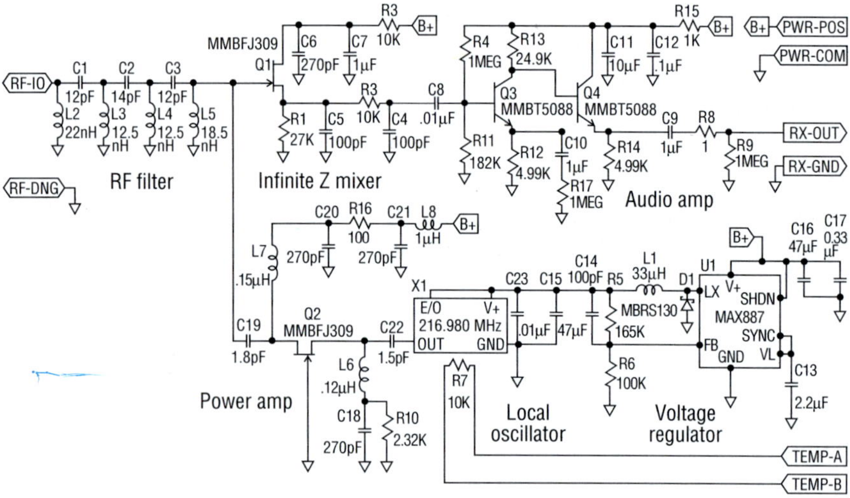

Fig. 8.29. XP217 Radar Fence transponder circuit schematic.

The MAX887 voltage regulator provides 3.3 V dc at up to 30 mA for the local oscillator. The local oscillator uses a Fox Electronics Just-in-Time-Oscillator module (JITO-2) operating at 216.980 MHz over a temperature range of –20°C to +70°C with a frequency tolerance of ±20 parts per million. The oscillator provides an output of 3 V peak-to-peak at a maximum current of ±2 mA and drives the single transistor grounded-source power amplifier Q2. R10 sets the output level. It has a 2 kΩ input impedance and 50 Ω output that brings the signal up to +8 dBm_0. The bidirectional 50 Ω RF filter provides nearly 40 dB rejection at 145 MHz to keep the downlink transmit signal from interfering with the receiver. The insertion loss is about 1 dB.

For receive, the same RF filter provides rejection of the 145 MHz transmitter, and the infinite-Z mixer mixes the received signal with the transmitter to produce an audio difference tone equal to the radar Doppler shift seen by the satellite. The low-noise audio stage has an adjustable gain via R17. With R17 open, the gain was about 4, providing about 100 mV rms output into a 1.5 kΩ load at the maximum receive signal level. The voltage divider, R8 and R9, set the output level as required by the downlink transmitter audio input.

8.6.10 RAFT Transponder Results

The output of the receiver is audio that contains the Doppler tone of the passage of the spacecraft through the Fence. Although the design assumed that only a 1 sec tone would be detected during the brief transition through the 0.2 deg fan beam, this assumption overlooked the sidelobes of the 6000 MW radar transmitter, which do produce some energy from horizon to horizon. A Doppler plot of the recording made by student Matt Bennett at University of California-Irvine is shown in Fig. 8.30. This data was taken four days before deorbit, and by chance,

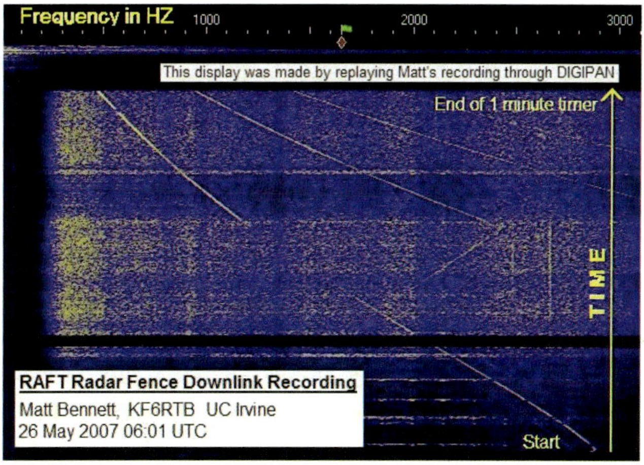

Fig. 8.30. Doppler plot of the XP-217 downlink showing the radar signal recorded over Southern California as RAFT approached the Space Surveillance System Fence during its final days on orbit. The second, third, and fourth harmonics are also visible.

Bennett turned on the transponder early. It turns out that the XP-217 receiver was sensitive enough to detect a Doppler, as long as the spacecraft was above the horizon to the Space Surveillance System radar transmitter as shown here. The Doppler is seen to be increasing in rate toward the upper left, but also getting stronger as it approaches the main lobe of the radar beam. Unfortunately, the 1 min failsafe timer ended at that point, but it did show that the XP-217 receiver was performing as designed.

8.6.11 RAFT Lessons Learned

8.6.11.1 Preflight Mating

A deployment anomaly resulted in an 80 rpm spin on MARScom and 40 rpm on RAFT. This was probably caused by the six-month preflight mating of the two spacecraft and the tip of one antenna spring perhaps being compressed all that time into a tiny amount of cured epoxy on the other spacecraft. Scotchweld 2216 B/A two-part epoxy remains slightly compressible after curing, and this may have captured the spring under that long duration pressure. The spin was created when the tip of this one antenna stuck to the other spacecraft during their separation. This high rate of spin probably prevented full deployment of the MARScom antenna and, therefore, its downlink. MARScom was never heard after deployment, although it did fulfill its primary mission, which was to be a passive reflector for the radar Fence beam. Similarly, the matching HF receive antenna for PSK-31 on RAFT probably did not deploy, and this prevented this secondary experiment from hearing any PSK-31 uplinks. These high spins had damped out completely by three months.

8.6.12 RAFT Successes

RAFT was successful in all primary missions and was used by students and experimenters around the world during its brief five-month mission.

8.7 Conclusion

The incentive for these five Naval Academy APRS educational satellite missions was the consideration that "everything is going wireless," and students need hands-on experience with wireless technology—including satellites, with their global distribution of users and applications. Amateur walkie-talkies and mobile radios can serve as inexpensive user terminals, and their hardware components can work effectively in space as well at very low cost compared with conventional materials. These inexpensive ground terminals give ham radio operators an educational, experimental, and emergency satellite communications capability in the palm of their hands or on their dashboards.

This APRS global digital communications capability can be a major driver for future amateur, educational, and emergency response missions. The Naval Academy satellites, PCSAT-1, PCSAT-2, ANDE, RAFT, and MARScom, have fulfilled this mission objective and will hopefully encourage other schools to contribute to this mission. Other schools are invited to either build additional satellites to join the constellation on 145.825 MHz, or to build permanent ground stations to contribute to global connectivity, or to develop remote environmental sensors or other communications devices and projects to inspire their students.

By encouraging these digital transponders as auxiliary payloads on many small satellites, the Amateur Satellite Service can bring all of these pieces together into the most powerful and far-reaching educational project to date. Student projects and educational institutions can easily contribute to this capability while also serving their own needs of viable payloads and ongoing operations training. Each university project contributes to a bigger mission, engendering a spirit of community among student experimenters.[29]

8.8 References

[1] R. Bruninga, "The APRS Satellite Mission and CubeSat Design Notes," *Proceedings of the AIAA/Utah State University Small Satellite Conference* (Logan, UT, 15 Aug. 2001).

[2] APRS web page, www.usna.edu/Users/aero/bruninga/aprs.html.

[3] ASTARS web page, www.usna.edu/Users/aero/bruninga/astars.html.

[4] TRAKNET web page, www.usna.edu/Users/aero/bruninga/traknet.html.

[5] MIR Packet Radio Test #2, www.usna.edu/Users/aero/bruninga/mirex.html.

[6] MIR Packet Radio Test, www.usna.edu/Users/aero/bruninga/mir-ht.txt.

[7] SAREX Packet Tests, www.usna.edu/Users/aero/bruninga/sarex.html.

[8] R. Bruninga, R. Waller, "Using GO-32 for Mobile APRS Satellite Operations," *Proceedings of the 2006 Annual AMSAT Space Symposium* (Pittsburg, PA, 26–28 Oct. 2007).

[9] R. Ross, "Return to Antarctica: A digital Satellite Journal," *The AMSAT Journal* (March/April 2000).

[10] PCSAT-1 web page, www.usna.edu/Users/aero/bruninga/pcsat.html.

[11] R. Bruninga, LtCol B.R.Smith, "PCsat and Follow-On Payloads," *Proceedings of the 16th Annual Small Satellite Conference* (AIAA/Utah State University, Logan, UT, 12–15 Aug. 2002).

[12] R. Bruninga, "The PCsat Mission," *Proceedings of the 19th AMSAT Space Symposium* (Decatur, GA, 5–8 Oct. 2001).

[13] "GPS Operations on the PCsat Microsatellite," *ION GPS 2002* (Portland, OR, 24–27 Sept. 2002).

[14] ITU Rules for the Amateur Satellite Service, www.iaru.org/satellite/prospective.html.

[15] Naval Academy ITU rules accommodation, www.usna.edu/Users/aero/bruninga/pec/pec-itu.txt.

[16] PCSAT-2 web page, www.usna.edu/Users/aero/bruninga/pcsat2.html.

[17] PSK-31 web page, www.usna.edu/Users/aero/bruninga/psk31uplink.html.

[18] ANDE web page, www.usna.edu/Users/aero/bruninga/ande.html.

[19] A. C. Nicholas, S. E. Thonnard, I. Galysh, P. Kalmanson, B. Bruninga, H. Kelley, S. Ritterhouse, J. Engelhardt, K. Doherty, J. McGuire, D. Niemi, H. Heidt, M. R. Hallada, D. Dayton, L.Ulibarri, R. Hill, M. Gaddis, and B. Cockreham, "An Overview of the ANDE Risk Reduction Flight," *Proceedings of the 2003 AMOS Technical Conference* (Maui, HI, Sept. 2002).

[20] R. Pohle, M. Kovacs, K. Ayers, "HI-CLASS/AEOS Optical System—A Versatile Tool For Space Surveillance and BMDO Applications," *Proceedings of the 2001 AMOS Technical Conference* (Maui, HI, Sept. 2001).

[21] D. Dayton, M. Elowitz, L. Ulibarri, "Using Laser Radar Polarization Signatures for Determining Properties of a Spinning Sphere," *SPIE Conference# 5158 Polarization Science and Remote Sensing,* paper #5158-25 (San Diego, CA, Aug. 2003).

[22] R. Bruninga, A. Monteiro, "An Experimental Space Surveillance Radar Transponder for the RAFT CubeSat," *Proceedings of the AMSAT 22nd Space Symposium* (Arlington, VA, 8–10 Oct., 2004).

[23] RAFT Satellite web site, www.usna.edu/Users/aero/bruninga/raft.html.

[24] RAFT operations page, www.usna.edu/Users/aero/bruninga/ande-raft-ops.html.

[25] Steven L. Berg, "Theoretical Radiation Patterns of NAVSPASUR Transmitter Antennas" (Interferometrics, Inc., 30 Nov. 1988).

[26] Tom Kneisel's NAVSPASUR web site, www.k4gfg.us/navspasur/index.html.

[27] P. A. Crossley's NAVSPASUR web site, www.jump.net/~crossley/NAVSPASUR.

[28] EZNEC Antenna Modeling Software, v. 2.0, by Roy Lewallen, www.eznec.com.

[29] R. Bruninga, "AMSAT Builders Builder's Channels for HT and Mobile Satellite Communications," *Proceedings of the 17th Annual AMSAT Symposium,* pp 191–195 (San Diego, CA, 8-10 Oct. 1999).

Chapter 9

The Naval Postgraduate School Small Satellites Program

Dan Sakoda,* Jim Horning,* and Rudy Panholzer*

9.1 Introduction

The Naval Postgraduate School's Space Systems Academic Group was formed in 1982 to develop a curriculum in space systems operations. At that time, it was a little-known fact that the United States Navy was the largest user of space. The space systems program now includes courses in both space systems operations and space systems engineering and supports men and women from all services in the Department of Defense (DOD). This chapter will discuss the Naval Postgraduate School space systems curricula and the small satellite program.

9.2 Space Systems Curricula

The space systems operations curriculum provides a broad technical introduction to the many issues related to a space system as well as topics important for Navy technical managers. Students study the technical aspects of space, such as orbital mechanics, launch vehicle systems, the space radiation environment, and spacecraft design while focusing on the operational aspects of space systems. Core courses also include studies in information technology, operations analysis, and acquisition management. In addition to the traditional coursework, students spend two weeks visiting a federal, academic, or commercial space program. Graduates receive a Master of Science in Space Systems Operations.

Students in the space systems engineering curriculum pursue one of four primary disciplines: electrical engineering, physics, mechanical and astronautical engineering, and computer science. The core courses cover astronautics, electrical and computer engineering, math, physics, management, and some specialized topics such as remote sensing and military applications of space. In addition, students elect specialization courses toward their respective Master of Science degree. The space systems engineering curriculum also requires a two-week tour, similar to that of the space systems operations curriculum.

A common theme in both programs is the study of a broad scope of disciplines important to the military space user and graduate-level thesis research. Each curriculum has a capstone course that requires students to work in teams to produce a conceptual design of a spacecraft. The project topic is often one of

*Space Systems Academic Group, Naval Postgraduate School
Monterey, California

interest to a Navy or other DOD sponsor, and the level of detail researched is similar to what a vendor would supply in response to a request for proposal. The end product is an educational experience not only for the students involved, but for the sponsor as well, who usually attains a better understanding of the system requirements. Students also have the opportunity to work on an actual space vehicle through the Small Satellite Design Program, which marries the educational objectives of the Naval Postgraduate School with militarily relevant research into spacecraft technology and missions amenable to small satellites.

9.3 University and Defense Efforts in Small Satellites

The watershed event that brought small satellites to the educational forefront was the launch of the Northern Utah Satellite (NUSAT), built jointly by Utah State University in Logan, Utah, and Weber State College in Ogden, Utah. NUSAT was launched in 1985 aboard the STS-51B mission using the shuttle's Get-Away Special (GAS) canister; the ejection mechanism was a spring-loaded plunger held in place by a marmon clamp with pyrotechnic bolt cutters. NUSAT was a tumbling spacecraft approximately 48 cm in diameter and 63.5 kg used for radar measurements of air-traffic-control radar beacons. It was soon followed by the Defense Advanced Research Projects Agency (DARPA) Global Low-Orbiting Message Relay (GLOMR) satellite, also using the GAS canister and ejection mechanism. Launched in 1985, GLOMR was the first satellite launched by DARPA under its Advanced Satellite Technology Program.[1] A method for space access for small satellites was thus established with involvement from universities and the DOD.

The first conference devoted entirely to small satellites was held at the Naval Postgraduate School in August 1987. It was sponsored by the American Institute of Aeronautics and Astronautics (AIAA) and DARPA,[2] which had begun investigating what were then called "lightweight satellites" for application to defense needs. DARPA's Advanced Satellite Technology Program had the goal of fielding space-based systems to support operational field commanders in force planning and execution by the early 1990s.

The DARPA conference was the impetus for the Small Satellite Design Program at the Naval Postgraduate School, under the direction of Allen E. Fuhs, then chair of the Space Systems Academic Group and Distinguished Professor.

9.4 Orion

The initial focus of the Small Satellite Design Program was the development of Orion,[a] a general-purpose satellite bus for small payloads (Fig. 9.1). Orion was conceived by Allen Fuhs as early as 1983. At the time, NASA had a very active

[a] Interestingly enough, the name Orion is not an acronym, but was selected by the lead engineer, Austin W. Boyd, who flew P-3 Orion maritime patrol aircraft prior to matriculating into the space system engineering curriculum.

Shuttle Small Payloads Project Office at Goddard Space Flight Center, where secondary payloads could be flown from the orbiter using the Spartan free-flying spacecraft (retrieved by the orbiter during the same mission), the GAS canister, or the HitchHiker secondary payload launcher. Other plans were also on the drawing boards to provide access to space for secondary payloads. In particular, an extended GAS canister[3] had been proposed that would accommodate a 48 cm diameter by 89 cm high payload with mass as high as 113 kg. Orion was designed to take advantage of the extended GAS canister. An initial mission concept was to take measurements in the Van Allen radiation belts. Orion was to use hydrazine thrusters for orbital maneuvering, attitude control, and stationkeeping. Starting at a 250 km insertion altitude, it would be capable of reaching a range of almost 1500 km circular or approximately 4000 km elliptical orbits. The spacecraft could accommodate 23 kg of payload mass with 60 W of electrical power at end of life.

Orion was canceled because of funding issues (although the extended GAS canister did successfully fly and deploy its payload on two shuttle missions); nonetheless, its impact on space systems education at the Naval Postgraduate School was profound, and laid the foundation for the Small Satellite Design Program. The iterative process of engineering design and analysis required in the development of Orion, and the associated design trades and their effects on operational concepts, proved the value of spacecraft mission analysis and design in the educational process.

After Orion, the Small Satellite Design Program adopted an additional focus area—spaceflight hardware development—as a means of exposing students to the issues and trades associated with systems engineering. As a DOD entity, the

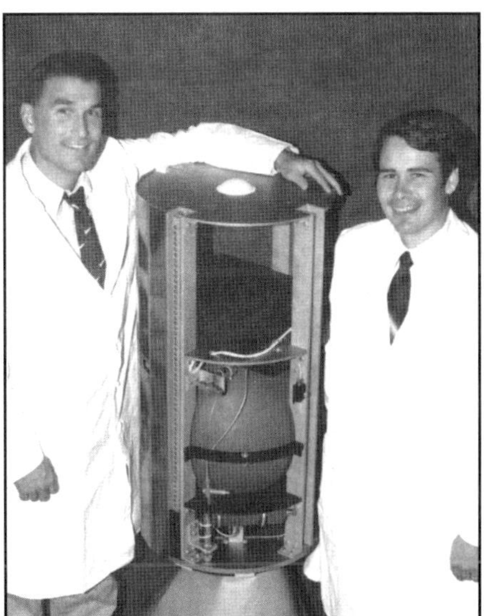

Fig. 9.1. Mockup of the Orion satellite with Austin Boyd (left) and Marty Mosier.

Naval Postgraduate School was able to secure spaceflight opportunities through the DOD Space Test Program. Spaceflight experiments included development of a solid-state data recorder using bubble memory,[4] a GAS canister payload to investigate thermoacoustic refrigeration for space applications,[5,6] and investigation of ferroelectric materials for space.[7] The bubble memory recorder never flew, and the technology was abandoned in favor of alternative technologies, such as ferroelectric memory and flash memory. The thermoacoustic refrigeration experiment flew aboard the STS-42 mission in January 1992. Data retrieved from the refrigerator included shuttle payload bay temperatures, although an electrical power anomaly prevented the refrigerator from operating nominally. Two spaceflight experiments were flown investigating ferroelectric materials as "piggyback" experiments, one on the APEX satellite and the other as a payload on the ORBCOMM-1 satellite. Unfortunately, data was not recovered from the APEX experiment, and the ORBCOMM-1 satellite failed on orbit.

9.5 PANSAT

The first fully autonomous satellite built by the Naval Postgraduate School was the Petite Amateur Navy Satellite (PANSAT). Rudy Panholzer, chair of the Space Systems Academic Group, served as the experiment's principal investigator. PANSAT was launched on the Discovery orbiter on the STS-95 mission on 29 October 1998. It was inserted into a circular, 550 km orbit at 28.5 deg inclination and had a design life of two years. PANSAT operated for almost four years, limited by the cycle life of its NiCd batteries.

PANSAT was designed to be ejected from the shuttle using a HitchHiker canister. The Hitchhiker canister allowed a payload envelope of 48 cm in diameter by 55.5 cm high, including the ejection system, and a total payload mass of 68 kg. Because PANSAT was designed as a tumbler, the optimal shape was that of a rhombicuboctahedron for best average solar panel illumination.

The PANSAT mission was to provide store-and-forward communications using spread-spectrum techniques in the UHF amateur radio frequency bands. As a tumbling spacecraft, the antennas were oriented in a tangential turnstile configuration to allow omnidirectional coverage. Although not all of the mission objectives were realized, the spacecraft hardware and ground segment worked successfully, with many lessons learned and design decisions validated.[8]

The PANSAT objective was to provide educational opportunities for the students in the space systems engineering and space systems operations curricula, and in this regard, it was highly successful. The first master's thesis on PANSAT, published in 1989, investigated a concept for a reliable computing system.[9] Through the nine years of development and four years of operation, more than 50 graduate theses were published related to the PANSAT project. Many other students of varied academic backgrounds were exposed to PANSAT through design projects, seminars, and subsystem case studies within their course work (see Fig. 9.2).

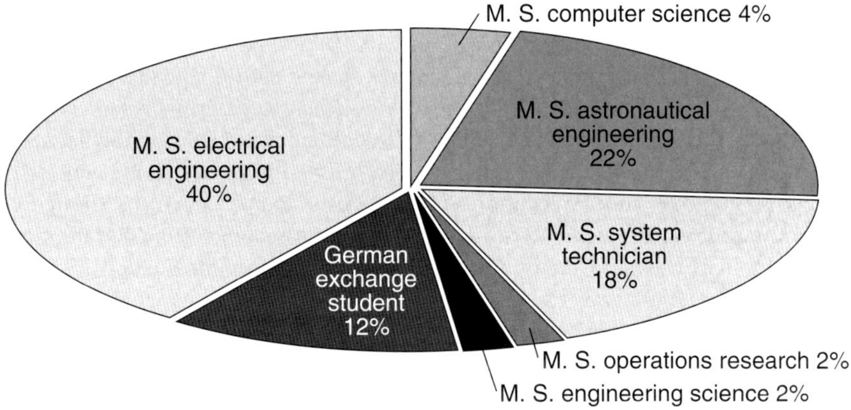

Fig. 9.2. PANSAT student participation by discipline.

9.5.1 PANSAT Design

The first design decisions made in any project have a profound effect on the overall configuration, and PANSAT was no different. Shuttle compatibility, a low-radiation low Earth orbit, and a "single-string" design were some initial design decisions that dictated much of the subsequent development. By "single-string," it is meant that single points of failure would be tolerated in the design. This did not mean that redundancy was not implemented; redundancy was used, but sparingly—where reliability was considered critical and where failures were likely to occur. Redundancy was also implemented where required by NASA for safety. The major design challenges were limited electrical power—roughly 12 W orbital average—and payload safety, which is much more complicated for shuttle missions than it is for unmanned expendable launch vehicles. It was decided early in the program that the spacecraft would have no moving parts, with the exception of microswitches at the launch interface. These microswitches were implemented as redundant safety inhibits, a trade in favor of payload safety versus reliability. When attached to the orbiter, the microswitches would be in the open position. After separation, they would close the circuit between the solar-panel power bus and the electrical power subsystem. Should a microswitch fail in the open position, the spacecraft would fail to operate. No moving parts also meant the solar panels would be body-mounted.

Redundancy was implemented in the RF chain from the antennas through the spacecraft's command and data-handling computer. Redundant digital memory storage units were also implemented. This was considered reasonable, given that the mission was digital communications. Redundancy in the communications path allowed for cross-strapping to bypass amplifiers, filters, and even command processors, in the event of a failure. Redundant paths also meant an increase in complexity for onboard decision-making. Switching between various permutations required implementing a state machine for the various cases, which were stepped through using software timers. The system would be considered in a normal state if there were communications with the ground. If there were no communications within a set time, software would change paths to the next state. The time interval coded into the onboard memory was therefore a critical element. Unfortunately, the time interval selected was shorter than the longest period between access times. For PANSAT's orbital geometry, with altitude of 550 km and inclination at 28.5 deg, and the location of Naval Postgraduate School in Monterey, California, there were generally four consecutive passes approximately 90 minutes apart. This was followed by about 18 hours without communications. Setting the time interval less than 18 hours meant that the state machine would advance even though the spacecraft was operating nominally. An interesting fact concerning this error is that the time interval was reduced late in the design because of a sense of urgency that communications be reestablished in case of failure. Although not a catastrophic error (because the state machine could be reset during a normal pass), it nonetheless shows how human action can force a mistake by reversing a good decision.

Reliability was improved by implementing a number of conventional design solutions—namely, a watchdog timer in the electrical power subsystem and error detection and correction (EDAC) memory in the command and data-handling subsystem. Component selection was also a consideration, although limited funds forced the use of commercial components. Radiation-tolerant devices were used where deemed necessary.

9.5.2 PANSAT Technical Details

9.5.2.1 Electrical Power Subsystem

The electrical power subsystem consisted of 17 silicon solar panels and one GaAs panel, two NiCd batteries, and power distribution electronics. The electrical power subsystem provided a range of 9–16 V. The bus voltage was clamped to a maximum of 16 V. The GaAs panel was added to the bottom of the spacecraft (where the spacecraft connected to the launch vehicle) to minimize nulls in solar power production when the bottom of the spacecraft pointed to the sun. Eight solar panels were equipped with current sensors, which provided information about spacecraft power production.

The NiCd battery cells were commercial off-the-shelf (COTS) Sony Cadnica D-size model KR 4400D. This is a 1.2 V cell of 4.4 Ahr capacity. Cells were matched based on charge and discharge profiles and placed in a serial string of nine cells. A sealed aluminum housing was designed with a safety factor of 2.5 times the maximum pressure requirement. The housing was also fitted with a pressure-relief valve set to 32 psig cracking pressure. Interior empty space of the housing was filled with glass wool to reduce the volume for gas to accumulate and to help contain electrolyte leakage. The interior of the housing was insulated with Teflon film, and thermal cut-outs were placed to take the batteries off line above 55°C. Each battery housing was purged with dry nitrogen at ambient pressure prior to integration with the spacecraft. The batteries were fully discharged as a final safety step during spacecraft integration at Goddard Space Flight Center.

Three microswitches were used in the design for shuttle safety: two on the power line and one on the return line between the solar panels and the electrical power subsystem electronics. Each microswitch constituted a single point of failure. Along with the discharged batteries, the microswitches ensured PANSAT could not operate while attached to the shuttle.

Battery maintenance included software control for efficient battery charging by minimizing the number of charge and discharge cycles. Batteries were nominally discharged to a depth of discharge of 40%. One battery would remain online, and thus discharging, while the other would be charging. The online battery would continue discharging even after the other was fully charged. PANSAT's mission life was two years, although nearly four years of cycle life was experienced with PANSAT's batteries.

9.5.2.2 Command and Data-Handling Subsystem

The PANSAT command and data handling was performed by the digital control subsystem, composed of the processor board or system controller, the peripheral control bus, the mass storage units, and temperature sensor multiplexer modules.

There were two redundant and mutually exclusive system controllers. If one failed to reset a watchdog timer, power would be switched to the other one. Each system controller[10] had an Intel M80C186XL 10 microprocessor operating at 7.3728 MHz. The system used an EDAC controller, the Harris ACS630MS, with static random-access memory (SRAM) for system RAM of 512 kb; the SRAM integrated circuits were Mosaic MSM8256VLMB 256 kb devices. The EDAC controller communicated with the microprocessor using discrete logic that minimized memory delays by only performing memory write-back correction when necessary, rather than during memory read cycles.[11] Two M27C256 UV erasable programmable read-only memories (PROMs) provide 64 kb of ROM. A National Semiconductor LM12H458 provided a four-channel analog-to-digital converter with 12 bits of precision.

A Harris 82C55A handled programmable peripheral interface functionality for the peripheral control bus. A serial communications controller handled data packet duties; as a two-channel device, it could provide synchronous receive and transmit data to the modem while providing standard asynchronous universal receiver and transmitter functionality. The universal receiver and transmitter was interfaced to a standard RS-232 transceiver, a Maxim MAX211E. This asynchronous channel was used for spacecraft testing during development and integration.

Telemetry was captured by two mass storage units, with a capacity of 4.5 Mb and 4 MB, implemented in SRAM. Four Am29F010 flash memory devices from Advanced Micro Devices made up the remaining 512 kb that PANSAT used for nonvolatile telemetry storage. Bit errors in the flash devices were rarely detected, despite extensive use for a period of nearly four years. The mass storage units could be operated at the same time to provide a larger total capacity; however, they were operated individually to provide data redundancy.

Two separate temperature multiplexers were located on the same printed circuit board (PCB) as mirror images. Each had 32 channels, one of which was used for calibration. Critical temperature measurements were redundant across the two units.

The spacecraft operated autonomously with the onboard program stored in ROM; however, it could also receive software updates from the ground that would operate in system RAM. Upon system reset, the spacecraft would revert to software stored in ROM.

Two errors were possible with a system RAM access. The first was a "soft" completely correctable one-bit error. The second was a "hard" error that could not be corrected, but could be detected. The system's software could distinguish between soft and hard errors. If a soft error were encountered, the software would note the time so that it could be used by ground controllers to correlate a position over Earth when the error was detected. A hard error would bring the operational system controller to a halt, which in turn, would force the electrical power subsystem to turn it off and switch to the alternate. This power cycling would clear the hard error. The software performed a regular "RAM wash" by reading data from RAM (although the data would not need to be written back). This ensured that all memory cells in the system RAM were accessed at a regular interval. PANSAT used an initial wash rate of once every 68 min, or about two-thirds of one orbit. This rate was insufficient for locating the position of the spacecraft over Earth; however, later experiments increased the wash rate to once every 2 min. An average error in distance of about 512 km was achieved for more precise location of soft errors. Logged bit errors allowed for a crude identification of the location of the single-event effects attributed to radiation. Analysis of the telemetry (see Fig. 9.3) showed what one would expect: that the vast majority of single-event effects occurred near the South Atlantic Anomaly (a local distortion of the geomagnetic field causing higher than average radiation levels).

9.5.2.3 Communications Subsystem

Digital communications operated in the UHF band at 436.5 MHz (an amateur radio frequency) and a bandwidth of 2.5 MHz using half-duplex. The communication system was designed with a bit error rate of 10^{-5}. It could operate in an experimental spread-spectrum mode at 9842 bps or a BPSK mode at 70 kbps. The spread-spectrum mode is direct sequence using a pseudonoise code of 127 chips, implemented with a 7-bit shift register with taps at bits 7 and 1. There were 127 chips per data bit.

The antenna consisted of four elements mounted in a tangential turnstile configuration on the spacecraft's upper body. Each antenna element was a monopole of approximately one-quarter wavelength. The antennas had to be bent to prevent contact with the HitchHiker canister (see section 9.5.3, below). Prior to antenna bending, analysis showed an omnidirectional radiation pattern with a gain from −8 dB (worst-case null) to +5 dB.[12]

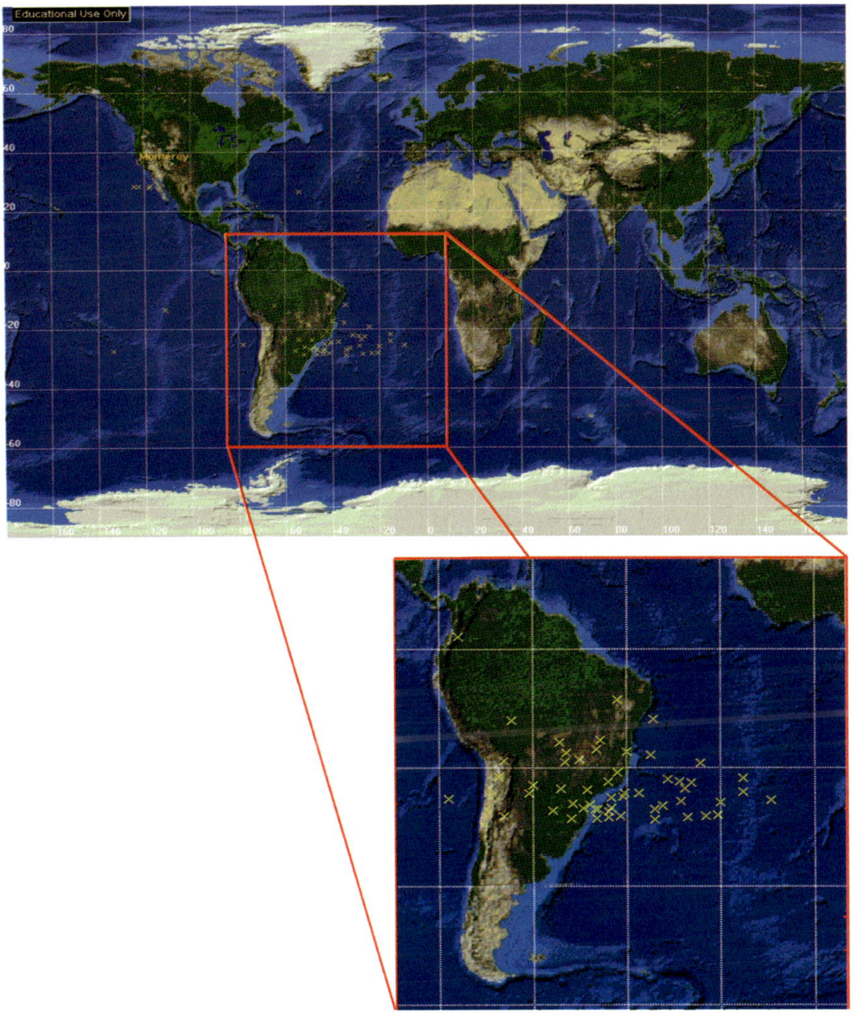

Fig. 9.3. Postprocessed telemetry of EDAC RAM errors logged.[13]

The RF section provides an up and down frequency conversion with an intermediate frequency of 70 MHz. Redundancy was built into the incoming and outgoing signals using switchable high-pass amplifiers, low-noise amplifiers, and local oscillators. The RF section is not fully redundant; pin-diode switches and mechanical relays constitute certain single points of failure. The half-duplex nature of PANSAT requires isolation switches in the RF section at both the antenna mixing point as well as the up and down conversion. In addition, another single point of failure, a 15-second timer, was added as a safety inhibit to disallow any inadvertent transmissions after separation from the shuttle.

A Lockheed Martin PA 100 application-specific integrated circuit (ASIC) was used to perform the acquisition, tracking, and demodulation of the incoming signal. The PA 100 transferred a demodulated signal to the Advanced Micro Devices 85C30 serial communications controller, which provided packet-level detection of data with cyclic redundancy checking of incoming data and also handled packet construction for outgoing data.

The redundant RF elements were software controlled using the spacecraft computer. Determination of a failure in the receive portion of the RF section would be inferred by a lack of communication from the ground, while a failure in the transmit portion would be inferred by repeated requests from the ground. The software could operate between eight different RF states, cycling through each possible RF configuration.

9.5.3 Integration and Testing

An issue uncovered during integration of PANSAT with the HitchHiker canister was that the antenna elements came in contact with the canister. Although there was no lid on the canister, the potential hazard proposed was that the spacecraft antennas could snag on the canister and hinder deployment from the shuttle. The only acceptable alternative at this late stage was to bend the antenna elements. This was clearly a design failure; it had been identified earlier in the program, but was not mitigated correctly, due largely to a combination of configuration changes and misinterpretation of some (two-dimensional) engineering drawings. Antenna deployment scenarios were proposed in early safety reviews, but rejected by NASA as being either too complex or unreliable. It could also be argued that the small project team had limited resources to thoroughly address such an issue. The bending of the antenna elements proved sufficient to meet NASA safety requirements, although the resulting antenna pattern was affected with unknown—but certainly less than optimal—results. No simulation was performed because of the lack of time, although local experts deemed the antenna pattern changes would be tolerable.

Spacecraft testing was divided into functional testing, environmental testing, and testing required for shuttle safety and interface compliance. The design philosophy emphasized designing to test. This was extremely important because the development took place in a university laboratory and was handed over from one student to the next. Continuity in testing from the board level to an integrated system served in minimizing the unnecessary redesign of tests and test procedures, which would prove costly in time. Testing for integration at Goddard Space Flight Center was done with a minimum of test equipment, including a laptop, a programmable power supply, and a briefcase-sized modem for testing communications through the air (Fig. 9.4).

Environmental testing was emphasized on the subsystem level to try to discover any workmanship problems early. Testing included random vibration to at least acceptance-testing levels, and thermal-vacuum cycling.[13] Because of the compressed schedule prior to delivery to NASA, very little system-level testing

Fig. 9.4. PANSAT engineers *(left to right)* D. Sakoda, J. Horning, R. Phelps, and D. Rigmaiden during environmental testing at NASA Goddard Space Flight Center.

was done on its own; rather, it was performed as part of the shuttle safety and interface requirements. Random vibration was done on the integrated spacecraft at Goddard. In addition, a mass measurement was performed to verify that the spacecraft center-of-gravity was within the prescribed envelope for the Hitch-Hiker ejectable payload.

9.5.4 PANSAT Ground Segment

The ground station at the Naval Postgraduate School was a low-cost setup. The majority of components were COTS units, the main exception being the modem and RF mixing, which were identical to those used within the spacecraft. The ground station provided for automatic update of NORAD two-line element sets, control of antenna rotors for azimuth and elevation pointing, automatic spacecraft access determination and scheduling, and manual or scripted communication with PANSAT. It was also responsible for Doppler compensation using programmable frequency synthesizers. The ground station controller was a PC running the Linux operating system; it provided communication to a plug-in PacketTwin serial communications controller board. The ground station antenna was a circularly polarized Yagi with 15.2 dB of gain and a beamwidth of 25 deg. Mounted near the antenna on the tower was a low-noise amplifier and bandpass filter. Azimuth and elevation pointing was performed with a Yaesu G-5400B controller and two antenna rotors connected to the PC via serial RS-232. Antenna pointing calibration was provided by a photoresistor. By tracking the sun using the same tracking software and measuring the resistance, the error in antenna pointing could be determined and corrected; this was performed automatically about once a month.

Software for the ground station controller consisted of PANSAT command and control software, drivers for control of peripheral equipment, spacecraft tracking software, and various customized scripts. The tracking software, SatTrack, was open source, which was amenable to customization for such things as updating two-line elements, exporting a schedule or PANSAT access times, and launching scripts based on those access times. SatTrack also provided an interface for azimuth and elevation control and Doppler compensation.

Different scripts were used for satellite command and control. These scripts allowed an automated flow of logic for the various tasks necessary for PANSAT operations, such as downloading telemetry or uploading new software. The scripts could determine whether the spacecraft was operating on uploaded software, or whether it had reset and was working from ROM software. Uploads and downloads could also be performed over multiple passes. Other ground station activities were also automated. The Naval Space Command provided daily updates for PANSAT ephemeris on their bulletin board system (a dial-up). The ground system would routinely download the two-line element, publish it on the Space Systems Applications Group website, and repropagate PANSAT's orbit to determine more accurate access times. The access times, along with downloaded telemetry information (numerical and graphical) and a recent log of satellite communication sessions, would be posted to the website.

Having a reliable mechanism of acquiring spacecraft telemetry and archiving the data allowed for some interesting questions to be resolved with surprising results. On 11 August 1999, a solar eclipse occurred, with the shadow of the moon crossing over central Europe, the Middle East, and India. On 12 August, PANSAT's principal investigator Rudy Panholzer, wondered, "What if PANSAT flew through the eclipse?" Within a short time, with some help from orbital analysis software, the times of PANSAT's location was narrowed down, and an analysis of the solar-panel currents showed that the spacecraft did indeed fly through the eclipse (Fig. 9.5). Although PANSAT did not fly through total eclipse, it was at least within the 90% shadow region, as evidenced by the slight power generated by the solar panels. Asked by a reporter, what would be the odds of this happening, one engineer responded, "astronomical," of course.

9.5.5 PANSAT Lessons Learned

The successful launch of PANSAT aboard the Discovery orbiter on STS-95 on 29 October 1998 (Fig. 9.6) marked a major milestone in the Naval Postgraduate School space education program. The successful downlink of the first telemetry packets from PANSAT was in some sense a validation of the program, and certainly of the more than 50 graduates who published their master's theses related to

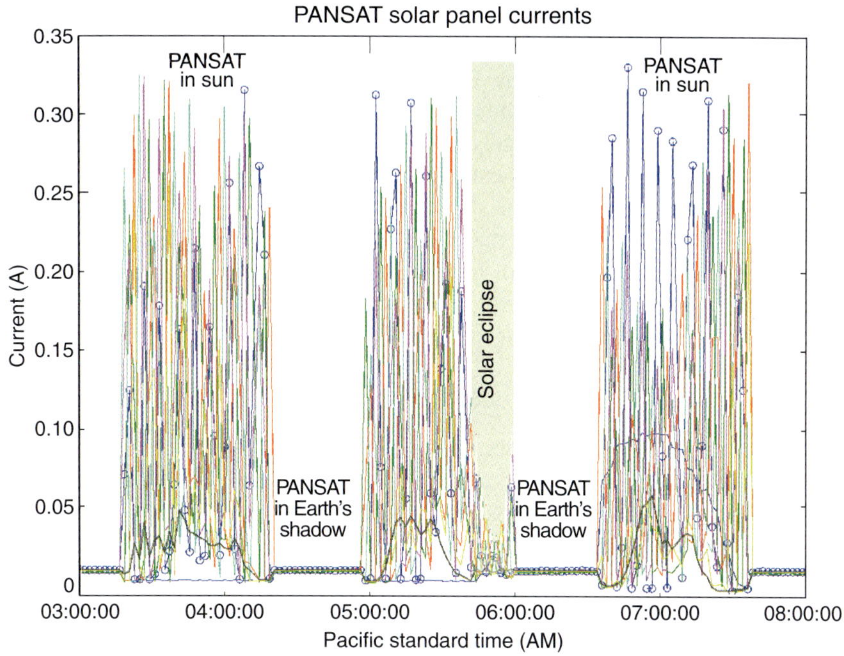

Fig. 9.5. PANSAT flies through the solar eclipse.

Fig. 9.6. PANSAT deployed from the Discovery shuttle orbiter on STS-95.

PANSAT. The satellite survived the launch, booted up properly after separation, and was operating nominally. This also meant that ground operations were occurring at regular intervals. As the low Earth orbit precessed about the globe, the four consecutive ground passes over the Naval Postgraduate School cycled through the hours of the day over approximately two weeks. Early-hour watches were strong incentives to automate ground operations.

Table 9.1 lists a few lessons learned from PANSAT. Some of these lessons validated design decisions that were made, while others identified design decisions that should be avoided. Included are programmatic decisions such as implementing a testing philosophy where resources are committed early in the program for potential future gains in the schedule and reliability.

9.6 NPSAT1

NPSAT1 is the follow-on to PANSAT. It began as a design project by students in the Fall 2000 Spacecraft Design course, who were looking into concepts of operations. A concerted effort was made to fold the Small Satellite Design Program into the course work within the space systems curricula. For example, the attitude control design was investigated as part of a Spacecraft Attitude, Dynamics, and Control course. Mission and orbital analysis was presented in an introductory

Table 9.1. Short List of Lessons Learned from PANSAT

Topic	Description
Spacecraft Design: Memory	EDAC RAM for spacecraft processor is necessary for continuous operations.
Spacecraft Design: Safety	Early design decision made with payload safety in mind, (e.g., battery selected & designed to meet shuttle payload safety requirements for integration and launch; battery able to be totally discharged prior to launch).
Spacecraft Design: Software	Unique spacecraft processor architecture delays software development. Standard architectures would allow early software development.
Spacecraft Design: Communications	Half-duplex communications required complex coordination of hardware interfaces, timings, and large delays between data throughput due to switching the channel.
Spacecraft Operations: Autonomy	PANSAT did not require a great deal of interaction from the ground under normal or anomalous conditions (e.g., initial startup after deploy allowed safe mode of operations until contacted from the ground).
Spacecraft Testing:	Emphasis on environmental testing early in the development cycle proved helpful in identifying problems prior to the fully integrated spacecraft.

orbital mechanics course. Other relevant courses used NPSAT1 as case studies, such as Spacecraft Power, Spacecraft Thermal, and Space Testing.

The major difference between PANSAT and NPSAT1 is in the level of complexity. Whereas PANSAT had a narrowly focused mission of digital communications, NPSAT1 will serve as a test bed for small satellite technology as well as an experiment platform. NPSAT1 communications will require three-axis attitude control, albeit with coarse ±10 deg pointing accuracy; in contrast, PANSAT was a tumbling spacecraft.

9.6.1 NPSAT1 Scientific Experiments and Technology Demonstrations

Spacecraft technology to be tested onboard NPSAT1 include a lithium-ion battery, experimental triple-junction solar cells, a Linux-based command and data-handling subsystem, and a three-axis MEMS rate sensor. Although these can be considered experiments in themselves, the distinction is that they are required bus subsystems or components, as opposed to standalone experiments. For example, the lithium-ion battery is experimental, but it's also the only battery onboard NPSAT1. Within the electrical power subsystem, ferroelectric RAM will be used as a nonvolatile memory device. This will be the first flight demonstration of ferroelectric RAM in a functional capacity. Even the three-axis attitude control subsystem is itself an experiment in implementing a low-cost control scheme.

Additional experiments on NPSAT1 include the configurable fault-tolerant processor (CFTP), two ionospheric physics experiments from the Naval Research Laboratory, and a solar-cell measurement system.

9.6.1.1 CFTP

The CFTP is an experiment in which field-programmable gate arrays (FPGAs) will be used in a triple-modular, redundant configuration. Within the FPGA, three processor cores are programmed along with the voting logic; if an error occurs from a single-event effect, it should be detected and corrected, and the CFTP should continue processing without having to reboot. Single-event effects will be logged and downloaded as telemetry. Having a reconfigurable device onboard allows for a myriad of research opportunities in spacecraft computing. The CFTP experiment is an example of the benefit of having an interdisciplinary framework in an academic setting. The principal investigators, Herschel Loomis and Alan Ross, are both members of the Space Systems Academic Group and have ties to the Department of Electrical and Computer Engineering. Students and engineers within the Space Systems Academic Group supported the design, development, and testing of the CFTP, which also flew as a standalone experiment on a separate spacecraft built by the U.S. Naval Academy, MidStar-1, which launched aboard an Atlas V on the DOD STP-1 mission, 8 March 2007.

9.6.1.2 Ionospheric Measurement

The two Naval Research Laboratory experiments will perform ionospheric physics measurements. The CERTO experiment is a radio beacon that transmits up to three separate frequencies: 150, 400, and 1067 MHz. Working with a network of ground receivers, the CERTO beacon is used to measure the integrated electron density of the ionosphere in the plane of observation. CERTO will also be used to develop and test tomographic algorithms for reconstruction of ionospheric irregularities, to provide a database for global models of the ionosphere, to characterize the ionosphere for geolocation, and to perform scintillation studies in the ionosphere. The Langmuir probe collects on four separate 12 bit analog/digital channels at rates up to 1000 samples/sec. The Langmuir probe provides *in situ* measurements at the spacecraft altitude, where the data can be processed for correlation with the ground observations of the CERTO beacon.

9.6.1.3 Solar-Cell Measurement System

The solar-cell measurement system is a late-generation implementation of research started in 1986 by students at the Naval Postgraduate School under Sherif Michael of the Department of Electrical and Computer Engineering.[14–16] It takes measurements of individual experimental triple-junction solar cells. These cells are not used for spacecraft power. Conversely, none of the solar panels used for spacecraft power are being measured for performance. Parameters measured by the solar-cell measurement system include the current vs. voltage, temperature, and sun angle. Microprocessor control of the circuit allows for timely and accurate data acquisition. Timeliness is important because of the rate of temperature change of the solar cells in low Earth orbit.

The solar-cell measurement system will investigate both the solar cells and the experiment control hardware, which will be used to construct the current vs. voltage plots used to characterize the solar cells. One of three solar panels on each of NPSAT1's 12 sides is made up of experimental solar cells. On each of these, two solar cells are exclusively used for current vs. voltage curve testing.

The sun sensors for the solar-cell measurement system are Goodrich 13-515 Sun Sensing Modules. These are single-axis sun sensors with a field of view of ±64 deg. Two sensors are collocated, with one rotated 90 deg to provide both the azimuth and elevation angles with respect to the spacecraft. With each sensor pair viewing 128 deg, a minimum of three pairs are required for a full 360 deg view in azimuth.

9.6.1.4 COTS Camera

Another experiment onboard is a COTS digital camera that will point in the Earth-viewing, or nadir, direction. The camera is intended to provide data generation for the scenario of using the CFTP as a hardware image compressor, testing various compression schemes. In addition, the camera will provide a means of educational outreach, where students from elementary through secondary schools can learn about spacecraft systems and task the satellite to take a picture of their home town.

9.6.2 NPSAT1 Technical Details

The NPSAT1 spacecraft is a 12-sided cylinder with overall dimensions of 55 cm in diameter by 86 cm high with a mass of 82 kg. The spacecraft has two booms, which, when deployed, point opposite each other, in the direction of the orbit normal. One of the booms is the antenna, and the other supports the Langmuir probe, both provided by the Naval Research Laboratory. The structure is aluminum 6061-T6 and consists of four equipment decks and three longerons. Side panels mount to the decks and longerons.

The NPSAT1 flight vehicle structure successfully completed qualification testing at the Naval Research Laboratory (Fig. 9.7). This included sine-burst testing to 15 g peak in each axis to simulate static loads, followed by random vibration testing to +6 dB above the maximum predicted environment of 14.2 g (rms). An engineering development unit was used for qualification testing; it was identical to the load-bearing flight structure, but with mass simulators of the electronics housings. The actual flight vehicle structure will undergo random vibration acceptance-level testing at +0 dB of the maximum predicted environment.

Fig. 9.7. NPSAT1 engineering development unit undergoing qualification testing at the Naval Research Laboratory.

9.6.2.1 Electrical Power Subsystem

The electrical power subsystem consists of triple-junction solar cells for energy conversion, an experimental lithium-ion battery for energy storage, and the electrical power distribution and control electronics, composed of a digital processor board and an analog/switching board.

The 12 sides of the spacecraft each have three solar panels. Two are made up of commercial triple-junction solar cells, and the third is made of experimental triple-junction solar cells. To reduce unwanted noise into the spacecraft, all solar-panel power lines are combined in the solar-panel junction box, which uses filtered connectors.

The flight battery is composed of 49 Sony US18650S lithium-ion hard carbon cells, extracted from Sony BP 945 camcorder battery packs. The flight battery housing is maintained at 1 atm and heated to maintain an optimal temperature for the battery. It is wired in a 7S-7P (seven cells in series and seven strings in parallel) configuration, providing a battery capacity of 225 Whr with a full charge voltage of 29.5 V. Each string has a current sensor to examine current sharing between strings as the cells age. One experimental aspect of the battery architecture is to show successful current sharing between the cell strings, eliminating the need for more complex charge circuits at the cell or string level (Fig. 9.8).

The electrical power subsystem electronics are designed around a radiation-tolerant UTMC80C196KD microcontroller on the digital board, which also hosts a real-time clock, radiation-tolerant RAM, a watchdog interface for the command and data-handling subsystem, serial port to the command and data-handling subsystem, and 8 kb of ferroelectric RAM. The analog/switching board hosts all of the power distribution to the other subsystems, providing a spacecraft power bus of 24–40 V. Each subsystem is responsible for regulating power from the bus. The analog/switching board is also responsible for measuring battery voltages and temperatures via 32 analog channels, which are multiplexed into a dual, 12-bit analog-to-digital converter.

9.6.2.2 Command and Data-Handling Subsystem

The command and data-handling subsystem contains several electronic boards attached to a PC-104 bus and housed within one box. The circuit boards are a motherboard, a mass storage and analog/digital and digital output board, the CFTP experiment board, a power-supply board, a modem, and RF components. The PC-104 bus is modified to carry alternate signals between the command and data-handling board stack to allow special operations; however, it still allows for use of PC-104 stacks during development. The command and data-handling subsystem also interfaces with the GPS receiver, providing command, data, and power.

The command and data-handling motherboard is built around a Sharp LH79520 microprocessor CPU and operates at 51 MHz. This is an ARM 720T core supporting a 32-bit low-power processor, memory management, watchdog timer, clock and power management, three universal asynchronous receiver/

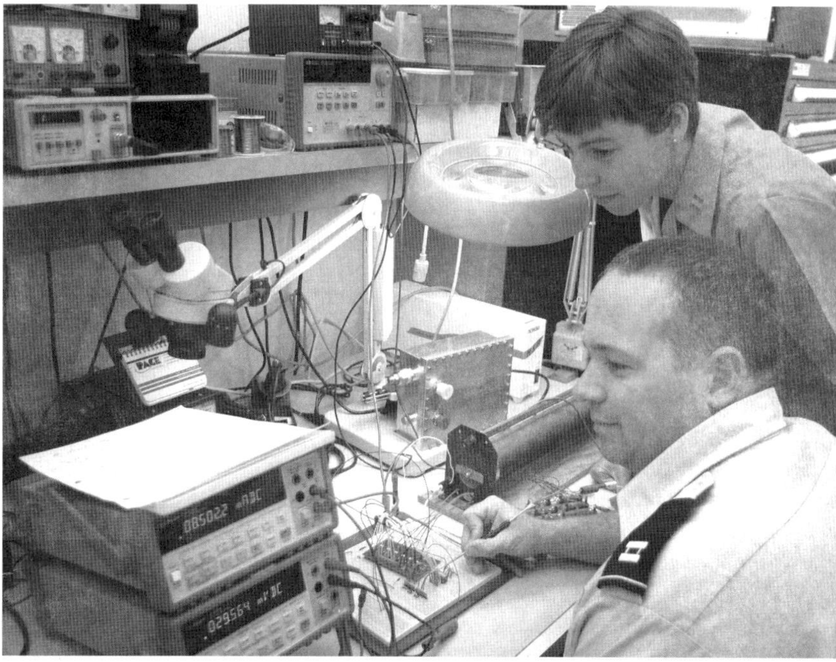

Fig. 9.8. Naval Postgraduate School students performing measurements on NPSAT1 prototype battery electronics.

transmitter units, programmable general-purpose input/output, and a JTAG (Joint Test Action Group) interface. The motherboard also contains a Xilinx XCV100 FPGA, which interfaces with nearly all of the ARM signals. This allows complete manipulation of the microprocessor signals between other devices within the command and data-handling subsystem, allowing modification via software reconfigurations (of the FPGA) during development. In addition to microprocessor signal manipulation, the motherboard FPGA implements the EDAC logic and microprocessor-to-SRAM control interface.

The motherboard memory consists of 64 kb of PROM (AT28C256: 256 kb EEPROM from Atmel) that contains the boot loader. There are four Sharp LH28F32 flash memory devices, 4 Mb each, which contain quadrupled versions of the operating system and a root disk. There are 16 Mb of static RAM with 1-bit error correction and 2-bit error detection.

The Linux 2.4.17 operating system has been ported to the motherboard, providing a mature and rich software platform. Linux is a POSIX-compliant operating system. It was chosen for several reasons: Linux is open source, robust, and highly configurable. It also accommodates multitasking and provides a Unix-like software environment. Plenty of powerful software development tools are available for free. Currently, the operating system is about 560 kb, compressed, and 890 kb uncompressed. The root disk is a compressed version of the system application software, which must always be present when the command and data-handling subsystem starts up. The compressed root disk is about 820 kb and about 1.8 Mb uncompressed. There is a limit of the compressed operating system and root disk sizes of 1 Mb and 3 Mb, respectively.

The three ARM universal asynchronous receiver/transmitter units are used for a low-level boot-loader interface (for initial development), a test port interface (used as a point-to-point protocol [PPP] link during development, test, and integration), and an interface to the GPS unit onboard the spacecraft. The motherboard also contains a Texas Instruments TL16C754B, a quad-port universal asynchronous receiver/transmitter. This device provides four additional serial ports, which are used for the attitude control subsystem, the electrical power subsystem, the solar-cell measurement system, and the COTS camera. A simple protocol operating at 9600 bps is used to communicate with the first three subsystems; the interface with the COTS camera is via a PPP link at 115.2 kbps. A complete network socket interface is available between the command and data-handling subsystem and the camera.

The GPS-12 receiver from SpaceQuest provides up to 12 channels of tracking within a cold start time of 120 sec. The unit consumes approximately 1.5 W of power, but will not be on continuously. Instead, it will power on every 60 min, and a spacecraft state vector will be obtained. This position information will be used with a simple Keplerian orbit propagator to predict the spacecraft's position within the next hour. In this duty cycle, the GPS will consume only about 4% of the power it would consume in continuous use. Orbit propagation with this technique provides positional information with less than 0.5 deg of error, which is more than sufficient for the spacecraft's navigational requirements. If the GPS fails, the spacecraft can perform orbit propagation using two-line element sets and an SGP4 (simplified general perturbations satellite orbit model 4) algorithm. Without navigational information, the spacecraft defaults to a 25% duty listen cycle, awaiting a connection with the NPSAT1 ground station in Monterey, California.

An array of eight Sharp LHF54 flash memory devices, each with 8 Mb, provides a total of 64 Mb of mass storage. Using flash chips compliant with common flash interface specifications, and the memory device support offered within the Linux kernel, results in a sophisticated mass storage for recording experiment data and other spacecraft telemetry.

Two National Semiconductor LM12H458 analog-to-digital devices digitize various analog signals used by the command and data-handling subsystem. One interfaces with signals within the command and data-handling subsystem, and the other will digitize the four analog channels from the Langmuir probe at rates up to 1 kHz.

The command and data-handling power supply provides all the power conditioning (1.8, 2.5, 3.0, 3.3, 5.0, and 15.0 V dc) for all the electronics within the subsystem housing. The ARM cores are known for very low power consumption while providing a powerful architecture for modern operating system support. The motherboard with the switching power supplies operate at less than 3 W of power.

The CFTP circuit board consists of two Xilinx XCV600 FPGAs, each of which is capable of implementing a stand-alone processing unit. Direct communication to the command and data-handling processor is via the 16-bit PC-104 interface. Upon reset, the CFTP automatically configures one of its FPGAs to implement a PC-104 controller. This device can be programmed from the command and data-handling subsystem via a JTAG interface (IEEE 1149.1 specification), allowing the FPGA to be reprogrammed from its original configuration.

9.6.2.3 Communications Subsystem

The RF components within the command and data-handling housing are isolated from the other electronics to reduce electromagnetic interference. The modem and RF subsystem are physically located in the bottom portion of the housing. The RF subsystem is a full duplex communication system providing 100 kbps data rate on both the uplink and downlink channels. The uplink operates at 1767.565 MHz, and the downlink operates at 2207.3 MHz. The RF subsystem uses patch antennas designed and built by the Naval Postgraduate School—one antenna for each frequency. The spacecraft uses one pair of transmit/receive antennas for nadir-pointing and another pair for zenith-pointing. This allows full 4π steradian coverage in case the spacecraft is tumbling. RF switches are used to alternate between nadir and zenith antenna pairs. Nominally, only the nadir-pointing antennas are used when the spacecraft is oriented in its correct attitude. An asynchronous, Gaussian-minimum-shift-keying frequency modulation is used. A maximum bit-error-rate of 10^{-5} was defined as a requirement.

The system uses a dual conversion to baseband, with the first conversion at 140 MHz and the second at 10.7 MHz for the uplink channel. A Philips SA639 mixer FM IF (intermediate frequency) system performs the IF-to-baseband conversion. The downlink channel uses an Analog Devices AD9852 direct digital synthesizer to modulate the serial data to an FM signal. There is a dual conversion to the downlink RF channel. Both the uplink and downlink channels use a common local oscillator to achieve the RF-to-IF and IF-to-RF conversions.

9.6.2.4 Attitude Control Subsystem

The pointing requirements for the spacecraft are defined by the Naval Research Laboratory experiments and the field of view of the onboard digital camera. Pointing accuracy is only required to be within ±10 deg. Three-axis stabilization is necessary, although the coarse pointing requirements allow a low-cost implementation.[17] The attitude control subsystem consists of three magnetic torquer coils for actuators, a three-axis magnetometer, a MEMS three-axis rate sensor, and the attitude control subsystem controller. The torquer coils were built at the Naval Postgraduate School and are basically aluminum wheels wound with 28 gauge wire. The attitude controller uses onboard orbit information to obtain the spacecraft's location to lookup the values for the local magnetic field vector at the latitude, longitude, and altitude. Magnetometer measurements are compared to the lookup values, and the control algorithm attempts to null the difference between the two. Although the magnetic field (B field) vector gives only one reference vector, because the spacecraft is orbiting and is continually chasing the B-field vector, three-axis stabilization can be achieved. Simulation results yield a pointing accuracy of better than 2 deg.

The attitude controller is nearly identical to the electrical power controller, and uses the same digital processor board and a similar analog/switching board. The analog/switching board drives the torquer coils and takes measurements for the magnetometer, MEMS rate sensor, and temperature sensors. It uses a 12-bit digital-to-analog converter to produce the precise currents needed to generate a specific magnetic moment, which can range from a few hundred $\mu A\text{-}m^2$ up to approximately 5 $A\text{-}m^2$. The analog board provides power for the MEMS rate sensor and magnetometer. Navigation information is provided by the onboard GPS receiver. Backup navigation can also be provided by the SGP4 orbit propagator running on the command and data-handling subsystem.

A design challenge is to ensure that the spacecraft has the proper mass distribution for a "gravity gradient friendly" configuration and to minimize the center-of-mass to center-of-pressure offset responsible for aerodynamic disturbance torques. Simulations showed that the center-of-mass to center-of-pressure offset needs to be within fractions of a centimeter. Under the direction of Barry S. Leonard, a ground-based three-axis air-bearing test platform was developed to validate many of the modeling techniques used in the computer simulations (Fig. 9.9). Although funding and time limited the scope of testing to a static magnetic field environment, analytical simulations performed solely on a computer were duplicated with great fidelity on the air-bearing platform. This served to validate the modeling techniques and to gain confidence.

Because of the experimental nature of the attitude control subsystem, some risk mitigation was taken by the addition of a momentum wheel. The required orientation for NPSAT1 is to have its body coordinates aligned with a local-horizontal/local-vertical orientation. This lends itself readily to a momentum-biased attitude control subsystem, where the wheel's momentum vector is aligned with the orbit normal. The major cost of adding a momentum wheel, in addition to the monetary cost, is the increased electrical load and the greater complexity in the integration and test. This means additional interfaces are needed for power, control, and software as well as a path for component, subsystem, and system-level testing. However, should the attitude control subsystem not meet the specified requirements, the momentum wheel will be commanded on so that the overall mission is not jeopardized. The momentum wheel is a Dynacon MW200 MicroWheel. It has a speed range of up to 10,0000 rpm with an angular momentum capacity from 0.1 to 1.0 N-m-s or a torque capacity greater than 30 mN-m. It has an RS-422 interface for command and telemetry and requires approximately 3 W of power

Fig. 9.9. ***From left to right:*** **Alex Schmidt (German Armed Forces student), Rudy Panholzer, and Barry Leonard with NPSAT1 air-bearing test platform.**

for steady-state operation at 10,000 rpm. The wheel is powered from the electrical power subsystem to accommodate the larger power requirements.

The MEMS rate sensor has three orthogonally mounted BEI Systron-Donner QRS-11 rate sensors, each approximately 3.78 cm in diameter and 1.35 cm thick, in a hermetically sealed housing 11.1 cm in diameter and 7.6 cm high (Fig. 9.10). Each device has seven pins for power, rate output, and built-in test. The sensor assembly fits within a sealed housing that is purged with dry nitrogen. Signal filters are incorporated into the housing, which ensures good signal quality of the low-voltage output. At the time these devices were selected for NPSAT1, they had not been flown in space. The smallest rate ranges advertised for the sensors was also ±5 deg/sec, well above orbital rates (i.e., 360 deg per orbital period). The rate range would appear to make them unattractive as an attitude control sensor, where rates need to be resolved on the order of 0.006 deg/sec (i.e., 1/10 orbital rates). However, research performed by Thomas Pugsley, a student working under James Newman, NASA visiting professor and astronaut, suggests that with addition of filters, careful calibration, and characterization of the sensor, performance is sufficient for attitude control.[18] Since the start of NPSAT1 program in 2000, these MEMS devices have flown on other satellites as well as the Mars Rovers.

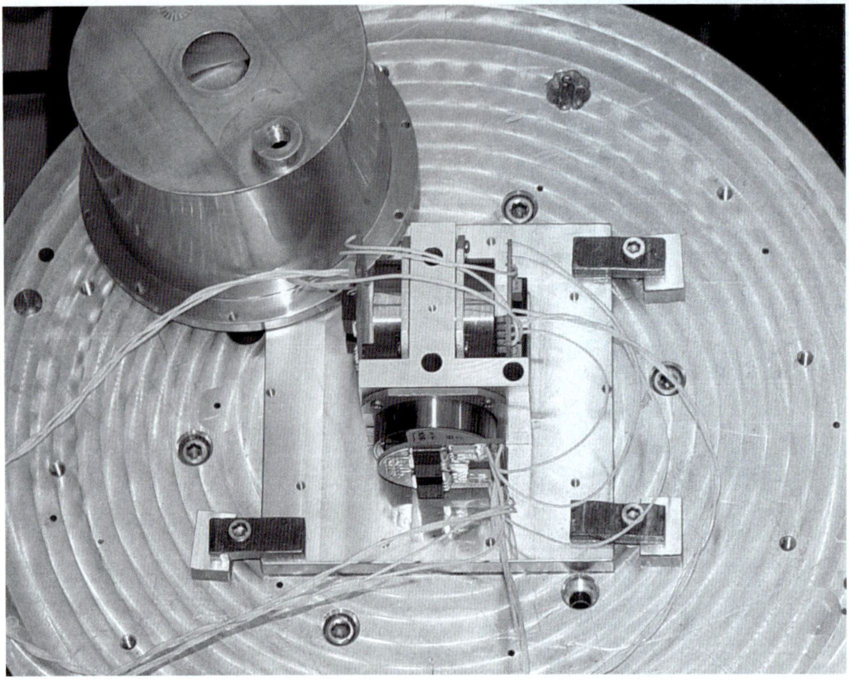

Fig. 9.10. NPSAT1 MEMS three-axis rate sensor.

9.6.3 NPSAT1 Ground Segment

The NPSAT1 ground station is a low-cost installation.[19] The majority of components are COTS. The ground station provides for automatic two-line element set updates, control of antenna rotors for azimuth and elevation pointing, automatic spacecraft access determination and scheduling, and the availability of manual or scripted communication with NPSAT1. It is also responsible for Doppler compensation using programmable frequency synthesizers. The ground station controller is a PC running Linux. The downlink frequency will be demodulated in an L3 Communications PCI-2070 telemetry receiver with tracking, a software radio card for a PCI slot within the ground station controller PC. The RC2800PX/AZ and RC2800PX/EL controllers are used for simultaneous control of the azimuth and elevation motors. The controllers interface with the PC via RS-232 using a simple ASCII-oriented command set. The antenna has a 3 m diameter parabolic dish with pointing accuracy of less than 2.5 deg. There is a calculated 6.0 dB of downlink margin and 10.7 dB of uplink margin in the link budget.

Software for the ground station controller consists of NPSAT1 command and control software, drivers from control of peripheral equipment, spacecraft tracking software, and various customized scripts. Different scripts are used for satellite command and control. These scripts allow automated NPSAT1 operations, such as downloading telemetry or uploading new software. Uploads and downloads can also be performed during multiple passes. The ground system routinely downloads the two-line elements, publishes them to a website, and re-propagates NPSAT1's orbit to update access times. The access times, along with downloaded telemetry information (numerical and graphical) and a recent log of satellite communication sessions, are posted to the website.

9.6.4 NPSAT1 Lessons Learned

The NPSAT1 spacecraft was presented to the DOD Space Test Program Space Experiments Review Board in November 2000, and placed on the manifest for launch on the STP-1 mission. However, schedule delays in the development of spacecraft electronics forced its removal from the manifest late in 2006. The STP-1 mission launched successfully on March 8, 2007, without NPSAT1; instead, it carried an aluminum simulator with identical mass and center of gravity so as not to affect the overall mission. The schedule delays can be attributed to a number of factors that might be common to any space program and that are especially possible within an academic environment. For example, limited resources in engineering staff, coupled with the uncertain student involvement, was one area of programmatic risk. The very nature of an university satellite program ensures that the project team will incur a high turnover rate. In addition, whereas PANSAT was flown on the shuttle, which had an established process for flying secondary payloads, NPSAT1 was slated for launch on the STP-1 mission, which was the first Air Force launch of the Atlas V Evolved Expendable Launch Vehicle (EELV) and the first flight of the EELV Secondary Payload Adapter, which carried four other secondary payloads, not including the NPSAT1 mass simulator. The complexity in coordinating the launch of six spacecraft on a single launch vehicle could not be circumvented, and naturally flowed down to the secondary payloads. The level of programmatic requirements was greater than what the NPSAT1 project team expected; however, all STP-1 programmatic deliverables in the form of analyses, reports, and documentation were met, but at the cost of progress in the hardware development. Given the number of experiments alone, NPSAT1 is a very ambitious satellite for a university program. Having failed to meet its launch opportunity, the NPSAT1 project is continuing development with hopes of another launch opportunity as a secondary payload through the Space Test Program, but with the previous development hurdles overcome and many program requirements completed.

9.7 CubeSats

The development time for PANSAT was approximately nine years from the first thesis until its launch. NPSAT1 development time will probably be similarly that long. One means of reducing this cycle time would be to adopt CubeSats in the Small Satellite Design Program. CubeSats are the invention of Prof. Bob Twiggs of Stanford University, who formerly led the NUSAT design team at Weber State College. CubeSats are 1 kg satellites in the shape of a cube, 10 cm on a side.

Depending on the definition used, CubeSats fall within the nanosatellite or picosatellite class of spacecraft. Many universities worldwide are using the CubeSat standard as a means of providing hands-on education in aerospace curricula.[20] In recent years, NASA,[21] Boeing,[22] and other entities have shown interest in CubeSats and have successfully flown them. One advantage of having a standard form factor for the satellite bus is the reduction in design and development. Design options are constrained, forcing students to focus on payload development rather than bus design trades. The small form factor also means a small area for solar panel energy conversion, and thus a minimal power budget. Design teams are therefore forced to either set a narrow scope on the mission objectives or grow beyond the single-unit (10 cm^3) size.

The Naval Postgraduate School is beginning work on conceptual designs and prototypes of CubeSat missions. These include funded research in an imaging mission and an advanced solar-cell testing mission. The bulk of work for the imaging mission is focused on achieving attitude control within the size of an extended CubeSat. The advanced solar-cell testing mission basically takes the Solar-Cell Measurement System experiment of NPSAT1 and fits it into a CubeSat.

Shorter development times for CubeSats would not necessarily guarantee more opportunities for students to operate an orbiting spacecraft. Access to space is a major hurdle for any secondary payload. The small size of CubeSats does offer some benefits because their interaction with the launch complement via coupled loads is minimized. Also, they are designed to be nonoperable until deployed from their launch carrier. Still, the ratio of active CubeSat projects to available launch opportunities is very small. The fact that NPSAT1 was replaced with a mass simulator rather than another secondary payload speaks to the difficulty in attempting changes in the manifest late in the process. A late substitution must maintain the same mass distribution, achieve flight qualification, and be ready to fly on short notice. CubeSats might be easier to substitute, given the number of active projects, the commonality of their configuration, and their low mass and size.

The Naval Postgraduate School is developing a means of launching multiple CubeSats to address the issue of space access not only for its own missions, but for other DOD, NASA, and university missions as well.[24] The successful launch of the EELV Secondary Payload Adapter (ESPA), although very complex and expensive, proved the technical feasibility of launching multimission EELVs. By maintaining the interface of the ESPA, the Naval Postgraduate School CubeSat Launcher (NPSCuL) can be manifested as a secondary EELV payload or serve as a standby, should a secondary payload be removed from the manifest for any reason. Rather than reinvent a deployment mechanism, the NPSCuL will incorporate the Cal Poly Picosat Orbital Dispenser (P-POD), designed to deploy up to three standard CubeSats. A standard electrical signal from a launch vehicle initiates a pyrotechnical reaction that releases a captured bolt attached to a hinged door. The CubeSats are deployed from the dispenser together. The NPSCuL is designed to be mass reconfigurable within a range of 45.4 to 181.4 kg, with the center of mass ranging between 25.4 and 50.8 cm above the ESPA mating surface. The larger mass allows for as many as 10 P-PODs or 30 CubeSats as well as ballast to match the upper mass value. The NPSCuL electronics will be required to interface with the input of the launch vehicle's single separation signal, and the deployment signals of each of the mounted P-PODs. As a secondary payload itself, the NPSCuL may be required to deploy from the launch vehicle, depending on mission planning and the requirements of the primary payload.

9.8 Conclusion

The value of small satellites in education is clear, especially where experiential education exposes students to a systems perspective. The various engineering trades necessary in working with competing requirements at all levels of design are much less forgiving when dealing with actual hardware, and the benefits of operating an orbiting spacecraft cannot be realized unless the space segment has been performed. The experience gained through spacecraft design, build, test, and operations in an academic setting is invaluable, especially for the graduates of the Naval Postgraduate School, who may eventually have responsibilities over larger programs with much more complexity. Small satellites typically encompass a narrower scope in mission requirements, funding levels, and programmatic overhead; however, university small satellites also offer an environment where risk is tolerated and failure can be the greatest teacher, albeit undesirable.

The vision for the Small Satellite Design Program is to engage in CubeSat development as a platform for technology demonstrations, narrowly focused missions, and mission concepts for clusters and constellations. In parallel, microsatellite projects will continue, allowing for a broader focus of mission concepts. The primary objective of educating military officer students is paramount and will continue to be augmented with "hardware-in-the-loop education." The development of a highly skilled space cadre with the depth and breadth of knowledge that can only be gained through hands-on experimentation is the Naval Postgraduate School's greatest contribution to U.S. national security space.

9.9 References

[1] *Affordable Spacecraft: Design and Launch Alternatives—Background Paper*, OTA-BP-ISC-60 (U.S. Congress, Office of Technology Assessment, Washington, DC, Jan. 1990).

[2] A. W. Boyd, Jr., "Design Considerations for the ORION Satellite: Structure, Propulsion and Attitude Control Subsystems for a Small, General Purpose Spacecraft," MSEE thesis, Naval Postgraduate School, Monterey, CA, March 1988.

[3] "Extended GAS Canister, Final Report, Contract Number 88-C-0015" (Ball Aerospace and Communications Group, December 1992).

[4] T. D'Ercole, "A Solid-State Data Recorder for Space-Based Applications Using Magnetic Bubble Memory," MSEE thesis, Naval Postgraduate School, 1986.

[5] M. P. Susalla, "Thermodynamic Improvements for the Space Thermoacoustic Refrigerator (STAR)," M.S. Physics thesis, Naval Postgraduate School, June 1988.

[6] J. A. Adeff, "Measurement of the Space Thermoacoustic Refrigerator Performance," M.S. Physics thesis, Naval Postgraduate School, Sept. 1990.

[7] C. E. Josefson, "Evaluation of Ferroelectric Materials for Memory Applications," MSEE thesis, Naval Postgraduate School, June 1990.

[8] D. Sakoda, R. Phelps, J. Horning, D. Rigmaiden, and J. Damerau, "Naval Postgraduate School PANSAT: Lessons Learned," Paper AIAA 2001-4556, Space 2001 Conference & Exposition (Albuquerque, NM, 28–30 Aug. 2001).

[9] J. K. Hiser, "Design of a Reliable Computing System for the Petite Amateur Navy Satellite (PANSAT)," MSEE thesis, Naval Postgraduate School, March 1989.

[10] J. A. Horning, "System Controller Hardware and Embedded Software for the Petite Amateur Navy Satellite (PANSAT)," MSEE thesis, Naval Postgraduate School, Sept. 1997.

[11] C. R. Oechsel, "Implementation of Error Detection and Correction (EDAC) in the Static Random Access Memory (SRAM) aboard Petite Amateur Navy Satellite (PANSAT)," MSEE thesis, Naval Postgraduate School, Sept. 1995.

[12] E. Karapinar, "Modification and Verification of an Antenna Design for the Petite Amateur Navy Satellite Using NEC," MSEE thesis, Naval Postgraduate School, June 1995.

[13] P. J. Overstreet, "Environmental Testing of the Petite Amateur Navy Satellite (PANSAT)," M.S. Astronautical Engineering thesis, Naval Postgraduate School, Sept. 1997.

[14] R. K. Callaway, "An Autonomous Circuit for Measurement of Photovoltaic Devices Parameters," MSEE thesis, Naval Postgraduate School, 1986.

[15] R. R. Oxborrow, "A Microprocessor-Based, Solar Cell Parameter Measurement System," MSEE Thesis, Naval Postgraduate School, June 1998.

[16] B. Lo, "Evaluation and Testing of the Naval Postgraduate School Satellite (NPSAT1) Solar Cell Measurement System," MSEE thesis, Naval Postgraduate School, Sept. 2004.

[17] B. S. Leonard, "NPSAT1 Magnetic Attitude Control System," Paper SSC02-V-7, 16th Annual AIAA/USU Conference on Small Satellites (Logan, UT, Aug. 2002).

[18] T. S. Pugsley, "Characterization, Optimization, and Test of the NPSAT1 MEMS 3-Axis Rate Sensor Suite for Use in Small Satellite Attitude Control," M.S. Space Systems Operations thesis, Naval Postgraduate School, Sept. 2007.

[19] L. E. Koerschner, "Ground Segment Preparation for NPSAT1," M.S. Space Systems Operations thesis, Naval Postgraduate School, Sept. 2007.

[20] L. Brooks, "CubeSats in Education and Industry," presentation, CubeSat Developers Workshop (California Polytechnic State University, San Luis Obispo, CA, April 2007).

[21] C. Kitts, K. Ronzano, R. Rasay, I. Mas, P. Williams, P. Mahacek, G. Minelli, J. Hines, E. Agasid, C. Friedericks, M. Piccini, M. Parra, L. Timucin, C. Beasley, M. Henschke, E. Luzzi, N. Mai, M. McIntyre, R. Ricks, D. Squires, C. Storment, J. Tucker, B. Yost, G. Defouw, and A. Ricco, "Flight Results from the GeneSat-1 Biological Microsatellite Mission," AIAA/USU 21st Annual Conference on Small Satellites (Logan, UT, 13–16 Aug. 2007).

[22] "Boeing Pico-Satellite Mission to Advance Miniature Satellite Technology," (Boeing Advanced Network and Space Systems, St. Louis, MO, 19 April, 2007).

[24] J. H. Newman, D. Sakoda, and R. Panholzer, "CubeSat Launchers, ESPA-rings, and Education at the Naval Postgraduate School," Paper SSC07-III-10, 21st Annual AIAA/USU Conference on Small Satellites (Logan, UT, 13–16 Aug. 2007).

Chapter 10

The UK's First Nanosatellite—SNAP-1

Craig I. Underwood*

10.1 Introduction

In October 1981, NASA launched the first University of Surrey satellite—the 52 kg UoSAT-1 (UoSAT-OSCAR-9), the ninth in a series of satellites designed to support amateur radio communications. The success of this mission led directly to Surrey's second satellite—the 60 kg UoSAT-2 (UoSAT-OSCAR-11), which was designed and built in the record time of just 6 months for launch in March 1984 (see Fig. 10.1). Built by university staff and students, these satellites were groundbreaking in that they were among the first to make extensive use of microprocessor technology in their orbital and payload operations and thus could be said to have ushered in the modern era of the microsatellite.

Both UoSATs were launched "piggy-back" alongside NASA spacecraft (Solar Mesosphere Explorer and LANDSAT-5) and thus, as free-flying but non-propelled spacecraft, ended up in the same sun-synchronous, low Earth orbits as the primary missions. UoSAT-1 continued to operate successfully until its reentry in 1989, while UoSAT-2 remained operational for more than 20 years.

Both UoSATs were designed primarily to support amateur-radio science and engineering experimentation, rather than real-time voice communication, and both played a major role in supporting space education in hundreds of schools, colleges, and universities world-wide.[1,2]

The 1980s saw a rapid increase in the availability of relatively powerful electronic devices, and this stimulated a boom in small satellites by enabling the construction of small, yet sophisticated payload instruments and bus subsystems. By the 1990s, such state-of-the-art technologies, coupled with increased access to space (initially via the European Ariane Structure for Auxiliary Payloads, or ASAP, and then via increased access to former Soviet launchers) enabled microsatellites to carry multiple payloads managed by highly automated onboard computers, supporting a wide range of technology demonstration, commercial, and scientific missions (see Fig. 10.2).

Surrey played a major role in these developments through the continuation of its UoSAT program and beyond. Consequently, Surrey began providing small-satellite platforms and expertise to emerging and established spacefaring nations

*Surrey Space Centre, University of Surrey
Guildford, Surrey, UK

Fig. 10.1. UoSAT-1 (left) and UoSAT-2 (right).

through its spin-off company, Surrey Satellite Technology Ltd. (SSTL).[3] By the end of 2007, SSTL and the university's Surrey Space Centre had launched a total of 28 satellites.

Today, the staggering developments in microelectronics continue apace. The semiconductor industry has integrated analog, digital, and even radio-frequency (RF) functions into single integrated circuits (ICs). Application-specific ICs (ASICs) and various forms of field-programmable gate-array (FPGA) devices are now widely available, significantly reducing the volume, mass, and power needed to support logic, communications, and processing functions. Memory densities have increased exponentially, and microcontrollers are now available that combine many peripheral functions with a central processing unit and high-speed data-bus interfaces.

These advances in commercial-off-the-shelf (COTS) electronic technologies now make it feasible to construct even smaller satellites in the 1–10 kg mass range (nanosatellites), with comparable functionality to the most advanced microsatellites of just a few years ago, but significantly cheaper. This opens up the possibility of designing and building satellites, commercially or as part of an educational exercise, at a price that is widely affordable. In many ways, the relationship of the modern "microsatellite" to "conventional" space technology can be seen as similar to that which existed between the personal computer (PC) and conventional computer technology in the 1980s—that is, the microsatellite is sufficiently low in cost and yet has sufficient utility to serve a much wider user community than would otherwise be possible using a conventional approach.

Fig. 10.2. The Ariane Structure for Auxiliary Payloads (ASAP).

10.1.1 SNAP Nanosat Program

Work began in this area at the University of Surrey in 1995, with a series of undergraduate student projects examining the feasibility of a soccer-ball–sized satellite for Earth imaging. This proved to be a highly successful educational exercise, and resulted in Surrey's first prototype nanosatellite concept: a multifaceted polyhedral satellite with a mass less than 10 kg. The students built prototype hardware, including an onboard computer based on a Motorola 68HC11, a power system with a 14 V NiCd battery providing a regulated 5 V and 9.5 V power supply, and a VHF transmitter. A CMOS miniature video camera payload was also produced, and this was actually incorporated into the TMSat (Thai Paht) microsatellite that was being built at Surrey at the time; it produced some of the first CMOS camera images from orbit when Thai Paht was launched in 1998.[4]

In 1997, work began on the topic of autonomous space vehicle inspection using nanosatellites, and as part of this work, staff and students at the Surrey Space Centre were brought together in a series of coordinated projects to help design a practical, modular, low-cost, multimission nanosatellite platform. Thus, the Surrey Nanosatellite Applications Program (SNAP) was born.[5]

In 1999, the program was adopted by SSTL as part of its research and development activities. This resulted in the final definition of the SNAP-1 mission, which was formally begun in October 1999, when a launch opportunity was identified alongside the SSTL-built Tsinghua-1 microsatellite. The fully tested SNAP-1 spacecraft was delivered to the launch site in early May 2000 for a June launch—thus, the total time from definition to orbit (for a brand new spacecraft design) was just 9 months!

To realize the project, a small team of academics and engineers was assembled, supported in full by SSTL's satellite manufacturing infrastructure. The primary objectives were to develop and prove a modular COTS-based nanosatellite bus, and in the process, to evaluate new manufacturing techniques. The spacecraft would also be used to photograph Tsinghua-1 during its deployment and demonstrate the systems required for future nanosatellite constellations—e.g., three-axis attitude control, precise GPS-based orbit determination, and automated orbital maneuvers. Finally, if propellant reserves allowed, SNAP-1 was to rendezvous with Tsinghua-1 and carry out formation-flying maneuvers. In actuality, when launched in June 2000, SNAP-1 met all its primary mission objectives.

10.2 SNAP Design Philosophy

10.2.1 Motivation—Why Build Nanosatellites?

The continuous drive to reduce size, mass, and power consumption of devices in consumer electronics means that nanosatellites (1–10 kg) can now be built in less than a year with capabilities that previously could only be supported by much larger platforms. This trend has a direct impact on nanosatellite cost, which, given the COTS-based approach advocated here, should be about a tenth of that for current microsatellite technologies. While the launch cost will not scale down in quite the same way, the ability to launch several nanosats together will effectively reduce launch costs on a per-spacecraft basis—and in any case, complete space missions costing $1 million or less are quite feasible.

This means that space missions become affordable to nontraditional space-faring nations—and indeed, even to universities. Nanosatellites are of particular benefit to those missions that require constellations or swarms of satellites or that require a rapid deployment of a space asset—for example, for multipoint remote sensing of Earth, pervasive monitoring for situational awareness, or for space environment monitoring.

Nanosatellites offer a relatively cost-effective way to demonstrate and test new technologies in space—particularly where such technologies would be considered too risky for more expensive conventional spacecraft. These new technologies—microelectromechanical systems (MEMS), low-voltage electronics, micropropulsion, active surfaces, etc.—may in turn greatly enhance the capabilities of future nanosatellites, leading to a virtuous circle of technological development. The short lead time from concept to mission means that such technological developments can be rapid, as demonstrated by Surrey in the case of microsatellites.

The low-cost approach to nanosatellites advocated here also leads to their direct applicability in education. They can become affordable within the context of "laboratory" equipment, and their relative simplicity (particularly with regard to interfaces) means that systems and principles can be clearly taught with direct relevance to systems engineering as applied to major aerospace projects.

The SNAP-1 spacecraft was built to practically demonstrate these advantages—i.e., it was a project to build a spacecraft weighing less than 10 kg for less than $1 million in less than a year to carry out a sophisticated task (remote inspection and orbital maneuvering), and in doing so, to support university space engineering education.

10.2.2 Modular or Integrated Design?

The approach adopted by Surrey for microsatellite missions is to design a modular spacecraft, where a common core bus with standardized electrical and mechanical interfaces can support many different space missions. This decreases cost, but adds some constraints—the bus may not provide exactly what the user wants.

An alternative approach is to integrate the spacecraft functions on to a single board (or even a single chip, as in the case of Surrey's satellite-on-a-chip program—see chapter 19). The spacecraft then becomes simply a set of interfaces, as far as the user is concerned—similar to the concept of plugging peripherals into a PC. It is likely that this single-board approach will become increasingly important in the future; however, at present, the best approach lies somewhere between these two methodologies. Thus, for a low-cost nanosatellite design, the tried-and-tested modular approach was favored—but with increasing levels of integration.

For SNAP-1, a simple standard electrical interface was prescribed for each module, consisting of a regulated 5 V and raw battery (~7.2 V) power connections, with a single bidirectional controller area network (CAN) bus for data transfer. The CAN bus was originally developed for the automotive industry and is a good example of applying COTS technology to space. These connections are provided via a 9-way D-type connector, which is standard to all modules. This simplified the testing and harness design. Indeed, all the SNAP modules (except the power system) can be powered and tested using just the standard 9-way connector. In addition, a single 44-way D-type connector was also allowed for each module to provide for specific point-to-point connections (where absolutely necessary).

All modules—except the onboard computer and machine vision system payload—contain a standard 8 bit CAN microcontroller (the Siemens C515), which provides telemetry and telecommand operations, data transfer, and a degree of subsystem autonomy. The onboard computer and machine vision system use 32-bit StrongARM 1100 RISC (reduced instruction set computer) processors, to which have been added external CAN interfaces operated via the StrongARM's built-in small peripherals interface bus. The decision to standardize on the use of the CAN system for onboard data handling followed its successful demonstration on Surrey's 50 kg FASAT-Bravo microsatellite and 312 kg UoSAT-12 missions launched in 1998 and 1999, respectively.

A standard equipment box was also defined at the beginning of the SNAP program; thus, every module on SNAP-1 has the same external dimensions. This approach allowed the mechanics, avionics, and payload design to occur in parallel, and largely in isolation. It also allowed procurement to start very early. This was essential, given the very tight schedule to meet the launch date.

With modularity also comes flexibility. The standardized modules can be put together to form a nanosatellite in several different configurations, and may even form core units in microsatellites and minisatellites. Indeed, subsequent to SNAP 1, SSTL's spacecraft have all made use of the "nanotray" technology derived from SNAP-1 to reduce the size of the avionics systems, thus leaving more capacity for payloads.

10.2.3 COTS or Advanced Technologies?

In the context of micro- and nanosatellites, much is made of the potential benefits of advanced technologies such as MEMS, multifunctional structures, system-on-chip, etc. However, if we focus on achieving a low-cost multifunctional nanosatellite, rather than getting hooked on "technology for technology's sake," it is clear that many of the engineering problems associated with nanosatellites can be answered by conventional surface-mount electronics and miniaturized but otherwise conventional mechanical systems—that is, through the use of present-day COTS technology. Indeed, on SNAP-1, we showed that, at the time, a miniaturized propulsion system was more economical than a MEMS-based one.[6]

Certainly, as spacecraft masses decrease—to the picosatellite (< 1 kg) size, for example—and performance requirements increase, technologies such as MEMS and highly-integrated electronics will become increasingly attractive, if not essential. However, for SNAP-1, we felt that the modular COTS-based approach was the best way to produce a truly low-cost nanosatellite.

10.2.4 Risk Management

Nanosatellites are by their very nature experimental. While we may strive to make maximum use of flight heritage and experience, there is of necessity much new technology. The risk is therefore potentially greater than one would normally accept, even for a microsatellite mission. It should be remembered, however, that the nanosatellite costs are considerably lower than for a microsatellite, and therefore, the degree of acceptable risk can be much greater (we can replace a failed mission quickly and relatively inexpensively, if necessary).

The nanosatellite, by nature, also has less capacity for redundant systems. We therefore accept that the spacecraft is basically single-string, containing a number of critical systems that could bring the mission to an end if they fail. However, to offset this, the systems often contain fewer components, and therefore should be less prone to failure.

In more than 25 years of designing, building, and operating largely COTS-based spacecraft, Surrey has developed a methodology that has proven successful in managing risk: We rely more on sound design practices, experience, and system-level testing, than on programs of testing individual devices—recognizing that this requires the design engineers to have a good knowledge of the operational environment and to use appropriate design techniques and margins. The COTS electronic devices flown so far have proven to be very robust, with just a few device failures attributable to ionizing radiation; all performed well beyond the planned mission lifetimes.

By minimizing the use of moving parts, using passive thermal-control techniques, employing body-mounted solar cells, and using near-omnidirectional antennas, we have designed and built demonstrably robust spacecraft.[7,8] Although we fly multiple onboard computers on our spacecraft, we do not rely on them for the successful operation. Instead, multiple processors and complex software is used to supply enhanced functionality. By routinely uploading new software, the missions are often upgraded—postlaunch. Indeed, SNAP-1 was the first Surrey spacecraft to fly with embedded software other than simple bootloaders (this was necessary for SNAP-1's machine vision system, which had to activate immediately upon deployment). The onboard computers also act as intelligent "operators in the sky," giving rise to a great degree of operational autonomy. This onboard autonomy is matched in the ground station, allowing multiple satellites to be routinely operated with minimal human intervention—and with all of the cost savings that implies. These principles of robustness and flexibility were all incorporated into the SNAP-1 nanosatellite design.

10.3 Design and Construction of SNAP-1

10.3.1 Methodology and Approach

The SNAP-1 project was highly constrained in terms of time and budget; nonetheless, the spacecraft had to be constructed to meet the rigors of the launch and space environment and meet our normal quality standards. This meant using SSTL's normal procurement and assembly and test procedures, and having a rigorous design review process—preliminary design review, critical design review, test-readiness design review, and flight-readiness design review. The only difference to SSTL's normal practice was that we were effectively our own customer—the spacecraft was funded internally. We therefore minimized bureaucracy as much as we could.

Because SNAP-1 was a new spacecraft, a structural qualification model was built and tested to flight-qualification levels; however, as is often the case with Surrey's microsatellites, we did not construct an engineering model. Instead, we moved straight from prototype hardware to the flight model (see Fig. 10.3). The fact that the interface was so tightly constrained meant that we were able to rapidly test the hardware once constructed, and the whole assembly, integration, and testing phase went remarkably smoothly.

Key to the rapid development of the spacecraft was ready access to effective computer-aided design and manufacture (CAD/CAM) software. For the printed circuit board (PCB) design, we used our standard CAD package, CADSTAR. For mechanical design, we used a new package, SolidEdge, which proved invaluable.

One of the manufacturing experiments conducted as part of SNAP-1 was to use, for the first time, flexi-rigid PCBs in all the subsystems. This did prove initially problematic because the flexible tracks were easily damaged during assembly. Moreover, if a new revision of a PCB had to be produced, the cost was much greater than it would have been for a conventional PCB. We did consider a flexi-rigid wiring harness to begin with, but we soon decided that, given the

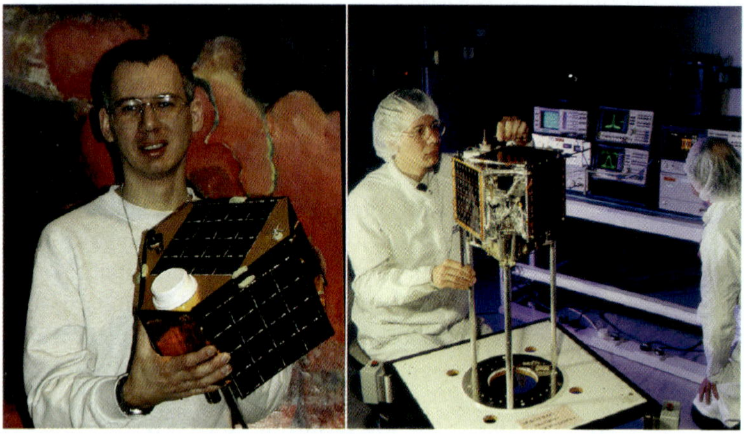

Fig. 10.3. Author with SNAP-1 design concept model (left) and flight model (right).

relatively few wires needed, a conventional harness would be more economical. The flexi-rigid PCBs proved effective in the end, and the lessons learned during the production of SNAP-1 were valuable for the adoption of this technology in subsequent spacecraft.

10.3.2 Mechanics and Thermal Design

Having decided to make SNAP a modular spacecraft, the principal design questions were, "How big should these modules be?" and "How should they be configured?"

Surrey's microsatellites make use of a standard "microtray" module format. A module has approximately 300 × 300 mm of usable PCB area and is usually 22 mm deep (though this varies with the individual module). Some subsystems occupy half a microtray. For SNAP-1, we wanted a module size that was challenging, and yet a practical step down in size—and thus settled on the "Eurocard" PCB size (approximately 160 × 100 mm). In terms of usable height, ~13 mm looked practical, which gave the module box a wall height of just over 20 mm. Thus, a fixed standard module box size of 168 × 122 × 20.7 mm was proposed.

As to layout, the Surrey microsatellites simply stack the modules one on top of another, and have the wiring harness run down one side. This arrangement could have been adopted, making SNAP-1 a rectangular prism—perhaps with body-mounted solar panels to make a miniaturized version of our microsatellites. Instead, we proposed a triangular prism (which is very strong and compact), with the modules forming the three sides of the satellite. Stacking three modules per side gave a nine-module configuration (about all that would be needed), which occupied an area of 341 mm in diameter—rather less than the cross-sectional area of one of Surrey's microsatellites. By arranging all the connectors on one side of the modules, a simple "ring" harness could be placed on the top of the structure to link the modules together.

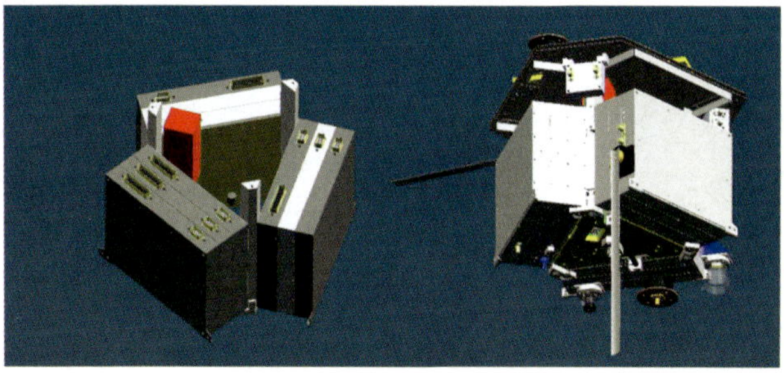

Fig. 10.4. SNAP-1 structural layout (SolidEdge CAD representation).

The final structural layout of SNAP-1 thus comprised standardized aluminum-alloy electronics boxes, arranged as three stacks of three modules in a triangular prism structure (Fig. 10.4). There was a 0.3 mm aluminum screening plate between each module and a 1 mm outer screen for radiation protection (in addition to the outer honeycomb panels). The modules were held together by four tie-bars, one in each corner.

The structure is closed with aluminum honeycomb panels. Three side panels and the top panel each house solar cells to form four fixed body-mounted solar panels. The bottom panel is actually part of the propulsion system, which is inserted into the central triangular void. This central volume is also used for some of the attitude determination and control sensors and actuators—notably, a three-axis flux-gate magnetometer, a pitch-axis momentum wheel, and two of the three magnetorquer rods. The height of the spacecraft, end-facet to end-facet, was approximately 200 mm (326 mm from the base of the launch vehicle's attachment fitting to the tip of fixed S-band antenna).

By design, the spacecraft geometry results in each module box having three sides with a good field of view to space or Earth. This simplifies the accommodation and integration of payloads that require external access for sensors. Additionally, it simplifies the RF harnessing, as the RF module boxes effectively act as their own EMC bulkhead. There is additional aluminum-foil EMC screening to close the internal volume to RF interference, and all antenna connections are external to this (Fig. 10.5a).

Thermal control is passive, achieved via thermo-optical tapes applied to the boxes and honeycomb panels. Typical solar panel temperatures ranged from –25°C to +65°C during the course of an orbit, while the internal temperatures ranged from approximately 10°C to 30°C, with 10–15°C of temperature cycling per orbit. The thermal design, as for all previous SSTL spacecraft, was carried out using proprietary software.

Standard aerospace metal alloys and aluminum-skinned honeycomb panels (in preference to composites) were used throughout the spacecraft. Although this results in some mass penalty, it greatly simplifies design, reduces manufacture time, and maximizes the ability to deal with design changes late in the program.

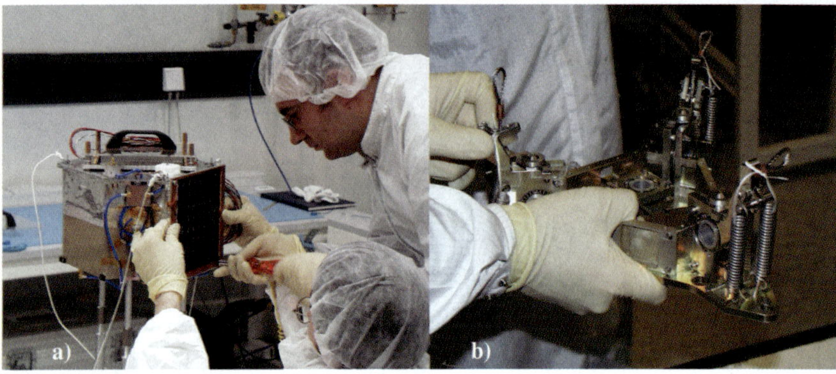

Fig. 10.5. SNAP-1 in assembly, integration, and test phase (a); SNAP-1 separation mechanism (b).

A low-shock three-point separation system was also designed for SNAP-1 (Fig. 10.5b). This employed three tensioned cables to latch the spacecraft down during launch with two (redundant) pyrotechnic guillotine-cutters to release the cables for separation. The spacecraft was held down by three "hooks" spring-loaded to snap open upon deployment. Three more separation springs, one next to each release clamp, gave SNAP-1 its correct push-off velocity (1 $msec^{-1}$). The whole system fit within the spacecraft envelope. As with Surrey's microsatellites, a separation switch was housed in SNAP-1's power system, which activates only after the spacecraft has physically separated. An engineering model of the separation system was constructed, and test separations were carried out on the ground prior to flight. The system proved to be robust and reliable, giving SNAP-1 a gentle push.

10.4 Avionics

The avionics systems of SNAP-1 filled six of the nine modules (see Fig. 10.6). The power system took up two modules, one for the batteries and battery charge regulators, and one for the power conditioning and power distribution systems. The RF system also took up two modules, one for the VHF receiver, and one for the S-band transmitter. The last two modules were filled by the onboard computer and the guidance, navigation, and control (GNC) system.

10.4.1 Power System

The key design principles behind the SNAP-1 power system were:

- Make efficient use of power at low voltage. Use the most efficient and (if possible) the largest solar cells available. Use the power system controller to implement maximum power-point tracking.

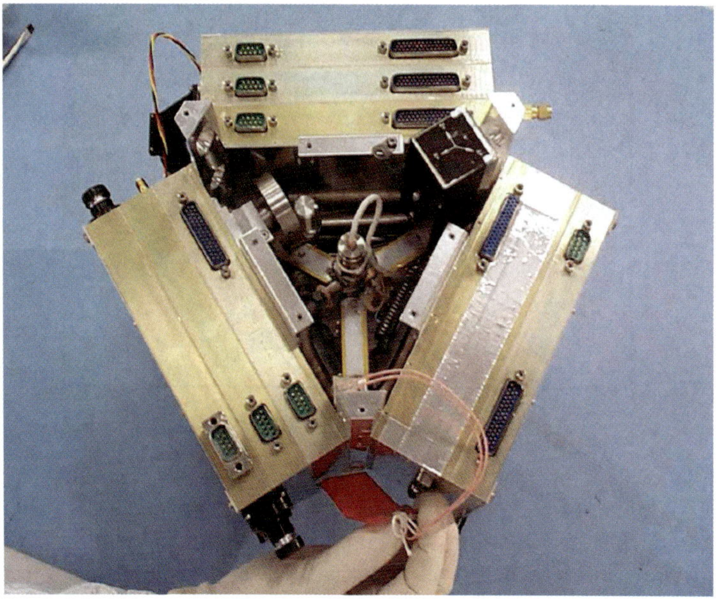

Fig. 10.6. SNAP-1 internal layout, showing the avionics stacks.

- Ensure safety via hardware design. Make sure that the power system (particularly the battery) is maintained in a safe state irrespective of any software in the controller.
- Strive for flexibility. Provide power rails that may be used directly, or that may be used by systems to generate their own power rails.
- Strive for modularity. Keep the battery/battery charge regulator and power conditioning and distribution modules separate so that multiple units can be flown on larger spacecraft, if required.

Given these principles, it was decided that the fastest and least risky approach was to use a smaller version of the power system used in Surrey's microsatellites at the time (Fig. 10.7).

10.4.1.1 Power Generation

SNAP-1 generates power through four body-mounted solar panels, each populated with 20% efficient GaAs/Ge cells producing, nominally, 0.5 A per panel at 12 V (~6 W) at 25°C. Because of the mechanical configuration of the satellite, the total orbit average power available is also approximately 6 W; however, the minimum required bus power is only 650 mW (which represents only the receiver and power system being on). Thus, under nominal power conditions, these systems, together with the onboard computer and the attitude control system, can be in continuous use. In addition, the GPS navigation system, payloads, and transmitter can be activated periodically as required.

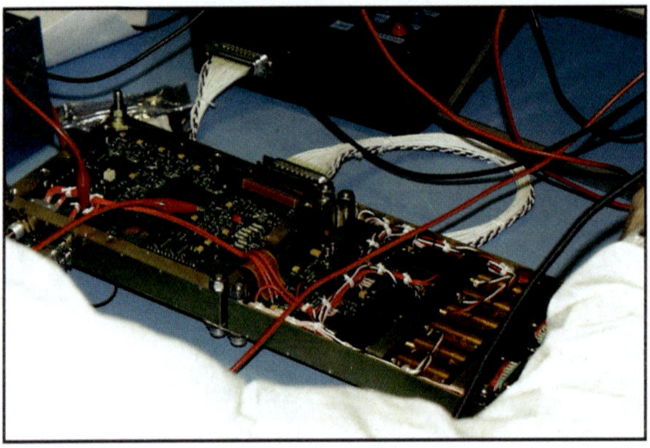

Fig. 10.7. SNAP-1 power system.

Solar-array temperatures, voltages, and currents are measured by the power system via its built-in CAN microcontroller. Each solar panel has it own independent battery charge regulator, which implements maximum-power-point tracking in hardware. This approach gives maximum power transfer efficiency and flexibility for future panel configurations. Each regulator is rated at 10 W, stepping down the 12 V from the solar arrays to the 7.2 V unregulated bus. The battery charge regulators are housed in the same module box as the battery for ease of connection; they are self sustaining and do not rely on the battery to power their operating circuits. This is a safety feature to prevent a complete discharge of the battery from killing the mission.

The operating frequencies of the battery charge regulators are synchronized together using a phase-locked loop. Synchronization is required to prevent beat frequencies interfering with sensitive RF systems such as the receiver.

The regulators charge a single 10 Whr battery consisting of six size-A Sanyo KR-1400AE Cadnica cells serially linked to give a nominal 7.2 V. The total mass of the battery pack is 270 g, which gives a relatively high energy density of 37 Whr/kg. Overcharging is prevented by a temperature-compensated end-of-charge voltage trigger to switch the battery charge regulators into a trickle-charging mode. These NiCd cells were selected over lithium-ion cells because they were deemed more economical at the time (1999), they could better withstand a degree of overcharging should a cell fail, and they could better supply large peak currents if necessary. We recognized that using these small cells, SNAP-1 would have a relatively large depth-of-discharge fraction, and this would limit the lifetime of the spacecraft (unmatched commercial cells operated to high depths of discharge typically give 800–1000 charge/discharge cycles). Given our previous flight experience with this cell manufacturer, we set the mission lifetime to be one year (i.e., approximately 5,500 charge/discharge cycles). In actuality, the battery operated well for more than 8,000 cycles, effectively giving SNAP-1 an 18-month operational lifetime.

10.4.1.2 Power Conditioning and Distribution

The power conditioning system provides a regulated 5 V and an unregulated 12 V supply for the spacecraft systems. The maximum total current that can be safely drawn from the power conditioning system is ~3 A. In total, the battery can sustain up to ~10 A (~60 W) output for a few minutes. This feature was designed to allow future SNAP vehicles to support thrusters or payloads with high power demands. The power conditioning system also provides deep-discharge protection by shutting off when the battery voltage falls below approximately 6.5 V.

The power distribution system uses miniaturized low on-resistance (RDS-on) field-effect transistor power switches, which are based on the design used in SSTL microsatellites at the time. There are eight switched 5 V lines and eight switched V_{batt} lines and a single switched 12 V line. All switches provide overcurrent protection; essential systems (e.g., the receiver) are connected directly via a fuse (for ground testing).

Power system telemetry is gathered via a built-in CAN microcontroller. Due to the high number of telemetry channels on the power module (32), there was a need to interface the signals to the CAN microcontroller via an analog multiplexer. The telemetry channels provide solar-array voltage, current, and temperature (for each of the four panels); battery voltage, current, and temperature; power-conditioning module input current; and a subsystem current monitor (one per power switch, for a total of 16). Telecommanding of the power system was also implemented via the CAN microcontroller. This allowed each battery charge regulator to be enabled or disabled or put under computer control and allowed each subsystem to have a regulated power line (5 V) or regulated power line and raw battery line (5 V, ~7.5 V) pair, switched on or off as necessary.

10.4.2 RF Systems

The key design principles behind the RF systems were:

- Maintain contact under all attitude scenarios. Use near-omnidirectional antennas, but tailor the pattern to give a uniform signal strength when the spacecraft is in its nominal attitude.
- Make it compatible. Use tried and tested frequencies and modulation schemes to make the satellite compatible with existing microsatellite ground stations, but also include enhanced communications capabilities such as basic Consultative Committee for Space Data Systems (CCSDS) compatibility—potentially enabling NASA or ESA ground stations to work with SNAP-1.
- Use VHF/S-band. The VHF uplink has a long established heritage. It is simple, reliable, and inexpensive and provides an adequate quality and data-rate. At the time, microsatellites typically used a UHF downlink, but this is now becoming a limiting factor in supporting the required downlink data rates at a sufficient quality, given the crowded spectrum

and the continuing buildup of interference on this frequency band. S-band offers a higher link speed in a more compact package and gives a clearly defined upgrade path; it is also relatively underutilized in the amateur radio context, so obtaining and coordinating frequency allocations is easier.

- Make it simple and modular. The receiver and transmitter functions should be separate so that, if required, multiple receivers and transmitters may be flown to give module-level redundancy. Use single fixed-frequency systems with minimum tuning-requirements to minimize testing and setup.

Given these design principles, the RF communications system architecture required two modules: a VHF receiver and an S-band transmitter, each with its own CAN-microcontroller-based telemetry and telecommand (TTC) node (Fig. 10.8). As with all the SNAP-1 spacecraft modules, the primary control and monitoring of these systems is achieved via the CAN bus; however, the RF systems also have dedicated links to the onboard computer to provide enhanced TTC and digital communications.

The receiver CAN node acts as the primary TTC link with the ground, and as such, it is permanently powered up on orbit. The transmitter node may be powered on or off at will, and the downlink can be switched between a number of data sources by means of CAN telecommands—issued from the ground or from any other intelligent system in the spacecraft. In larger spacecraft, multiple receivers and transmitters may be flown, allowing for module-level redundancy.

10.4.2.1 Antennas

A design criterion applied to all Surrey spacecraft is that they must remain safe and controllable in any attitude. This greatly simplifies the launch interface, as we can deploy the satellite in a passive mode, eliminating any possible interference with the launch vehicle. This also allows the spacecraft to be operated from the ground in case of a failure in the attitude determination and control system. Omnidirectional (or near omnidirectional) antennas are therefore necessary (Fig. 10.9).

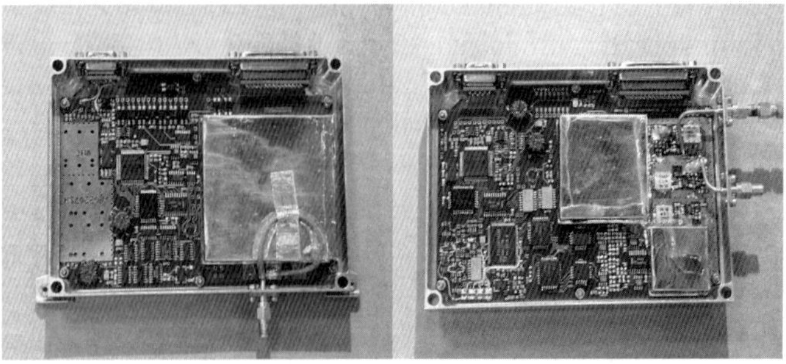

Fig. 10.8. SNAP-1 VHF receiver (left) and S-band transmitter (right).

The VHF receive antenna is a simple quarter-wave monopole, which has an impedance set to allow a physically short antenna ~20 cm (8 in.) long. The antenna is a sprung blade ~1 cm (0.5 in.) wide. This had been used successfully on all the previous Surrey spacecraft. At 2 m wavelength, the nanosatellite body does not act as a simple ground plane, and so the precise placing and matching of the antenna is a function of the mechanical size and configuration of the satellite. This was modeled using an RF finite-element modeling package and measured using RF representative physical models once the SNAP-1 configuration had been defined.

The S-band transmit antenna comprises a pair of quarter-wave ground-plane antennas. The high operating frequency (~2.4 GHz) means that the antennas are physically small, < 5 cm (2 in.) long. Each has a solid disc ground-plane, which is used to define a reference plane independent of the satellite body. As with the VHF antenna, the nanosatellite body will have an impact on the precise antenna pattern, and so this was also modeled and measured prior to flight.

The antennas' sensitivity patterns were tailored to provide a nearly constant power profile over a ground station when the spacecraft is in its nominal attitude, while also allowing a usable link in case it's not.

10.4.2.2 VHF Receiver

The VHF receiver is a crystal-controlled device operating at a single selected frequency in the 140–150 MHz band. The uplink modulation scheme is frequency-shift keying (FSK), which ensures that the receiver is compatible with existing SSTL microsatellite ground-station uplink facilities.

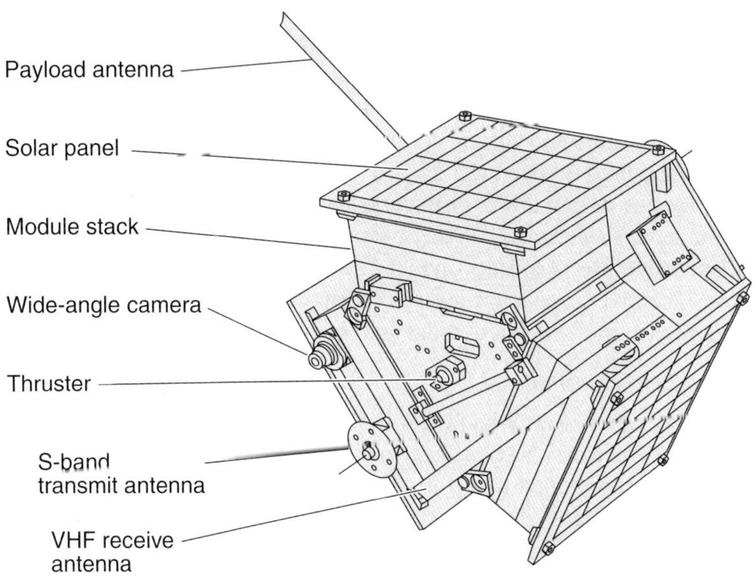

Fig. 10.9. SNAP-1 external layout, showing the antenna configuration.

The nominal uplink data rate is 9600 bps, which ensures that the signal bandwidth remains within the limits specified for amateur radio communications. For non-amateur use, the receiver can be operated up to a maximum of 76.8 kbps. The data rates and operating frequencies were selected prior to flight.

The receiver is controlled via a CAN microcontroller, which provides the module's TTC functions and acts as the primary TTC link for the spacecraft.

In the primary mode of operation, digital data are uplinked to the spacecraft asynchronously using a CAN-based packet communications protocol. The receiver's CAN microprocessor passes these packets to the spacecraft's CAN TTC bus. All spacecraft systems are linked to this bus and will obey any CAN telecommands addressed to them, if they are powered up. The CAN command system is directly compatible with previous Surrey spacecraft CAN TTC systems and makes use of existing flight software. The CAN microcontroller is programmed prior to flight to perform these functions, with the code held in read-only memory.

The receiver also sends the uplink asynchronous digital data stream directly to the onboard computer, which may then exercise more complex control functions such as operations scheduling and special data formatting. This bypass link also allows program code and data to be uploaded to the onboard computer efficiently using Surrey's existing "bootloader" protocol.

As an added feature, the receiver sends "received data" and "received data clock" signals to the onboard computer so that synchronous digital communications can be supported. In this way, the satellite can be made directly compatible with the existing systems at Surrey's Mission Operations Centre (and compatible stations worldwide) using the AX.25 packet communications protocol; basic compatibility with CCSDS TTC formats were also supported.

The receiver has two special telemetry channels, receiver signal strength and discriminator output, which help characterize the performance of the receiver. Both of these values may be requested via the CAN TTC system. As with other modules, temperature and current consumption are also made available—the current being measured in the power system.

The receiver design draws on experience gained from existing Surrey microsatellite receivers (UoSAT-5, FASAT-Bravo, UoSAT-12, and Tsinghua-1) and makes extensive use of surface-mount COTS technology to reduce size, mass, and cost. The fixed-frequency design means that minimal tuning is required, which greatly simplifies system setup and testing.

10.4.2.3 S-Band Transmitter

The transmitter operates at S-band (2.4–2.5 GHz) and, like the receiver, is a single-channel fixed-frequency device. The transmitter may operate on any frequency within this band, provided that it is set on a 100 kHz boundary. The frequency was fixed prior to flight.

The transmitter delivers between 100 and 200 mW of RF power from ~3.3 W. A power splitter feeds two external quarter-wave monopole antennas—one on the top, and one on the bottom of the spacecraft.

The downlink modulation scheme is nominally binary phase-shift keying (BPSK), but quadrature phase-shift keying (QPSK) and offset quadrature phase-shift keying (OQPSK) are also supported. The nominal data rate is fixed at 38.4 kbps to ensure compatibility with commercially available S-band ground-station equipment; however, the transmitter is capable of operating at data rates from 10 kbps to 10 Mbps. The downlink is nominally a 38.4 kbps asynchronous digital link; however, through the use of the onboard computer, it can also be made compatible with AX.25 and CCSDS protocols.

The transmitter power supply is switched, and thus the transmitter may be operated as needed to make maximum use of the available power budget.

The transmitter contains a CAN microcontroller, which provides the necessary TTC functions as well as the packet formatting (i.e., the synchronization, spacecraft ID, and cyclic-redundancy checks (CRCs) necessary to transport CAN telemetry to the ground).

Through this node, the transmitter performance parameters are made available (voltage standing-wave ratio [VSWR], temperature, etc.), and control is exercised over the downlink multiplexer. This may be switched to the CAN bus (via the transmitter's CAN microcontroller), the onboard computer's asynchronous digital serial output, or the onboard computer's synchronous digital output. In this way, the basic health of the spacecraft can be ascertained directly through the CAN TTC, or the onboard computer may be used to provide enhanced functionality such as summarized health status checks.

By having both a synchronous and asynchronous digital connection to the downlink multiplexer, the onboard computer can be used to support a wide range of specialized communications and payload operations. To aid this, special communications functions such as Viterbi and differential encoding, V.35 CCITT and Intelsat scrambling/descrambling, etc., are performed by the logic in the field-programmable gate-array (FPGA) in the transmitter.

The transmitter output power is, in itself, sufficient to be picked up by a moderate-gain (30 dBi, 2.3 m diameter) steerable antenna on the ground.

10.4.3 Onboard Computer and Data Handling

None of the existing SSTL flight computers would fit in the SNAP-1 nanotray, so we needed to develop a completely new onboard computer—and we viewed this as an opportunity to implement state-of-the-art technology. However, flight experience from UoSAT-12 had shown just how effective the CAN bus was for handling onboard data. Thus, the key design principles for the SNAP-1 onboard computer and data-handling system were:

- Make maximum use of existing software and hardware support. The CAN TTC bus was well established and well supported by existing Surrey ground and flight data-handling systems.
- Use simple interconnections. Essentially, CAN is a two-wire interface. Bus contention and priority is dealt with by the CAN hardware itself.

- Achieve flexibility and upgradability via the onboard computer. The onboard computer is the "operator in the sky." It should ease and enhance spacecraft and payload operations, but should not be relied on for basic spacecraft functionality.
- Use the onboard computer to achieve autonomy. In normal operations, the ground station and onboard computer arrange spacecraft operations and data transfer without operator intervention. This is the approach taken at Surrey, where some 16 or more spacecraft are typically operated by a single person (part time!).
- Minimize software in the loop. Some basic software is needed, such as bootloaders and TTC handling, but no system should depend upon complex software being ready and debugged prior to flight (though in fact, the machine vision system payload did need preloaded flight code because it had to activate immediately upon spacecraft separation without ground support).

Using these design principles, we adopted the CAN-bus architecture to link all the subsystems of SNAP-1. However, we did introduce new technology in terms of the processors used in the onboard computer and machine vision system (Fig. 10.10). Both these systems used the 220 MHz StrongARM 32-bit SA1100 RISC processor (running nominally at 88 MHz for the onboard computer).

The onboard computer had a 4M × 8 bit program memory, protected against single-event upset by a double-bit correcting code implemented in hardware. Low-level programs and a bootloader were stored in a 1M × 16 bit flash memory. Unfortunately, the move away from the Intel chip architectures used on Surrey's microsatellite meant that all the SNAP-1 flight code had to be developed anew.

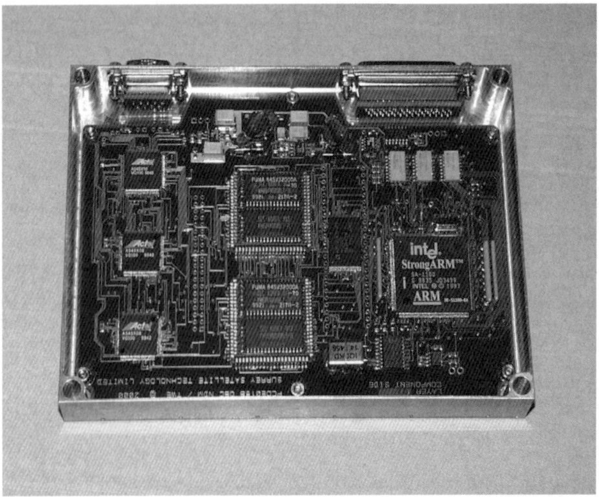

Fig. 10.10. SNAP-1 StrongARM high-speed RISC processor-based onboard computer.

The Surrey microsatellites all run an operating system that allows multitasking. With SNAP-1, there was no time to develop or port an operating system to the new hardware. Thus, a multitasking executive was written, programmed in C, which allowed the spacecraft tasks to be run as subroutines of the main code. This proved to be highly effective in practice.

The onboard computer greatly enhances the spacecraft's capabilities, providing automatic control—but, as with all of our spacecraft, it is not relied upon for basic operations.

10.4.4 Guidance, Navigation, and Control

The GNC system is designed to allow three-axis momentum-biased operation with autonomous orbit maneuvers (Fig. 10.11). Attitude knowledge is provided via a compact internal three-axis flux-gate magnetometer and a Kalman-filter attitude estimator. Attitude stabilization and control is provided by a Surrey-designed miniature pitch-axis momentum wheel, running nominally at 2000 rpm, together with three miniature magnetorquer rods for momentum dumping and magnetic attitude control (if required). Full three-axis stabilization (to better than 1 deg in roll and yaw and 0.2 deg in pitch) and orbit maneuvers are controlled via the onboard computer.[9] A 12-channel GPS navigation system, based on the GEC-Plessey Orion GPS receiver connected to a single patch antenna, provides accurate orbit knowledge (~15 m position uncertainty).[10]

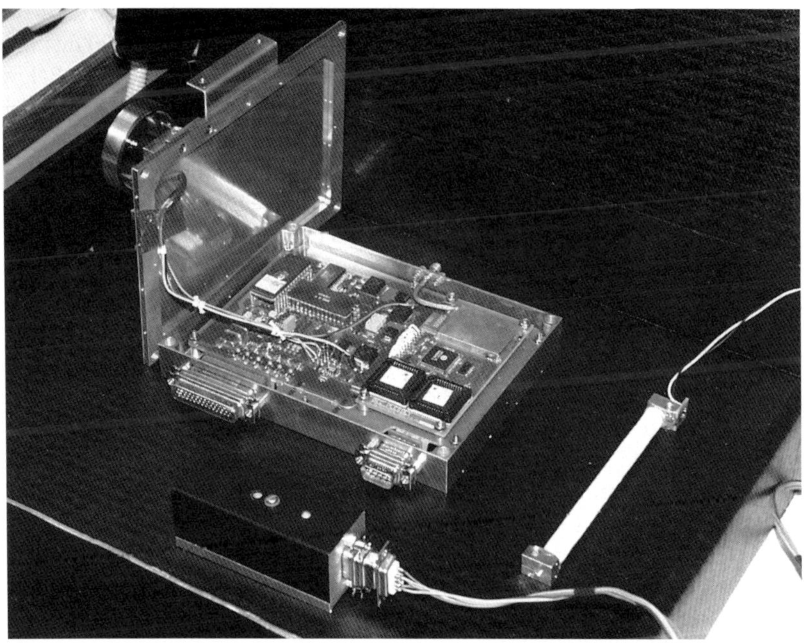

Fig. 10.11. SNAP-1 GNC module, three-axis magnetometer, and one of three magnetorquers.

10.5 Propulsion System

The role of the propulsion system was to demonstrate, for the first time, orbital maneuvers by a nanosatellite. The plan was to bring SNAP-1 to rendezvous with another Surrey-built spacecraft, Tsinghua-1, launched at the same time. The calculated change in velocity (ΔV) requirement was 1 msec^{-1}, using a single axial thruster, firing as closely as possible through the center of gravity of the spacecraft. The mission imposed other requirements as well: the propulsion system would need to withstand a temperature range of 0–40°C, perform multiple firings, generate a maximum thrust of 100 mN, and operate from a supply voltage of 7.2–9 V dc with low power consumption. Needless to say, it would also have to be inexpensive and accommodate a very tight delivery schedule.

We determined that these requirements were best met by a butane cold-gas system (Fig. 10.12). The system had to fit within the central triangular void of SNAP-1, with a single thruster positioned along the central axis. This meant that there was no room for a single, centrally mounted propellant tank. The system also had to be designed so that SNAP-1 could fly without it, if it wasn't ready in time!

To meet the tight deadline, we sought existing COTS designs and hardware, where possible from UK or European sources so as to avoid any possible delays associated with obtaining the necessary export licenses.

Thus, a British company, Polyflex Aerospace Ltd., was selected as the valve supplier. They had already developed a cold-gas thruster, in conjunction with SSTL, under a program sponsored by the British National Space Centre. The thruster valve was designed for use with regulated nitrogen, giving 100 mN thrust at 4 bar chamber pressure; hence, the use of butane at 2 bar for SNAP-1 was well within its capabilities. The most obvious feature of the complete propulsion pipework assembly is that there is no propellant tank, as such. Instead, the propellant is stored in 1.1 m of coiled titanium tubing, providing 65 cm^3 of storage volume. This has a number of advantages over a conventional tank. For example, there is easily verified compliance with MIL-STD-1522A (and subsequent regulations), as the system does not contain a pressure vessel—only pipework and fittings. Compliance with the standard merely requires a minimum burst of 4× maximum operating pressure, which was demonstrated in the system proof test. The material costs are low—standard Airbus titanium tubing was used. There is an even distribution of mass along the tube. A fill valve is welded directly to one end of the coiled tube assembly. The other end is connected to a titanium manifold. This manifold has a pressure transducer and temperature sensors for system monitoring and contains stainless-steel mesh discs that act both as filters and heat transfer elements. It has an external heater (a 15 W commercially available resistor), which was intended to ensure propellant vaporization during firings, and includes an isolation valve and a thruster valve. As the system contains only a small amount of propellant, it is sensitive to leakage. Consequently, all joints are welded or contain double seals. The isolation valve protects against leakage from the thruster valve.

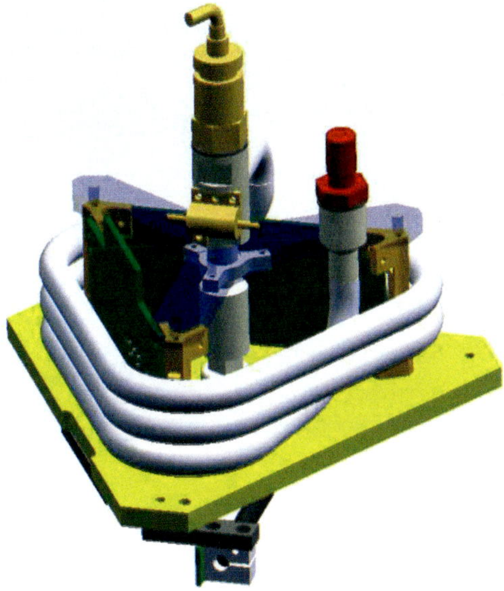

Fig. 10.12. SNAP-1 propulsion system (SolidEdge CAD model).

A total of 32.6 g of butane was loaded into the system three weeks before the spacecraft was shipped to the launch site (Plesetsk in northern Russia). This was done in a standard laboratory fume cupboard at the Surrey Space Centre and took two engineers three hours to perform. Due to the low toxicity of butane, no additional safety equipment was required—the major precaution was simply to avoid any potential ignition source. Once completed, no further propulsion system operations were required at the launch site. We estimated that this quantity of propellant would give ~3.5 $msec^{-1}$ ΔV with ~50 mN of thrust.[11]

10.6 Payloads

Three payloads were flown on SNAP-1: a VHF spread-spectrum communications package for a commercial customer; a UHF intersatellite link aimed at deriving relative position via differential GPS to aid rendezvous with Tsinghua-1; and a machine vision system for remote inspection of Tsinghua-1, imaging deployment, and Earth imaging at meteorological resolution.[12]

The machine vision system (Fig. 10.13) consists of four ultraminiature COTS monochrome CMOS video cameras—three with wide-angle lenses (90 deg field of view) and one with a narrow-angle lens (20 deg field of view). The lenses were standard closed-circuit TV-type lenses, stripped down and modified for use in vacuum. The narrow-angle camera had a COTS photographic near-infrared filter to give good differentiation between land, sea, and cloud and to give a clear view through the atmosphere. The other cameras were left unfiltered to maximize

Fig. 10.13. SNAP-1 machine vision system payload with CMOS cameras.

sensitivity for low-light imaging. From a 700 km altitude, ground resolution was ~500 m for the narrow-angle camera and ~3 km for the wide-angle cameras.

Each CMOS imager array was organized as 350 × 288 pixels and produced a standard 1Vp-p composite video output. The four video streams were digitized by the machine vision system using a software-based method—i.e., the entire video stream was recorded, including sync pulses. This meant that we could simplify the hardware, requiring only fast video analog-to-digital converters, a processor, and fast-access memory. Sync separation and data formatting could then be left to software. This proved to be a good design decision, as the hardware, thus simplified, worked the first time.

Processing and storage of images was carried out by a 220 MHz StrongARM SA1100 processor equipped with 2 Mb of flash memory and 8 Mb of SRAM.

The machine vision system also carried 10 kb of flash RAM-based operating code. This comprised an asynchronous bootloader for changing the machine vision system mission code once in orbit and an "autograb" sequencer designed to grab a time-lapse video sequence during SNAP-1's and Tsinghua-1's deployment.

As the CMOS cameras only had a three-wire interface—power, ground, and composite video output—there was no mechanism for controlling the exposure level. This was done by automatic on-chip circuitry in the cameras themselves. One consequence was that the cameras would tend to overcompensate for the blackness of space, giving and overexposed image. These were primitive devices compared to today's CMOS webcams! A useful additional feature of the machine vision system is that it could act as back-up onboard computer, if required.

10.7 Flight Results

After completing a successful full environmental verification test program (vibration, spin-balance, thermal-vacuum, and EMC), SNAP-1 was delivered to the launch site in May, 2000 (Fig. 10.14).

On 28 June 2000, it was launched together with the Tsinghua-1 microsatellite and a Russian COSPAS-SARSAT satellite, Nadezhda, on a Kosmos launch vehicle into a 700 km sun-synchronous (near polar) orbit.

SNAP-1 and Tsinghua-1 were both carried into orbit by Nadezhda. After Nadezhda had been released from the launch vehicle's upper stage and the orbit established, SNAP-1 was commanded to separate from Nadezhda, and five seconds later, Tsinghua-1 was similarly commanded. Immediately, SNAP-1 powered up its machine vision system and began to capture images of the deployment.

The launch and early operations proceeded smoothly and successfully. Receiving data proved relatively easy once SNAP-1 had risen above the radio horizon. The relatively simplified TTC operations proved to be surprisingly effective.

10.7.1 Machine Vision System Operations

The spacecraft was acquired by Surrey's Mission Control Centre on the first pass, and the data received showed that, as planned, the machine vision system payload had successfully acquired a sequence of video images showing the deployment of Tsinghua-1 from Nadezhda (Fig. 10.15). From this video sequence, we could accurately estimate the deployment tumble rate (later confirmed by Tsinghua-1's telemetry data).

Analysis of the video sequence also confirmed initial data from the SNAP-1's GNC system indicating that SNAP-1 was tumbling at 26 deg/sec. This was not expected, as during all ground tests, the deployment worked very smoothly. We concluded that something unexpected had occurred immediately (or within a second or so) after deployment. Nonetheless, this attitude rate was within the range that the attitude control system could accommodate, and a couple of days after deployment, the GNC system's automatic magnetorquer-based rate-damping controller was activated. Within a day, the high initial rate of tumbling was

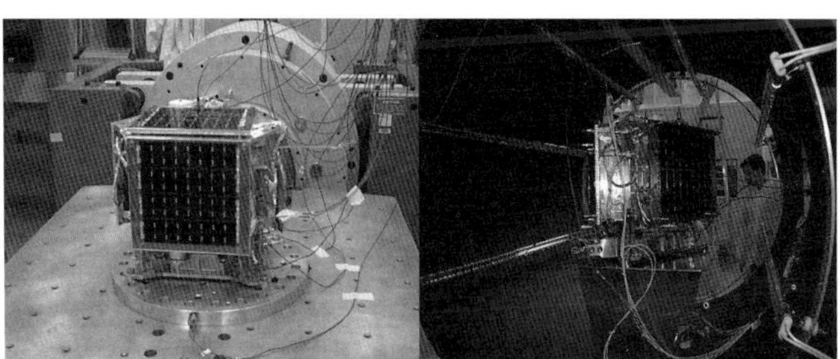

Fig. 10.14. SNAP-1 in environmental verification testing—vibration test (left) and thermal vacuum test (right).

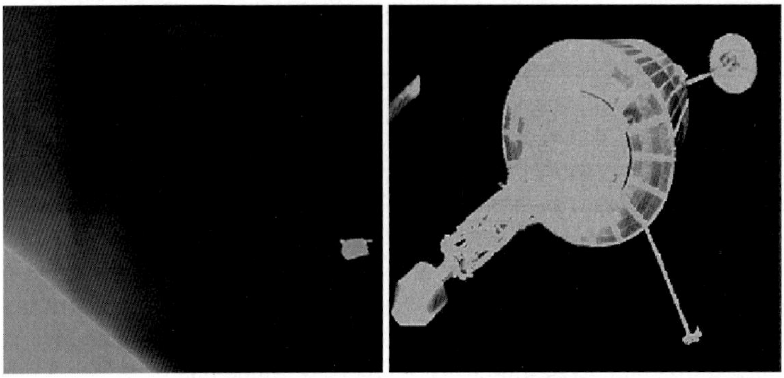

Fig. 10.15. SNAP-1 machine vision system image of Tsinghua-1 and Earth 10.5 sec after SNAP-1 separation (left); SNAP-1 machine vision system image of Nadezhda 2 sec after SNAP-1 separation (right).

completely damped, and the spacecraft was placed into a slow, stable rotation about its pitch axis.

Once we had finished with attitude control experiments and SNAP-1 was fully controlled by its pitch momentum-wheel system, we were able to use the machine vision system in a series of imaging experiments—acquiring still images and video sequences of Earth under differing lighting conditions and putting SNAP-1 into (ground) target tracking mode so that we could view the same target from multiple angles (Fig. 10.16). Unfortunately, we were not able to get SNAP-1 back sufficiently closely to Tsinghua-1 to reacquire it after the initial deployment.

10.7.2 Attitude and Orbit Control System Operations

Once the initial tumble-rate had been damped down, it was discovered that, left to its own devices, SNAP-1's thruster-axis would tend to track Earth's magnetic field vector. This was unexpected. Initially, we thought this might be due to current loops caused by ground-return currents in the spacecraft structure or in the harness. However, we had experienced this effect in ground testing, and had taken steps to reduce them. In any case, the harness was designed using good EMC principles, and current loop areas were minimized by the proper routing of power and ground returns. Calculations showed that any residual current would be insufficient, and further investigations determined that the probable cause was the small amount of residual magnetism in the propulsion system's solenoid valves.

The degree of the disturbance was significant, but fortunately, SNAP-1 had enough torque capability to break the magnetic lock. So, by August, the attitude control system was sufficiently functional for orbital maneuvers to be started.

To improve stabilization, accurate knowledge of the magnitude and direction of the magnetic disturbance was needed. Thus, a Kalman-like recursive estimator was used to determine the disturbance moment components from in-orbit attitude and rate estimates provided by the GNC system. Over the course of the following

Fig. 10.16. SNAP-1 machine vision system image of Greek islands (near-infrared narrow-angle camera—500 m ground sample distance) on 31 March 2001.

months, the attitude control algorithms were refined, and new software uploaded to the spacecraft, so that by late November, full-accuracy nadir-pointing momentum-biased operation had been achieved using the pitch-axis momentum wheel with rms error residuals over 24 hours being less than 3 deg in roll and yaw and 0.3 deg in pitch (1σ).[13]

During this period, we also checked out the onboard systems and payloads, including the spread-spectrum payload and the intersatellite link, and found these to be working satisfactorily.

One of the more ambitious elements of the operations plan was to try to maneuver SNAP-1 back into the vicinity of Tsinghua-1 in the months following orbital injection. The ability to do this depends critically upon the initial relative orbits of the two spacecraft, which in turn depends upon the orientation of the carrying vehicle during the separation sequence—a factor outside our control.

Because of the low mass of SNAP-1 (6.5 kg) compared to Tsinghua-1 (~50 kg), atmospheric drag would cause SNAP-1 to lose altitude faster than Tsinghua-1; thus, it was hoped that SNAP-1 would be deployed into a slightly higher initial orbit. Unfortunately, the deployment was such that SNAP-1 ended up with a semimajor axis around 2 km lower than that of Tsinghua-1. This significantly increased the ΔV required to bring the two back together for a rendezvous, as we first had to overcome the altitude difference and then boost SNAP-1 to an even higher altitude to phase them correctly.

Once SNAP-1's attitude control system had been commissioned and verified, the propulsion system could be tested. This began with the first firing on 15 August 2000. Initial tests showed that the propulsion system functioned correctly, and so a series of firings were begun to raise SNAP-1's semimajor axis above that of Tsinghua-1. SNAP-1 would automatically fire its thruster several times a day. Each firing would lead to an attitude disturbance, and so the spacecraft would automatically damp this out before resuming the next firing.

By September 19, SNAP-1 had gained 2.6 km in altitude and was ~1 km higher than Tsinghua-1. A total of 80 firings had been performed, mostly of 3 sec duration, under the full automatic control of the onboard computer. Unfortunately, the mission was carried out at a period close to solar maximum, and the high levels of solar activity meant that the atmospheric density was particularly high. SNAP-1 continued to fall rapidly—up to 20 m per day in terms of absolute height, and ~10 m per day relative to Tsinghua-1, as recorded by their GPS receivers (Fig. 10.17).

Because SNAP-1 was initially at a lower altitude that Tsinghua-1, its orbital velocity was greater; thus, by the time SNAP-1 had climbed to the same altitude as Tsinghua-1, it was a considerable distance ahead—approximately 15,000 km!

To allow Tsinghua-1 to catch up, SNAP-1 had to be sent to an even higher orbit. Thus, a further series of firings was undertaken from 20–29 October 2000, which boosted SNAP-1's altitude another 350 m before the propellant was finally depleted. In total, 98 firings were made, with a total firing duration of 297.1 sec.

Taking atmospheric drag effects into account, the propulsion system raised the altitude of SNAP-1 by the equivalent of ~4 km (with a corresponding total ΔV of 2.1 m/sec). The ΔV would have been greater, were it not for the fact that the early firings definitely vented liquid rather than gaseous butane. It was determined that this was because there was insufficient ullage built into the very compact thruster.[14] This was a valuable lesson, and subsequent thrusters built by SSTL were given greater ullage, more heater power, and a "sponge" to hold the liquefied propellant away from the propellant tank outlet. These measures proved entirely successful.

By means of these maneuvers, SNAP-1 dropped back to Tsinghua-1's orbital altitude on 18 March 2001, at a separation distance of approximately 2,000 km. Thus, while true rendezvous was not achieved, the agility and maneuverability of SNAP-1 under automatic control was clearly demonstrated.

10.8 SNAP-1's Legacy

SNAP-1 was an internally funded research and development experiment. Although we carried out the mission for around $1 million—a low cost in space terms—this was a significant investment of our own money, and one which we would now find difficult in an increasingly competitive commercial environment. So, as yet, there has been no SNAP-2.

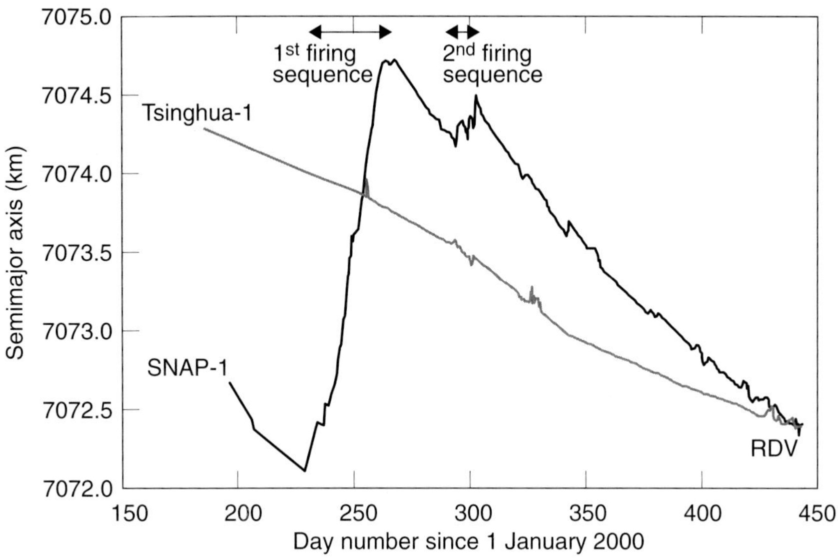

Fig. 10.17. SNAP-1 and Tsinghua-1 semimajor axis from onboard GPS data.

However, the SNAP-1 experience lives on in other ways. The technology produced has enabled the avionic suites of all the subsequent SSTL spacecraft to be much smaller than before. As a result, SSTL's spacecraft have been able to carry payloads of greater mass and volume on small satellite platforms, producing, for example, the world's first operational international Disaster Monitoring Constellation (DMC) and the first of the European Galileo Satellite Navigation Test-Bed satellites (Giove-A). These missions have truly brought small satellite technology into the domain of operational spacecraft meeting real user needs. The DMC spacecraft use a cold-gas butane-based propellant system to great effect—indeed, the constellation control requires this to work effectively over their seven-year planned mission lifetime. This would not have been possible without the flight experience gained with SNAP-1's propulsion system.

A second SNAP may not have been built, but SNAP-1 avionics have been used by the U.S. Air Force Academy to form the core systems of their educational FalconSAT program.[15] SNAP-1 has also provided the core spacecraft design in a number of European space system studies.

Finally, the modular COTS-technology design approach of SNAP-1 has been continued in our PalmSat concept:[16] In 2001, we challenged Surrey students to build a "coffee-cup" sized spacecraft that could perform a significant space mission. The development of PalmSat is now at a stage similar to that of SNAP-1 just before it was adopted by SSTL. In less than a year from kick-off, the SNAP-1 mission could now be carried out (or indeed exceeded) with a ~2 kg satellite using "credit-card" sized avionic modules.

10.9 Conclusions

Following its pioneering work with the UoSAT microsatellite program of the 1980s and 1990s, the University of Surrey and its spin-off company SSTL have continued to push at the boundaries of small satellite technology through the development of the SNAP-1 nanosatellite. The 6.5 kg spacecraft demonstrated that sophisticated, agile nanosatellites can be built in rapid time scales (< 9 months) and at very low cost (~$1 million), with all the technology needed to achieve formation flying—i.e., orbital position determination and orbital maneuverability, full attitude determination and control, intersatellite communications, and a high degree of onboard autonomy and intelligence through the use of powerful onboard computers.

SNAP-1's legacy lives on in the enhanced capabilities in Surrey's current generation of small satellites, and itself paves the way for the deployment of (potentially) large numbers of nanosatellites to synthesize functions previously unobtainable from single higher-cost spacecraft.

10.10 Acknowledgments

The author wishes to thank all those at SSC/SSTL and other external companies who have contributed to the success of the SNAP-1 mission and legacy.

10.11 References

[1] M. N. Sweeting, "The UoSAT-1: A Review of Orbital Operations and Results," *J. IERE*, **57** (5) (Supplement), pp. S184-S194 (Sept./Oct. 1987).

[2] M. N. Sweeting, "The University of Surrey UoSAT-2 Spacecraft Mission," *J. IERE*, **57** (5) (Supplement), pp. S99-S115, (1987).

[3] M. N. Sweeting, "Microsatellite and Minisatellite Programmes at the University of Surrey for LEO Applications and Effective Technology Transfer," *International Symposium on Satellite Communications and Remote Sensing* (Xi'an, China, 20–22 Sept. 1995).

[4] M. N. Sweeting and Sitthichai Pookyaudom, "TMSAT: Thailand's First Microsatellite for Communications and Earth Observation," IAA-96-IAA.11.1.02, *Proceedings of the 47th International Astronautical Congress* (Beijing, China, Oct. 1996).

[5] C. I. Underwood, M. Crawford and J. W. Ward, "A Low Cost Modular Nanosatellite Based on Commercial Technology," SSC98-V-4, *Proceedings of the 12th AIAA/USU Conference on Small Satellites* (Logan, UT, Aug. 1998).

[6] A. M. Baker, D. Gibbon, and C. Underwood, "Micropropulsion from SNAP to PALMSAT: When Does MEMS Become the Way Forward?" AIAA 2002-5759, *Proceedings of the IAA Nanotech 2002 Conference* (Houston, TX, Sept. 2002).

[7] M. N. Sweeting and C. I. Underwood, "Small Satellite Engineering and Application," in *Spacecraft Systems Engineering* (3rd Ed.), edited by P. Fortescue, J. Stark, and G. Swinerd (John Wiley & Sons, Chichester, UK, 2003).

[8] J. W. Ward and C. I. Underwood, "Microsatellite Mission Operations," in *Cost-Effective Space Missions*, edited by D. G. Boden and W. J. Larson (McGraw-Hill, New York, 2003).

[9] W. H. Steyn, Y. Hashida and V. Lappas, "An Attitude Control System and Commissioning Results of the SNAP-1 Nanosatellite," SSC00-VIII-8, *Proceedings of the 14th AIAA/USU Conference on Small Satellites* (Logan, UT, Aug. 2000).

[10] M. Unwin, P. Palmer, C. Underwood and M. Oldfield, "The SNAP-1 and Tsinghua-1 GPS Formation Flying Experiment," *Proceedings of the ION GPS-2000 Conference* (Salt Lake City, UT, Sept. 2004).

[11] D. Gibbon, J. Ward and N. Kay, "The Design, Development and Testing of a Propulsion System for the SNAP-1 Nanosatellite," SSC00-I-3, *Proceedings of the 14th AIAA/USU Conference on Small Satellites* (Logan, UT, Aug. 2000).

[12] R. Lancaster and C.I. Underwood, "The SNAP-1 Machine Vision System," SSC00-III-6, *Proceedings of the 14th AIAA/USU Conference on Small Satellites* (Logan, UT, Aug. 2000).

[13] W. H. Steyn and Y. Hashida, "In-Orbit Attitude Performance of the 3-Axis Stabilised SNAP-1 Nanosatellite," SSC01-V-1b, *Proceedings of the 15th AIAA/USU Conference on Small Satellites* (Logan, UT, Aug. 2001).

[14] D. Gibbon, C. Underwood, M. Sweeting and R. Amri, "Cost Effective Propulsion Systems for Small Satellites Using Butane Propellant," IAF-01-S.1.11, *Proceedings of the 52nd International Astronautical Congress* (Toulouse, France, Oct. 2001).

[15] J. Sellers, L. Sauter, C. Underwood and J. Ward, "Bounding The Problem: Microsatellite Design Using Commercial-Off-The-Shelf Architecture," *Proceedings of the 2002 IEEE Aerospace Conference* (Big Sky, MT, Mar. 2002).

[16] C. Underwood, V. Lappas, A. da Silva Curiel, M. Unwin, A. Baker, and M. Sweeting, "Using 'PalmSat' Pico-Satellite Technologies to Meet Mission Scenarios," IAC-04-P.5-A.01, *Proceedings of the 55th International Astronautical Congress* (Vancouver, Canada, Oct. 2004).

Chapter 11

Microvariability and Oscillations of STars (MOST) Mission: Canada's First Microsatellite and Space Telescope

Robert E. Zee* and Jaymie M. Matthews†

11.1 Introduction

The microsatellite Microvariability and Oscillations of STars (MOST) continues to exceed expectations several years after its launch in June 2003. Continuously observing up to 30 stars at a time for up to seven weeks, the microsatellite measures oscillations in brightness to deduce their ages, core compositions, and sizes. MOST also has expanded its science mission to encompass exoplanetary science, including the hunt for exoplanets as small as Earth, studies of the interactions of exoplanets and their parent stars, and measurements of the properties of their atmospheres. Considering its original planned lifetime of one year to observe only 10–15 stellar targets, MOST has far exceeded both its scientific and technical objectives many times over.

The 53 kg suitcase-sized satellite was developed in less than four years for a mere 9 million Canadian dollars. To put this in context, when MOST was conceived in 1997, microsatellites had been used mainly for simple technology demonstration and store-and-forward and bent-pipe communication missions. Three-axis attitude control was not readily available for satellites under 100 kg because of the lack of components that could meet the mass, volume, and power restrictions. MOST's science requirements, on the other hand, imposed a pointing requirement that demanded pitch and yaw stability to within 25 arcsec. No one believed that such control was possible, or that Canada—a country that had not built a science satellite in more than 35 years—could build a barrier-breaking microsatellite on such a tight schedule and an even tighter budget.

Kieran A. Carroll (then of Dynacon Enterprises Ltd., Toronto) and Slavek Rucinski (then of the Centre for Research in Earth and Space Technologies, CRESTech, Toronto) had hypothesized that a highly capable microsatellite could be developed with the availability of key enabling technologies and that the microsatellite could perform a demanding space astronomy mission. When Rucinski joined the Canada-France-Hawaii Telescope in 1997, Jaymie Matthews (University of British Columbia, Vancouver) became the principal investigator

*Space Flight Laboratory, University of Toronto Institute for Aerospace Studies
Toronto, Ontario, Canada
†Department of Physics and Astronomy, University of British Columbia
Vancouver, British Columbia, Canada

for the mission. At the end of 1997, a Phase-A study was completed for the Canadian Space Agency (CSA) under the Small Payloads Program. On the basis of the Phase-A report, MOST was awarded funding for the following year.

That same year, the Space Flight Laboratory (SFL) was founded at the University of Toronto Institute for Aerospace Studies. SFL established a full-time staff, acquired training from the Radio Amateur Satellite Corporation (AMSAT), and installed facilities to develop substantial portions of the MOST satellite. MOST, depicted in Fig. 11.1, was a significant accomplishment for a team and a country that had never built a microsatellite. SFL continues to build inexpensive, capable satellites, and operates them (including MOST) through its mission control center.

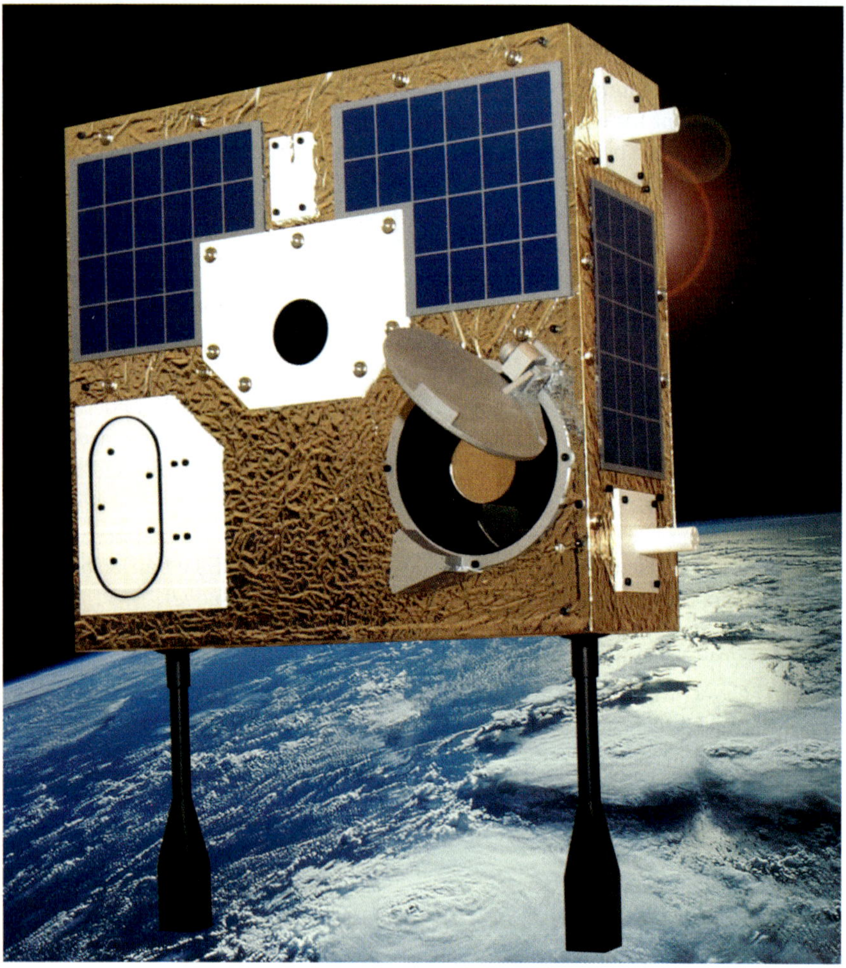

Fig. 11.1. A depiction of MOST on orbit.

11.2 The Science

The primary science objectives of MOST include measuring subtle oscillations in the brightness of the sun and other stars to perform seismic analyses of their interiors; determining the age of nearby "metal-poor, subdwarf" stars (very old intruders from the galaxy's halo passing through the solar neighborhood), which will in turn allow an independent lower limit to be set on the age of the universe; and detecting the first reflected light from orbiting exoplanets to determine the composition of their atmospheres, clouds, and even weather patterns.

Stellar seismologists exploit extremely tiny surface vibrations of stars—detected through brightness oscillations with amplitudes of only a few parts per million—to probe stars' hidden interiors and ages and address big questions such as, "What is the age of the universe?" A big telescope is not needed to study many of the sun's nearer neighbors in the galaxy, which are bright enough to provide large photon fluxes and high signal-to-noise. However, atmospheric turbulence—plus the day-and-night cycles inherent to single-site, ground-based observation—mean that the telescope must be placed in space. Even there, a big budget is not necessary, if proven optical and detector technology and advances in microsatellite attitude control are made.

MOST features a small optical telescope (15 cm aperture) equipped with a CCD (charge-coupled device) photometer designed to return unprecedented photometric precision (light intensity changes of 1 ppm) and frequency resolution (~0.1 Hz) on stars other than the sun. Given the fact that this instrument (see Fig. 11.2) is carried aboard a microsat bus about the size and mass of a suitcase, the Canadian public has come to know MOST as the "Humble Space Telescope."

Fig. 11.2. The MOST telescope (15 cm aperture).

11.2.1 Probing Mysterious Planets: Following in Galileo's Footsteps

MOST was originally designed to detect rapid brightness oscillations (with periods of minutes to days) in sun-like stars to seismically probe their interiors. However, once the project had passed the critical design phase, the MOST instrument proved to be more sensitive and versatile than originally conceived. It also had the potential to detect reflected light from some of the giant planets recently discovered around other nearby stars. The amount of light reflected and scattered back to Earth by such an exoplanet varies as the planet goes through illumination phases such as those of the moon or of Venus, as first observed telescopically by Galileo in the early 1600s. Therefore, in the detailed design phase of the project, the MOST team added an exciting new science application without changing the hardware or software design or the selected orbit—and at no added cost.

The amplitude and shape of the reflected light curve of an exoplanet depends on the planet's size, its orbital inclination and eccentricity, and most importantly, its atmospheric composition (which determines its albedo). However, the exoplanet signal can be very subtle—about 1 part in 10^5 relative to the brightness of its parent star—with orbital periods of only a few days for the exoplanets with the smallest orbits. Simulations[1] indicated that MOST should be able to detect this signal for exoplanet systems such as 51 Pegasi (the first planetary system to be discovered around a sun-like star, with planetary mass at least half that of Jupiter and an orbital period of 4.2 days) and tau Boötis (see Fig. 11.3 for a simulated plot of electrons versus time illustrating an orbital period of 3.3 days).

MOST is the first operational instrument in history capable of detecting these signals and giving direct information about the atmospheres of these mysterious worlds. At the same time, data on the oscillations of the parent stars would specify the ages of these stellar and planetary systems—an important test of models of how these planets formed and evolved and a new insight into how our own solar system formed and evolved.

11.2.2 Why Go to Space?

It is not currently possible or practical to perform photometry from Earth's surface at the precision of 1–10 ppm required for solar-type seismology or exoplanet light curves. For example, if the sun were observed as a point source (like a distant star), then the net amplitude of its integrated oscillations would be only a few centimeters per second in velocity and a few micromagnitudes in brightness. Such levels lie about two orders of magnitude below present detection limits for ground-based measurements (although efforts to improve the spectroscopic limits are under way). In fact, even the solar oscillations have been detected in brightness only from space, by the ACRIM (Active Cavity Radiometer Irradiance Monitor) bolometer aboard the Solar Maximum Mission, the IPHIR (InterPlanetary Helioseismology by IRradiance measurements) experiment aboard the Phobos probe, and the VIRGO (Variability of solar IRradiance and Gravity Oscillations) instrument on the SOHO (Solar and Heliospheric Observatory) satellite.

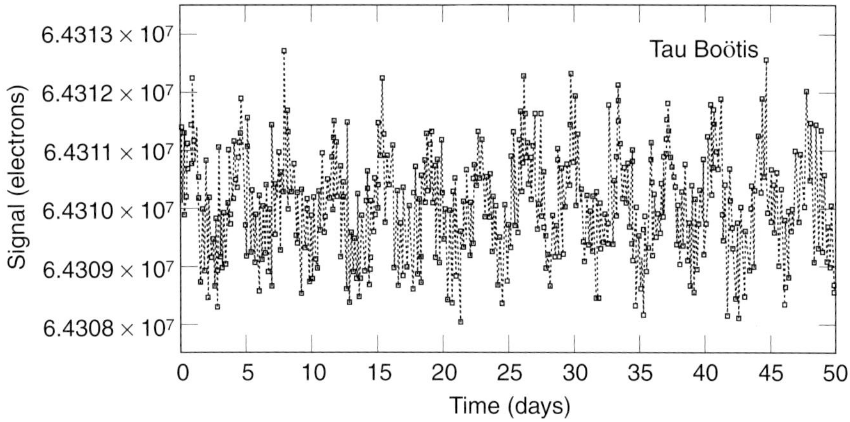

Fig. 11.3. Simulated MOST "observations" of a star with an exoplanet orbiting every 3.3 days.

Performing seismology of sun-like stars and the detection of reflected light from exoplanets requires precise detections of photometric signals as low as a few parts per million and long, nearly uninterrupted monitoring of each star for weeks at a time. Neither is easy to achieve, but getting both simultaneously is virtually impossible from the surface of Earth.

11.2.2.1 Scintillation Noise

Turbulent cells of air at altitudes of about 10 km in Earth's atmosphere have slightly different temperatures that change their refractive indices. As these cells, on the order of 10 cm across, are carried over the beam of a telescope, they modulate the intensity of starlight randomly and rapidly. This is the cause of star "twinkling," that is, atmospheric scintillation. Scintillation is the dominant source of photometric noise in measurements of oscillation amplitudes below about 0.1% (1000 ppm) and at timescales of only a few minutes. This noise term depends on the position of the star relative to the horizon, the exposure time, and the altitude and aperture of the telescope. Scintillation noise decreases with increasing aperture. As the light collecting area becomes larger, the incoming stellar signal is integrated over a larger number of turbulent cells, and their effects begin to average out more effectively. To avoid an impractically large telescope on Earth and succeed with a small aperture, the telescope has to be placed above Earth's atmosphere.

11.2.2.2 Continuous Monitoring

From Earth, an astronomer has only limited options for monitoring a star continuously for times much longer than half a day. The astronomer can either observe the star from a site near one of the poles during prolonged night, or operate a network of telescopes spread in longitude around the globe so that at least one of the sites can observe the star in darkness at all times. Both options require clear weather everywhere for weeks at a time.

The first approach is frustrated by the less-than-ideal conditions for stellar photometry in the Arctic and Antarctic (although a South Pole stellar observatory is planned at Concordia Dome C that may achieve near-space-like photometric conditions). The latter approach has been used successfully by the Global Oscillation Network Group (GONG) to study the low-amplitude velocity oscillations of a very bright source (the sun), and by the Whole Earth Telescope (WET) to study the relatively high-amplitude (10%) oscillations of faint sources (pulsating white dwarfs). Both approaches are expensive in terms of money, facilities, and human resources; but neither can avoid the scintillation limit. Again, a telescope placed above the atmosphere seems to be the logical choice, if it can be done cheaply in comparison to a ground-based alternative.

11.2.2.3 The "Humble Space Telescope"

The MOST instrument is a 15 cm aperture Rumak-Maksutov telescope (see Figs. 11.2 and 11.4) fed by a flat "periscope" mirror that allows it to sit in the allowed volume of the satellite's instrument bay. Maksutov optics gives MOST a wide field of view (about 2 deg diameter) so that it can serve as a guiding star sensor for the microsat attitude control as well as a science instrument. The telescope feeds starlight to a pair of frame-transfer CCDs—one for guiding, one for science. Through frame transfer, the integrations can be stopped without the need for a mechanical shutter. In fact, the MOST instrument is designed to have no moving parts (to reduce cost and increase reliability). The structure is athermal, so that the optics maintain focus across the wide temperature range between alignment on the ground and the orbital environment.

11.3 The Satellite

The MOST satellite design originates from trying to fit several items—as large a telescope as possible, all the electronics to interface with the telescope, the satellite bus equipment that allows the storage and transmission of the science data to the ground, the attitude control system (ACS) hardware that is required to maintain better than 25 arcsec pointing accuracy for the telescope, and a power system to provide enough power to the satellite—into a package that can meet the requirements for launch as a secondary payload on a Delta II launch vehicle (originally, MOST was planned to be launched with Radarsat-2 on this vehicle). MOST probably bears little resemblance to the satellite design that would have resulted had the actual launch vehicle—the "Rockot" out of Plesetsk, Russia—been identified from the start of the project, given that the available volume was not as restrictive as the Delta II secondary allocation, which forced MOST to be more flat than cubical. However, a microsatellite is often subject to some launch uncertainty, especially as a secondary payload on someone else's launch.

Fig. 11.4. The MOST tray stack and telescope.

A periscope mirror allows the long axis of the telescope to lie perpendicular to the aperture of the telescope and therefore fit into the Delta II secondary payload physical constraints. Attached to the telescope, separate from the satellite bus structure, is a two-stage passive cryocooler capable of maintaining the focal plane of the telescope at a temperature of –40°C within 0.1°C. The detector is biased cold, and a trim heater is used to achieve this temperature stability.

The satellite structure is based on a tray stack design. The structure consists of aluminum trays that house the satellite's electronics, battery, radios, and attitude actuators. These trays are stacked (see Fig. 11.4), forming the structural backbone of the satellite. To this backbone, the telescope is mounted with its barrel parallel to the stacking axis. Six aluminum honeycomb panels, acting as substrates for solar cells and carriers for attitude sensors, enclose the tray stack and telescope assembly, forming the box seen in Fig. 11.1. An actuated telescope door mounted on the star facing side of the satellite protects the telescope focal plane from direct stares at the sun should the satellite tumble or lose attitude lock.

11.3.1 Satellite Electronic Architecture

The satellite electronic architecture is shown in Fig. 11.5. The housekeeping computer, which is central to the design (and the figure), is an off-the-shelf product from the late 1990s (when the project began) that has been modified to meet

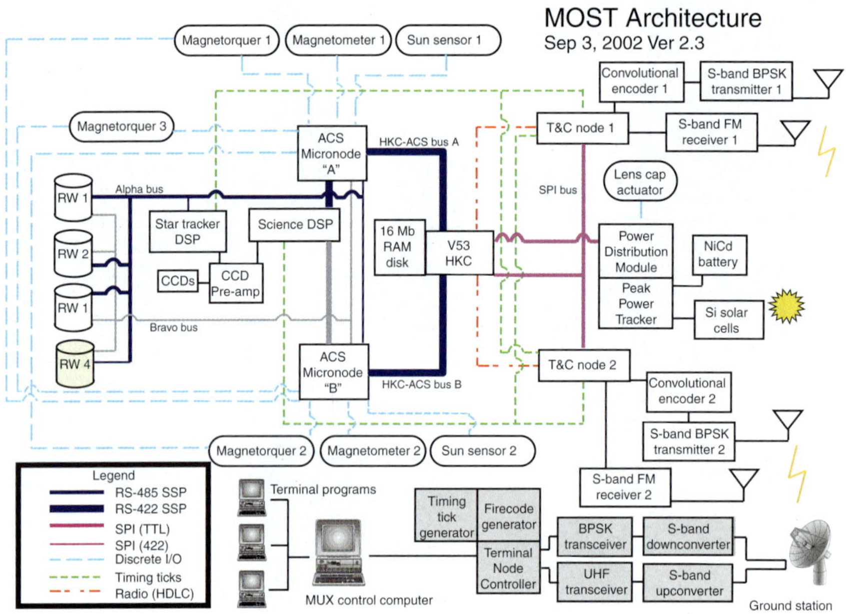

Fig. 11.5. The MOST architecture.

MOST requirements. Based on a NEC V-53 processor, the computer's crystal frequency has been increased from 9 MHz to 29 MHz to accommodate the processing demands of the mission. It interfaces with the rest of the satellite through a custom interface card that provides power, serial, and digital I/O connections. The housekeeping computer's main tasks include receiving, executing, and distributing commands and/or files uploaded from the ground, and collecting and transmitting engineering and science data to the ground.

In the figure, roughly from the V53 to the right, the satellite core design is typical of AMSAT-based designs[2] of the late 1990s in terms of processing power, communication over the AX.25 protocol, silicon solar cells, and nickel-cadmium battery. This part of the figure includes the main housekeeping computer, radio transmitters and receivers (including support electronics), and the power system for the satellite.

MOST employs two 0.5 W RF output BPSK transmitters and two 2 W FM receivers. All radios operate at S-band frequencies. Sufficient downlink margin is maintained by using a 0.5 rate convolutional code, implemented on a custom board. On the uplink, FM receivers provide a simple, robust, and low-cost means to talk to the satellite. Both receivers and transmitters connect to custom telemetry and command nodes that serve as modems and telemetry collection devices. To maintain omnidirectional coverage, one receiver/transmitter pair is located on either side of the satellite, connected to quadrifilar antennas. Each radio operates on its own frequency. Thus, the appropriate transmitter is selected based on the receiver being used.

The power subsystem is based on centralized switching and decentralized regulation feeding from an unregulated bus that varies between 9 and 14 V. Power regulation occurs through switching power supplies to maximize conversion efficiency (power is very limited in a satellite of this size—35 W in fine pointing operations and only 9 W in safe-hold or tumbling operations). While this poses EMC/EMI challenges for the science digital signal processor (DSP) that must read its CCD array with almost zero noise, these challenges have been met.

The power system switches are controlled via the housekeeping computer. Two levels of load shed (a system to turn off nonessential loads if the battery is deeply depleted) protect the satellite from unrecoverable battery drainage, allowing contingency operations to resume in safe-hold mode. All power lines have overcurrent protection.

For energy storage, a NiCd battery provides power during eclipses and supports peak power draws from equipment such as the transmitters. High-efficiency (17%) silicon solar cells on all sides of the satellite generate energy to recharge the battery and provide power for fine pointing and safe-hold operations. Peak-power tracking hardware and software (run by the housekeeping computer) maximize the available power to the satellite subsystems.

To the left of the V53 computer in Fig. 11.5 is the equipment that enables MOST to provide scientific contributions, making it unique for a microsatellite. These are the electronics that support the telescope, and the ACS hardware and electronics. The ACS equipment consists of magnetometers, sun sensors, a star tracker, and magnetorquers and reaction wheels. The key developments here have been the use of reaction wheels for three-axis attitude control and the development of a star tracker that is a fundamental part of the science telescope. Together, these enable the satellite bus to maintain pointing accuracy of better than a few arcseconds.

Science and star tracker images are taken on dual 1024 × 1024 CCD arrays that share the focal plane of the telescope. Each CCD is connected to a preamplifier, and to analog and digital electronics boards (Fig. 11.6). These boards are based around a Motorola 56303 DSP and provide digital control and analog-to-digital conversion of the signals from the CCDs. The instrument computers are designed to provide CCD readings while minimizing disturbances from switching power supplies.

There are four attitude control modes for the satellite.

- **Safe hold:** The satellite power balance is essentially positive in all practical orientations. Therefore, this is an uncontrolled state in which there is no active attitude control. In this mode, the focus is nominally on commissioning or recovery operations.
- **Detumbling:** This mode involves using the magnetometers and magnetorquers to implement B-dot control ("B-dot" referring to the use of the

Fig. 11.6. The MOST instrument computer.

rate of change of Earth's magnetic, or "B," field in the feedback control law) to slow the tumble rate of the satellite so that coarse pointing control can be executed. Normally, this is used after kick off from the launch vehicle.

- **Coarse pointing:** After the satellite is detumbled, the ACS uses sun sensors and magnetometers to determine the spacecraft attitude while using reaction wheels to orient the main solar array toward the sun and to point roughly in the direction of science interest. The magnetorquers are used to desaturate the reaction wheels.
- **Fine pointing:** The ACS uses the star tracker to determine spacecraft attitude to an accuracy of 3 arcsec. The reaction wheels control attitude, while the magnetorquers desaturate the reaction wheels. (Normally, it takes only a few minutes to transition from one target

field to the next—a "coarse slew." The transition from coarse to fine pointing is accomplished within a few seconds once the star maps have been uploaded to the spacecraft.)

The attitude control computers (ACS nodes) are also based on the Motorola 56303 DSP. The DSP acts as the fundamental processing unit that runs the ACS software. The computers provide analog control of the magnetorquers, power, and analog-to-digital conversion of the magnetometer and sun sensor signals, as well as RS-485 connections to the main housekeeping computer, the reaction wheels (which contain their own microcontroller), the star tracker, and science DSP boards. Nominally, only one ACS node is operational. The second is designed as a cold spare to add redundancy while not incurring significant additional testing requirements.

All computers have error detection and correction (EDAC) hardware and software to correct for bit errors induced by radiation. Single-event latch-ups are corrected by power cycling the affected device.

To ensure that components within the satellite operate at suitable temperatures, a combination of passive surface treatments are used, including aluminum, gold, and silver Teflon tapes. In the event that the satellite enters a cold state caused by a disadvantageous attitude relative to the sun, resistive heaters are used to keep the battery and trays sufficiently warm. During fine pointing operations, a passive radiator cools the telescope focal plane so as to minimize thermal noise in the CCD readout.

11.3.2 Ground Stations

Three ground stations in Toronto, Vancouver, and Vienna are used to download data from MOST. The primary control station is located at SFL in Toronto (Fig. 11.7), while the secondary stations (Vancouver and Vienna) are controlled and coordinated over the Internet. Although the basic mission can be accomplished with just one ground station, additional science data can be acquired using the secondary stations.

The stations used for MOST communications are based on an amateur radio core station operating at VHF and UHF frequencies. They are upgraded with S-band transverters and BPSK transceivers connected to a 2.1 m parabolic antenna (downlink) and a 45-element loop Yagi (uplink). The antennas are mounted on a heavy-duty, precisely controlled rotator located atop a 20 ft (6.1 m) tower (Fig. 11.7b).

The ground station radios are connected to a custom terminal node controller (combination modem and serial communications controller), which is in turn connected to a computer that coordinates multiple terminals each running interface software for specific components on the satellite (see Fig. 11.5). Through this system, terminal users have a virtual link to their satellite hardware of interest. The terminal node controller also generates "firecodes" or emergency commands to reset satellite hardware. A "timing tick" generator also implemented in the terminal node controller is used to maintain knowledge of clock drift in the instrument computers so as to accurately time tag science observations.

11.3.3 Commissioning

The commissioning of the satellite consisted of the following eight steps: health monitoring, spacecraft software upload, ACS equipment checkout, detumble maneuver, coarse pointing, star tracker checkout, fine pointing, and science commissioning.

When the satellite was ejected from the launch vehicle, it began its mission in safe-hold mode, with no attitude control. When first contact was made, the satellite's health was monitored and found to be in excellent condition. The data downloaded included the power state, voltages, currents, and temperatures. The V53 housekeeping computer did not have prepositioned code, so much of the housekeeping software had to be uploaded from the ground.

After the upload was successfully completed, the ACS equipment was turned on in sequence (except for the star tracker) to ensure that it was safe to proceed to coarse pointing operations. The checkout included turning on and querying the ACS computer, magnetometer, sun sensors, magnetorquers, reaction wheels, and rate sensors. Through B-dot feedback control, a detumble maneuver was performed to reduce rotation rates from 3 deg per sec to less than 0.1 deg per sec. This was then followed by coarse-pointing mode.

In coarse-pointing mode, the initial objective was to point the main solar panel to the sun to generate sufficient power for the upcoming mode transitions (although all faces of the satellite contain solar cells for safe-hold power, one face had the most number of cells for science operations). Coarse-pointing accuracy was on the order of 2–3 deg. Once coarse-pointing mode was entered and its performance evaluated, the star tracker was commissioned.

Commissioning the star tracker was accomplished in four steps. First, images were taken to ensure that the CCDs were functioning properly. Then, the attitude of the satellite was determined from sample images. The polarity of the star tracker was then verified before using the star tracker for continuous attitude estimation. This commissioning was done in coarse-pointing mode—that is, without feeding back the star tracker readings in the pointing control law.

After the star tracker was commissioned, it was put into the control loop, and fine pointing was entered. MOST successfully achieved the required pointing stability of better than 25 arcsec rms. (When switching between target fields, MOST enters coarse pointing, slews, acquires coarse pointing again, and then finally begins fine pointing. Fine slews can be executed to position target stars in appropriate areas of the CCDs as desired by the science team.)

In fine pointing mode, science commissioning was accomplished using a similar approach to star tracker commissioning. Test images were first taken with the telescope door closed to measure dark current before the door was opened and sample science data were collected. Preliminary science targets were observed to verify the operation of the instrument.

Fig. 11.7. The MOST ground station (a) and antenna system at UTIAS/SFL (b).

- Commissioning was not without its challenges. Several hurdles were overcome in the process. For example, the V53 housekeeping computer exhibited behavior on orbit that had not been observed in ground-based testing. The computer would enter a state in which it no longer responded to packets from the ground. The MOST team hypothesized that the noise environment on orbit was considerably different, creating significantly more interrupts than the computer could handle. To deal with this issue, a software workaround—simply to periodically reset the elements of the V53 that were connected to the receivers—was implemented.
- Also, stray light from Earth entering the telescope was a function of roll angle around the boresight of the telescope. To remedy this, the MOST engineering team modified the satellite's star tracking algorithm to remove the background signal before identifying stars. Once this was done, MOST was able to achieve fine pointing to an rms stability of 4.6 arcsec in pitch and 4.2 arcsec in yaw. The MOST science team was also able to adapt the data processing software to subtract out the effect of this stray light.

11.3.4 Postcommissioning Improvements

Following the completion of satellite commissioning, additional improvements were made that enabled MOST to outperform its originally defined mission:

- **Pointing stability.** Upgrades to the star tracking correlation approach to compare star images to star maps improved the pointing stability. The star tracking algorithm originally had a resolution of 3 arcsec, equal to the CCD pixel size. Resolution was first improved through averaging over many correlation solutions rather than just taking the best pixel fit. It was improved yet again by implementing a centroiding algorithm that allowed subpixel interpolation. This modification allowed a factor of 5–10 improvement over the 25 arcsec stability goal. The final pointing stability was on the order of 1 arcsec rms (some variation was present from target field to target field depending on the availability of guide stars).
- **Simultaneous photometry.** The improved pointing allowed MOST to perform simultaneous photometry of multiple targets, including guide stars—up to 30 stars at a time. This greatly increased the scientific return of the mission.
- **Full automation of data downloads.** MOST observed stars for up to seven weeks continuously and collected science and engineering telemetry over these long observation campaigns. During these periods, the ground station was able to request data from the satellite autonomously while being monitored over the Internet. This approach greatly reduced the cost of operations.
- **Implementation of a three-ground-station network.** During commissioning, the SFL station was primarily used. Subsequently, both the Vancouver and Vienna stations came online, increasing the data download capacity by a factor of approximately 2.5.
- **Power generation.** It was discovered soon after launch that MOST was generating more power than expected because Earth albedo was intentionally left out of power budget calculations. This allowed the satellite to operate at a fixed solar-array voltage without having to use its peak power trackers. In fact, at the beginning of the mission, so much power was generated that the battery was not needed for peak transient loads (such as transmitters turning on) while in sunlight. The abundance of power also meant that MOST would have sufficient power for many more years of scientific operations than originally foreseen (beyond its original one-year design lifetime and beyond the current fifth year of operations, at the time of writing) if the other equipment onboard the satellite continued to function.
- **Multiple target fields.** The ability to slew to multiple target fields over the course of an orbit was added, greatly increasing the number of targets that could be observed over a given period of time. This in turn increased the rate of scientific discovery.

11.3.5 Operations

The satellite was commissioned with two teams of three people. Each team was responsible for either early morning or evening passes. After commissioning, much of the standard operations involving the downloading of telemetry and science data were automated. (Although some issues were encountered during commissioning, all have been overcome through software improvements. In this way, the satellite's performance has, in fact, been enhanced, with an achieved pointing stability of as low as 1–2 arcsec and streamlined tasking capabilities.) At present, human intervention is only required when acquiring new target fields or for anomaly resolution. The primary ground station at SFL is fully automated with Internet connectivity. Operators check in periodically to ensure the satellite is healthy and that the expected data have been downloaded. Furthermore, the ground station network can be reconfigured in case one ground station is offline for repairs or maintenance. Large margins in the data transfer budget accommodate occasional station dropouts (the mission was originally designed around 80% throughput of a single ground station, and not the three ground stations that are currently online). The operation of MOST is therefore inexpensive and fully consistent with the overall cost model.

11.4 The Approach

The key to the success of MOST was the ability of the team to follow the "microspace philosophy," which makes satellite development significantly cheaper than the traditional approach. Credit also goes to the CSA for allowing the MOST team to implement (or sometimes modify) the recommendations of AMSAT without being tied to the documentation and management approaches typical of most space programs. Had CSA not granted this freedom, it is unlikely that MOST would have been as successful—at least not within the same tiny budget. It is ironic and perhaps counterintuitive that in the case of a microsatellite, less oversight by the customer—in this case, CSA—proved to be better and more effective. This sort of freedom can result in a resounding success, but it could also lead to failure, depending upon the competency of the developers. It is a risk that some customers refuse to take. Space agencies typically want to be involved in a significant way and have been known to impose policies and programmatic requirements based on formulas that they believe ensure quality and reliability. However, these preconceptions have practically no place in the development of a microsatellite. MOST represents a success, not just scientifically or technologically, but also programmatically. Fortunately, CSA had no preconceptions of how a microsatellite should be built. It was thus able to accept the microspace approach without needing to tinker with it.

The microspace approach is not about cutting corners. It is about building a satellite around a focused mission typically involving a single instrument for a very specific environment. Understanding the environment and designing to it is what drives the development and test philosophies. In every stage of the program, every action taken must contribute to good and robust design. Tests are selected on the basis of what needs to be verified and what contributes most directly to reliability, rather than on adhering to a recipe whose ingredients are not understood or well justified by members on the team.

A small team (10–15 people) located in the same facility is ideal for microsatellite development. This allows for reduced formality and minimal documentation to help keep costs down. This is not to say that a microsatellite program is devoid of documentation, only that the essential documents are written to capture the requirements, design, test approaches, and test results. Having a team with diverse skills and multidisciplinary understanding is also essential. In a microsatellite program, there is no time or money for team members to be insulated from the work of others through narrow work assignments and narrow understanding.

In the development of MOST, the microspace quality assurance approach was inherited from AMSAT, but several challenges were faced that required the MOST team to extrapolate beyond past experience. Specific challenges included noise reduction in the instrument electronics, the need to create software workarounds because of the limitations of the off-the-shelf housekeeping computer, modifications to the spacecraft acceptance campaign to accommodate issues discovered with the instrument and housekeeping computer, and the use of space research frequencies instead of amateur frequencies.

Moreover, the development of enabling technologies—reaction wheels and star tracker software—in parallel with the development of the satellite was an additional risk that had to be managed. In all cases, the microspace extrapolations were similar:

1. Negotiate requirements by descoping if possible or reorganize functional architectures.
2. Rearrange schedule to give uncertain efforts more time while not affecting the critical path—this may involve segmentation of the test plan, but in such a way as to assure the original intent of testing is satisfied.
3. Employ "cold-spare" redundancy where easy to do so, without adding onerous testing or analysis.
4. Ensure robustness and maximum flexibility in the programming of the onboard computers to work around problems as needed.

Given these challenges and the extensions to the microspace approach that were employed out of necessity, MOST represented—and may still represent—one of the most challenging microsatellite missions ever developed. Put simply, MOST was a significant advance in the microsatellite state of the art.

11.5 The Outcome

MOST has produced scientific data and discoveries well in excess of initial expectations. The mission has allowed astronomers to gain a better understanding of solar-type stars and to confirm or challenge previously held theories on stellar pulsation based on ground observations.[3–6] Recently, in celebration of the fourth anniversary of MOST on orbit, Jaymie Matthews published a "Top Ten List" of MOST's accomplishments to date.[7] These accomplishments highlight the satellite's barrier-breaking achievement and represent a definitive milestone in what microsatellites can accomplish.

11.5.1 The MOST Top 10 List

10. Stars as extreme physics labs.

MOST has observed pulsating stars with magnetic fields thousands of times more intense than the global field of Earth or the sun. The resulting data make these stars unique remote laboratories of how magnetic fields and chemical processes operate in hot charged gas. The MOST team has been able to make the first direct comparisons of the star's pulsation frequencies to models of the internal magnetic fields of stars other than the sun.

9. Testing the "Earthiness" of an alien planet.

MOST has monitored the Gliese 581 system, which contains a planet that may be able to support life. MOST has shown that the red dwarf star is sufficiently calm and old, that the climate on that planet would not be a temperature rollercoaster, and that life would have at least a few billion years to evolve into something complex. Maybe as complex as humans.

8. Discovering totally new classes of pulsating stars.[8]

MOST measurements of a massive star approaching a spectacular death as a supernova has revealed a 10-hour rhythm that may be pulsation at the base of the star's hot high-speed wind. The light variability of Wolf-Rayet stars seen from earthbound observatories only accentuated the mystery of the true nature of the stars and the origin of their winds, which inject gas into the interstellar medium to contribute to future generations of star and planet formation. MOST has revealed for the first time the true complexity of the variations of such stars, and hidden within the complex light curve is a periodic signal that theorists believe is the pulsation of the underlying star. This signal can test models of the massive star as a diagnostic of the conditions prior to its supernova explosion.

MOST has also discovered new classes of hot pulsating stars that represent an important step in understanding the structure and evolution of these stars, whose rapid rotation presents many challenges to theorists. The discovery of buoyancy waves in a hot supergiant overturned the original theoretical expectations that this type of pulsation would be suppressed in such massive, evolved stars. New models now show that in fact the opposite is true, in agreement with the MOST observations. So, MOST has discovered yet another previously unknown class of variable star—the slowly pulsating B supergiant (SPBsg)—charting new territory in the path of understanding the life histories of massive stars.

7. Finding that an exoplanetary "tail" wags its stellar "dog."[9]

A planet orbiting very close to its star is expected to be tidally locked by the star's gravity, keeping one face to the star at all times as the moon does to Earth. However, nobody expected that an exoplanet might force the star—more than 1000 times more massive—to stay in synch with the planetary orbit. MOST has seen this happening in the tau Boötis system, where the variability of the star is regulated by the planet. They have stayed in lockstep for at least two years, or about 200 planetary orbits. The planet is orbiting so close to the star—only a few stellar diameters—that its magnetic field is actually tangled with the star's field. Flares in a star occur when the magnetic field lines are twisted and disconnect; spots occur where those field lines emerge from the gas in the star's upper atmosphere. The close proximity of the planet and the distorting influence of its magnetic field generate excess flares and spots in a concentrated area of the star's surface roughly beneath the position of the planet in its orbit.

The MOST finding motivated other astronomers to investigate the star tau Boötis more closely, and they measured the magnetic field of the star with the Canada-France-Hawaii Telescope—the first time a field has been detected in a star with an exoplanet. MOST is showing that much can be learned about these mysterious exoplanets by their influences on the stars they orbit.

6. Measuring the surface rotation of a young sun.[10]

The MOST team has found that a nearby young star spins in the same complex way that powers the giant dynamo responsible for the magnetic field of Earth's sun. The star, kappa 1 Ceti, is less than 20% of the age of Earth's sun, somewhat cooler and smaller, and spinning three times faster. And yet its spin pattern slows away from its equator in a carbon copy of the pattern for the sun's surface. It is like witnessing what the sun may have been like in its childhood, at a time when life was just appearing on Earth.

5. Stabilizing a "suitcase in space" thousands of times better than ever achieved for a microsatellite.[11]

One of MOST's greatest accomplishments is not a scientific one, but a technical one. The very fact that MOST even exists and operates successfully is a milestone in aerospace technology—made in Canada. MOST has a mass of only 53 kg and dimensions of 65 × 65 × 30 cm. Nobody had ever considered performing astronomy from such a small microsatellite. Why? Because the best pointing accuracy of microsats was about ±2 deg (four times the angular diameter of the full moon in the night sky). The MOST ACS regularly achieves pointing of about ±1 arcsec—about 7000 times better! This has paved the way for other extremely cost-effective Canadian and international space astronomy missions such as NEOSSat and BRITE.

4. "Ultrasound" of stellar embryos.

In the stellar cluster officially designated NGC 2264 but known to amateur astronomers as the "Christmas Tree Cluster," there are many "baby" stars and even stellar "embryos." Astronomers would designate these as pre-main-sequence (PMS) stars. Two of these stars in NGC 2264 pulsate, and MOST has observed literally hundreds of oscillation frequencies in them caused by sound waves bouncing around in their gas. It's as though the MOST team is performing a remote form of "ultrasound" on these embryos of stars to probe their hidden internal physiology. The data are giving insights into the theories of how stars are born and how Earth's sun and planets came to be.

3. Much ado about nothing—the pulsations of Procyon?[12]

MOST observed its first primary science target, the bright star Procyon, early in 2004 to detect the oscillations in light caused by pulsating ripples in its surface. The result, which appeared in the journal *Nature*—that no such oscillations were detectable in the data down to levels of only a few parts per million—generated not just ripples but seismic waves of reaction in the astroseismic community.

MOST's null detection was controversial, but inspired two teams of theorists to check their models of the turbulent motions expected in the outer envelope of a star like Procyon. The outcome has been an improved understanding of the physics of convection and pulsational mode excitation.

Initial MOST results on Procyon also inspired astronomers to reexamine the star more closely. In early 2007, a network of spectrographs around the globe were trained on Procyon to search for its oscillations in velocity through Doppler shifts. At the same time, 820 km above the globe, MOST was obtaining the best photometry ever obtained of Procyon (indeed, of any star) with precision better than 1 ppm. The results of these two concerted campaigns should settle some of the controversies, but are likely to raise new questions about the nature of a star that—prior to MOST—astronomers thought they understood well.

2. Exoplanetary meteorology.[13,14]

The exoplanet HD 209458b, like the other gas giants found through Doppler surveys of sun-like stars, is too faint to be seen in any image. But its orbit carries it in front of its parent star, causing the starlight to dim slightly but measurably in a transit once per orbit. That orbit also takes the exoplanet behind the star. The dip in reflected light during each eclipse is so small that it is equivalent to the change in light you would see if a mosquito flew behind a city streetlamp 1000 km away. MOST can measure changes this small, and its photometry of the HD 209458 system has shown that the close-in planet reflects only a few percent of the light it receives from its sun. Combined with NASA Spitzer space telescope infrared data, MOST data tell us that this "hot Jupiter" has no bright cloud cover. The data serve as a thermometer to measure the atmospheric temperature.

1. The search for Earths around other stars.

The fact that MOST can put stars under a stakeout for up to two months at a time without interruption, and achieve unprecedented light-measuring (photometric) precision, has enabled it to hunt for planets approaching the mass and size of Earth around some other sun-like stars. In the transiting system HD 209458, the transit times of the known giant planet are the "ticks" of an orbital clock whose pace can be changed by the gravities of other unseen planets in the system. MOST has been able to measure how accurately that clock runs, and the results demonstrate that there are no Earth-mass planets orbiting close to the star. The same data have been searched for transits at other orbital periods and eliminate the presence of planets as small as twice the radius of Earth in tight orbits. The gravitational influence of possible small planets was the best theory to explain the fact that the giant planet HD 209458b is too large for its mass. MOST has forced the theorists to reconsider and is testing this theory in other exoplanet systems. The search continues.

11.6 Low-Cost Satellite Technologies and Approaches: The Space Flight Laboratory

As a result of the opportunity and challenges met by the MOST development team, SFL was established with a full-time staff and facilities to continue to push the state of the art in low-cost nano and micro missions. The Canadian Advanced Nanospace eXperiment (CanX) program was initiated in 2001 based on the desire to continue to develop leading-edge missions subsequent to the launch of MOST in 2003. The CanX program has created versatile generic technologies that can be used to implement a wide variety of missions. SFL's three-axis-controlled Generic Nanosatellite Bus—another innovative, barrier-breaking platform—is being used to support the following missions.

- The BRITE constellation,[15] a mission involving up to four 7 kg, 20 cm cube nanosatellites each using 3 cm aperture optical telescopes to photometrically observe the most luminous stars in Earth's galaxy of apparent visual magnitude +3.5 or brighter with oscillation periods ranging from hours to months.
- CanX-4&5,[16–18] a precise, autonomous formation-flying demonstration mission targeting relative position determination to within a few centimeters and relative control to an accuracy of less than 1 m.
- AISSat-1, a mission to detect and decode Automatic Identification System (AIS) messages from ships.
- Lunette, a moon mission concept for the European Space Agency to accurately map the far-side gravity field of the moon.
- MOMENT, a Mars mission concept for the CSA to measure the remnant magnetic field of Mars to nano-Tesla accuracy and 100 km spatial resolution.

This multimission platform is scalable to larger missions, including its reaction wheels developed jointly between SFL and Sinclair Interplanetary (Toronto, Canada), high-data-rate radios (~1 Mbps), and low-power onboard computers (~200 mW, several MIPS). The outcome of MOST is not only a highly successful space astronomy mission, but also a continuing capability for barrier breaking missions developed at SFL at extremely low cost.

11.7 Acknowledgments

The authors would like to acknowledge the following: The members of the MOST development team (in addition to Robert E. Zee): Kieran A. Carroll, Glen B. Sincarsin, Simon C. O. Grocott, Peter Stibrany, Wayne G. Sincarsin, Ron Wessels, Noah H. Hansen, Douglas N. Sinclair, Alexander M. Beattie, Hugh G. Chesser, Daniel G. Foisy, Andre E. Laurent, Freddy M. Pranajaya, Henry Spencer, Andrew Walker, Ron Johnson, Greg Burley, and John Pazder. The members of the MOST Science Team (in addition to Jaymie Matthews): Slavek M. Rucinski (University of Toronto), Rainer Kuschnig (UBC), Anthony F. J. Moffat (Universite de Montreal), Dimitar Sasselov (Harvard-Smithsonian Center for Astrophysics), Gordon Walker (UBC), David Guenther (Saint Mary's University), Werner W. Weiss (University of Vienna). The Canadian Space Agency, especially Glen Campbell. The Ontario Research and Development Challenge Fund. The Ontario Centres of Excellence Inc. (Earth and Environmental Technologies, formerly the Centre for Research in Earth and Space Technologies). The Natural Sciences and Engineering Research Council. The Canada Foundation for Innovation. The University of Toronto. The University of British Columbia. The University of Vienna (Prof. Werner Weiss). The Vienna University of Technology (TU Vienna). The Spectral Applied Research. Ceravolo Optical Systems. AeroAstro Inc. Burley Scientific. Routes AstroEngineering.

11.8 References

[1] D. Green, J. M. Matthews, S. Seager, and R. Kuschnig, "Scattered Light from Close-In Extrasolar Planets: Prospects of Detection with the MOST Satellite," *The Astrophysical Journal*, **597**, 590–601 (1 November 2003).

[2] M. Davidoff, *The Radio Amateur's Satellite Handbook* (The American Radio Relay League, 1997).

[3] G. Walker, J. Matthews, R. Kuschnig, R. Johnson, S. Rucinski, J. Pazder, G. Burley, A. Walker, K. Skaret, R. Zee, S. Grocott, K. Carroll, P. Sinclair, D. Sturgeon, and J. Harron, "The MOST Astroseismology Mission: Ultraprecise Photometry from Space," *Publications of the Astronomical Society of the Pacific,* **115**, 1023–1035 (September 2003).

[4] S.C. O. Grocott, R. E. Zee and J. M. Matthews, "Exploring the Mysteries of the Cosmos on the MOST Microsatellite Mission," *Proceedings of the 17th AIAA/USU Conference on Small Satellites* (August 2003).

[5] R. E. Zee, S. C. O. Grocott and J. M. Matthews, "The MOST Microsatellite Mission: All Systems Go for Launch," *Proceedings of the 12th CASI (Canadian Aeronautics and*

Space Institute) Conference on Astronautics (November 2002).

[6] R. E. Zee and P. Stibrany, "Canada's First Microsatellite: An Enabling Low-Cost Technology for Future Space Science and Technology Missions," *CASI (Canadian Aeronautics and Space Institute) Journal,* **48** (1), (March 2002).

[7] L. Brian, "Canada's Space Telescope Celebrates Fourth Birthday: Canadians Invited to Join the Party." Media release, University of British Columbia Public Affairs (Vancouver, 29 June 2007).

[8] L. Lefevre, S. V. Marchenko, A. F. J. Moffat, A. N. Chene, S. R. Smith, N. St-Louis, J. M. Matthews, R. Kuschnig, D. B. Guenther, C. A. Poteet, S. M. Rucinski, D. Sasselov, G. A. H. Walker, and W. W. Weiss, "Oscillations in the Massive Wolf-Rayet Star WR123 with the MOST Satellite," *The Astrophysical Journal Letters,* **634**, 109 (2006).

[9] J. M. Matthews, "MOST Space Telescope Sees 'Tail Wagging Dog' in Star-Exoplanet System." Press release, Canadian Astronomical Society (University of Montreal, 2005).

[10] G. A. H. Walker, B. Croll, R. Kuschnig, A. Walker, S. M. Rucinski, J. M. Matthews, D. B. Guenther, A. F. J. Moffat, D. Sasselov, and W. W. Weiss, "The Differential Rotation of kappa 1 Ceti as observed by MOST—II," *The Astrophysical Journal,* preprint doi:10:1086/'511851', (The American Astronomical Society, 2006).

[11] S. C. O. Grocott, R. E. Zee, and J. M. Matthews, "The MOST Microsatellite Mission: One Year in Orbit," *Proceedings of the 18th Annual Conference on Small Satellites* (August 2004).

[12] J. Matthews, R. Kuschnig, D. B. Guenther, G. A. H. Walker, A. F. J. Moffat, S. M. Rucinski, D. Sasselov, and W. W. Weiss, W.W. "No Stellar p-mode Oscillations in Space-Based Photometry of Procyon," *Nature,* **430**, 51–53 (1 July 2004).

[13] B. Croll, J. M. Matthews, J. F. Rowe, R. Kuschnig, A. Walker, B. Gladman, D. Sasselov, C. Cameron, G. A. H. Walker, D. N. C. Lin, D. B. Guenther, A. F. J. Moffat, S. M. Rucinski, and W. W. Weiss, "Looking for Giant Earths in the HD 209458 System: A Search for Transits in MOST Space-Based Photometry," Accepted by *The Astrophysical Journal* (2007).

[14] E. Miller-Ricci, J. F. Rowe, D. Sasselov, J. M. Matthews, D. B. Guenther, R. Kuschnig, A. F. M. Moffat, S. M. Rucinski, G. A. Walker, "MOST Spacebased Photometry of the Transiting Exoplanet System HD 209458: Transit Timing to Search for Additional Planets," *The Astrophysical Journal,* submitted 2007.

[15] A. F. J. Moffat, W. W. Weiss, S. M. Rucinski, R. E. Zee, M. H. van Kerkwijk, S. W. Mochnacki, J. M. Matthews, J. R. Percy, P. Ceravolo and C. C. Grant, "The Canadian BRITE NanoSatellite Mission," *ASTRO 2006—13th CASI Canadian Astronautics Conference* (Montreal, Canada, April 2006).

[16] S. Eagleson, K. Sarda, S. Mauthe, T. Tuli and R. E. Zee, "Adaptable, Multi-Mission Design of CanX Nanosatellites," *Proceedings of the 20th Annual AIAA/USU Conference on Small Satellites* (August 2006).

[17] K. Sarda, S. Eagleson, E. P. Caillibot, C. Grant, D. D. Kekez, F. Pranajaya and R. E. Zee, "Canadian Advanced Nanospace Experiment 2: Scientific and Technological Innovation on a Three-Kilogram Satellite," *Acta Astronautica,* **59**, 236–245 (2006).

[18] N. G. Orr, J. K. Eyer, B. P. Larouche and R. E. Zee, "Precision Formation Flight: The CanX-4 and CanX-5 Dual Nanosatellite Mission," *Proceedings of the 21st Annual AIAA/USU Conference on Small Satellites* (August 2007).

Chapter 12

TUBSAT: The Technical University of Berlin Satellite Program

Matthias Buhl*

12.1 Introduction

TUBSAT is a microsatellite development program sponsored by the Institute of Aeronautics and Astronautics at the Technical University of Berlin (TU Berlin), Germany. Initiated in 1985, TUBSAT is an educational program in step with actual practice, including the design, manufacture, testing, launching, and operation of microsatellites. The design is deliberately kept simple, so that advanced students can build and operate the satellite, the ground station, and the experiments while less experienced students can still understand the functionality of the complete system. Over time and after several successful missions, the program has expanded, and cooperative programs with national space agencies have commenced. Engineers from the German, Moroccan, and Indonesian Space Agencies have participated in performing experiments, training personnel, and conducting joint satellite design, manufacture, and operations.

Seven TUBSATs have been successfully launched and operated to date. Table 12.1 provides a launch summary, including the operational status of each satellite and its primary mission goals (apart from student training).

12.1.1 History and Overview

In 1985, Prof. U. Renner was called to head the new Department for Spacecraft and Space Equipment at TU Berlin. At the time, NASA was offering launch opportunities using the Get-Away Special space shuttle payload containers. Utah State University and Weber State University deployed their NUSAT from a modified Get-Away Special canister in April 1985. Renner started an educational program that would include the design and manufacture of a satellite to be launched with the shuttle. The result was TUBSAT (Fig. 12.1), designed to perform communication experiments and video recording in space. A primary objective was to build and test a momentum-bias attitude control system for microsatellites; the design included a Teldix momentum wheel.

When the space shuttle Challenger was destroyed in January 1986, the shuttle fleet was grounded, and Prof. Renner's team began looking for new launch opportunities. In 1989, a new possibility emerged. Arianespace offered to launch microsatellites as piggyback payloads with their Ariane 4 rocket. The German Space

*Technical University of Berlin, Germany

Table 12.1. TUBSAT Launches

Satellite	Launch Date	Mission	Operational Status
TUBSAT-A	17 July 1991	Store-and-forward communication	Operational up to date
TUBSAT-B	25 Jan. 1997	Earth observation	Lost contact after 39 days
TUBSAT-N/N1	7 July 1998	Store-and-forward communication	Mission ended (reentry)
DLR-TUBSAT	26 May 1999	Earth observation	Operational to date
MAROC-TUBSAT	10 Dec. 2001	Earth observation	Operational to date
LAPAN-TUBSAT	10 Jan. 2007	Earth observation	Operational to date

Agency offered to finance a TUBSAT launch. Because the spacecraft specification of Arianespace differed significantly from that of NASA, a completely new satellite had to be developed and built within two years to meet the launch deadline.

Working day and night, the team at TU Berlin managed to prepare a satellite for launch in July 1991. TUBSAT-A was only the fifth of the so-called university-class spacecraft,[1] following the two UoSATs of the University of Surrey, NUSAT of Utah State University, and WeberSat of Weber State University. Remarkably, TUBSAT-A is still operational. The satellite performed communication experiments with great success. For example, ultramobile communications were demonstrated by researchers of the Alfred Wegener Institute in Antarctica and by Arved Fuchs on his sailing trip around the North Pole.

After the launch of TUBSAT-A, a new satellite for Earth observation was planned. TUBSAT-B was launched in 1997. Unfortunately, controllers lost contact after 39 days of successful operation. Nonetheless, the attitude control system was successfully tested. Experiences with TUBSAT-A and TUBSAT-B led to a standard microsatellite bus named TUBSAT-C,[2] which formed the core of subsequent microsatellite projects.

Following TUBSAT-A and TUBSAT-B, a new nanosatellite research project was initiated by Renner and his colleague Robert Schulte. This project, named TUBSAT-N (for "nano"), would demonstrate that fast, ultralow-cost space access was feasible. Using design information from TUBSAT-A and TUBSAT-B, Renner and Schulte calculated that two satellites could be designed, manufactured, and launched for less than US$350,000 each. The double launch took place in July 1998, and communication experiments have been successfully carried out.

Following TUBSAT-N, the microsatellite program expanded as the university began looking for new partners to promote its TUBSAT-C microsatellite bus. Joint ventures with national space agencies were initiated, beginning with the German Aerospace Center, DLR (successor to the German Space Agency). An Earth observation mission was planned and implemented, resulting in the launch

Fig. 12.1. TUBSAT.

of DLR-TUBSAT. The payload consisted of three video cameras, with the biggest one having a focal length of 1 m. Because DLR-TUBSAT was designed to transmit live video, new interactive attitude control strategies had to be developed.

In 1994, Karim Taha, a graduate of TU Berlin, convinced the Moroccan government to engage in a joint venture. The objective was to build an Earth observation satellite: TU Berlin would be responsible for the satellite bus, and the Moroccan Institute would be responsible for the payload. The contract also included the teaching of Moroccan engineers. The resulting MAROC-TUBSAT was launched in 2001, taking images for cartography, irrigation observation, and ice monitoring. The development of the MAROC-TUBSAT was a major step in the creation of the Moroccan Space Agency.

The Indonesian space agency, LAPAN, was looking for partners to build up a center for satellite design and manufacturing. The Indonesian government wanted access to experienced spacecraft developers who could not only help build a satellite but train future engineers as well. LAPAN selected TU Berlin, and signed a contract in 2003 that included support up to the early stages of on-orbit operation. Indonesian engineers were actively included in the design and manufacturing process. LAPAN-TUBSAT was launched in January 2007 and has been taking breathtaking color videos ever since.

LAPAN-TUBSAT was the last TU Berlin microsatellite launched so far, but a new picosat, BeeSat, is nearly complete and should be ready for launch soon.

The TUBSAT design philosophy can be summarized as a direct implementation of the KISS principle: Keep It Simple, Stupid. TUBSAT missions often lack extensive safety protocols during development, and redundancy of the satellite components is deemphasized in favor of an ambitious attitude control system, payload, or experiment module. Over time, TU Berlin has acquired and built facilities and developed protocols to do the flight qualification, including radiation and thermal vacuum testing. Shake tests are done at the German Aerospace Center in Berlin. Communication tests have been carried out with sounding rockets in Kiruna, Sweden. A self-designed air bearing simulator is used to verify the attitude control system design and to implement new strategies. But to control costs, flight qualification testing has been limited to the essentials, and proven hardware and software is used whenever possible—including standardized subsystems, as was the case with the TUBSAT-C microsatellite bus.

The following sections present each mission in more detail, with particular focus on the first launched satellite TUBSAT-A and the most recent LAPAN-TUBSAT.

12.2 TUBSAT-A

TUBSAT-A[3] as seen in Fig. 12.2 is a low-cost university satellite designed for store-and-forward communications. Developed by Prof. Renner and the TUBSAT-A team, it has been one of the most successful microsatellite missions. The satellite is still operational after 16 years in orbit. It was launched as a secondary payload to ERS-1 (Earth observation) by an Ariane V44 from Kourou, French Guiana, on 17 July 1991, becoming one of the first university-class spacecraft ever. The orbit was defined by the primary payload and is sun-synchronous with an altitude of 780 km.

The launch contract offered by Arianespace posed a difficult challenge because the primary payload dictated the launch date, and the secondary payload had to be ready by then—otherwise, it would lose the launch opportunity and still have to pay the launch cost. The contract was signed in February 1990 with a designated launch date of July 1991. The German Space Agency financed the launch to boost the satellite design industry in Germany. After many night shifts, the satellite was completed and transported to the launch site in Kourou.

Fig. 12.2. TUBSAT-A.

Sometimes there is an advantage to being an inexperienced university group, rather than a professional agency, launching your first self-made satellite. You get a lot of indulgence, and Arianespace bent the rules more than once. The documentation of TUBSAT-A—such as the shake test results or the flight readiness reviews—was rather small compared with the stacks of paper for the other passengers. The satellite also came without a mass dummy, which is necessary in case the satellite can not fly. But Arianespace helped out, and even built that mass dummy. On 17 July 1991, the rocket finally launched with TUBSAT-A onboard. An overview of the mission is given in Table 12.2.

12.2.1 Design

12.2.1.1 Mechanical Design and Ejection Mechanism

The dimensions of secondary payloads are limited by the launcher. Essentially, these payloads fly in place of inert mass, so mass restrictions are not as tight as volume restrictions. In the case of TUBSAT-A, the volume was restricted to 45 × 45 × 60 cm and mass to about 50 kg. The final mass was actually closer to 35 kg, even though the team did not focus on saving mass. A cube was chosen as the basic configuration, with edges measuring 38 cm (defined by need of solar cell panels) made with square aluminium bars. Relatively thick (3 to 5 mm) aluminum plates were connected to this framework. They are heavy and stable and provide good surfaces for ease of mounting. Aluminum is also a good heat conductor, has a high heat capacity, provides good radiation shielding, and serves as a good

Table 12.2. TUBSAT-A Mission Overview

Mission	Store-and-forward communications (animal tracking, arctic expeditions); Technology demonstration (VHF communication, star sensor, GaAs solar cells)
Launch/Status	17 July, 1991 (Ariane 4); Operational
Orbit	Sun-synchronous, altitude 780 km, inclination 98.4 deg
Mass/Volume	35 kg, 38 × 38 × 38 cm
Structure	Framework of aluminum bars with aluminum plates constituting the hull
Thermal Control	Passive
Power	Ten NiCd battery cells, 7 Ahr, 12 V; Four silicon solar panels (six parallel strings of 36 cells, max. 17 W)
Communication	VHF (143.075 MHz, FFSK, 1200 baud, 2 W)
Attitude Control	Three-axis magnetic field sensor, three-axis magnetorquer, sun sensor, star sensor
Payload	Store-and-forward communication; Radiation exposure experiments (transputer measuring single-event upsets); Long-duration GaAs solar cell tests

electric grounding potential. Figure 12.3 shows the TUBSAT-A design as well the Arianespace adaptor that carries the secondary payloads. It is a sandwich platform, which can be seen at the bottom. Between the satellite and the platform is a bed plate manufactured by TU Berlin that is screwed to the Arianespace platform and remains there. At that time, every secondary payload had to bring its own ejection mechanism (Fig. 12.4). TUBSAT-A was ejected by a prestressed spring kept in place by a copper-beryllium bolt of 6 mm diameter. This bolt was severed by a pyrotechnically driven bolt cutter after the primary payload ERS-1 was separated. The ejection velocity was specified to 1.4 m/sec.

12.2.1.2 Thermal Control and Power Supply

Only passive thermal control is used. The inside of the satellite is anodized black to ensure good thermal balance; the outside of the satellite is also anodized black, although superisolation or white coating could have been used to improve thermal control. A simple one-node model incorporating albedo and eclipse fractions yielded an average temperature of −5°C, which is a very good operating temperature for most electronic devices. This was substantiated by measurements in orbit. Figure 12.5 presents temperatures of one orbit of two outer plates and the ground plate. For the solar panels, the extreme temperatures are relatively close, about

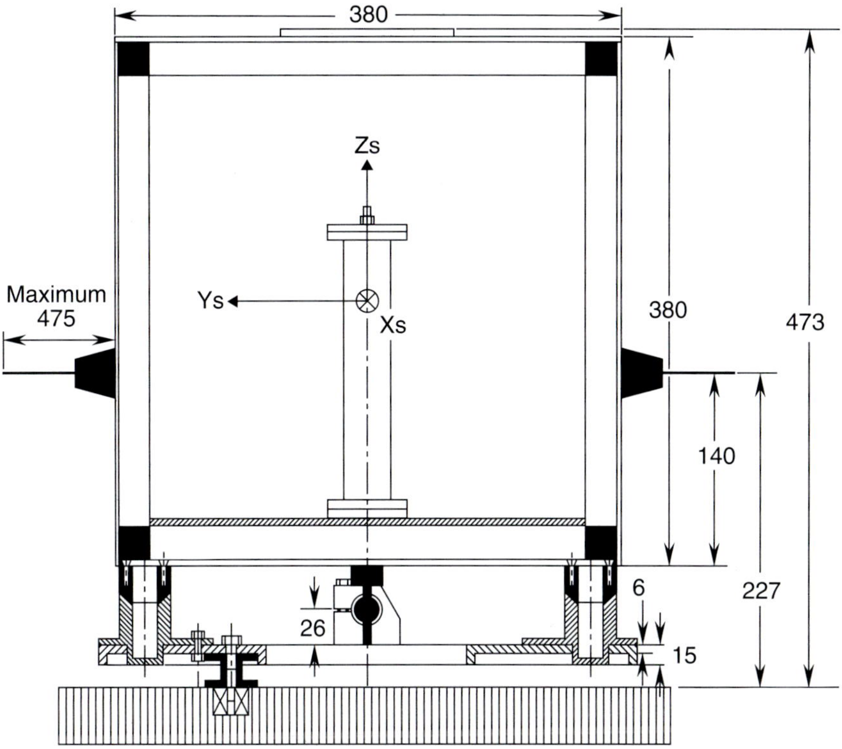

Fig. 12.3. TUBSAT-A design and Arianespace interface platform (all dimensions are in millimeters).

±20°C. The battery block is sandwiched between two 5 mm aluminum plates with high thermal capacity and conductivity to maintain equal temperatures among the batteries, resulting in maximum temperatures of only −8°C to 2°C.

The power supply consists of 10 NiCd batteries with 7 Ahr (SAFT Vo7s) with 12 V nominal voltage. The batteries are charged by silicon solar cells (36 cells of 2 × 2 cm connected in series) with off-load voltage of 22 V. Charging starts when the voltage drops below 12.3 V and is cut off when it reaches 14.3 V. Four of the six surfaces are covered with six parallel solar-cell strings, yielding a maximum current of 800 mA (in inclined attitude, a total of 1.4 A can be generated, as seen in Fig. 12.6). There are also GaAs solar cells sponsored by the Italian company Fiar within the scope of a European Space Agency technology demonstration program (20 cells of 2 × 4 cm with an off-load voltage of 22 V and 230 mA). The power supply provides more than the TUBSAT-A off-load current of about 280 mA. Long-duration solar-cell degradation tests revealed less degradation than predicted. Furthermore, tests on NiCd battery cells show that the degradation due to the memory effect is fast in the beginning of a mission, but stabilizes after some years.

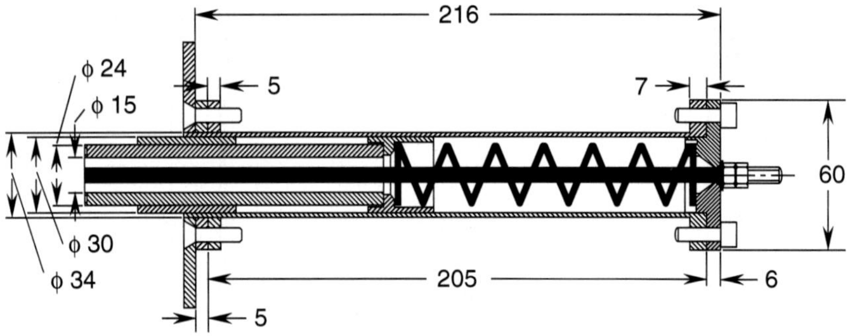

Fig. 12.4. TUBSAT-A ejection mechanism (all dimensions are in millimeters).

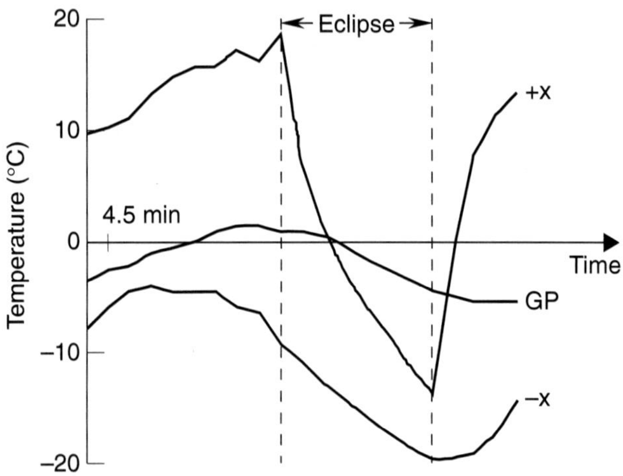

Fig. 12.5. TUBSAT-A measurements of the temperature of +X, −X, and ground plate (GP) over one orbit.

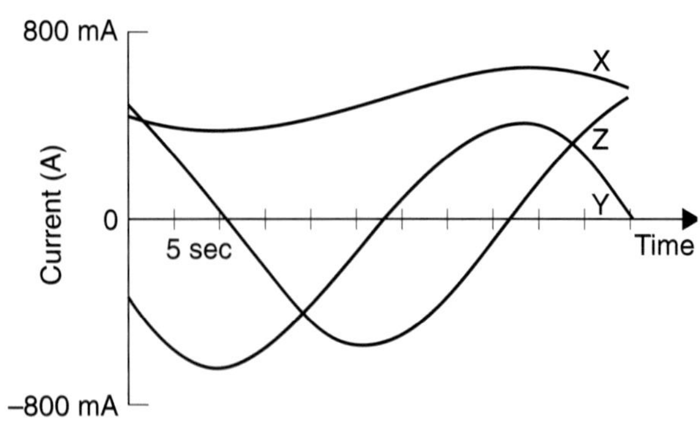

Fig. 12.6. TUBSAT-A on-orbit measurements of the solar panel currents over one nutation period.

12.2.1.3 Attitude Control System

The TUBSAT-A is a three-axis stabilized vehicle. TU Berlin developed a star sensor based on CCD technology as well as a sun sensor. Further sensors include a three-axis magnetic-field sensor and the solar panels themselves.

The actuator of the attitude control system is a three-axis magnetorquer. The magnetic coils are embedded into the framework and are made of 100 windings with 0.7 mm copper wire. Maximum current that can be applied is 2 A.

The star sensor provides a "Top-10 list" containing the coordinates of the 10 brightest pixels in the image. Figure 12.7 shows 15 successive Top-10 lists of the star sensor. The attitude is determined from the first one with a star-pattern detection algorithm on ground. Because data are taken every 0.5 sec, one can also determine the angular velocity about the star sensor axis. In the example shown in Fig. 12.7, the angular velocity is given by 23 deg/7.5 sec. One can also observe static objects, which are easily identified as hot spots on the image sensor.

12.2.1.4 Communication and Data Handling

TUBSAT-A features an L-band experiment to test communication with patch antennas, but the standard communication is realized in the 2 m band. The transceiver is the standard radio amateur transceiver Yaesu-FT-23R; it features small dimensions while keeping good transmission characteristics. It is developed for voice communication and modulates the signals from the modem onto the carrier of 143.075 MHz using frequency modulation. A single-chip modem is used (FX439 of CML) with FFSK modulation at 1200 bps (logic one is realized as 1200 Hz oscillation and a logic zero as 1800 Hz). The satellite sends the signals via a $\lambda/4$ monopole antenna using about 2 W of power. Every 4.5 minutes, TUBSAT-A sends an identification followed by the total content of the RAM. It takes about 8 sec to send 1 kb. The telemetry data contains temperatures, voltages, star sensor images, and uploaded text. After sending the data, TUBSAT-A is ready to receive and execute commands for a short time; if it receives no commands, it falls back into hibernation mode.

The main onboard computer is composed of Hitachi H8/532 single-chip microcontrollers. These feature 32 kb EPROM, 1 kb RAM, and 8 channels of A/D conversion. The main onboard computer is connected to a second microcontroller with bigger memory capacity to save messages and telemetry. TUBSAT-A also carries a transputer (parallel working computer) from the German company Kayser-Threde to assess radiation sensitivity; single-event upsets are counted by checking patterns in the memory for any variance.

The main ground station is a crossed-Yagi antenna installed on the roof of the TU Berlin Institute building. TU Berlin also designed a hand-held mobile ground station weighing 300 g with 2 W transmitting power, designed to maintain communications with explorers in the arctic region. TUBSAT-A is also capable of recording and transmitting voice messages 2 minutes long; this experiment was designed by Prof. P. Noll of TU Berlin Institute of Telecommunication.

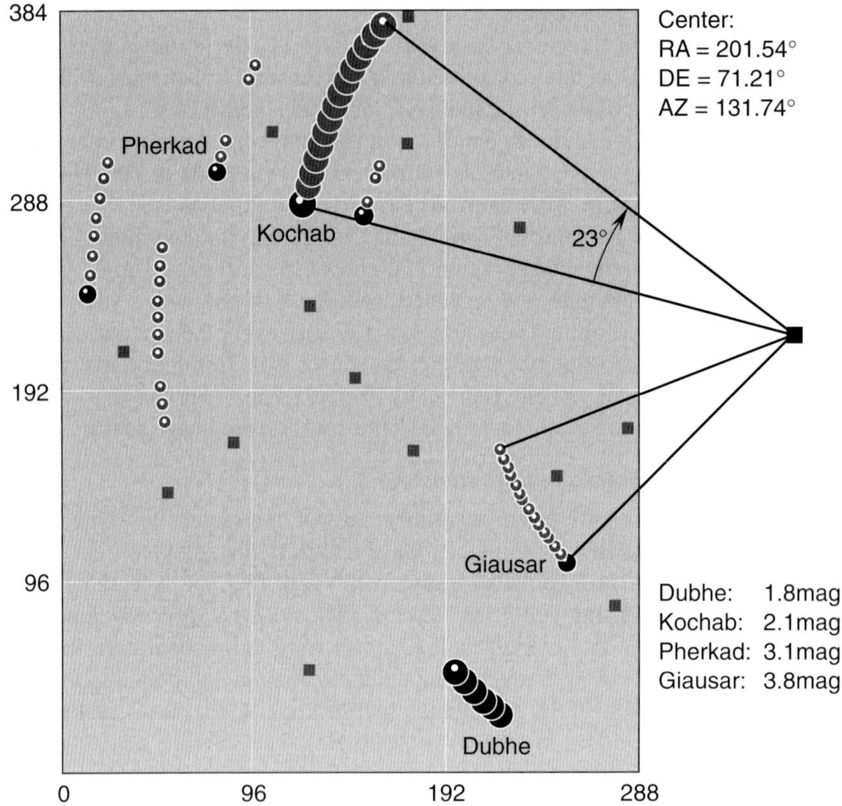

Fig. 12.7. Star sensor data of TUBSAT-A.

12.2.2.1 Communication Experiments

TUBSAT-A was used extensively for communication experiments. One of the many experiments was a pilot project[4] with the University of Göttingen to investigate the spatial distribution and migration of red deer in the Harz mountain range in northern Germany. The measurements were conducted from February to August 1995. Two male and one female deer were each outfitted with an electronic collar that recorded its location at specific times and sent the data to the satellite. These collars were developed and built at TU Berlin; they contained a GPS receiver, microcontroller, modem, transceiver, batteries, and antenna. Total weight was about 1.6 kg. The data were then transferred via TUBSAT-A to the ground station in Berlin. It was not only a test of the applicability of the satellite for wildlife research purposes, but also a study to provide data on the size of the deer's home ranges and seasonal migrations in this area. The success of this experiment engendered other animal tracking experiments with TUBSAT-A in the Harz forest. More experiments have been conducted with the Finnish Institute for

Wildlife and Fish Research and the Institute of Geography of the University of Oulu in Finnish Lappland.

TUBSAT-A also became an important tool for arctic expeditions, because of its near-polar orbit. Several arctic and antarctic expeditions were equipped with handheld ground stations. Adventurers like Arved Fuchs on his sailing trip around the North Pole and the Weber-Malakhov expedition to the North Pole used communication with TUBSAT-A extensively. The Alfred Wegener Institut used TUBSAT-A to recover data from unmanned gauging stations in Antarctica and to transmit the data to headquarters in Bremerhaven, Germany.

Even today TUBSAT-A is used for education at TU Berlin. For example, students enrolled in satellite technologies classes are expected to build up the ground segment of the UHF communication link. They get first-hand experience with spacecraft development by implementing the communication protocol and recovering contact to TUBSAT-A for operation as an exercise.

12.3 TUBSAT-B

TUBSAT-B (Fig. 12.8) was the first Earth observation satellite of TU Berlin. Launched as a secondary payload aboard a Tsyklon 3 on 25 January 1994 from Plesetsk, Russia, it reached a medium orbit of 1250 km altitude at 82 deg inclination. The main payload was the Meteor 3-7 Russian weather satellite. The launch was sponsored by Kayser-Threde. Unfortunately, the satellite was lost after 39 days in orbit. The cause for the failure could never be determined definitively. The most probable cause was radiation damage, due to the satellite's position in the inner Van Allen belt. Nevertheless, most experiments onboard could be performed.

Fig. 12.8. TUBSAT-B.

The principal objective of TUBSAT-B was space technology experimentation, with a focus on the demonstration of a high-resolution attitude control system using a star sensor and three reaction wheels. The satellite was a cuboid measuring 38 × 38 × 50 cm. Total mass was about 40 kg. The power system was based on four solar panels directly attached to the main body, which charged four NiH_2 batteries. The structure was dictated by the largest component—a 1 m telescope for Earth observation at about 20 m resolution. A Zeiss Jena telescope with 1 m focal length (folded optics) and aperture of f/5.6 was used together with a Thomson-CSF CCD array (288 × 384 pixels), resulting in a field of view of 0.35 × 0.5 deg.

The star sensor onboard was derived from the TUBSAT-A star sensor. A complete image from the unit can be seen in Fig. 12.9. As a secondary sensor, a new sun sensor was tested in orbit. The three reaction wheels in the attitude control system were designed at TU Berlin.

Despite the early loss of TUBSAT-B, it was an important step in the evolution of the TUBSAT family. The power system and attitude control system became the basis of the new standardized microsatellite bus, TUBSAT-C. This satellite bus was used with DLR-TUBSAT, MAROC-TUBSAT, and LAPAN-TUBSAT as well as other external projects.

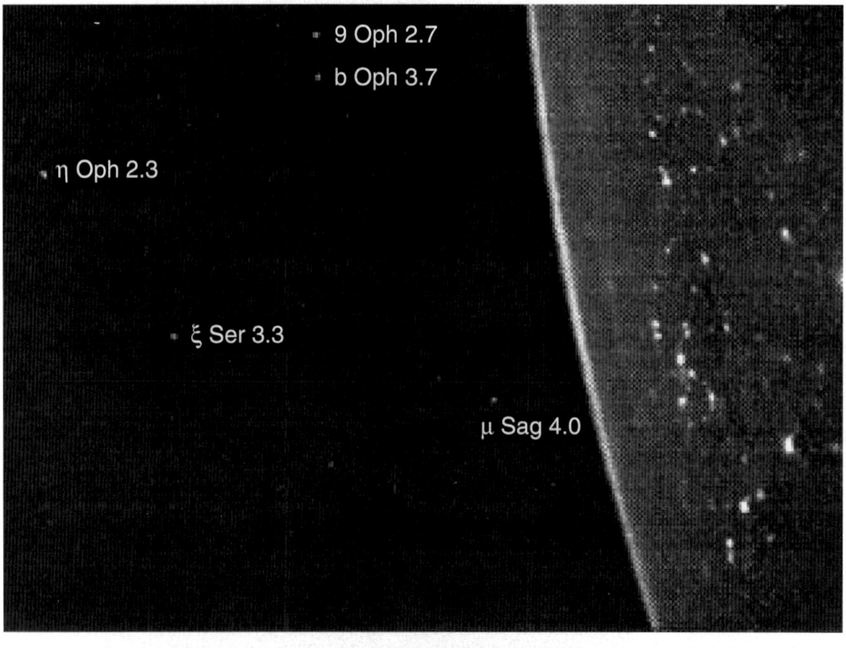

Fig. 12.9. Star sensor image of TUBSAT-B: constellation Ophiuchus and night view of Ural area in Russia.

12.4 TUBSAT-N/N1

TUBSAT-N[5] and TUBSAT-N1 (Fig. 12.10) were the first nanosatellites to be developed at TU Berlin. They were launched together on 7 July 1998 from a Russian submarine with a converted SS-23 SHTIL military rocket. The satellites only separated when the designated orbit was reached. The launch by a K-407 Novomoskovsk submarine in the Barents Sea was the first time a satellite had been launched into orbit from a submarine.

The satellites were brought to Russia for integration by airplane (Schulte actually took the two satellites as hand luggage) before they were transported to the submarine harbor station. The submarine itself was closed to outside scientists, but the complete operation proceeded without any serious problems or delays. Reentry from the highly elliptical orbit of 770 × 400 km occurred one year later, as predicted. The prime mission was the demonstration of a low-cost global communication system.[6] TUBSAT-N and N1 were capable of the same communication

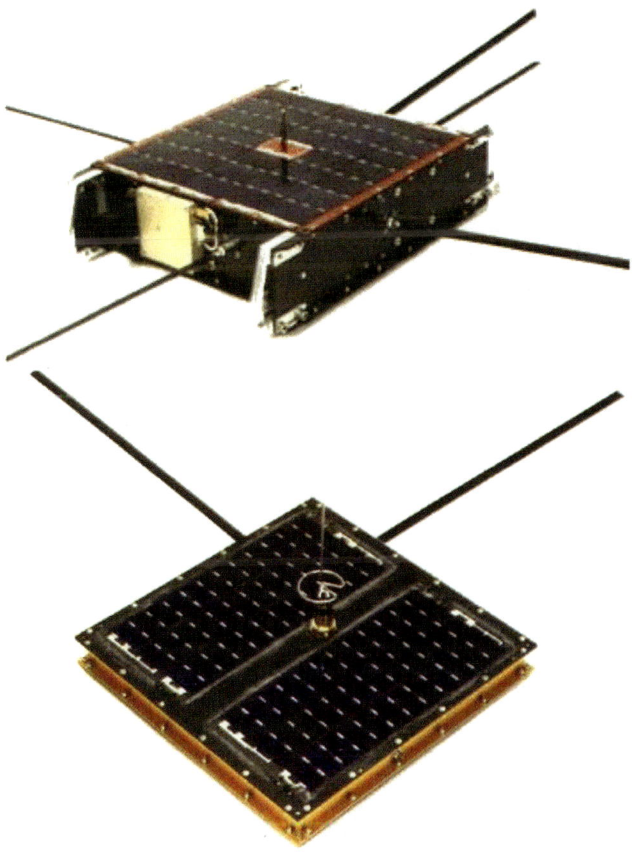

Fig. 12.10. TUBSAT-N (top) and TUBSAT-N1 (bottom).

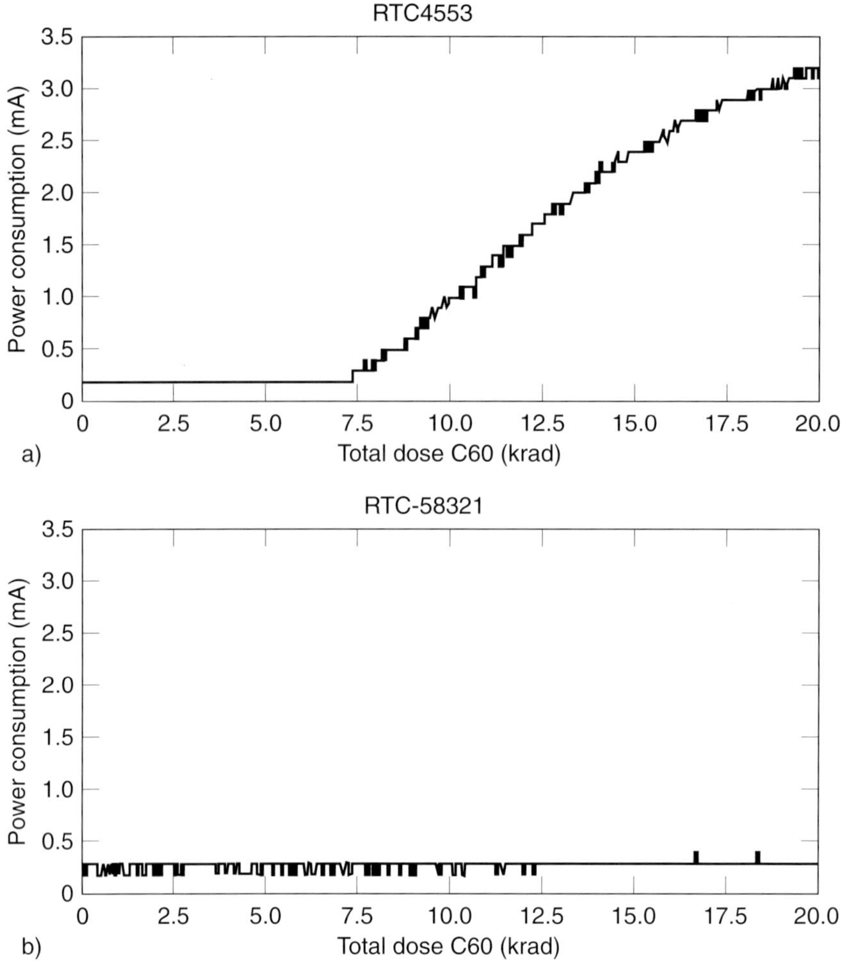

Fig. 12.11. Radiation test results for real-time units RTC4553 (a) and RTC58321 (b).

experiments as TUBSAT-A, but with notably lower mass and volume requirements. TUBSAT-N had a mass of about 8.5 kg, and N1 of about 3 kg. Both could be designed, manufactured, and launched for about US$350,000 per satellite.

12.4.1 Design

One of the major problems with nanosatellites is the lack of space for the installation of the solar cells. All electronic subsystems have to be designed with low power consumption. For TUBSAT-N, designers sought to reduce cost and size by omitting mechanical and electrical separation of subsystems (they were not self-contained), using low-power devices (but capable of withstanding radiation of 20 krad), designing for operational temperatures in the range of –40ºC to 80ºC,

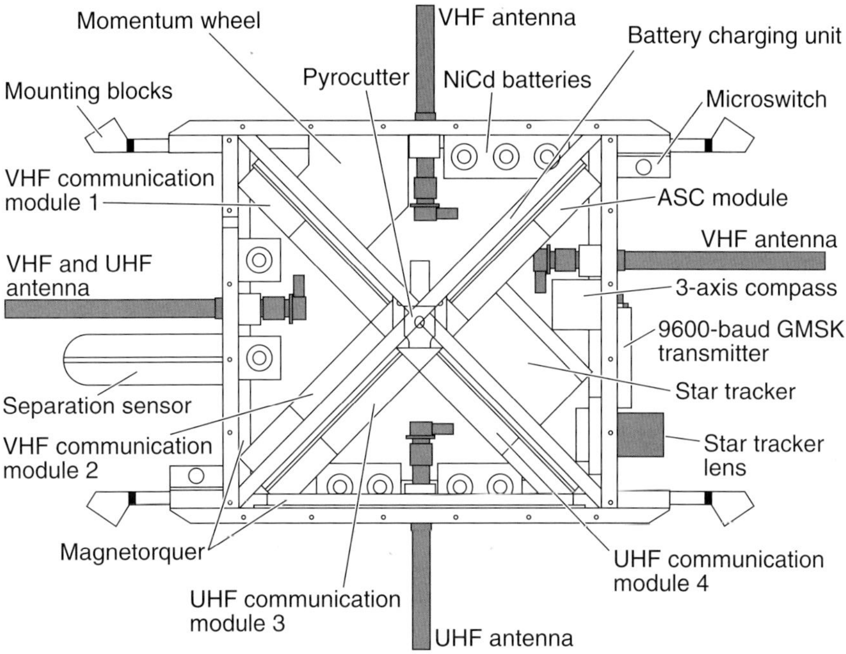

Fig. 12.12. Schematics of TUBSAT-N.

and ensuring short delivery time for components. The communication system would need to function without an attitude and orbit control system; therefore, the antenna radiation pattern needed to be sufficiently omnidirectional. This was accomplished by using two orthogonal linearly polarized antennas.

During the design phase, many low-power semiconductors were tested for radiation hardness. In fact, the real-time clock RTC4553 had to be replaced by RTC58321 as a result of such a test (Fig. 12.11). Each power consumer was configured with its own fuse that would monitor the current and shut down in case of overcurrent. The solar cells were mounted on the top and bottom surfaces.

High-energy-density NiCd cells from Sanyo were chosen for the battery: TUBSAT-N had nine cells with 5 Ahr (KR-500DEL), and TUBSAT-N1 had nine cells with 2.8 Ahr (KR-2800CE). The satellites used passive thermal control, and predictions from a 20-node thermal model were confirmed. The structure was made from carbon fiber (Fig. 12.12).

TUBSAT-N had four communication units for telemetry and commanding: two VHF units, and two UHF units, all with FFSK modulation (GMSK optional). Each communication module consumed less than 60 mA while receiving and about 1.5 A while transmitting. It had dimensions of 8 × 16 × 2 cm and a mass of 260 g. Special linear ground insensitive antennas were developed, because the nanosatellites were too small to be an effective RF ground plane.

12.4.2 Communication Experiments

The store-and-forward communications payloads were used to keep track of transmitters placed on vehicles, migrating animals, and marine buoys. For example, the Institute of Applied Physics in Kiel, Germany, used TUBSAT-N to retrieve data (water temperature, salt content) from buoys. The user communication access to the satellite was organized for modified preassignment time-division multiple access (MP-TDMA). Each mobile communication terminal (Fig. 12.13) had a GPS receiver; with knowledge of position, time, and orbital parameters, it could calculate the contact times with the satellite. After the satellite emerged above the horizon, the terminal would wait for the defined time slot for transmitting data.

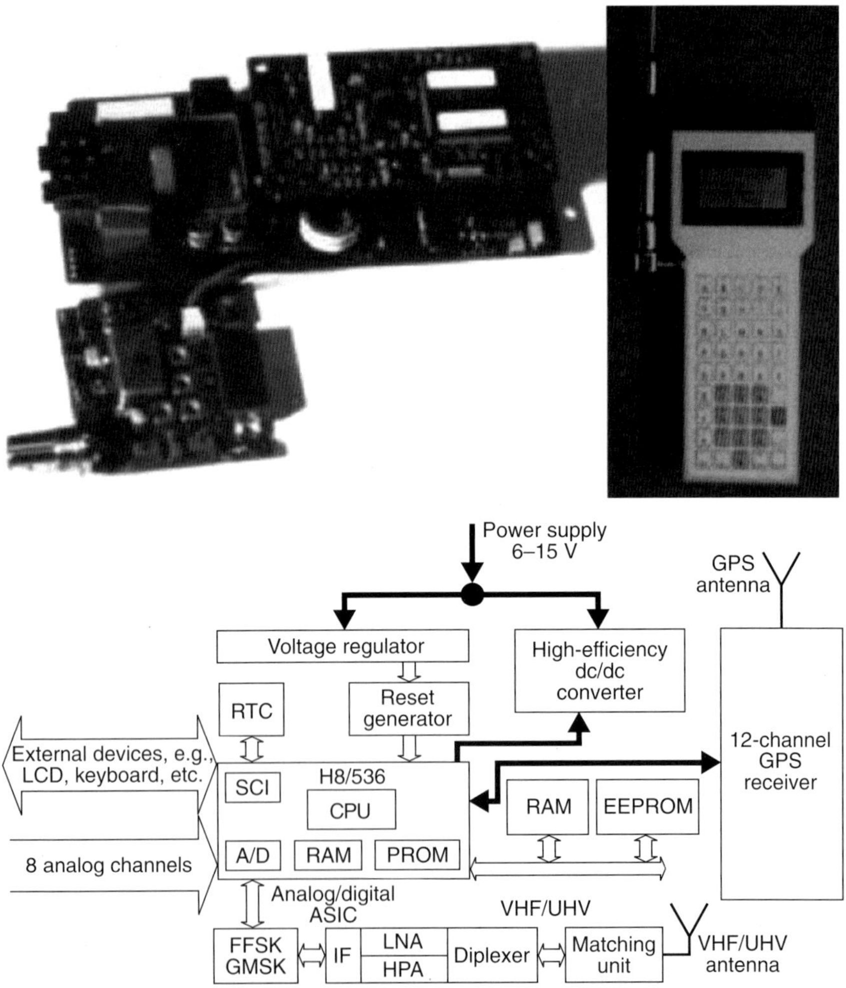

Fig. 12.13. Mobile communication terminal with integrated GPS receiver.

12.5 DLR-TUBSAT

DLR-TUBSAT[7] (Fig. 12.14) was launched on 26 May 1998 on a Indian PSLV-C2 rocket as a secondary payload into a sun-synchronous orbit (altitude 726 km, inclination 98.3811 deg) together with KITSAT-3. The primary payload was the IRS-P4 satellite. The development of the DLR-TUBSAT was a joint project of TU Berlin and the German Aerospace Center, DLR (Deutsches Zentrum für Luft- und Raumfahrt). The satellite was designed for interactive Earth observation,[8] where the target is not identified in advance and must be located, or it has to be visually followed. A project overview is given in Table 12.3.

Most of the satellite remote sensing programs of that time were huge government projects. These systems commonly used only nadir pointing with repeat cycles of 3 to 26 days, delivered only photos of predefined targets, and were in general heavy systems with huge development and launch costs.

DLR-TUBSAT tried to exploit the characteristics of being a microsatellite. It would be smaller and cheaper and could be offered to customers unwilling

Fig. 12.14. DLR-TUBSAT.

Table 12.3. DLR-TUBSAT Mission Overview

Mission	Earth observation (interactive attitude control); Technology demonstration (wheel/gyros, low-cost video imaging)
Launch/Status	26 May 1999 (Indian PSLV); Operational
Orbit	Sun-synchronous, altitude 726 km, inclination 98.4 deg
Design	TUBSAT-C microsatellite bus (TUB); Optical Payload (DLR)
Mass/Volume	44.8 kg, 32 × 32 × 32 cm
Structure	Four aluminum compartments
Thermal Control	Passive
Power	Four NiH_2 battery cells, 12 Ahr, 10 V 4 silicon solar panels (single string of 34 cells, max. 14 W)
Communication	TTC: VHF (FFSK, 1200 baud, 5 W), UHF (FFSK, 1200 baud, 3.5 W); Payload: S-band (125 kbaud BPL coded or analog video)
Attitude Control	Three wheel/gyro pairs, magnetorquer, sun sensor
Payload	Focal length 16 mm, pixel resolution 375 m (f/0.95, bw video); Focal length 50 mm, pixel resolution 120 m (f/1.80, bw video); Focal length 1000 mm, pixel resolution 6 m (f/11, bw video)

or unable to pay for a larger satellite. One customer could buy a small system designed for special needs and avoid sharing it with others. Even launch could become affordable, as many new small launchers would be viable (mostly converted ICBMs). The disadvantages were the small payload volume and limited power supply. Radar observation would not be possible, but visible and near-infrared viewing with resolutions of 3 to 6 m would be possible for a microsatellite in low Earth orbit. The primary mission of DLR-TUBSAT was to investigate these new possibilities while providing video surveillance of Earth.

12.5.1 Design

DLR-TUBSAT is cube shaped with dimensions of 32 × 32 × 32 cm and weighs 44.8 kg. It contains three modules: housekeeping, attitude control, and payload (Fig. 12.15). The housekeeping and attitude control modules represent the TUBSAT-C bus.[2] The structure comprises four compartments and is made of aluminum. Thermal control is passive, and stable temperatures between −8°C and +4°C are maintained in orbit.

The housekeeping module contains batteries, the power control unit, the air core magnetic torque coil, and the two telemetry and telecommand units in VHF

Fig. 12.15. Exploded view of DLR-TUBSAT.

(2 m) and UHF (70 cm) band, as well as the two antennas. It uses four duplex NiH_2 battery cells from EaglePicher, each with a capacity of 12 Ahr supporting an unregulated 10 V bus (no charge regulator). Four solar panels, each containing a single string of 34 silicon cells, charge the batteries. The short-circuit current of each panel is 960 mA. The power control unit contains the dc-to-dc converter and the power distribution device. The UHF/VHF telemetry and telecommand units receive and transmit data via FFSK modulation at a rate of 1200 bps (similar to TUBSAT-A). Both transceivers nominally operate in parallel in a listening mode. Both antennas are omnidirectional dipoles. As long as no command is received from the ground, the satellite is silent.

The attitude control module contains three reaction wheels (RW202 by TU Berlin) with integrated wheel drive electronics and three fiber-optic gyros (Litef μFORS-6). The wheels weigh less than 1 kg and consumes about 1 W in steady state. Furthermore, the module contains two different magnet torquers with electronics and the onboard data-handling system. A single string of solar cells is attached at the surface in the z-axis and is used as a sun acquisition sensor. The magnet torquers are only used to reduce or to increase the angular momentum.

The payload module contains three TV cameras: two wide-field cameras with 16 mm focal length (f/0.95), and one camera with 50 mm focal length (f/1.8). There is also a high-resolution telescope with 1 m focal length (f/11) and a ground pixel resolution of 6 m. Each CCD chip (ICX039DLA) contains 752 × 582 pixels, and each camera can transmit video images in CCIR-standard and single digital pictures. The analog video signal is transmitted via S band with a bandwidth of 8 MHz, and the beamwidth of the helix antenna is 70 deg.

12.5.2 Interactive Attitude Control

DLR-TUBSAT is commanded by the TU Berlin VHF/UHF ground station in Berlin. For payload data reception, several cooperative ground stations have been used, such as the German Remote Sensing Data Center (DFD) S-band receiving station in Neustrelitz, north of Berlin. In the course of the satellite operational phase, a complete satellite control station in Tromsø, Norway, was developed, and a cooperative satellite control station in Rabat, Morocco, was used. In addition, tests with a ground station in South Africa and a radio amateur ground station in Kiel have also been performed.

With the DLR-TUBSAT microsatellite, TU Berlin developed the interactive attitude control strategy. The satellite is controlled almost like a helicopter hovering over a target of interest on Earth, and is completely rotated as it follows the target. For example, one experimental objective was to take a video of Capetown

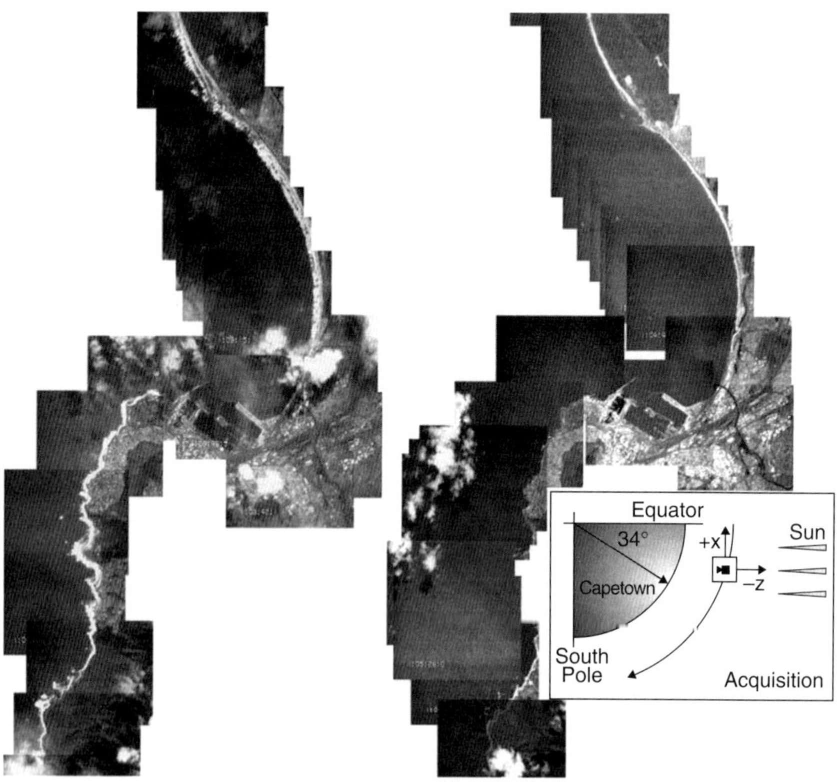

Fig. 12.16. Interactive attitude control in Capetown, South Africa.

harbor, South Africa.[9] The satellite starts out in a hibernation mode with low power consumption. The attitude control system is switched off, and the satellite is completely passive and self contained (tumbling mode). To begin the targeting event, the three reaction wheels and the three fiber-optic gyros are switched on. Now, the satellite needs to be prestabilized (acquisition). Sun pointing modes are used depending on the season and geographical position. The satellite attitude is fixed using the reaction wheels and gyros in all three axes. Now, one axis of the satellite is aligned with the sun vector. Using the solar cells as a sun sensor, the $-z$ axis is brought into the sun (maximum solar cell current). In this position, the satellite is kept inertially stabilized. Orientation accuracy is about ±5 deg. Boot software was developed for the satellite to make this acquisition automatically in advance of the pass. Other acquisition using the camera as an Earth horizon sensor have also been developed.

When the satellite comes into sight of the ground station, the operator starts receiving the video signal of the cameras. As a result of the acquisition, the satellite is almost looking nadir and ahead. The satellite now sets up the azimuth angle by rotating about the $+z$ axis until the image runs from top to bottom. Because the orbit is nearly polar, the north direction is now at the top of the image. The operator now follows the satellite until it has reached the coastline of South Africa. Then, a counterrotation about the y axis is induced to keep the visual focus at the coastline. From here, the operator follows the coastline to the south until Capetown appears. The operator conducts this interactive session using a joystick to change the satellite attitude by commanding angular velocities in the two axes perpendicular to the line of sight. The results[10] can be seen in Fig. 12.16, which presents two distinct approaches to finding Capetown harbor.

Other scenarios include searching an area for a specific target with one of the wide-field cameras and, after locating it, switching to the high-resolution camera. It is also possible to follow a moving target. With interactive attitude control strategies, no star sensor or GPS system is required. That keeps the system as simple as possible and cheap as well. Nevertheless, using a star sensor for the acquisition maneuver makes it faster and more exact. In the exercise,[10] the payload cameras were programmed to send down star images. These were processed in real time on the ground, and from the data, the attitude could be determined. This procedure only worked in the eclipse portion of the orbit at night. It was used to take night images of bright cities in Europe.

DLR-TUBSAT is still operational. The attitude control system is working without any problems. Apart from the 50 mm camera, the optical payloads are not useful anymore. After nine years in orbit, radiation damage to the CCDs and defocusing of the optics rendered the system nearly inoperative. The flight experiences[11] of DLR-TUBSAT and the interactive attitude control have been incorporated in the attitude control strategies of MAROC-TUBSAT and the design and operation of LAPAN-TUBSAT.

12.6 MAROC-TUBSAT

MAROC-TUBSAT (Fig. 12.17) is a cooperative effort of TU Berlin and CRTS (Centre Royal de Teledetection Spatial) of Morocco. It is a microsatellite for Earth observation. It allows observation of any place in the world within 24 hours and can document events that change with time between 12 hours and 2 weeks. The camera has a focal length of 72 mm and a field of view of about 8 deg. Typical applications include ice movement, cartography, or vegetation development. A mission overview is presented in Table 12.4.

The satellite was designed at TU Berlin and is based on the microsatellite bus TUBSAT-C. The payload camera was developed in Morocco. The cooperation also included teaching satellite technology and design to Moroccan engineers. MAROC-TUBSAT was launched with the Russian Zenit rocket. After the transport to Russia in December 2000, a final test revealed that one of the power amplifiers was broken and had to be replaced. It was probably damaged during transportation. There was some excitement and plenty of discussion on how to proceed, as the launch was scheduled for the beginning of the new year. Nobody knew at the time that the launch would be delayed by almost a year, leaving plenty of time to repair the satellite without the rush. So, it was decided to replace

Fig. 12.17. MAROC-TUBSAT.

Table 12.4. MAROC-TUBSAT Mission Overview

Mission	Earth observation (ice movement, cartography, vegetation); Technology Demonstration (star sensor, low-cost imaging mission)
Launch/Status	10 December 2001 (Russian Zenit); Operational
Orbit	Sun-synchronous, altitude 1000 km, inclination 99.4 deg
Design	TUBSAT-C microsatellite bus (TUB); Optical Payload (Morocco, CRTS)
Mass/Volume	47 kg, 32 × 34 × 36 cm
Structure	Four aluminum compartments
Thermal Control	Passive
Power	Four NiH_2 battery cells, 12 Ahr, 10 V 4 silicon solar panels (single string of 34 cells, max. 14 W)
Communication	TTC: VHF (FFSK, 1200 baud, 5 W), UHF (FFSK, 1200 baud, 3.5 W); Payload: S-band (max. 250 kbaud, 2 W RF, BPSK modulation)
Attitude Control	Three wheel/gyros pairs, magnetorquer three-axis magnetic field sensor, star sensor, sun sensor
Payload	Focal length 72 mm, pixel resolution 32 m (f/6, b/w image)

the amplifier. The satellite was opened, and the broken device was sent back to Berlin. Just before Christmas of 2000, the satellite was repaired and made ready for launch.

On 10 December 2001, the satellite was finally launched from the Baikonur Cosmodrome into a sun-synchronous orbit with an altitude of 1000 km.

The satellite was used extensively by TU Berlin until the launch of the LAPAN-TUBSAT in January 2007. TU Berlin is now focusing on the operation of the LAPAN-TUBSAT and has left the operation of the MAROC-TUBSAT to the Moroccan operators.

12.6.1 Design

In the design process, TU Berlin was responsible for the satellite bus, and the Moroccan partners for the payload. Figure 12.18 shows an exploded view of the MAROC-TUBSAT space segment. The satellite uses the TUBSAT-C bus, which already had flight heritage through DLR-TUBSAT.

Power is generated by four solar panels of 320 × 320 mm (34 cells in series) with a maximum generated power of 14 W each. The energy is stored in four NiH_2 batteries with 10 V nominal voltage and 12 Ahr of energy.

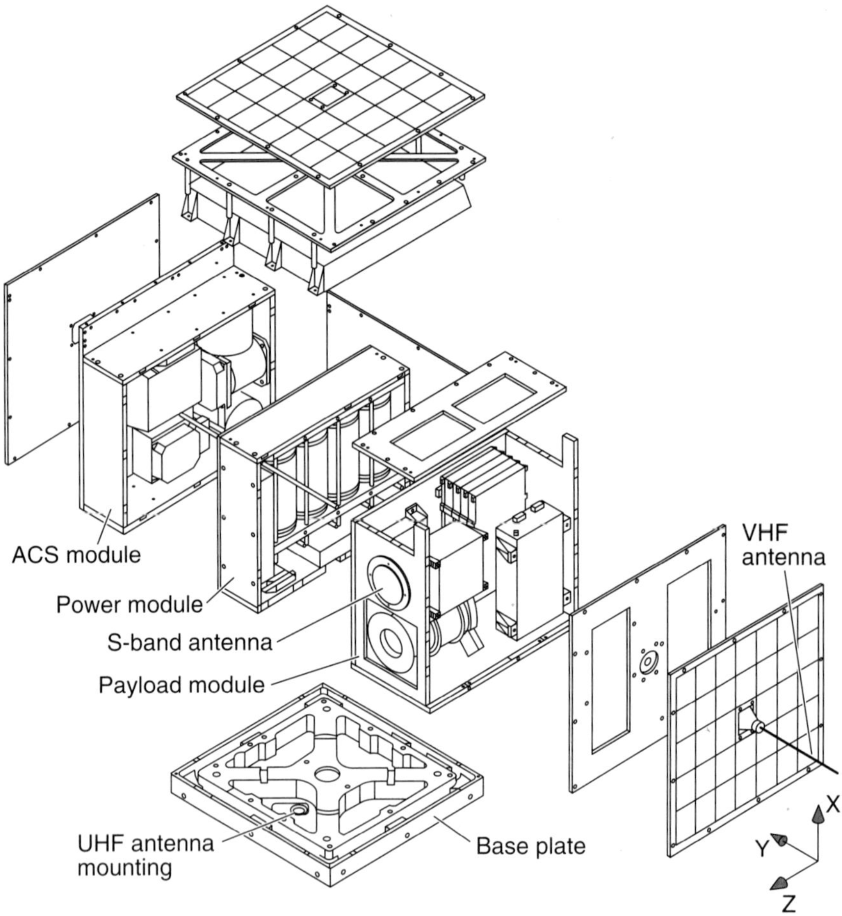

Fig. 12.18. Exploded view of MAROC-TUBSAT.

There are two telemetry and telecommand units for the communication, both of which employ FFSK modulation. The first operates in UHF (1200 baud, 3.5 W RF output), and the second in VHF (1200 baud, 5.0 W RF output). The attitude control system features three wheel/gyro pairs in orthogonal axes (similar to the DLR-TUBSAT), a star sensor for three-axis inertial attitude determination, a three-axis magnetic field sensor, a magnetic coil in one axis, and single solar cells as sun sensor.

The star sensor KM1301 was developed by TU Berlin and manufactured by Kayser-Threde. It has a mass of 0.55 kg and consumes 4.2 W of power. It features a 288 × 384 pixel CCD matrix from Thomson. Star pattern recognition is done with a carefully selected star catalog, and the three-axis attitude information is given in the form of Euler angles or quaternions. With a focal length of 16 mm,

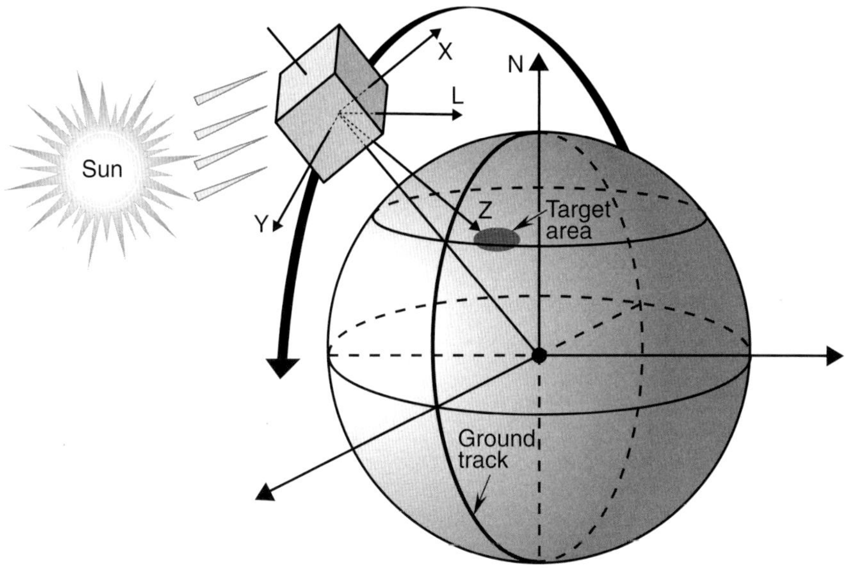

Fig. 12.19. Simplified geometry of MAROC-TUBSAT acquisition maneuver.

the theoretical accuracy is about ±0.02 deg. It is designed for an update period of 250 msec after an initial acquisition of about 0.3 sec.

The payload CCD-sensor contains 770 × 576 pixels. Focal length of the optics is 72 mm, f/6, and the field of view is 8 deg. The nominal integration time is 28 msec. The camera can only deliver photos. An S-band transmitter with patch antenna, BPSK modulation (max 250 kbaud), and 2 W RF power is used to send down the digital image information.

121.6.2 Flight Results

MAROC-TUBSAT is operated by TU Berlin from ground stations in Berlin and Longyearbyen, Norway. Image download is managed in conjunction with mission partners such as DFD in Neustrelitz or Svalsat in Longyearbyen. The satellite is also operated by Morocco, which has a ground station in Rabat for commanding and image reception.

The satellite is kept in a momentum bias mode: The angular momentum vector L is put perpendicular to the orbital plane. It is controlled with the attitude determination by the star sensor and realigned with the help of the magnetorquer. The star sensor only operates in eclipse because it is blinded by the light of the Earth and sun. Furthermore, the gyro drift is too big to hold an attitude with acceptable precision. Direct targeting using the star sensor and wheel/gyro pairs is not precise enough. On the other hand, the momentum-axis pointing direction is known by 0.2 deg. It is very stable, and can therefore provide a very precise reference for attitude control.

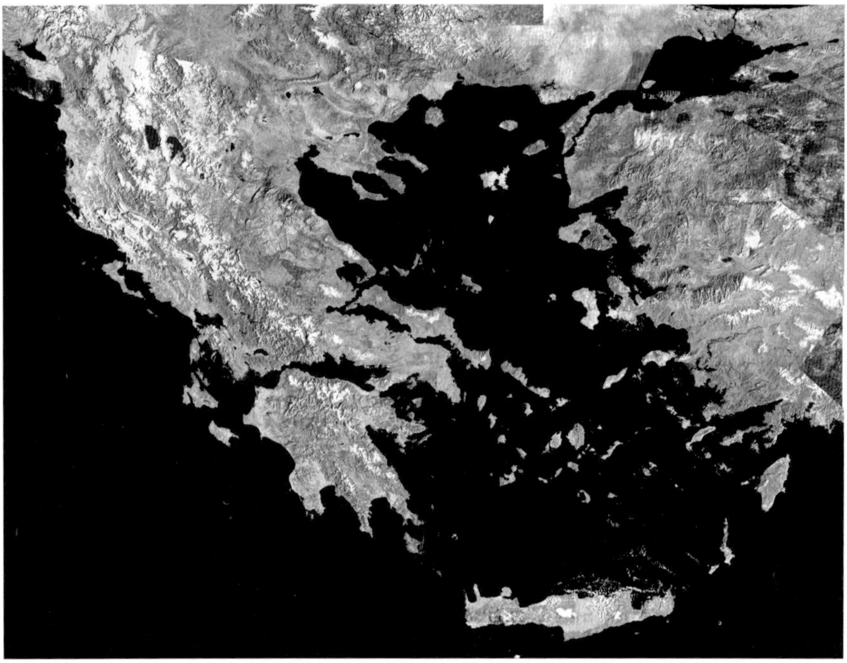

Fig. 12.20. Stitched image of Greece and Turkey (images collected between December 2002 and March 2003) by MAROC-TUBSAT.

The satellite starts out in sleep mode. All angular momentum is in the x axis (camera line of sight is perpendicular in the $+z$ axis). First, attitude acquisition is done by splitting the momentum vector L between z and x axes. This defines the offset from the nadir. (Doing nothing results in nadir pointing, with the x axis and momentum vector L perpendicular to the orbit.) Since the orbit is almost polar, this defines the east-west direction of the target. The rotation about the x axis is determined with the sun. This defines the north-south direction of the target. Figure 12.19 shows a simplified schematic of the geometry in question.

Figure 12.20 shows a collection of images of Greece taken by the MAROC-TUBSAT. It took several days to collect all the images for such a big area. There is also the possibility for stereoscopic pictures because the satellite can capture one image of a target (such as a mountain) before passing over and a second image after rotating back on the falling branch of the pass. Ice monitoring is one of the many applications, and has been done for the Icesail Expedition of Arved Fuchs in the polar region.

MAROC-TUBSAT was a very successful microsatellite mission, especially for the Moroccan government. The cooperation and promising mission results have been a major step in the subsequent creation of a National Moroccan Space Program.

12.7 LAPAN-TUBSAT

LAPAN-TUBSAT[12] (Fig. 12.21) is a cooperative endeavor between TU Berlin and the National Institute of Aeronautics and Space of Indonesia (LAPAN). One of LAPAN's primary objectives is to master satellite technology and its applications, such as remote sensing and data collecting. The agency set up a program with three phases. The first phase entails partnering with an established satellite manufacturer to build a satellite and train Indonesian engineers. The second phase includes operation of the satellite and mission analysis. The third step is to build a satellite of its own in Indonesia, where the cooperation is reduced to the role of consulting and advising.

LAPAN did a survey of satellite manufacturers that also provide training, and ultimately appointed TU Berlin. A fixed contract of 1 million euro was signed in 2003 to build an Earth observation satellite based on the DLR-TUBSAT and MAROC-TUBSAT architectures by the end of 2005. During the design and manufacturing process, four Indonesian engineers resided in Germany for more than a year and half. Another 12 engineers have been trained over a three-month period. The satellite was ready in time, but the launch was postponed because of investigations following the destruction of a Indian GSLV rocket on 10 July 2006. LAPAN-TUBSAT was finally launched on an Indian PSLV on 10 January 2007. The camera, with focal length of 1000 mm, allows video surveillance with up to 6 m pixel ground resolution. A mission overview is given in Table 12.5.

Fig. 12.21. LAPAN-TUBSAT.

Table 12.5. LAPAN-TUBSAT Mission Overview

Mission	Earth observation (traffic, agriculture, disaster monitoring); Technology demonstration (star sensor, refocusing mechanism); Training of Indonesian engineers (LAPAN)
Launch/Status	10 January 2007 (Indian PSLV); Operational
Orbit	Sun-synchronous, altitude 625 km, inclination 98.7 deg
Design	Advanced TUBSAT-C microsatellite bus (TUB)
Mass/Volume	56 kg, 45 × 45 × 27 cm
Structure	Aluminum plates (divided into two shelves)
Thermal Control	Passive
Power	Five NiH_2 battery cells, 8Ahr, 12.5 V; Four silicon solar panels (single string of 34 cells, max. 14 W)
Communication	TTC: 2× UHF (FFSK, 1200 baud, 3.5 W); Payload: S-band (2220 MHz, 5 W RF)
Attitude Control	Three wheel/gyro pairs, magnetorquer, star sensor, sun sensor
Payload	Focal length 35 mm, 80 km swath (color video); Focal length 1000 mm, pixel resolution 6 m (f/11, color video)

12.7.1 Design

LAPAN-TUBSAT's design[13] follows the TUBSAT family, with dimensions of 45 × 45 × 27 cm and a mass of about 56 kg. The satellite bus is a consequent development of TUBSAT-C. The structure of the satellite is made of aluminum alloy, and is black anodized for thermal properties. The satellite is divided into two shelves (Fig. 12.22) to provide maximum access while harnessing and testing.

The power generator in the satellite consists of four silicon solar panels measuring 432 × 243 mm. Each panel has 34 cells in series, which provides a maximum power of 14 W. The power storage system in the satellite consists of five NiH_2 batteries, configured in series (nominal voltage 12.5 V, capacity 8 Ahr). The power charging is done by hardwiring the batteries, through diodes, parallel to the solar panels (direct energy transfer/unregulated bus).

There are two identical telemetry and telecommand systems in the satellite for redundancy. The system uses a frequency of 436.075 MHz, with FFSK modulation (half duplex); the data rate is 1200 bps, and the RF output is 3.5 W. The half-dipole antennas are placed orthogonally to guarantee an adequate link budget at any attitude. The onboard data-handling system uses a Hitachi SH7045 processor with external memory for store-and-forward communication experiments.

The attitude system is based on three wheel/gyro pairs, a three-axis magnetorquer (air coil), sun sensor (using four solar panels and two solar cells to cover

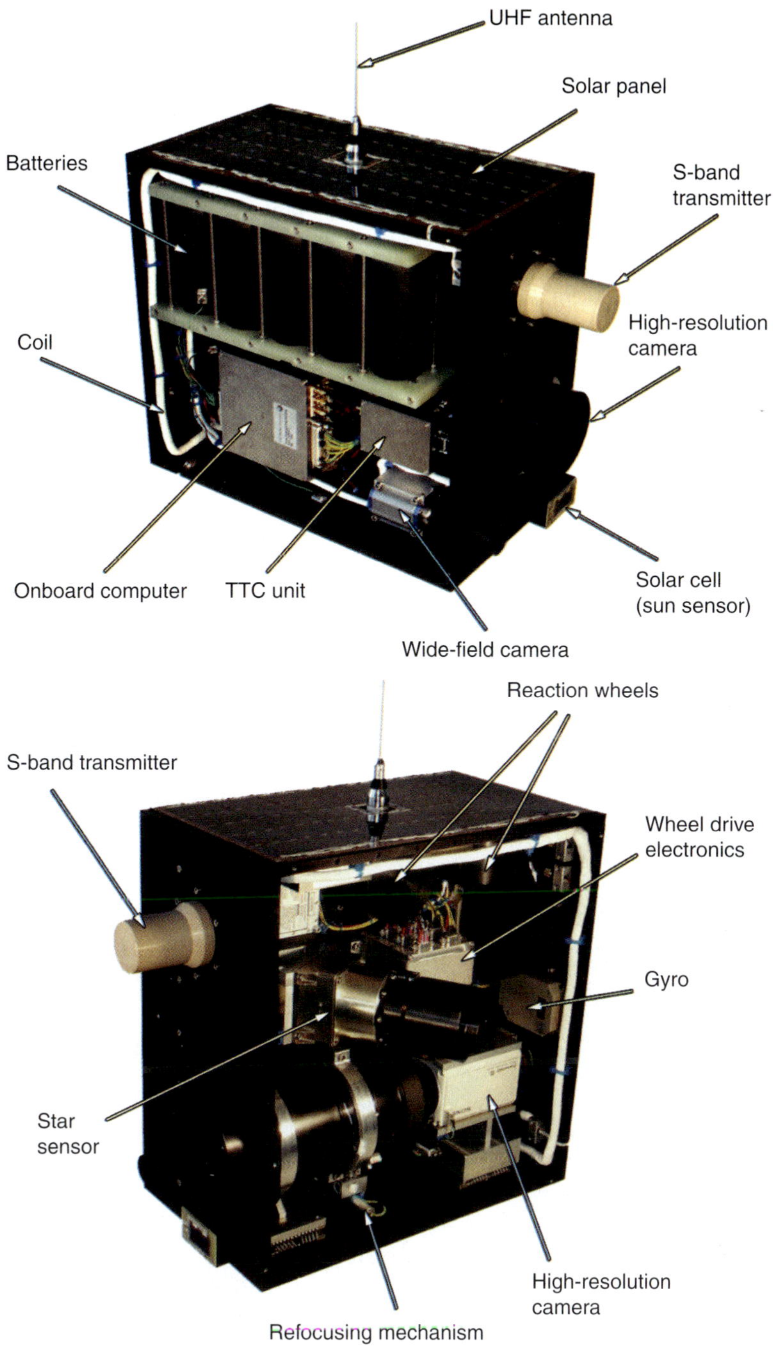

Fig. 12.22. LAPAN-TUBSAT design, front (top) and aft (bottom).

all six surfaces), and a star sensor. The reaction wheels, fiber-optic laser gyros (Litef μFORS-6), and corresponding electronics were developed by TU Berlin and based on the TUBSAT-C microsatellite bus. The star sensor (developed by Vectronic Aerospace GmbH) uses a 512 × 512 radiation-tolerant CMOS image sensor and 50 mm optics. The data processor for star recognition uses a 32 bit Hitachi SH7045; it is based on the star sensors of previous TUBSAT missions.

The payload analog video signal is frequency modulated onto a carrier of 2220 MHz. The output is 5 W RF. One can switch between a wide-field camera and a high-resolution camera. Both are off the shelf. The wide-field camera is a color CCD video recorder with an effective imager size of 752 × 582 pixels and a 50 mm lens (built by KAPPA Opto-Electronics GmbH); it produces an image with an 80 km swath and a ground resolution of 120 m. The high-resolution camera (Fig. 12.23) is a three-CCD video recorder with a color splitter prism (Sony DXC-990P). Each CCD has an effective imager size of 752 × 582 mm pixels. A reflex telescope with focal length of 1000 mm (Nikon f/11) is mounted on the camera, resulting in images with 4 km swath and 6 m ground resolution from a 625 km altitude. TU Berlin developed a mechanism to allow refocusing of the high-resolution camera. This is necessary to counter the heat expansion of the telescope.

12.7.2 Flight Results

LAPAN-TUBSAT[14] is operated from Indonesia as well as from Germany. LAPAN has one ground station in Rumpin (close to Jakarta) and is building a second one

Fig. 12.23. LAPAN-TUBSAT high-resolution camera design.

in Biak (Papua) to cover all of Indonesia. Both have UHF antennas for commanding and an S-band dish to receive the payload signal. TU Berlin has command stations in Berlin and in Longyearbyen. The payload signal can be received with a 3 m dish in Berlin. TU Berlin also has access to several professional satellite ground stations—namely, with the German Remote Data Sensing Center (DFD) and Svalsat—as well as with a radio amateur in Kiel, Germany.

The satellite is operated in a similar manner to DLR-TUBSAT. It transmits live video, and the interactive attitude-control strategies developed with DLR-TUBSAT can be applied. The big advantage of LAPAN-TUBSAT is the star sensor, which allows an acquisition with very high precision. It can be used to pinpoint a target with an uncertainty of only 0.1 deg. The precision is high enough that the target is visible in the 3 × 4 km high-resolution image before the operator has to intervene. Once the target comes into view, the satellite can be "stopped" by means of a counterrotation (to compensate for the satellite movement). Therefore, it is possible to hover over the target or to scan the area around the target. Furthermore, by looking off nadir, the ground repeat time can be reduced.

LAPAN-TUBSAT is operated as a momentum-bias system. The momentum vector is not necessary for targeting because operators can get the actual attitude from the star sensor and rotate the satellite with the wheel/gyro pairs—in fact, using the momentum vector can have negative effects, because it induces momentum cross coupling when maneuvering and can disturb the maneuver because the control loops for the three reaction wheels are separated. Nevertheless, there are some advantages to using it. The satellite can be set into a hibernation mode either as single or dual spinner. This reduces costs because few personnel are required to operate the satellite. The momentum vector is maintained perpendicular to the orbital plane. The constant angular momentum vector stabilizes the satellite, resulting in constant angle to the sun. This, in turn, stabilizes the temperatures, and operators can ensure that the cameras are not exposed to direct sunlight. Because the momentum vector is known, the sun vector can be used as a second reference to guarantee that the star sensor is not blinded by the sun or Earth, resulting in a very fast acquisition. A target can be pinpointed from hibernation mode in just a few minutes.

Figure 12.24 shows images from LAPAN-TUBSAT, most of which are stitched together from video frames. The wide-field camera can not only be used for orientation in the interactive attitude maneuvers, but also for ice monitoring and fire detection (12.24a). The high-resolution camera can be used for surveillance in maritime areas (12.24b) or disaster monitoring (12.24c). It can help environmental organizations detect illegal lumbering or monitor disasters such as oil pollution of the sea caused by tanker accidents. Further applications include agriculture, traffic, and cartography. Operators have also investigated new ways of using the satellite—for example, taking pictures of the European metropolis at night (12.24d) or high-resolution images of the moon (12.24e).

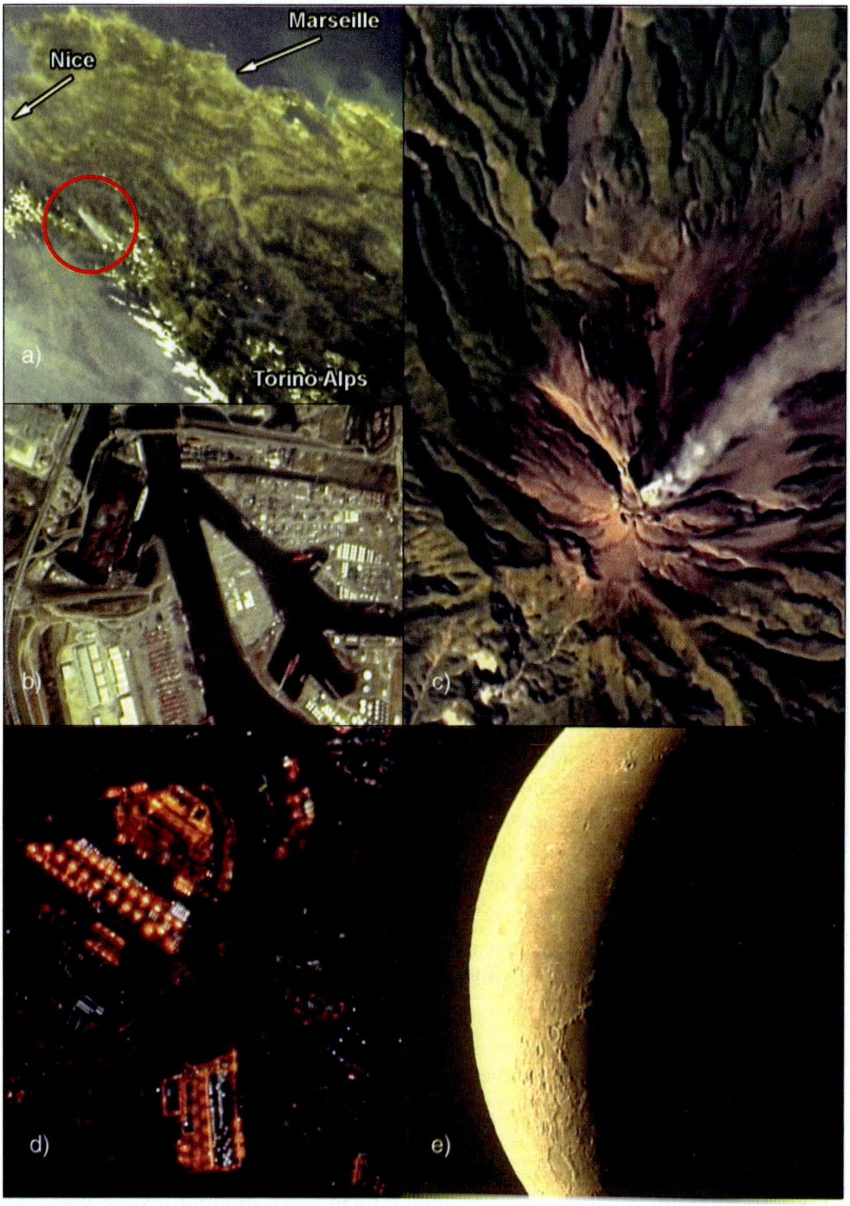

Fig. 12.24. Images from LAPAN-TUBSAT. Smoke of a forest fire close to the Italian-France border (image center: 44.00 N, 6.06 E) on 26 July 2007 looking about 30 deg off nadir (a); high-resolution image of Hamburg harbor, Germany (image center: 53.516 N, 9.941 E; image size: 3.6 × 2.5 km), taken on 26 March 2007 (b); Indonesian Vulcan Merapi (7.543 S, 110.444 E; 5.5 × 6.8 km) taken on 24 May 2007 (c); high-resolution image of Hamburg harbor (53.517 N, 9.931 E) at night from 11 March 2007 (d); high-resolution images of the moon taken on 11 June 2007 (e).

12.8 Future Developments at TU Berlin

Space programs everywhere are facing continuous budget cuts and looking for an easier, cheaper, faster way to exploit space. The microsatellite technologies developed at TU Berlin provide an answer to the problem of increasing budget pressure. Though developed primarily as an educational exercise, these technologies have been exported for use in national space programs. For example, the attitude control system of the TUBSAT-C bus was used for DLR-TUBSAT, MAROC-TUBSAT, and LAPAN-TUBSAT; it is also being used by DASA for INSPECTOR and by KAIST (SaTReC) for the Korean KITSAT-3 program. The LAPAN-TUBSAT microsatellite proved that high-resolution color images with 6 m ground resolution could be delivered with high reliability and for a total project cost of only 1 million euro (not including launch, but including the education and teaching of support engineers).

TU Berlin is still involved in microsatellite design besides operating the LAPAN-TUBSAT. For example, the Indonesian second-generation microsatellite will be built completely in Indonesia, but will be a further development of the LAPAN-TUBSAT, with a TU Berlin microsatellite bus. TU Berlin is advising LAPAN not only on the design of the satellite bus, but also on the choice and design of the optical payloads and payload data transmitters. The launch of the second-generation satellite featuring higher ground resolution in four spectral bands is scheduled for 2010. Experiences of the LAPAN-TUBSAT operation will not only enrich TU Berlin's consulting activities but also enhance student education. Technical difficulties and solutions of actual problems arising during operation of the TUBSAT fleet, as well as further developments such as implementing an automatic targeting mode on the LAPAN-TUBSAT or improving subsystem design, are discussed with students in class.

While development continues on the TUB microsatellite fleet, Prof. Brieß of TU Berlin and his team are developing the picosatellite BeeSat[15] (Berlin Experimental and Educational Satellite) to be launched at the end of April 2009—which also simultaneously initiates the development of an in-house nanosatellite program. These projects try to carry the new developments in the areas of miniaturization and MEMS up to space. The primary mission of BeeSat is the on-orbit verification of the newly developed microreaction wheels for picosatellite applications. BeeSat is built according to CubeSat specifications (10 × 10 × 10 cm and a maximum mass of 1 kg) and is now under assembly at TU Berlin (Fig. 12.25). It features a UHF amateur radio communication system and a color camera as payload.

The coin-sized microwheel system (total 150 g) can be used to set up a precise attitude control system[16] and open the door to new applications for picosatellites in the areas of Earth observation, astronomy, and on-orbit verification of new technologies. Other space-related projects at TU Berlin include development of a foldable telescope for microsatellite application (the Dobson Space Telescope[17]), design of miniaturized planetary rovers, and development of hot water propulsion systems (Aquarius[18]).

The extensive satellite program allows students at TU Berlin to enjoy first-hand experience in satellite design and manufacture. The space flight community is being enriched not only because of the TU Berlin's satellite projects, but also as a result of the well trained students and instructors. Partnerships all over the world helps to boost space travel and to expand the exploration of Earth and outer space.

Fig. 12.25. BeeSat.

12.9 References

[1] M. Swartwout, "Twenty (plus) Years of University-Class Spacecraft: A Review of What Was, An Understanding of What Is, And a Look at What Should Be Next," *20th Annual AIAA/USU Conference on Small Satellites* (21 Aug. 2006).

[2] P. Butz and U. Renner, "TUBSAT-C, A Microsat Bus for Earth Observation Payloads," *3rd International Symposium, Small Satellites Systems and Services* (Annecy, France, June 1996).

[3] U. Renner, B. Luebke-Ossenbeck, and P. Butz, "TUBSAT: Low-Cost Access to Space Technology," *Small Satellites Systems and Services* (SEE N94-17131 03-18) (Jan. 1993).

[4] U. Fielitz, U. Renner, R. Schulte, and H. Wölfel, "Satellitentelemetrie an Rothirschen im Harz—Eine Pilotstudie," *Zeitschrift für Jagdwissenschaft,* **42** (1) (March 1996).

[5] R. Schulte, "TUBSAT-N, A Global Communication Satellite System, based on Nanosatellites," *Small Satellite Systems and Services, 4th International Symposium* (Antibes, France, 1998).

[6] R. Schulte, "TUBSAT-N, An Experimental Global Communication Satellite System, based on Nanosatellites," *IAF Specialists Symposium: Novel Concepts for Smaller, Faster and Better Space Missions* (Redondo Beach, CA, 1999).

[7] S. Schulz and U. Renner, "DLR-TUBSAT: A Microsatellite for Interactive Earth Observation," *Small Satellite Systems and Services 5th International Symposium* (La Baule, France, 2000).

[8] S. Schulz, "Interaktive Lageregelung zur Erdbeobachtung mit Mikrosatelliten am Beispiel DLR-TUBSAT," Ph.D. thesis, Technische Universität Berlin (2001).

[9] U. Renner, J. Bleif, and S. Roemer, "The TUBSAT Attitude Control System: Flight Experience with DLR-TUBSAT and MAROC-TUBSAT," *4th IAA Symposium on Small Satellites for Earth Observation* (Berlin, 7–11 April 2003).

[10] J.-H. Bleif, "Interaktive Lageregelung bei Nacht mit dem Mikrosatelliten DLR-TUBSAT," Ph.D. thesis, Technische Universität Berlin (2004).

[11] S. Roemer and U. Renner, "Flight Experiences with DLR-TUBSAT," *Acta Astronautica,* **52** (9), pp. 733-737 (5 May 2003).

[12] Triharjanto, Hasbi, Widipaminto, Mukhayadi, and Renner, "LAPAN-TUBSAT: Micro-Satellite Platform for Surveillance and Remote Sensing," *Proceedings of the 4S Symposium: Small Satellites, Systems and Services (ESA SP-571)* (La Rochelle, France, 20–24 Sept. 2004).

[13] S. Hardhienata, A. Nuryanto, R. H. Triharjanto, and U. Renner, "Technical Aspects and Attitude Control Strategy of LAPAN-TUBSAT Micro Satellite," *5th IAA Symposium on Small Satellites for Earth Observation* (Berlin, Germany, 4-8 April 2005).

[14] U. Renner and M. Buhl, "High Precision Interactive Earth Observation with LAPAN-TUBSAT," *Proceedings of the 4S Symposium Small Satellites, Systems and Services* (Rhodos, Greece, 26–30 May 2008).

[15] H. Kayal, F. Baumann, K. Briess, and S. Montenegro, "BEESAT: A Pico Satellite for the On Orbit Verification of Micro Wheels," *3rd International Conference on Recent Advances in Space Technologies* (Istanbul, Turkey, 14–16 June, 2007).

[16] M. Herfort, M. Berlin, H.-P. Geile, and Z. Yoon, "BeeSat Attitude Determination and Control System," *Proceedings of the 6th IAA Symposium on Small Satellites for Earth Observation* (Berlin, Germany, 23–26 April 2007).

[17] T. Segert, B. Danziger, D. Gork, and M. Lieder, "Dobson Space Telescope—Development of an Optical Payload of the Next Generation," *Proceedings of the 6th International Conference on Space Optics, ESTEC* (Noordwijk, The Netherlands, 27–30 June 2006).

[18] H. Adirim, R. E. Lo, N. Pilz, and M. Kreil, "Hot Water Propulsion Development Status For Earth and Space Applications," *42nd AIAA/ASME/SAE/ASEE Joint Propulsion Conference and Exhibit* (Sacramento, CA, 9–12 July 2006).

Chapter 13

Proba: ESA's Series of Small Satellite Testbeds

Frederic Teston,* Dirk Bernaerts,* David Hardy*

13.1 Background

Two major factors have influenced the development of the European Space Agency (ESA) series of small satellites: the need for an affordable development environment in which to prove new ideas and technologies, and the need to reduce ground intervention after a spacecraft reaches orbit. Proba (Project for On Board Autonomy) was an idea first put forward in 1995 to address the second of these needs. Until then, most spacecraft depended upon micromanagement from ground controllers. The plan was to develop a spacecraft that could be sent a list of broad-brush tasks and a list of periods when it could downlink its data and get on with the job, possibly for several days. Proba would be a small and relatively low-cost mission to fly the experimental software needed to transfer decision making to the onboard system. It would also make it possible to fly a number of experimental technologies to address the first of the target requirements (affordable development environment for new ideas and technologies). A target was set of less than 20 million euros total for development, launch, and first-year operating costs.

A call for ideas for experimental technologies went out to the manufacturers, and at the same time an announcement of opportunity for science experiments was circulated among the science community (with the clear understanding that the technologies and experiments would be flying on an experimental platform). ESA was inundated with responses and could easily have produced proposals for a whole series of spacecraft to be developed in parallel. Funding and labor concerns led to the practical decision to start a series of small satellites that would each flight-test new technologies while fulfilling a practical science need. These have all been named "Proba" so far, but future spacecraft in the series may well acquire more descriptive names. At the time of writing (mid 2008), Proba-1 has been successfully operating in orbit for more than six years, Proba-2 is preparing for launch in 2009, and Proba-3 (actually two satellites) is in phase B of the development process, with a launch foreseen in 2012.

What is presented here is an overview of the Proba program. Many of the technological and scientific results have been presented at conferences or published in journals, and a list (not exhaustive) of publications and Web links where more detailed information can be found is given at the end of the chapter.

*European Space Agency Technology Centre
Noordwijk, The Netherlands

13.2 Ground Station and Operations Developments

In the early days of space exploration, most spacecraft were in low Earth orbits (LEO), where they were visible only to a particular ground station for short periods (typically only 7 to 12 min per pass). They were unsophisticated, and had no means of onboard data storage. Ground stations were installed all over the globe to get maximum data return from the spacecraft, which frequently transmitted continuously. The stations were staffed by teams of operators 24 hours per day, every day of the year. The station at Redu, Belgium, for example, opened in 1966 and was at the forefront of these activities (see Fig. 13.1). It is still used, among other activities, as the control center for Proba-1 and will be used in a similar capacity for Proba-2.

Over time, however, the number of active ground stations declined. With the rapid progress of technology, onboard storage of data became the norm, so satellites no longer required constant contact with a ground station. Moreover, after the initial flood of scientific data in the 1960s and 1970s, the number of LEO missions declined, and the majority of spacecraft moved to higher orbits—many to geosynchronous Earth orbit (GEO) and to non-Earth centered orbits. Data-relay satellites also grew more prevalent, considerably reducing the number of operational ground stations around the globe.

As a result, mission planners for polar-orbiting LEO spacecraft faced a challenge. Either the number of ground stations had to be expanded to maintain sufficient contact to control complex maneuvers in real time (with coordinated control distributed between those stations), or the spacecraft must be given the autonomy to perform those maneuvers without reference to a ground controller. The latter was the approach taken by Proba.

The Proba concept of transferring autonomy to the onboard system was further developed into the area of ground operations automation, which was still in the early development stage when Proba-1 was launched. The current situation allows a team of two, working normal office hours, to support the mission planning, spacecraft operations, and data dissemination for continuous operation of the satellite. The ground-station hardware comprises a number of computers, a

Fig. 13.1. Redu ground station, Belgium, housing the Proba mission control.

fully steerable antenna system, and specialized hardware for controlling the antenna and encoding and decoding the uplink and downlink data.

Triggered by a master schedule, all the other components of the station can be automatically controlled such that the system will start up 15 min prior to the appearance of the spacecraft above the horizon. The system will automatically check its own operability and then acquire the spacecraft signal, check its status, proceed with uplink of any new onboard schedule, and at the same time monitor the spacecraft housekeeping and downlink and record all the science data. Once the spacecraft has disappeared over the horizon, the system will automatically replay the recorded data and distribute the information to users via a Web server. In this way, the station supports up to six passes per day, seven days a week. Assuming no failures, it can operate for four or five days without manual interaction. If the spacecraft signal is not acquired, or if a status check indicates a spacecraft anomaly, an email can be sent to alert mission managers of the need for manual intervention. For the scientists particularly interested in the image data from the spacecraft, this is generally available to them in raw form, together with a quick-look product to allow rapid assessment of data quality, less than two hours after being downlinked from the spacecraft.

13.3 Space-Segment Quality and Reliability Considerations

With space-qualified components becoming either exorbitantly expensive or unavailable, the Proba series is proving to be an excellent testbed for component evaluation. Many commercial, off-the-shelf (COTS) components and even complete units are included, mixing, where necessary, fault-tolerant design, redundancy for critical units, and mitigation techniques for radiation effects. For those unfamiliar with these techniques, redundancy generally implies two identical units, either of which can fulfill the mission requirements. There are two approaches to redundancy—hot and cold. Hot redundant units are permanently powered (such as the uplink receivers); arbitration logic in the spacecraft determines which one is used at a given time. In a cold redundant setup, one unit is switched on and normally active while its copy is powered off. If the primary unit experiences an anomaly, the fault diagnosis, isolation, and recovery (FDIR) process within the onboard software will switch it off and switch on the redundant unit. Generally, only a short time is needed to recover from such a failure—much less time than waiting for a ground contact to detect the problem and recover the mission operations. Fault-tolerant designs generally operate several (usually three) identical circuits in parallel—with a voting mechanism to choose which of the circuits is currently active—on the basis that errors generally occur one by one, and if two of the three circuits agree, they are likely to give the correct result. Proba-2 uses this latter technology in a number of its subsystems.

Both Proba-1 and Proba-2 have error detection and correction (EDAC) mechanisms for their main memories, with software-scrubbing algorithms to correct any radiation-induced errors. The extended operational life of Proba-1 has allowed analysis of memory-error rates, and since all the software elements can be reprogrammed in flight, changes have been made to these algorithms to optimize performance and provide feedback for the next spacecraft design. In this way, major failures are minimized, as are unwanted reboots of the onboard software caused by multibit memory errors. For Proba-1, the software was upgraded after 15 months in orbit, and since that time, an unplanned software reboot has happened on average every six months. Indeed, the flight software is sophisticated enough to switch to a redundant unit or a backup mode of operation without losing control of the spacecraft, and generally without modification of the schedule. It is possible to reboot the software in many of the attitude control system units—such as the star tracker, reaction wheels, and GPS receiver—without having to restart the overall spacecraft.

In the Proba series, accent has been on robustness of the onboard systems. For Proba-1, the possibility of ground intervention is less than one hour per day. The Redu ground station is the only one with control authority, and on average, it has only five visible passes of 9 to 10 min each day. Thus, the spacecraft has to autonomously carry out its complex imaging maneuvers and must also detect, isolate, and recover from any faults that may occur.

Proba-1 relied primarily on cold-redundant units in key areas such as the data-handling system, star trackers, telecommunications, and some parts of the power system; Proba-2 has in addition fault-tolerant designs for many of these systems. Where possible, critical components are specified to MIL-883 standard, but almost none has full space qualification. As far as possible, latch-up free electronics are used, but many of the boards also include latch-up protection and recovery. The spacecraft are specifically designed to recover automatically from a complete power-down and return to a safe mode when power is restored. Proba-2 is designed to function on its solar arrays alone, even if the battery fails, to the extent that if sufficient power is available, the full mission can still be achieved.

Both individual units and the full spacecraft are subjected to full environmental testing (thermal-vacuum, vibration, acoustic, and shock), and many critical items for which no radiation data were available have been tested—particularly for heavy-ion damage, which is most prevalent in LEO. By analysis of expected environmental effects in the proposed orbit and within the operational constraints, it has been possible to reduce the risks as far as possible. Nonetheless, in a few cases, mission developers have had to rely on engineering judgment and say, "This is an experimental spacecraft—the risk is acceptable."

13.4 Proba-1

Proba-1, built in Belgium with contributions from several other European countries, was launched 22 October 2001 aboard a Polar Satellite Launch Vehicle (PSLV) from the Indian Space Research Organization's launch site in southeast India (see Fig. 13.2). It was injected directly into its final polar, sun-synchronous, slightly elliptic 620 km orbit with a right ascension of 10 hr, 30 min. Originally

Fig. 13.2. Proba-1 launch, Shriharikota, India.

planned for up to two years of science operations, the spacecraft has operated almost flawlessly and has completed seven years of operation, supplying data to more than 90 scientists in 27 countries. If all subsystems continue to function correctly, the orbit drift should allow at least two more years of continued science exploitation. Proba-1 has no propulsion system, so orbital corrections cannot be introduced. Ground operations are controlled from Redu ground station with additional downlink capacity provided from Kiruna ground station in Sweden.

13.4.1 Proba-1 Mission Objectives

The primary objectives were always intended to be the flight testing of various components and technologies, including the autonomy software. The attitude control and navigation system software, which provides the core of the autonomy, used autocoding technology for its implementation and validation. Four science experiments were included to provide added value. It was agreed that the technology goals would be fully evaluated within the first three to six months in orbit, and the emphasis would then transfer to the scientific results for any continued exploitation.

The technological packages being flight-tested fit into two groups: those that formed an integral part of the spacecraft infrastructure, and those that were simply flown as passengers. The first group included novel GaAs solar cells, a new lithium-ion battery, a dual-head star tracker, new GPS receiver, and significant use of COTS items—indeed the GPS receiver is a COTS item. Among the COTS components were some high-density three-dimensional memory cubes, each having 160 Mb of storage. Proba-1 also provided the first flight of the ERC-32 processor—a three-chip implementation of the SPARC-V7 processor in radiation-hardened space-qualified ASICs. The second group included another design of star tracker, a miniature radiation monitor measuring electron and proton fluxes (scintillation type sensor) and a small monitoring device that measures both temperature and total-dose radiation (radfet type sensor). A number of these latter devices are attached at various points throughout the spacecraft.

Four scientific payloads were included, two of which were specifically chosen to make use of the attitude agility of the spacecraft. What has become the primary payload as the mission has developed is the Compact High-Resolution Imaging Spectrometer (CHRIS). This operates in the visible and near-infrared bands between 415 and 1050 nm and can select from up to 200 separate spectral bands (with a maximum of 62 in any one image). Operating in a push-broom scanning mode, CHRIS sweeps a path across track about 15 km wide (at nadir) and can, in principal, extend the image along track until the memory is full.

The second imaging payload actually comprises two similar microcameras—a wide-angle lens camera, and a high-resolution camera. Each produces monochrome images. The high-resolution camera is in a Cassegrain-type telescope; it produces a ground image of approximately 4 × 4 km with a resolution of 5 m. It has also become a major source of science data.

The remaining two payloads are for space environment monitoring. The Standard Radiation Environment Monitor is also flying on the Integral (ESA space observatory) and Rosetta (ESA-led robotic spacecraft mission to study the comet 67P/Churyumov-Gerasimenko) missions and will fly on Herschel-Planck (two-satellite ESA space science astronomy project) and Giove-B (ESA Galileo In-Orbit Validation Element), allowing various radiation measurements to be coordinated with results from the other spacecraft. The Debris In-orbit Evaluator is an experimental payload, which measures submillimeter particulate fluxes and which can derive the mass and velocity of impacting particles. These environment experiments can operate in parallel with either of the imagers, and Standard Radiation Environment Monitor in particular has benefited from the much extended operational lifetime of the spacecraft.

13.4.2 Proba-1 System Description

Proba-1 has a mass of nearly 100 kg and is compatible with Ariane ASAP5 (Ariane 5 Structure for Auxiliary Payloads) and PSLV launcher requirements as a secondary payload. Its structure is built using aluminium honeycomb panels and is approximately 60 × 60 × 80 cm. Body-mounted solar panels provide power to the spacecraft, and a Li-ion battery is used for energy storage. A centrally switched regulated bus distributes the power to the units and instruments. An ERC-32 (SPARC-V7-based) processor provides the computing power for all the platform functions, and a TSC21020 digital signal processor provides the processing power to the imaging payloads. All software components, either in the central computer or embedded in the spacecraft subsystems, can be reprogrammed in flight. The S-band telecommunications subsystem has omnidirectional coverage, and the uplinks and downlinks utilize the data structures and coding defined by the Consultative Committee on Space Data Standards (CCSDS)—an international body supported by many of the space agencies, including both NASA and ESA. The imaging payloads are rigidly mounted on the structure and are pointed at their intended targets by slewing the whole spacecraft. Thermal control of the spacecraft is completely passive. Figure 13.3 shows Proba-1 ready for flight, and Fig. 13.4 shows an exploded view of the spacecraft.

Fig. 13.3. Proba-1 prior to launch.

Fig. 13.4. Exploded view of Proba-1.

All essential platform subsystems are dual and in cold-redundant configuration with the exception of the uplink receivers, which are hot redundant—both are always on—and the GPS receiver, for which no redundant unit is included. In the event of a GPS failure, the attitude control and navigation system can function by uplinking two-line elements from the ground. The onboard timing and orbit stability is sufficient for normal operations to be maintained for several days on one set of elements. To date, none of the primary systems has failed. The experiments are nonredundant.

Table 13.1. Proba-1 Attitude Control and Navigation System Performance

Attitude control	Three-axis stabilized providing high-accuracy nadir and off-nadir pointing capabilities
Sensors	Cold-redundant, dual-head advanced star trackers, redundant three-axis magnetometers, GPS receiver
Actuators	Four magneto-torquers, four reaction wheels.
Absolute pointing accuracy	Better than 360 arcsec
Absolute pointing knowledge	Better than 125 arcsec

13.4.3 Proba-1 Attitude Control and Navigation

The attitude control and navigation system is the heart of the spacecraft autonomy (see Table 13.1). To fulfill the science mission requirements, the spacecraft attitude must be agile but accurate, stable, and reproducible. A high-level command can be sent to the spacecraft, which simply specifies which imager to use and requests an image of a point on the Earth's surface defined by latitude, longitude, and height above sea level. The spacecraft itself then calculates when it is within sight of its target and within the default constraints (in daylight and less than 25 deg to right or left of the vertical track) and records a set of images. The completion of requested tasks is flagged in the housekeeping telemetry, and the spacecraft will continue with the next target until either the memory is filled or becomes available for reuse as a result of downlinking the data to a ground station.

The system can sort the requests and process them in the order that the targets become visible and memory is available, and not necessarily in the order in which they were initiated. However, this mechanism can be overridden by optional request parameters; in practice, the scientists seem to prefer to do their own planning, and the onboard planner is rarely used. Some presorting of the requests is carried out at the ground station prior to uplink by merging with weather forecast information to try to minimize loss of image data due to cloud cover. The default requirement for imaging is less than 30% cloud cover. Since the maneuvering process for target imaging begins when the spacecraft is many hundreds of kilometers away from the target area, no attempt has been made to estimate cloud cover in real-time—indeed, none of the imagers is suited to this function as both CHRIS and high-resolution camera have too narrow a field of view, while the wide-angle lens camera has too wide a view.

The sensors provided for orbit and attitude determination comprise a dual-head star tracker, three-axis magnetometer, and GPS receiver. No Earth or sun sensors or gyroscope are included, as the imaging requirements with rigidly mounted cameras would conflict with the use of these—either the cameras or the sensors would have to be independently steerable. For attitude control, four reaction wheels and two redundant pairs of magneto-torquers are used. Dependent upon the pointing mode, either two or three of the reaction wheels are actually used at any given time, and different combinations can be used to ensure redundancy in case of failure. There is no propulsion system, and orbit control is not possible.

The CHRIS experiment (shown in Fig. 13.5) and the high-resolution camera both require accurate and stable pointing of the spacecraft to achieve their goals. Both operate in the visible part of the spectrum, and, therefore, can only take images when the ground targets are in sunlight. Hardware limitations on the spacecraft mean that the imagers cannot be simultaneously active. Also, it is not possible to record a new image while downlinking an existing image.

The attitude control and navigation system has a number of operating modes, three of which are most important for normal operation of the satellite.

The first is magnetic mode, or B-dot. This is a minimum-power mode using the minimum set of sensors and actuators whereby the spacecraft remains in an attitude locked to Earth's magnetic field. It was the mode used to detumble the spacecraft immediately after launch and is used as a "safe mode" in the event of any onboard anomalies. It relies primarily on the magneto-torquers to generate a small field, which then reacts with Earth's field and, given time, will align with it. The magnetometers are used to monitor that alignment. From an initial tumbling spacecraft following separation from the launcher, it can take several orbits to dampen the motion and align to the field. This mode does not require either the

Fig. 13.5. The CHRIS instrument on Proba-1.

star tracker or GPS receiver and provides a very stable behavior from which to recover in the event of problems.

Imaging mode (Fig. 13.6) is the prime operating mode for using the CHRIS imager. The imager has a single line of detectors scanning a distance across track of about 15 km, and a pushbroom technique is used to build up a picture covering 15 × 15 km with a pixel size of about 18 m at nadir, in much the same way as a television picture is built up from scan lines. The orbital speed of about 7 km/sec is too fast to give good resolution in this mode, and the effective ground speed is reduced by a factor between 3 and 5 by slewing the body of the spacecraft. In this way, a sequence of five images is built up as the spacecraft flies over the target through a total angle of 110 deg as shown. The different viewing angles for each image permit the study of the reflectance variation as a function of the viewing geometry—that is, the analysis of the bidirectional reflectance distribution function. By default, the images are recorded in 19 spectral bands. It is possible by extending the request command parameters to change the default settings—more or fewer spectral bands and different image formats, for example, a single image 75 × 15 km rather than five individual images of the same 15 km square. The default ±25 deg off-track limit for the target position can also be tightened or relaxed at the request of the experimenter.

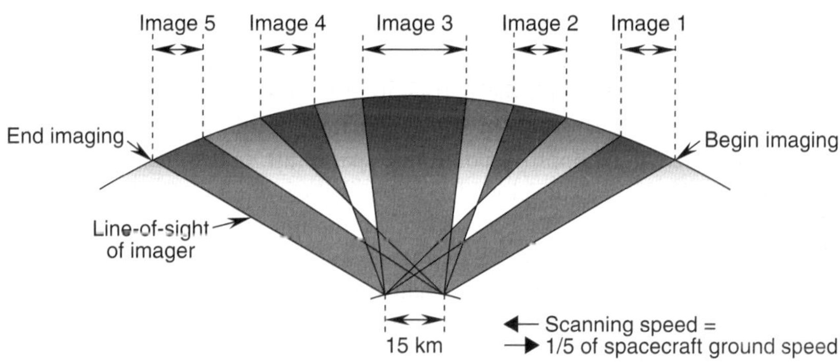

Fig. 13.6. Imaging mode maneuver for the CHRIS instrument.

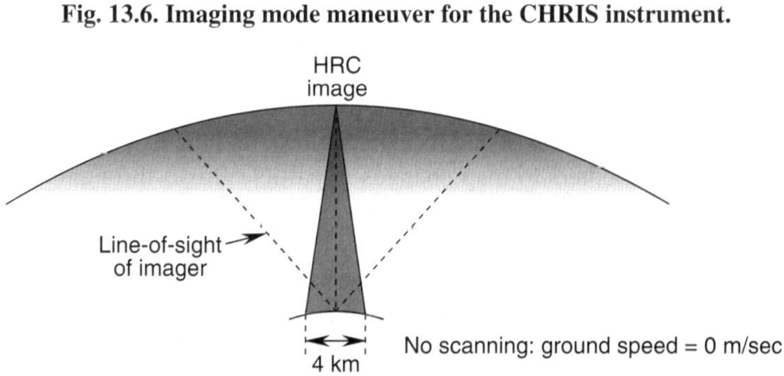

Fig. 13.7. Fixed target mode maneuver for the HRC instrument.

Fixed-target mode (Fig. 13.7) is used for the high-resolution camera. This is similar in operation to a standard digital camera, and the target needs to be held still for the duration of the exposure. In this mode, the effective ground speed of the orbiting spacecraft is reduced to zero as the structure is slewed to "point and stare" at the target as shown, which in this case is a 4 × 4 km square with a pixel size of 6 m. Once again by default five images are taken, and as with the CHRIS imager these defaults can be overridden by request.

Figures 13.8 and 13.9 show typical results obtained from the two imagers. The four high-resolution camera pictures were taken at 30-second intervals, and between the first and last of these, the spacecraft actually traveled 600 km along the ground track.

13.4.4 Proba-1 Data Results and Usage

Many of the Proba-1 images have been published in the media and displayed on posters at exhibitions. Such exposure heightens the general public's familiarity with ESA space exploration and provides good publicity for space research. This may help to generate sufficient interest to encourage future funding, although public relations is not the primary purpose of the mission.

All software components are reprogrammable in flight. Developments made to the main system software have improved the accuracy and stability of the attitude control and navigation system and have fixed some inevitable "bugs." The star tracker, too, has been improved over the period of operation. Because of a particular problem found with the baffles used to protect the sensitive charge-coupled device (CCD) heads from stray light from Earth, moon, or sun, CCD optical properties will be improved on future spacecraft. In the meantime, special filter algorithms have been developed to improve the overall performance even in the presence of stray light. With the extended mission proving so successful for the science community and with the spacecraft still operating on its prime units—none of the subsystems has yet failed—many of the now proven technologies will find flight opportunities on major new spacecraft. Not all of the items flown as demonstrators have been successful, but these are commercial developments, and the companies concerned will be reviewing their results for future developments. Fortunately, all those integral to the spacecraft infrastructure were successful.

Of the science experiments flown, the stars of the show have undoubtedly been the CHRIS and high-resolution cameras, but the Standard Radiation Environment Monitor has also exceeded its expectations, and the extended mission is enabling the experimenters to map changes in particle fluxes throughout a large portion of a solar cycle. Figure 13.10 shows an example of a particle flux map obtained from the instrument, showing clearly the increased flux above the South Atlantic anomaly. Data from this experiment have enabled mission operators to tailor the memory-scrubbing algorithm for the main processor, and the number of unexpected reboots of the software is now very low (about twice a year on average). It has also enabled users to analyze the effects of solar flares during the life of the spacecraft, including a strong flare in December 2006, which caused

Fig. 13.8. CHRIS image of San Francisco, California, USA.

a software reboot of the main processor and burned out a few pixels in the star-tracker heads.

The two primary imagers have provided many thousands of images during the flight to date. Their primary purpose is to supply data for monitoring forests, agricultural land use, soil biochemical and biophysical properties, and suspended particulate matter in coastal waters. The Belgian government, one of the major funding bodies for Proba-1, has produced an educational program for Belgian

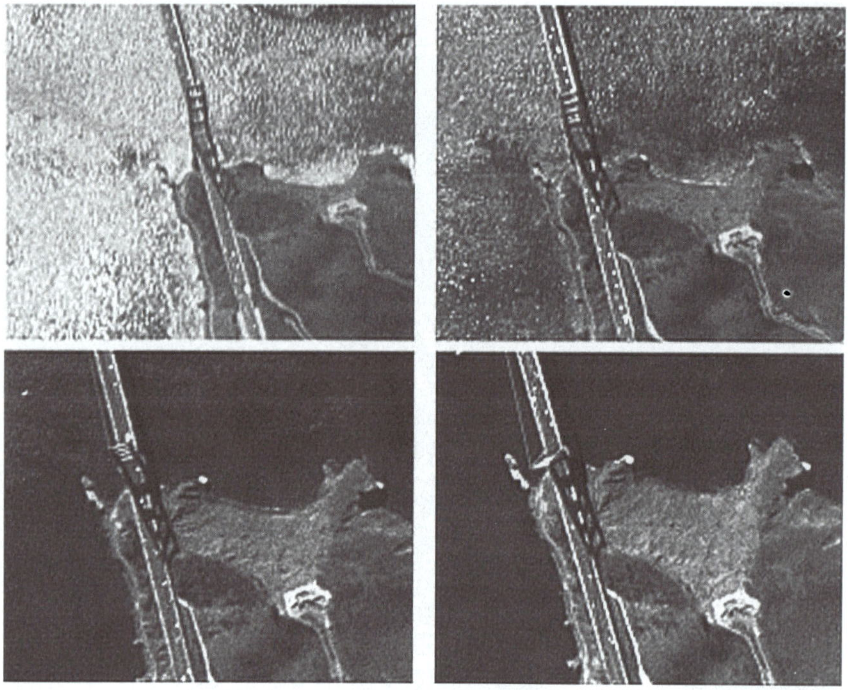

Fig. 13.9. High-resolution camera images of northern tower of Golden Gate Bridge, San Francisco, California, USA.

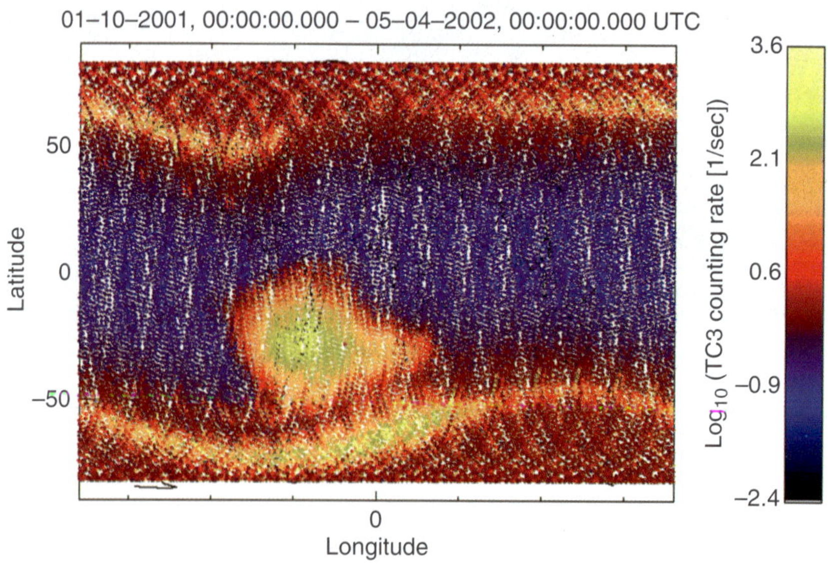

Fig. 13.10. Standard Radiation Environment Monitor map of proton flux around the globe, clearly showing the South Atlantic anomaly.

schools to request and use Proba images, mainly using the high-resolution camera. The imagers have also been used to monitor natural disasters such as volcanic eruptions, forest fires, and floods. Since the end of 2003, Proba-1 has been an invited guest in the International Charter "Space and Major Disasters"—an agreement to make space resources available to civil protection agencies responding to natural or manmade disasters—and can be called upon to provide monitoring of such events at very short notice. Although the revisit time of any given point is about a week, the ability to take images off track gives a much quicker effective response time

Researchers hope that the spacecraft will remain operational for a few more years until the orbit degrades and marginalizes the science return. The performance of the solar cells and battery remain good, and if the reaction wheels remain operational, there is every chance that many more images will yet be taken.

13.5 Proba-2

The second spacecraft in the series is once again being built in Belgium and is now approaching flight readiness, as shown in Fig. 13.11. Proba-2 is expected to be launched in the second quarter of 2009. This spacecraft, at 130 kg, is still small, and some effort was required to find a suitable partner spacecraft to mitigate launch costs. The primary science payloads are for solar observation, and a sun-synchronous dawn/dusk polar orbit with a right ascension of 06:00 is being sought to minimize eclipse periods. Eventually, a joint launch with the ESA Soil Moisture and Ocean Salinity (SMOS) spacecraft was approved, as both spacecraft can benefit from such an orbit. The agreed orbit is at a height of 750 km and should be eclipse-free for nine months of the year. Proba-2 is currently in the final stages of integration and test.

By using a high-latitude ground station (Svalbard, Norway), the mission is expecting a considerable volume of science data to be dumped daily because almost all orbits pass over that station. Redu will remain the control center for the mission.

13.5.1 Proba-2 Mission Objectives

Once again, the primary objective is to test a number of technologies. Four science experiments will also be flown, two of which are for solar observation. SWAP (Sun Watcher using Active Pixel system detectors and image processing) is an extreme ultraviolet (EUV) telescope for studying the solar corona at temperatures near one million degrees. It is a much-enhanced version of the instrument carried on the SOHO (Solar and Heliospheric Observatory) spacecraft, which has now been operational for more than 10 years. It is an off-axis Ritchey-Chrétien telescope that allows for simpler baffling and a smaller aperture than the old design and uses a new CMOS-APS (active pixel sensor) detector, rather than the older CCD technology, covered with an EUV-sensitive scintillation layer. LYRA (Lyman-alpha radiometer) is a solar radiometer with four UV bands (1–20 nm,

Fig. 13.11. Proba-2 during thermal vacuum tests.

17–31 nm, 115–125 nm, and 200–220 nm) chosen for their relevance to solar physics, aeronomy, and space weather. The experiment uses wide-bandgap detectors based on diamond, which is resistant to the radiation encountered in the chosen orbit. The instrument has been calibrated against a synchrotron source and carries two onboard calibration sources (visible and near UV) to monitor any degradation of the detectors and filters.

The remaining two science payloads are space weather experiments. The first is a thermal plasma measurement unit to measure electron temperature, total ion density, ion composition, ion temperature, and the floating potential of the satellite body. The second is a dual-segmented Langmuir probe, an updated version of the experiment flown on the French Demeter spacecraft. Its objectives are to measure space plasma density in the range 100 particles/cm up to 5 million particles/cm, electron temperature in the 500–3000 K range, and the spacecraft potential. Because these two space weather experiments use different sensor types, it should be interesting to compare results of the overlapping measurements.

Once again, the technology testbeds are in two groups: those integrated as part of the spacecraft infrastructure, and those being flown for test and demonstration only. The infrastructure items include triple-junction GaAs solar cells with integrated diodes; an updated star tracker with enhanced electronics, modified baffles, and improved software; a new Li-ion battery; and a new combined data-handling and power subsystem, with the main processor being the LEON-FT fault-tolerant single-chip version of the SPARC-V8. The system also extensively utilizes the new range of Actel RTAX4000 field-programmable gate array (FPGA) devices, which provide the bulk of the digital logic for each board in a single chip—for example, the telecommand decoder with two virtual channels and telemetry encoder with five virtual channels all fit into a single FPGA.

A total of 12 items are being flown to gain flight experience for new developments. These include no fewer than three triple-axis magnetometers from different manufacturers, each using a different sensor technology, as well as a digital sun sensor; a demonstration model of a new star tracker, which will subsequently be flown as a primary attitude control and navigation system component on the future Bepi-Colombo mission to Mercury; and a GPS receiver operating at both L1 and L2C frequencies. Additionally, the system will have a xenon gas propulsion system using a resistojet thruster and a cool gas generator to repressurize the xenon propellant tank with nitrogen generated from solid-state cartridges once the xenon has been exhausted. This complete subsystem is shown in Fig. 13.12. Propulsion is not necessary for the science mission, but these items are being characterized for use on other missions and can be used to initially mitigate orbital decay or at a later stage to accelerate end-of-life deorbiting. The choice of the gas propellants is to minimize any contamination of the UV optics of the solar experiments. The other demonstrators being flown are an Earth-pointing experimental camera; a fibre sensor demonstrator for measuring temperatures and pressures, which will include six sensors for measurements on the propulsion system; a small experimental solar panel with a concentrator lensing system; and a laser reflector cube. This latter is a passive device to support a ground-based experiment.

13.5.2 Proba-2 System Description

Proba-2 is very similar in dimensions to Proba-1, but whereas the first was an Earth-pointing mission with different facets pointing sunward at different times,

Fig. 13.12. Propulsion subsystem for Proba-2.

Proba-2 will be sun pointing. A single side panel, with space required for experiment apertures, gives insufficient space for the required power generation, even from the high-efficiency solar cells being used, and Proba-2 is equipped with two deployable solar panels made from carbon fiber. These are folded flat against two opposite sides of the main body of the spacecraft for launch, and their deployment, which should occur automatically a few minutes after separation from the launcher, will also help to despin the spacecraft. Additionally, solar cells are mounted on one side panel, which will normally be facing away from the sun. This is to act as a backup supply in the event of loss of attitude control. Unlike Proba-1, the new power system uses an unregulated bus in the nominal range of 25–29 V dc, although in emergency, some subsystems can remain operable down to 21 V dc; the system is specifically designed to be able to operate from the solar arrays alone—so, in sunlight, the spacecraft can continue to function normally even with the battery disconnected (or failed). Overall, the solar arrays are electrically split into six parallel sections, and the battery has three parallel strings of cells.

Communications is once again based on CCSDS uplinks and downlinks in S-band, reusing the Proba-1 frequencies. The telemetry downlink is also at the same rate as Proba-1, but the two spacecraft will be in significantly different orbits and they only transmit when commanded, so interference between the two is not a problem. Proba-2 will be the first ESA spacecraft to utilize the "medium rate" telecommand mode, with an uplink of 64 kbps, which should permit uplink of sizeable blocks of test software in the short passes available in LEO. Its purpose is also to validate the newly upgraded ground systems to support this increased uplink rate.

The spacecraft relies on a single LEON-FT processor for both spacecraft infrastructure and image processing of the SWAP data. Nearly 600 Mb of memory are available for data storage. Once again, all essential platform units are dual and in cold-redundant configuration except for the hot-redundant telecommand receivers. Many of the subsystems and experiments contain their own processors, and most can be reprogrammed in flight if necessary.

13.5.3 Proba-2 Attitude Control and Navigation

Once again, all experiments are rigidly fixed to the spacecraft structure, and pointing them at their targets relies on the agility of the attitude control and navigation system to slew the whole spacecraft. In this instance, the two major experiments requiring accurate pointing are both solar experiments, and they have been mechanically aligned during integration so that both can be operated simultaneously. The originally specified and expected performance of the system is shown in Table 2. The improvements in pointing accuracy compared with Proba-1 largely result from improvements in the star tracker sensors and software and the use of physically larger reaction wheels for finer control. The spacecraft once again has a low-power "safe mode" of operation locked to Earth's magnetic field to be used in emergencies and from which transition to any other mode can be made if sufficient power is available.

Table 13.2. Proba-2 Attitude Control and Navigation System Performance

	Specification	Expected
Absolute measurement error	10 arcsec	19.9 arcsec
Absolute pointing error without compensation	100 arcsec	184 arcsec
Absolute pointing error with compensation for thermal deformation	100 arcsec	38 arcsec
Relative pointing error (5 sec)	5 arcsec/5 sec	1.7 arcsec/5 sec
Relative pointing error (60 sec)	5 arcsec/60 sec	1.7 arcsec/60 sec

The main science mode will be sun pointing, and for much of the time, there will be no eclipses. The sun-pointing face will thus be considerably hotter than the opposite face and the attitude control and navigation system software will have to take thermoelastic deformation of the structure into consideration. Additionally, to avoid blinding of the star trackers by Earth light, the spacecraft will be rotated through 90 deg in the plane of the velocity vector four times per orbit, as shown in Fig. 13.13, to keep the star trackers pointing away from the body of Earth. The moon does not usually cause blinding problems, and in any case, will not be visible to both heads at the same time.

Some additional modes are also being prepared for occasional use to optimize results from the technology experiments. These will allow for pointing the experimental camera at Earth targets, thruster pointing for the propulsion system, and fine positioning of the thermal plasma measurement unit and dual-segmented Langmuir probe to investigate the plasma field around the spacecraft.

13.6 Proba-3

Unlike its predecessors, Proba-3 will actually be two spacecraft with a combined mass close to 400 kg in highly elliptical orbits, with an orbital period approximately 24 hours. The primary purpose of the mission is to test technologies and develop software algorithms to maintain multiple spacecraft in tight formation. Once again, the science benefit will be a solar observation mission with the two spacecraft, when not close to Earth, acting as a coronagraph (Fig. 13.14). The occulting spacecraft, roughly circular in the aspect presented to the other, will maintain a position, tightly controlled in three dimensions, between its sister craft and the sun. In this situation, either one or both of the spacecraft are flying in non-Keplerian orbits and need propulsion systems to maintain their course. It is expected that the system will operate in the primary coronagraph mode for more than half of the orbit period. For perigee passes close to Earth, the formation control will be released, and the two spacecraft will return to free-flying Keplerian

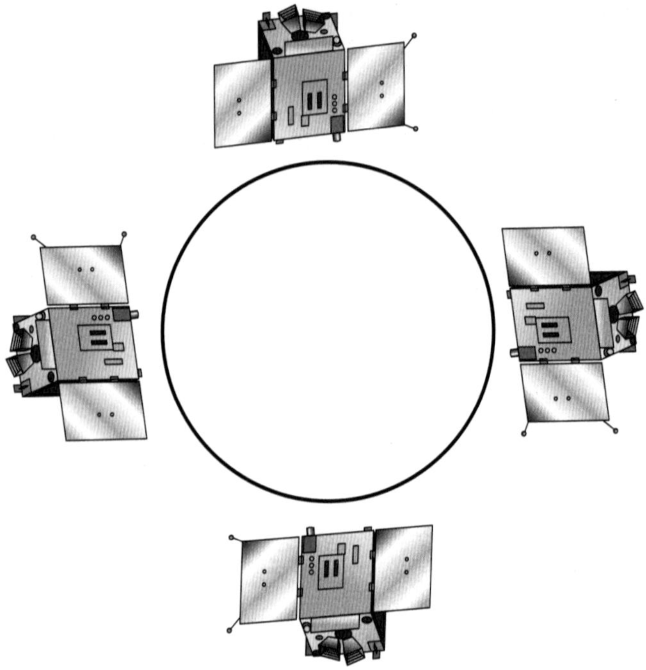

Fig. 13.13. Proba-2 maneuvers (as seen from the sun) to avoid blinding of the star trackers.

orbits. This dual-spacecraft mission is still in the development stage, and no flight hardware has yet been built. The final configuration is still being analyzed.

The present baseline is for both spacecraft to have communications links with the ground, and both require propulsion systems, which may be cold-gas thrusters or electric propulsion or a mix. Four measurement systems are proposed for station keeping: coarse and fine RF systems and coarse and fine laser systems. For the coronagraph experiment, the two spacecraft must be positioned relative to each other with a lateral accuracy of 3 mm and a separation accuracy of 60 mm over a baseline of 150 m. Pointing accuracy for the main spacecraft is 19 arcsec. These performances can be met with the first three metrology systems—the fine laser system will be for experimental tests of even finer control. In addition to station keeping, the coarse RF metrology system will also be used for collision avoidance when the spacecraft are free flying close to perigee and not functioning in the coronagraph mode.

13.7 The Future

The experience of the Proba projects to date as well as similar space projects in Europe have demonstrated the interest and benefits of small spacecraft missions combining advanced technology and user payload. However, for the Proba

Fig. 13.14. Proba-3 coronagraph concept.

projects in ESA, the one-at-a-time approach followed does not maximize the benefits obtainable from such projects. Indeed, when a project is decided, then several conditions have to be fulfilled:

- A set of technologies at the right development level must be available for demonstration.
- User payloads at the proper maturity and appropriate for the mission are required.
- A launch opportunity into a compatible orbit, generally as a secondary payload in view of the relatively small mass, is necessary.
- Recurrent spacecraft elements must be available at low cost.

The whole process from mission decision to orbit takes several years and is barely compatible with the timely demonstration of new space technology. A much quicker turnaround time would be welcomed by the space industry suppliers.

Considering both the interest of such missions reflected by the number of proposals for demonstration, the increasing capability of small spacecraft with advanced miniaturization, and the current drawback of ad-hoc missions, ESA is investigating a possible in-orbit demonstration program that will support a mission every two and a half years, thereby providing timely access to space for new technologies, more opportunities for user payloads compatible with small missions, and reduced cost resulting from recurrent missions.

13.8 Bibliography

6th Small Satellite Systems and Services Symposium, The CHRIS instrument expanded article, http://earth.esa.int/pub/ESA_DOC/PROBA/Barnsley_Smallsat_2004.pdf (2004).

J. de Lafontaine, J.-R. Lafleur, I. Jean, P. Van der Braembussche, and P. Vuilleumier, "In-flight Performance and Calibration of the Autonomous Minisatellite Proba," *American Astronomical Society,* AAS 03-152 (2003).

J. L. Jorgensen, T. Denver, M. Betto, and P. Van der Braembussche, "The Proba Satellite Star Tracker Performance," *Acta Astronautica,* Vol. 56 (2005).

Autocoding information:

W. I. Lundgren, "Code Generation and the Complexity of Application," DSP-FPGA. com Product Resource Guide (2005).

NASA Ames Research Center website: www.arc.nasa.gov.

A. Mohammadzadeh, H. Evans, P. Nieminen, E. Daly, P. Vuilleumier, P. Bühler, C. Eggel, W. Hajdas, N. Schlumpf, A. Zehnder, J. Schneider, and R. Fear, "The ESA Standard Radiation Environment Monitor Program: First Results from Proba-1 and Integral," *IEEE Transactions on Nuclear Science,* **50** (6), (Dec. 2003).

A. L. R. Pouponnot, "Strategic Use of SEE Mitigation Techniques for the Development of the ESA Microprocessors, Past, Present and Future," *Proceedings of the 11th IEEE International On-line Testing Symposium* (2005).

F. Teston, D. Bernaerts, and K. Gantois, "Proba—an ESA Technology Demonstration Mission: Results After 3 Years in Orbit," *6th Small Satellite Systems and Services Symposium* (2004).

http://earth.esa.int/proba links to many of the science papers published about the use of the Proba-1 data.

http://www.esa.int/proba links to information about all three present Proba missions.

Chapter 14

A Revolutionary Approach to Spacecraft Design, Production, and Operations: Orbital's MicroStar Satellite

Dean Hawes* and Srimal W. Choi*

14.1 Introduction: The MicroStar Spacecraft

The MicroStar spacecraft bus was developed by Orbital Sciences Corporation in the early 1990s as a low-cost point design to meet the requirements of the ORBCOMM satellite-based communications network (Fig. 14.1). The intent was not to reduce the price of a satellite by a few million dollars, but to reduce it by an order of magnitude. The MicroStar team achieved that goal by developing an economical spacecraft and a production process that could deliver it within 24 months. Additionally, a stackable design allowed multiple spacecraft to launch from a single vehicle, reducing mission costs even further.

This chapter will summarize the history of the MicroStar spacecraft. It will explore the concepts behind the design and production of the spacecraft for the ORBCOMM program, subsequent missions that used derivatives of the original architecture, and future prospects.

14.2 What is ORBCOMM?

ORBCOMM is a machine-to-machine communications network designed to send short data messages in near real time between any two terminals on Earth. These terminals may be connected to an infrastructure (such as the Internet), or they may be remotely located. Messages from remote terminals are relayed via satellite to a ground-based network control center in Dulles, Virginia, where they are processed and forwarded to their destination. The ORBCOMM system allows users to track mobile assets (such as trucks, ships, and rail cars), monitor assets located in remote regions, or remotely control unmanned equipment.

The requirement for near real-time communications means that a user terminal must constantly be in view of at least one ORBCOMM satellite. To accomplish this on a global scale, the constellation requires about 35 low Earth orbit (LEO) satellites in six orbital planes (four planes at 45 deg and two near polar) at altitudes ranging from 740 to 825 km. Coverage is optimized for mid latitudes

*Orbital Sciences Corporation
Dulles, Virginia

Fig. 14.1. ORBCOMM: The first MicroStar mission.

ranging from 20 to 60 deg because the largest population resides within this belt. Several ground stations located around the world and the network control center are used to control the constellation. An overview of the system is shown in Fig. 14.2.

For the system to be financially viable, the constellation had to meet the following key requirements:

- Eight satellites would have to be launched from a single Pegasus XL launch vehicle, which was also in the early stages of its life cycle.
- Each satellite needed to have low mass (approximately 46 kg), occupy a small volume (0.21 m^3), and have a high power-to-mass ratio (5.5 W/kg). This required the satellite communications and power subsystems to be highly efficient to support the 16 dBW downlinks in VHF frequencies.
- The communications subsystem had to be frequency-agile and support real-time packet and frequency switching to avoid region-specific in-band noise.
- Each satellite had to have a five-year operational design life.
- The operation of the satellites had to be highly autonomous to allow the entire constellation to be managed by only four satellite and ground network controllers per 12-hour shift.

The development of the system began in 1993, at a time when a typical space system was a very expensive venture and usually involved many years of development and significant launch costs. If ORBCOMM were to become a reality, all of this would have to change.

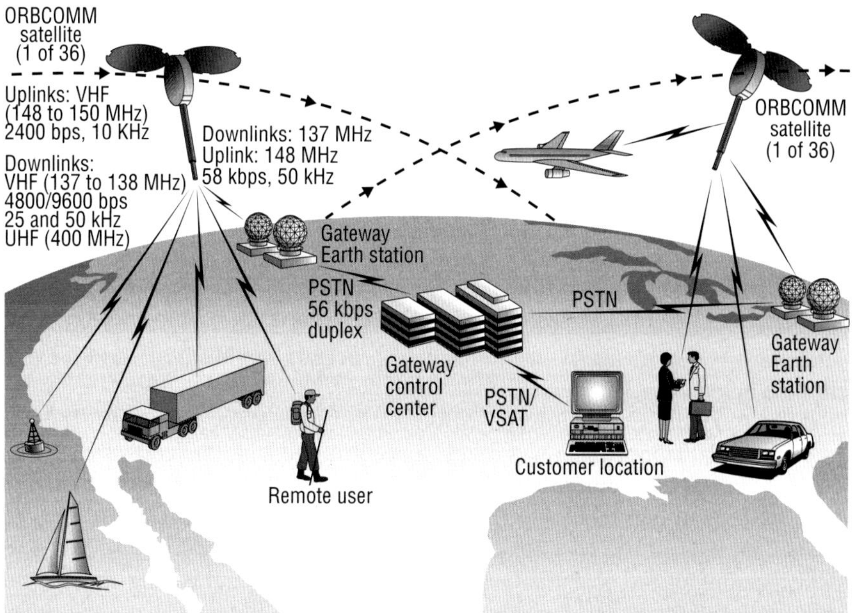

Fig. 14.2. ORBCOMM system overview.

14.3 Development of the MicroStar Spacecraft for ORBCOMM

The implementation of the ORBCOMM constellation began with the development of the first two flight models, FM1 and FM2. The idea was to use FM1 and FM2 as pathfinders, not prototypes. In other words, these satellites would not only serve as technology demonstrators for the full constellation that would follow, but would also be placed into commercial service as soon as they were launched and commissioned.

The tight financial constraints of the program required a new approach. The cost of the satellites was of course critical, but the cost of deploying and operating the constellation also drove many decisions behind the design. Probably the most significant ideas involved the stacking of multiple satellites on a single launch vehicle and design choices that would allow highly autonomous operation, with minimal interaction from ground controllers. ORBCOMM's customers were commercial, and this allowed the development team to adopt new paradigms, as opposed to those that were more common for NASA or DOD customers.

Another key concept in the development of the FM1 and FM2 satellites was the idea of a simplified spacecraft design, which would ultimately translate into fewer components onboard, reducing both cost and mass. The performance requirements of the ORBCOMM payload meant that this spacecraft would have to adequately point the antenna and convert solar power into the RF power needed to support the mission. The idea of a magnetically based attitude control system had been studied and discussed in conference papers, but to Orbital's knowledge, no one had actually flown such a system. Developing one would require significant effort, but would ultimately help the development team achieve the goal of a simplified attitude control subsystem.

A single-string architecture, as opposed to an architecture with redundant avionics, would further simplify the spacecraft design. However, because revenues depended on satellite coverage, consideration had to be given to the possibility of a spacecraft going off-line, for whatever reason, and how that would affect the revenue stream. To address this, the system reliability was not evaluated at the spacecraft level; rather, it was evaluated at the constellation level. In other words, there are eight spacecraft in each plane. If one of these spacecraft went off-line, temporarily or permanently, the remaining spacecraft would still need to provide the coverage to continue service.

Finally, parts selection would also prove to be a major challenge. The use of commercial components was a key concept in reducing the cost of each satellite. Commercial components are not space-qualified and therefore do not have the established reliability that typical space-qualified parts have. This, of course, introduced risk, but some of that risk could be mitigated by gaining an understanding of the potential component failure modes and adding either extra protection against environmental effects in the form of epitaxial layers and external shielding, or by simply allowing the components to experience an upset and building a solid recovery strategy. Orbital's in-house test program also played an important part in this philosophy. Orbital exerted a significant effort in radiation-testing specific components, establishing unique qualification and acceptance test programs, and analyzing the performance of commercial parts in the anticipated environments. In the end, this philosophy paid off—but many hard-won lessons were learned along the way.

14.3.1 Spacecraft System Overview

The ORBCOMM satellite is a single-string, distributed-architecture spacecraft that weighs approximately 46 kg. It is a compact cylinder measuring about 1 m in diameter and 16.5 cm high. When stowed, all of the avionics and deployable structures reside within this volume. Once launched, the spacecraft deploys its two circular solar array panels and a 3 m boom that supports the antennas for the payload and for the RF communications subsystem.

Figure 14.3 shows the ORBCOMM spacecraft in its stowed and deployed configuration, along with its components. (Note that the reaction wheel did not exist on the FM1 and FM2 spacecraft but was added for later production models.) Each spacecraft consists of a circular structure and a single aluminum equipment shelf holding the avionics, batteries, and cold-gas propulsion system. The antenna boom assembly (which also contains the magnetometer and gravity-gradient mass) folds up for stowage across the center of the equipment bay. Two ultrathin solar arrays fold 90 deg from their zenith position to stow against the equipment shelf and across the top of the spacecraft body, enclosing it like two halves of a

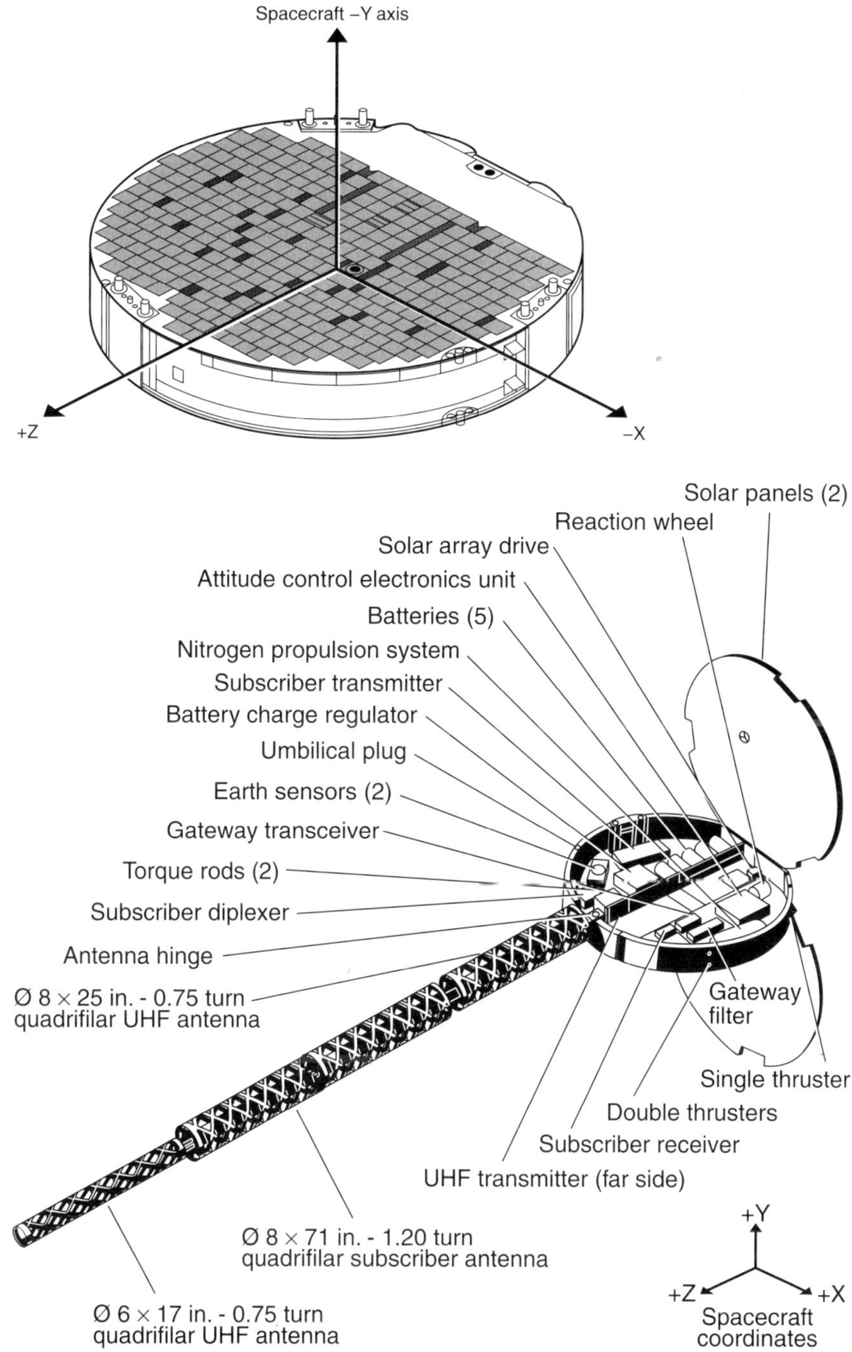

Fig. 14.3. ORBCOMM spacecraft before (top) and after (bottom) deployment.

clam shell. Following launch, the deployed arrays are positioned in elevation by a single-axis drive and in azimuth by the spacecraft's yaw motion. The attitude control system also controls the two-thruster cold-gas reaction-control subsystem, which is primarily used for orbit maintenance.

The spacecraft's avionics architecture (shown in Fig. 14.4) is built around three serial buses (one RS-485 and two RS-422) radiating from a central flight computer. This computer handles telemetry, command, and message storage and routing and also hosts the master health and maintenance task for the spacecraft and payloads. The first serial bus connects to the attitude control electronics unit, which autonomously runs all attitude control system functions and software, and to the battery charge regulator, which autonomously runs all electrical power system functions. The second serial bus connects to the gateway transceiver, which handles uplinks and downlinks to and from the Gateway Earth Stations. The third bus connects the computer to the payload suite, which consists of a dual-channel subscriber transmitter (downlink) and seven-channel subscriber receiver (uplink). An additional UHF beacon is also provided. Each major avionics unit contains a Motorola 68302 microprocessor.

The architecture of the spacecraft is such that each subsystem operates independently of the others; however, each subsystem monitors adjacent subsystems to verify that they are online and operational. This concept supports the layered failure detection and correction strategy employed onboard. When an anomaly is detected, each subsystem attempts to resolve the anomaly at the component level. If the anomaly persists, attempts are made for resolution at the subsystem level, and then at the system level. Should these efforts fail, the spacecraft can be safely "rebooted" (shut down and immediately powered back up) and brought back online fully configured, either autonomously or through ground intervention. In fact, all autonomous spacecraft functions can be augmented or overridden through ground intervention. The distributed architecture also provides some limited overlap of critical functions, which improves the reliability of this single-string spacecraft.

In the launch configuration, up to eight ORBCOMM spacecraft can be stacked on their matching fittings on a single Pegasus XL or Taurus launch vehicle, allowing an entire orbit plane to be deployed with a single launch. Each spacecraft can carry the weight and bending moment of seven others stacked on top of it, even when exposed to the significant lateral transient that the Pegasus XL experiences as it is dropped from its carrier airplane. Once the launch vehicle has reached its insertion target, commands are sent to individual spacecraft to eject themselves from the units below. Each spacecraft then separates from the launch vehicle and autonomously deploys its solar arrays and communications antenna. By design, each spacecraft will most likely be in a tumbling state after deployment and will autonomously null out its rotation rate, usually within two orbits.

The onboard GPS receiver is used for both navigation and time management. The time-management function supports the timing requirements of the payload,

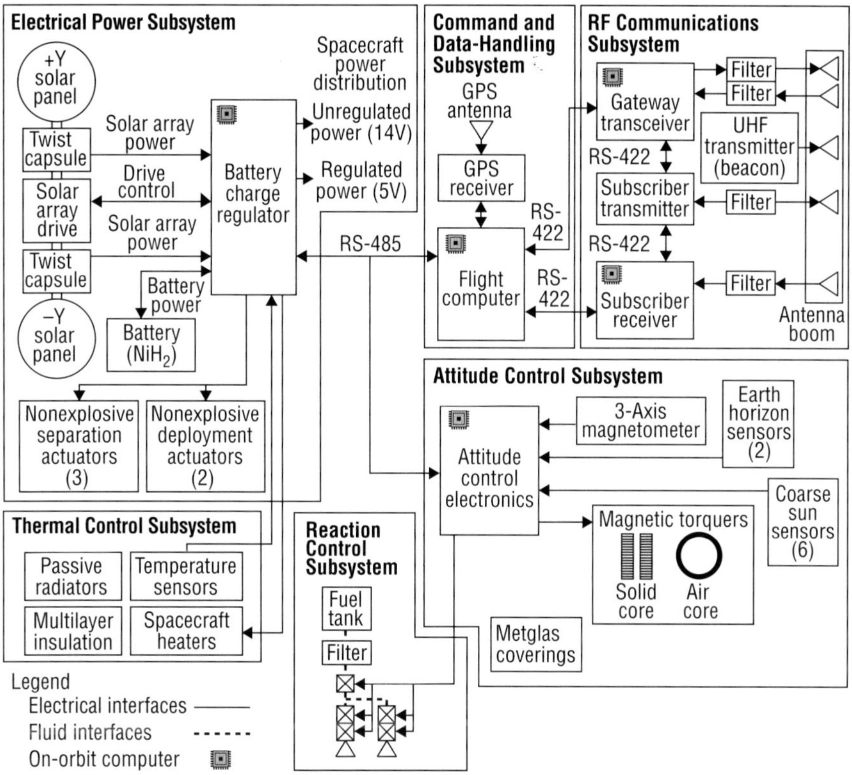

Fig. 14.4. ORBCOMM FM1 and FM2 architecture.

while the navigation function supports normal operations of the attitude control system propagator.

14.3.2 Structures and Mechanisms

The stackable design and the use of materials such as AlBeMet, an aluminum/beryllium alloy, helped to significantly reduce mass of the ORBCOMM spacecraft. In addition, the use of nonexplosive actuators allowed the spacecraft to initiate multiple deployment events while protecting the sensitive payload from excessive shock.

Each MicroStar satellite consists of a honeycomb ring with AlBeMet facesheets, which is about 1 m in diameter and 16.5 cm high, and a single equipment shelf. The 3 m antenna boom assembly supports three cylindrical antennas: a UHF quadrifilar antenna that measures 20.3 cm in diameter by 63.5 cm in length, a second UHF antenna that measures 15.2 cm in diameter by 43.2 cm in length, and a VHF antenna that measures 20.3 cm in diameter by 180.3 cm in length. This antenna boom folds in four pieces for stowage across the center of the equipment bay. Two circular solar-array panels fold over either end of the spacecraft structure.

The satellite must initiate three separate mechanical events during launch and deployment: separation from the launch vehicle, deployment of the two solar-array panels, and the release and deployment of the antenna boom. All of these events rely on nonexplosive actuator mechanisms, which are electromechanical devices that use miniature wire-wrapped spools to hold and release applied loads. The spools are made of two matched halves that are held together by a circumferential wire wrap. The mechanical advantage allows the spool to restrain external loading on the actuator. To actuate, a current of 4.5 A is placed across a retention wire on the spool, causing the wire to break. This allows the spools to separate and the load to be released.

When stacked on the launch vehicle, the bottom spacecraft is mated to the launch vehicle avionics section at three load-bearing points. The remaining seven spacecraft are stacked to the bottom spacecraft, and each other, at these same points, with the bottom satellite in the stack carrying the maximum load. The Pegasus XL launch vehicle is an air-launched vehicle and therefore imparts large lateral and bending loads when it is dropped and ignited. The maximum bending load can exceed 15,000 N-m at the joint where the bottom spacecraft is mated to the launch vehicle. The satellite-to-satellite interface must therefore tolerate these loads. Each spacecraft has three bracket assemblies evenly spaced around its circular structure. The top and bottom of these brackets have a cup-and-cone shear interface, and within each bracket is a pair of compression springs and a nonexplosive separation nut actuator that receives a captive bolt from the spacecraft below it. When all eight spacecraft are stacked, these brackets line up to make three solid columns up the stack, providing the needed rigidity. The cup-and-cone features accommodate the shear loads. Tensioning the three separation bolts at each joint provides the preload necessary to maintain stiffness across the interface. All three actuators are initiated simultaneously to prevent excessive tip-off during spacecraft separation. Both solar array panels are tied to the same nonexplosive release actuator, which uses lanyards instead of bolts to hold the panels in place. When initiated, the actuator releases the lanyards, allowing each panel to open 90 deg.

The antenna boom is a four-segment assembly. The four segments are hinged so that when they stow, the assembly fits into a trough that is less than 100 cm long, 12 cm high, and 26 cm wide. The antenna elements are embedded within a flexible Kapton membrane, which can be folded when stowed, returning to its original shape when the antenna is deployed. When deployed, the antenna springs into its 3 m flight configuration, and the elements resume their cylindrical shape around the boom segments. The stowed antenna is held in place by one of the closed solar-array panels. The antenna deploys in two phases. When the arrays are released, the entire stowed antenna bundle deploys 180 deg out of its trough, completing Phase 1. A single nonexplosive actuator is used to release the antenna for its Phase 2 deployment. Total deployment time of the spacecraft is less than 5 minutes.

14.3.3 Avionics

The MicroStar avionics suite is a low-cost, single-string design that is capable of recovering from single-event upsets. The avionics are configured in a distributed architecture based on the concept of independent software processes running on independent processors connected by message-passing interfaces. In the event of a single-event upset, this separation prevents one process from corrupting another. This makes the system very robust and supports the implementation of autonomous fault detection and correction.

The avionics suite consists of three major components: the flight computer, the battery charge regulator, and the attitude control electronics. Each of the components supports a major subsystem—command and data handling, electrical power, and attitude control, respectively. They communicate over an RS-485 serial bus using a token protocol (see Fig. 14.5).

An innovative method of analysis and test was developed to select parts for radiation tolerance. Extensive environmental screening allowed the use of commercial off-the-shelf (COTS) plastic encapsulated modules while maintaining reliability.

The flight computer is the heart of the command and data-handling subsystem. It is based on special epitaxial-wafer Motorola 68302 microprocessors operating at 12 MHz. The epitaxial processing makes the processors resistant to latch-up and increases their tolerance to radiation as high as 15 krad. On FM1 and FM2, static random-access memory (SRAM) data was protected by a COTS error detection and correction (EDAC) device. On FM3 through FM36, 3 Mb of SRAM are protected by EDAC implemented in a field-programmable gate array

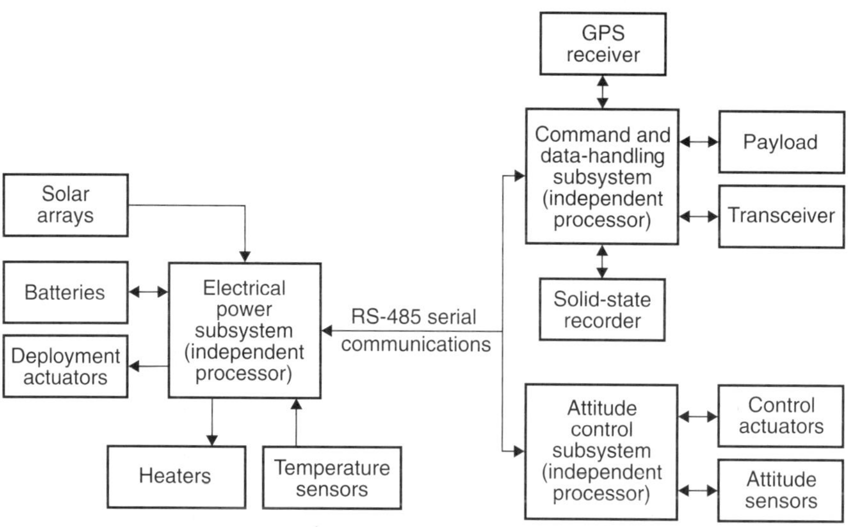

Fig. 14.5. MicroStar distributed architecture.

(FPGA). Flight code is executed out of EEPROM. The EEPROM memory space is physically segmented into 256 kb of fixed code, writable only during ground processing, and 256 kb of code that is writable on orbit. The flight computer has four RS-422 ports and two RS-485 ports for communicating with the other subsystem components and the launch vehicle.

The battery charge regulator is the heart of the electric power subsystem. It is based on the same epitaxial Motorola 68302 microprocessor used in the flight computer, but has 128 kb of fixed EEPROM and 128 kb of on-orbit writable EEPROM. On FM3 through FM36, this was updated to include 256 kb of fixed EEPROM, 256 kb of on-orbit writable EEPROM, and 256 kb of SRAM. It has two RS-422 ports and one RS-485 port to communicate with other subsystems and the launch vehicle.

The attitude control electronics unit is the heart of the attitude control subsystem. It is also based on the same epitaxial Motorola 68302 microprocessor with 256 kb of fixed EEPROM, 256 kb of on-orbit writable EEPROM, and 256 kb of SRAM. It has two RS-422 ports and one RS-485 port. It collects data from the attitude control system sensors and executes the attitude control algorithms. It also provides the drive commands for the onboard actuators.

14.3.4 Flight Software

The unique avionics architecture of the FM1 and FM2 spacecraft, and the fact that some of the avionics onboard had limited processing capabilities, precluded Orbital from using one of the more powerful real-time operating systems that were available at the time. Instead, a "home-grown" solution was required. The result was a custom operating system, dubbed "OSX." It was derived from a commercial real-time operating system, but carried much less overhead. It gave Orbital the facilities to develop a multitasking architecture and enabled the implementation of a lightweight message-based communication protocol between avionics boxes and tasks. This protocol, named OX25, has no relationship to the industrial protocol X.25.

The OX25 protocol is used throughout the spacecraft, the ground station, and the electrical ground support equipment. This approach provides a common development, integration, testing, and operating environment for flight software, system, subsystem, and ground system engineers.

14.3.4.1 Operating System Changes for FM3 through FM36

Small improvements in the architecture and system design used for FM3 through FM36 increased the processing capabilities compared to FM1 and FM2. This opened up the possibility of using a more powerful COTS real-time operating system that could support:

- onboard serial debugging, eliminating the need for in-circuit emulators.
- the ability to trim the real-time operating system to only incorporate the features required, thus reducing overhead.

- the ability to dynamically download and execute software during development, even (in rare cases) on flight boards.

This more powerful operating system significantly increased the overall robustness of the system.

14.3.4.2 Software Features

The MicroStar software architecture is extremely flexible, thanks to its modular message-based approach. Components and interfaces can be added and removed with little impact to the overall system. This is largely made possible by the universal messaging system that provides communications between software tasks, and even between subsystems.

A unique concept within the MicroStar software architecture is the self-monitoring software, which is embodied in autonomous, event-based software modules. Events include a wide variety of states, trends, and actions that may be detected within the hardware or software. In turn, these events may trigger one of any number of actions. All actions and events are fully configurable on-orbit.

This event system forms the basis of the MicroStar health and maintenance system. This feature is a key element in the layered reset strategy of the spacecraft because the lower levels of protection require the software tasks to monitor themselves for faults.

The development of the MicroStar software platform for a constellation of satellites meant that code configuration, control, and maintenance were extremely important. To that end, coding standards, styles, and templates were developed, and their usage was strictly enforced across the entire software baseline. A complex release system enables a constellation of similar but not identical satellites to reuse as much common software as possible. These efforts reduce both the cost and risk associated with the development and management of the constellation and follow-on MicroStar-based programs.

14.3.5 Attitude Control Subsystem

The attitude control subsystem uses an innovative design that enabled a 46 kg microsatellite to provide three-axis pointing using a minimal sensor and actuator suite. The primary requirement of the MicroStar attitude control system is to autonomously point a communications antenna nadir to within 5 deg CEP (circular error probable—the median of all error samples) while maintaining continuous sun tracking via spacecraft yaw maneuvers. This was achieved using an attitude control system that primarily uses magnetic control, which even by today's standards is a unique approach.

The attitude control subsystem sensor and actuator suite consists of a combination of magnetic torquers and passive gravity-gradient stabilization for control authority. A three-axis magnetometer and two static horizon sensors are used to determine the attitude of the spacecraft. Additionally, six coarse sun sensors located around the spacecraft are used to determine the sun vector. The attitude control electronics is the avionics unit that provides the interfaces to all of the sensors and actuators, as well as the processing required to execute the attitude control algorithms. The attitude control system also controls the two-thruster cold-gas reaction control system used for orbit maintenance.

The attitude control system propagates an orbit ephemeris based on information from the onboard GPS receiver or from an uploaded state vector, and this serves as the primary input for the orbit determination algorithm. Fairly accurate estimates of the spacecraft's position were required to support the magnetometer inputs into the estimator. A geomagnetic model is stored onboard the attitude control electronics and provides an expected magnetic field for the given spacecraft position, which, when compared to the magnetometer measurements, provides attitude error estimates. A magnetic torquer bar drive uses a proportion drive pulse-width modulation scheme, which minimizes torque disturbances.

To remedy spacecraft sensitivities to internally generated magnetic dipoles, extremely careful attention was paid to harness routing, making sure that all power lines and returns were routed in such away to cancel out current-induced fields. To address residual fields around the spacecraft, nonmagnetic materials were used as much as possible, and in the case of the battery, which induces the highest concentration of magnetic fields, the cells were wrapped in a MetGlas material.

14.3.6 GPS Subsystem

An innovative approach would be required to find a solution that would meet the navigation and time synchronization requirements of the ORBCOMM mission, even though these requirements were not as stringent as those of more common scientific or remote sensing applications. Early trade studies showed that the benefit of using a GPS-based navigational system outweighed the perceived risks of what was considered at the time to be an unreliable approach. The most significant benefits of a GPS-based navigation system were the ability to use the time output from the GPS receiver to help with autonomous onboard time synchronization and the ability to use a low-cost magnetometer-based attitude determination system, given the navigational data available from the GPS. Additionally, the output of the GPS receiver would provide spacecraft ephemerides that could be used for orbit determination. The autonomy provided by the GPS-based system helped minimize operational costs by greatly reducing the amount of support required from the ground.

Although orbit determination using GPS receivers on small satellites is now well demonstrated, at the time of the FM1 and FM2 missions, it was a novel concept. The Trimble TANS II six-channel receivers were chosen for MicroStar. These receivers were equipped with specialized firmware and a Motorola 68000 processor to allow them to operate in the highly dynamic environment of space with high velocity and high velocity derivatives. The receiver's four antennas were mounted on the solar arrays. Controlled by the spacecraft flight computer, the receiver provided navigation fixes to the attitude control subsystem and

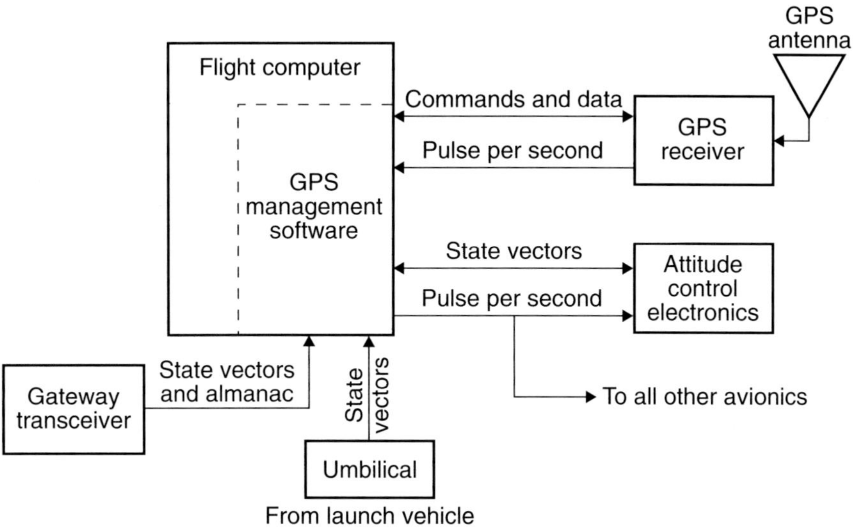

Fig. 14.6. MicroStar GPS subsystem.

provided absolute and relative time synchronization inputs to the spacecraft time control loops in the form of a time transfer message and pulse-per-second signal, respectively. GPS data were also used on the ground for orbit determination and constellation maintenance. Figure 14.6 shows the components and interfaces associated with the GPS subsystem.

14.3.7 Electrical Power Subsystem

The electrical power subsystem provides and controls electrical power to operate the spacecraft. Its functions include solar power control, battery charge management, power distribution and switching, initiation of separation and deployment actuators, and monitor and control of the thermal control subsystem. The electrical power subsystem is comprised of the battery charge regulator, five common-pressure-vessel (CPV) NiH_2 batteries, two solar arrays, one solar-array drive, and two twist capsules, which are used to manage the solar-array wires when the panels rotate. The system can provide a nominal orbit average power of approximately 150 W (beginning of life) and 125 W (end of life) when flying in a circular orbit at an altitude of about 750 km. A block diagram of the electrical power subsystem topology is shown in Fig. 14.7.

During spacecraft anomalies, the electrical power subsystem performs load shedding (turning off unnecessary loads) to provide the necessary power to operate the spacecraft and to recover from the anomalies.

The electrical power subsystem architecture uses a peak-power tracking topology, meaning that there is a current regulator between the solar arrays and the battery—an innovative approach, as opposed to the more common direct energy

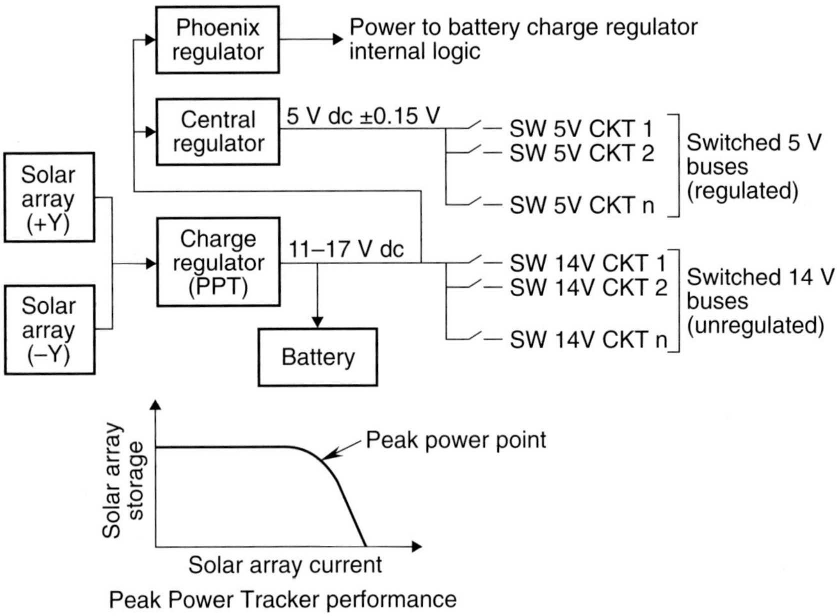

Fig. 14.7. MicroStar electrical power subsystem block diagram.

transfer systems of that time. Peak-power tracking was required because of the solar cells' inherent nonlinear characteristics and the resultant variance in the output power curve. The system continuously computes the power slope, determines whether it is positive or negative, and compensates by controlling the current draw from the solar arrays to keep the power slope as close to zero as possible. The use of a peak-power tracking system allows the size of the electrical power subsystem to be optimized by providing more efficient control of the available power on the solar arrays, especially during periods where the power demand from the bus is greater than the available power on the arrays.

The solar cells on FM1 and FM2 were made of silicon, used a back surface field/back surface reflector (BSF/R) technology, and had an efficiency of approximately 17% (not very impressive by today's standards). The cells on each solar array panel were configured into 10 strings, with 27 cells on each string for a total available power of 231 W (beginning of life).

Both solar array panels articulate on a common shaft that is driven by the solar-array drive. This drive articulates the panels in a single axis, and "yawing" the spacecraft provides the other axis of articulation. The drive electronics for the solar-array drive reside in the battery charge regulator. The desired solar-array angle is determined by the attitude control system, based on the attitude and position in orbit at that time, and sent to the battery charge regulator, which then commands the solar-array drive to the desired angle. The attitude control system commands the spacecraft yaw angle directly.

The use of NiH_2 CPVs was also a groundbreaking decision at that time, because NiH_2 batteries had just come on the market and did not have much flight heritage. But the increase in energy density and the improved performance across temperature over the more common NiCd batteries of that day made the benefit outweigh the risk. Five CPVs were used to give a bus voltage that ranged from 11–17 V dc and a capacity of 10 Ahr. Use of the standard 28 V ±6 V battery voltage was considered, but a trade study revealed that a significant mass reduction could be gained by reducing the number of CPVs from ten to five. Of course, this meant that all of the avionics units would have to be designed to accommodate this input voltage, but the battery was the single heaviest component on the spacecraft, so when it came time to look for ways to reduce mass, that proved to be very low hanging fruit.

Battery-charge management is straightforward. Pressure and temperature sensors located on the CPVs are used to monitor the pressure of the battery (normalized for temperature), which is then used to calculate the state of charge of the batteries. The charge current requirement is then determined based on that state of charge.

Another unique aspect of the MicroStar electrical power subsystem is the use of centralized regulators. When developing the MicroStar architecture for ORBCOMM, it was found that eliminating dc-to-dc converters in each of the avionics units could further reduce mass. This led to the concept of providing 5 V for digital electronics from a central regulator within the battery charge regulator and distributing that, along with the 14 V ±3 V battery bus, to each of the avionics on independently switched buses. This concept did introduce challenges concerning electromagnetic interference, but in the end worked well and is still in use today.

The battery charge regulator also accommodates the interfaces and the processing required for the thermal control subsystem. Temperature sensors located throughout the spacecraft are connected to the battery charge regulator. Likewise, the battery charge regulator provides individually switched heater circuits for each of the heaters throughout the spacecraft. All heater control is software based, meaning there are no mechanical thermostats. This allowed adjustments to be made on-orbit, and even allowed heaters to be assigned or reassigned via ground command to any temperature sensor on the spacecraft.

During launch, the battery charge regulator interfaces with the launch vehicle through an RS-422 serial port. At the appropriate time, the launch vehicle sends a message to the battery charge regulator indicating it is time to separate from the launch vehicle. Upon receipt and verification of this message, the battery charge regulator will initiate the three redundant nonexplosive separation actuators—the primary side first, and then the secondary. A continuity loop between the spacecraft and the structure below provides an indication to the battery charge regulator that the spacecraft has separated. Approximately one minute after separation, the battery charge regulator initiates the nonexplosive actuators that release the two solar array panels, followed by the actuator that releases the antenna boom, allowing the spacecraft to deploy itself.

14.3.8 RF Communications

The ORBCOMM constellation provides global narrowband two-way digital messaging, data communications, and geopositioning services. The following description of the RF system is not meant to be an in-depth explanation, but rather a brief overview, because the design is proprietary.

The communications suite consists of a gateway transceiver, which handles uplinks from and downlinks to the Gateway Earth Stations, a dual subscriber (downlink) transmitter, and the seven-channel subscriber (uplink) receiver. An additional UHF beacon is provided to help locate the satellite during initial acquisition. The VHF and UHF communication payload is capable of operation in the 137.0–150.05 MHz and the 400.075–400.125 MHz bands. The use of the system uplink (Earth to space) spectrum is managed by an onboard computer that employs an algorithm called DCAAS (Dynamic Channel Activity Assignment System). DCAAS continuously scans the authorized spectrum, identifies frequencies in use by other users of the band, and assigns subscriber communication uplink channels to minimize interference. DCAAS changes the uplink frequency at least every 15 seconds, which allows the system to coexist with current terrestrial users of the VHF band and limits interference to acceptable levels. The Gateway Earth Stations and the subscriber terminal communicate with the satellites in the same VHF band, thus eliminating the design complexity—as well as the associated bulk, power, and cost—of supporting multiple communication equipment on a single satellite. The satellites also contain packet-routing communications capability, including a limited store-and-forward capability.

The gateway transceiver serves as the primary communications path with the ground controllers and is primarily used for satellite telemetry and control, but also has the capability of handling limited amounts of subscriber traffic.

14.3.9 Thermal Control Subsystem

The thermal control subsystem design is an essentially passive system that primarily uses radiators to maintain the spacecraft temperature within acceptable limits. Software-controlled heaters help maintain battery temperatures between –5°C and 0°C. See Table 14.1 for a summary of the key features of the subsystem.

Design of the thermal control subsystem focused on minimizing risk and mass. To reduce risk, flight-proven components—including heritage temperature sensors and heaters, multilayer insulation blanket designs, and surface coatings—were used. Given the performance capability of Pegasus XL, minimizing mass was imperative for all subsystems. The thermal control subsystem mass was minimized by careful analysis of thermal requirements and by limiting the use of thermal tools to only those that were absolutely necessary. Overall, the thermal control subsystem mass was only 2% of the total spacecraft mass.

Table 14.1. Thermal Control System Design Summary

Feature	Design
Overall design approach	Passive
Radiators	ITO-coated silver-Teflon
Multilayer insulation outer layer	Al Kapton
Cylinder, nonradiator payload shelf area	Multilayer insulation
Cylinder open end	Multilayer insulation with EMI protection
Box-mounting thermal enhancement	Electrically conductive adhesive
Heat spreader (batteries)	K1100 composite material
Heater control	Batteries, solar array drive, antenna hinge

14.3.10 Launch, Deployment, and Acquisition

Launch of the FM1 and FM2 spacecraft had been scheduled for 17 March 1995, but the first two attempts were foiled—first by fog, and second by the debonding of a section of thermal protection material on the launch vehicle fairing.

The Pegasus XL rocket carrying FM1, FM2, and MicroLab I (Fig. 14.8) finally launched 3 April 1995 from Vandenberg Air Force Base into an orbit 741 × 759 km inclined at 70.03 deg (the target orbit was 739 × 743 km). The New York Gateway Earth Station connected to both FM1 and FM2 about 8 hours after launch. Initial spacecraft telemetry indicated that attitudes were all nominal, and all subsystems were behaving as expected. The investigation of downloaded back-orbit telemetry from launch through acquisition also confirmed that all activities had occurred nominally, including separation and deployment of arrays and antennas.

14.3.11 On-Orbit Anomalies

Three major spacecraft anomalies occurred during the early operations of the FM1 and FM2 spacecraft. The first was an uplink loss to FM2. The second was a gateway oscillation problem with FM2. The third involved difficulties in acquiring GPS signals. The anomalies were all successfully resolved through a variety of approaches, including fault-tree analysis, testing on spare hardware, and system analysis.

14.3.11.2 Uplink Loss to FM2

The Gateway Earth Station failed to connect to FM2 on its sixth pass, but experienced no problems connecting to FM1. Basic telemetry was available, thanks to the telemetry mode in which the spacecraft was in, but soon after the failure to connect the uplink, this basic telemetry stream ceased. An initial attempt to resolve the problem included a reset of the gateway transceiver processor, but no commands were passing through to the transceiver's processor at all. The design

Fig. 14.8. ORBCOMM FM1, FM2, and MicroLab mating to Pegasus XL.

of FM2 included automatic processor resets to unresponsive units, but unfortunately this feature had been disabled prior to launch because of concerns about the unknown effects of potential resets during captive carry and launch. Another option to address the problem involved stored command resets, which were based on a timer that executes if the timer itself is not reset by ground command (i.e., if communication is not available). However, these stored reset commands had not yet been loaded onto the spacecraft.

The only remaining option to restore communication with FM2 was the master, or "fire-code," reset. Implementation of this option was made difficult by the fact that the two spacecraft had not separated enough to ensure that the sent command would not be received and executed by FM1. Eventually, however, the command was sent, the uplink was swept in frequency, and different modulation schemes were attempted. Unfortunately all these attempts failed. It was not until 13 May 1995 that FM2 was heard from when it automatically connected to a Gateway Earth Station. After dumping the available back-orbit telemetry, it was discovered that the spacecraft had "browned out," and as a result, all avionics boxes had been reset. All FM2 hardware appeared to be healthy, allowing the team to narrow the list of likely causes to a latch-up or a single-event upset in the receiver DSP. To prevent similar situations in the future, stored watchdog reset timers were loaded and automatic reset capabilities were enabled.

14.3.11.3 GPS Anomaly

Throughout early operations, the GPS receivers on both FM1 and FM2 occasionally failed to provide navigation fixes for periods of time. Testing showed that resetting the receivers could remedy the problem, as could uploading a new GPS ephemeris. The outages were found to be caused by the orientation of the articulated solar arrays upon which the four GPS antennas were mounted: apparently, the spacecraft itself would occasionally block the antennas as they rotated. The GPS software neglected to consider the geometry of the spacecraft, and would not command the receiver to look for a new GPS vehicle in the event of a spacecraft blockage. The solution to this problem involved moving the solar arrays to a particular position and then sending a reset command to the GPS receiver, which then tracked the correct GPS vehicles.

14.3.11.4 Software Anomalies

Several minor software anomalies were detected and fixed throughout early operations, but one somewhat more significant bug involved processor overloading. The first symptom was a series of unexpected resets from the health and maintenance task of the flight computer. It was later found that the processor was being momentarily overloaded, so responses to the health and maintenance polls were not heard by the flight computer, causing it to reset the supposedly offending avionics box. This behavior was primarily limited to periods of rapid telemetry dumping in a particular mode, but was also seen during processor-intensive GPS resets and high communications traffic. The simplest fix was an extension of the timeout period that the health and maintenance task waited for a response. Since the operations that overloaded the processors were short in duration, the longer timeout period was all that was needed to solve the problem.

14.3.12 Initial Operations

Following the resolution of the initial anomalies during launch and early operations, the mission continued with great success. Planning, analysis, and other support processes such as telemetry logging, pass-plan generation, and orbit determination were shown to be easily handled by a single operator in the control center without shift work.

Calibration firings of the thrusters soon began, with the goal to correct for thruster misalignments. This process went smoothly on FM1, but the mass properties or thruster misalignments of FM2 proved to be much more difficult to overcome. In fact, thruster firings often caused the attitude control system to lose sun-tracking, and sometimes created significant pitch and roll excursions. Eventually, the calibration firings of the thrusters were able to account for the inherent misalignment/misbalance of the FM2 spacecraft, and the two spacecraft were placed into service on 3 August 1995. FM2 remained in service until 2000, and FM1 remained in service until 2006.

14.4 The Constellation—A Full Production Program

The development team had a brief period of time to monitor the performance of FM1 and FM2 before starting on the production program. Many lessons had been learned, and the team was now charged with moving ahead to produce the balance of the constellation—34 spacecraft! The two pathfinder spacecraft had proven that most of the innovative concepts were working as intended. However, the team realized that certain aspects of the spacecraft design needed to be revisited before beginning a full-scale production effort. At the same time, the team had to develop a methodology for this unique production process.

As with FM1 and FM2, schedule and cost would be very large constraints, but the designs and processes that were developed not only overcame those challenges but defined the MicroStar product line as it is today—not just from a design perspective, but from a process perspective as well. The FM1 and FM2 spacecraft provided an innovative design. This next phase of the program would give MicroStar something that no other company had at that time: a production line that by its end turned out about five spacecraft per month at an average cost of less than $5 million each.

14.4.1 Changes to the Spacecraft Design

The on-orbit experience with FM1 and FM2 had revealed aspects of the design that needed to be improved. Other changes had also been identified before the launch as a result of the intense effort to integrate and qualify the two satellites. Table 14.2 summarizes the most significant design changes that would be incorporated into the production units FM3 through FM36. Other changes included reducing mass, enhancing performance, and extending the satellite design life to five years from the original four. The following sections highlight some of the more significant changes.

14.4.1.1 Reset Strategy

A key design change identified as a result of an on-orbit anomaly experienced by FM1 and FM2 involved the reset strategy. The new approach used four levels of monitoring to recover portions of the spacecraft. The new reset strategy was particularly designed to deal with single-event upsets. The first layer monitors the performance of the components—in particular, the Motorola 68302 microprocessor—through the use of watchdog timers. The second level resides within the software of each avionics unit: If a task is not performing, then monitors within the software detect it and reset the individual task. The third level is implemented through the interfaces between subsystems: each subsystem monitors adjacent subsystems, and when an anomaly is detected—usually through a loss of communications on the serial interface—the subsystem that detected it can request a power cycle of the offending unit. Finally, the fourth level is a system-level reset that can be implemented by the ground through a "fire-code" command, or through the master timer located in the battery charge regulator. The master timer must receive a strobe signal from the ground once within a 37-hour period; otherwise, the spacecraft master reset will be initiated. In both cases, the entire spacecraft will be effectively shut down and then autonomously return to life. The diagram in Fig. 14.9 summarizes the reset strategy employed for FM3 through FM36.

Table 14.2. Summary of Design Changes for FM3 Through FM36

Component	FM1 & FM2	FM3–FM36	Reason
Reset strategy	Very simple	More comprehensive layered approach	Increased reliability
Parts selection	High use of commercial parts	More stringent selection/testing	Increased reliability
Battery charge regulator	Original design	Upgraded switching	Improved latch-up protection, increased operational flexibility
Battery charging	Absolute pressure readings	dm/dC charging algorithm	Improved performance, increased battery life
Solar arrays	Silicon	GaAs	Increased efficiency and power, extended life
GPS	Trimble TANS (4 array-mounted antennas)	Rockwell MPE (Single body-mounted antenna)	Reduced size, mass, power, and cost; improved performance
Attitude control system	Gravity gradient/ magnetic control	Added reaction wheel, better attitude reference system	Improved pointing performance
Software	MPX/OSX operating system	VxWorks	Improved development support, increased reliability
Reaction control system	2 thrusters	3 thrusters	Relaxed alignment requirements, longer burns

14.4.1.2 Piece-Part Components

One of the originating philosophies for the MicroStar spacecraft development was to maximize the use of commercial piece-part components. FM1 and FM2 revealed some areas that required higher reliability parts. Therefore, more stringent requirements regarding piece-part selection, testing, and protection were instituted as part of the updated design. The substitution of parts with better radiation characteristics (and more available data) was emphasized, along with improved test and parts screening programs. Lastly, the constellation design stressed improved protection against damage from single-event latch-up current in critical parts.

14.4.1.3 Battery Charging Algorithm

The MicroStar satellites employ a patented battery charge algorithm that uses battery pressure to provide autonomous adaptation to battery capacity aging and temperature effects. This algorithm monitors charging receptivity in a closed loop,

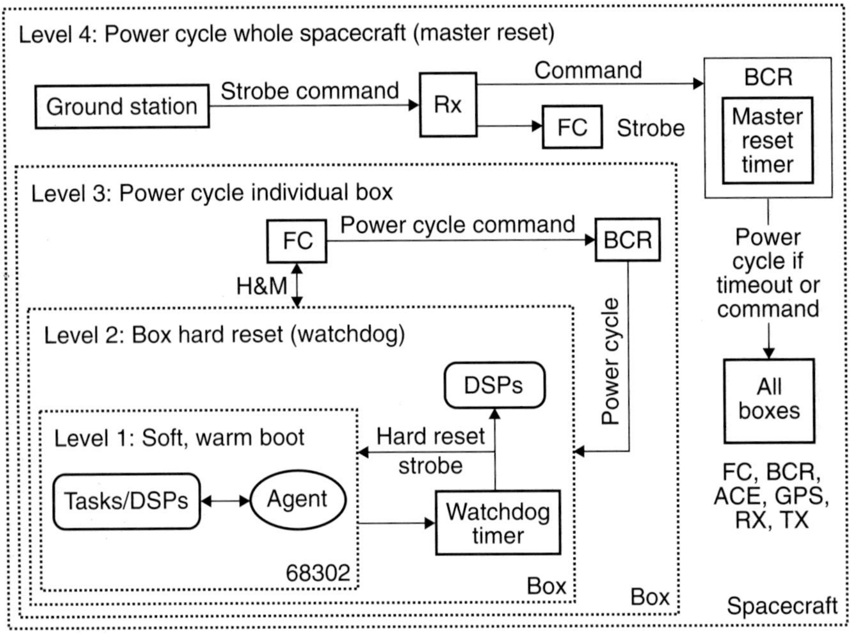

Fig. 14.9. FM3 through FM36 reset strategy.

as opposed to the more common practice of charging a battery to a fixed point, or "name plate" capacity, regardless of environmental factors. A key behavior of NiH_2 batteries is the way their actual capacity varies across their operating temperature range and life. When the batteries are colder, the capacity rises significantly, and conversely, when the batteries are warmer, the capacity decreases. This means that at different temperatures, the NiH_2 batteries can be safely charged to different points. The algorithm uses the Ideal Gas Law (taking pressure P and temperature T as inputs) in terms of moles of hydrogen gas produced, m, for a given input charge, C. The derivative, when expressed as dm/dC, is a function of changing battery pressure, temperature, charging current I_{bat}, and time t. This relationship is shown in Eq. (14.1).

$$\frac{dm}{dC} = \frac{d\left(\frac{P}{T}\right)}{I_{bat}dt} \tag{14.1}$$

The charging behavior of the dm/dC controller is shown in Fig. 14.10. Since the algorithm relies on a derivative, noise rejection is very important, so a two-state derived-rate estimator was developed to cope with any noise issues. In the end, a novel system was developed that reliably charges each battery relative to its real,

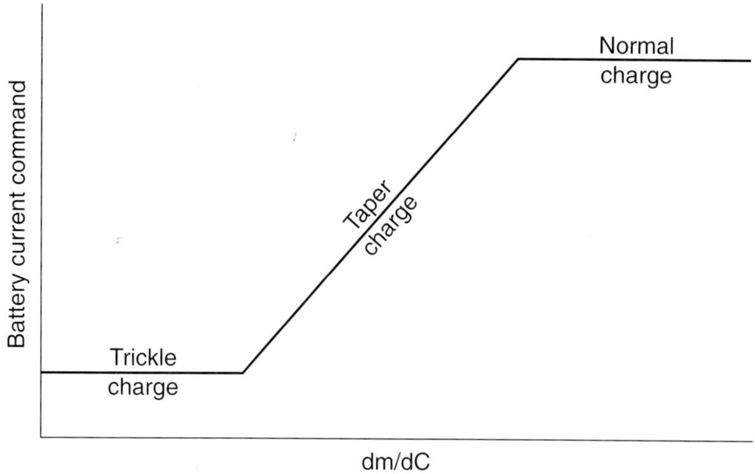

Fig. 14.10. Battery charging behavior (dm/dC).

environmentally dependent capacity rather than a predefined limit, which often leads to over- or undercharging.

There are two benefits of this algorithm. First, the spacecraft can take advantage of the increased battery capacity in colder temperatures (the MicroStar spacecraft uses a cold-biased thermal control system). Second, the life of the batteries is extended by charging the batteries in a more desirable charge cycle.

14.4.1.4 Global Positioning System

The Trimble receiver performed well on FM1 and FM2, but a cheaper solution was needed. Therefore, the inexpensive Rockwell MPE-I five-channel GPS receiver was selected as the core component for the GPS subsystem on FM3 through FM36. Although this Rockwell receiver had been used for many years on the ground, many modifications were needed before it could be used in space. A hybrid solution that included a commercial ground-based GPS receiver along with some in-house development provided the most economical solution, even at the expense of increased program risk. This additional risk was mitigated through the use of a design and development approach that required additional testing of the commercial hardware for spaceflight, extensive ground-based testing of the subsystem hardware and software, and an increased flexibility to configure the receiver through management software that was developed in-house.

The main software interface between the GPS receiver and the rest of the spacecraft is the GPS task, which resides on the flight computer. The design of the GPS task focused on three priorities: providing a command and telemetry capability that maximized the ability to reconfigure the receiver in the event of on-orbit anomalies; providing a self-contained health and maintenance strategy that would catch receiver anomalies that went undetected during ground testing;

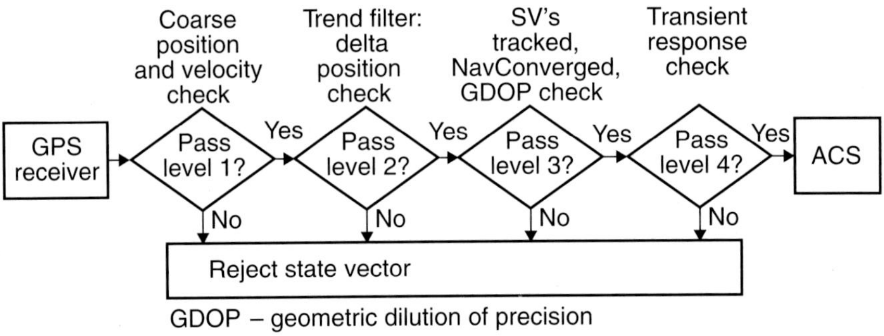

Fig. 14.11. State vector filtering scheme.

and providing a simple yet effective strategy to prevent erroneous navigation-state vectors from being passed to the attitude control subsystem. The onboard attitude control subsystem is sensitive to bad state vectors, so a state vector "filtering" scheme is employed within the GPS task. As shown in Fig. 14.11, the filtering scheme consists of four main steps. A state vector from the GPS receiver has to pass every check before being made available to the attitude control subsystem; failure to pass any one results in rejection of the state vector and an increment in the counter of the time since the last good navigation fix from the receiver.

14.4.1.5 Attitude Control Subsystem

For the constellation, the attitude control subsystem design was improved in several ways. A reaction wheel oriented in the yaw axis with precise torque-command tracking capabilities was added to the design along with a least-squares control algorithm to provide three-dimensional control capability. The total mass budget for the wheel (including controller electronics) was just 0.63 kg. Teldix GmbH of Heidelberg, Germany, developed a compliant wheel that was used for all constellation spacecraft. The additional mass of the reaction wheel was offset by a reduction in the gravity gradient tip mass. In addition, the attitude reference system estimator was improved in three ways: first, the dynamics introduced by the reaction wheel were accounted for; second, given the spacecraft's magnetic sensitivity, additional states to estimate the components of the spacecraft's residual magnetic dipole were incorporated into the attitude estimator; and finally, the sun vector provided by the coarse sun sensors was provided to the attitude estimator, thus adding robustness to sensor outages.

Use of a single vertically oriented reaction wheel, along with mild gravity gradient assist, has been shown to provide good three-axis pointing in a LEO spacecraft. These changes allowed a threefold increase in MicroStar's tolerance to magnetic disturbances, as compared with FM1 and FM2. The addition of the reaction wheel also improved yaw maneuverability, which enhanced the spacecraft's sun-tracking capabilities.

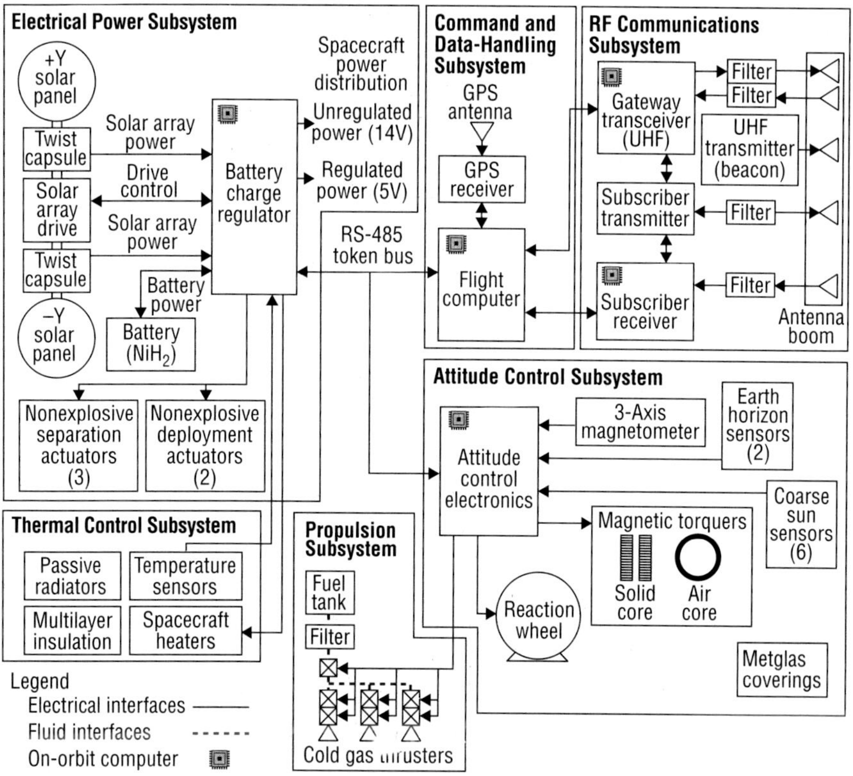

Fig. 14.12. ORBCOMM FM3–FM36 block diagram.

14.4.1.6 The Updated Architecture

The block diagram in Fig. 14.12 reflects the hardware changes made to the architecture. This new architecture also provided the baseline for MicroStar spacecraft flown on the other missions not associated with ORBCOMM. Throughout the years, the architecture has been updated to address issues with parts obsolescence and to add mission-specific features that make sense for the product line as a whole, but the overall architecture is the same as used today.

14.4.2 The Satellite Production Line

When producing a single spacecraft, the integration and test flow is fairly standardized among most satellite producers. A production run of 34 spacecraft is too large to use standard industry practices, but not large enough to initiate a full-blown production line. In addition, an aggressive, market-driven schedule forced the overlap of development, qualification, and early production phases and imposed a production cycle that required the team to produce satellites at a rate of four to five per month. The ORBCOMM production program would require an innovative qualification and acceptance test approach, a highly motivated team, a

Table 14.3. MicroStar Spacecraft Resource Plan

Spacecraft	Description	Application
FlatSat #1	Controller cards harnessed together for software development	Test software, develop flight test scripts, and simulation development
FlatSat #2	Controller cards and partial communications platform	Software and test procedure development
Engineering Development Unit	Complete and fully functional spacecraft using preliminary units	Functional and environmental test pathfinder
Qualification Spacecraft	Complete spacecraft (qualification units)	Design qualification
Flight Models 1 & 2 (FM1 & FM2)	First two operating spacecraft, launched 3 April 1995	Initial operations pathfinder, also used for on-orbit test development and testing of attitude control system upgrades

high level of test automation, and a creative and flexible flow that could accommodate potentially late deliveries and conflicts with test facilities.

When the production plan was being developed, it became apparent that two items were critical. First, a detailed plan was needed that accommodated all the complex operations on multiple spacecraft. A conventional integration and test flow, with tasks in series, would not support the schedule. Second, any such plan would be ruined if development problems lingered into production, or unit flaws repeatedly emerged at the system level. To prevent this, the program allocated a large number of flight and nonflight spacecraft assets for development and flight testing. Table 14.3 summarizes those assets and how they were used.

This resource plan allowed the qualification program to start before the development program was complete. The development hardware test requirements were as severe as those for the qualification hardware, but the testing flow differed to target those tests most likely to stress the design first. For example, the engineering development spacecraft was subjected to extensive functional testing over temperature, but only a couple of cycles were run. Subsequently, the qualification spacecraft was put through a conventional thermal-vacuum test; however, with a significant part of the risk retired by the previous thermal cycle test, this was considered unlikely to uncover any unit design changes.

By using a prototype "FlatSat" (a flight-like copy of the spacecraft) for software and test development, and the engineering development unit for test refinement and selected environmental tests, the development process stayed ahead of the qualification program. When the test program found a design flaw or a required hardware change, the change was immediately implemented in the

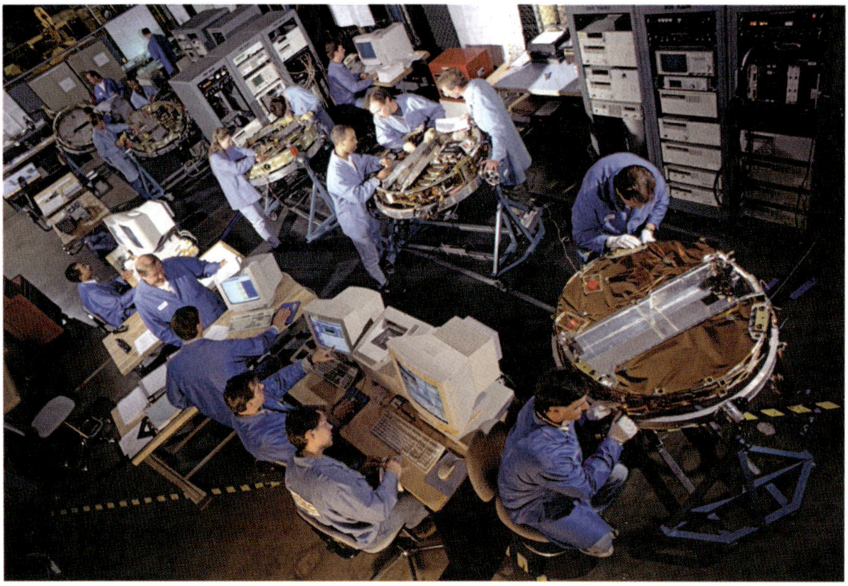

Fig. 14.13. ORBCOMM MicroStar satellites in production.

qualification hardware. This caused some rework to the qualification hardware; however, because the development program requirements were so stringent and the early development work was complete before the start of the qualification program, very few of these changes occurred.

The production team determined that the most efficient way to meet the production schedule was to set up a system that could produce and test four spacecraft at a time. Later, as the team became more proficient, the number was raised to five (Fig. 14.13). The spacecraft system production flow includes integration, functional testing, and environmental testing. The production cycle for each spacecraft started with preprocessing of the spacecraft structures to install many of the small items that required bonding, installation of the propulsion subsystem, and measurement of critical alignments. The structures would then proceed to the unit integration phase. The flat configuration of the MicroStar satellites simplified the mechanical integration, allowing most units or subsystems to be installed in less than two weeks from the start of the flow. In addition, the components used in the spacecraft design did not warrant any unique cleanliness requirements, which simplified the production, integration, and test process.

Because of the physical requirements of the stowed communications antenna, a large portion of the spacecraft's volume was reserved for the antenna "trough." This necessitated crowding the other units into a small area. However, careful placement used the allowable space efficiently. Units could be installed in almost any order, though in some cases, physical limitations, unit availability, and the system functionality needed for certain tests dictated the preferred order of buildup. As unit production crept ahead of system production on later spacecraft, most of the integration was done, as preferred, before any testing.

The subsequent functional and environmental test flow was much like traditional programs for the initial several spacecraft. However, test durations and the type of environmental and functional tests required decreased as the program progressed, and so the process was retired. In addition, functional tests were highly automated to minimize human error and reduce test time. After eliminating design flaws and refining the manufacturing processes, the resultant test flow was efficient and lean. Table 14.4 summarizes the changes in testing as production progressed.

Following mechanical integration, the spacecraft underwent functional testing. Testing of the power subsystem started when the installation for the power system components was complete. The remaining bus tests—including attitude control, flight computer, and GPS testing—followed power subsystem testing. The communications systems were tested last. This is because certain RF units were typically available last and because the complicated nature of the RF tests initially required extra time and refinement. Extensive use of automated test software and scripts made the functional tests fast and repeatable and codified the detailed knowledge that would otherwise have required experts to run many of the tests.

After initial functional tests, the spacecraft underwent environmental tests. The first few spacecraft (including the development and qualification vehicles) saw all of the major environmental tests usually associated with a production spacecraft program. These tests included shock, vibration, thermal balance, thermal vacuum, thermal cycle, EMC/EMI, and critical clearance and deployment checks. In addition to these standard tests, each MicroStar spacecraft underwent a magnetic calibration procedure that is essential to the proper functioning of the attitude control system. As additional spacecraft completed their mechanical integration and started environmental testing, environmental tests were either decreased in duration or eliminated altogether. This reduction was predicated on a minimum number of previous spacecraft completing the same test successfully. By the time the final sets of spacecraft went through environmental test, the test sequence was very efficient. This sequence included functional testing along with a vibration test to verify workmanship, the required magnetic calibration, and the electromagnetic interference test to verify EMI close-out workmanship of each spacecraft. The final functional test included a verification of the final flight software load. Following successful completion of the test sequence, each spacecraft went to storage to await shipment to the launch site.

14.4.3 ORBCOMM Flight Data Performance Analysis

The production program produced 34 satellites (in addition to FM1 and FM2); however, only 33 of these were launched and commissioned. One spacecraft, FM29, was not launched because the stack of eight spacecraft violated the mass

Table 14.4. MicroStar Test Matrix

	Functional	Vibration	Shock	Deployment	S/A Verification	Thermal Balance	Thermal Vacuum	Thermal Cycle	Magnetic Calibration	Radiated Comm/ EMC	Performance	Comm Validation	Multi S/C Testing
Eng. Dvp. Unit	X	X		X	X			X	X	X	X	X	
Qualification	X	X	X	X	X	X	X		X	X	X	X	
FM3	X	X		X	X		X		X	X	X	X	X
FM4	X	X		X	X			X	X	X	X	X	X
FM5	X	X		X	X			X	X	X	X	X	X
FM6	X	X		X	X		X	X	X	X	X	X	X
FM7	X	X		X	X			X	X	X	X	X	X
FM8	X	X		X	X			X	X	X	X	X	X
FM9 and up	X	X		X	X			X	X	X	X	X	X

allocation, resulting in that particular launch only deploying seven satellites. This spacecraft was eventually recommissioned by the Naval Research Laboratory for the TacSat-1 program.

By the end of 1999, ORBCOMM had successfully deployed and commissioned a 35-satellite constellation, a space-based network that continues to provide service to this day. The following sections summarize aggregate hardware performance data by subsystem of the constellation and assess the success of the mission, based on a sampling of very consistent data obtained as of August 1999.

14.4.3.1 Electrical Power Subsystem

The hardware components of interest in the electrical power subsystem are the battery and solar array. The performances of the batteries, on average, have exceeded specifications. Although the nameplate capacity of each battery is 10 Ahr at 10°C, the observed capacity at beginning of life is significantly higher, sometimes reaching 14 Ahr, even when taking temperature into account. This was found to be generally true for all MicroStar spacecraft; however, the variability observed between them is on the order of 25%. Even accounting for this variability, all battery capacities have exceeded their nameplate capacities of 10 Ahr. Additionally, the discharge/charge cycle for each satellite, depending on orbit, is about 35 min for maximum eclipse discharge and about 45 min to charge up to full capacity again. Figure 14.14 shows the trend across four satellites in the C-plane.

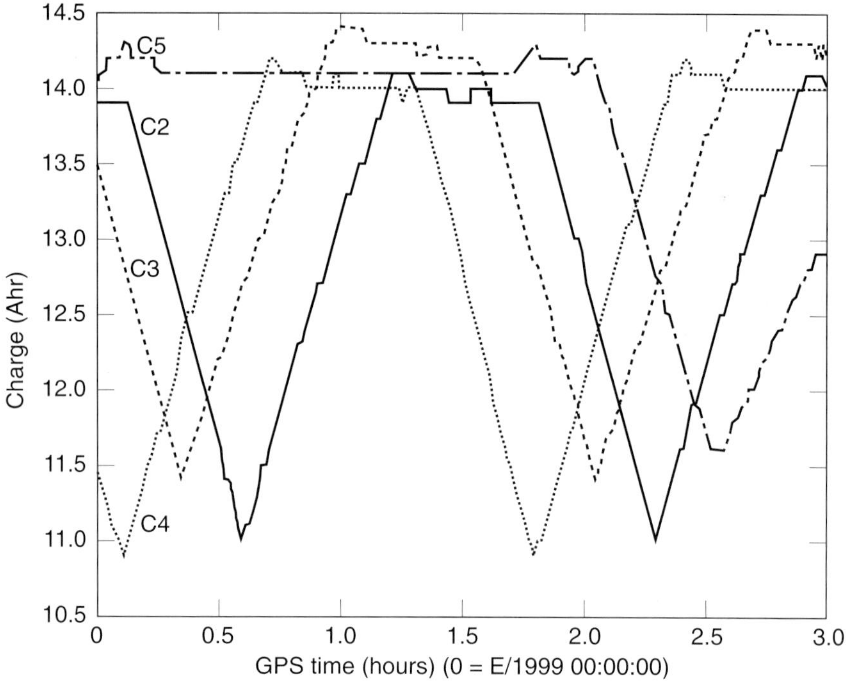

Fig. 14.14. Battery state-of-charge trending for four ORBCOMM spacecraft.

The solar-array control algorithm provides fine control of both the solar-array voltage and current operating point in response to spacecraft loads. The array displays the classic inverse relationship of solar-array voltage and current, with the expected variability due to fluctuations of temperature across the solar arrays. All MicroStar spacecraft have displayed similar behavior.

14.4.3.2 Attitude Control Subsystem

There are two main attitude control subsystem pointing requirements—antenna nadir pointing to within 5 deg CEP and within 10 deg 90% of the time. The plot shown in Fig. 14.15 illustrates the nadir pointing trajectory for one spacecraft. Typical of all the spacecraft in the constellation, the nadir error is less than 5 deg 90% of the time. All spacecraft exceed requirements, within 5 deg more than 90% of the time and better than 10 deg all the time. The CEP performance of the spacecraft is about 1 deg, in comparison with the 5 deg requirement.

The attitude control subsystem is also required to provide yaw pointing commensurate with the needs of the electrical power subsystem for solar-array power generation. The yaw pointing requirement is 18 deg to meet the 95% power generation requirement. All spacecraft with the exception of four had yaw error rates that stayed below 5 deg at least 90% of the time, easily meeting the requirement.

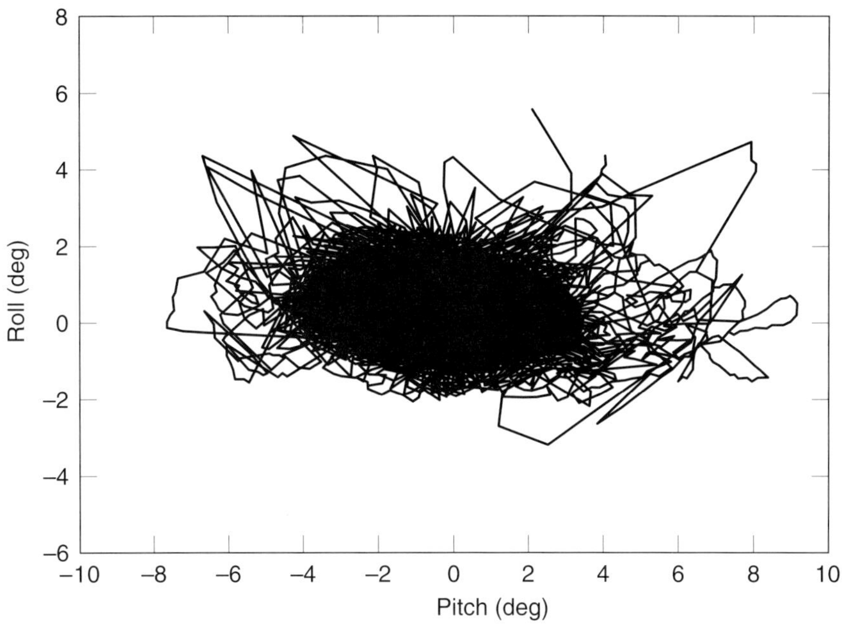

Fig. 14.15. Spacecraft pointing performance for one ORBCOMM spacecraft.

The four spacecraft that failed to meet the requirement did so because of scheduled commanded maneuvers, which took the arrays away from the sun.

14.4.3.3 Global Positioning System

Performance of the GPS system can be quantified in three ways: tracking performance, accuracy of the fixes (position figure of merit), and percentage of time the receiver provides fixes. Data from four spacecraft that are representative of the constellation show that, despite occasional loss-of-track events, the five-channel receivers track four or five GPS spacecraft most of the time, and navigation fixes are still provided more than 99% of the time. The position figure of merit for all spacecraft has shown to be in the 75–125 m range, thanks to the newly developed GPS task in the flight computer and functions within the attitude control subsystem that filter out large errors when processing GPS navigation data. The constellation average for fix percentage is 90%.

14.4.3.4 Thermal Control Subsystem

Thermal control subsystem performance is best characterized by the most important thermal requirement for each spacecraft: that the battery be kept in a temperature range of –5°C to 10°C. The data shown in Fig. 14.16 confirms that the battery temperature, although varying between spacecraft, was maintained in the range –2°C to 7°C over two orbits, thereby complying with the requirement.

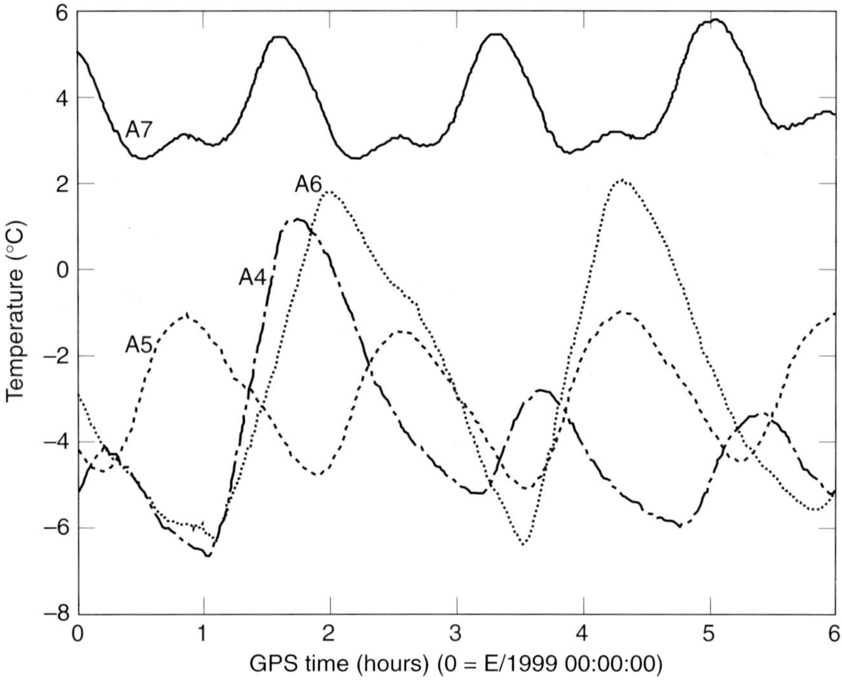

Fig. 14.16. Battery temperature across two orbits for four ORBCOMM spacecraft.

14.4.3.5 On-Orbit Performance Conclusions

After producing satellites at an astounding throughput of one per week, Orbital successfully launched 33 ORBCOMM satellites in 23 months from late 1997 to late 1999 (FM1 and FM2 were already up and operating). The ORBCOMM space-based communication network could now be fully commissioned. This was the world's first commercial LEO data constellation in service. It has already amassed more than 300 spacecraft-years of on-orbit experience, meeting and in many areas exceeding performance requirements. Unlike a single spacecraft mission, the ORBCOMM constellation provides the rare opportunity to assess the performance and amass statistics of various hardware components across numerous spacecraft simultaneously and over longer periods of time. This is especially valuable when attempting to gain a thorough understanding of the differences in performance between the standard practices of spacecraft development and the unique philosophies adopted for the MicroStar design.

It is evident from the sampling of the initial years of service that performance is consistent across spacecraft and over time. The variability that is apparent is generally due to three factors: hardware peculiarities, orbital position and attitude, and atmospheric variations. The hardware peculiarities particularly manifest themselves in the plots of battery capacity. The variability caused by orbital location

Table 14.5. MicroStar Mission Summary

	ORBCOMM	MICROLAB	BATSAT	MUBLCOM	QuikTOMS	ROCSAT-3	IBEX
Primary Mission	Global Data Gram Communications	Remote Lightning Detection and Mapping	Communications Demonstration	Military Communications Demonstration	Total Ozone Mapping System	GPS Occultation Experiment	Interstellar Boundary Characterization
Secondary Mission	None	GPS Occultation Experiment	None	Demonstration of Autonomous Rendezvous Technology	None	Atmospheric Scintilation Measurements	None
Launch Vehicle	Pegasus/Taurus	Pegasus	Pegasus	Pegasus w/HAPS	Taurus	Minotaur	Pegasus XL
Launch Date	Apr-95	Apr-95	Feb-98	Apr-99	Sep-01	4th Quarter 2005	4th Quarter 2008
Orbit	750 km	750 km	580 km	750 km	800 km	700 - 800 km	7000 km x 50 RE
Inclination			97.8 Degrees	98 Degrees	97.3 Degrees	72 Degrees	11 Degrees (Initial)
Attitude Control	Magnetic 3-Axis, Gravity Gradient	Magnetic 3-Axis, Gravity Gradient	1 Degree, 3-Axis	Magnetic 3-Axis, Gravity Gradient	Momentum Biased	Magnetic 3-Axis	Spinner
Knowledge	4 ° CEP,	4 ° CEP, 7.5 °	0.8 °	4 ° CEP, 7.5 °	0.25 °	2 °	0.2 °
Control	2 °	2 °	1.0 °	2 °	0.5 °	2 °	2.5 °
Sensor Suite	EHS, TAM, CSS	EHS, TAM, CSS	HCI, TAM, IMU DSS, CSS	EHS, TAM, CSS	DES, TAM, IMU, D SS, CSS	EHS, TAM, CSS	ST, CSS
Actuator Suite	Torquers (x3) RWA (x1)	Torquers x 3	Torquers (x3) RWA (x3)	Torquers (x3) RWA (x1)	Torquers x 3 RWA (x1)	Torquers (x3) RWA (x1)	None
Dry Mass	47 kg	75 kg	85 kg	46 kg		54 kg	77 kg
Deployables	Solar Arrays, Antenna Boom	Solar Arrays, Gravity Gradient Boom	Non-Articulated Solar Arrays	Solar Arrays, Antenna	Solar Arrays	Solar Arrays, Antenna Boom	None
Bus Power	27 W	31 W	64 W	27 W	96 W	45 W	48 W
Payload Power	64 W	40 W	14 W	73 W	28 W	46 W	36 W
Available Power	165 W	137 W	165 W	165 W	169 W	169 W	114 W
Type of Architecture	Distributed Intelligence	Distributed Intelligence	Distributed Intelligence	Distributed Intelligence	Distributed Intelligence	Distributed Intelligence	Distributed Intelligence
Science Data Storage	Stored w/SC SOH	45 MB	Stored w/SC SOH	Stored w/SC SOH	64 MB	128 MB	156 MB
SOH Data Storage	1.3 MB	1.3 MB	1.3 MB	1.3 MB	1.3 MB	1.3 MB	1.3 MB
Communications	VHF	S-Band	UHF	UHF	S-Band	S-Band	S-Band
Uplink Frequency	149 MHz	2.090 GHz	450 MHz	450 MHz	2.095 GHz	2039.5 MHz	2.038 MHz
Uplink Data Rate	32 Kbps (HDLC)	19.2 Kbps	32 Kbps (HDLC)	32 Kbps (HDLC)	2 Kbps	32 Kbps (HDLC)	2 Kbps
Propulsion System	Cold Gas	None	None	None	Mono Propellant	Mono Propellant	Mono Propellant
Fuel Capacity	145 cu inches @ 6000 psi	None	None	None	24 kg	6.65 kg	26.25 kg
Thrusters		None	None	None	4 x 1N (0.2 lbs)	4 x 1N (0.2 lbs)	4 x 5N (1.1 lb) 2 x 22N (4.9 lbs)
Delta V		None	None	None	360 m/sec	190 m/sec	455 m/sec

is expected, and in fact verifies that the spacecraft are behaving as intended. The atmospheric variations are obviously uncontrollable, but it appears that adequate margins were adopted in the communication subsystems such that the spacecraft meet all requirements within the specified elevation ranges.

14.5 Adapting MicroStar to Other Missions

The spacecraft developed for the ORBCOMM constellation was originally intended as a point design to meet a specific set of requirements—which, it should be noted, were not too stringent by any measure. As the development of FM1 and FM2 progressed, it became apparent that other missions without stringent requirements could benefit from the low-cost approach. This realization first occurred with the MicroLab lightning-mapping mission, which effectively spurred the transition from a point design for a specific mission to a product line. Table 14.5 summarizes some of the key missions that have flown the MicroStar bus.

In addition to the versatile design, the innovative production process and existing production line greatly helped to establish MicroStar as a turnkey system that could be developed, manufactured, tested, and delivered in less than 24 months. The ability to use aspects of the ORBCOMM ground station protocols and hardware added further benefit.

As more missions came on line, it also became apparent that the MicroStar architecture was flexible enough to add components necessary to significantly improve the pointing performance without significantly increasing the price. What started out as a bus designed to provide 4.5 deg of pointing accuracy is now capable of providing subdegree accuracy. While this may not be good enough for sophisticated optical payloads, it is good enough to support a wide variety of science and communications missions. The use of standardized serial interfaces between components and subsystems allows new components to be added quickly, with only software modifications in most cases. So, if a particular mission has a large enough memory requirement to warrant the addition of a solid-state recorder, one can be added by connecting it to an available serial port on the flight computer. The standard MicroStar block diagram is shown in Fig. 14.17 and illustrates the modularity as well as the high level of heritage within the design.

There are a limited number of serial interfaces in the MicroStar architecture, so if additional interfaces are required for a specific mission, a component known as the estimator computer, which is a clone of the flight computer, can be added to provide six more serial interfaces. This computer is typically used for additional attitude control system sensors and actuators, and may or may not be used for its processing capabilities.

There are cases, however, where hardware modifications are required. As stated earlier, the bus voltage is a nonstandard 11–17 V dc. The first modification that most missions require is the addition of a dc-to-dc converter that steps the 11–17 V dc bus up to the standard 28 V system that most off-the-shelf components and payloads require. This is accomplished through the addition of a power

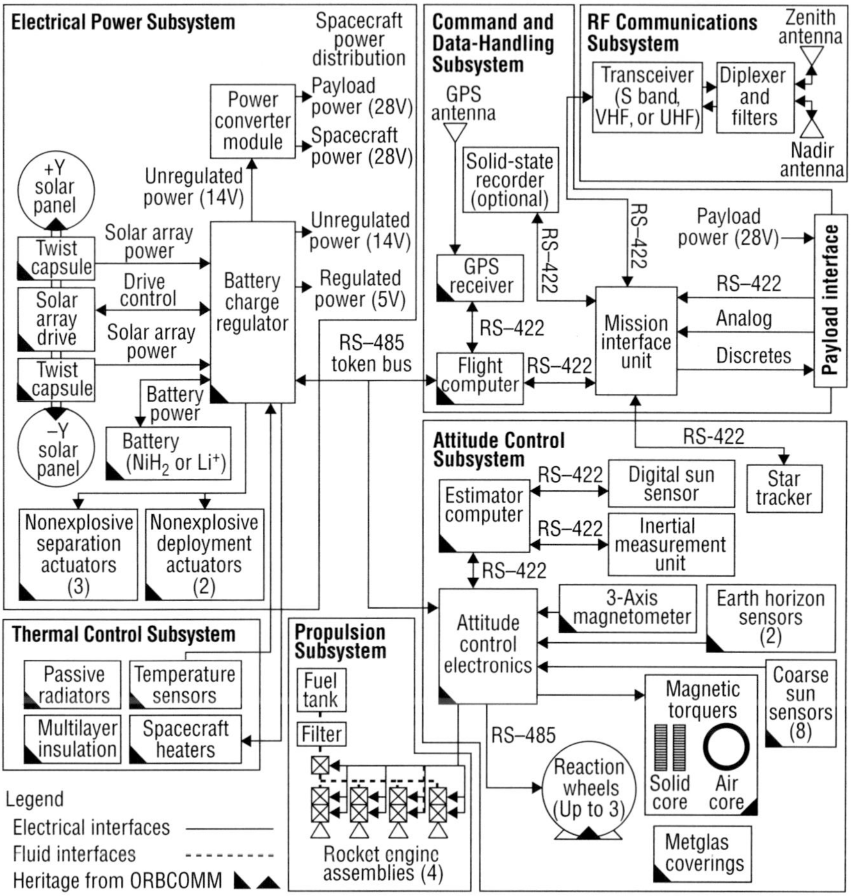

Fig. 14.17. Standard MicroStar block diagram.

converter module. The basic spacecraft form remains the same; however, several may be combined when additional payload volume is required.

14.5.1 MicroLab

The MicroLab architecture started with the ORBCOMM spacecraft bus avionics. Then, a payload interface computer, a solid-state recorder, and an S-band transceiver were added. The payload interface computer was required to provide the interfaces to the payload and the solid-state recorder. Additionally, because the standard ORBCOMM battery bus voltage was 14 V ± 3 V, a power converter was added to step that up to the 28 V input required by the payload, solid-state recorder, and S-band radio.

This spacecraft was developed in parallel with FM1 and 2 and was launched on the same vehicle on 3 April 1995. The launch on that day not only initiated the

Fig. 14.18. BATSAT.

deployment of the ORBCOMM constellation, but also served as the genesis for the MicroStar product line.

14.5.2 BATSAT

Teledesic's Broadband Advanced Technology (BATSAT) satellite (Fig. 14.18) was one of the first uses of MicroStar for scientific or technology demonstration. The BATSAT mission was an experimental satellite designed and built by an Orbital, Teledesic, and Boeing team. It was the world's first commercial Ka-band LEO spacecraft. Orbital provided the satellite bus, and Boeing supplied the payload. BATSAT was successfully launched in 1998 and operated for more than three years.

14.5.3 MUBLCOM

The Space and Terrestrial Communications Directorate (S&TCD) of the Army Communications—Electronics Command (CECOM) was the program manager for the MUBLCOM (Multiple Path Beyond Line of Sight Communication) program (Fig. 14.19). The MicroStar satellite bus served as the platform to host the MUBLCOM communications payload. In fact, the MUBLCOM bus was identical to ORBCOMM, with the payload modified to accommodate spread-spectrum communications.

Using a LEO satellite as an extremely high relay tower in the sky, the project's goal was to demonstrate mobile, over-the-horizon, networked communications among military forces in their theater of operations. In May 1999, the MUBLCOM satellite was launched from Vandenberg Air Force Base on a

Fig. 14.19. MUBLCOM.

Pegasus launch vehicle. In October 2004, it was recommissioned to serve as a rendezvous target for Orbital's DART mission.

14.5.4 FORMOSAT-3

The FORMOSAT-3/COSMIC program is an international collaboration between Taiwan and the United States that uses a constellation of six remote-sensing micro-satellites to collect atmospheric data for weather prediction and for ionosphere, climate, and gravity research. In early 2001, the National Space Organization of Taiwan selected Orbital as their partner for this mission. Orbital was responsible for constellation design and analysis, development of the spacecraft bus, payload instrument development, and oversight of components made by vendors in Taiwan. Orbital also provided assistance with system integration, satellite integration and testing in Taiwan, early on-orbit checkout, satellite positioning, and mission operations.

FORMOSAT-3/COSMIC (Fig. 14.20) consists of six planes of one spacecraft each, spaced 24 deg apart, in a 800 km circular orbit with 72 deg inclination. Mission life for the spacecraft is two years, with expendables sized for five years. FORMOSAT-3/COSMIC was launched aboard a Minotaur I launch vehicle on 14 April 2006 and is in operation today.

14.5.5 TacSat-1

TacSat-1 is a joint Naval Research Laboratory (NRL) and Dept. of Defense Office of Force Transformation experiment that combines existing military and commercial technologies to provide cheaper and faster access to space. NRL is

Fig. 14.20. FORMOSAT-3/COSMIC.

the program manager for this spacecraft. TacSat-1 uses a modified ORBCOMM spacecraft bus combined with payloads from the Predator UAV to meet the program goal of designing, building, and launching a functional experimental LEO satellite in less than 12 months with a total mission cost (including launch vehicle) not exceeding $15 million. A joint Orbital and NRL team completed the spacecraft integration in 10 months. TacSat-1 was to be launched in early 2005, but due to launch vehicle issues, the mission was canceled.

14.5.6 Interstellar Boundary Explorer (IBEX)

As part of a Southwest Research Institute team, Orbital was selected in January 2005 by NASA's Small Explorer Program to develop, build, and launch a small sun-pointing major axis spinning spacecraft (the first spinner for the MicroStar product line) on a Pegasus rocket for the Interstellar Boundary Explorer (IBEX) mission.

Successfully launched in October 2008 with a 24-month mission life, the MicroStar-based IBEX spacecraft (Fig. 14.21) was boosted into orbit by a Pegasus launch vehicle from the Kwajalein launch site in the Pacific Ocean. The spacecraft incorporated an onboard solid rocket motor to propel it to the final high-altitude orbit beyond Earth's magnetosphere, as required by IBEX's scientific instruments. The IBEX satellite will orbit the Earth every five days on a highly-elliptical path that takes it from a perigee of 7000 km to an apogee of 320,000 km or 50 Earth radii (approximately 83% of the distance to the moon), and will use its instruments to remotely detect and map the boundary between the solar system and interstellar space.

Fig. 14.21. IBEX.

14.6 The Future of MicroStar

Although the marketplace today looks quite different than it did in the early 1990s, it is still driven by the same basic need for capable microsatellites that can meet performance requirements while maintaining schedule and controlling costs. Today, customers for MicroStar satellites fall into three categories: operationally responsive military missions, small science and technology demonstrations, and small LEO commercial satellites for communications. An updated, next-generation MicroStar bus has been developed to serve these markets. The baseline bus is comprised of one or more modules stacked together vertically, with additional solar panels as needed. Figure 14.22 shows some example MicroStar configurations.

To develop an essentially "off-the-shelf" microsatellite, mission constraints were gathered, categorized, and traded with an eye toward maximizing performance while minimizing cost. The design takes advantage of the essential strengths of the original ORBCOMM architecture—low cost and mass, modularity, stackability, and innovative manufacturing methodologies to minimize development time. In addition, the design and development approach considered new technologies and changes in the availability of components in the marketplace. For example, the next-generation MicroStar takes advantage of the increased availability and commoditization of spacecraft avionics by using an integrated avionics unit from a third-party vendor.

The modern MicroStar offers high performance, multiple payload and spacecraft interfaces that are compliant with key standards, quick turnaround, and a

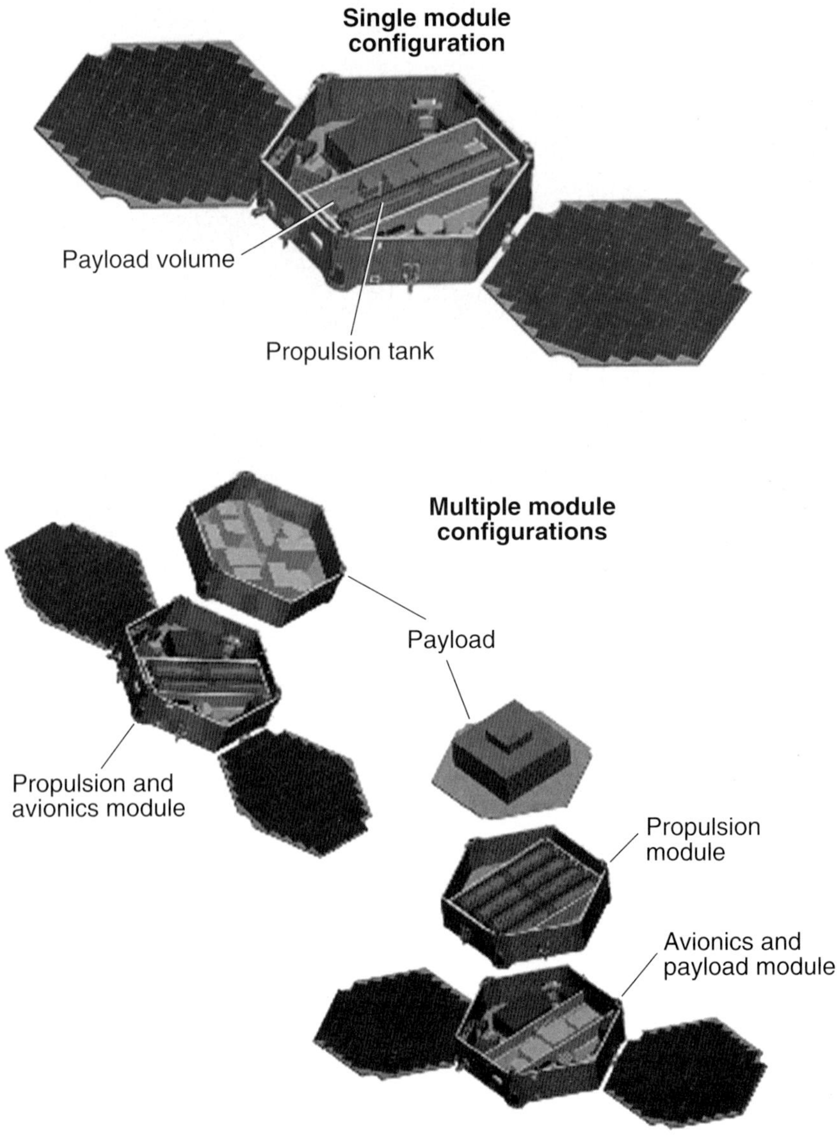

Fig. 14.22. Possible MicroStar configurations.

substantially lower bus cost. It combines heritage technology and state-of-the-art hardware to support advanced payloads.

14.7 Conclusion

Full-scale production of the ORBCOMM spacecraft ended in 1999, but the evolution of the MicroStar product has continued and evolved into Orbital's smallest spacecraft product line. The spacecraft being produced for current programs are being designed and built using the same methods and processes that were originally developed for the ORBCOMM program. The high level of heritage, accumulated on-orbit experience, and demonstrated performance substantiates the utility of this inexpensive spacecraft bus for future commercial communications and remote sensing applications.

Through the joint efforts of the ORBCOMM customer and the MicroStar engineering team at Orbital, remarkable feats were accomplished. In an industry where change happens slowly and innovation is achieved in small, easily-managed increments, the MicroStar story proves that when the right people come together with a common goal and an original vision, the rules can be broken—and innovators can succeed.

14.8 Acknowledgments

The authors would like to thank the following people for helping us to document the MicroStar story and assisting with the production of this chapter: Michael Balling, Ed Belte, Melinda Biegon, David Bryan (The PTR Group), Jennifer Campbell, Melissa Carpenter, William Ducas, Rob Fulton, Bret Hautamaki, Craig Lalley, Elisabeth Palyi, Bhavesh Patel, John Scott, David Steffy, and Jason Westphal. We would also like to acknowledge the innovation and hard work of the ORBCOMM development team, ORBCOMM Inc., and the MicroStar product team who have served as an inspiration to "out of the box" thinkers in the space industry. Finally, a special thank you to David W. Thompson not only for his in-depth review of our final draft, but also for providing an environment that fosters the kind of creativity that made the development of the MicroStar spacecraft possible.

14.9 Bibliography

F. DeMauro and W. Ducas, "Thermal Design of the ORBCOMM Constellation Spacecraft," *SAE 26th International Conference on Environmental Systems* (Monterey, CA, 1996).

G. Dorsey, *Silicon Sky* (Westview Press, Boulder, CO, 1999).

M. R. Krebs, "A New Attitude Control Mechanism for LEO Satellites," *10th AIAA/USU Conference on Small Satellites* (Logan, UT, 1996).

A. W. Lewin, "A Maintenance-Free Battery Charge Control Approach for Nickel-Hydrogen Batteries," *13th AIAA/USU Conference on Small Satellites* (Logan, UT, 1999).

ORBCOMM, Inc. "ORBCOMM: Space Segment," http://www.orbcomm.com (as of 24 Sept. 2007).

B. Patel, S. Schroll, and A. Lewin, "On-orbit Performance of the ORBCOMM Spacecraft Constellation," *13th AIAA/USU Conference on Small Satellites* (Logan, UT, 1999).

B. Patel, W. Walter, and D. Patton, "An Innovative GPS Solution for the ORBCOMM Constellation Spacecraft," Institute of Navigation National Technical Meeting (San Diego, CA, 1999).

A. Robinson, Craig Courtney, and Thomas Moran, "Non-explosive Actuation for the ORBCOMM Satellite," *29th Aerospace Mechanisms Symposium* (Houston, TX, 1995).

D. A. Steffy, "ORBCOMM Satellites Launch and Initial Flight Operations," *10th AIAA/USU Conference on Small Satellites* (Logan, UT, 1996).

P. M. Stoltz, M. Krebs, and Rick Baltman, "ORBCOMM Attitude Determination and Control," *AIAA/AAS Astrodynamics Conference* (San Diego, CA, 1996).

Chapter 15

Japan's Small Scientific Satellite, "INDEX" (Reimei)

Hirobumi Saito,* Masafumi Hirahara,** and the INDEX Project Team

15.1 Introduction

The Institute of Space and Astronautical Science, Japan Aerospace Exploration Agency (ISAS/JAXA) has launched a series of scientific satellites including planetary spacecraft and astronomical observation satellites. Although the missions have achieved fruitful scientific results, these satellites cost nearly $160 million each, including the use of the M-V launch vehicle, and take longer than seven years to be developed.

To differentiate from these "big, expensive, slow" missions, ISAS/JAXA also planned to launch "small, inexpensive, fast" piggyback satellites to demonstrate new technologies and to perform science observations. Thus, the INDEX (INnovative Technology Demonstration EXperiment) satellite began development in 2002 for aurora observations and small satellite technologies[1,2] and launched in August 2005. INDEX has been functioning for 38 months as of November 2008. This paper summarizes the satellite overview, some of the on-orbit results, and lessons learned from developing INDEX.[3–5]

15.2 Mission and System Description

15.2.1 Aurora Observation

Auroral phenomena in the polar regions of Earth are characterized by emissions over a broad range of wavelengths as well as electron and ion distributions over a wide range of energy and pitch angles. The main scientific purpose of INDEX was to investigate small-scale dynamics of terrestrial auroras, particularly the spatial distributions and time variations and the relations between spectral properties and the spatial distributions of the charged particles that generate these emissions. On INDEX, one of the observation instruments is a multispectral aurora camera using three CCD imagers, and another is an electron and ion energy spectrum analyzer using top-hat type electrostatic optics.

*Institute of Space and Astronautical Science,
Japan Aerospace Exploration Agency (ISAS/JAXA)
Sagamihara-shi, Japan
**Tokyo University
Bunkyou-ku, Tokyo, Japan

15.2.2 Demonstration of Advanced Satellite Technologies

In 1995, ISAS organized a technology program known as STRAIGHT (STudy on Reduction of Advanced Instruments weiGHT) to develop onboard instruments with reduced mass but high performance for planetary exploration. The STRAIGHT program produced various instruments, including sensors, processors, semiconductor devices, and batteries, most of which were actually applied to the ISAS satellites. The goal of the first INDEX engineering mission was to demonstrate these advanced spacecraft technologies on orbit. Table 15.1 summarizes the items and the on-orbit results from INDEX.

The second engineering mission was to demonstrate that a small satellite can perform a scientific mission. ISAS/JAXA has launched scientific satellites in various fields, including x-ray astronomy, infrared astronomy, solar physics, and magnetosphere physics. Using INDEX technology, missions such as these can now be realized with relatively small payload mass. To perform a mission based on scientific requirements using a small satellite, limited resources have to be used to achieve high performance for attitude control, power generation, computation capability, and downlink capacity. We selected the auroral observation mission because the payload is relatively small and requires only moderate attitude pointing accuracy of several arcminutes.

15.2.3 Development Strategy

The development of INDEX focused on two points—namely, to train young engineers and scientists, and to serve as an incubator for new space technologies and systems. The INDEX satellite bus concept was therefore defined as follows:

- It would be a piggyback satellite with 50–70 kg mass to save on the launch cost.
- The cost goal would be around $4 million, excluding ISAS/JAXA personnel expenses.
- The development period would be about three years to accommodate students and young personnel.
- The satellite design would be optimized within available resources to meet the scientific requirements. The mission scientists would be asked to join the INDEX satellite working team.

To meet these constraints, we set the work breakdown as follows:

- Without direct outside contractor assistance, ISAS/JAXA performed the satellite design, development of onboard software, development of mission instruments, satellite integration, subsystem and system test, integration of the ground station, and the satellite operation. The INDEX effort was mentored by retired experts from a leading satellite manufacturer in Japan.
- Graduate and undergraduate students were involved in developing INDEX. Their contributions included onboard software development,

Table 15.1. Technologies Demonstrated in INDEX

Demonstration Technology	Feature	On-Orbit Results
Fine 3-axis attitude control of 70 kg satellite[6–9]	• Simple hardware (3xMTQ, 1xRW) • Estimation of residual magnetic moment on orbit and feed-forward control	• 0.05 deg (1σ) accuracy • 0.004 deg/sec stability
Integrated control system with triple-voting commercial proceessor[10]	• Hitachi SH3, 60 MHz • Reconfiguration at voter error	• Satellite keeps operating through CPUs reconfiguration against 28 SEUs of non-rad-hard devices for 27 months
Li-ion battery with pouch cell[11]	• Potted with resin for vacuum	• Works more than 16,500 cycles • Coincidence with simulation tests
Solar-concentrator paddle with thin-film reflector[12,13]	• Specific power 94 W/kg for reflector contribution • Robust design against reflector contamination	• Concentration factor 1.25 • No power degradation for 38 months
Flexible variable-emittance radiator[14]	• Flexible thick film of $LaSrCaMnO_3$	• Confirm performance
Miniature space GPS receiver[15–18]	• Converted from car-navigation receiver • 200 g including RF hybrid	• Cold-start acquisition in less than 28 min • 1.5 m (1σ) random error
Commercial fiber-optic gyroscope with temperature control[19]	• Temperature control (±0.2°C) improves bias stability 0.05 deg/hr	• 0.05 deg/hr (1σ) on-orbit
Compact star tracker	• Medium accuracy (0.02 deg) • Lightweight (1.3 kg)	• Works well
Commercial high-temperature SOI-SRAM	• High-temperature application • 100 krad tolerance, single-event latch-up free	• No radiation damage
β-titanium bolt without creep[20]	• Solve creep issue of conventional α-titanium bolt	• Survived launch, environment

advanced instrument development, system tests, flight operations, and satellite operations. INDEX could not have been launched without their contributions.

- The advanced onboard instruments developed in the STRAIGHT program (see Table 15.1) were manufactured as components by industrial partners. Other conventional onboard instruments (e.g., the power control unit, power converter, and input/output circuits) were manufactured as components by small companies; we took responsibility for unique manufacturing processes such as black painting and coating for the space environment.

15.2.4 Cost, Parts, and Components

The cost of INDEX, including development and flight hardware, was about $4 million, excluding ISAS/JAXA personnel expenses. If personnel expenses were included, the total cost would be about $6.5 million. The distribution of the total cost is shown in Fig. 15.1. The cost of INDEX is rather low compared to a typical satellite of the same mass class.[21] A few guidelines for parts and components helped ensure a cost-effective product.

For example, high-reliability parts, which are supported by certification documents, seem not to be cost-effective. On-orbit failure analysis[22] reveals that most satellite malfunctions are not random failures governed by constant failure rates, but early failures caused by inadequate design or manufacturing. In fact, 41% of satellite malfunctions occur during the first year of operation. Power (48%), data-handling, and communication (46%) subsystems have very high rate of failures during the first year. These subsystems are mostly made of electronic components, which should be reliable for at least 3–5 years of operation. INDEX used Japanese industrial-grade electronic parts, MIL-grade discreet active electronic parts, radiation-tolerant field-programmable gate arrays (FPGAs), and commercial high-performance logic devices with triple voting or error correction codes. Active devices were confirmed to tolerate a total ionizing dose of 10–20 krad. Components that contain parts with a risk of single-event latch-up were protected by circuit breakers. Intensive ground testing served to eliminate initial parts failures and design problems. Circuit boards were also designed to enable easy modification and rework.

We purchased conventional instruments and sensors as components from manufacturers, either domestic or overseas, based on our in-house system design. We used conventional instruments manufactured by small companies for system integration functions such as input/output circuits, a power control unit, cable harness, mechanical instrumentation, and thermal instrumentation. We selected small companies with high quality levels and provided them with unique manufacturing processes/services for flight instruments such as black paint and coatings.

We used commercial components when they were significantly superior to space components in terms of cost, performance, mass, and procurement,

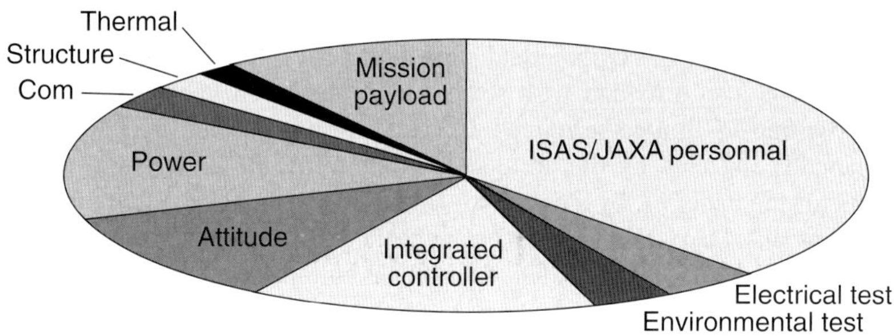

Fig. 15.1. Cost structure of INDEX satellite (total about $6.5 million).

as described in Sec. 15.4. Examples include the GPS receiver, the fiber-optic gyroscope, and the lithium-ion secondary battery. In most cases, commercial components without modification can not meet the requirements of a high-performance satellite such as INDEX; however, by understanding their operation, we were able to modify them for space applications. For example, software in the GPS receiver was modified to allow a large Doppler frequency shift, and the fiber-optic gyroscope was temperature controlled to improve its bias stability. Based on their structures, materials, and ground tests, these modified commercial components were deemed compatible with space conditions and could meet requirements concerning mechanical durability, thermal performance, outgasing in vacuum, and susceptibility to space radiation. The use of outgas-free materials was strictly enforced on the INDEX satellite because it is equipped with particle analyzers that are extremely sensitive to outgasing.

Integration of similar functions into one instrument is economical. We developed an integrated control unit to oversee the data handling, attitude control processing, heater control processing, and actuator driving (e.g., magnetic torque rods and heaters). Ideally, the design of such an integrated system should have an internal structure with clear interfaces at the circuit-board, circuit-block, and FPGA levels; however, this policy was not perfectly observed due to resource limitations.

15.2.5 Reliability and Product Assurance

The reliability and product assurance of small satellites such as INDEX can be maintained without implementing a formal document-based procedure, but by the talent and enthusiasm of the project members and company engineers who take pride in their work. We tried to keep the amount of documentation as low as possible. The only effective document of INDEX is the System Design Document, which describes all key performance parameters for the satellite. Because satellite tests are very important and help to detect possible defects, procedural documents for ground testing should be prepared with great detail.

As mentioned, reliability for INDEX was ensured through design strategies such as failure isolation and safe-mode recovery as well as intensive ground testing. For example, satellite system and subsystems are isolated from each other by circuit breakers and current limiters; this approach is especially useful for mission instrumentation that uses commercial parts that are at risk of a single-event latch-up. INDEX is also provided with several error detection, recovery, and safety features within its limited resources. Critical on-orbit problems for INDEX are automatically detected by the attitude subsystem, by the undervoltage sensing systems for the batteries, by a processor subsystem (e.g., a watchdog timer and a voting error checker), and by a 2-bit error checker of the data recorder. Depending on how serious the failure is, a safety operation is automatically initiated, including the reconfiguration of triple voting processors and activation of a safe-hold mode with spin stabilization and a minimum number of instruments in operation.

Ground tests are significantly useful to detect possible defects of the satellite.[22] Design defects and initial defects of hardware are detected through long-duration tests and environmental testing with satellite functions being monitored. We found many defects of the satellite during these ground tests and solved all of them before launch.

15.2.6 System Description

INDEX is a small scientific satellite with size of 72 × 62 × 62 cm and a mass of 72 kg. It has two deployable solar paddles, equipped with solar-concentrators, with a power capability of 150 W. Figure 15.2 shows INDEX with the solar paddles deployed. Table 15.2 lists the main specifications for the satellite. The configuration of the satellite is shown in Fig. 15.3. Table 15.3 lists the mass budget.

A system block diagram of the satellite is shown in Fig. 15.4. INDEX was designed with an integrated computer system[10] that executes almost all satellite control tasks. As depicted in Fig. 15.4, the integrated control unit is directly connected with almost all instruments. These include sensors (star tracker, spin/nonspin-type sun aspect sensor, geomagnetic aspect sensor, and fiber-optic gyroscopes), actuators

Fig. 15.2. INDEX satellite with solar paddles deployed.

Table 15.2. Main Specifications of INDEX

Mission	Observation of aurora Demonstration of small science satellite and advanced technologies
Mass	72 kg (bus 61 kg, payload 11 kg)
Size	72 × 62 × 62 cm
Cost	$4 million $2.5 million (ISAS/JAXA personnel)
Project work breakdown	Design, integration, and test by ISAS/JAXA Components by industry
Orbit	Nearly sun-synchronous orbit 608 × 655 km, noon
Launch	Dnepr rocket from Baikonur, 23 August 2005
Life	Has been working for 38 months as of November 2008
Power	150 W solar-concentrator with thin-film reflector, triple-junction cell 3.0 Ahr × 2 lithium ion battery 3-sequential partial shunt
Attitude	Bias-momentum three-axis stabilization Pointing error: 0.05–0.1 deg Determination error: 0.05 deg Attitude stability: 0.004 deg/sec 1 reaction wheel (0.5 N-m-s), 3 magnetic torquers (6 A-m^2)
Communications	S-band USB Uplink bit rate: 1 kbps Downlink bit rate: 132/64/32/16/8 kbps

for attitude control (reaction wheel and magnetic torquers), and the S-band receiver and transmitter. In other words, the integrated control unit combines the role of the data-handling, attitude-control, and heater-control electronics that are found in conventional satellite design. For the sake of system compactness, there is no data bus outside of the integrated control unit. All instruments are controlled directly in the address space of the processor. Input/output ports for the sensors and the actuators are developed with space-qualified antifuse FPGAs.

The INDEX satellite achieves attitude control via a bias-momentum three-axis stabilization system.[6–9,19] Details are described in Sec. 15.4.2.

Commands are sent to the satellite through an S-band link at 1 kbps, while telemetry is sent at 132, 64, 32, 16, and 8 kbps. The main control station of INDEX is the newly developed Sagamihara small satellite station, which uses a 3 m diameter antenna.[10] For auroral observation, telemetry is also received at the Norwegian Svalbard satellite station and at the Japanese Showa Base at the South Pole.

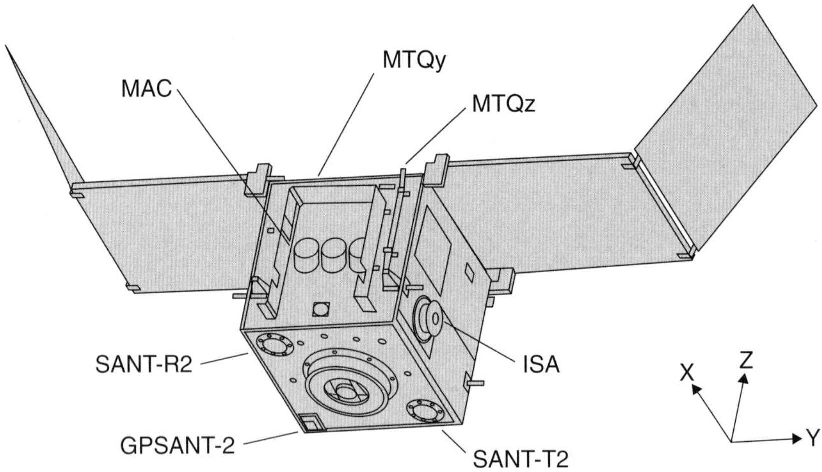

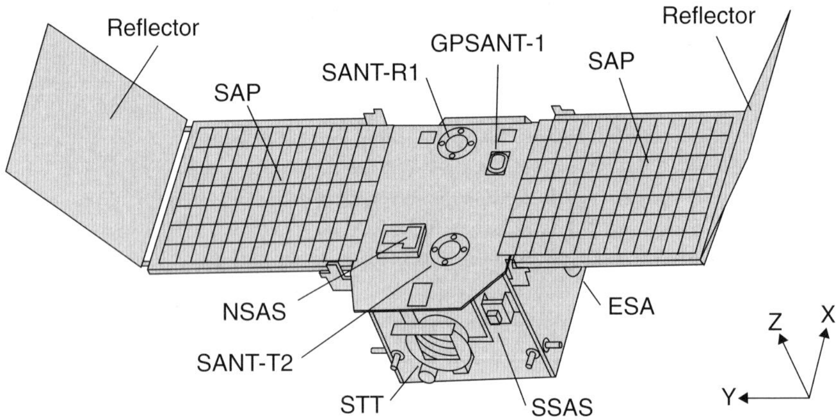

Fig. 15.3. Instrument layout of INDEX (see Fig. 15.4 for acronym key).

15.3 Launch and Tracking Operations

15.3.1 Launch and Initial Attitude Acquisition

INDEX was launched by a Dnepr rocket from the Baikonur launch site in Kazakhstan on 23 August 2005. The main payload was the JAXA satellite OICETS, which carried an intersatellite optical communication experiment. Apogee and perigee altitude for INDEX were 654.866 km and 608.731 km, respectively. The orbit is nearly sun-synchronous with the local time of about 12:00, which is adequate for auroral observations.

Because INDEX has only one reaction wheel and no thrusters, quick acquisition of the sun is impossible. Libration dumping control was the first stage of the

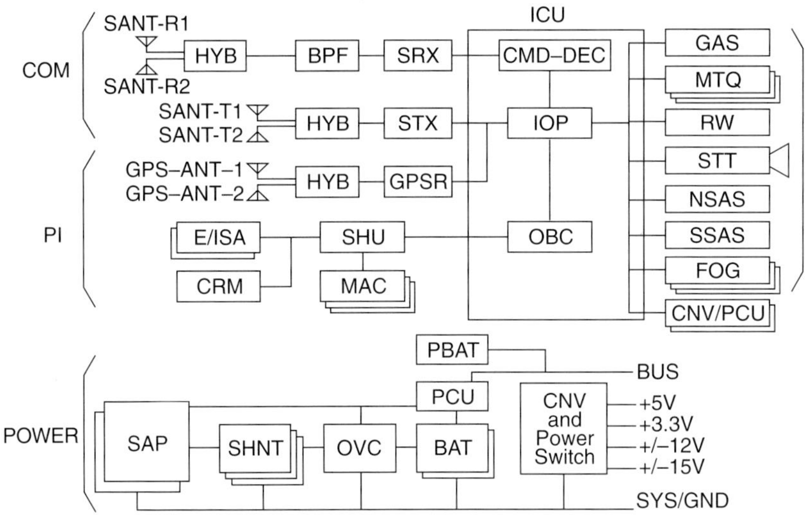

ATT	Attitude control subsystem	NSAS	Non–spin sun aspect sensor
BAT	Secondary battery	OBC	Onboard computer
BPF	Bandpass filter	OVC	Overvoltage control
BUS	Bus power line (hot)	PBAT	Primary battery
CNV	Converter	PCU	Power control unit
COM	Communications subsystem	PI	Payload instruments
COM–DEC	Command decoder	RW	Reaction wheel
CRM	(Particle) current monitor	SANT	S–band antenna
E/ISA	Electron/Ion energy spectrum analyzer	SAP	Solar paddle
FOG	Fiber-optic gyro	SHNT	Shunt circuit
GAS	Geomagnetic aspect sensor	SHU	Science data handling unit
GPSR	GPS receiver	SRX	S–band receiver
HYB	RF hybrid circuit	SSAS	Spin sun aspect sensor
ICU	Integrated control unit	STT	Star tracker
IOP	Input/output port circuit	STX	S–band transmitter
MAC	Multispectral aurora camera	SYS/GND	System ground
MTQ	Magnetic torquer		

Fig. 15.4. Block diagram of INDEX.

initial startup sequence, and was finished at 1.9 hr as scheduled. The actual libration at separation was less than 0.05 rpm (0.3 deg/sec), much less than the Dnepr specification (7 deg/sec). In the second stage, spin-up control was started. The onboard spin-rate estimator, based on the geomagnetic aspect sensor, terminated the spin-up at 2.4 hr from separation. A nominal spin rate of 1 rpm was achieved. Then, the sun acquisition procedure was started, with initial sun angle at 110 deg with respect to the satellite's z axis. Sun acquisition was performed at a slow rate using the magnetic torquers. The sun angle was reduced by the control process, and it reached 60 deg in 5.6 hr from separation. At this point, the attitude control system automatically deployed the solar-array paddle. The output power of the solar-array paddle was balanced with onboard power consumption at a sun angle

Table 15.3. Mass Budget of INDEX (in kilograms)

Payload	11.5	
Multispectral aurora camera		4.6
Electron/ion spectral analyzer		2.8
Science data-handling unit		3.5
GPS receiver		0.2
Others		0.4
Structure	15.3	
Power Subsystem	16.7	
Secondary battery (2)		2.3
Primary battery (6)		4.5
Solar paddle		3.0
Converter		3.2
Power control unit		2.3
Others		1.5
Communications Subsystem	2.4	
S-band receiver		0.5
S-band transmitter		1.0
S-band antennas (4)		0.7
Hybrid/filter		0.2
Integrated Control Unit	4.1	
Attitude Control Subsystem	8.4	
Fiber-optic gyro		1.0
Geomagnetic aspect sensor		0.5
Magnetic torquer (3)		0.9
Non-spin sun aspect sensor		0.9
Spin sun aspect sensor		0.2
Reaction wheel		2.6
Star tracker		2.4
Harness	5.1	
Balance weight	6.9	
Others	2.3	
Total:	**72.8**	

of 40 deg after 6 hr from separation. Monte Carlo simulations predicted that this condition should be achieved in the worst case within 11.1 hr after separation, and the battery capability was designed to survive for this duration. For the initial startup, primary batteries were used to enhance the survivability during the initial sun acquisition. Finally, the satellite acquired stable sun-pointing attitude with a spin rate of 1 rpm. As the sun angle decreased down to 8 deg, only nutation dumping control was applied. If the sun angle exceeds 13 deg, INDEX is designed to switch back to sun acquisition mode again.

INDEX's first pass over a Japanese ground station was 6 hr after launch. The main ground station used for the INDEX launch phase (Uchinoura, with a 20 m diameter antenna) received the telemetry signal on 24 August, which filled us with great joy.

15.3.2 Tracking Operation

Daily operation of INDEX is performed with a dedicated ground station at ISAS/JAXA Sagamihara campus[10] using a 3 m diameter antenna on the roof of the main building. If the onboard S-band transmitter is in the high-power mode, a bit transfer rate of 132 kbps is available with the 3 m antenna.

To download all the scientific and engineering data, the Norwegian Svalbard satellite station and the Japanese Showa Base at the South Pole are also used. The acquired data is made available via FTP (File Transfer Protocol) within a few minutes after INDEX goes over the horizon.

The operation team consists of young researchers, engineers, and students—who are not necessarily familiar with all the details of INDEX functions. Daily satellite operation is performed by one or two people. Most operational software and tools in the ground station were developed in-house and at a small software house to allow some autonomy and flexibility of operation.

The automatic generation of a satellite operation procedure is a unique process that we developed to enhance the robustness of satellite operation. It consists of merging "procedure libraries" of individual subsystems. These libraries, prepared by the people in charge of the individual subsystems, are written in XML (Extensible Markup Language) and are managed with a concurrent version control system. The satellite operation procedure is compiled from these libraries and automatically generated in two formats: a human-friendly TeX-style document, and an uplink command plan. Some of these plans contain programmed command sequences. The generated commands are checked with a prototype model of the integrated control unit before sending them to the real satellite. This procedure is used because it is impossible to test all combinations of software tasks running on the integrated computer before launch.

This satellite operational environment provides a convenient command planning and checking system for distributed users, as shown in Fig. 15.5. Auroral scientists, who are at ISAS/JAXA at Sagamihara as well as in Tohoku University at Sendai, individually choose attitude-control modes and decide the reference values. They describe their desired operational sequences in procedure library

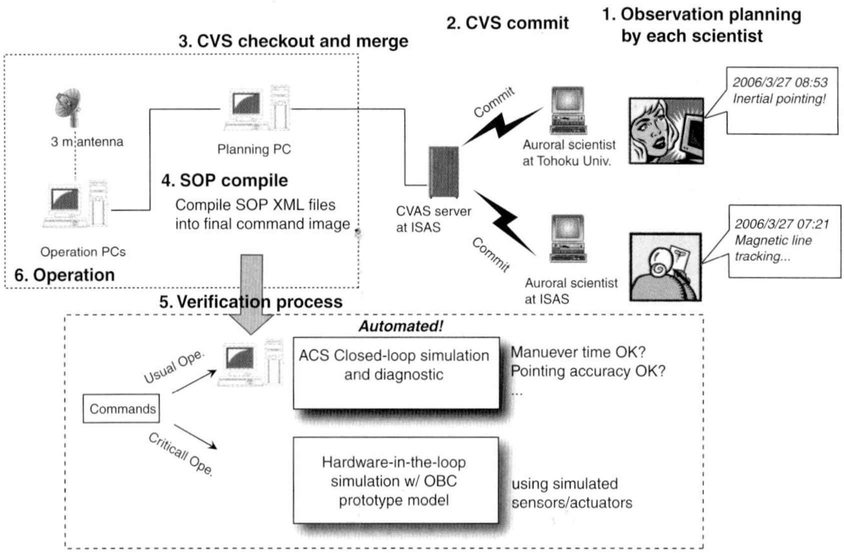

Fig. 15.5. Operation flow of auroral observation and attitude control.

files with the XML format. Then, they upload their library files to the version control server via the Internet. The merged top-level procedure is very carefully checked. A semiautomatic attitude-simulation tool is applied as a diagnostic tool. This reads the operational procedure and automatically generates simulation configuration files. If something suspicious is found, the operator is advised how to modify the procedure. When the problem seems to be serious, the tool warns the operator not to use the procedure and to contact the attitude-control engineers.

15.4 Satellite Performance and On-Orbit Results

15.4.1 Integrated Control Unit

Almost all tasks for satellite control are executed on the integrated control unit, which is based on the Hitachi SH-3 (SH7708, 60 MIPS) RISC processor. Since the SH-3 is not radiation-hardened, three processors with DRAMs and EEPROMs compose a triple voting system to ensure fault tolerance. If a voting error or DRAM 2-bit errors have occurred due to a single-event upset, the cell in error is automatically reconfigured by copying DRAMs and register data from a different cell. It takes about two seconds to execute the reconfiguration process. Proton/heavy-ion irradiation tests were conducted along with a radiation-effect analyzing tool (CREME96). From these, it was estimated that a proton-induced upset of the SH-3 processor (i.e., the cache memory and the registers) in the South Atlantic Anomaly would be observed every 5 days; a cosmic-ray-induced upset over the polar regions would be observed every 25 days.

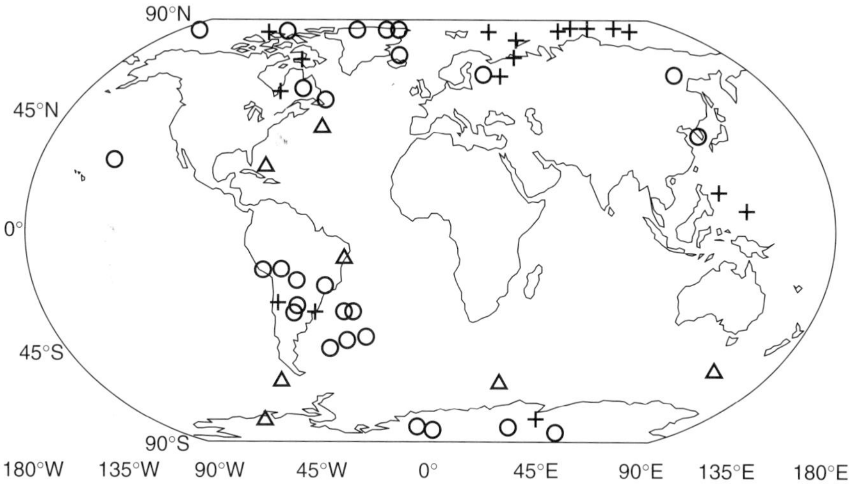

Fig. 15.6. Error map of integrated control unit onboard processor for two years from launch. Pluses denote cell reconfiguration (total 19 events) due to voter errors (3 events), synchronous errors (2 events), and DRAM 2-bit errors (14 events). Circles (28 events) and triangles (7 events) denote DRAM 1-bit errors detected by direct access and memory patrol at 30 min intervals, respectively. DRAM 1-bit errors are corrected by error correcting code.

The actual number of cell reconfigurations measured in orbit was 19 in two years. This number is much smaller than expected. Furthermore, 14 of the 19 cell reconfigurations were caused by 2-bit errors in peripheral DRAMs. The reconfigurations originating from the CPU itself, which are voting and synchronization errors, were observed only five times. Figure 15.6 is a ground map showing where the errors were detected. The pluses (+) denote the ground positions where the reconfiguration process was activated. Most of the errors certainly occurred over the polar regions and in the South Atlantic Anomaly. The DRAM 1-bit errors, which are corrected by the error correcting code, are also plotted in Fig. 15.6. Since the 1-bit errors are detected not only by direct access (marked by circles, 28 events) of the user's program but also by memory patrol (marked by triangles, 7 events) of the error correcting code every 30 min, there is a little ambiguity in the ground position.

Software modules corresponding to each task are executed in a cyclic manner at 1, 8, and 32 Hz using external interruption signal cues. The modules are attitude control and determination, command reception and distribution, telemetry formatting and downlink control, collection of housekeeping data, and mission modules (i.e., scientific observation and GPS). The whole onboard software is designed and implemented in-house, while the ground station software is coded by a small company as described in Sec. 15.3.2. The project members interacted closely to perform the system software test with very long running time.

Table 15.4. Requirements and Achievements of INDEX Attitude Control System

	Requirement	Achievement
Absolute pointing error	0.5 deg	0.05–0.1 deg
Determination error	0.06 deg	0.05 deg
Control error		0.05 deg
Attitude stability	0.03 deg/sec	0.004 deg/sec

During the software development effort, it was discovered that the computation capability of the integrated control unit, especially in the case of three-axis attitude control, would not be sufficient, even though relatively high-speed processors were used. The problem was caused by software bottlenecks from the voting system and the fact that the SH-3 has no floating-point number processing unit function.[10]

15.4.2 Attitude Control

The attitude of INDEX is controlled so that the field of view of the multispectral aurora camera covers the INDEX footprint, which is the projection of the satellite at an altitude of 110 km along a magnetic field line. The particle analyzers measure the energy spectra at the satellite, and the aurora camera captures images at the foot of the magnetic field line. Such observation requires three-axis stabilization and inertial pointing capability. Table 15.4 shows the requirements for the mission as well as the achievements of the attitude control system.[6–9, 19]

Table 15.5 shows the specifications of the attitude control sensors and actuators. The sensors are a spin-type sun sensor, a two-dimensional sun sensor (accuracy 0.05 deg), a star tracker (accuracy 1 arcmin in the $-x$ direction), a three-axis geomagnetic-field aspect sensor, and a commercial three-axis fiber-optic gyroscope. The field of view of the aurora camera is in the x direction. The attitude-determination algorithm is based upon an extended Kalman filter, which estimates the bias rates of the gyroscope and the attitude of INDEX.

The attitude-control actuators are a small reaction wheel (0.5 N-m-s) and a three-axis magnetic torquer (6 A-m^2). The reaction wheel provides agility and rotation control around the z axis, and the magnetic torquers generate attitude control torque around the x and y axes.

15.4.2.1 Temperature Control of a Commercial Gyro

The attitude of INDEX can be determined by the star tracker with an accuracy of 1 arcmin, as well as with an extended Kalman filter and fiber-optic gyroscope. The Kalman filter is used when the sun and Earth are within the star tracker's field of view. Since INDEX is pointed toward a specific direction in inertial coordinates, the star tracker is available for only 30 min during an orbital period of about 100 min. For the remaining 70 min, the attitude of the satellite is propagated

Table 15.5. Specifications of INDEX Attitude Components

Reaction wheel	
Mass	2.6 kg
Angular momentum (max.)	1.0 N-m-s
Output torque	>0.01 N-m
Magnetic torquer (3)	
Mass	0.3 kg at 1 axis
Linear dipole	>6.0 Am^2 at 1 axis
Star tracker	
Mass	2.4 kg
Field of view	30 × 40 deg
Random noise	<3 arcmin (3σ)
Bias error	<1 arcmin (3σ)
Sampling rate	1 Hz
Fiber-optic gyro (3)	
Mass	0.14 kg at 1 axis
Pulse weight	4.5 arcsec/pulse
Bias stability	<3 deg/hr (1σ) 0.05 deg/hr (1σ) with temperature control
Non-spin sun aspect sensor	
Mass	0.9 kg
Field of view	50 × 50 deg
Accuracy	0.05 deg
Random noise	<0.01 deg
Spin sun aspect sensor	
Mass	0.1 kg
Field of view	120 deg
Accuracy	<0.3 deg
Geomagnetic aspect sensor	
Mass	0.5 kg
Range	±1000 mG

with angular rate data from the gyroscope. The extended Kalman filter estimates the gyroscope bias rate and the attitude itself. The requirement for the gyroscope bias stability is less than 0.1 deg/hr.

The gyros used in INDEX are the JG34-FP14 manufactured by Japan Aviation Electronics Industry, Ltd. It is a commercial product consisting of a closed-loop interferometric fiber-optic gyroscope with a mass of 140 g for one axis. Bias stability is 3 deg/hr, which does not meet the mission requirement. The dominant factors that decrease bias stability are day-to-day instability and temperature dependence. The gyroscope, as mounted on INDEX, is expected to experience a temperature variation of several degrees during solar illumination (65 min) and eclipse (35 min). Unfortunately, this cycle time coincides with the time duration for propagating the gyroscope rate output. Therefore, we applied a temperature-control system using heaters, temperature sensors, and multilayer insulation to stabilize the gyroscope bias rate during propagation.[19] Temperature-control software is implemented on the integrated control unit's onboard computer. This software determines the heater's on/off operation, based on the measured temperatures with a sampling period of about 2.7 sec. The achieved temperature stability is about 0.2ºC.

The gyros are connected to the baseplate through thermal insulator spacers, and their temperature is about 20ºC higher than the baseplate temperature due to self power dissipation (2.2 W per one axis). To simulate the space thermal environment, we gave to the baseplate a periodic temperature variation of 25°C ±2.5°C with a 100 min period. Then, the gyroscope bias rates were measured for both cases—with and without temperature control (Fig. 15.7a and 15.7b). Without temperature control, the bias rate varied by about 0.7 deg/hr with a period of 100 min. With temperature control, the bias rate is almost stabilized, with a random noise of about 0.2 deg/hr.

The Allan variance is a convenient method to evaluate drift errors of gyros as a function of time.[23] The Allan variance $\sigma_\Omega(\tau)$ is defined as a difference of a drift rate averaged over time τ in a root-mean-square (rms) value. A variance σ_θ of the attitude error propagated only with gyro sensors during loss of star tracker output is given as

$$\sigma_\theta^2 = (\sigma_\Omega \tau)^2 + (\sigma_\omega \tau)^2 \qquad (15.1)$$

where τ is the propagating time during star tracker's blinding and σ_ω is the bias-rate estimation error of the Kalman filter.[24]

To evaluate the effect of temperature control of the commercial fiber-optic gyroscope on the ground, we measured the Allan variance in vacuum for 40 hours. The variance over time is shown in Fig. 15.8. Without temperature control, the Allan variance is roughly proportional to $\tau^{-1/2}$ in the region of $\tau < 1000$ sec, where gyroscope-angle random walk is dominant.[24] However, in the region of $\tau > 2000$ sec, the variance degraded as a result of the day-to-night temperature change (gyroscope temperature varied from 47°C to 54°C within a period of 24 hours). If there were also a periodic temperature variation within the period of an

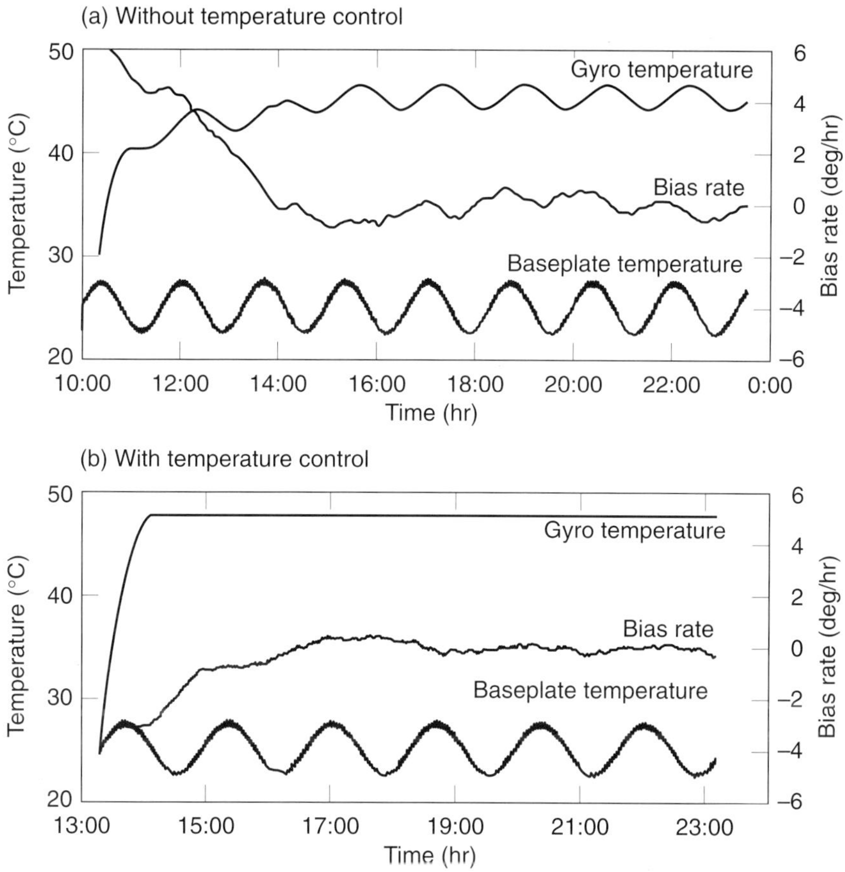

Fig. 15.7. Temporal variation of gyro bias rate when temperature of baseplate varies periodically: without temperature control (a), and with temperature control (b).

orbit (~6000 sec), the Allan variance could degrade to several tenths of a degree per hour at several thousand seconds, as shown in Fig. 15.7a. With temperature control (using a sampling period of 8 sec and a reference temperature of 58°C), the Allan variance decreases further to 0.05 deg/hr around τ = 4000 sec (67 min), which is coincident with the time scale of interest. On the other hand, the Allan variance increased around τ of several tens of seconds, which is caused by the heater control. However, in this short time scale, the increase in the Allan variance due to heater control deteriorates the propagation error only by 10 arcsec, as can be derived from Eq. (15.1).

In summary, temperature control improves the fiber-optic gyroscope bias-rate stability over long durations, while it deteriorates the angle random walk during the short heater-control periods. For INDEX, temperature control was adopted

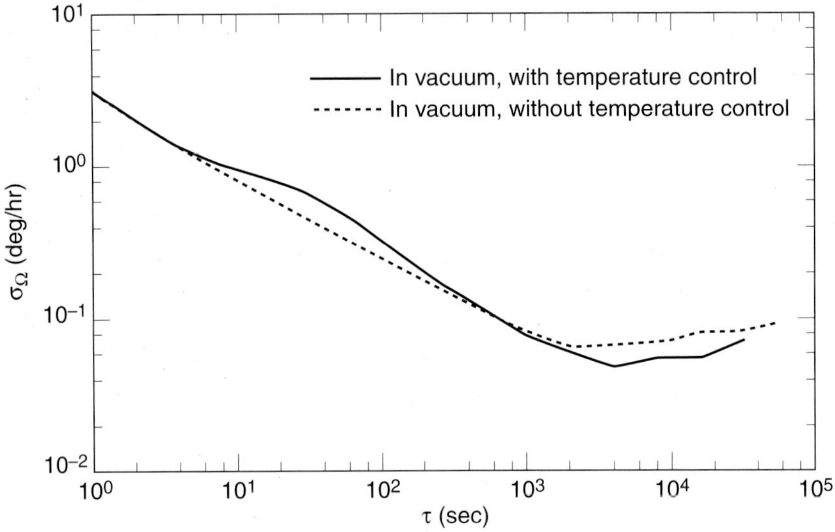

Fig. 15.8. Allan variance of gyro bias rate in vacuum with and without temperature control.

because improvement of the attitude propagation performance is important during the 70 min loss of star tracker output.

It is possible to estimate the total attitude propagation error on orbit for 70 min during blinding of the star tracker by the sun or Earth. If the Allan variance around 70 min is $\sigma_\Omega = 0.05$ deg/hr (from Fig. 15.8) and the bias-rate estimation error of the Kalman filter is $\sigma_\omega = 0.04$ deg/hr, then, the total attitude propagation error on orbit for 70 min is calculated from Eq. (15.1) as $\sigma_\theta = 0.075$ deg = 269 arcsec.

15.4.2.2 Attitude Determination Algorithm and On-Orbit Results

Attitude determination in INDEX uses standard extended Kalman filter methods, as shown in Fig. 15.9. It estimates the attitude error and the attitude-rate bias error (gyroscope-rate bias error) as state vector elements for the Kalman filter. Estimation is performed through two processes: The time-update process for attitude propagation is based on the three fiber-optic gyroscopes, and the observation-update process for correction of propagating attitude is based on the star tracker data. The cycle frequencies are 8 Hz for attitude propagation and 1 Hz for correction, determined by the performance of the onboard computer and data-output cycle of the star tracker.

Figure 15.10 shows attitude determination errors measured by the angle difference between the star vectors observed by the star tracker and those predicted from the Kalman filter. INDEX performed rotation maneuvers by 160–170 deg with the reaction wheel around the *z* axis at 12:00, 18:00, and 20:30 on 5 Mar 2006 to avoid blinding of the star tracker caused by the sun or Earth being near its

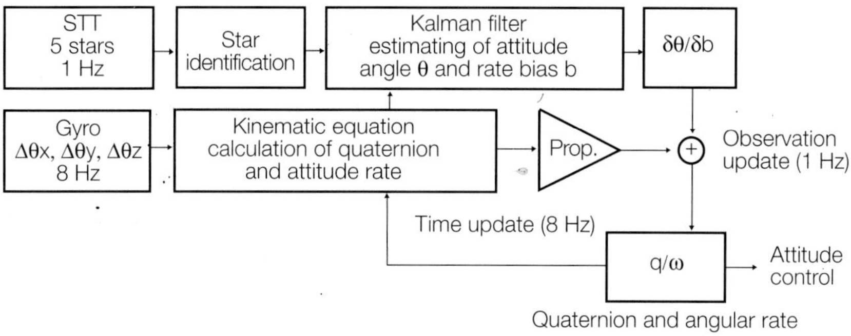

Fig. 15.9. Block diagram of attitude determination.

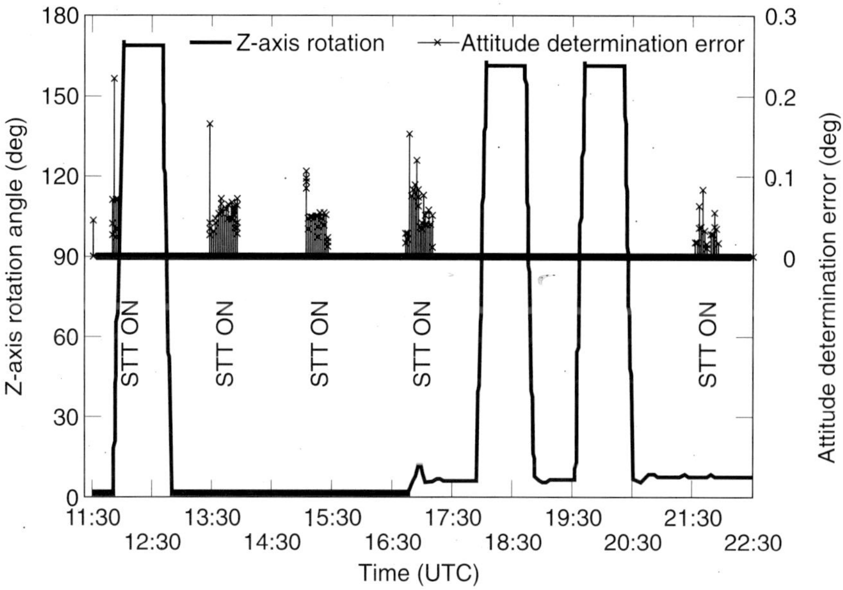

Fig. 15.10. On-orbit data of attitude determination error. There are three large rotation maneuvers and five periods with star tracker on.

field of view. The rotation maneuvers took several minutes. Figure 15.10 shows that the attitude determination error angle is as great as 0.05 deg in most cases. We can see that the first data of the attitude error angle in each period of star tracker operation has a spike of 0.05–0.2 deg, which is the attitude propagation error. These are mainly caused by the rate bias of the fiber-optic gyroscopes.

Figure 15.11 is the histogram of the attitude propagation errors during the 65 periods of star tracker blindings between 25 February and 1 March 2006. These on-orbit results are in relatively good agreement with the predicted value $\sigma_\theta = 0.075$ deg = 269 arcsec based on the ground measurement as shown in Sec. 15.4.2.1.

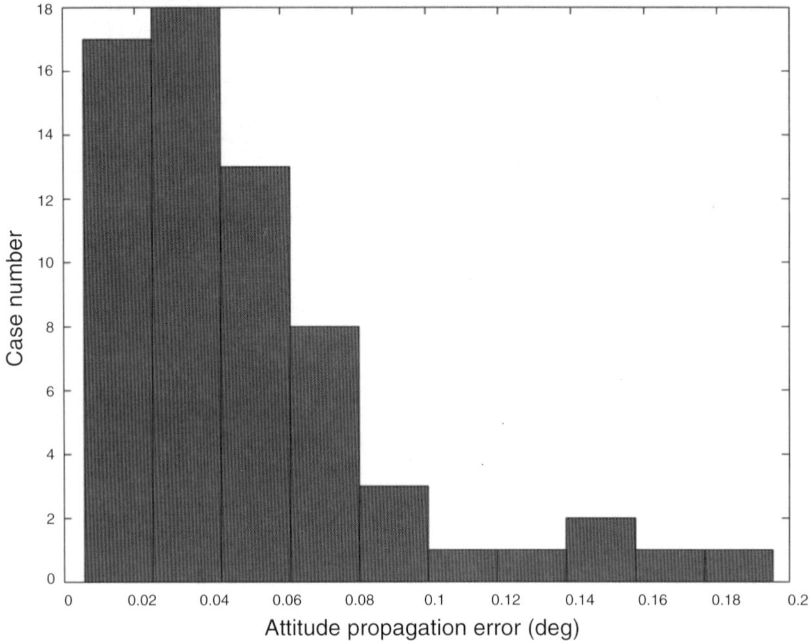

Fig. 15.11. Histogram of attitude propagation errors during star tracker blinding of 70 min. During this period, gyro rate output is integrated to determine attitude.

15.4.2.3 Magnetic Torque Control Law

Torques around the *z* axis are generated by the reaction wheel. Torques around the *x* and *y* axes are generated by the magnetic torquers. The problem with magnetic torque control is to find the magnetic moment **M** that generates the required torque on two axes in the geomagnetic field **B**. Since **M** generates a magnetic torque $\mathbf{T} = \mathbf{M} \times \mathbf{B}$, an equation of the reference torque $\mathbf{T}^*$ and **M** can be described as

$$\mathbf{T}^* = \mathbf{B}_{m2}\,\mathbf{M} \tag{15.2}$$

where $\mathbf{T}^* = [T_x^*, T_y^*]^t$ and

$$\mathbf{B}_{m2} = \begin{bmatrix} 0 & B_Z & -B_y \\ -B_Z & 0 & B_X \end{bmatrix} \tag{15.3}$$

Usually, $\mathbf{B}_{m2}$ is rank = 2 matrix; therefore, the magnetic moment **M** for $\mathbf{T}^*$ can be calculated using a pseudoinverse matrix

$$\mathbf{M} = \mathbf{B}_{m2}{}^{\#}\,\mathbf{T}^* \tag{15.4}$$

where $\mathbf{B}_{m2}{}^{\#}$ is the pseudoinverse matrix of $\mathbf{B}_{m2}$.

However, a problem with this calculation occurs when $B_z = 0$, since it is a singular point of Eq. (15.2). M_x and M_y have no contribution on magnetic torque—i.e., Eq. (15.4) becomes overdetermined with two equations for one variable. The magnetic torquers are saturated with this solution and generate a large disturbance torque on the *z* axis.

To improve this behavior, a novel algorithm was proposed in the INDEX project.[6–9] It is based on the singularity robust inverse (SR-inverse) method, which was first proposed in the field of robotics and is also used in control-moment gyro torque steering laws. The SR-inverse matrix $\mathbf{B}_{m2}^{+}$ of $\mathbf{B}_{m2}$ is

$$\mathbf{B}_{m2}^{+} = \mathbf{B}_{m2}^{t}(\mathbf{B}_{m2}\,\mathbf{B}_{m2}^{t} + k\mathbf{I})^{-1} \tag{15.5}$$

where k is a parameter called a scale factor and **I** is the unity matrix. The solution of the SR-inverse was determined by evaluating the exactness and the feasibility simultaneously, and k is a weighting factor to tune this balance. The SR-inverse method provides the same **M** as the pseudoinverse method when k = 0.

In INDEX, k is dynamically adjusted according to the magnitude $|B_z|$ of B_z to limit the magnetic disturbance torque T_z near the singularity at $B_z = 0$. The on-orbit performance of INDEX reveals[8] that the SR-inverse method is very effective in reducing the disturbance torque near the singularity at $B_z = 0$.

15.4.2.4 Residual Magnetic Moment Observer

Generally, the dominant attitude disturbance source for small satellites in low Earth orbit is the residual spacecraft magnetic moment. Other disturbance sources—such as gravity-gradient, solar-radiation pressure, or air drag—have relatively small effect. An orthodox approach to suppress attitude disturbances is to enhance the feedback controller. This approach, however, depends on several conditions such as sensor noise, computational performance, and the accuracy of the dynamics model.

Another approach is to apply feed-forward cancellation of the residual magnetic moment. In this case, performance depends on the accuracy of residual magnetic moment estimation. To estimate it with on-orbit attitude data, a residual magnetic moment observer was applied to INDEX. The estimated disturbance torque, which is the output of disturbance observer, was calculated on the ground. The period of the disturbance torque is found to be about half of the orbital period. With estimated disturbance torques and measured geomagnetic field **B**, the residual magnetic moment $\mathbf{M}_r$ is estimated, using the least-square method, to be $\mathbf{M}_r = [0.51, 0.042, 0.11]$ A-m^2.

15.4.2.5 Performance on Orbit

The estimated residual magnetic moment $\mathbf{M}_r$ is uploaded to the spacecraft from a ground station. This value is used in INDEX as a feed-forward value to cancel the residual magnetic moment. It means that this estimated residual moment $\mathbf{M}_r$ is subtracted from the magnetic moment **M**, which is the output from the magnetic torque control law. Figure 15.12 shows the effect of feed-forward cancellation. The upper, middle, and lower graphs are the attitude-control angle errors of the *x*, *y*, and *z* axes, respectively. Each graph consists of four time periods. In the first, labeled "w/o FF," attitude was controlled by only the feedback controller, without any feed-forward cancellation. In the second, labeled "w/ 50% FF," feed-forward cancellation was applied, but with only 50% of the estimated value. In the third, the value for feed-forward cancellation was increased to 75% of $\mathbf{M}_r$. Finally, in the fourth period, 100% feed-forward cancellation was applied. These gradual steps clearly show the effect of $\mathbf{M}_r$ feed-forward cancellation. It indicated the validity of the proposed $\mathbf{M}_r$ estimation and cancellation method. The on-orbit attitude control error is about ±0.2 deg with no feed-forward cancellation. With full $\mathbf{M}_r$ feed-forward cancellation, the pointing error decreased to 25%, to less than ±0.05 deg.

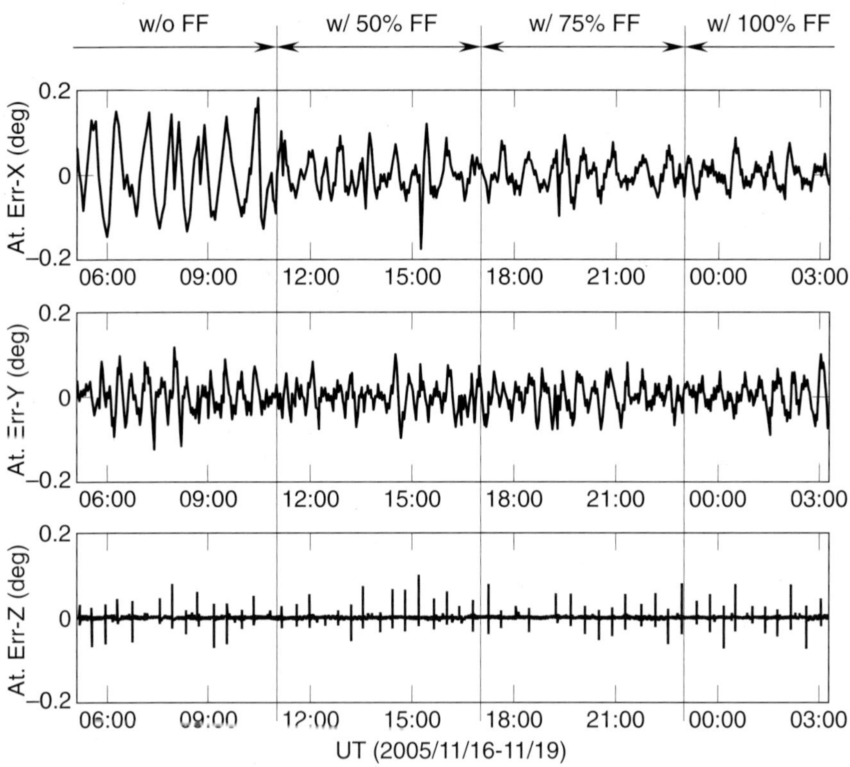

Fig. 15.12. Attitude control errors of *x*, *y*, and *z* axes as functions of time. Feed-forward cancellation of $\mathbf{M}_r$ (residual magnetic moment) is added with various intensity to conventional feedback control. The notation "w/ 50% FF" means that 50% of output from feed-forward control law is added to feedback controller. Feed-forward cancellation with 100% intensity is applied in normal aurora observations and achieves control accuracy of 0.05 deg (1σ).

Just before launch, the residual magnetic moment for INDEX was measured in a magnetically shielded chamber and determined to be $\mathbf{M}_r$ = [–0.054, –0.039, –0.069] A-m^2. The attitude error was as large as ±0.2 deg when this measured value was applied for feed-forward cancellation. The estimated value on orbit was much more effective for feed-forward control than the measured one. It suggests that the residual magnetic moment was changed after the measurement. The reason, however, is still a mystery.

Figure 15.13 shows a histogram of attitude control errors during one month, July 2006, when INDEX carried out about 90 auroral imaging observations. It indicates that the worst-of-the-worst control error is about 0.1 deg. The attitude determination error has been evaluated to be less than 0.05 deg (Fig. 15.10); therefore, pointing accuracy was estimated to be 0.1 deg. Total pointing accuracy was also confirmed by data from the aurora imager, and the end-to-end absolute pointing accuracy was revealed to be between 0.05 and 0.1 deg. These evaluations clearly indicated that the achieved performance met the mission requirements, as listed in Table 15.4.

15.4.3 Solar-Concentrator Paddle

15.4.3.1 Merit of Solar-Concentrator Paddle

The performance of the solar-concentrator paddles on INDEX was evaluated in terms of specific power (W/kg), which is defined as ratio of the generated power to the solar-paddle mass. The two paddles weigh 2.6 kg and generate about 120 W of power, providing specific power of 46 W/kg. The structure of the paddles is shown in Fig. 15.14. There are two wings that have thin-film reflectors. These

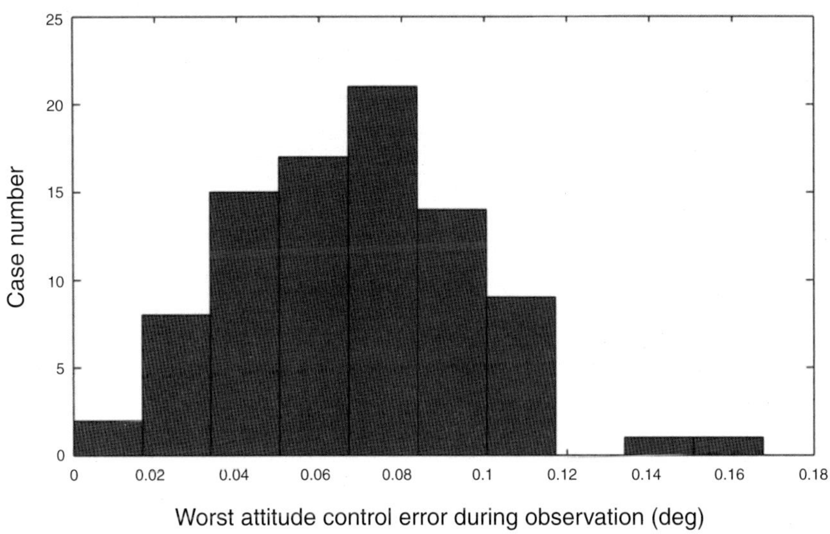

Fig. 15.13. Worst attitude control errors during auroral observations.

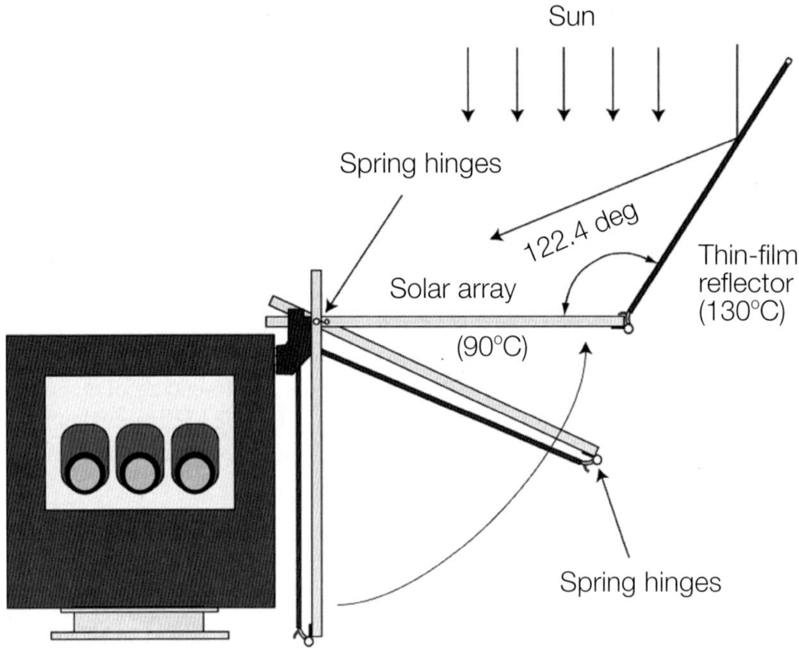

Fig. 15.14. Structure of solar paddles and reflectors.

have a total of mass of 0.32 kg and enhance solar intensity by a factor of 1.25, which translates to a power increase of about 30 W. The specific power of the solar-concentrator paddles with the reflectors is 51 W/kg. The specific power of the reflectors, which is defined as the ratio of the power increment and the reflector mass, is 94 W/kg.

In INDEX, a single reflector is attached to the end of a solar paddle to enable use of a single-stage deployment mechanism. If a double-stage deployment were used, two reflectors could be attached at both side facets of the paddle, doubling the effect of solar concentration.

15.4.3.2 Avoidance of Power Degradation

The Boeing BSS-702 satellite bus launched around 2000–2001 had been equipped with double-sided solar-concentrator paddles with thin-film reflectors.[25] However, they suffered from power loss[26,27] that exceeded the prediction, which was for a nearly linear drop in power as a function of time. It has been reported[26] that the root cause of the power loss was contamination coupled with UV/radiation exposure and shrinkage. The maximum temperature of the solar cell is around 120°C, while the concentrator film is around 95°C. This temperature difference may induce contamination: Outgasing from the cell panel may condense on the reflector surface, which is at a lower temperature. It has been confirmed[26] that UV/radiation exposure causes loss of surface reflectance of the contaminated surface and also causes the concentrators to shrink. For the BSS-702 reflector geometry, the shrinkage of the film caused the deployed angle of the reflector to increase, thereby reducing the solar illumination on the cells.

To avoid the power degradation observed in the BSS-702 satellite, special care was devoted to the design and manufacturing of the INDEX solar concentrator. One modification was to ensure that the deployment angle would be independent of the shape of the reflector film, as described below. The other modification was that the thermal design of the thin-film reflector was improved to prevent possible contamination.

For INDEX, the maximum temperature of the cell panel was estimated to be 100°C. The solar cells are triple-junction InGaP/GaAs/Ge cells with 27% efficiency, which is very effective in reducing the cell temperature even for solar concentration ratios of 1.3. To avoid possible contamination, the reflector temperature should be higher than the cell temperature. Thin-film polyimide (25 μm) aluminized on both sides (UTC-025S-AANN manufactured by Ube Industries, Ltd.) was selected as a reflector. The aluminized backing has low infrared emission, which is effective in keeping the film temperature high. The solar absorptance and infrared emissivity were measured to be about 0.08–0.09 and 0.01–0.02, respectively. This emissivity is found to be rather lower than the one on BSS-702 (~0.04).[25] Consequently, the temperature of the reflector was estimated to be 130°C in sunshine, given an incident angle of 60 deg, which corresponds to a sun angle of 0 deg with respect to the z axis. During eclipse, the temperature was estimated to be –70°C. Flight data show a cell temperature of 90°C in sunshine with 1.3× solar concentration and –60°C during eclipse. The assessment of possible power degradation was performed by Office National d'Etudes et Recherches Aerospatiales (ONERA/DESP) in France. Their result indicates that, due to the higher temperature of the reflector and the relatively low temperature of the cell, the predicted degradation due to contamination is very small for the INDEX satellite—less than 1% per year.

15.4.3.3 Structure of Reflector and Prediction of Solar-Concentration Effect

The mechanical configuration of the thin-film reflector was carefully designed to minimize wrinkling and maximize reflection. The film is wrapped around a rectangular fiber-reinforced plastic frame and tensioned in the longitudinal direction by five springs on the reverse side.

The reflector is connected to the end of the solar array with hinge springs. In the stowed configuration, it is folded between the solar array and the satellite panel, as shown in Fig. 15.14. In the deployed configuration, it is unfolded at an angle of 57.5 deg from the solar array. This deployment angle is determined only by the hinge mechanism and is independent of the shape of the film, which is different from the BSS-702. As a figure of merit, the paddle concentration ratio is defined as the ratio between generating power with and without the reflector. If the thin-film reflector were a perfect mirror, the paddle concentration ratio would be 1.426, which is geometrically determined.

Table 15.6. Paddle Concentration Ratios between Generating Power with/without Reflectors

	Reflector (+Y)	Reflector (-Y)	Average
Configuration A	1.26	1.25	1.24
Configuration B	1.22	1.23	

Surface shape measurements of the thin-film reflectors by laser ranging during ground tests were performed for the vertical configuration of the reflector surface.[13] The tip of the reflector is down in configuration A and up in configuration B, respectively. The maximum out-of-plane displacement was less than 2.5 mm. On orbit, the exact shapes of the thin-film reflectors might change slightly due to launch vibration, zero gravity, and thermal deformation; however, the wrinkling mainly depends on tension applied by the springs.

Ray racing calculations of the reflected solar light on the thin-film surface were performed based on the surface shape measurements. First, concentration ratios of all cells were individually calculated based on the illumination intensity distributions and the real configuration of solar cells on the paddle. Second, concentration ratios of individual cell series were calculated under the assumption that a cell with the lowest concentration ratio governs the electric power of the series. Finally, paddle concentration ratios were calculated. The result was an average concentration ratio of 1.24 with an assumed solar reflectance of the thin-film reflector of 0.9 (see Table 15.6).

15.4.3.4 Performance on Orbit

The performance of the solar concentrator was evaluated on orbit. At a constant bus voltage, the solar current at the moment of zero shunt current should be proportional to the solar intensity on the solar-array paddles. Figure 15.15 shows the solar current at the moment of zero shunt current as a function of the sun angle. If there were no solar concentrators at the solar paddle, the solar current would have a cosine dependence on the sun angle, as shown by the broken line. The solid line shows the predicted value for the real case with the solar concentrator. The solar-concentration effect is enhanced in the sun-angle range between –15 and 15 deg. The solid circles indicate the measured solar current on orbit for a sun angle of 0 deg as well as 25 deg. The solar concentration vanishes at sun angles greater than 25 deg due to the reflection geometry. The predicted solar concentration at a sun angle of 0 deg was 1.25. The measured values (solid circles) are in good agreement with the predicted value.

To verify our improvement in concentrator design, we monitored the trends in power generation for INDEX at a sun angle of 0 deg since launch. Figure 15.16 shows two years of measured solar current at a sun angle of 0 deg. The values are normalized to the initial values just after launch. Solid circles are shunt currents for a safe-hold mode, which depend linearly on the solar current. The open circles

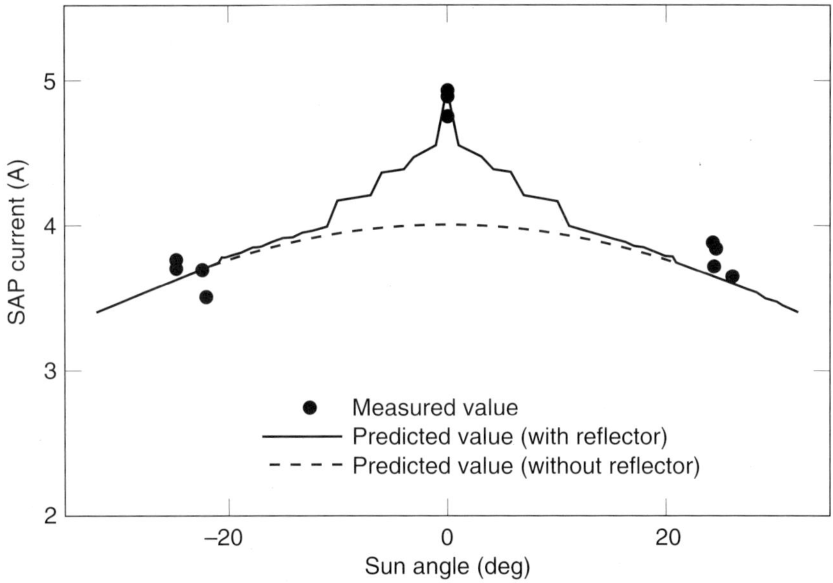

Fig. 15.15. Solar current as function of solar angle. The solid and broken lines show predicted values with and without solar concentrator, respectively. Solid circles are measured values on orbit.

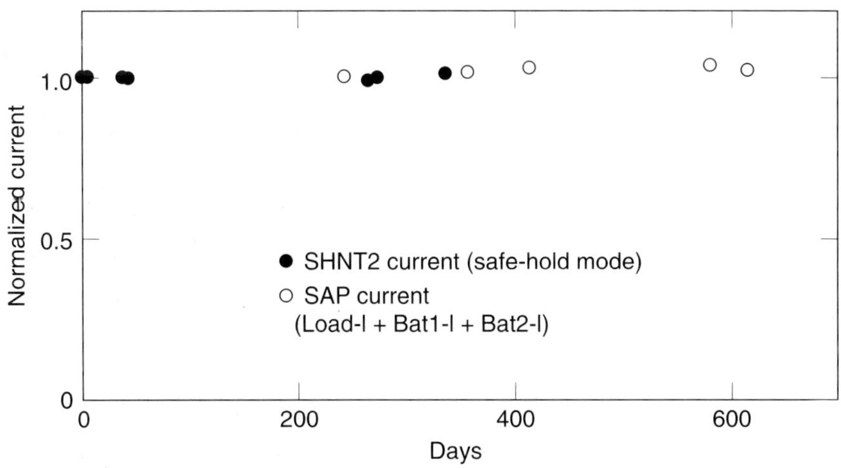

Fig. 15.16. Two years' on-orbit trend of solar current with solar-concentrator paddles. Sun angle is 0 deg.

are the sum of the bus load current and the battery charge currents, which is equal to the solar current. It is observed that there is no degradation of the solar current, or solar power. At present, improvements to counter contamination-induced degradation appear to have been effective.

15.4.4 Battery

A lithium-ion secondary battery based on "pouch" cells (NEC/TOKIN, IML 8204/150) was used on INDEX. The rated capacity of the cells is 3.0 Ahr. Pouch cells are potted with resin and reinforced by the aluminum housing to enhance tolerance to the vacuum of space. Seven cells were connected in series to create a 28 V bus battery. The energy density of the battery is about 70 Whr/kg. The battery has been used satisfactorily for 16,500 cycles on orbit as of November 2008.

Seasonal aurora observations around the South Pole each winter require a discharge current of more than 2 A at the 3 Ahr rated capacity at the end of the eclipse period. It was feared that this operation might trigger a low-voltage condition, in which case satellite operation would shift to a safe-hold mode. To predict the precise performance of the batteries on orbit, ground tests have been conducted using batteries with the same specification as the batteries in space.

When auroral observations are performed on orbit around the South Pole, a 0.78 A discharge current is first applied for 25 min, followed by a 0.96 A discharge for 5 min, and finally a 1.63 A discharge for 3 min. Ground tests were conducted, ahead of on orbit operations, to predict the operability of INDEX on orbit. The ground tests used a 1 A constant discharge current for 35 min at 25°C.

Orbital performance of the batteries were calibrated based on the discharge rate, which differs in the dc impedance of the battery between the ground test and the flight operation. Figure 15.17 shows the trend of end-of-discharge voltage for the ground test and for the two years' on-orbit data after calibration. A good agreement between the ground test and the on-orbit data is obtained, indicating satisfactory operation of the satellite and its scientific observations.

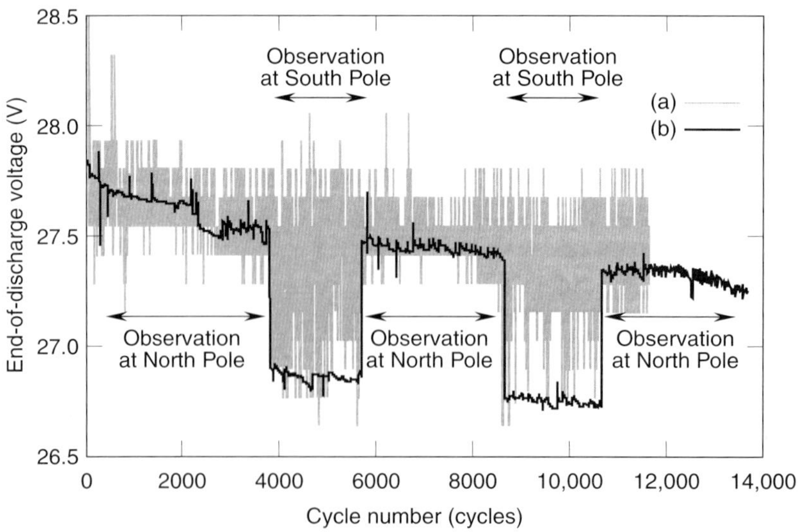

Fig. 15.17. Two years' trend of battery end-of-discharge voltage for on-orbit data (a) and ground simulation results (b).

15.4.5 Miniature GPS Receiver

A miniature GPS receiver was developed for INDEX based on the automobile-navigation receiver (Japan Radio Corporation, CCA-370 HJ).[15–18] It is an eight-channel receiver with frequency (L1) with code ranging. The frequency sweep range was expanded to ±57 kHz to cover the large Doppler shift (±45 kHz) on orbit. Performance for low Earth satellite applications was tested by means of a GPS simulator, which predicted a range error of 0.9 m rms. The receiver can tolerate 20 krad of radiation and resists latch-up for 200 MeV protons.

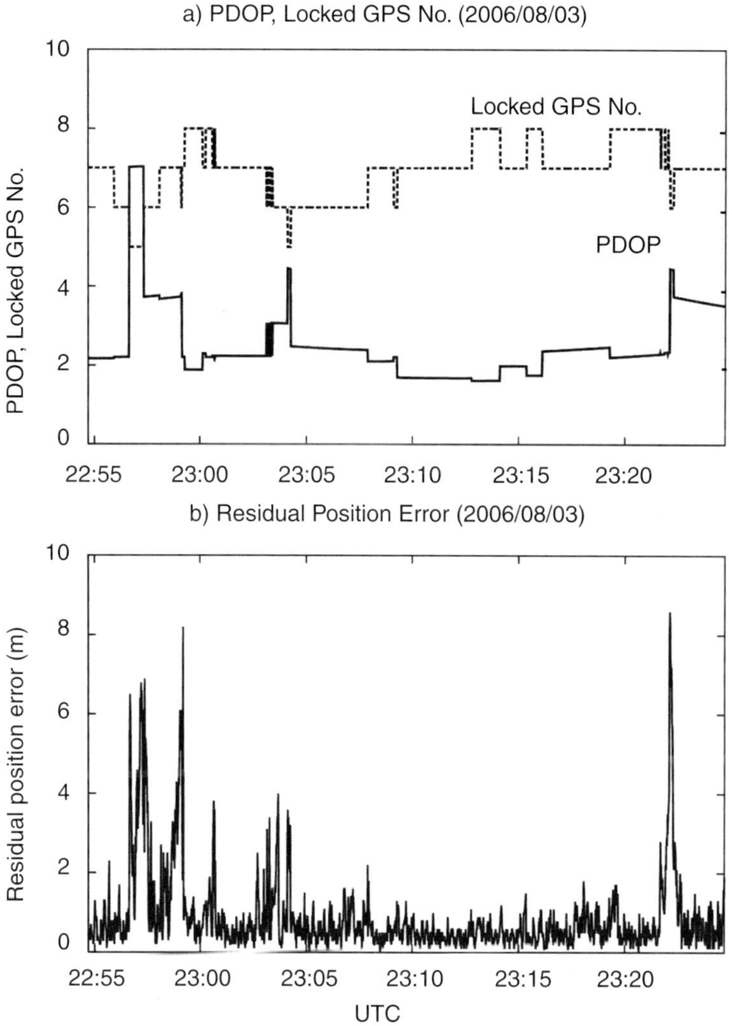

Fig. 15.18. On-orbit data of GPS receiver: position dilution-of-precision (PDOP) and number of locked GPS satellites (a), and residual position error (random position error) (b). Interval of orbit determination is 60 sec.

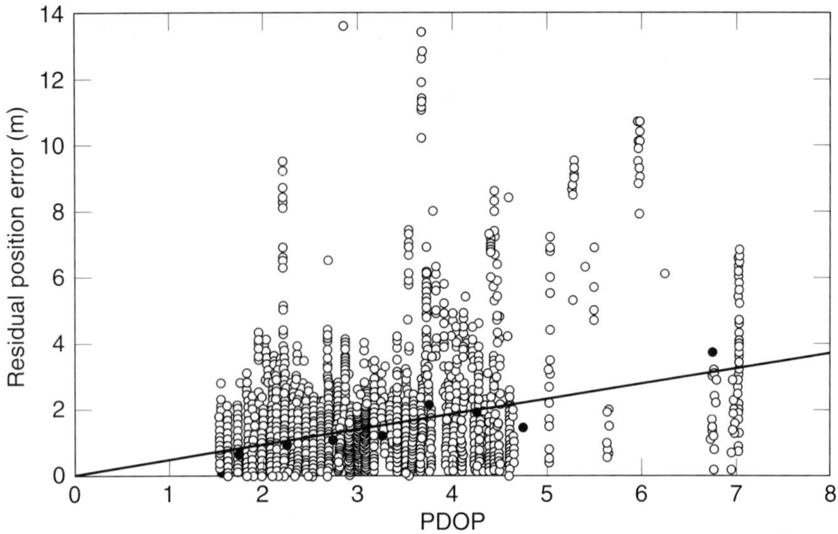

Fig. 15.19. Relation between residual position errors and position dilution-of-precision (PDOP) values. Solid circles are rms values for each bin of PDOP value. Straight fit is least-square-error line of solid circles.

The modified GPS receiver in INDEX works perfectly on orbit. Cold-start positioning has been confirmed repeatedly to finish within 28 minutes of start. Orbit determinations were performed with GPS position data for 60 seconds, and the residual position error between the orbit determination and the GPS position data was evaluated. Figure 15.18 shows the flight data, where (a) is the number of visible GPS satellites and position dilution-of-precision (PDOP) and (b) is the residual position error. The short-term random error of GPS positioning is as large as 1.5 m for a PDOP of 2.7 on orbit, which is as good as on the ground.

Figure 15.19 shows the relation between PDOP values and residual position errors: The PDOP axis is divided into nine bins, and the rms values of the residual position errors in each bin are indicated by solid circles. As expected, the rms values of residual position errors are nearly proportional to the PDOP values. The straight line in the figure is the least-square error fit of the solid circles, and its slope is 0.46 m, which corresponds to the standard deviation of range error of the receiver. This miniature space GPS receiver is now on the market[28] as a promising receiver with low cost and medium performance.

15.4.6 Auroral Observation

The scientific purpose of INDEX is to investigate fine-scale structures and dynamics of auroral emissions and particles. The most advanced approach in the auroral observation by INDEX is the simultaneous measurement of auroral emissions and particles using the satellite attitude control and the satellite orbit. Figure 15.20 explains details of the observation. The simultaneous measurement

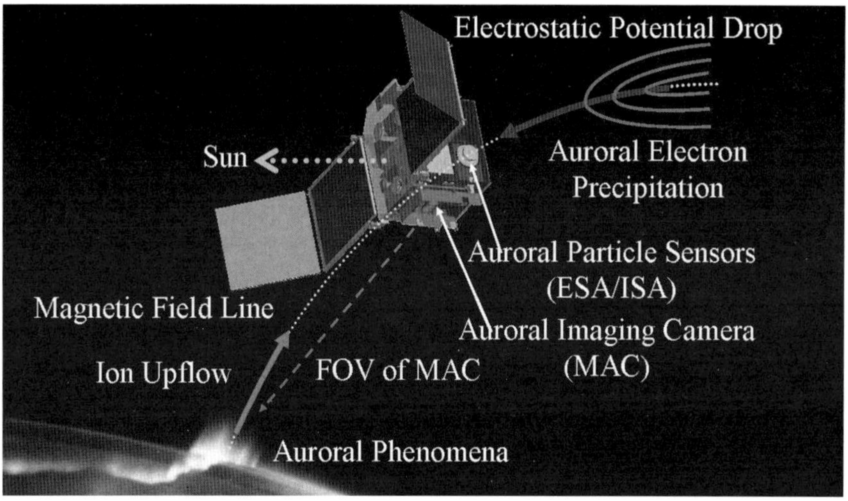

Fig. 15.20. Configuration of simultaneous observations with auroral imaging and particle measurement.

provides 64 × 64 pixels of two-dimensional aurora images covering about 70 × 70 km area at 110 km altitude at 428, 558, and 670 nm wavelengths every 120 msec, and energy-pitch angle distribution functions of aurora particles with energies of 12 eV to 12 keV for electrons and 10 eV/q to 12 keV/q for ions every 20 msec. The specifications of the auroral camera and the particle analyzers are shown in Table 15.7.

Table 15.7. Scientific Instruments Onboard INDEX

Scientific Instruments	Specification	
Electron energy spectrum analyzer	Electrostatic electron/ion energy analyzer	
Ion energy spectrum analyzer (no mass resolution)	Energy	10 eV/q–12 keV/q
	Number of steps	16(32) logarithmic
	Pitch angle	0–180 deg
	Time resolution	20(40) msec
Multispectral aurora camera	CCD camera with monochromatic filters	
	Wavelength	428, 558, 670 nm
	Field of view	7.6 × 7.6 deg
	Spatial resolution	1–4 km, 0.2 km (cal. mode)
	Time resolution	120–1000 msec, 4040 msec (cal. mode)
Current monitor	Time resolution	5 msec
Geomagnetic aspect sensor	Time resolution	5 msec

The attitude of INDEX is controlled so that the camera's field of view tracks the projection point of the satellite at an altitude of 110 km (auroral region) along a magnetic field line. This means that the particle analyzers could measure aurora particle properties on the same geomagnetic field line on which the aurora emissions could be observed by the multispectral camera. This simultaneous observation procedure is planned by the science team on the ground before the actual observation, and is uplinked by stored commands.

An interesting example from the simultaneous observation results is shown in Fig. 15.21. This event on 29 December 2005 is valuable for discussion of the auroral emission-particle correlation in the midnight aurora region at an altitude of 640 km. In the energy-time spectrogram of precipitating electrons (top image), a clear structure of the accelerated electrons is observed from 00:21:23 to 00:22:35 UTC. This is displayed mainly in red. It is remarkable that the energy of the accelerated electrons drastically varies in the range from 1–10 keV. It would not be obvious that these energy variations are due to the spatial distribution or time

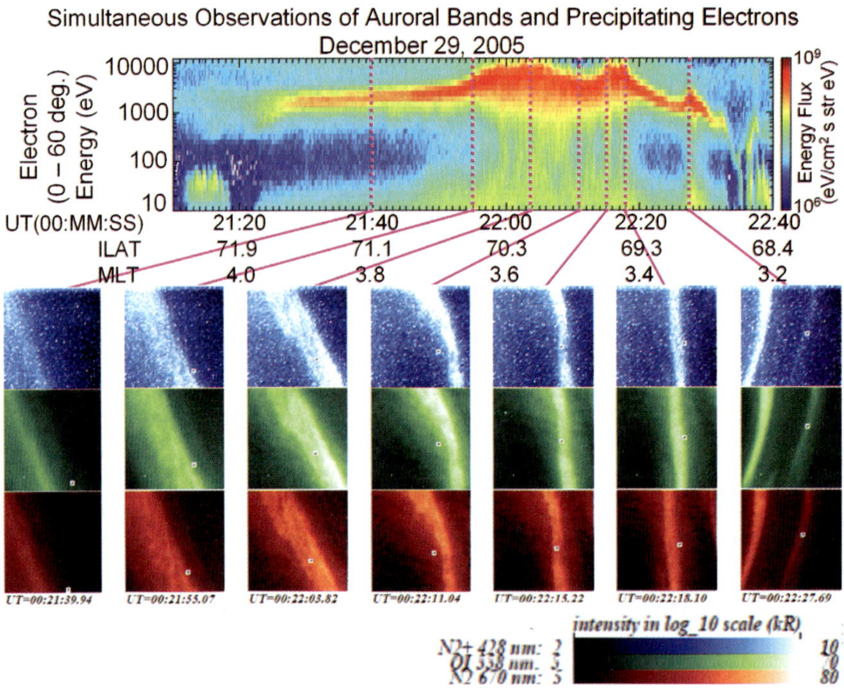

Fig. 15.21. An example of simultaneous observational results of the particle counter and multispectral auroral camera for the event on 29 December 2005. The top panel shows an energy (vertical axis) and observational time (horizontal axis) spectrogram for the electrons of a pitch angle range from 0 to 60 deg. The electron energy flux is displayed by color code. The lower panels with blue, green, and red are auroral images at 427.8, 557.7, and 670 nm, respectively. A dot in each image indicates the footprint of INDEX mapped onto the ionospheric altitude of 110 km along the local geomagnetic field line.

variation of the acceleration mechanisms using only the *in situ* electron measurements from the particle sensor.

The comparison with the aurora images presented in the lower panels of Fig. 15.21 is crucial in identifying the cause of the energy variations of precipitating electrons. It can be clearly recognized that the footprint of INDEX crossed a large aurora band twice: the first time from right to left, and the second time from left to right. The rightmost panels of the aurora images, moreover, indicate that the INDEX footprint crossed a smaller aurora band. It should be noted that the global change of the aurora form in each image is due to the orbital motion of the satellite with a speed of about 7.5 km/sec. Dotted and solid red lines connecting the electron energy-time spectrogram and the aurora images depict these correspondences. Three peaks of the electron energy in the energy-time spectrogram correspond to the bright areas of the auroral bands. Although the electron acceleration region associated with the larger auroral band may change the peak energy of auroral electrons and the area of aurora emissions, one can conclude that the electron energy variations observed in the energy-time spectrogram are caused by the spatial distribution of two aurora bands.

A number of the "bursty" electron signatures could also be noticed at the lower energy range of 10 eV to 1 keV in a main auroral emission region. These electron components are considered to be accelerated by active plasma waves excited in bright auroral emission regions. The aurora motions would also be related with the activity of the plasma waves and bursty electron precipitations.

The advanced auroral observations[30–34] by the INDEX satellite are summarized as follows:

- INDEX has successfully been observing the fine-scale properties on aurora emissions and particles through well planned satellite operation sequences in conjunction with the fine satellite attitude control.
- The simultaneous observations of aurora emissions and particles show the fine correlation between them in terms of intensities of aurora emission and downward energy fluxes of aurora electrons.
- Pulsating auroras are often observed by INDEX and are well correlated with the energy-dispersed precipitating components of the electrons.

15.5 Lessons Learned

The INDEX satellite project was a great success. A small science mission served as a testbed for technology demonstrations using a three-axis controlled satellite of 0.05 deg accuracy under the constraint of low mass and cost. The main driver of success was the enthusiasm of capable researchers and engineers. The new advanced technologies listed in Table 15.1, some of which were developed in the STRAIGHT technology development program, enabled completion of the INDEX satellite. The high-performance, lightweight onboard instruments, including fiber-optic gyroscopes and the GPS receiver, were developed with innovative approaches rather than conventional approaches.

The design of INDEX was optimized in a practical manner under a number of constraints during the ~3.5 year project period. INDEX may be considered a success story of a single small satellite with low cost and high performance. It would be difficult to generalize our experiences and apply them throughout the space community because the project was highly dependent on the personal talents and enthusiasm of everyone involved. Regretfully, development of INDEX was not performed in a way that other people and companies could follow.

The next step is to create a framework and a generalized method within ISAS/JAXA to develop many small satellites with low cost and high performance in short timeframes. An important lesson learned is that we will proceed from an excellent single satellite to a series of good satellites.[29]

15.5.1 Configuration of Bus and Mission Instrument

Conventional piggyback satellite dimensions are about 50 × 50 × 50 cm; however, this small size makes it hard to design the instrument layout with high reusability. To generate a small satellite design that can be repeatedly used for following missions, the bus segment and the mission segment should be separated. Scientific instruments may have various sizes and shapes—e.g., a telescope, a wire antenna, and a particle measurement instrument. There may even be an inherent inconsistency between developing a piggyback satellite and one that uses a standard bus layout.

15.5.2 Integrated Data Processing System

An integrated data processing instrument can significantly reduce satellite mass, volume, and cost. On the negative side, several subsystems (attitude control, command and telemetry, housekeeping, science) must share a computer, FPGAs, and digital circuit boards. Another demerit is that an integrated instrument is not adequate for reuse with a just slight modification. The development processes of several subsystems can interfere with each other. It is important to implement a modular approach in the integrated data processing system. For example, the subsystems should use a separate FPGA, a separate circuit board, and an isolated software group with a layered structure. This policy was not perfectly observed in INDEX development due to limited resources. Many simulators of the integrated data processing instrument should be prepared for parallel development of subsystems.

15.5.3 Project Team Members

It's important to organize a highly capable project team and to set interesting goals to keep enthusiasm high. Generally, the manufacturing team for a small satellite consists of a small number of people who work closely together. Each subsystem member collaborates closely with other members, depending on how closely each subsystem links with another. However, each subsystem member should work independently through a clear interface with others, if the subsystem does not closely couple with others.

15.5.4 Small Companies

Small satellite projects must have the freedom to collaborate with capable, small companies. Project members as well as small companies can learn something new interactively. It is important for the project to give a new challenging task to the small companies. There is, however, a risk that a small company can not complete the work if an unexpected critical situation happens. A small company does not have any margin for staffing or funding.

15.6 Conclusion

INDEX is a small scientific satellite for auroral observation and on-orbit testing of advanced satellite technologies. It was launched into a nearly sun synchronous polar orbit on 23 August 2005 and has been working satisfactorily for 38 months as of November 2008. Three-axis attitude control has achieved a control accuracy of 0.05 deg (1σ). INDEX is simultaneously observing auroral images and measuring particle energies, demonstrating that even a small satellite with 72 kg mass can successfully perform unique scientific missions.

15.7 References

[1] H. Saito, Y. Masumoto, T. Mizuno, A. Miura, M. Hashimoto, H. Ogawa, S. Tachikawa, T. Oshima, A. Choki, and H. Fukuda, M. Hirahara, and S. Okano, "INDEX Piggy-Back Satellite for Aurora Observation and Technology Demonstration," *Acta Astronantica,* **48**, 723 (2001).

[2] H. Saito, "Small, Advanced Scientific Satellite 'INDEX,' Aurora Observation and Advanced Satellite Technology," *Inst. Elect. Info. Comm. Engineering*, **J88-B**, p. 58 (2005) (in Japanese).

[3] H. Saito, T. Mizuno, K. Tanaka, Y. Sone, S. Fukuda, S. Sakai, N. Okuizumi, M. Mita, Y. Fukushima, M. Hirahara, K. Asamura, T. Sakanoi, A. Miura, T. Ikenaga, and Y. Masumoto, "An Overview and Initial In-Orbit Status of 'INDEX' Satellite," Paper keynote-04v, *25th International Symposium on Space Technology and Science* (June 2006).

[4] H. Saito, T. Mizuno, K. Tanaka, Y. Sone, S. Fukuda, S. Sakai, N. Okuizumi, M. Mita, Y. Fukushima, M. Hirahara, K. Asamura, T. Sakanoi, A. Miura, T. Ikenaga, and Y. Masumoto, "An Overview and Initial In-Orbit Status of 'INDEX' Satellite," Paper SSC06-IV-1, *20th AIAA/USU Conference on Small Satellites* (Aug. 2006).

[5] H. Saito, T. Mizuno, K. Tanaka, Y. Sone, S. Fukuda, S. Sakai, N. Okuizumi, M. Mita, Y. Fukushima, M. Uno, Y. Yanagawa, T. Takahara, R. Kaneda, T. Honma, M. Hirahara, K. Asamura, T. Sakanoi, A. Miura, T. Ikenaga, H. Nagamatsu, K. Ogawa, and Y. Masumoto, "An Overview and Lessons Learned of Small Scientific Satellite 'INDEX' (REIMEI)," Paper 07-B4.4.05, *58th International Astronautical Congress* (Sept. 2007).

[6] S. Sakai, Y. Fukushima, Y. Sone, T. Honma, R. Kaneda, and T. Nagao, "Design and Evaluation of Accurate Attitude Control System for Small Satellite INDEX," *Inst. Elect. Info. Comm. Engineering,* **J88 B**, p. 69 (2005) (in Japanese).

[7] S. Sakai, Y. Fukushima, and H. Saito, "Design and On-orbit Evaluation of Attitude Control System for Small Scientific Satellite 'REIMEI'," Paper 2006-d-84, *25th International Symposium on Space Technology and Science* (June 2006).

[8] S. Sakai, Y. Fukushima, H. Saito, and R. Kaneda, "Studies on Magnetic Attitude Control System for the REIMEI Microsatellite," *AIAA Guidance, Navigation, and Control Conference* (Keystone, CO, USA, 2006).

[9] Y. Fukushima, S. Sakai, and H. Saito, "Flight Performance of the REIMEI Microsatellite Attitude Determination System," *4S Symposium: Small Satellites Systems and Services* (Chia Laguna, Sardinia, Italy, Sept. 2006); printed in ESA SP-625 (Nov. 2006).

[10] S. Fukuda, T. Mizuno, S. Sakai, Y. Fukushima, K. Asamura, T. Homma, T. Takahara, R. Kaneda, Y. Yanagawa, and H. Saito, "On-Orbit Performance of the REIMEI Satellite with Integrated Computer System and Introduction to Its Flexible Operation," Paper 2006-f-01, *25th International Symposium on Space Technology and Science* (June 2006).

[11] Y. Sone, M. Uno, Y. Takeda, K. Tanaka, H. Saito, and M. Tajima, "The Flight and Ground Test Data of the Lithium-Ion Secondary Battery of 'REIMEI'," Paper 2006-f-04, *25th International Symposium on Space Technology and Science* (June 2006).

[12] H. Saito, N. Okuizumi, K. Tanaka, M. Mita, J. Roussel, and B. Boulanger, "Development and On-orbit Results of Solar Concentrated Paddles with Thin Film Reflector," Paper 2006-f-03, *25th International Symposium on Space Technology and Science* (June 2006).

[13] N. Okuizumi, K. Higuchi, H. Saito, and K. Tanaka, "Wrinkling and Reflection Characteristics of Membrane Solar Reflectors of REIMEI Spacecraft," Paper 2006-c-14, *25th International Symposium on Space Technology and Science* (June 2006).

[14] Y. Nakamura, A. Okamoto, S. Tachikawa, A. Ohnishi, J. Kawaguchi, H. Saito, and K. Shimazaki, "In Orbit Test Result of Smart Radiation Device," Paper 2006-c-20, *25th International Symposium on Space Technology and Science* (June 2006).

[15] H. Saito, T. Mizuno, K. Kawahara,T. Saiki, Y. Tuda,Y. Fukushima,Y. Hamada, H. Sasaki, S. Katumoto, S. Kuroki, and Y. Kajikawa, "Tiny GPS Receiver for Space Application," *Inst. Elect. Info. Comm. Engineering,* **J88-B**, p. 79 (2005) (in Japanese).

[16] H. Saito, T. Mizuno, K. Kawahara, K. Shinkai, T. Saiki,Y. Hamada, and H. Sasaki, "Development and On-Orbit Results of Miniature Space GPS Receiver by Means of Automobile-Navigation Technology," Paper 2006-f-02, *25th International Symposium on Space Technology and Science* (June 2006).

[17] H. Saito, T. Mizuno, K. Kawahara, K. Shinkai, T. Saiki,Y. Fukushima, Y. Hamada, H. Sasaki, S. Katumoto, and Y. Kajikawa, "Miniature Space GPS Receiver by Means of Automobile-Navigation Technology," Paper SSC07-VIII-1, *20th Annual AIAA/USU Conference on Small Satellites* (Aug. 2007).

[18] H. Saito, T. Mizuno, K. Kawahara, K. Shinkai, T. Saiki,Y. Fukushima, Y. Hamada, H. Sasaki, S. Katumoto, and Y. Kajikawa, "Miniature Space GPS Receiver by Means of Automobile-Navigation Technology," *J. Jpn. Soc. Aeronautical Space Sci.,* **56**, p. 123 (2008) (in Japanese).

[19] S. Sakai, Y. Fukushima, A. Ohno, and H. Saito, "In-Orbit Performance Evaluation of Temperature Controlled Small Fiber Gyro on Microsatellite 'REIMEI,'" *18th International Conference on Optical Fiber Sensors,* Topical Meeting (Mexico, Nov. 2006).

[20] E. Sato, T. Yamada, H. Tanaka, and I. Jimbo, "Categorization of Ambient-Temperature

Creep Behavior of Metals and Alloys on their Crystallographic Structures," *Materials Transactions,* **47**, 1121 (2006).

[21] D. Bearden, "Small-Satellite Costs," *Crosslink* (Winter, 2001); available at http://www.aero.org/publications/crosslink/winter2001/04.html.

[22] M. Tafazoli, "A Study of On-Orbit Spacecraft Failures," Paper D1.5.02, *58th International Astronautical Congress* (Sept. 2007).

[23] IEEE Aerospace and Electronic System Society, "IEEE Standard Specification Format Guide and Test Procedure for Single Axis Interferometric Fiber Optical Gyros," IEEE Std 952-1997 (Feb. 1998).

[24] R. L. Farrenkopf, "Analytic Steady-State Accuracy Solutions for Two Common Space Attitude Estimators," *J. Guidance Control,* **1**, 282 (1978).

[25] R. Stribling, "Hughes 702 Concentrator Solar Array," *28th IEEE Photovoltaic Specialists Conference* (2000).

[26] J. M. Bodeu, "Root Cause of the BSS702 Concentrator Array Anomaly," *Space Power Workshop* (April, 2003).

[27] M. Caceres, "Satellites and Launches Trend Down," *Aerospace America,* **18** (Jan. 2004).

[28] http://spacelink.biz/.

[29] H. Saito, S. Sawai, S. Sakai, S. Fukuda, and K. Kitade, "A Series of Small Scientific Satellite with Flexible Standard Bus," paper B4.7.02, *58th International Astronautical Conference* (Sept. 2007).

[30] Y. Obuchi, T. Sakanoi, A. Yamazaki, T. Ino, S. Okano, Y. Kasaba, M. Hirahara, Y. Kanai, and N. Takeyama, "Initial Observations of Auroras by the Multi-Spectral Aurural Camera on board the Reimei Satellite," *Earth, Planets and Space* (2008).

[31] J. Lunde, S. C. Buchert, Y. Ogawa, M. Hirahara, K. Seki, Y. Ebihara, T. Sakanoi, K. Asamura, M. Okada, T. Raita, and Häggström, "Ion Dispersion and Rapid Electron Fluctuations in the Cusp: A Case Study," *Annales Geophysicae*, **26**, 2485–2502 (2008).

[32] K. Sakaguchi, Y. Miyoshi, Y. Otsuka, T. Ogawa, K. Asamura, and M. Connors, "Simultaneous Appearance of Isolated Auroral Arcs and Pc1 Geomagnetic Pulsations at Subauroral Latitudes," *Journal of Geophysical Research*, **113**, A05201 (2008).

[33] Y. Ogawa, K. Seki, M. Hirahara, K. Asamura, Kanagawa, Japan, T. Sakanoi, S. C. Buchert, Y. Ebihara, Y. Obuchi, TMiyagi, A. Yamazaki, I. Sandahl, S. Nozawa, and R. Fujii, "Coordinated EISCAT Svalbard Radar and Reimi Satellite Observations of Ion Upflows and Suprathermal Ions," *Journal of Geophysical Research*, **113**, A05306, doi:10.1029/2007JA012791 (2008).

[34] Y. Ebihara, Y. Miyoshi, K. Asamura, and M. Hirahara, "Microburst Cusp Ion Precipitation Observed with Reimi," *Journal of Geophysical Research*, **113**, A03201 (2008).

Chapter 16

Mini AERCam: The Miniature Autonomous Extravehicular Robotic Camera

Steven Fredrickson*

16.1 Introduction

To meet the remote viewing and inspection needs foreseen for future human spaceflight missions, NASA Johnson Space Center (JSC) has developed the Miniature Autonomous Extravehicular Robotic Camera (Mini AERCam), a nanosatellite-class free-flying vehicle. The Mini AERCam technology demonstration unit, which has been integrated into the approximate form and function of a flight system, represents a significant technology breakthrough in the field of free-flying robotic space vehicles.

The goal of the Mini AERCam project is to improve the performance and safety of human space operations by providing useful on-orbit views that cannot be obtained from fixed cameras, cameras on robotic manipulators, or cameras carried by crewmembers during an extravehicular activity (EVA). The Mini AERCam could be used to support human and robotic EVAs by supplying orthogonal views to the robotic operators or crewmembers inside the spacecraft or on the ground and by conducting independent visual inspections of areas of interest around the spacecraft.

The potential applications of Mini AERCam extend beyond low Earth orbit (LEO) to future long-duration human exploration missions, for which intelligent mobile inspection agents are likely to become essential crew helpers. A free-flyer could be used to perform damage inspection of future translunar or transmartian vehicles using either piloted teleoperation or autonomous control. Assembly in space could also be greatly enhanced with capability to obtain an orthogonal viewpoint of mating interfaces during construction. Close-up views of maintenance and repair sites could provide valuable insight in preparing and conducting servicing tasks. Finally, the Mini AERCam platform itself, or a close derivative, could be used as a forerunner to test other free-flying technologies required for exploration.

The spherical Mini AERCam is 19 cm (7.5 in.) in diameter and weighs approximately 5 kg (11 lb.). The vehicle is designed for either remotely piloted operations or supervised autonomous operations, including automatic stationkeeping

*NASA Johnson Space Center
Houston, Texas

Fig. 16.1. AERCam Sprint on STS-87.

and point-to-point maneuvering. The Mini AERCam is accompanied by a control station for command and control as well as a docking system for automated deployment, docking, and recharge at a parent spacecraft. Functional testing has been conducted successfully on an air-bearing table and in a 6-degree-of-freedom closed-loop orbital simulation with avionics hardware in the loop. This chapter provides the history of the AERCam program and the technical details of the Mini AERCam technology demonstrator.

16.2 AERCam Project History

Interest in external remote inspection of human spacecraft extends back more than two decades, but only in recent years has the miniaturization of spacecraft components made such free-flyers practical. The first successful free-flying inspector built specifically for human space flight was AERCam "Sprint," a 16 kg (35 lb), 35 cm (14 in.) unit that flew as a remotely piloted shuttle flight experiment. Sprint was developed in house at Johnson Space Center with NASA Headquarters Code X sponsorship for a flight experiment on Space Transportation System (STS)-87.[1]

On 3 December 1997, Sprint was released and retrieved in the space shuttle orbiter payload bay by astronaut Winston Scott during EVA. Figure 16.1 shows views of AERCam Sprint during the STS-87 flight. Astronaut Steve Lindsey controlled the free-flyer, which was piloted from the orbiter aft flight deck using out-of-the-window views for navigation and a single, multifunction hand controller for remote control. Lindsey was able to observe Sprint at all times through the orbiter windows, and also monitored the Sprint video on an internal display. The Sprint propulsion system consisted of 12 cold-gas thrusters with nitrogen as the propellant. Sprint used angular-rate gyros for angular-rate feedback and automatic attitude hold. The video stream could be switched between two cameras with different focal lengths for a regular or telephoto view. A light was available for illumination during orbital night. Sprint had a soft covering for impact protection. The flight test on STS-87 lasted approximately 1 hr 15 min and was highly successful in demonstrating the feasibility of free-flyer inspection and remote viewing. The controllability and flying characteristics of Sprint were confirmed by the crew, and the precision and stability of the in-space camera platform resulted in excellent video imagery.

Following the flight of STS-87, a crew evaluation was conducted in Johnson Space Center's Virtual Reality Laboratory to evaluate the suitability of Sprint for use as an operational vehicle outside the International Space Station (ISS). A series of crew-in-the-loop evaluations were conducted with a high-fidelity Sprint and station simulation and state-of-the-art graphical scenes. The study by nine astronaut participants yielded a set of recommendations for improving situational awareness, safety, and handling qualities.[2] The results of these evaluations were incorporated into requirements for additional capability in future AERCams. These requirements included providing:

- Standard hand-controllers for teleoperation.
- Situational awareness to the pilot.
- AERCam in-space position information (not just visual).
- Stationkeeping capability without pilot action.
- Point-to-point navigation capability without crew teleoperation.
- Orthogonal camera views to allow scanning during a fly-by.

NASA Headquarters Code M sponsored an advanced development initiative to develop and demonstrate the technologies necessary to build an operational version of AERCam that addressed the new requirements for enhanced capabilities. This effort, known as the AERCam Integrated Ground Demonstration, culminated in 1998 with a series of air-bearing demonstrations of autonomous technologies that could serve a dual role by enabling intuitive pilot situational awareness in a teleoperated mode.

In 2000, the AERCam team began efforts to reduce free-flyer size (compared with Sprint) while simultaneously adding new capabilities that would contribute to overall system controllability, reliability, and utility. The result was Mini AERCam. The motivation for miniaturization stemmed from advantages in safety and operational logistics. Drawing on the findings of the Sprint experiment, the crew evaluations, and the Integrated Ground Demonstration, the team succeeded in developing and demonstrating a free-flyer with vastly expanded capabilities, compared with the AERCam Sprint, in just 20% of the volume. Technical requirements included all capabilities anticipated for a flight system: relative navigation so that crewmembers would know the location of the free-flyer with respect to the parent spacecraft structure; automatic stationkeeping and point-to-point maneuvering to improve the free-flyer handling qualities and reduce crew workload; and an orthogonal-view camera for crew situational awareness and fly-by scanning.

The development team accepted three associated challenges for the development of Mini AERCam:

- Embrace high-risk technology to achieve a compact integrated package.
- Build an approximately flight-like technology demonstration unit that is "one step from flight."
- Demonstrate all planned free-flyer capabilities in relevant environments (NASA Technology Readiness Level 6).

By 2002, these challenges had been met: A fully functional prototype had been designed, developed, integrated, and tested. The nanosatellite-class, spherical Mini AERCam free-flyer (Fig. 16.2) is 19 cm (7.5 in.) in diameter and weighs only 5 kg (11 lb), yet it incorporates significant additional capabilities compared with the 16 kg (35 lb), 35 cm (14 in.) AERCam Sprint.

After initial testing in 2002, the Mini AERCam technology demonstrator was used as a prototype for various engineering studies and analyses during

Fig. 16.2. The Mini AERCam technology demonstrator (top) and Mini AERCam air-bearing testing (bottom), circa 2002.

2003–2005. During that period, the technology demonstration capability was also enhanced in various ways—most notably, by the addition of a prototype electromagnetic docking system. On the free-flyer side, the technology demonstrator was updated with a docking test interface and a vision-based automatic docking system. Each of these enhancements followed the original philosophy of keeping the technology demonstrator "one step from flight" and avoiding any "cheating" that would preclude a straightforward engineering development of a flight system based on the technology demonstrator design. That discipline was crucial because during the same period (2003–2005), the Mini AERCam was being considered for application as a second-generation inspection tool for on-orbit imaging following the loss of the space shuttle Columbia in early 2003. Although the space-shuttle program ultimately decided to continue using the orbiter boom sensor system for inspection, the "one step from flight" doctrine was tested during initial flight-system development of a Mini AERCam for shuttle inspection.

16.3 Mini AERCam Technology Demonstrator

The Mini AERCam vehicle is designed for both remotely piloted operations and supervised autonomous operations such as stationkeeping, point-to-point maneuvering, and docking. The Mini AERCam hosts a full suite of miniaturized avionics, instrumentation, communications, navigation, video, power, and propulsion subsystems, including two video imagers and one higher resolution still-frame imager. Technology innovations include a rechargeable xenon-gas propulsion design, a rechargeable lithium-ion battery, custom avionics based on the PowerPC 740/750 microprocessor series, "camera on a chip" CMOS imagers with wavelet video compression, MEMS gyros, GPS relative navigation, digital RF communications, micropatch antennas, a digital instrumentation network, and compact mechanical packaging.

16.3.1 Mechanical Design

The spherical Mini AERCam free-flyer achieves its nanosatellite size by a combination of dense packaging and miniaturized components. Figure 16.3 shows a size comparison between the AERCam Sprint and the Mini AERCam.

Fig. 16.3. AERCam Sprint and Mini AERCam size comparison.

Figure 16.4 shows the external features of the Mini AERCam prototype. The dual cameras are coaligned: The lower one provides color video, and the upper one generates high-resolution still images when selected. A third camera points 90 deg offset from the other two and provides color video for an orthogonal view.

The Mini AERCam prototype was designed with two custom GPS antennas, one on the top and one on the bottom, for full spherical coverage. A communications antenna, which is not visible in Fig. 16.4, is mounted near the top GPS antenna. An LED array provides illumination. Figure 16.4 depicts two different LED arrangements. The left image shows an earlier 2002 configuration; the right image shows the LEDs embedded with the camera cluster.

16.3.2 Power and Propulsion

As shown in Fig. 16.5, the Mini AERCam vehicle is designed with a central ring that houses the power and propulsion systems. The battery consists of rechargeable lithium-ion cells. The propulsion system is designed for cold xenon gas (which packs more densely than nitrogen) but is compatible with low-cost nitrogen in the current ground-test configuration. Attitude and position control are achieved with the use of 12 thrusters, distributed in four clusters around the central ring.

The rechargeable lithium-ion battery is composed of seven cells in series, each with an independent safety circuit and cell bypass capability. The safety circuits sense and prevent overdischarge and underdischarge (automatically reset by discharge and charge, respectively), overtemperature, overcurrent, and short-circuit conditions. The battery is packaged as four modular subassemblies; three are dual-cell modules, and one is a single-cell module. The battery assembly is sized to supply a nominal load of 20 W for 4 h, although significantly higher peak loads can be supported. Mini AERCam power usage is typically closer to 15 W, enabling run times of up to 6 hr.

The propulsion system design is a semiconventional regulated cold-gas supply. High-pressure xenon is stored at up to 3000 psi (20,685 kPa) in a small

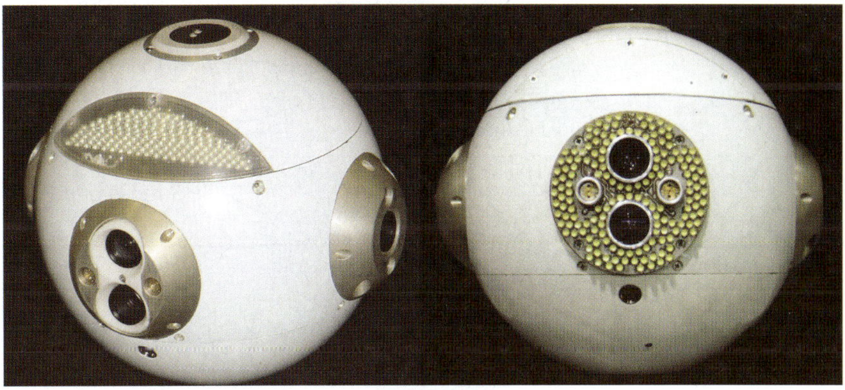

Fig. 16.4. Mini AERCam external features (LED configuration in 2002 at left and 2005 configuration at right).

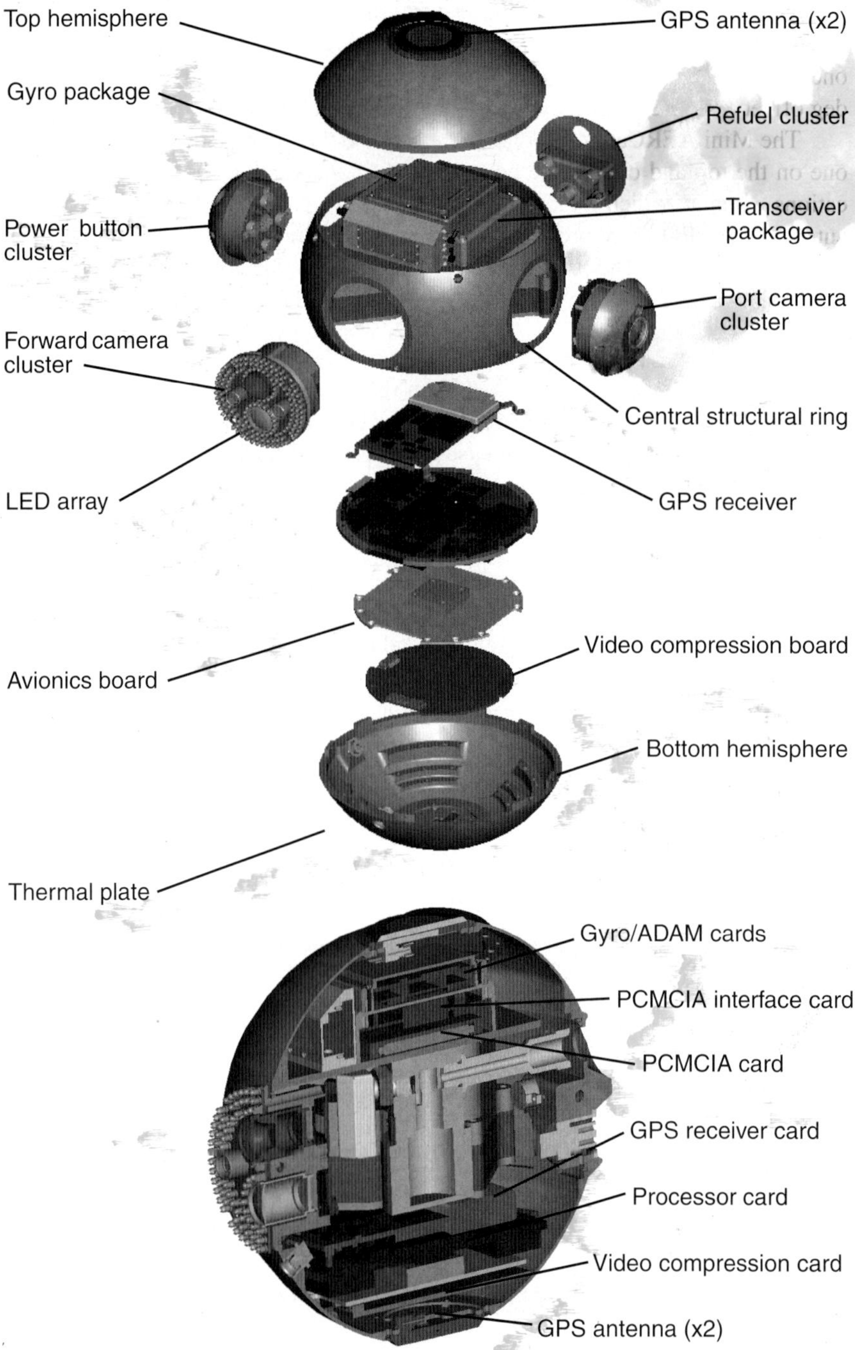

Fig. 16.5. Mini AERCam exploded view (top) and cutaway view (bottom).

internal tank at the vehicle center of mass (flight design) or in external run tanks to support extended operations using nitrogen (technology demonstration unit). The flight-unit design scavenges heat from the avionics subsystem to maintain the thermodynamic availability of the xenon. The ground unit supplies high-pressure nitrogen to the free-flying vehicle through a flexhose, where a small commercial regulator provides low-pressure gas to the four thruster clusters. Distribution is via polymer tubes that engage in semiquick-disconnect fittings at each cluster. These clusters have integrally machined manifolds to supply the cartridge-style mounted thruster valves. Small, replaceable thruster nozzles are installed near the cluster surface, reducing impingement issues and allowing accurate thrust balance. The designed thrust level for in-space operations is 0.025 lb (0.11 N) for each thruster.

Figure 16.6 shows the 12 thrusters and the plume direction for each. The input to the thruster selection logic is a set of commanded thrust durations (in integer milliseconds) for all rotation and translation axes that controls all six degrees of freedom.

16.3.3 Avionics

The Mini AERCam avionics system[3] includes an avionics processor board, a video compression and imaging system, wireless Ethernet for communications, and LED illuminators. The avionics processor board, which is shown in Fig. 16.7, is a 22-layer board that constitutes the hub-of-a-star architecture. The board houses a PowerPC 740 that runs at 266 MHz, with 8 Mb of flash memory, 64 Mb of RAM, and an additional 8 Mb for error detection and correction (EDAC). The

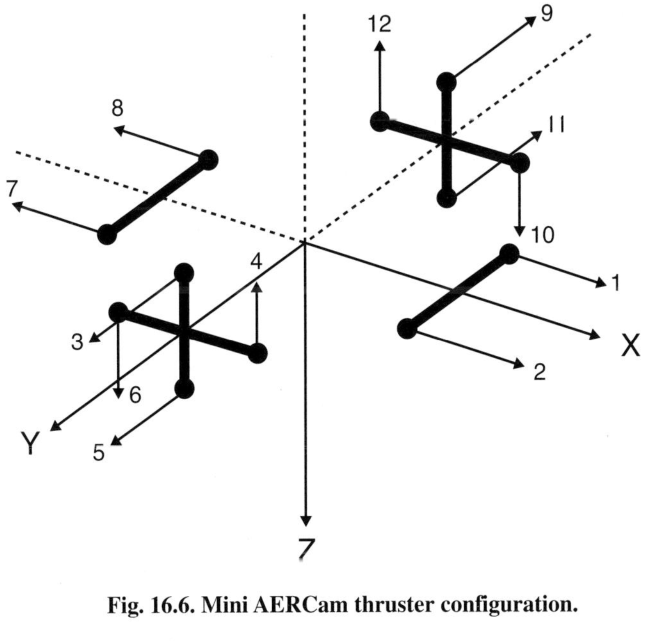

Fig. 16.6. Mini AERCam thruster configuration.

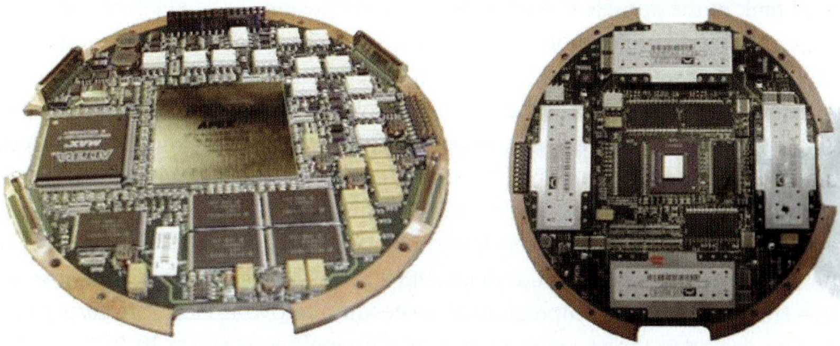

Fig. 16.7. Avionics processor board (front and back views).

board includes a field-programmable gate array (FPGA) that implements most of the hardware interface functions. Data buses include RS-232, I2C for instrumentation, RS-422, and low-voltage differential signaling (LVDS). The flight design follows this technology demonstration architecture closely but includes components with higher radiation tolerance, including a PowerPC 750FX.

The onboard video subsystem consists of three CMOS imagers and a compression board. Two of the imagers provide a 640 × 480 color pixel, motion image. Integrated circuits compress the full-motion images using wavelet compression algorithms. The third color imager captures higher resolution (1.3 megapixels) color still images, which are compressed by software with an onboard digital signal processor (DSP). (The later flight design incorporated higher-resolution multimegapixel imagers to meet anticipated shuttle inspection criteria.) The off-board video components of the control station consist of a video decompression and display system. The video is transmitted as part of a single multiplexed data stream.

The technology demonstration communications system is implemented with a bridge that allows a commercial PCMCIA 802.11b wireless Ethernet card to interface with an LVDS spoke-of-the-star architecture. The output of the PCMCIA 802.11b is transmitted to an omnidirectional linearly polarized microstrip patch antenna. The uplinked commands operate in reverse to the downlink. In the flight design, the 802.11b is replaced with customized hardware that includes an onboard Ethernet switch and dual antennas.

16.3.4 Guidance, Navigation, and Control

The Mini AERCam navigation system[4] includes a GPS receiver for position and velocity determination and MEMS gyros for angular rate sensing. The position and velocity are processed in a Kalman filter, and the attitude and attitude rate are processed separately. The navigation system provides data for operator situational awareness around the parent vehicle. Additionally, precise navigation allows for automated functions such as automatic stationkeeping, which is helpful as an operator aid during piloted flight, and "hands-off" point-to-point maneuvering.

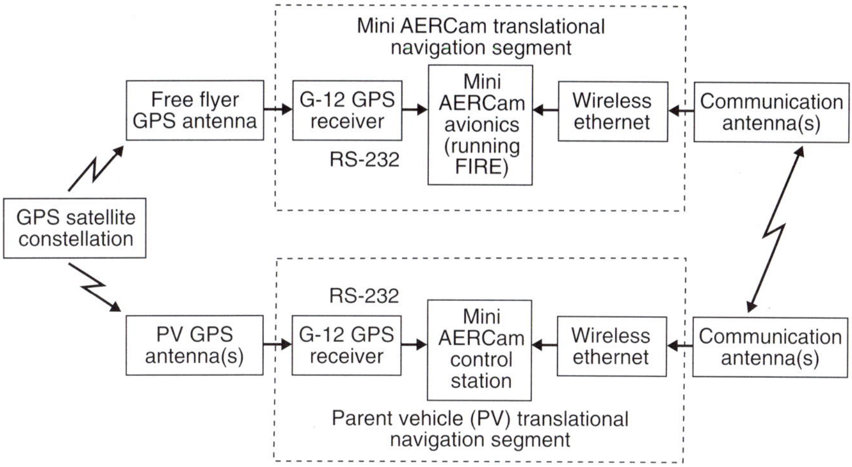

Fig. 16.8. Translational navigation system components.

Translational navigation uses precise relative GPS for estimating relative position and velocity between the free-flyer and parent vehicle. Precise relative GPS, which is based on reconciling the pseudorange and carrier-phase measurements between the GPS receivers of two vehicles, can achieve accuracies typically on the order of 10 cm or better. For the technology demonstration implementation, a commercial off-the-shelf (COTS) GPS receiver, the Ashtech G-12, was selected for both the free-flyer and the parent vehicle.

Figure 16.8 shows the Mini AERCam translational navigation system components. In the current prototype configuration, the parent vehicle GPS data are collected at the Mini AERCam control station and transmitted to the free-flyer using a wireless Ethernet transceiver. The data are then processed onboard the free-flyer, along with the free-flyer GPS measurements, by the avionics software.

The GPS measurements are processed and filtered using Johnson Space Center's Flexible Inertial/Relative Estimator (FIRE) software,[5] which includes a Kalman filter. FIRE was originally based on the generalized Kalman filter algorithms that are used for many NASA navigation applications. FIRE was initiated in 1998 in cooperation with the U.S. Air Force Research Laboratory as part of a risk mitigation activity for the XSS-11 microsatellite program. FIRE development continued when the Mini AERCam project was initiated in 2000. In addition to the core Kalman filter, FIRE software includes real-time data collection and synchronization software.

FIRE samples the GPS receiver data and processes them onboard the free-flyer. Three Kalman filters run simultaneously: one estimates the World Geodetic System (WGS)-84 position and velocity of the parent vehicle based on its GPS data; another estimates the WGS-84 position and velocity of the free-flyer based on its GPS data; and the third estimates the relative position and velocity of the free-flyer in the WGS-84 frame. FIRE provides the position and velocity estimates at 1 Hz. It also generates detailed performance information for display and recording for postflight analysis.

The Mini AERCam attitude navigation system uses angular rate measurements from a three-axis Draper MEMS gyro system to estimate inertial attitude and attitude rate. The MEMS gyro package outputs digital data at a rate of 300 Hz, which is averaged down to a 25 Hz rate. The body axis roll, pitch, and yaw attitude rate measurements are converted to quaternions and then are integrated to maintain an estimate of inertial attitude. To estimate the relative attitude of the free-flyer with respect to the parent vehicle, the inertial attitude is transformed to other reference frames, including the Mini AERCam local vertical/local horizontal (LVLH)-to-body frame, J2000 inertial-to-Mini AERCam LVLH, body-to-J2000, body-to-ISS LVLH, J2000-to-ISS LVLH, and body-to-an-arbitrary inertial frame. The flight-version gyro hardware has a maximum drift-rate specification of 1 deg per hour per axis. For short sorties, gyro reinitialization is not necessary, but could be performed using either docking target pose data (see below) or GPS-based techniques.

The attitude guidance subsystem on the Mini AERCam is designed to accommodate either manual (crew-piloted) or autonomous attitude and attitude-rate commands. The attitude guidance system accepts inputs from either source and computes errors that are then passed on to the attitude control system, which uses a phase plane to control each independent body axis. It is designed to rapidly center the vehicle in the phase plane space (target angle and angular rate). Achieving and remaining in the center of the attitude phase plane ensures that the vehicle is aligned as commanded.

Like the attitude guidance subsystem, the translational guidance subsystem on the Mini AERCam is designed to accommodate either manual (crew-piloted) or autonomous commands. It accepts inputs from either source and computes errors that are then passed on to the translational control subsystem. The translational control subsystem accepts as inputs the errors from the translational guidance subsystem and the LVLH-to-body quaternion estimate from the rotational navigation subsystem. The translational errors are then converted from the LVLH frame to the body frame. Translational control uses a phase plane controller similar to the attitude controller on the space shuttle orbiter. As with the Mini AERCam attitude phase plane, the translation phase plane uses the anticipated thruster accelerations to compute a burn time that forces the vehicle back toward the center of the phase plane in a fuel-efficient manner. The center of the phase plane represents the commanded position and velocity.

16.3.5 System Software

The Mini AERCam avionics software architecture consists of a central database and a task for each subsystem. Two task rates are used: a high rate of 25 Hz and a low rate of 1 Hz. The backbone of the Mini AERCam software is the common data area, which manages all data passed between software subsystems. Rate-sensor data are read directly from the FPGA and exported to the common data area for debugging and recording, while raw GPS data are queued in the common data area for use by the navigation task. All commands and telemetry pass through the common data area. The only exceptions are the thruster commands, which are directly issued by guidance, navigation, and control (GNC) to the FPGA, and real-time video data that are sent directly to the communications interface. Figure 16.9 shows the Mini AERCam avionics software data flow diagram.

The Mini AERCam video system software executes on a dedicated DSP. It has three primary functions, which are to capture and compress still frame images from the megapixel imager, capture and packetize the hardware compressed video streams from the two NTSC resolution imagers, and multiplex and send all imagery data to the avionics system for transmission to the control station. To maintain a fixed output video data rate, the software also provides feedback compression control for the hardware compressors in the NTSC resolution imagers.

The Mini AERCam is controlled remotely from the control station—a laptop computer running Microsoft Windows. This computer interfaces to two hand

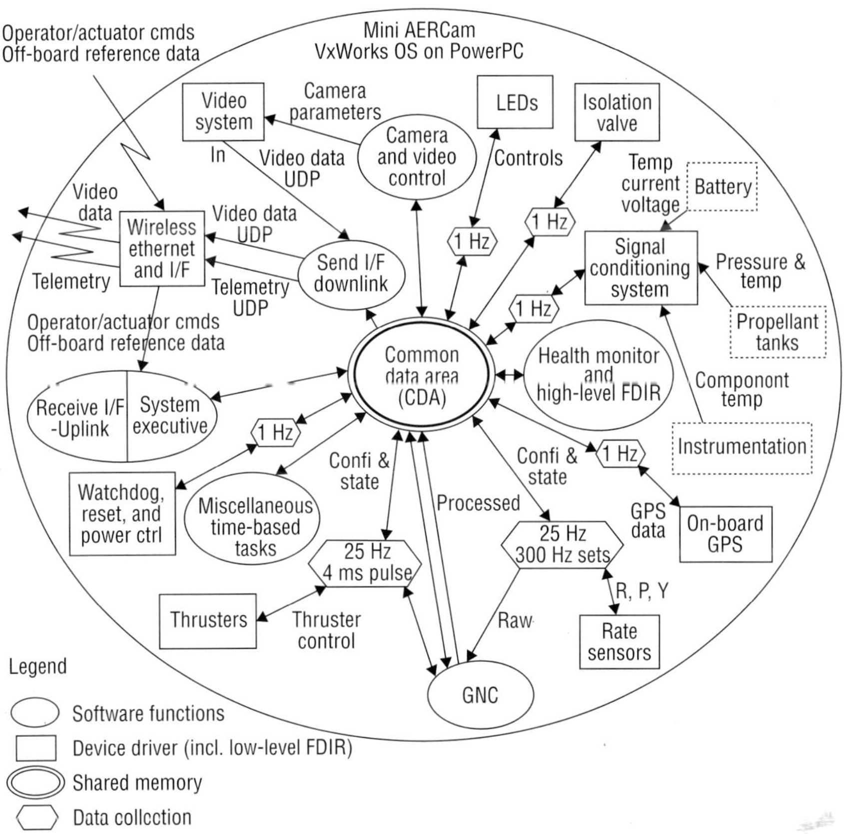

Fig. 16.9. Mini AERCam avionics software data flow diagram.

controllers (separate rotational and translational) and to the parent vehicle GPS receiver. The control station software architecture is similar to the avionics architecture. It uses a common data area and has tasks for GPS reference receiver data processing, command and telemetry processing, and telemetry display. Commands, telemetry, and video are transmitted between the control station and the free-flyer using a wireless Ethernet transceiver.

Several COTS software packages, as well as custom software developed at Johnson Space Center, are used on the Mini AERCam project. The software packages and products are listed in Table 16.1. Trick is a C-language simulation development environment that makes it easy to "plug-and-play" with an extensive set of models and algorithms. The Dynamic Onboard Ubiquitous Graphics (DOUG) software package provides high-fidelity graphical models that are updated for each International Space Station increment and shuttle flight.

16.3.6 Autonomous Control Software

The Mini AERCam system supports autonomous capabilities via an intelligent systems architecture[6] that has three major components: A robust command sequencer enables task-level scripting of operations such as automatic inspection scans and automated docking; a system manager built on the same expert-system software as the command sequencer provides detection and smart-response capability for potential system-level anomalies such as loss of communications between the free-flyer and the control station; and an AERCam dynamics manager provides nominal and off-nominal management of GNC[7] functions.

Table 16.1. Software Development and Simulation Tools

Function	Product
Operating system	WindRiver VxWorks
Compiler	WindRiver Diab
GNC algorithms and software	MathWorks Matrixx
Data processing and plotting	MathWorks MATLAB
Real-time engineering data displays	National Instruments LabWindows
Orbital simulation development environment	JSC Trick
High-fidelity orbital simulation graphics	JSC Virtual Reality Laboratory's DOUG
Video System Software Development	Analog Devices VDSP++ 3.0
Command Script Processor	C Language Integrated Production System (CLIPS)
Control Station Display	Microsoft Visual C++ 7.0

The autonomy software architecture that was developed for the Mini AERCam was influenced by several practical goals for embedded flight software design. Foremost was that all components needed to operate within the available memory and processor utilization constraints of an embedded single-processor running the flight software under VxWorks real-time operating system. The software development team implemented a rule-based expert system for conditional sequencing and system management because of advantages in representation, flexibility, and ease of maintenance compared with traditional hand-coded methods.

The following sections will describe the architectural components of Mini AERCam autonomy, as well as the interaction of these elements with a human operator during supervised autonomous control.

16.3.6.1 Command Sequencer

The Command Script Processor is a robust command executor based on CLIPS[8] computer software that automates free-flyer trajectory waypoint guidance commands. The command sequencing and execution is based on rules that model the way in which an operator manually executes a paper procedure. Figure 16.10 illustrates the parallel between operator functions and Command Script Processor functions: Both agents monitor system feedback through telemetry, follow a procedure, decide which command to issue next based on the current system state (reflected in telemetry), and then issue the commands.

The Command Script Processor resides on both the control station and the free-flyer. On the control station, it allows the operator to develop, load, and run command scripts that automate free-flyer tasks in a closed-loop fashion, as if the operator were directly controlling the free-flyer. From the perspective of the free-flyer, there is no difference between a command issued directly by the Command Script Processor and a command issued by the operator using the control-station graphical user interface.

The sequencing and execution rules implement the following logic.

- Monitor specified telemetry (system state) to determine readiness to issue a command to the system.
- Verify communications between the free-flyer and the control station.
- Verify that the free-flyer navigation state is valid before executing motion control commands.
- Verify successful completion of the previous command.

If all checks are true for a given command, the command is issued.

The user interface is designed for supervised autonomy. The operator can monitor autonomous operation and quickly take over control at any point. Also, if monitored checks fail during autonomous execution (e.g., communication between the control station and the free-flyer is lost or a command fails to execute onboard), command script processing stops, an operator alert is issued, and control is handed back to the operator. If the communications link is severed, control

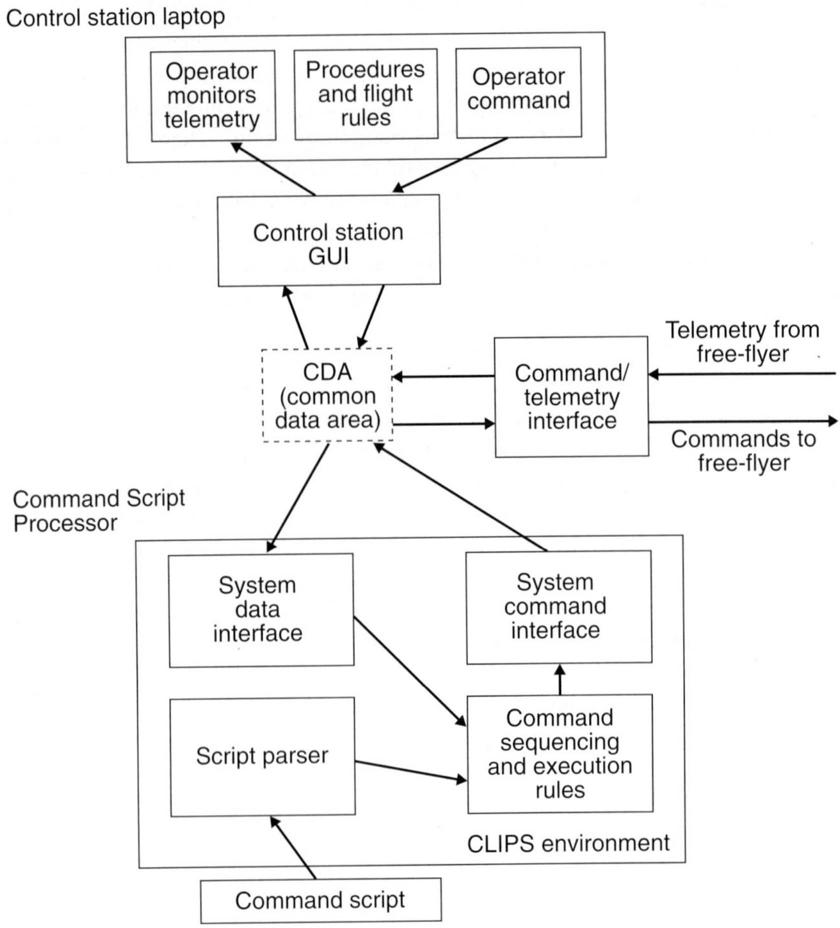

Fig. 16.10. Mini AERCam command scripting architecture.

reverts to the Command Script Processor onboard the free-flyer, which follows a predefined set of contingency scripts executed by the system manager as described in the next section.

As shown in Fig. 16.11, the user interface allows the operator to load, execute, pause, and/or resume a script, which is stored as a text file. Scripts can be created for configuring the free flyer, performing automated scanning operations, performing automated docking (which involves both maneuvering the free-flyer and reconfiguring onboard navigation sensors), and virtually any sequence of commands that the operator might issue. New script files could even be uploaded from the ground if supported by the communications infrastructure of the parent vehicle. Script entries consist of a command keyword set followed by a set of

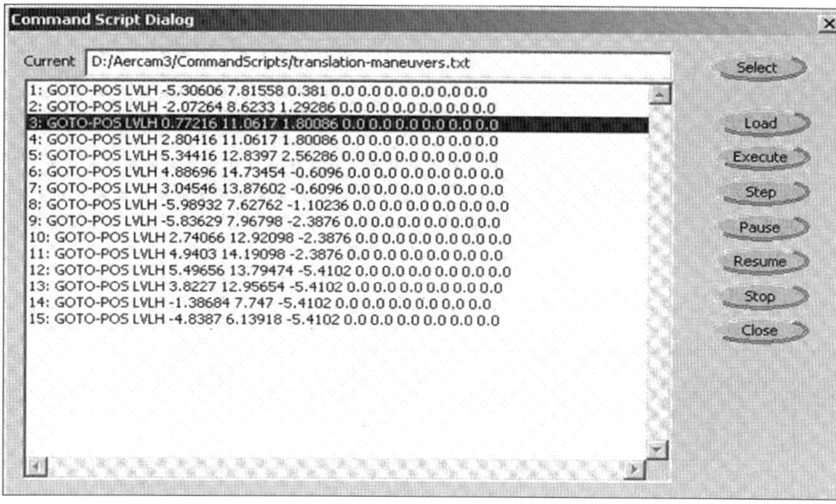

Fig. 16.11. Command script processor user interface (waypoint script example).

numerical indicators. The operator can also manually select the starting point in the script. The script shown in Fig. 16.11 consists of a simple series of waypoints.

The simple handover capability between autonomous operation and manual operation is particularly useful during a scanning mission. The operator can stop the autonomous scan, take over manual control for a detailed inspection at a point of interest, and resume the scan where it left off.

16.3.6.2 System Manager

The primary functions of the system manager are fault detection, isolation, and recovery and resource management. The system manager employs a copy of the same Command Script Processor that runs on the control station for handling the execution of onboard closed-loop recovery scripts and resource management scripts.

The system manager is built from a rule-based, expert system in which the rules "look for" predefined data signatures to detect when an event of interest (e.g., possible failure, anomaly, or safety condition violation) has occurred, then automatically loads and executes a recovery script. An important consideration for a teleoperated free-flyer is what action to take if a loss of communications occurs between the control station and the free-flyer. The specific action to take depends on several factors, which correspond to specific data signatures. Another common consideration involves close-proximity operations, for which several rules are implemented to protect the free-flyer and parent vehicle. Figure 16.12 highlights the end-to-end control architecture, including the parallels between control-station control and onboard control using the system manager.

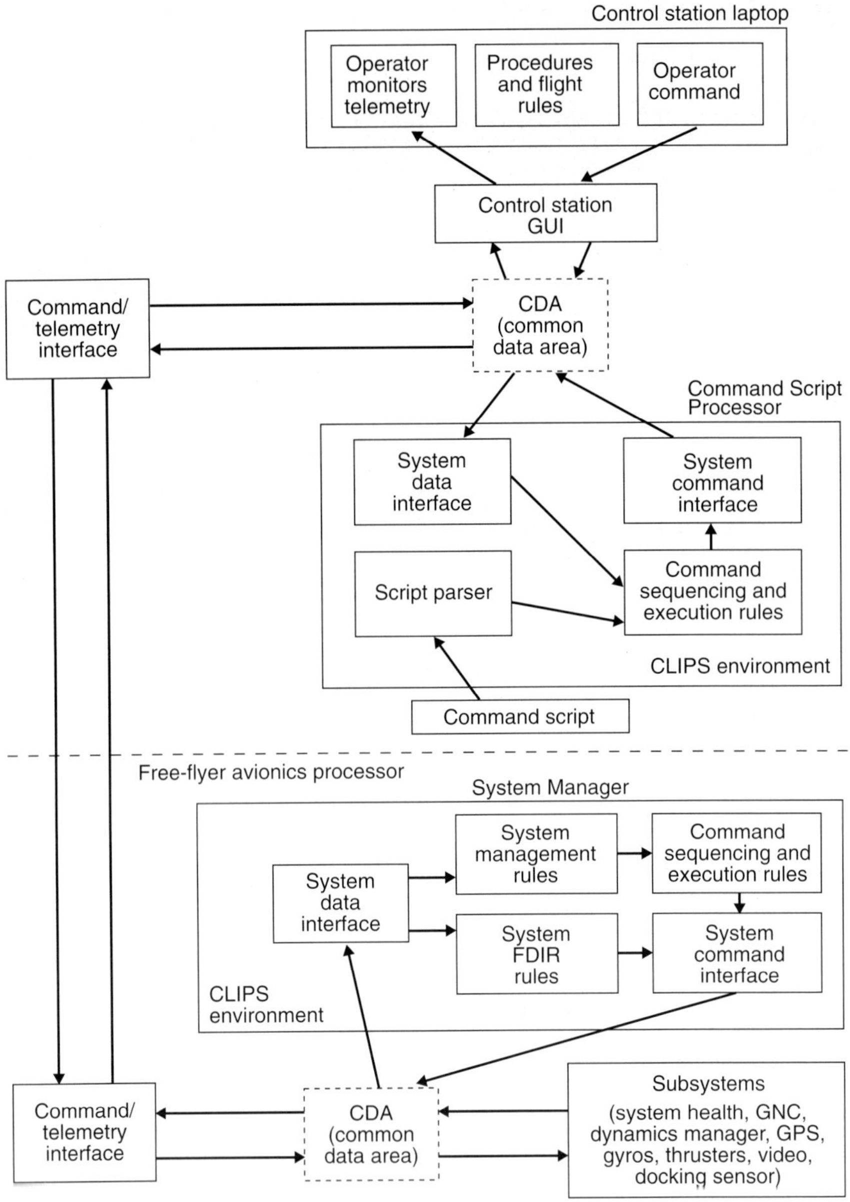

- Rule-based autonomy, based on operator knowledge, procedure, flight rules
- GUI, Command Script Processor, and system manager all share same command and data handling interface
- Command sequencer/executor is the same on the control station and free-flyer

Fig. 16.12. End-to-end control architecture.

16.3.6.3 Dynamics Manager

The AERCam Dynamics Manager[9] (ADM) software is a set of logic tied in to the system manager and tightly coupled with the core GNC software that commands free-flyer trajectory maneuvers in response to system failures. The ADM manages free-flyer maneuvering primarily during periods of communication loss, but also in situations where automatic safe maneuvering is required (e.g., when self-imposed minimum standoff distance or speed-limit violations occur). All ADM commands are aimed at either recovering the mission or safing the free-flyer in the event of a hardware or software failure. The system manager is responsible for choosing the appropriate command based on the current vehicle state. Command requests are then sent through the common data area to the ADM, which calculates the appropriate parameters, including trajectory targets, and performs direct commanding of the GNC system. Figure 16.13 shows the commanding relationship between these different components for different situations.

Certain commands may be passed from the system manager to the ADM. Each command has a list of flags and necessary conditions associated with it that must be satisfied for that command to execute. For example, communication must be lost for a "reacquire communications" command to be valid, and the free-flyer must have good attitude-state information to perform automatic translation maneuvers. The system manager is constantly monitoring system-state information and will issue to the ADM only the command available for a given system state. The ADM commands have a priority order based on mission preservation

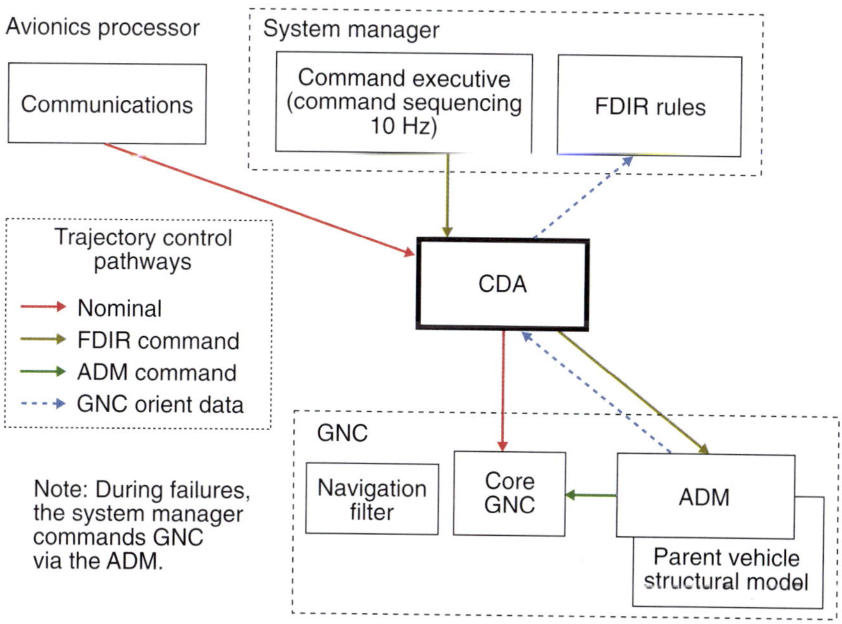

Fig. 16.13. ADM/system manager commanding interface.

likelihood, subject to the specifics of a particular mission. Using the combination of command availability based on vehicle state and priority order, the system manager can choose the best ADM command in any situation by selecting the highest priority command that is available at any given time.

The ADM uses "waypoints" and "breadcrumbs" as steering targets. Waypoints are predefined position and attitude values, relative to the parent vehicle, that are used to perform routine functions such as scanning the exterior of the parent spacecraft or traveling from the free-flyer hangar to a region of interest around the parent vehicle. Between waypoints, or when the free-flyer is in manual mode, breadcrumbs are defined at user-defined distance intervals that allow the free-flyer to retrace its steps. Breadcrumbs are a record of position, attitude, and creation time. The time stamp is recorded to ensure that breadcrumbs are not "stale" when being used to retrace the path of the free-flyer around the parent vehicle. A breadcrumb can be marked as a point of interest by the crew. Once a waypoint is marked as a point of interest, the crew is able to use a "return" command to automatically revisit that location.

16.3.7 Control Station Hardware

The Mini AERCam is controlled remotely from the control station. Data are transmitted between the control station and the free-flyer using a wireless Ethernet transceiver. The technology demonstration used commercially available 802.11b/g wireless Ethernet routers with the 802.11b PCMCIA card on the free-flyer described earlier. The control station laptop has serial interfaces to the rotational and translational hand controllers and to the parent vehicle GPS receiver. The control station also contains the video decompression and display system.

The rotational and translational hand controllers are used to operate the vehicle manually. The operator may input commands through the control station laptop, which also displays system status and engineering data. Forward- and side-facing camera views as well as still-frame snapshots are available to the operator. In addition to the live imagery, a situational awareness display, which can be shown on a second laptop, provides a high-fidelity graphical representation of the Mini AERCam location with respect to the parent vehicle. The situational awareness display uses the DOUG program to render synthetic views from any desired viewpoint. Figure 16.14 shows the control station hand controllers and displays.

During planning for potential Mini AERCam integration into the shuttle orbiter for on-orbit inspection, an alternate hand controller concept was developed and prototyped. This alternate concept is a single control pad with an ergonomic design—similar to commercial game controllers, but with radiation-tolerant avionics. The primary advantage of the control pad is that it requires no structural mounting in the parent spacecraft, thereby eliminating the associated mechanical integration costs. The prototype design, shown in Fig. 16.14, was evaluated by a group of astronauts and found to represent a suitable replacement concept relative to the traditional hand controller pair.

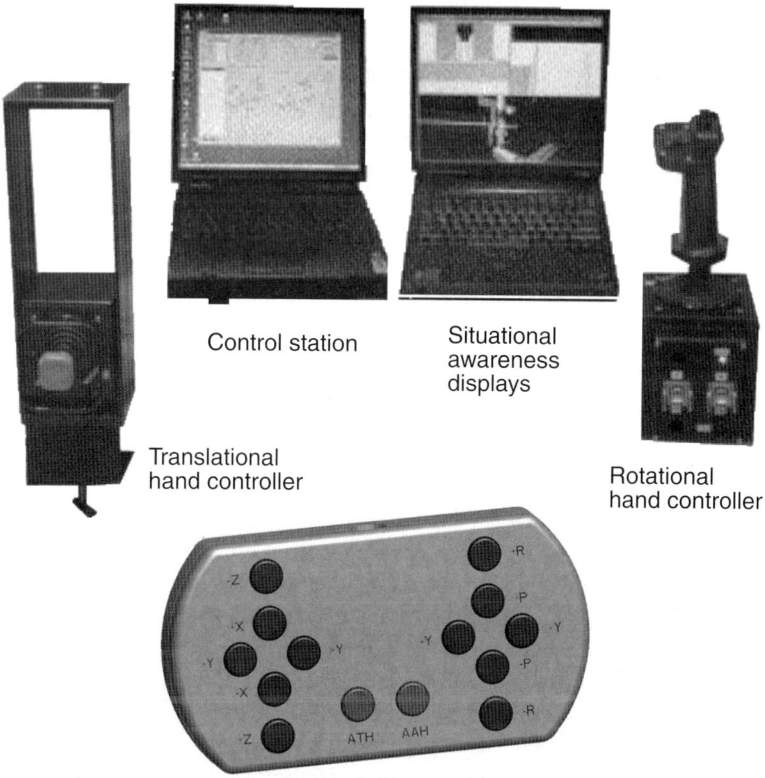

Fig. 16.14. Operator controls and displays (top) and alternate "control pad" controller concept (bottom).

16.3.8 Docking System

The Mini AERCam docking system[10] consists of sensor and mechanical components. The free-flyer video system determines the pose of the free-flyer with respect to a docking target to supply docking-approach navigation for onboard maneuvering of the free-flyer into close proximity with a magnetic docking mechanism. Magnetic attraction between the free-flyer and the docking system performs the final alignment and precision capture of the free-flyer. The docking system then engages a hard-dock capture that is suitable for connecting propulsion and electrical recharge elements. After hard dock is achieved, the electromagnet can be deenergized.

16.3.8.1 Magnetic Docking System

The Mini AERCam free-flyer was designed to be small enough to reside on almost any space vehicle and to inspect the parent spacecraft on a routine basis or on demand. Realizing the goal of a highly versatile inspection tool requires a means to deploy, retrieve, and recharge the free-flyer without human intervention.

The Mini AERCam docking system was developed to provide automatic deployment and retrieval. The two primary elements are magnetic docking and vehicle retention. The magnetic docking element aligns and captures the vehicle, while the retention element supports the vehicle during heavy loading conditions (e.g., launch and landing). Hard-dock retention is achieved when the drive motor secures the ball-lock mechanism in alignment with the ball-lock detents on the free-flyer. The docking element is compatible with hardware for refueling the propulsion system and recharging the onboard batteries. Both of these services use an axial access port in the refuel cluster. The design of the docking element is shown in Figs. 16.15 and 16.16. To show the available axial access port, the refueling and recharging components have been omitted. The figures show the configuration tested in 2003.

Lab tests of the prototype docking hardware have been successful. The system that was developed is able to both align and draw in the vehicle before any physical contact is made. The ability to align before contact helps ensure successful docking. The capture envelope extends approximately 13 cm (5 in.) from the front center of the electromagnet, and misalignments on the order of ±30 deg can be corrected; thus, the system can capture an almost completely passive vehicle.

To provide initial alignment, the system uses magnetic fields generated by passive permanent magnets or active electromagnets. The magnets begin to interact during the early stages of docking approach, providing initial axial and rotational alignment as well as an attractive force that helps draw the vehicle in. Once the vehicle is within the capture envelope and relative velocity tolerances, no further maneuvering by the vehicle is required. For low-mass vehicles like Mini AERCam, the magnetic force is enough to completely draw the vehicle in to soft-dock. The mechanical rotational alignment guides provide the final rotational positioning, and a ball-lock mechanism is used to secure the vehicle. If an electromagnet is used on either the vehicle or dock, an additional active release is available. Reversing the polarity of the electromagnet and controlling the current allows the vehicle to be released with a tunable parting velocity. Figure 16.17 shows the prototype docking hardware used for air-bearing testing. The structure on the left shows the modified thruster cluster that held a permanent magnet; the electromagnetic dock is on the right.

A vehicle retention element or "hangar" is intended to complement the magnetic docking system by providing a second level of capture during loading conditions such as launch. The retention element could also be used in a stand-alone capacity (no magnetic docking) to capture the vehicle when precision docking is not required.

Figure 16.18a portrays one hangar concept. The petals of the spherical enclosure are operated by a linear actuator or jackscrew located in the base. Lined with foam, the petals allow the vehicle to be misaligned and do not require precise

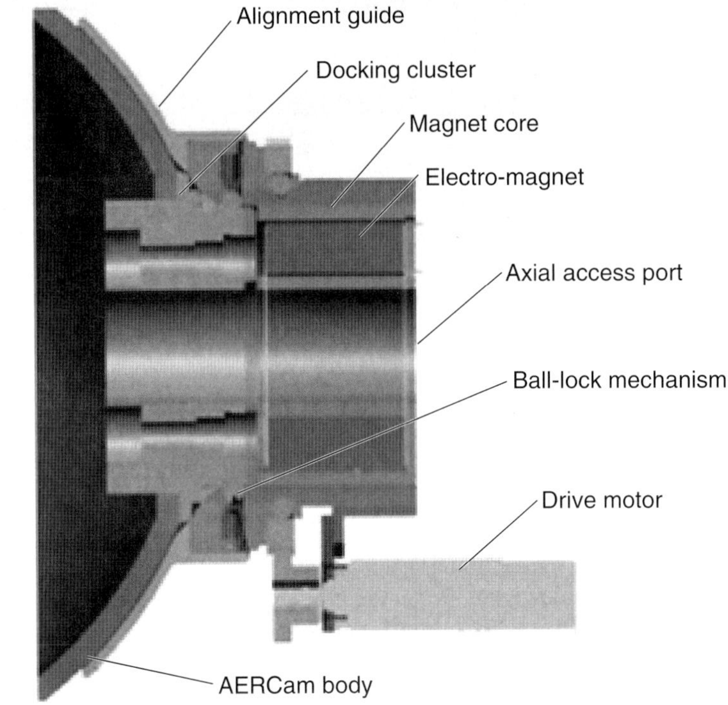

Fig. 16.15. Magnetic docking prototype cross section (free-flyer docked).

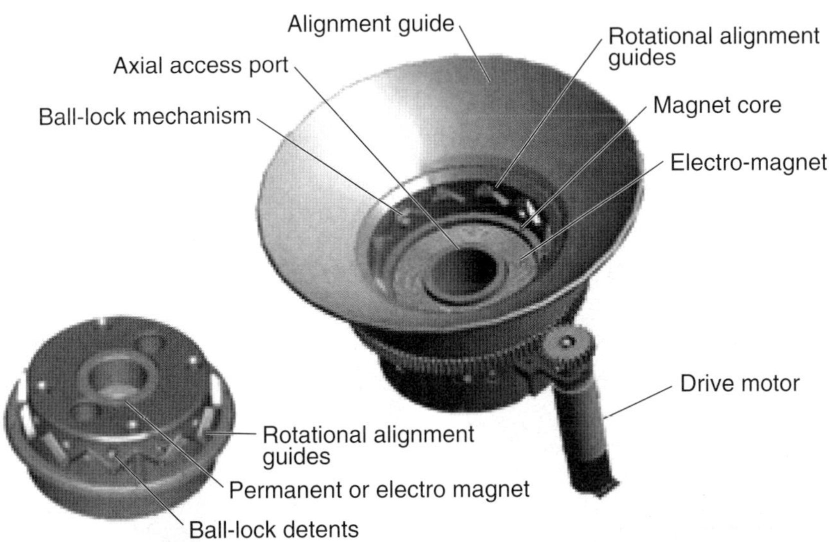

Fig. 16.16. Magnetic docking prototype interfaces (the free-flyer docking cluster is shown on the left; the parent vehicle docking system is shown on the right).

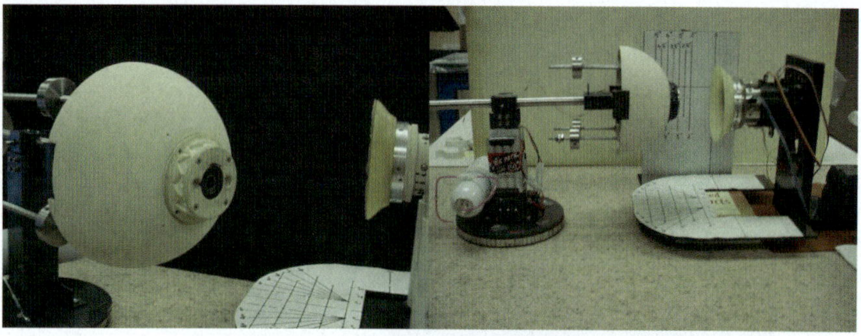

Fig. 16.17. Magnetic docking prototype testing.

positioning. The linear actuator drives the petals open or closed through a four-bar mechanism. The mechanism is driven over center in the closed position. By driving the mechanism over center, the petals are effectively locked and no additional brake is needed.

Figure 16.18b depicts an alternate hangar configuration developed during planning for potential shuttle orbiter inspection. This configuration is customized for mounting on the sidewall of the shuttle orbiter payload bay using an adaptive payload carrier.

16.3.8.2 Docking Navigation Sensor

Mini AERCam employs a docking sensor system that uses the onboard video system and requires only a passive target on the host spacecraft. The AutoTRAC Computer Vision System determines the 6-degree-of-freedom pose of a planar docking target with respect to a camera. It uses active lighting via LEDs in a ring around a camera lens to illuminate retroreflectors located on the target object in predetermined locations. The target is shown in Fig. 16.19. These retroreflective features are brighter in the images with the LEDs on and dimmer in the images

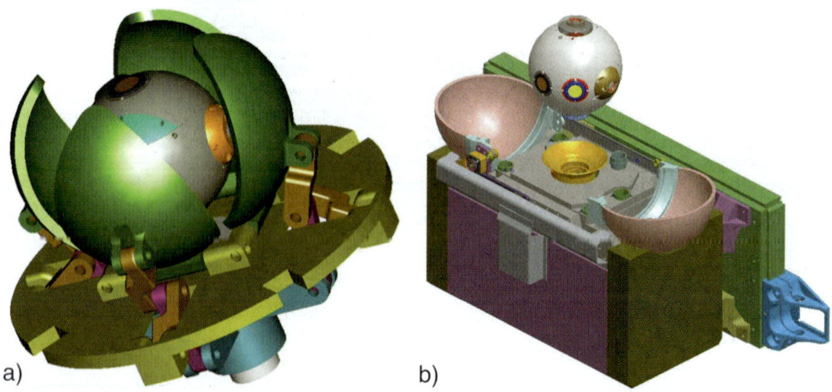

Fig. 16.18. Vehicle retention element (a) and alternate (b).

Fig. 16.19. Mini AERCam hangar base target with five retroreflectors.

with the LEDs off. When the on/off images are compared, the strongest features in the difference image are the retroreflectors. The difference image is binarized using a threshold value, and then the centroids of the connected components of the binary image (blobs) are determined. These are the potential target features. The correspondence between the potential target features and the known target model is determined. Then a nonlinear least-squares fit of the image data is performed to determine the pose of the target object with respect to the camera. The resulting pose (x, y, z, pitch, yaw, and roll) is then converted from the camera reference frame to the vehicle body frame and input into the navigation system (Fig. 16.20.) All image processing and pose estimation calculations are performed onboard the free-flyer. The AutoTRAC Computer Vision System provides a precise relative navigation signal that supports automated deployment and docking. Navigation using the system has been integrated into the overall navigation filter and has been demonstrated in the air-bearing table test environment.

16.3.9 Free-flyer Specifications Summary

The success of the Mini AERCam technology demonstration system was fostered by the successful integration of the free-flyer hardware and software into an approximate flight-like configuration. A summary of Mini AERCam specifications is provided in Table 16.2.

16.4 Mini AERCam Testing and Analysis

Free-flyer testing has been conducted on an air-bearing table and in a 6-degree-of-freedom closed-loop orbital simulation. In the air-bearing table environment, the

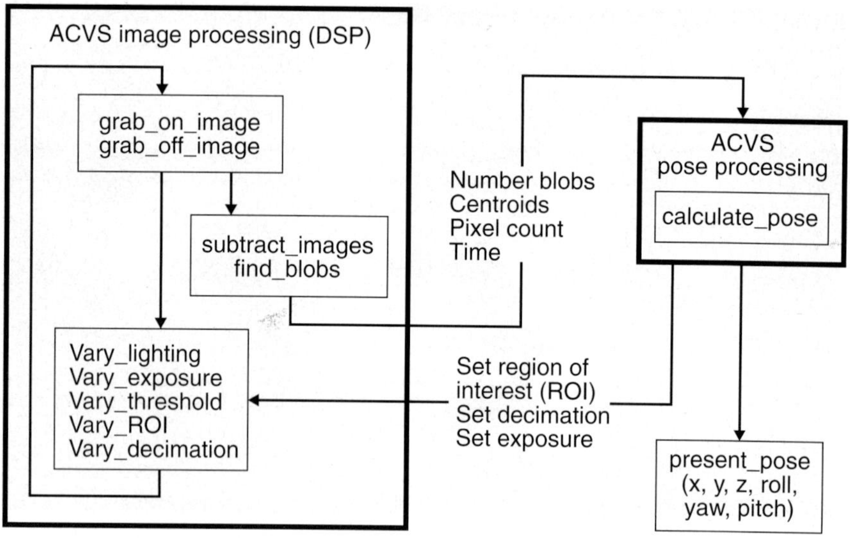

Fig. 16.20. AutoTRAC Computer Vision System (ACVS) processing and control.

free-flyer hardware is placed in the cradle of a tetherless air-bearing sled, which produces a nearly frictionless environment. The free-flyer fires its thrusters to maneuver along the air-bearing surface while it transmits video and telemetry to the pilot control and display station. The orbital simulation models the 3-D dynamics of the free-flyer in proximity to the parent vehicle, and produces corresponding situational awareness views and simulated free-flyer camera views. A high-fidelity simulation is achieved using free-flyer avionics hardware in the loop.

16.4.1 Air-Bearing Table Test Environment

The air-bearing table test facility is used for testing free-flyer hardware in two translational directions and about one axis of rotation. An air-bearing sled produces a cushion of air to float the free-flyer on an air-bearing table, which simulates microgravity. The air-bearing test configuration is shown in Fig. 16.21. The air-bearing sled is located below the rectangular sled cradle. The cradle enclosure contains nitrogen-run tanks that enable low-cost ground testing of Mini AERCam in lieu of a xenon flight tank. The run tanks feed the internal free-flyer regulator through the free-flyer propulsion recharge port.

The free-flyer can perform rotational and translational maneuvers, including automatic attitude hold, automatic translation hold, and point to point maneuvering, on the air-bearing table. Since GPS is not available inside the laboratory, a COTS image-based positioning system called Visualeyez is used for navigation.

16.4.2 Orbital Simulation Test Environment

The orbital simulation models the 3-D dynamics of the free-flyer in proximity to a parent vehicle, such as the shuttle or International Space Station, and produces

Table 16.2. Mini AERCam Technology Demonstrator Specifications

Size	19 cm diameter sphere
Mass	5 kg
Power (average)	15W (6 hr run time)
Attitude and position control	12 thrusters mounted around central ring
Propellant	Xenon gas (12.2 m/sec [40 ft/sec] ΔV) (nitrogen gas used for ground test)
Attitude rate sensors	MEMS gyros
Relative position sensors	Precise relative GPS
Avionics processor	Custom-built 22-layer circular PowerPC board
Instrumentation	I^2C network
Communications	802.11b wireless Ethernet; micropatch antenna
Lighting	LED array
Video	Two digital video cameras One digital inspection camera for still images

corresponding "God's-eye" views and simulated free-flyer camera views. The orbital simulation testbed, shown in Fig. 16.22, is designed for testing GNC capabilities in a high-fidelity simulation with much of the Mini AERCam avionics in the loop. The simulation is driven directly by thruster driver signals from a free-flyer avionics board. MEMS gyro responses are emulated in specialized hardware, and the position "truth" state from the simulation drives a GPS signal generator that provides realistic RF signals to the GPS receivers.

As on the air-bearing table, the free-flyer is controlled in the orbital simulation using separate rotational and translational hand controllers and a control station. In place of live video, the images as seen from the free-flyer are simulated with high-fidelity graphics. Engineering data and state information are displayed on the control station. In addition, a "God's eye" view is generated for the situational awareness display, which provides the operator with a graphical view of the Mini AERCam location with respect to host vehicle structure. Figure 16.23 shows an example scene from the high-fidelity graphics used for the orbital simulation.

In the orbital simulation, the free-flyer can perform rotation and translation maneuvers, including automatic attitude hold, automatic translation hold, and point-to-point maneuvering.

A duplicate of the avionics hardware is used during orbital simulations to ensure full-fidelity avionics processor testing. For the orbital simulation, the Trick

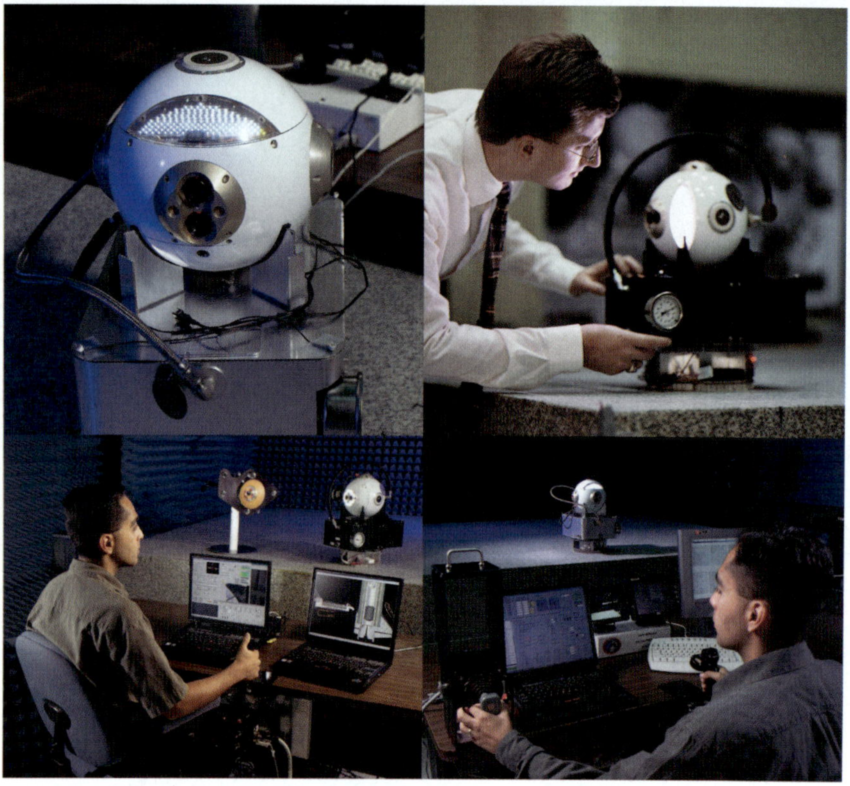

Fig. 16.21. Mini AERCam testing on an air-bearing table.

simulation communicates with a Spirent Federal Systems GPS simulator through a user datagram protocol socket connection to an interface personal computer (PC). The Trick simulation sends vehicle state information for both the free-flyer and the parent vehicle to the interface PC, which packages the data in a format that the GPS signal generator can interpret and transfers the information over an IEEE-488 connection. The signal generator then uses that information to generate the RF signals, which are transmitted via RF cables to both the Mini AERCam and the parent vehicle GPS receivers. The Mini AERCam avionics processor board is connected serially to the G-12 GPS receiver. The control-station laptop provides an interface to the parent vehicle GPS receiver and packages the receiver data, along with operator commands, to send across the wireless Ethernet interface to the free-flyer.

The two high-fidelity test environments described in this section have been established to demonstrate the capabilities of Mini AERCam. Table 16.3 summarizes the featured elements for each environment, all of which have been accomplished.

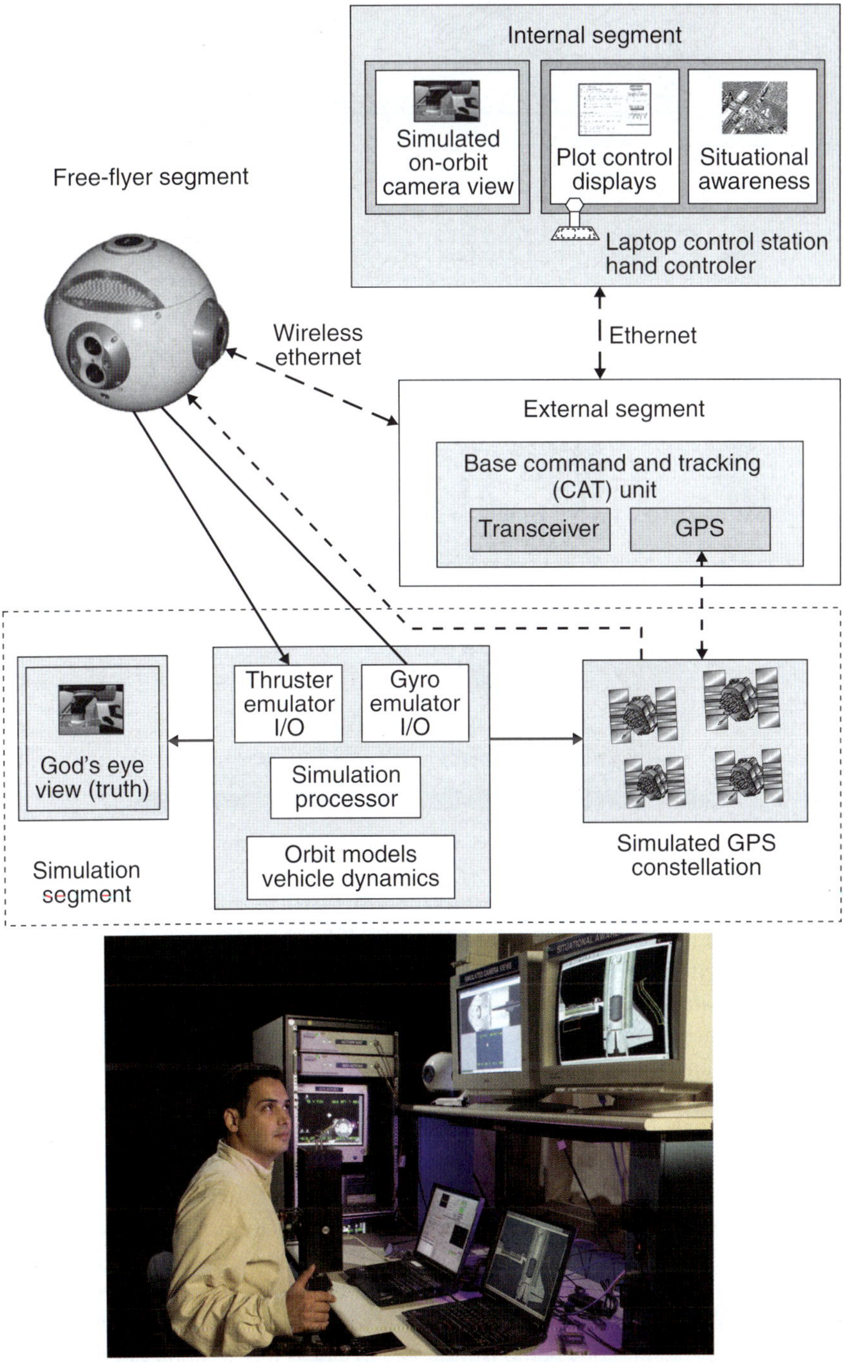

Fig. 16.22. Orbital simulation testbed architecture (top) and lab configuration (bottom).

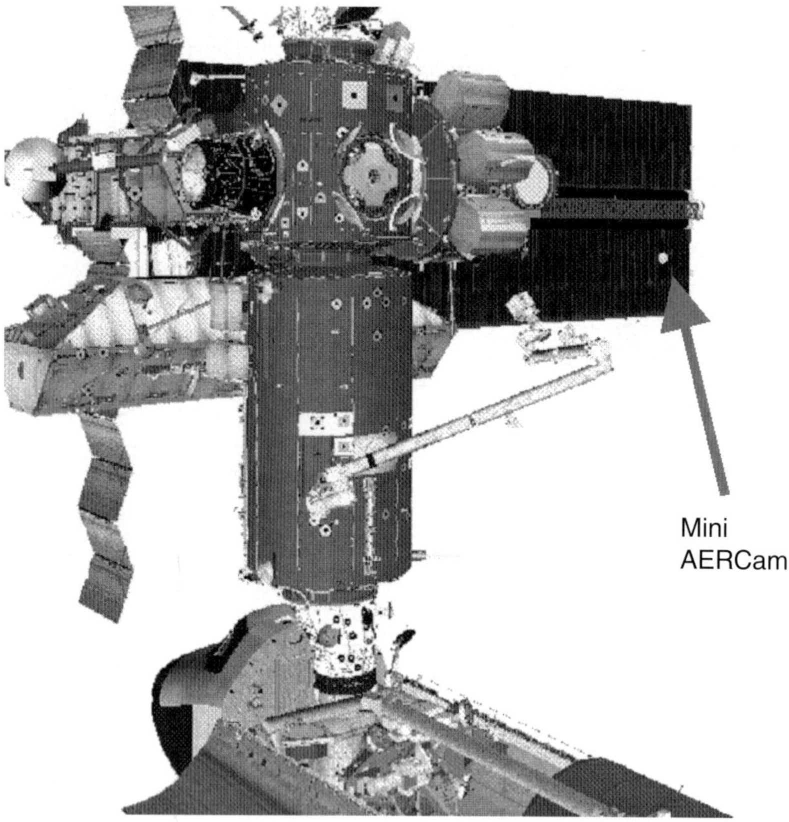

Fig. 16.23. "God's-eye" view for the orbital simulation.

16.4.3 AERCam Environments Analyses and Testing

In addition to functional testing, the Mini AERCam technology demonstrator has been subjected to space-environment analyses and testing to ensure that its design provides a suitable basis for a flight system. Environmental testing with the technology demonstrator has included thermal vacuum, radiation, communications link margin, and solar lighting.

Early analyses focused on RF coverage and the space thermal vacuum environment. As a quick look at RF coverage, simulations were used to analyze line-of-sight blockage around the International Space Station.[11] The composite line-of-sight blockage, from six hypothetical antenna locations analyzed, obscured only 2.3% of the total coverage volume surrounding the Space Station. Moreover, the associated link margin analysis shows that 77% of the coverage volume has a 22 dB forward link margin with a minimum forward link margin of 13.8 dB.

Thermal analysis[12] was performed both to support the development of the Mini AERCam technology demonstrator and to assess the feasibility of flying the

Table 16.3. Demonstrated Mini AERCam Capabilities

Orbital simulation demonstration elements (with simulated video)	Air-bearing table demonstration elements (with actual free-flyer video)
1. Manual 6-DOF control	1. Manual 3-DOF control
2. Automatic attitude hold (inertial and relative)	2. Automatic attitude hold
3. Relative navigation with GPS hardware in the loop	3. Relative navigation with indoor navigation system
4. Automatic translation hold with GPS hardware in the loop	4. Automatic translation hold with indoor navigation system
5. Automatic point-to-point maneuvering in 6-DOF	5. Automatic point-to-point maneuvering in 3-DOF
6. Automatic docking with dynamic docking model	6. Automatic docking with magnetic docking hardware
7. Command scripting	7. Command scripting

design in space. A thermal math model of the Mini AERCam vehicle was developed. It includes the structure and electronic board level models.[13] The electronic board models were developed based on designer input, ground bench-top testing correlation, and infrared camera testing. Figure 16.24 shows the Mini AERCam thermal math model.

The on-orbit thermal environment is complex because the heat incident on a Mini AERCam is a function of not only orbit angle and beta angle, but also the location of the Mini AERCam with respect to the parent vehicle. The problem is further complicated given that a parent vehicle such as the International Space Station is an articulating structure, changing configuration with both orbit and beta angle. A sample environment of the Space Station is presented here. A simplified model was obtained and populated with Mini AERCam-size spheres to sample the environment at locations around the station. This analysis was later extended to the case of the shuttle in free flight. Figure 16.25 shows the International Space Station screening model. This model computes 700 "effective environment" temperatures that take into account the direct and reflected solar and infrared energy exchanges in the complex environment. The detailed thermal math model was assessed at these effective environment temperatures to quantify the feasibility of the Mini AERCam design in space.

The resulting effective environment temperatures ranged from –83°C to 22°C when all points and all beta angles were considered. If the points at the more extreme locations (located directly over the solar arrays) and the more extreme beta angles (±75 deg) are excluded, the resulting range is –57°C to –18°C. The

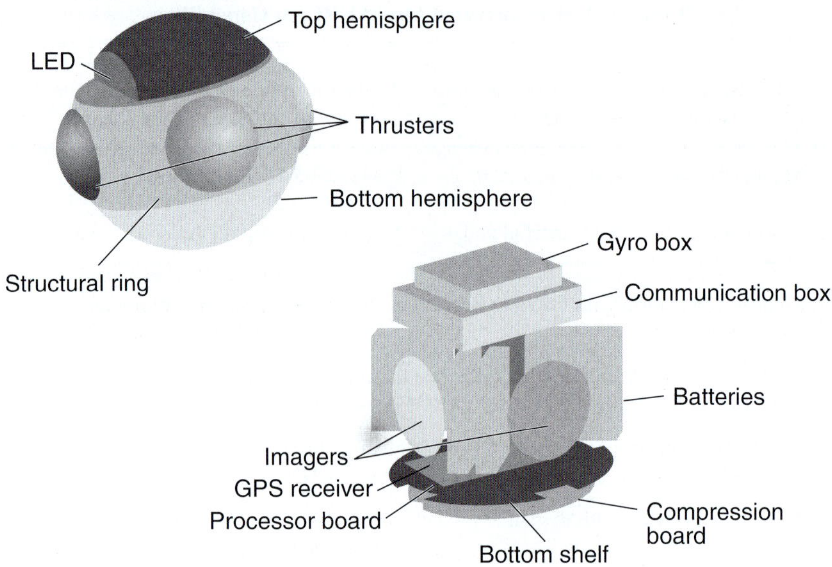

Fig. 16.24. Mini AERCam thermal math model.

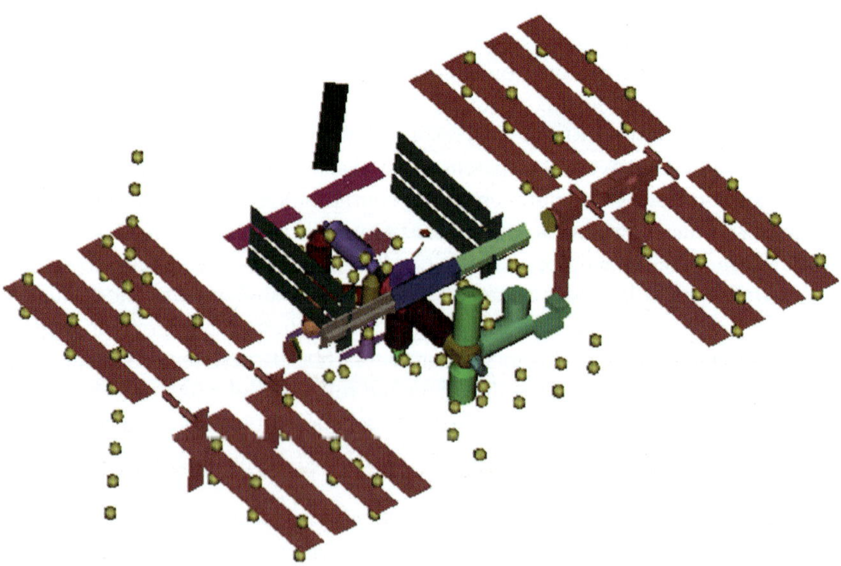

Fig. 16.25. International Space Station screening thermal math model.

flight design of Mini AERCam, incorporating only minor design changes from the technology demonstrator to enhance local thermal connectivity, could operate at steady state in effective environments ranging from –83°C to –18°C, and could operate in an effective environment temperature outside this range by limiting the exposure duration and/or reducing the power of the free-flyer. In fact, the technology demonstrator hardware functioned successfully during a 36 hour thermal vacuum test conducted at temperature settings of –45°C, –18°C, and 10°C in Johnson Space Center's Chamber N during 2003. That thermal vacuum test of the free-flyer (shown in Fig. 16.26) also served to validate the detailed thermal analysis math model. Overall, the size, power level, and general design of the Mini AERCam ground demonstration vehicle are thermally feasible in a space environment, based on the modeling.

16.5 Mini AERCam Applications

Because the AERCam was designed to assist human space flight efforts, it is inherently applicable to future exploration missions. For example, the Mini AERCam could potential be used to:

- Inspect the descent vehicle's thermal protection system prior to entry and landing.
- Provide an alternative to inspection of EVAs during lunar cruise or interplanetary flight.
- Provide a remote view for in-space assembly and maintenance.[14]
- Provide an orthogonal view of rendezvous and docking or other berthing/mating operation.

Fig. 16.26. Free-flyer thermal vacuum testing at NASA Johnson Space Center.

Inspection represents a stand-alone capability in which the free-flyer is used for either routine or contingency scanning. Such an application was considered for second-generation shuttle inspection. Figure 16.27 depicts the orbital simulation of the Mini AERCam inspecting the leading edge of the orbiter wing. The left pane shows a situational awareness view of the relative free-flyer at 4.6 m (15 ft) from the starboard wing (shown to scale). The right pane shows the corresponding simulated field of view of the wing as seen from the free-flyer. Figure 16.28 shows the trace of the free-flyer flight path resulting from a full scan of the starboard wing. Due to the curvature of the wing leading edge, this scan pattern involves four passes and takes 30 min to complete.

Stand-alone inspection is clearly a potential application for exploration. The Mini AERCam might serve as a reconnaissance agent to help plan a complex maintenance task; it might be used to provide supplemental views (e.g., orthogonal to a mating interface) during a complex task; or, it might be called upon to record "as-built" documentation of a completed task. Having an extra set of "eyes" (or other desired sensor) on a mobile platform is clearly applicable to a large range of tasks.

Even before a system like Mini AERCam is deployed operationally for inspection or viewing tasks on a crewed exploration mission, it could be used as a platform for testing other advanced capabilities such as autonomous rendezvous and docking technology. A flight demonstration of Mini AERCam conducted for this purpose would produce:

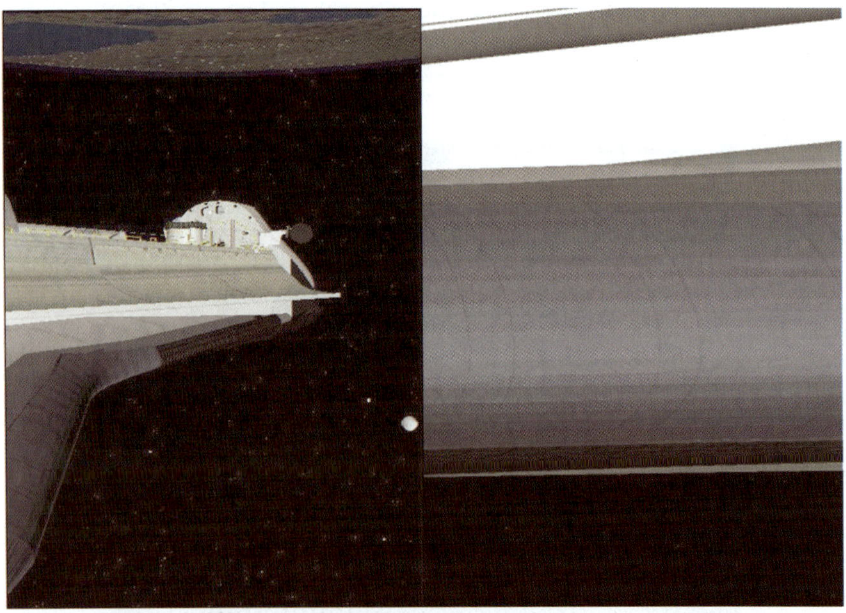

Fig. 16.27. Mini AERCam shuttle inspection simulations. The left pane shows a situational awareness view at 4.6 m; the right shows the wing as seen from the free-flyer.

- Demonstrated teleoperated and automatic maneuvers, including docking.
- A verified relative navigation capability (precise relative GPS) for use during LEO in-space assembly flights and for use as a "truth" sensor during tests of non-GPS relative navigation.
- A flight-proven candidate docking navigation sensor system, which could also serve as the non-GPS relative navigation system on exploration missions.
- A flight-tested spacecraft platform (avionics core, propulsion, video imaging) for integration with future sensor payloads.
- Launch-vehicle integration completed to support future flight tests.

16.6 Conclusions

Engineers at NASA Johnson Space Center have designed, developed, and tested a nanosatellite-class free-flyer intended for future external inspection and remote viewing of human space-flight activities. The Mini AERCam technology demonstration system has been integrated into the approximate form and function of a flight system and demonstrated in two high-fidelity simulation environments.

In a future operational scenario, system checkout is conducted with the free-flyer at its docking location. Free-flyer deployment is commanded from

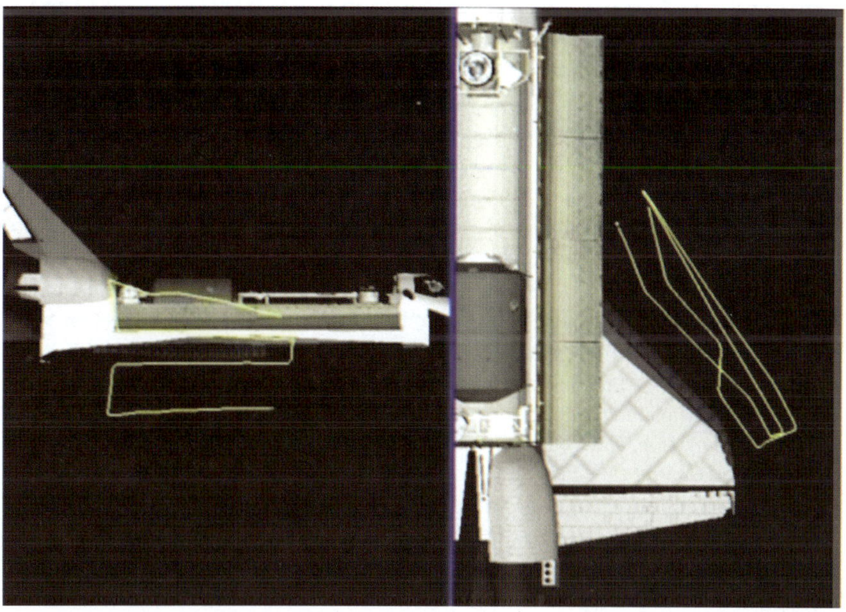

Fig. 16.28. Mini AERCam shuttle inspection trace. The two panes show the trace of the free-flyer flight path.

an astronaut at the control station. The free-flyer is released from the docking mechanism. Inspections are conducted using either teleoperation or automatic maneuvering. After the inspection mission is complete, the Mini AERCam returns to the docking station. After docking, the free-flyer is recharged and refueled to support additional inspection missions as needed.

The current Mini AERCam prototype has demonstrated capabilities that can be of significant benefit to NASA's human space-flight program for external visual inspection of space vehicles. Potential Mini AERCam applications range from stand-alone remote viewing to complex multiagent cooperative tasks. Mini AERCam can also serve as a test platform for evaluating algorithms and relative navigation for autonomous proximity operations and docking, evaluating other sensors and technologies, and conducting formation-flying experiments with multiple free-flyers. In the future, small free-flyers such as the Mini AERCam may serve many mission functions such as visual or nonvisual inspections, communications relay, or viewing for human and robotic EVAs, both in LEO and during longer exploration missions.

16.7 References

[1] AERCam Sprint website, http://AERCam.jsc.nasa.gov/sprint.

[2] S. E. Fredrickson, P. S. Lockhart, and J. D. Wagenknecht, "Autonomous Extravehicular Robotic Camera (AERCam) for Remote Viewing," *Proceedings of the AIAA ISS Service Vehicles Conference* (26–29 April 1999).

[3] L. Abbott, "Mini AERCam Avionics," *2003 IEEE Aerospace Conference* (8–15 March 2003).

[4] J. Wagenknecht, S. Fredrickson, T. Manning, and B. Jones, "Design, Development and Testing of the Miniature Autonomous Extravehicular Robotic Camera (Mini AERCam) Guidance, Navigation, and Control System," *26th Annual American Astronautical Society Guidance and Control Conference* (5–9 Feb. 2003).

[5] K. W. Key and S. Vitalpur, "Performance of the Flexible Inertial Relative Estimator (FIRE) for Terrestrial and On-Orbit Relative Navigation," *Institute of Navigation Global Positioning Systems Conference* (Sept. 2002).

[6] S. Fredrickson, S. Duran, A. Braun, T. Straube, and J. Mitchell, "AERCam Autonomy:, Intelligent Software Architecture for Robotic Free Flying Nanosatellite Inspection Vehicles," *Proceedings of AIAA Space 2006* (San Jose, CA, 19–21 Sept. 2006).

[7] J. Wagenknecht, S. Fredrickson, T. Manning, and B. Jones, "Design, Development and Testing of the Miniature Autonomous Extravehicular Robotic Camera (Mini AERCam) Guidance, Navigation, and Control Systems," *26th Annual American Astronautical Society Guidance and Control* (5–9 Feb. 2003).

[8] CLIPS website, http://clipsrules.sourceforge.net.

[9] R. Proud, J. Hart, "FLOAAT, A Tool for Determining Levels of Autonomy and Automation, Applied to Human-Rated Space Systems," *Proceedings of InfoTech @ Aerospace* (Arlington, VA, 26–29 Sept. 2005).

[10] S. Fredrickson, S. Duran, N. Howard, and J. Wagenknecht, "Application of the Mini AERCam Free Flyer for Orbital Inspection," *Proceedings of SPIE Vol. 5419, Spacecraft Platforms and Infrastructure,* Peter Tchoryk, Jr., Melissa Wright, Eds. (April 2004).

[11] S. Bolen, L. Vazquez, S. Fredrickson, P. Fink, C. Sham, J. Cox, C. Phan, and B. Panneton, "ISS MAC Radio Frequency (RF) Coverage Analysis Using iCAT Development Tool," *21st AIAA International Communications Satellite Systems Conference* (15–19 April 2003).

[12] N. M. Williams, "Mini Autonomous Extravehicular Activity Robotic Camera (AERCam) Thermal Analysis Report," JSC-29894 (2002).

[13] S. W. Miller and E. Marotta, "Thermal Modeling and Testing of a Nanosatellite's Avionics Board," *Journal of Thermophysics and Heat Transfer,* **21** (3), pp. 496–504 (2007).

[14] W. Whittaker, P. Staritz, R. Ambrose, B. Kennedy, S. Fredrickson, J. Parrish, and C. Urmson, "Robotic Assembly of Space Solar-Power Facilities," *American Society of Civil Engineers Journal of Aerospace Engineering,* **14** (2), pp. 59–64 (April 2001).

Chapter 17

Developing and Maturing Microsatellite Formation Flight Technology

Alvar Saenz-Otero,* Simon Nolet,* Raymond Sedwick,† David W. Miller*

17.1 Introduction

Synthesizing the functionality of larger satellites through the formation flight of multiple microsatellites carries great promise for future space exploration. But while the motivation for satellite formation flight can be quite compelling, the challenges of developing and maturing the technology can be equally daunting. This chapter focuses on the efforts of the MIT Space Systems Laboratory (MITSSL) to address these challenges and to answer the questions central to their resolution:

- What current and future missions will require formation flight?
- What are the challenges in formation flight?
- How does one bring formation-flight technology to maturity?
- Can propellantless techniques be employed to reduce fuel consumption during formation flight?

The following sections address the motivation and challenges concerning satellite formation flight. First, a technique for propellantless satellite formation flight—"electromagnetic formation flight"—is introduced and applied to the Terrestrial Planet Finder Interferometer (TPF-I) mission. Next, the philosophy behind the SPHERES facility is introduced; the hardware is described, and sample results of its use to mature microsatellite technologies—including estimation, docking, and formation flight—are presented.

17.2 Motivation for Satellite Formation Flight

Astronomy has entered a golden age. We are starting to answer the age-old questions: How did it all begin? How will it end? Is there life beyond Earth? Discovering the answers to these questions raises daunting engineering challenges. Space telescopes—our premier investigatory tools—are becoming ever larger, exceedingly precise, and more exotic. Whether it is through the Mount Wilson telescope in California or the Hubble Space Telescope in Earth orbit, dramatic vistas of the universe have been opened to us. The next generation of astronomical telescopes, such as the Space Interferometry Mission[1] and TPF-I,[2] will operate

*MIT Space Systems Laboratory, Cambridge, Massachussetts
†University of Maryland, Dept. of Aerospace Engineering, College Park, Maryland

10 million miles from Earth. To improve angular resolution (the ability to distinguish between two closely spaced objects), they will employ multiple telescopes that are spread apart. In the case of TPF-I, these telescopes will lie on separate spacecraft that are flown in a formation that must be maintained to a precision less than the diameter of a hydrogen atom.

This space system—perhaps the most complex ever built—will be expected to work in an environment in which it has never run. It is not even possible to have an end-to-end functional test of such a space telescope on Earth. Furthermore, since launch costs are so high and opportunities so rare, it is only natural for program managers to include as much functionality as possible in each spacecraft, increasing the difficulty of testing such a system. The result is a development environment characterized by high design costs, complexity of system integration and validation, risk of large deployment, low levels of maturation because of design customization, and limited design heritage and legacy.

One innovative approach to satellite design that addresses these issues is to modularize satellite functions (e.g., propulsion, power, attitude control) and achieve subsystem interconnectivity through genderless docking ports and wireless command and data handling. The lab expects to simplify assembly and test prior to launch while enabling self-assembly and reconfiguration on orbit. The goal is to reduce the cost, risk, and time required for the deployment of new spacecraft by changing the fundamental methodology used in their development.

Coordinating the use of multiple satellites to facilitate assembly, servicing, upgrade, and operation of future space-based telescopes is an emerging field. Distributing mission functionality across multiple satellites has the promise of revolutionizing space exploration in general while also presenting unique challenges. These challenges are addressed below.

17.3 Electromagnetic Formation Flight

A cluster of spacecraft in Earth orbit will experience geopotential disturbances that will cause them to drift apart without constant formation keeping. For non-optimized cluster configurations, the ΔV required in LEO (low Earth orbit) is about one centimeter per second per meter of separation per orbit (scaling linearly with orbital period and separation distance), causing the total ΔV over the mission lifetime to be substantial.

Most proposed methods of actuating spacecraft in sparse-aperture arrays use propellant as a reaction mass. For formation-flying systems, propellant can become a critical consumable that is quickly exhausted while maintaining relative separation and orientation. Furthermore, the total required propellant mass increases exponentially with ΔV. Additional problems posed by propellant include optical contamination, plume impingement, thermal emission, and vibration excitation. An alternative is to use a system of solar-powered electromagnets, in concert with reaction wheels, to replace consumables such as propellant.[3,4] This is the principle of electromagnetic formation flight (EMFF).

EMFF can keep formation indefinitely as well as dump any accumulated angular momentum into Earth's magnetic field.[5,6] EMFF is particularly beneficial during close-proximity operations (within 10 m), where thruster firing directions are severely constrained. This is particularly true when positive separation rates for collision avoidance are being created. If thruster firing directions are not properly constrained, plume impingement can pose a contamination and damage threat to neighboring spacecraft. EMFF does not have this problem. In fact, EMFF thrives in close proximity where the coupling between magnetic fields on neighboring spacecraft can be stronger, increasing control authority.

17.3.1 How Does EMFF Work?

EMFF replaces thruster-based propulsion with electromagnetic forces for controlling the relative degrees of separation of a spacecraft cluster.[7] By creating attractive and repulsive forces between spacecraft (Fig. 17.1a), as well as torques and shear (off-axis) forces (Fig. 17.1b), EMFF coils and reaction wheels can control all degrees of freedom (DOF) except the three translations of the center of mass of the entire cluster.

EMFFs may come in two basic flavors: HTS and micro. The first uses high-temperature superconducting (HTS) wire to greatly amplify the interspacecraft force, or extend it over much further distances. A typical HTS-EMFF system could produce an electromagnetic dipole with micropropulsion level thrust (tens of millinewtons) at a hundred meters of separation using commercially available superconducting wire. The prototype creates this force by driving up to 100 A of current on each spacecraft, at a power consumption of about 10 W. As of 2007, the state of the art in commercial HTS wire carries about 150 A through a cross section of 0.5 mm^2. Micro-EMFF uses conventional conductors to provide close-proximity (up to 1 m in ground operations) perturbation compensation, collision avoidance, or slow maneuvering. Micro-EMFF is still in the early stages of development.

Force is the weakest interaction between the spacecraft. So, to illustrate the effectiveness of EMFF in controlling the geometry of an array, force is analyzed. The far-field force (F) between two identical electromagnetic coils is given by the

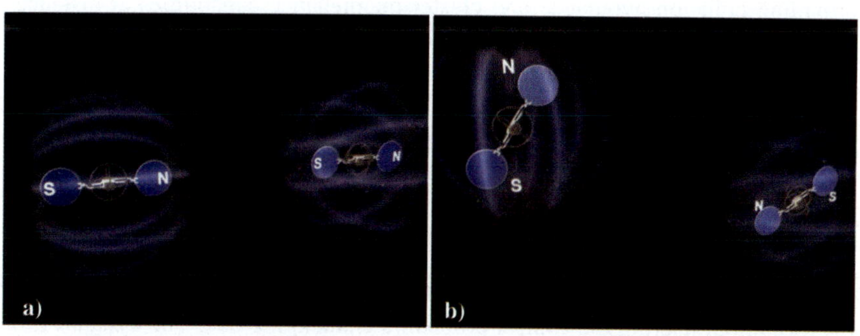

Fig. 17.1. EMFF attraction and repulsion (a) and shear and torque (b).

product of the permeability of free space (μ_0), the square of their total amp-turns (I_T), and the 4th power of the ratio of coil radius to their separation (*R/d*), as shown in Eq. (17.1). The key limitation of EMFF is the rapid reduction of the dipole force with distance. However, the key strength is the fact that the dipole nature of the field permits both force and torque to be created, allowing control of not only separation but also shear and rotation between the spacecraft.

$$F \sim \frac{3\pi}{2}\mu_o I_T^2\left(\frac{R}{d}\right)^4 \sim \frac{3}{2}\left(10^{-7}\right)\left(\frac{I_c}{\rho}\right)^2\left(MR\right)^2\frac{1}{d^4} \tag{17.1}$$

Substituting for the properties of the superconducting wire, this force relationship can be rewritten, revealing two important terms. The first (I_c/ρ) is a technology term associated with the HTS. The maximum, or critical, current density (I_c) supported by the HTS limits the achievable force. However, this critical amperage density can be increased by lowering the temperature of the superconductor. The volumetric mass density (ρ) accounts for the density of the coil and its insulation. This technology term can be altered by selecting different HTS products, changing the operating temperature, and modifying the insulation. The second term is a resource allocation that shows the impact of adding coil mass (*M*) with a major radius (*R*). The equation shows that improvements in HTS amperage density (increase) and insulation mass (decrease) can increase performance substantially, as they have a quadratic effect on the resulting force. Figure 17.2 plots force versus separation for different products of commercially available coil mass and radius (*MR*).

Because power consumption is not an issue, HTS-EMFF allows continuous precision positioning of spacecraft cluster elements. Using reaction wheels can independently control all relative DOF within a cluster of spacecraft. The challenges are thermal control to the required temperature of commercial HTS (~77 K) and the coupled, nonlinear dynamics, both of which have been experimentally studied in depth.[8,9]

Micro-EMFF produces an electromagnetic dipole using conventional wire, providing collision avoidance and cluster maintenance capabilities at distances up to 30 m. The HTS-EMFF system is dominated by the power dissipation in the current controller, while the wire dissipates minimal power. Micro-EMFF does not require thermal control; however, substantial power is dissipated in the wire, in addition to that used by the current controller. Therefore, pulsed operation is used. The challenge is to synchronize the pulses to ensure adequate electromagnetic coupling between the spacecraft.

17.3.2 Dynamics and Controls

The benefits of EMFF come at the cost of coupled and nonlinear dynamics of the formation, making the control problem a challenging one. The dynamics for a general N-satellite electromagnetic formation have been derived for both deep-

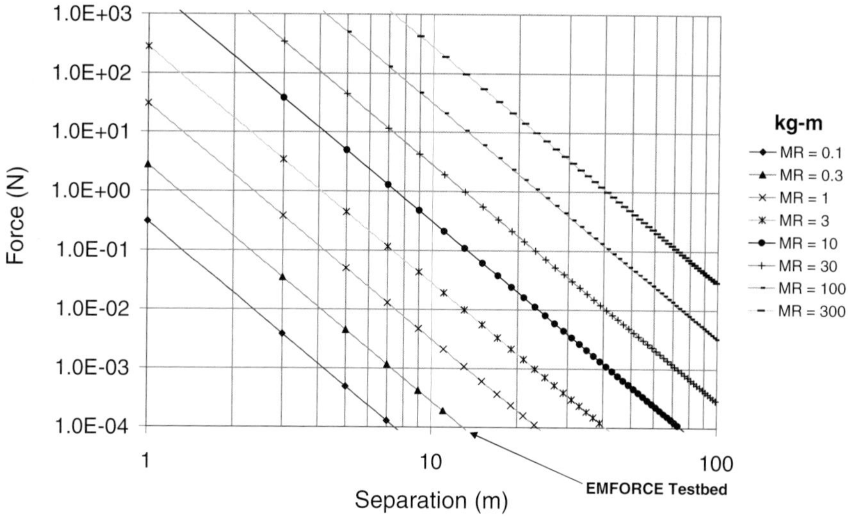

Fig. 17.2. HTS-EMFF can provide precise control at distances up to 100 m.

space missions and LEO formations.[10] Nonlinear control laws using adaptive techniques have been derived for general formations in LEO.[11] Angular momentum management in LEO is a problem for EMFF because of interaction of the magnetic dipoles with the Earth's magnetic field. A solution to this problem for general electromagnetic formations takes the form of a dipole polarity switching control law.[12,13] This switching law preserves the forces between spacecraft, but periodically reverses the torques induced by Earth's magnetic field. This periodic switching allows accumulated angular momentum stored in the reaction wheels to be drained into Earth's magnetic field, thereby avoiding saturation of the wheels.

For EMFF, the formation reconfiguration problem is a nonlinear, constrained optimal time-control problem, as fuel cost for EMFF is zero. Two different methods of trajectory generation, namely feedback motion planning using the Artificial Potential Function Method (APFM) and optimal trajectory generation using the Legendre Pseudospectral method, have been derived for general electromagnetic formations.[12] The results of these methods have been compared for up to 10 spacecraft electromagnetic formations. This comparison shows that the APFM is a promising technique for solving the real-time motion-planning problem for nonlinear and constrained systems, such as EMFF, with low computational cost. Specifically, a fully actuated N-satellite electromagnetic formation can be stabilized and controlled under fairly general assumptions, therefore showing the viability of this novel approach for satellite formation flight from a dynamics and controls perspective.[12,14]

A two-vehicle HTS-EMFF controls testbed is operational on the MITSSL's flat floor or optical table (Fig 17.3a) facilities. Each vehicle has two HTS coils, cooled by liquid nitrogen, that allow the dipole to be steered in the horizontal plane. The coils, designed for close-proximity operations in the laboratory, carry up to 95 A and produce up to 0.2 tesla. The vehicles, which float on air carriages, also contain one reaction wheel that provides torque about the vertical axis. This testbed has been used to test position regulation in the presence of perturbations, as well as trajectory following. In addition, processors, power systems (alkaline and rechargeable batteries), RF communication channels, reaction-wheel controllers, and GPS receivers have been exposed to the magnetic fields. No interference or degradation in performance, caused by the magnetic field, has been detected. While liquid-nitrogen boil-off is used to cool the HTS in the laboratory, it is not a good choice for space. Therefore, there also exists a thermal-vacuum chamber that is used to test thermal-control techniques for the coils.

A micro-EMFF testbed is also operational (Fig. 17.3b). A single circular coil is attached to the end of a swing arm that is mounted to an air bearing. A second coil is mounted on a pivot. When current is simultaneously pulsed through both coils (at up to 30 A), this swing-arm rotates due to the generated force. This testbed has been used to demonstrate axial forcing, shearing, collision avoidance, and deadband control in open-loop and closed-loop configurations.

17.3.3 TPF-I Mission Application

The TPF-I mission, a five-vehicle rotating interferometer at the second Earth-Sun LaGrange point under development by NASA, was studied to understand the reductions in spacecraft mass that can be realized using EMFF. The goal was to analyze the design of TPF-I with sufficient subsystem detail to fairly compare EMFF with micropropulsion systems. The micropropulsion techniques analyzed included pulsed plasma thrusters, field-emission electric propulsion, and two models of HTS-EMFF (named EM1x and EM3x). The five-vehicle TPF-I system consists of two outer collectors, two inner collectors, and one central combiner. The mass breakdown for each is shown in Fig. 17.4. The additional mass for the EMFF subsystem due to reaction wheels and power storage is captured either

Fig. 17.3. HTS-EMFF (a) and micro-EMFF (b) testbeds.

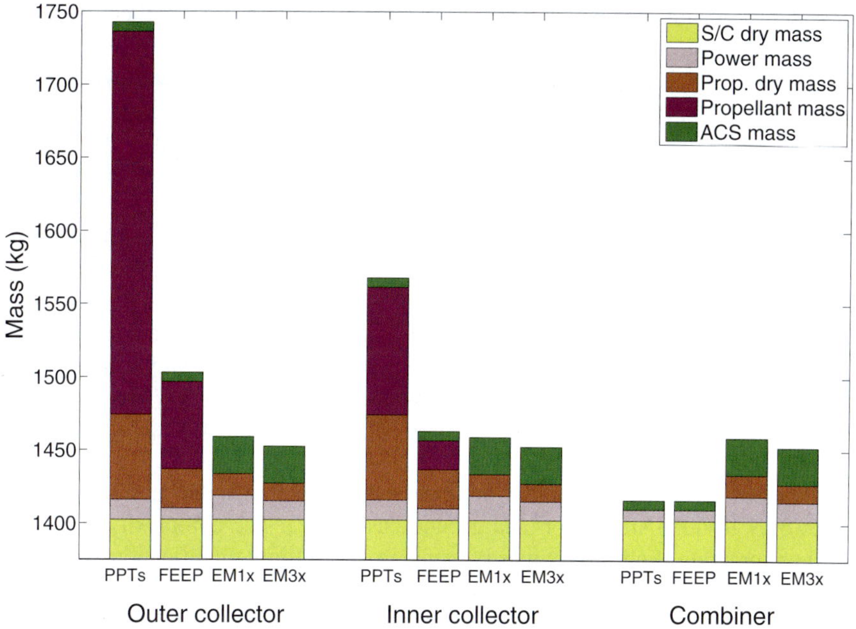

Fig. 17.4. Mass breakdown for various propulsion systems (4 hr rotation period, 75 m baseline). PPT: pulsed plasma thruster; FEEP: field-emission electric propulsion.

within the attitude-control system or the power subsystem. Note that for the propellant-based options, the outer collector is the most massive because it has the highest centripetal load, while the combiner does not have any propulsion mass because it only needs to rotate in place. The propellant-based options also have a relatively high amount of mass (excluding the dry mass) allocated for propellant. This shows that while the attitude-control system mass increases when EMFF is used, the dramatic reduction in propellant mass makes EMFF quite favorable.

Developing formation-flight technology (such as EMFF), demonstrating it in the laboratory, and analyzing its benefits in the context of future missions is only part of the challenge. Maturing the technology so that it can be used in operational systems can be equally difficult. The next section describes one approach to address technology maturation.

17.4 Maturing Formation-Flight GNC

Space-technology maturation is a challenging process. Substantial amounts of money, time, and human resources go into the development of new spacecraft. At every point in the design life of a new spacecraft, substantial risks are involved, especially as the complexity of a new design increases. More than a decade ago, NASA developed the Technology Readiness Levels[15] to determine where in the design process a specific technology stands. The levels attempt to divide the design process into incremental steps. As a technology gets to a higher level, the requirements for its demonstration environment increases in complexity, requiring the use of special testing facilities and eventually testing in an environment representative of space.

Microgravity experimental facilities vary in their fidelity, cost, and operational limitations. While not necessarily exhaustive, the list presented in Table 17.1 shows a wide range of facilities that can provide an environment to reproduce or simulate microgravity conditions in support of research. The table lists 15 different environments under which microgravity research can be emulated and conducted. The first column shows facilities that can be housed within the individual researcher's institution, but that don't necessarily simulate full 6-DOF microgravity. The second column lists facilities that have full 6-DOF capabilities, but which are usually managed by a third party. The third column lists the existing facilities that provide full microgravity conditions, but that present the largest development challenges.

Traditional space-system development, especially for single-satellite systems, goes only through simulation and ground-based tests. The complexity and high costs of creating flight prototypes prevents in-space flight qualification and demonstration of most high-value science missions; many *demonstration* missions turn into high-cost multiyear missions themselves (\$100 million+[16] such as DART[17]—Demonstration of Autonomous Rendezvous Technology—and Orbital Express[18]). Ground-based tests and simulations have important shortcomings for technology maturation: limited degrees of freedom, nonrepresentative dynamics (e.g., gravity and suspension systems change the dynamics), limited operation time, and so on.

An innovative solution that bridges the gap is to use the existing microgravity environment aboard the International Space Station: that is, to use *small satellites* in conjunction with the station to create a laboratory environment for space technology maturation. The use of microsatellites—and careful use of the resources that have already been built by others to accommodate research on the space station (such as NASA, DOD, and the space station partners)—overcomes the majority of the limitations of full space demonstrations: the facility is low cost

Table 17.1. Sample of Available Facilities for Microgravity Research

In-house	Third-Party/Full Microgravity	Space
Robot helicopters	Reduced gravity aircraft	Free-flyer
6-DOF robot arms	Neutral buoyancy tank	ISS
Helium balloons	Drop towers	Shuttle payload
Robot cars		Shuttle middeck
Flat floor		Sounding rocket
Air table		
Simulation		

(<$2 million), low risk, flexible, enables human intervention and supervision, and reduces the profile of the mission to that of standard laboratory research. At the same time, the facility allows operations in a 6-DOF microgravity environment with highly representative dynamics. The use of multiple small satellites, which emulate standard satellite-bus craft, enables testing of multiple research areas including formation flight and docking.

The small satellites are known as SPHERES (Synchronized Position Hold, Engage, Reorient Experimental Satellites). They are substantially less expensive than any traditional demonstration mission. By operating in the risk-tolerant environment of the International Space Station (Fig. 17.5), they can demonstrate that satellite formation flight is not only feasible but also robust. Further, by exploiting platforming concepts, where a common chassis with standardized interfaces allows modular components to be added, they are extensible in both hardware and software to accommodate a myriad of diverse research objectives. Uplinking software, downlinking data, and attaching payloads to the SPHERES expansion ports facilitates spiral algorithm development and hardware extensibility.

17.4.1 The Philosophy Behind SPHERES

The SPHERES formation-flight and docking laboratory is the culmination of a decade of dynamics and controls research conducted by MITSSL and flown on the shuttle, Mir, and International Space Station. The MODE (STS-48 and STS-62), DSL (STS-62, MIR), and MACE (STS-67, ISS Expedition 1 and 2) experiments took place from 1991 through 2001. Based on those experiences, the group developed a design philosophy[19] to aid the development of SPHERES. The philosophy formalized the lessons learned during those short and long-term missions and allowed a structured approach to implement a true laboratory environment for guidance, navigation, and conrol (GNC) aboard the International Space Station.

The first step in maturing a technology through hardware-in-the-loop testing is to clearly identify the objectives. The following hardware-in-the-loop testing objectives specifically address the GNC technology challenges studied at the laboratory:

- Demonstration and validation in the correct environment such that the results clearly show that the technology achieves the required performance.
- Repeatability and reliability of each test by ensuring that they provide the same results more than once under similar operating conditions with representative system noise.
- Determination of simulation accuracy by providing results to calibrate and validate simulations.
- Identification of performance limitations by pushing the limits of new technologies or algorithms until failure, that is, learning the limitations of an algorithm in a relevant environment before deployment in a full mission.

Photo courtesy of NASA

Fig. 17.5. U.S. astronaut Michael Lopez-Alegria operates SPHERES aboard the International Space Station in March 2007.

- Identification of operational drivers, or design parameters, that enable engineers to create design trade studies and system designs based on the most relevant parameters for GNC.
- Identification of new physical phenomena should be possible by allowing observation of the physical system behavior; the phenomena must be captured well enough to allow creation of models for future designs.

Based on these objectives and the unique resources made available by the International Space Station, the following design principles are proposed for designers of microgravity experiments for technology maturation to use in the development of their facilities to be operated aboard the space station. By generalizing the GNC objectives, these principles encompass a wider range of technology maturation experiments, beyond the dynamics and controls scope discussed above.

17.4.1.1 Principle of Iterative Research

A successful laboratory environment supports the established scientific process that a hypothesis be tested and modified as experiments are performed. A designer should consider the following steps:

- Conception: initial hypothesis definition and scientific objectives of the laboratory.
- Facility development: engineering design and manufacturing of the facility.
- Technology maturation tests:
  - Individual test development.
  - Experiment run and data capture.
  - Data analysis versus expected results.
  - Hypothesis reformulation as needed.

The principle of iterative research separates these steps into "science time" (hypothesis formulation and modification, test runs, and data analysis) and "overhead time" (facility design, implementation of tests to operate in the specific facility, and data collection and transfer). While overhead time should clearly be minimized, it is not as simple for science time. Science time must be flexible: sufficient to allow ample time for data analysis and hypothesis (re)formulation, while not so long that a scientist loses focus on the task at hand.

17.4.1.2 Principle of Enabling a Field of Study

To provide experimentation in a field of study, a laboratory must allow for experiments within the different research areas of the field. Given the intent to use the International Space Station (a highly valuable resource) and the large overhead in the development of facilities to operate there, scientists must use the facility as efficiently as possible. This principle calls for each program aboard the space station to develop laboratories that aid a field of study. To do so, a facility must support multiple investigators to work on individual topics to cover the whole field of study, and facilitate the collection of knowledge from the individual topics to enhance understanding of the field as a whole.

17.4.1.3 Principle of Optimized Utilization

The International Space Station offers a wide range of unique resources that make it ideal for the maturation of space technologies. Rather than thinking about using the least resources possible, this principle guides the researchers to maximize the value of their use with respect to the research goals. The following are specific resources to consider.

- **Crew:** The crew can provide valuable feedback, even interpretation of results, based on observations. Further, the crew can supervise experiments to minimize risks as scientists push the limits of algorithms.
- **Consumables:** Electrical power, pressurized gases, and liquids provided by the International Space Station can greatly reduce the cost of a facility and increase the researcher's focus on the technology instead of hardware development and maintenance.

- **Data telemetry:** The International Space Station communications and data management system allows scientists to obtain their data within hours of the experiments and enables upload of new software. In the future, it is expected to enable real-time video and other teleconferencing options as part of daily research operations.
- **Long-term experimentation:** Long-term microgravity experimentation enables flexible operating schedules, iterative research, controlled test conditions, and ample science time.
- **Risk-tolerant environment and benign atmosphere:** The availability of a risk-tolerant environment enables tests that push the limits of the technology while, at the same time, reducing complexity and costs.

17.4.1.4 Principle of Focused Modularity

Because experiments almost always contain basic elements that can support other similar experiments (e.g., power, communications, data storage, processing), the design phase of a facility should identify these common elements. These generic parts should be made available for future experiments as long as the practice does not compromise the objectives of the original experiment. Just as with any terrestrial laboratory, there should exist a suite of generic capabilities that can be used (and possibly reconfigured) by future researchers. The call for *focused* modularity is to prevent a "do-everything" system. The modular part on the design should focus on those common features identified as generic after the initial facility design.

17.4.1.5 Principle of Remote Operation and Usability

Remote laboratories are based in remote locations because they offer a scarce environment that researchers cannot emulate in their home institutions. The International Space Station offers a unique microgravity environment. The design of remotely operated laboratories for the space station must account for the following facts about the operation:

- Operators are usually not experts in the specific field and are a limited resource. Their time spent supervising a test must provide useful feedback to the scientist.
- Research scientists have limited or no experience in microgravity operations and the space station environment; therefore, what may seem easy to them on the ground might not be easy for the operators. While the scientists can modify experiments in their facility, it is not possible to do so remotely. Therefore, scientists must constantly be made aware of the differences between their operations and the remote facility operations.

The goal of a remote facility is to allow for a virtual presence of the research scientist in the operational environment by combining the abilities of the operator (astronaut) and scientist (on the ground).

17.4.1.6 Principle of Incremental Technology Maturation

The goal of incremental technology maturation is to make the complexity, risk, and cost increase smoothly as one climbs through the Technology Readiness Levels while cognizant of the changes in the environment required. The International Space Station provides an environment that can closely satisfy the requirements for a space environment (space radiation, illumination, vacuum, and thermal conditions can be replicated in terrestrial facilities); yet the presence of humans in the space station can greatly reduce the risks involved, and the existence of the station itself can reduce the costs. Further, successful tests in the space station may lead to less complexity when moving to even higher Technology Readiness Levels by providing researchers with a better understanding of the system and its true risks.

17.4.1.7 Principle of Requirements Balance

By the time of the Critical Design Review, hard requirements are usually set to determine the goals that must be met; they are mostly quantitative. Soft requirements are features desired by the researchers but that do not necessarily have a specific value or are not essential for the success of the mission (and usually creep up after reviews). A successful design creates a realistic set of requirements while taking into account the other principles presented herein. This principle does not call for all the requirements to be perfectly balanced or to necessarily eliminate the soft requirements; rather, it calls for the scientist to proactively pursue a realistic justification for each requirement and to ensure that a substantial part of the effort spent on the development of the facility goes toward clearly defined needs.

17.4.2 SPHERES Design

The SPHERES[20] laboratory for distributed satellite systems consists of three nanosatellites, a custom metrology system, communications hardware, consumables (propellant tanks and batteries), and an astronaut interface aboard the International Space Station. Figure 17.6 shows the SPHERES satellites being operated aboard the space station and identifies the different elements of the facility. The ground-based setup consists of another set of microsatellites, a more streamlined interface, and a SPHERES Guest Scientist Program to allow multiple researchers to use the facility (in keeping with the principle of enabling a field of study).

The SPHERES satellites were designed to provide the best traceability to upcoming formation-flight missions by implementing all the features of a standard thruster-based satellite bus. The satellites have fully functional propulsion, guidance, communications, and power subsystems. These enable the satellites to maneuver in 6 DOF, to communicate with each other and with the laptop control station, and to identify their position with respect to each other and to the experiment reference frame (2 mm position and 1 deg attitude resolution). The laptop control station is used to collect and store data as well as to upload control algorithms to the satellites. Figure 17.7 shows a CAD drawing of the open satellites as

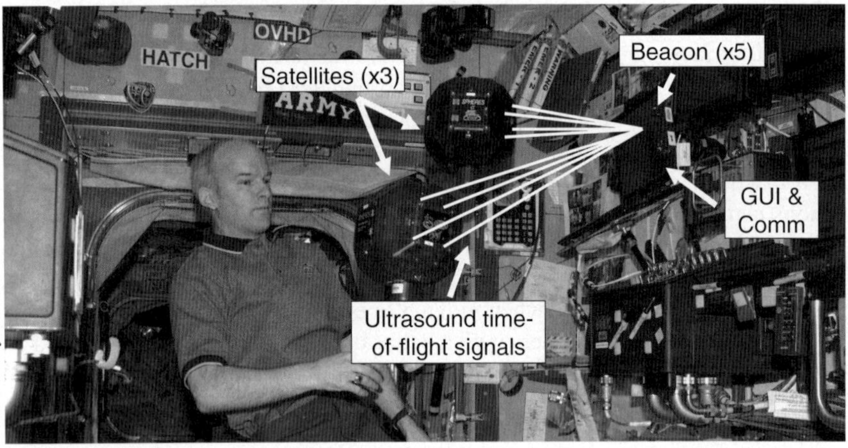

Photo courtesy of NASA

Fig. 17.6. SPHERES hardware aboard the International Space Station with U.S. astronaut Jeff Williams in November 2006.

well as a picture of an assembled SPHERES satellite and identifies its main features. Physical properties of the satellites are listed in Table 17.2. The following sections present basic descriptions of the subsystems; further details are available in the referenced publications and the SPHERES Critical Design Review.[21]

17.4.2.1 Propulsion

The satellites are propelled by pressurized carbon-dioxide propellant. The selection of the propellant was part of a large trade study between multiple inert cold gases and onboard air compressors. CO_2 provided the best performance, as it exists in liquid form at room temperature and is reasonably stable without the need for cryogenics or other power-consuming support systems. The trade-off was the requirement to operate only when the CO_2 removal assembly of the International Space Station is active. A manually adjusted regulator sets the pressure to 25 psig. Twelve thruster assemblies, distributed among the three orthogonal axes (four per axis, in opposite pairs), allow 6-DOF maneuvers, enabling both attitude and stationkeeping control. Each thruster provides approximately 0.10 N of force. The system is primarily used with pulse-width modulation (at up to 20 Hz) to achieve variable forces and torques. The propulsion system may be easily replenished; 96 gas tanks were delivered to the space station for SPHERES operations.

17.4.2.2 Guidance

The SPHERES metrology system closely resembles a GPS system, but it operates inside the space station (where GPS does not) at resolutions of up to ±2 mm. The goal of the metrology system was to create a system that can use software signal processing to provide state information equivalent of a wide range of space systems. For example, scientists can be provided only with attitude information to simulate a star tracker. The metrology system consists of a local 6-DOF inertial

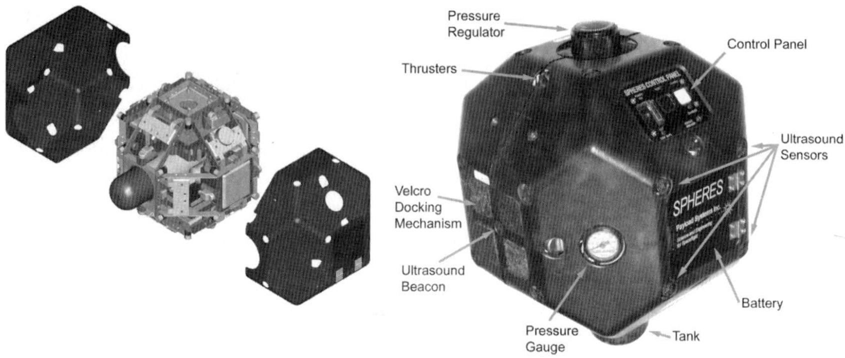

Fig. 17.7. SPHERES CAD view and assembled satellite with subsystems.

measurement unit (three accelerometers and three gyroscopes operating at 50 Hz) and a global pseudo-GPS system that uses ultrasonic time-of-flight measurements from transmitters (equivalent of GPS satellites) placed at known locations in the space station (or MITSSL laboratory). These time-of-flight measurements are converted to ranges and used to derive position and attitude at a frequency up to 5 Hz.

17.4.2.3 Communications

Each SPHERES unit uses two separate frequency communications channels to simulate the expected operations of future formation flying missions: one channel is used for satellite-to-satellite communications; the other channel enables satellite-to-ground communications. The system uses COTS (commercial off-the-shelf) 916 MHz and 868 MHz close-range ASK (amplitude shift keying) type RF transceivers. The effective data rate of each channel, after removing the bandwidth needed for packet management and error checking, is 20 kbps. The system does not use the higher bandwidth 802.11 standard because the space station has its own wireless local area network; use of the frequency would have held SPHERES to safety reviews of any communications between the satellites and the network. The goal of effective iterative research took precedence in this design choice.

Table 17.2. SPHERES Satellite Properties

Diameter	0.22 m
Mass (w/tank and batteries)	4.3 kg
Maximum linear acceleration	0.051 m/sec^2
Maximum angular acceleration	0.944 rad/sec^2
Power consumption	13 W
Battery lifetime	2 hr
Communications data rate	20 kbps

17.4.2.4 Power

Each SPHERES satellite is powered by two packs of eight AA alkaline batteries each. The packs provide up to two hours of operations. The power system regulates the battery voltage to six voltages used by the internal electronics. The choice of AA batteries was driven by requirements balance: while the better implementation would have been to select rechargeable batteries (and therefore use space station resources better), the substantial amount of effort required to approve rechargeable batteries for the space station would have delayed the project substantially and consumed too many funds. Therefore, while rechargeable batteries are under development in parallel with operations, the use of AA batteries allowed faster and cheaper deployment to the International Space Station.

17.4.2.5 Avionics and Software

A Texas Instruments C6701 digital signal processor (DSP) provides the computational power. These DSPs have been used in other space missions and ensure real-time operations required for GNC. Further, the DSP processors include all support functions of a standard processor, allowing it to control the whole unit. The primary driver in the selection of a DSP was the high ratio of processing to energy consumption; the selected DSP carrier board provides between 166–1000 MFLOPS (depending on optimization) and requires less than 7 W. The processor is supported by 16 Mb of RAM and 256 kb of flash memory (the avionics' most stringent limitation). The software is based on the TI DSP/BIOS Kernel, with a custom configuration of the real-time elements to allow easy access to the metrology (inertial and global) data, control interrupt (up to 1 kHz), propulsion driver, and communications data. The software enables access to high-priority periodic processes (e.g. data capture, control) and low-priority aperiodic background processes (e.g., high-level estimation routines).

17.4.2.6 Crew Interface

The graphical user interface for use aboard the space station was developed with the idea that the crewmembers may not have any contact with the SPHERES researchers while they run tests. Therefore, the interface not only guides the crew on how to run tests, but also to capture the observations of the crew—the crew becomes an extension of the scientist by providing interpretation of the test runs in presence of the hardware. By doing so, the interface becomes an electronic lab notebook.[22] Together with multiple angles of video capture and the possibility of live audio and video use during test sessions, SPHERES creates a virtual presence of the scientist aboard the space station.

17.4.2.7 Guest Scientist Program

The Guest Scientist Program was an integral part of the development of SPHERES. The high-priority objective to enable several scientists to use the laboratory heavily guided the development of the software and a set of simulations. The main SPHERES software, called SPHERES Core, creates a generic GNC software environment with high-level interfaces to the hardware (all other subsystems and the microprocessor) as well as generic algorithms developed at MIT for estimation, control, and utility functions. This allows scientists to concentrate on their science, rather than the specific implementation on SPHERES or science not in their field (e.g., estimation researchers do not need to develop their own control laws—they can use existing ones developed and tested by the SPHERES team). This required careful balance between the development of generic tools (which adds overhead to the software and DSP requirements) and the level at which scientists prefer to interface to the software. To achieve this balance, the SPHERES team held a science Critical Design Review.[23]

The Guest Scientist Program also included the development of a simulation for the use of the hardware. This enables scientists to create initial implementations of their algorithms; however, because of resource limitations, the simulations are not end-to-end. Therefore, the program also joins an MIT graduate research assistant with remote scientists so that tests can be conducted with ground-based satellites.

17.4.3 Verification and Validation of the GNC Flight Software

As any other flight software, the SPHERES software undergoes an extensive verification and validation process to increase the level of confidence prior to flight. This includes both simulations and hardware-in-the-loop tests at various levels of software integration. By being smaller and cheaper, microsatellites can have the advantage of being more accessible for hardware-in-the-loop tests for the integrated GNC system, which further increase the level of confidence. This section describes the different components of the verification and validation process used when preparing experiments on the SPHERES microsatellites.

17.4.3.1 Verification and Validation Process Overview

The verification and validation process is illustrated in Fig. 17.8. It involves software-in-the-loop simulations (right column) and hardware-in-the-loop experiments (left column). The simulations are performed in MATLAB. Hardware-in-the-loop experiments are typically performed at the ground two-dimensional (2-D) testing facilities (air table and/or flat floor). A series of five gates (diamond shaped boxes) are used to make a decision to move on to the next testing phase (the space station) or to iterate in case the systems engineer's expectations (requirements) are not met.

The process is initiated with the guest scientist integrating a GNC software in the MATLAB simulation. While integrating the software, the guest scientist ensures that all parameters proper to the simulation environment (mass properties, controller gains, dynamics coefficients, etc.) are contained in one file that later can easily be replaced by another one with flight parameters. Once the software simulation ensures that all expectations are met (gate 1), the guest scientist integrates the flight software in the C programming language, using as many validated modules as possible to benefit from code heritage. With all parameters set for 2-D

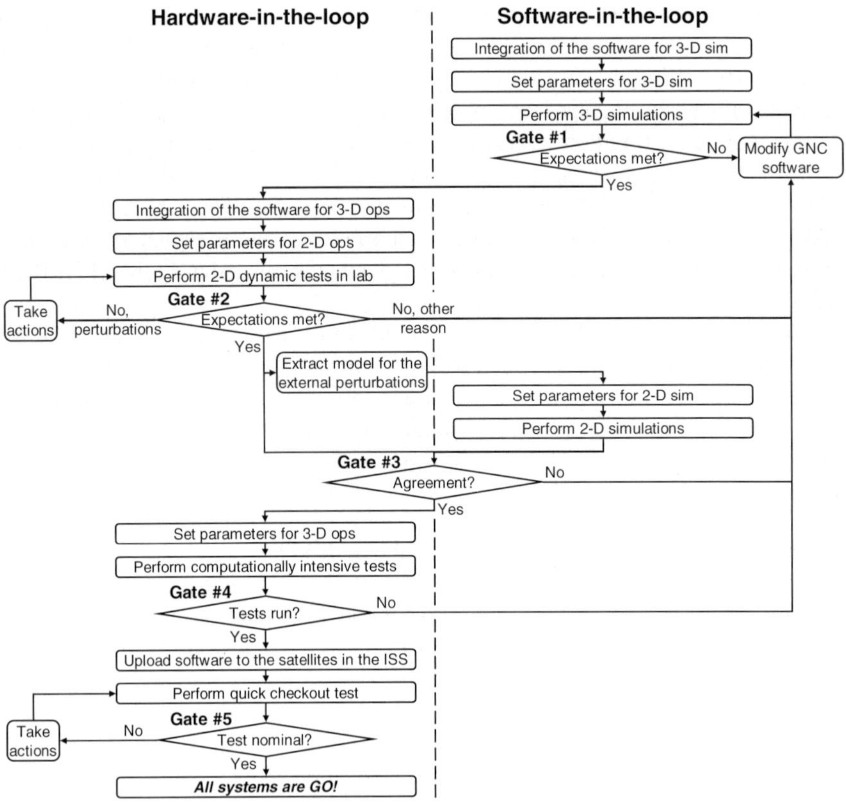

Fig. 17.8. SPHERES flight software verification process.

(ground) testing, experiments are performed using the ground 2-D facilities. Those perturbed by extensive ground effects (gravity, bumps, drag) have to be restarted. Once the expectations are met (gate 2), a model of the external perturbations is extracted whenever possible. Implementing that model in the simulation environment, along with the parameters used for 2-D testing, allows a direct comparison between the simulations with 2-D parameters and the experiments at the MIT ground facilities. An agreement between the two (gate 3) validates the simulation with the 2-D parameters, which increases the confidence that the first simulation with the 3-D (microgravity) parameters is representative of reality.

The last two gates are a means to further increase the level of confidence in the hardware prior to the flight experiments. After reinserting the parameters for 3-D experimentation in the same flight software used at gates 2 and 3, more ground testing is performed at the ground facilities. Although the ground dynamics and perturbations are likely to prevent the tests from completing successfully, the goal of gate 4 is to ensure that the onboard computer can support the flight software running in real time, with all sensors and thrusters being activated and with the 3-D parameters being used. After successfully going through gate 4, the software is uploaded to the space station. At the beginning of every space station test session, a checkout test (gate 5) is performed to ensure the hardware operates nominally and all subsystems are go.

The verification process presented here has been developed through multiple test sessions. As in any flight program, it played a crucial role in increasing the confidence of the onboard software to a level acceptable for flight. By test sessions 5 and 6, the process described above allowed for 100% success of the 15 experiments, from a total of seven guest scientists who participated in these two test sessions.

17.4.3.2 Software Simulation

The initial development and verification of a software module is performed in a MATLAB-based simulation. Created for the SPHERES Guest Scientist Program, the simulation facilitates the development of GNC software from a distant site. It also benefits from a large array of visualization tools made available by MATLAB, which facilitates the verification of the algorithms embedded in the software.

To create a realistic simulation of the dynamic response of the system, the simulation engine replicates the interrupts that occur in the hardware, executes the different primary interface functions in the same order as they occur on the satellites, and propagates the states of the satellites between the interrupts. The hardware models used in the simulation are based on data acquired from calibration of the SPHERES hardware on the ground and data measured during microgravity operations aboard the space station. Sensor noise, biases, and blackout behavior has been captured; the actuator models include thruster strength and pressure drop as multiple thrusters open. This simulation does not model communications losses.

17.4.3.3 Hardware-in-the-Loop Ground Experiments

Three testing facilities were used to mature the software modules that populate the GNC architecture. The 2-D facilities have enabled repeatable testing conditions. NASA's reduced-gravity aircraft allowed short duration testing in a microgravity environment. The first test sessions in the International Space Station, even with only a reduced set of hardware, were used to incrementally test the different GNC modules in the final operational environment.

Hardware-in-the-loop experiments typically lead to the discovery of problems at an early stage of the software development process, when it is relatively easy and inexpensive to make changes. Remedies can be designed early, thereby improving the reliability of subsequent experiments. Important lessons were learned by the SPHERES team after they discovered the nature of each problem encountered through hardware-in-the-loop experiments:

- When verifying a GNC software module, it is important to experimentally check every assumption associated with the algorithm. Numerical errors caused by the software implementation of the algorithm might invalidate some assumptions.

- Regular jumps in the state estimates at the measurement update phase of the navigation filter are likely to be caused by unmodeled biases in sensor measurements. Knowing that the hardware is identical between all SPHERES microsatellites, the scientist is aware that these unmodeled biases are likely to be common to all of them. If not, the problem can be quickly traced to a specific sensor onboard the satellite.
- Although hardware-in-the-loop experiments at a component level are invaluable, it is crucial to understand clearly the behavior of the navigation sensors after they are integrated into the system. The integrated sensors might show a behavior that was not observed by the manufacturer in stand-alone tests, and therefore not documented in the specifications.

17.4.3.4 Software Integration

Because verification and validation can be performed at various levels of software integration, it is important to understand the role and the effect the integration process can have on the overall functionality of the system. The integration process must ensure that the correct information is properly communicated between the software modules as well as the peripheral devices such as sensors and thrusters, and that the processor and memory are not overloaded by the various software processes. Because of the complexity of the integrated software, software integration can lead to undesired behaviors. Therefore, it is crucial not to overlook it, and to perform verifications whenever possible.

A modular GNC architecture greatly facilitates the integration process. GNC modules with different capabilities can be quickly integrated to allow an investigator to test different combinations. The selection of the right combination of modules is largely dependent on the available hardware and on the experiment to be performed. For example, the use of a certain module can be restricted when the hardware providing essential inputs is unavailable, or when the required resources exceed the hardware capability (e.g., computational processing, sensing, and communication bandwidth).

Real-time hardware-in-the-loop experimentation was key in identifying problems related to system integration on the SPHERES facility. It allowed the team to:

- Identify a proper sequence of maneuvers and GNC modes leading to close-proximity operations such as docking with cooperative and tumbling targets.
- Establish requirements regarding the transition between the maneuvers.
- Confirm the capability of each software module to fulfill its corresponding requirements when integrated into a GNC architecture.

17.5 Technology Maturation in Action Using Small Satellites

The use of small satellites is not limited to simple missions or Earth observations. By conforming to the space-technology maturation design philosophy presented above, small satellites provide a unique ability to help develop and mature technologies for future missions that may use small or large satellites. Since May 2006, the SPHERES facility has demonstrated these concepts by allowing scientists at MIT and elsewhere to mature a large array of algorithmic technologies for future formation flight and docking missions. Table 17.3 presents a summary of the research iterations that have taken place for three main areas of study: metrology, docking, and formation flight. These iterations started to occur as early as the software development phase, between 2003 and 2006, as well as during the first five SPHERES test sessions, between May 2006 and November 2006. The following subsections provide more details on the research iterations and technology maturation for each of these areas.

17.5.1 Metrology and State Estimation

The core metrology estimation algorithm has seen important iterative improvements to its design. The low-risk, reconfigurable, and long-duration nature of SPHERES enabled the research team to address filtering of noise in raw data and tuning of the extended Kalman filter during the core software development process and between space station test sessions. As the estimator base code became stable, the team continued incremental development with the introduction of fault detection and isolation algorithms to improve estimator convergence.

Table 17.3. Research Iterations from Test Sessions (TS) 1 to 5

Area	TS	Iterations	Description
Metrology	1, 2, 3, 4, 5	4	The metrology system began with a limited set of beacons, which resulted in the development of innovative filters. The lessons from that development and the capture of data in TS 4 showed no need for further development starting with TS 5.
Docking	2, 3, 4, 5	3	The basic maneuvers for docking began development during TS 2. As each session demonstrated individual parts of a full docking mission, these parts were put together to achieve the final demonstrations of TS 5. More iterations follow in future sessions.
Formation flight	3, 4, 5	2	Starting with attitude-only formation flight, the latest iteration of demonstrations aboard the ISS shows the basic maneuvers that are essential for missions such as TPF-I. Further iterations will occur.

The most common navigation problem encountered during the hardware-in-the-loop experiments was instability in the state estimates (position, velocity, attitude, and rotational rate) output by the extended Kalman filter.[24] The different forms of instabilities included jumps of the state estimates and temporary divergence (Fig. 17.9). The jumps, typically occurring in the measurement update phase of the extended Kalman filter, were caused by multipath effects (reflections of the ultrasonic signals on the walls and surrounding objects) and unmodeled biases in the ultrasonic sensor measurements. The temporary divergence was caused by the covariance matrix of the navigation filter being driven nonpositive definite during the state propagation phase of the extended Kalman filter. In response to these instabilities, the controllers produced erratic behavior. This proved to be problematic during periods of close proximity. Moreover, problems with the navigation system occurred frequently in recent uncrewed missions involving close-proximity operations (listed in Table 17.4). Consequently, the resolution of the instability problems with the state estimators quickly became a high priority. The solution, refined through multiple static experiments where the satellites were constrained to remain stationary, included:

- A prefilter to remove outliers in the ultrasonic measurements before they are passed to the navigation filter.
- An online fault detection algorithm preventing multipath from affecting the state estimates.
- A reduced covariance matrix improving the robustness of the navigation filter.

17.5.2 Docking

Close-proximity operations, such as docking, require multiple advancements of current GNC technologies. Because of the communication delays with the ground controllers and the amount of information to be monitored, the onboard computers have to be granted significantly more autonomy than in common satellite missions. In the past decade, multiple missions were launched to test autonomous close-proximity operations, as illustrated in Table 17.4. They all involved either autonomous docking or inspection. It is important to emphasize that anomalies occurred for each of these missions, with the possible exception of the XSS-11 mission, although no public report has been issued on its results. At least one of these missions failed, and all others were faced with anomalies that required ground intervention. Therefore, there is a need to mature technologies for autonomous docking.

In response to this need, the team initiated research into autonomous close-proximity operations through the SPHERES project. The first step in this research was to analyze the state of the art in GNC architectures used in past uncrewed missions involving close-proximity operations, and to identify the needed improvements. A GNC architecture is defined here as an abstract description of the entities of a GNC system and the relationships among them. This definition is

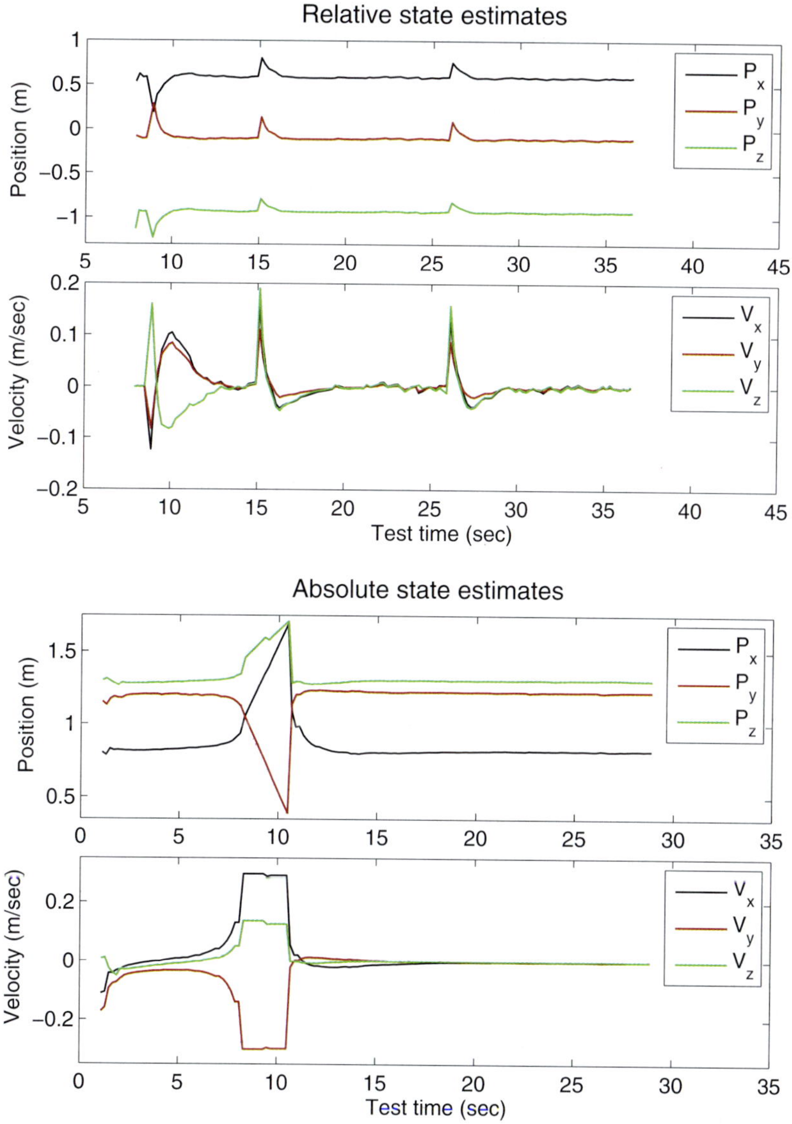

Fig. 17.9. Representative results of the SPHERES extended Kalman filter: jumps in the state estimates (top) and temporary divergence of the state estimates (bottom).

adapted from a broader system architecture definition proposed by the MIT Engineering Systems Division Committee.[37] GNC architectures found in the literature have common characteristics supporting the above definition:[38–43] a decomposition in simple entities, a hierarchy between the entities, and some interactions between the entities.

Table 17.4. Review of Automatic and Autonomous Missions Involving Close-Proximity Operations

Mission	Year	Status	Type	Anomalies
Cosmos-186, Cosmos-188[25,26]	1967	Success	Docking	- Misaligned capture
Progress[27]	1978–present	Success (multiple)	Docking	- On Progress 18, ground station problems prevented the command uplink for automatic docking
ETS-VII[28,29]	1998	Success	Docking	- Thruster anomaly
SNAP-1[30]	2000	Partial	Inspection	- Atmospheric drag larger than expected - Ran out of fuel before completing the rendezvous
XSS-10[31]	2003	Success	Inspection	- Navigation problem (confused star tracker) - Communication interruptions
XSS-11[32,33]	2005	Unknown	Inspection	- Unknown
DART[17,33,35]	2005	Failed	Inspection	- Navigation problem (biased velocity measurements) - Collision
Orbital Express[18,36]	2007	Success	Docking	- Navigation problem (GPS system initialization problem) - Inverted reaction wheel polarity - Navigation software error

In his book entitled *Automated Rendezvous and Docking of Spacecraft,*[44] Fehse presents a typical GNC architecture traditionally used for automated rendezvous and docking. The schematic he provides is simplified and shows only the levels of authority, not the actual functional relations within the system. This GNC architecture involves humans performing high-level monitoring and making mission-critical decisions, such as the triggering of a collision-avoidance maneuver when anomalies occur. However, for complex operations, such as docking to a tumbling target (especially if it were to be in lunar or Martian orbit), the short reaction time required in the event of anomalies needs an increased level of autonomy, which is not supported by the GNC architecture portrayed by Fehse.

Figure 17.10 shows a hierarchical view of the baseline GNC architecture used on the SPHERES microsatellites for autonomous docking. It is largely based on Fehse's architecture, but has been extended to enable complex autonomous operations. The components common to Fehse's architecture are grayed. The box labeled "autonomous onboard docking control systems" contains the software modules that run on a satellite. Each software module is an implementation of an

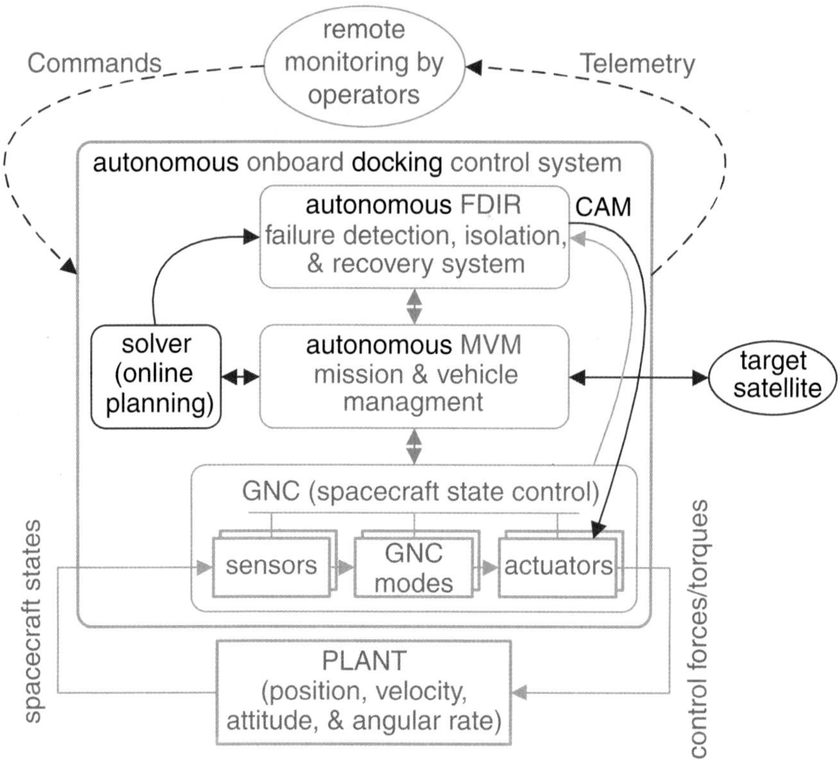

Fig. 17.10. SPHERES GNC architecture for autonomous close-proximity operations.

algorithm accomplishing a certain task (e.g., state estimation, trajectory control, thruster actuation). External interfaces include the plant (satellite), the operators, and the other satellite. The software modules are grouped into FDIR (fault detection, identification, and recovery), solver, mission and vehicle management, and GNC. The extended architecture accounts for the idea that communications with human operators may be intermittent or nonexistent (illustrated by dashed lines), potentially preventing ground operators from taking direct control of the satellite. This forces the integration of FDIR capability at all levels.[45,46] In this extended model, the onboard computer is granted the authority, through the main FDIR module, to trigger a collision-avoidance maneuver if a problem is detected. Furthermore, a solver module allows for online planning and scheduling. The solver is monitored by the FDIR module to prevent infeasible solutions or other failures. Because knowledge of the target's states is required, remote sensing and/or communications between the satellites are needed, as shown by the link with the target satellite.

So far, the team has focused on the implementation of low-level GNC modules[47,48] on the SPHERES hardware. The navigation extended Kalman filter processes the ultrasonic time-of-flight signals to compute the position, velocity, attitude, and rotational rate of the satellites. Controllers—based on PID (proportional integral derivative), phase-plane, and glideslope algorithms—compute desired forces and torques from the tracking error at 1 Hz when following a trajectory. Thruster management modules transform the desired forces and torques into thruster on/off times using a pulse-width modulation scheme. Each module has the potential to be used in either docking or formation flight experiments, and can be called at various frequencies.

The higher-level GNC modules are still at an early stage of development. The mission and vehicle management module consists of the partition of a sequence (e.g., a docking sequence) into multiple short closed-loop maneuvers.[50] Each maneuver is validated by steps and then carefully put together until a complex demonstration can be achieved. The current fault-detection system rejects ultrasonic time-of-flight measurements incoherent with the state estimates and throws a fault if the residual of the extended Kalman filter exceeds a threshold determined through simulations. A solver for typical linear-quadratic regulator problems has also been recently implemented.

On 19 August 2006 (test session 4), all the necessary SPHERES hardware for performing autonomous docking maneuvers between two satellites was confirmed ready for experimentation onboard the International Space Station. A series of four autonomous docking experiments, with cooperative and uncooperative (drifting) targets, took place that same date.[49,50] Experiments with fixed and drifting targets were successful. Following this test session, the addition of a fault-detection module using the navigation filter innovation (residual) significantly improved the robustness of the global estimator to bad measurements caused by multipath.

On the following test session (test session 5 on 11 November 2006), five docking experiments were attempted.[48–50] They were all successful and resulted in closely aligned contact (the Velcro patches on satellites—see Fig. 17.7—were joined). Two of these were performed using a safe approach trajectory, which is robust to failures occurring a few seconds prior to contact, when safety is critical (passive abort).[51] Two others were performed with a tumbling target, resulting in the first autonomous docking with a tumbling target ever achieved in microgravity (Fig. 17.11). All docking experiments were successful in spite of unexpected measurement errors, induced by the camera flash interfering with the infrared components of the navigation system when taking pictures of the satellites. These measurement errors were successfully detected and rejected online by the autonomous fault-detection system. Therefore, these experiments went beyond their intended objectives and demonstrated the capability of the GNC architecture implemented on SPHERES to perform well even in the presence of navigation sensor errors. These accomplishments illustrate the benefits of maturing GNC technologies on orbit in a risk-tolerant environment. They were possible due to the use of small satellites.

Photos courtesy of NASA

Fig. 17.11. Footage of the first autonomous docking to a tumbling target in microgravity.

Many iterations were performed to obtain an appropriate sequence of GNC modes (Fig. 17.12). The exit conditions for each mode were the result of multiple iterations to ensure timely transitions while maintaining the satellites along an appropriate trajectory. This experimentation permitted the tuning of different parameters in each GNC module (controller gains, estimator process noise, minimum impulse bit) to ensure good performance. The following lessons were learned:

- A holding point (berthing) is necessary at a safe but close proximity to the target to make precise alignment corrections prior to capture.
- The closer to the target, the more constraining the exit conditions must be on each GNC mode, to ensure that the tracking errors are within the tolerance of the capture mechanism prior to contact.
- The controllers must be tuned together with the exit conditions to ensure a smooth and timely transition between modes at a reasonable fuel cost.

17.5.3 Formation Flight

The prohibitive costs have prevented any three-satellite formation-flight demonstrations in space, but the availability of small satellites in the risk-tolerant

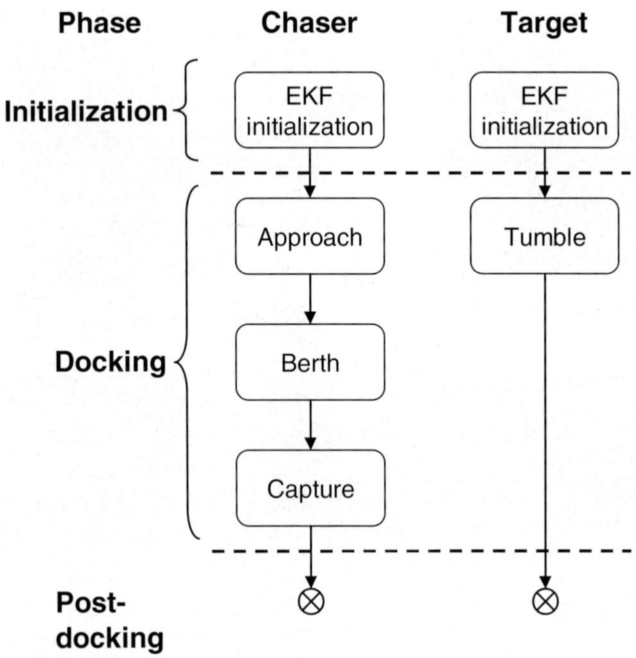

Fig. 17.12. Mode sequencing for autonomous docking to a tumbling target.

environment of the International Space Station now enables such tests in microgravity. Formation-flight algorithms have been demonstrated with two and three satellites.[52] The iterative nature of SPHERES algorithm development has been exploited at each step. Initial formation flight tests controlled the 3-DOF attitude between two satellites. These tests allowed the team to identify the operational drivers for formation flight with SPHERES: communications and synchronization. Two-satellite 6-DOF formation control followed with tests to demonstrate the use of acknowledgement packets with timing information to synchronize the satellites. The next incremental tests demonstrated formation flight of three satellites. The tests demonstrated triangular formations prescribing a circle. These initial tests used preplanned trajectories with classical PID-style, low-level control laws to maintain the formation; each satellite operated independently.

The next maneuver, illustrated in Fig. 17.13, involved changing the plane orientation of the three satellites; that is, while the satellites initially prescribed a circle parallel to the deck and overhead walls, the formation reconfigured to a plane perpendicular to the deck and overhead walls of the space station. This "plane change" maneuver was demonstrated with multiple high-level algorithms: simulation of a rigid body (resulted in low fuel efficiency), optimal fuel maneuver (resulted in a high-risk path), and a safer nonoptimal transfer. Figure 17.13a shows the optimal fuel maneuver (a 6 minute experiment); Figs. 17.13b and 17.13c compare the desired and measured path, respectively. The tests were performed to

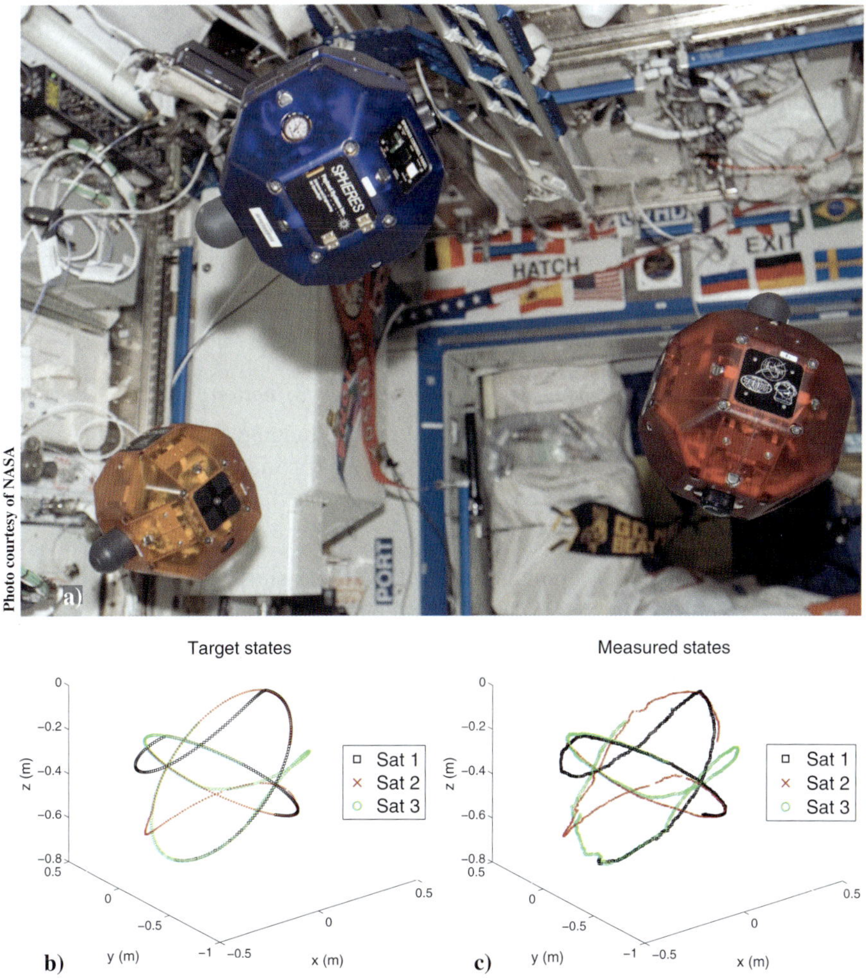

Fig. 17.13. The three SPHERES satellites during a formation flight initialization and plane-reconfiguration test inside the International Space Station: (a) NASA photograph. (b) desired path, and (c) and telemetry data.

begin research on the high-level commands of formation-flight systems. These tests already give a new insight into the differences in high-level control maneuvers between large rigid-body satellites and distributed formation-flight satellites; while the rigid-body fuel-optimal maneuver is to rotate about a primary axis (usually using a reaction wheel or gyroscope), a formation-flight system's optimal fuel maneuver involves complex paths that exit and reenter the formation. Future tests will involve the use of relative sensing among the satellites, collision-avoidance maneuvers, and online path planning, until SPHERES has been used to mature an essential range of GNC algorithms for future space telescopes.

17.6 Summary

This chapter provides an overview of some of the research that MITSSL has been performing in the development and maturation of satellite formation-flight technology. These activities span the satellite formation-flight engineering development life cycle—from mission architecture, to development of enabling technology, to on-orbit technology maturation. Through the use of microsatellites in the International Space Station, modular software, and a modular GNC architecture, the group is bringing to maturity the technologies necessary for autonomous formation flight and close-proximity operations. The results provide some answers to the questions posed in the introduction, which in turn open new perspectives on challenges of space technologies:

- *What current and future missions will require formation flight?*

These mission include sparse-aperture arrays, ranging from radio to x-ray wavelengths. In addition, formation flight is a key component of inspection, docking, assembly, and repair missions.

- *What are the challenges in formation flight?*

While the challenges are numerous, MITSSL has chosen to specifically address propellantless formation flight, autonomous docking, and cost-effective technology maturation. As others work to develop new hardware (sensors, computer, etc), MITSSL has started work to mature robust estimation, control, and autonomy algorithms that will enable microsatellites to synthesize large spacecraft, counter to the normal process where these algorithms are the last to be tested.

- *How does one mature formation-flight technology?*

One identified solution is the use of the International Space Station to provide a risk-tolerant, long-duration, microgravity research environment that permits the maturing of formation-flight technology at a cost comparable to ground-based facilities. Scientists can change the notion that space research cannot take risks by continuing to push the envelope in the use of this important new facility.

- *Can propellantless techniques be employed to reduce fuel consumption during formation flight?*

The research suggests that they can. EMFF is definitely a promising method, although it does remain a technology challenge. Other methods also being studied include tethers, electrostatics, and the efficient use of natural orbital motion.

The goal of the satellite formation-flight program at MITSSL is to develop new engineering practices that help the next generation of space systems keep pace with emerging needs of new missions. If developed and matured properly, these technologies will shape future space missions that will revolutionize the way we explore our universe.

17.7 Acknowledgments

MITSSL and the authors acknowledge the support of the DOD and U.S. Air Force throughout the development of the SPHERES and EMFF programs. We thank the DOD Space Test Program, which is the SPHERES payload integration manager with NASA payload operations. Multiple research centers, including Goddard Space Flight Center, Marshall Space Flight Center, Ames Research Center, the Jet Propulsion Laboratory, and the Naval Postgraduate School have contributed as part of the SPHERES Guest Scientist Program and as funding agents. We acknowledge the support of the Lockheed Martin Advanced Technology Center in the development of EMFF and docking technologies. We also thank the NASA International Space Station support teams at Johnson Space Center, Kennedy Space Center, Marshall Space Flight Center, and the many crew members who have and will operate SPHERES. Lastly, we thank Payload Systems Inc., now part of Aurora Flight Sciences Corp., for being our industry partner throughout the SPHERES program, including the production of the flight satellites.

17.8 References

[1] S. Unwin, "Searching for Planets with the Space Interferometry Mission," NASA report, Jet Propulsion Laboratory, Doc ID 20000056082 (2000).

[2] C. A. Beichman, N. J. Woolf, and C. A. Lindensmith, ed., "The Terrestrial Planet Finder (TPF): A NASA Origins Program to Search for Habitable Planets," Jet Propulsion Laboratory Publication 99-003 (1999).

[3] R. J. Sedwick and S. A. Schweighart, "Propellantless Spin-Up and Reorientation of Tethered or Electromagnetically Coupled Spacecraft," paper 4849-26 presented at the SPIE Highly Innovative Space Telescope Concepts Conference (Aug. 2002).

[4] S. A. Schweighart and R. J. Sedwick, "Propellantless Formation Flight Operations in Low Earth Orbit Using EMFF," paper AIAA-2006-5579, *Space Ops 2006 Conference* (June 2006).

[5] R. J. Sedwick and S.A. Schweighart, "Electromagnetic Formation Flight," paper AAS 03-005, *26th Rocky Mountain Guidance and Control Conference* (Feb. 2003).

[6] D. W. Kwon and D. W. Miller, "Electromagnetic Formation Flight for the Terrestrial Planet Finder," paper 5905-41, *SPIE International Symposium on Optics & Photonics* (July 2005).

[7] E. M. C Kong, D. W. Kwon, S. A. Schweighart, L. M. Elias, R. J. Sedwick, and D. W. Miller, "Electromagnetic Formation Flight for Multi-Satellite Arrays," *J. of Spacecraft and Rockets*, **41** (4), (2004).

[8] S. A. Schweighart, "Electromagnetic Formation Flight Dipole Solution Planning," MIT Department of Aeronautics and Astronautics, Ph.D. thesis, SSL Report #09-05 (June 2005).

[9] D. W. Kwon, "Electromagnetic Formation Flight System Design," *6th IAA Symposium on Small Satellites for Earth Observation*, IAA-B6-0602 (Berlin, 23–26 April 2007).

[10] L. M. Elias, D. W. Miller, and D. W. Kwon, "Electromagnetic Formation Flight Dynamics including Reaction Wheel Gyroscopic Stiffening Effects," accepted to *J. of Guidance, Control and Dynamics* (2006).

[11] U. Ahsun and D. W. Miller, "Dynamics and Control of Electromagnetic Satellite Formations," *2006 American Control Conference* (June 2006).

[12] U. Ahsun, "Dynamics and Control of Electromagnetic Satellite Formations," MIT Department of Aeronautics and Astronautics, Ph.D. thesis, SSL Report #12-07 (June 2007).

[13] T. Hashimoto, S. Sakai, K. Ninomiya, K. Maeda, T. Saitoh, "Formation Flight using Super-Conducting Magnets," *International Symposium of Formation Flying Missions and Technologies* (Toulouse, France, 2002).

[14] U. Ahsun, "Dynamics and Control of Electromagnetic Satellite Formations in Low Earth Orbits," paper AIAA-2006-6590, *AIAA Guidance, Navigation, and Control Conference and Exhibit* (Keystone, Colorado, Aug. 2006).

[15] J. Mankins, "Technology Readiness Levels," Advanced Concepts Office, Office of Space Access and Technology, NASA (April 1995).

[16] M. Matossian, "Design for Success: Optimizing the Earth Observing System for Performance and Cost with Managed Risk," paper IAF-95-Q.4.05, *46th International Astronautical Congress* (1995).

[17] T. E. Rumford, "Demonstration of Autonomous Rendezvous Technology (DART) Project Summary," *Space Systems Technology and Operations,* vol. 5088, edited by P. Tchoryk, Jr. and J. Shoemaker (SPIE, 2003).

[18] J. Shoemaker and M. Wright, "Orbital Express Space Operations Architecture Program," *Spacecraft Platforms and Infrastructure,* vol. 5419, edited by P. Tchoryk, Jr. and M. Wright (SPIE, 2004).

[19] A. Saenz-Otero, "Design Principles for the Development of Space Technology Maturation Laboratories Aboard the International Space Station," MIT Department of Aeronautics and Astronautics, Ph.D. thesis, SSL Report #16-05 (June 2005).

[20] A. Saenz-Otero, A. Chen, D. Miller, and M. Hilstad, "SPHERES: Development of an SS Laboratory for Formation Flight and Docking Research," paper 081, *2002 IEEE Aerospace Conference* (March 2002).

[21] "SPHERES Critical Design Review," MIT Space Systems Laboratory, http://ssl.mit.edu/spheres (Feb. 2002).

[22] J. Myers, E. Mendoza, and B. Hoopes, "A Collaborative Electronic Notebook," *IASTED International Conference on Internet and Multimedia Systems and Applications (IMSA 2001)* (Honolulu, HI, Aug. 2001).

[23] "SPHERES Science Critical Design Review," MIT Space Systems Laboratory, http://ssl.mit.edu/spheres (Feb. 2002).

[24] S. Nolet, "The SPHERES Navigation System: from Early Development to On-Orbit Testing," number AIAA-2007-6354, in *AIAA Guidance, Navigation, and Control Conference and Exhibit* (Hilton Head, SC, August 2007).

[25] V. P. Legostaev, and B. V. Raushenbach, "Automatic Assembly in Space," *Cosmic Res,* **7** (6), (1969).

[26] B. V. Rauschenbakh, M. Y. Ovchinnikov, and S. McKenna-Lawlor, *Essential Space Flight Dynamics and Magnetospherics* (The Space Technology Library, Kluwer Academic Publishers and Microcosm Press, Boston, MA, 2003).

[27] R. Godwin, *Rocket and Space Corporation Energia* (Apogee Books, Burlington, Ontario, Canada, 2001).

[28] I. Kawano, M. Mokuno, T. Kasai, and T. Suzuki, "Result and Evaluation of Autonomous Rendezvous Docking Experiments of ETS-VII," paper AIAA-1999-4073, *AIAA Guidance, Navigation, and Control Conference and Exhibit* (Aug. 1999).

[29] M. Oda, "Experiences and Lessons Learned from the ETS-VII Robot Satellite," *IEEE International Conference on Robotics and Automation (ICRA '00)* (April 2000).

[30] C. I. Underwood, G. Richardson, and J. Savignol, "In-orbit Results from the SNAP-1 Nanosatellite and its Future Potential," *Philosophical Transactions: Mathematical, Physical and Engineering Sciences,* **361** (1802), (2003).

[31] T. M. Davis and D. Melanson, "XSS-10 Micro-Satellite Flight Demonstration Program Results," *Spacecraft Platforms and Infrastructure,* vol. 5419, edited by P. Tchoryk, Jr. and M. Wright (SPIE, 2004).

[32] D. Zimpfer, P. Kachmar, and S. Tuohy, "Autonomous Rendezvous, Capture, and In-Space Assembly: Past, Present, and Future," paper AIAA-2005-2523, *1st Space Exploration Conference: Continuing the Voyage of Discovery* (Jan. 2005).

[33] M. A. Dornheim, "Rendezvous trials," *Aviation Week & Space Technology* (The McGraw-Hill Companies, Washington, DC, 18 April 2005).

[34] "Overview of the DART Mishap Investigation Results," May 2006, available at http://www.nasa.gov/pdf/148072main_DART_mishap_overview.pdf as of May 2007.

[35] M. Ruth and C. Tracy, "Video-Guidance Design for the DART Rendezvous Mission," *Spacecraft Platforms and Infrastructure*, vol. 5419, edited by P. Tchoryk, Jr. and M. Wright (SPIE, 2004).

[36] J. Morris, "Full Service: Pioneering Orbital Express Mission Offers Many Lessons for Future Satellite Servicing," *Aviation Week & Space Technology* (The McGraw-Hill Companies, Washington, DC, 23 July 2007).

[37] E. Crawley, O. de Weck, S. Eppinger, C. Magee, J. Moses, W. Seering, J. Schindall, D. Wallace, and D. Whitney, "The Influence of Architecture in Engineering Systems," technical report, Massachusetts Institute of Technology (March 2004).

[38] J. P. How and S. R. Hall, "Local Control Design Methodologies for a Hierarchic Control Architecture," *J. of Guidance, Control, and Dynamics,* **15** (3), (1992).

[39] N. Pedreiro, "Spacecraft Architecture for Disturbance-Free Payload," *J. of Guidance, Control, and Dynamics,* **26** (5) (2003).

[40] S. R. Hall, E. F. Crawley, and J. P. How, "Hierarchic Control Architecture for Intelligent Structures," *J. of Guidance, Control, and Dynamics,* **14** (3), (1991).

[41] E. Frazzoli, M.A. Dahleh, and E. Feron, "A Hybrid Control Architecture for Aggressive Maneuvering of Autonomous Helicopters," *38th IEEE Conference on Decision and Control* (Dec. 1999).

[42] M. Fujita, Y. Kuroki, T. Ishida, and T.T. Doi, "Autonomous Behavior Control Architecture of Entertainment Humanoid Robot SDR-4X," *2003 IEEE/RSJ International Conference on Intelligent Robots and Systems (IROS 2003)* (Oct. 2003).

[43] J-M. Park, I. Song, Y.-J. Cho, and S.-R. Oh, "A Hybrid Control Architecture using a Reactive Sequencing Strategy for Mobile Robot Navigation," *1999 IEEE/RSJ International Conference on Intelligent Robots and Systems (IROS '99)* (Oct. 1999).

[44] W. Fehse, *Automated Rendezvous and Docking of Spacecraft* (No. 16 in Cambridge Aerospace Series, Cambridge University Press, Cambridge, UK, 2003).

[45] M. D. Ingham, R. D. Rasmussen, M. B. Bennett, and A. C. Moncada, "Engineering Complex Embedded Systems with State Analysis and the Mission Data System," *J. of Aerospace Computing, Information, and Communication,* **2** (12), (2005).

[46] D. Dvorak, R. Rasmussen, G. Reeves, and A. Sacks, "Software Architecture Themes in JPL's Mission Data System," *2000 IEEE Aerospace Conference* (March 2000).

[47] S. Nolet, E. Kong, and D. W. Miller, "Autonomous Docking Algorithm Development and Experimentation using the SPHERES Testbed," *Spacecraft Platforms and Infrastructure*, vol. 5419, edited by P. Tchoryk, Jr. and M. Wright (SPIE, 2004).

[48] S. Nolet, A. Saenz-Otero, D. W. Miller, and A. Fejzic, "SPHERES Operations Aboard the ISS: Maturation of GNC Algorithms in Microgravity," paper AAS 07-042, *30th Annual AAS Guidance and Control Conference* (Feb. 2007).

[49] S. Nolet and D. W. Miller, "Autonomous Docking Experiments using the SPHERES Testbed inside the ISS," *Sensors and Systems for Space Applications,* vol. 6555, edited by P. Tchoryk, Jr. and Roger Hall (SPIE, 2007).

[50] S. Nolet, "Development of a Guidance, Navigation and Control Architecture and Validation Process Enabling Autonomous Docking to a Tumbling Satellite," MIT Department of Aeronautics and Astronautics, Ph.D. thesis, SSL Report #08-07 (June 2007).

[51] L. Breger and J. P. How, "Safe Trajectories for Autonomous Rendezvous of Spacecraft," paper AIAA-2006-6584, *AIAA Guidance, Navigation and Control Conference and Exhibit* (Aug. 2006).

[52] C. Mandy, A. Saenz-Otero, and D. W. Miller, "Satellite Formation Flight and Realignment Maneuver Demonstration aboard the International Space Station," *2007 SPIE Optics & Photonics Conference* (Aug. 2007).

Chapter 18

A Mass-Producible Glass/Ceramic Micropropulsion Unit for a Co-Orbiting Satellite Assistant (COSA) Mission

Adam Huang,* Siegfried W. Janson,† William W. Hansen,† Lee F. Steffeney,† and Henry Helvajian†

18.1 Overview

Traditional manufacturing technologies can be used to fabricate spacecraft as small as picosatellites and CubeSats. However, the benefit of using smaller and more numerous satellites hinges heavily on the possibility of economical mass production. Our goal was to demonstrate, for the first time, a mass-producible miniature space system prototype based on photostructurable glass/ceramic (PSGC) as the structural material, as opposed to metal (see chapter 24), and the use of commercial-off-the-shelf (COTS) components. We achieved this goal by designing, fabricating, assembling, and testing an inexpensive PSGC cold-gas propulsion system with onboard sensing and processing for a short-duration nanosatellite or picosatellite mission.

This propulsion system is designed to support a kilogram-class satellite known as COSA, the Co-Orbiting Satellite Assistant,[1] which would orbit about a host satellite to provide mobile external inspections. Two propulsion modules would be needed in a single COSA vehicle to provide six degrees of freedom (i.e., three axes of translation and three axes of rotation).

Three major generations of the COSA propulsion module (COSA-PM) have successively been developed. The first-generation design (Gen 1), as shown in Fig. 18.1 (left), began in 1999. It incorporated the basic cold-gas thruster system using machine-milled Lucite parts and wire-wrapped electronics mounted on stacked vector boards. The second-generation (Gen 2) design, as shown in Fig. 18.1 (center), started in 2001; it incorporated photo-patterned PSGC to fashion the thruster fluidic stacks and replaced the electronics with hybrid components on a PCB (printed circuit board). The biggest difference between Gen 1 and Gen 2 was the design process. Gen 1 was completely hand-built with limited use of computer-aided design and manufacturing (CAD/CAM) methodologies and electronic layout tools. Gen 2 represented the first attempt in integrated design, with emphasis in CAD/CAM and electronic layout software for the PCB design. In both Gen 1 and Gen 2, onboard components included COTS temperature, pressure, and

*University of Arkansas, Fayetteville, Arkansas
†The Aerospace Corporation, El Segundo, California

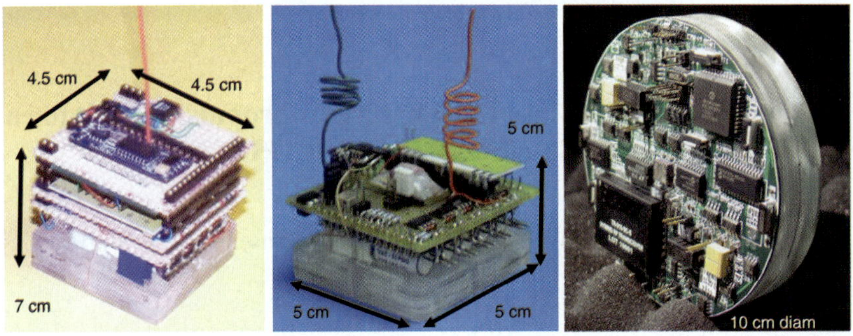

Fig. 18.1. The COSA-PM prototypes. Generation 1 is based on machined Lucite stacks for the cold-gas thruster and has a mass of 70 g without batteries (left). Generation 2 uses PSGC fluidic layers for the cold-gas thruster and has a mass of 84 g without batteries (center). Generation 3, or PSGC COSA-PM, uses surface-mount PCB electronics and has a mass of 365 g with batteries (right).

rate gyro sensors; a propellant heater; wireless telemetry; and microcontrollers. The third-generation (Gen 3) design, as shown in Fig. 18.1 (right), began in 2004 and is discussed in detail below.[2,3] While Gen 1 and Gen 2 focused on specific technology developments, they were not designed for the COSA mission. Gen 3 was designed for the COSA mission; it carries sufficient fuel and battery power and has the necessary thrust capability.

The COSA-PMs have been tested on a custom-built air table. Subsystems onboard include a complete two-axis cold-gas system with eight nozzles, a MEMS rate gyro and magnetometer as rotation sensors, distributed microcontrollers for command and control, propellant temperature and pressure monitors, and RF signal-strength indicators. All are terrestrial COTS components and were not designed for radiation hardness or tolerance. This is considered a reasonable design choice, given the short duration of the mission. The laboratory COSA-PM experiments included both onboard and offboard computer-controlled "flight" in addition to man-in-the-loop control via RF telemetry. We expect that wireless RF and/or optical (using a photonic bus through the clear PSGC structure) communication will be used for linking the COSA to the host spacecraft. This module provides a propulsion capability to small satellites with minimal interface hardware. In addition, numerous units can be mounted and controlled through multiplexing schemes to offer a truly modular and scalable propulsion system that can be tailored toward specific missions. Finally, establishing an approach for wireless connectivity will enable a block-modular approach to satellite manufacturing.

The entire Gen 3 structure, except for the main circuit board, is composed of PSGC, with fuel tank, fuel lines, filters, and nozzles integral to the structure. In the future, we will replace the standard glass-epoxy circuit board with a low-temperature cofired ceramic (LTCC) that can also be patterned using lasers. LTCC technology enables the three-dimensional (3-D) packing of both dc and RF electronics, and given that the material is a "cousin" of PSGC, integration of

PSGC with electronics becomes possible. As the name implies, PSGC materials are photosensitive. A sequence of processing steps that include lithography or direct-write patterning, baking, and chemical etching can produce high-aspect-ratio (~50:1) structures with high fidelity (<3 μm minimum resolution). The Gen 3 wafers were patterned by pulsed UV lasers and direct-write processing. The processing of PSGC merges a serial direct-write lithography step with a subsequent bulk chemical etching step that provides a fast and economical technique to pattern 3-D structures on the wafer scale. Direct-write patterning is maskless; all designs are directly translated from a solid model, generated by CAD software, to a CAM program that controls the laser patterning platform. Consequently, there is a high degree of flexibility to quickly develop micro- and mesoscale structures that is not present in other microfabrication processes. This chapter addresses the design and development of the COSA-PM, including the lessons learned and recommendations for eventual space applications.

18.2 The COSA Mission

The Aerospace Corporation and NASA-Houston held a series of workshops at the First International Conference on Integrated Micro/Nanotechnology for Space Applications in 1995.[4] One of these workshops focused on the concept of a picosatellite inspector for imaging the entire exterior of a host spacecraft at 3 mm resolution to resolve on-orbit anomalies. Key requirements included a mass of less than 1 kg, a volume of less than 350 cm^3, a 48 hour minimum operational lifetime, a maneuvering thruster system capable of 10 m/sec ΔV, and a deorbit thruster. Three-axis stabilization, a position-determination system, and an optical imaging system were also required. Examples of possible missions included inspections for improper array deployment, improper antenna deployment, micrometeoroid damage, surface damage due to space debris, and decreased solar cell output due to surface contamination. The workshop asserted that 1 kg picosatellite inspectors could accomplish the mission based on 1995 technology, but it did not specify how they would be constructed.

In 1999, DARPA (Defense Advanced Research Projects Agency) funded The Aerospace Corporation and Rockwell Scientific to develop two ~360 g picosatellites to provide low-cost on-orbit demonstrations of MEMS. This evolved over the next few years into the MEMS-based PicoSatellite Inspector (MEPSI) program; it used traditional spacecraft materials and fabrication techniques (see chapter 20).

Between 2000 and 2003, The Aerospace Corporation provided internal R&D funding to demonstrate the fabrication of mass-producible, cold-gas propulsion systems for nanosatellites. The Gen 1 and Gen 2 systems were developed under this effort. In 2003, DARPA funded the development of Gen 3—a small, laboratory-class, low-cost, expendable, mass-producible PSGC propulsion module for a two-week COSA mission.[5] An aspect of the project was to also demonstrate that the prototype could be produced via direct digital development (D^3) methodology. The goal of D^3 is to show a possible manufacturing path that implements aspects of mass-production. Initially, plastics and bulk silicon were also proposed as a structural material, but the degradation of plastics in the space environment becomes an issue when the dormant period, before the actual two-week mission, extends to many months or years. The benefits of using silicon as the structural material are obvious, since it provides the intrinsic capability of integrating microelectronics and MEMS devices *in situ*. However, a serious challenge exists in realizing the various integrated COSA-PM designs, given that the physical shaping of silicon structures remains predominantly a two-dimensional (i.e., mask) process, and the cofabrication of structures with varied aspect ratio becomes proportionally more expensive with the number of masking layers. Between the plastics and bulk silicon are the glass/ceramic materials, when considered with regard to material property strength, fabrication flexibility, and space environment durability. And within the glass/ceramic family, the PSGC subclass allows implementation of optical lithography, one of the hallmarks of the microelectronics industry's success.

In our basic COSA mission scenario, an inspection spacecraft is ejected from the host spacecraft to provide mobile external inspections for two weeks at 100 m constant range, as shown in Fig. 18.2. Both low Earth orbit (LEO) and geosynchronous (GEO) mission models were analyzed; Table 18.1 summarizes the key propulsion specifications. The mission scenarios include ejection from the host, maneuvers to co-orbit, drag make-up, north-south station-keeping (for GEO missions), and final disposal.[2]

Disposal of satellites in LEO is based on propulsive maneuvers aiding the eventual burn-up in the atmosphere, while satellites in GEO are required to transfer into disposal orbits either higher or lower than GEO as defined by the U.S. Space Command (minimum perigee of 36,100 km and maximum apogee of 35,300 km). Our COSA-PM is compatible with efficient disposal methods for both LEO and GEO missions, albeit via different means. For LEO, disposal could

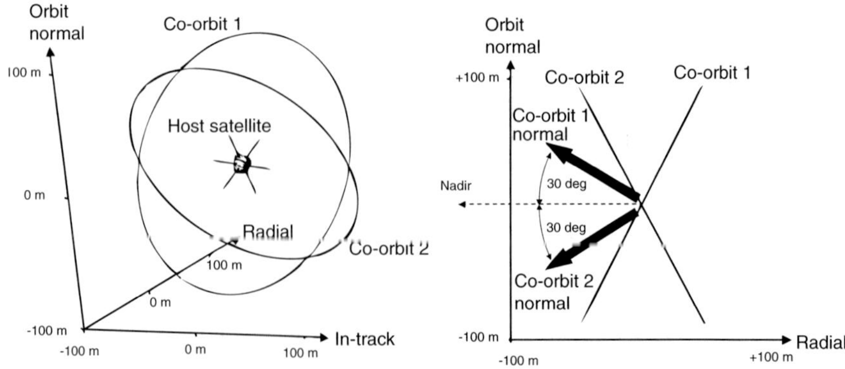

Fig. 18.2. Configuration of constant range co-orbits about a host vehicle as seen in the host vehicle's reference frame.

Table 18.1. Summary of COSA Propulsion Requirements

Mission	Mission ΔV (m/sec)	Thrust Level (mN)	Minimum Impulse Bit (mN-sec)	Disposal ΔV (m/sec)	Disposal Options
LEO	10–20	20	0.1	50–350	Solid rocket or balloon
GEO	0.15–7	1	0.005	11–18	Cold-gas thruster

be accomplished by the inflation of a 1 m balloon using the gas supply from the cold-gas system. In GEO, the thrusters have sufficient ΔV to transfer to the required disposal orbits. Thus, the PSGC COSA is potentially capable of both LEO and GEO missions with a required minimum ΔV on the order of 10 m/sec and a required impulse bit on the order of 1 μN-sec.

18.3 COSA-PM Design

The COSA-PM design is an attempt to integrate into the structure of the vehicle the propellant tank, liquid-to-gas phase separators, propellant lines, thruster nozzles, battery and valve compartments, and electric wiring. The ultimate goal is to use spacecraft structure as more than a simple scaffold. This concept has become feasible using a nonablative 3-D laser fabrication process developed at The Aerospace Corporation and the unique properties offered by PSGC materials. Unlike metals, which serve primarily as structural material, PSGCs are multifunctional and can simultaneously serve as support structure, radiation shield, optical element where necessary, chamber wall for propellant tank, surface for electrical lines and vias, and walls for fuel and thrust channels.

The ability to establish multifunctionality reduces the number of materials in a spacecraft and enables the design of highly integrated systems. This, in turn, decreases the number of interfaces and parts count, and can thereby reduce system mass. This project has demonstrated that it is possible to develop highly integrated space subsystems out of PSGC and that the fabrication method is amenable to mass-production, with feature-size control ranging from centimeters down to near micron scale. The hope is that multiple COSA vehicles can be manufactured economically enough to be placed onboard every large spacecraft.

18.3.1 PSGC Patterning

PSGC is a subset of the family of materials called glass/ceramics.[6] Nearly all glass/ceramics are manufactured in the amorphous glass state to enable shaping and molding via low-cost techniques. After the desired shape is fashioned, a thermal treatment is applied that converts the glass into the ceramic phase to obtain the desired properties. Common kitchen flatware (i.e., CorningWare) was revolutionized in the 1950s by the successful development of glass/ceramics by Corning

Corporation. The PSGCs differ from the common glass/ceramics in that while the common glass/ceramics are designed to phase transform in a single processing step, PSGC is designed to transform in two steps.[7] The extra step allows the material to also be shaped by optical lithographic patterning. The process is enabled by the inclusion of photosensitizers in the glass matrix so that only areas exposed to UV light undergo phase transformation during the thermal processing step.[8] In the early 1950s, Corning scientists realized that a particular formulation of glass/ceramic, when thermally processed with a particular bake protocol, resulted in the growth of a crystalline phase that is chemically soluble in dilute hydrofluoric acid. This discovery allowed the development of a chemically "machinable" glass/ceramic, or PSGC.[9] Consequently, it was realized that micro and macro structures could be fabricated in glass/ceramic. The processing of PSGC mimics that of a photoresist, a material that is routinely used in the microelectronics industry. The three basic processing steps can be broadly defined as exposure, phase transformation, and chemical etching. For the COSA-PM project, the commercially available PSGC material trademarked as Foturan (by Schott Corporation) was used.[10] Foturan is a lithium aluminosilicate glass/ceramic; by applying a specific bake temperature protocol after exposure, a lithium metasilicate crystal (Li_2SiO_3) grows that is soluble in dilute hydrofluoric acid (5% in water) at room temperature. However, recent research has shown that other properties of the processed PSGC material can be altered by simply controlling the initial exposure.[11] These affect the aspect ratio of the structures that can be fabricated, the colors (i.e., transparency) that can be infused, and the types of shapes that can be fashioned (e.g., 2-D, 3-D, embedded).[12] These results further enhance the inventory of multifunctional attributes possible with PSGC.

The traditional method of patterning PSGC involves lithographic techniques that are identical to those used in the integrated-circuit and MEMS microfabrication process. However, mask-based lithography is inherently a planar process, where materials are structured one layer at a time for the entire mask field, as illustrated in Fig. 18.3 (left). All exposed areas generally get the same photon flux. The Aerospace Corporation approach instead uses laser direct-write patterning without the use of masks.[13] Consequently, the incident laser photon flux can be locally varied, as shown in Fig. 18.3 (right). This approach enables tailoring of material properties on a local scale and in three dimensions. There are two fundamental approaches to generate 3-D exposure patterns. The first exploits the fact that for a given wavelength there is a corresponding practical optical penetration depth defined by the material absorbtivity. Thus, some control in exposure depth can be set. The second approach capitalizes on the photon density threshold to induce an interconnected network of etchable phase crystals. This technique is far more versatile, allowing true 3-D patterning. The Aerospace Corporation approach uses two laser wavelengths that can be derived from commercially available UV laser sources. At 266 nm, Foturan is strongly absorbing (absorbtivity = 1.45), with structures taller or deeper than 0.75 mm not practically feasible. On

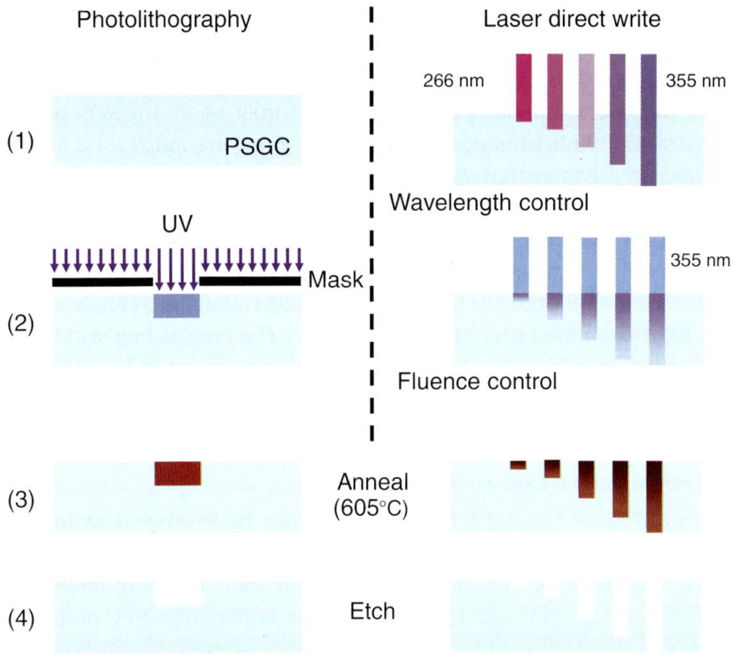

Fig. 18.3. Generalized schematics showing the photolithographic (left) and direct-write laser (right) patterning of PSGC materials.

the other hand, at 355 nm, Foturan is initially only weakly absorbing (absorbtivity = 0.09), and it is theoretically possible to fabricate structures many centimeters deep.[14] The second approach uses the measured exposure threshold value of the material to advantage. For any exposure wavelength, the local photon flux density can be increased by focusing the laser beam into a small focal volume. By then metering the dose as defined by the applied number of laser shots and the number of photons within a laser shot, the integrated exposure level in the focal volume can be controllably varied.[14] A 3-D structure is fabricated by parsing the total material volume into discrete volume elements (voxels) where the exposure in each voxel is prescribed.[15] During the laser direct-write patterning process, which uses a three-axis motion platform, each addressable voxel element is exposed to the prescribed level. Because this is a patterning operation where exposure doses are much lower than required for material ablation, processing speeds can exceed many meters per second. In addition, this technique enables fabrication of embedded patterns.[14] This is accomplished by carefully metering the laser photon density and dose *above* the threshold value for exposure[14] for each voxel that will be removed, while simultaneously keeping the exposure of all other voxels below threshold values. This is typically accomplished using a focused laser beam with a well-defined, high-intensity waist. The embedded patterns, when connected to conduits that feed a chemical etchant, result in the fabrication of channels (i.e., propellant lines) and cavities (i.e., propellant tanks) inside the structure.

At the time of the COSA-PM development, the capabilities of the Aerospace patterning tool were limited to motions on the order of millimeters per second. The practical resolution limit of a UV laser exposure patterning tool is on the order of a few hundred nanometers for far-field exposures[16] and tens of nanometers[17] for near-field exposures. Both of these exposure resolutions are far smaller than that which can be achieved by most thermally generated cutting tools (i.e., laser milling) and by most practical material additive tools. In the case of PSGC, the exposure resolution is actually defined by the material doping concentration, and for Foturan, it is on the order of a few microns. The typical line-width achieved using 355 nm radiation with a 5 μm laser spot size is shown in Fig. 18.4 after the bake step where the phase transformation to form the acid-soluble lithium metasilicate crystal has occurred.

18.3.2 Stacked-Layer Construction

Glass forming is an age-old technology that can be leveraged to form PSGC substrates of almost any size and shape. Commercially available Foturan is manufactured as a continuous slab that is roughly 1 m wide and 1 cm thick. It is then cut and polished into 100 and 150 mm diameter wafers of 1 to 6 mm thickness. The wafer size is determined by the wafer polish tooling. In the unexposed state, Foturan resembles optical clear glass (as shown in the right inset of Fig. 18.5).

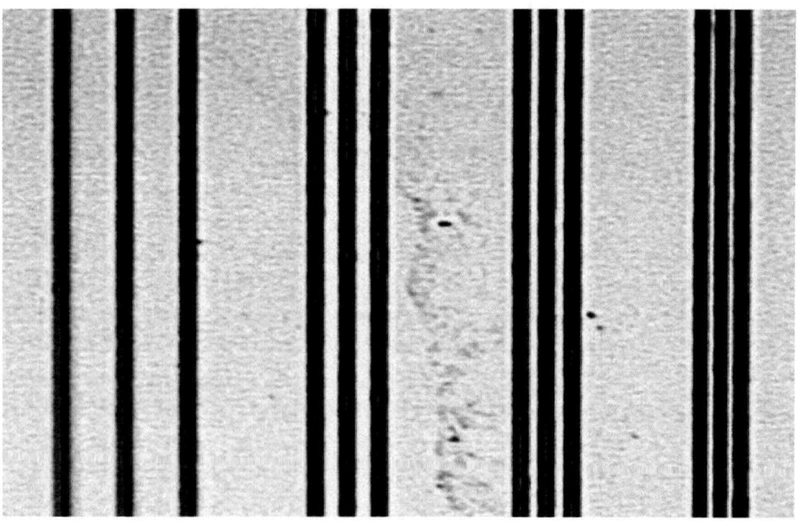

Fig. 18.4. An optical microscope image of exposed and baked lines in Foturan using a 5 μm laser-beam spot size. The four test patterns are shown as a function of line separation distance.

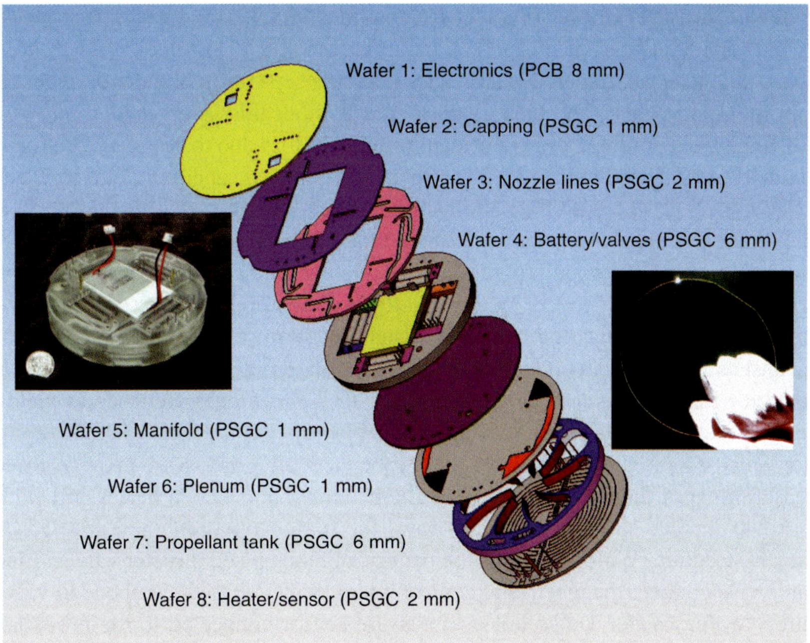

Fig. 18.5. Exploded schematic diagram of the PSGC COSA-PM.

This fact permits the development of an optical bus within the structure, which is being pursued at The Aerospace Corporation for future COSA designs.

The COSA propulsion module uses seven stacked layers of 100 mm diameter PSGC wafers. Figure 18.5 illustrates the thicknesses of the individual PSGC wafer layers, which together yield a total thickness of 19 mm for the PSGC stack. The electronics board with the mounted COTS chips adds another 8 mm in thickness, bringing the overall dimensions of the disk-shaped COSA-PM to 100 mm in diameter by 27 mm thick. It was designed as a flat puck to allow thrust testing on an air table. Other overall shapes could also have been fabricated. For example, making the COSA-PM resemble an antenna dish requires preshaping the flat wafers by standard molding techniques (e.g., softening over a mandrill) followed by the 3-D laser fabrication.

The key components of the propulsion system are the layers that include the micronozzles and propellant lines (wafer 3), the gas/liquid separator screen in the plenum layer (wafer 6), a scalable propellant tank (wafer 7), and a heater coil (wafer 8). Furthermore, there are electronic and fluidic vias that pass through specific wafers. Wafer 3 contains eight 15 deg half-angle converging micronozzles. The design throat area is 500 μm × 1 mm, and thrust per nozzle is 20 mN, reflecting the required thrust level for terrestrial experiments on an air table. A converging/diverging nozzle designed for on-orbit use, where the propellant expands into a vacuum rather than the ambient pressure in a terrestrial laboratory, would produce

45 mN of thrust. Wafer 6 contains two plenum cavities, each serving four thrusters and separated from the propellant tank (wafer 7) by 1315 microcapillary holes (50 μm diameter) and microstructures for ensuring separation of the propellant gas from its liquid phase. The Gen 1 prototypes without this layer yielded unsteady thrusting, rapid cooling, and significant propellant loss due to massive ejection of liquid droplets. The plenum with the associated liquid-gas separator holes for Gen 2 is shown in Fig. 18.6 with the patterns before (left) and after etching (right).

Wafer 7 is the scalable propellant wafer shown in Fig. 18.5 with antisloshing supporting ribs. The propellant tank is scalable because additional propellant wafers can be added as needed. Wafer 8 is the bottom-capping layer of the COSA, which also functions as the heater and fuel-temperature sensor layer. The heater consists of two independent coils of electroplated nickel grown on top of an electron-beam evaporated nickel seed layer in 700 μm deep trenches to yield a nominal resistance of 3 Ω (providing 4 W of heat per coil at 3.5 V). Two thermistor chips (for redundancy) are also embedded on wafer 8 to provide direct measurement of the fuel temperature. The patterned and etched heater coil wafer is shown in Fig. 18.7, which also shows the bottom view of the entire PSGC stacked structure. Wafer 4 houses the battery and the pulsed thruster valves. Electronic vias connect the electronic control board (wafer 1) to the embedded valve/battery wafer. Wafer 1 also holds a pressure sensor connected to the propellant tank (wafer 7) by a fluidic via. At the time of fabrication, a workable pressure sensor that could be embedded in the fuel tank was not available; we used an external pressure sensor connected to the fuel tank. The "dry" (no propellant) weight of the COSA-PM is 325 g.

18.3.3 COTS Components

Keeping with the goal of supporting mass-producibility, we emphasize COTS components. Major COTS components include the processors (Microchip Technology, PIC16F), rate gyro (Analog Devices, ADXRS150), RF transceiver (Linx, TR-916-SC-S), magnetometer (Honeywell, HMC1022), analog-to-digital

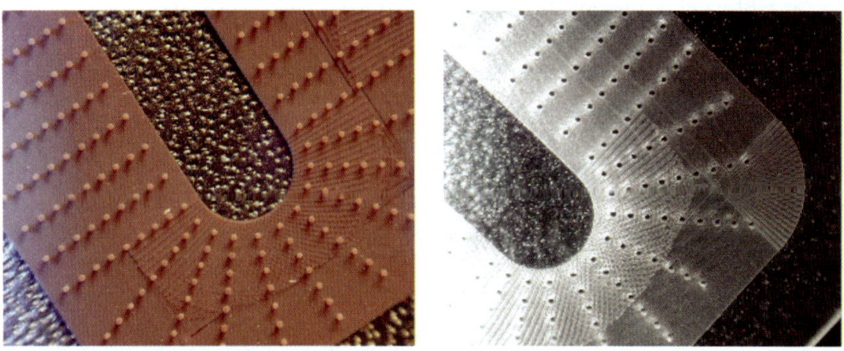

Fig. 18.6. Two photographs of a PSGC vapor-liquid separator showing the patterned PSGC material after baking (left) and the final product after chemical etching (right).

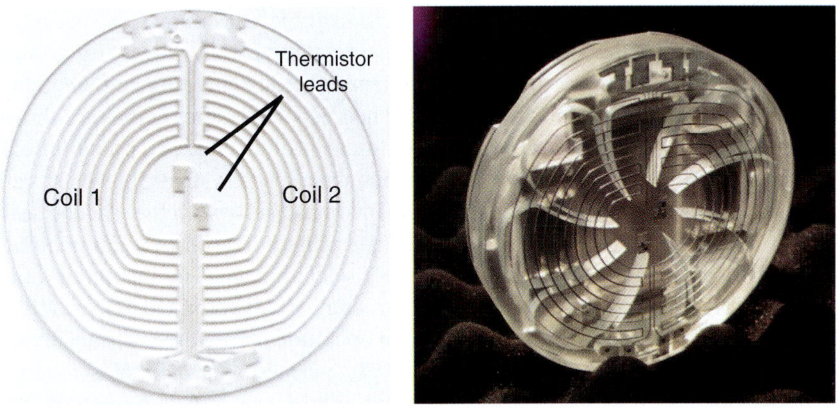

Fig. 18.7. Photograph of the PSGC heater layer (left) and the assembled PSGC structure viewed from the bottom (right).

converter (Linear Technology, LTC1290), voltage step-up regulator (Maxim, MAX1771), microdispensing solenoid valve (The Lee Co., INKA1222210H), lithium-ion polymer batteries (Plantraco Co., LP170, LP550, and LP850), pressure sensors (Measurement Specialties, 1451-100G-T), and temperature sensors (Panasonic, ERT-J1VR103J). These represent what was commercially available circa 2004. Components with higher performance and functionality are available today; all have to be evaluated for outgassing, thermal stability under vacuum, and radiation hardness before use on an actual flight vehicle.

18.3.4 Cold-Gas Thruster System

The COSA-PM was designed with a 20 mN thrust level and minimum impulse bit of better than 0.005 mN-sec, which is sufficient to satisfy both LEO and GEO missions. Up to 30 cm^3 of propellant can be stored in one propellant wafer. This can provide 15 m/sec ΔV for a 1 kg COSA based on an estimated on-orbit specific impulse of 38 sec for the $C_3H_2F_6$ 1,1,1,3,3,3-hexafluoropropane (ANSUL FE-36 or HFC-236fa) propellant. Two modules would be required for full six-axis control of an orbiting COSA, providing a combined ΔV of 30 m/sec, which would easily meet the propulsion requirements for a typical LEO mission.

The choice of the HFC-236fa over traditional cold-gas propellants was due to the requirement for low toxicity, low pressure, low ignition potential, and modest pressure at room temperature for ground-based tests. It may also be carried onboard the space shuttle, where propellant safety plays a key role. The specific impulse for the HFC-236fa is low at 38 sec (relative to ~300 sec for H_2 and ~80 sec for air) due to its high molecular weight. Since the fuel is stored in the liquid phase, the total ΔV is not as low as the specific impulse and propellant tank volume would otherwise suggest. The densities of HFC 236fa are 1371.58 kg/m^3 and 11.57 kg/m^3 in liquid and vapor phase at 293.186 K, respectively—which is equivalent to roughly 120 times expansion in volume.

Low propellant pressure is also more compatible with miniature space systems, as high-pressure components (such as pressure vessels, valves, fluid lines) are often bulky. Other low-pressure propellants were also investigated, including butane and octafluorocyclobutane (C318). Their vapor pressures curves are shown in Fig. 18.8 alongside that for HFC-236fa. Butane is often cited as a possible cold-gas (and potential hot-gas) propellant for miniature space systems; however, care must be taken with its application due to its incompatibility with the silicone seals that are often found in valve seats. This is the primary reason (not flammability concerns) why this project tested but did not select butane. The maximum burst pressure for the PSGC propulsion system structure was determined empirically to be >690 kPa (100 psia), while the maximum operational pressure is limited by the differential valve pressure of ~350 kPa (50 psi) when the normally closed valves automatically open and relieve the differential pressure across the valves (this mechanism actually provides a failsafe feature for burst protection).

18.3.5 Decoupled Thrust-Axis Control

Since the air-table COSA-PM demonstration is inherently planar, only two translational (x, y) and one rotational (yaw or θ) axes can be controlled. Four thrusters could provide this functionality (diagonally opposed thrust lines with equal offsets to the center of gravity), but eight thrusters were used to decouple translation from rotation to simplify the control protocol and provide simultaneous thrusting of all three axes. The general layout of the thruster arrangement and the corresponding wafer is shown in Fig. 18.9. The nozzle pairs for rotation control are placed 3.75 cm from the centerline. Although the nozzles used for the laboratory demonstration are convergent, converging-diverging nozzle designs would be used in space, where the background pressure is effectively zero. PSGC thruster nozzles with converging-diverging nozzles have also been fabricated and tested as part of the Gen 2 effort (Fig. 18.10).

18.3.6 Thrust Control Algorithm and Simulation

A generic, yet simple, control algorithm was used for controlling all three axes of the COSA:

$$\text{Thrust}_i = \alpha_i(\delta\Phi_i) - \beta_i\frac{d\Phi_i}{dt}, \text{where } \Phi = \text{displacement and } i = X, Y, \theta \quad (18.1)$$

This equation is essentially a proportional controller based on the positional error, $\delta\Phi_i$, compensated by the instantaneous momentum term, $d\Phi_i/dt$. In effect, the coefficient α_i dictates the response time, and β_i provides the artificial damping. Experimentally, the displacement data were provided by a vision system and the onboard rate gyro. A magnetometer was mounted on the COSA-PM, but the data were redundant and thus not required in these tests. A spreadsheet program was then used to find the six coefficients that provided critically damped control. The simulation was based on the propulsion module with 300 g mass, 20 mN

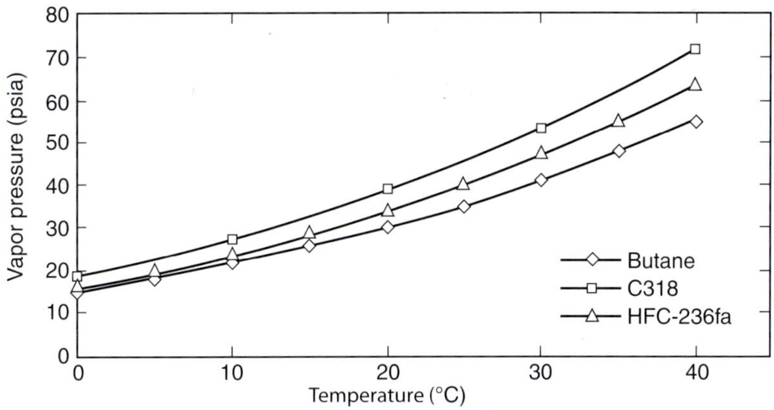

Fig. 18.8. Vapor pressures of the propellants investigated for the COSA-PM prototypes.

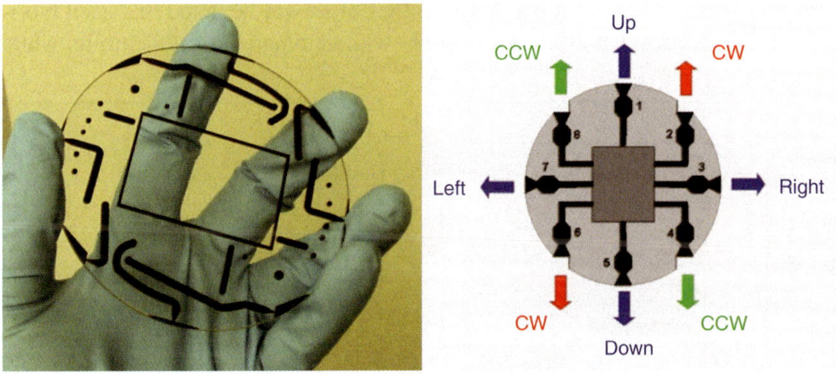

Fig. 18.9. The eight-nozzle thruster PSGC layer after exposed and baked (left) and the functional layout of the thrusters (right) for the COSA-PM.

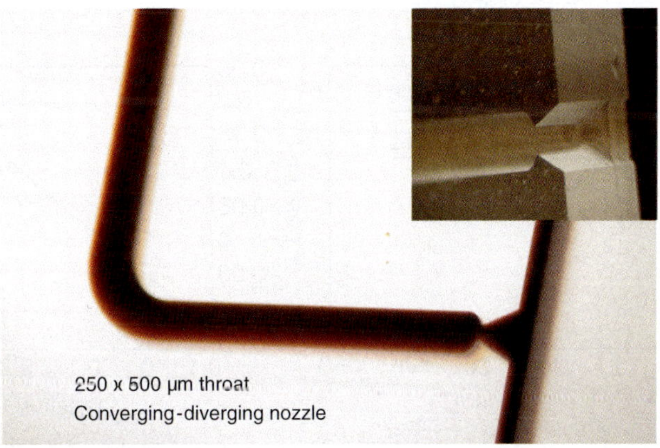

Fig. 18.10. The converging-diverging nozzle design used on the Gen 2.

maximum thrust nozzles, and 3.75×10^{-4} kg m^2 moment of inertia (based on a circular disk). The results of the simulations for rotation and translational control are shown in Figs. 18.11 and 18.12. The determined coefficients were $\alpha_X = \alpha_Y =$ 0.05 N/m, $\alpha_\theta = 0.0064$ N/rad, $\beta_X = \beta_Y = 0.2$ N-sec/m, and $\beta_\theta = 0.01$ N-sec/rad. It should be noted that control schemes utilizing constant impulse bits (i.e., "bang-bang" controls with fixed duty cycle) can not be adopted in this control scheme since thrusting will be needed even at $\delta\Phi_i = 0$. This is due to the inability to exactly cancel out a subimpulse bit-size $d\Phi_i/dt$ term in a real-world environment. The experimental results are discussed below in section 18.6.

18.3.7 Fabrication Time

As noted, this project viewed mass-producibility as the key technology for the eventual proliferation of miniature spacecraft. During the development and fabrication of the COSA-PM, detailed records of manufacturing time where made. The time required to fabricate each propulsion unit in a serial fashion (one part at a time) was 207.5 hr, as shown in Table 18.2. It should be noted that each fabrication step was designed to be compatible with automated manufacturing, while

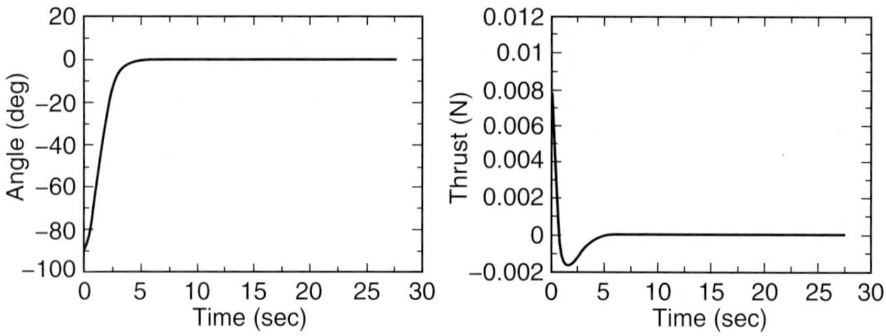

Fig. 18.11. The simulated response of the PSGC COSA-PM with 90 deg rotation (left) and the required thrust history (right).

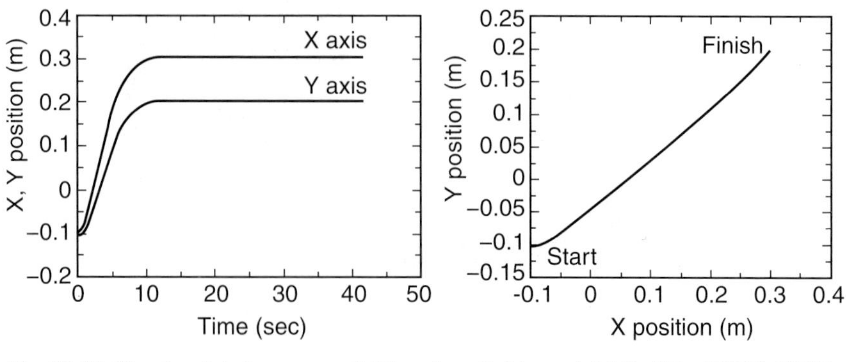

Fig. 18.12. The simulated response (left) and spatial trace (right) of the PSGC COSA-PM with translations of 40 cm and 30 cm in *x* and *y* axis, respectively.

Table 18.2. Time (hr) Needed for Fabricating Components of the PSGC COSA-PM

	Patterning	Baking	Polishing	Etching
Is the process automated? Can it run continuously?	Yes	Yes	Yes	Yes
Wafer				
Heater	10	4	12	2
Fuel	0.25	0	0	0
Plenum	19	4	12	3
Spacer	2	4	12	1
Battery	0.25	0	0	0
Nozzle	17	4	12	3
Capping	7	4	12	1
Total	55.5	20	60	10
Total time to fabricate COSA wafers			145.5	
Total time to bond all wafers			2	
Total time to bake bonded wafers			24	
Estimated total time for electronic wafer assembly			36	
Total time for COSA vehicle manufacture			**207.5**	

some of the steps allow batch processing with the use of multiple equipment (i.e., etching of multiple PSGC wafers in hydrofluoric-acid spray baths). In addition, the polishing step can be eliminated using recently developed baking and etching steps that can retain high surface smoothness. Analyses have shown that with a high-power laser (600 W UV)[18] and high-speed traverses (10 m/sec), the PSGC patterning time would take less than 15 min. It is then estimated that a complete PSGC COSA-PM can be produced within 15 hours.

18.4 Hardware Testing

Component tests documented in this section include evaluation of thruster-valve pulse-width modulation, characterization of the propellant monitoring system, characterization of the charge/discharge curves for the lithium-ion batteries, evaluation of Foturan wafer bonding, and measurement of MEMS rate gyro performance. Pulse-width modulation of the valves and propellant-gas state control (pressure and temperature) were used as throttling parameters for the cold-gas thruster. The former approach provides high-speed throttling (~1% thrust resolution), while the latter provides expanded thrust ranges to provide the required minimum impulse bit capabilities. The charge, discharge, and life-cycle characterization of the lithium-ion polymer battery was needed to determine its compatibility with the COSA mission. Because the propulsion module prototype was fabricated using layers of PSGC wafers, sample bonding experiments were performed to ascertain the proper protocol for final assembly of the stack. Lastly, the MEMS rate gyro was characterized to ensure that it could be used for autonomous stabilization and operations.

18.4.1 Valve Pulse-Width Modulation Characterization

Cold-gas thrusters are the simplest form of rocket engine and can be easily described by well known mass, momentum, and energy conservation equations. Their general systems design and performance are well understood; however, they need further refinement to operate effectively on pico- and nanosatellites. The main issue is the extremely low inertias for these miniature satellites, which require superior throttling range and minimum impulse control if the same thruster is to be used for translational and rotational control. This can be easily seen from scaling laws based on a characteristic length l, where thrust scales with the nozzle area as l^2, mass (inertia) scales as l^3, and moment of inertia scales as l^5. First, the scaling laws for smaller satellites favor high angular accelerations because moment of inertia decreases at a much faster rate than thrust. Second, a smaller satellite would require much finer thruster control as compared to a larger satellite to perform the same atttitude control maneuvers. For example, the required minimum thrust resolution for equivalent rotational control of a 1 kg picosatellite compared with a 10 kg nanosatellite is about 20 times smaller for a 10× reduction in mass.

After numerous efforts to identify various miniature COTS valve actuation schemes (shape memory alloy,[19] solenoid,[20] thermal,[21] piezoelectric,[22] and electrostatic[22]) and packaging/integration technologies (conventional, hybrid, and

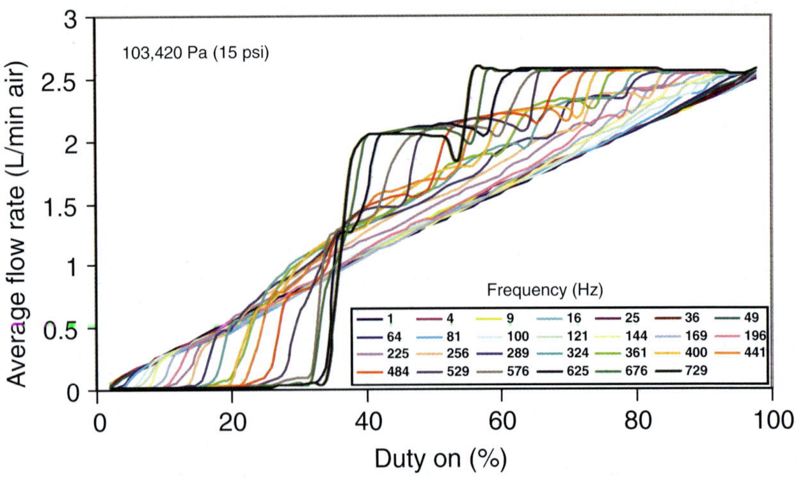

Fig. 18.13. The cold gas thruster valve (INKA1222210H) dynamic characteristics from 1–729 Hz and from 2–98% "on" duty cycle.

MEMS), nanoliter dispensing solenoid valves[20] were selected because of their commercial availability, size, ease of use, and proven reliability (more than 2×10^9 switching cycles). These valves are typically designed for liquid dispensing, but gas versions are available and were used on the thruster system. The valves can operate up to a 1 kHz pulsed repetition rate. The dimensions of the valve body are 5.72 mm in diameter by 21.1 mm long. We used valves rated at 12 V with a nominal coil resistance of 10.7 Ω. The rated voltage refers to the peak voltage required to turn on the valve (i.e., push the solenoid), while the maximum continuous voltage is 2.3 V to keep the valve open. Thus, a spike-and-hold controller is ideally needed for valve operations, where it is energized with a 250 μsec spike at the rated voltage followed by the continuous voltage to hold the valve open.

Valve characterization tests were performed with compressed air discharged directly into the environment. The valve was energized by an NPN Darlington transistor using a typical solenoid driver configuration. This simple on/off valve driver design was selected due to its minimal parts count (spike-and-hold circuits need three transistors for each valve) at the expense of reducing on/off repetition rate to ~800 Hz. The collector voltage of 8.5 V was experimentally determined to be the minimal voltage required to switch on the valve. Valve life is expected to decrease due to the excess power, but this can be alleviated by controlling the duty cycle. Testing parameters included open/close frequencies from 1 to 729 Hz with pulse-width modulation to vary the "on" duty time from 2% to 98% at gage pressures ranging from 34.47 to 206.84 kPa (5 to 30 psi). Typical valve response is shown in Fig. 18.13, where open/close frequencies above 100 Hz start to show dramatic nonlinearities. At frequencies above 256 Hz, the valve response becomes almost tri-state with on, intermediate, and off flow rates. The pressure dependence of the valve characteristics can be seen in Fig. 18.14 (at

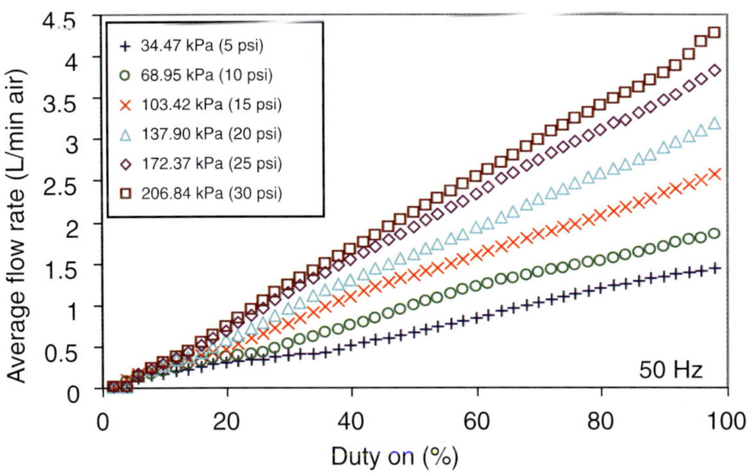

Fig. 18.14. The cold-gas thruster valve (INKA1222210H) volume flow-rate linearity at 50 Hz switching rate, 34.47–206.84 kPa pressure, and from 2–98% "on" duty cycle.

50 Hz). Notice that pressures above 68.95 kPa (10 psi) are considerably more linear. These valve data are used to provide the COSA-PM cold-gas thruster design parameters, such as limiting valve operations below 100 Hz and keeping fuel pressures above 68.95 kPa.

18.4.2 Propellant-State Control

A cold-gas thruster capable of high pulse rates is needed for the COSA to provide the required minimum impulse, especially in GEO missions. At GEO, the reduced thrust levels will require low propellant pressures, a propellant heating/cooling system, and higher switching frequencies. For the laboratory demonstrations, however, the minimum impulse requirement does not play a factor due to the relatively large damping capacity of the air table. Because the propellant used for the terrestrial demonstration had a significant heat of vaporization, we had to add a heating system to maintain propellant temperature, and hence pressure, during thrusting. Ideally, only a simple temperature sensor would be necessary to monitor the heating of the propellant because the vapor pressure can be directly determined by its temperature; however, this assumes that the heater and vapor are at the same temperature. In actuality, simple heat-transfer calculations, as shown in Fig. 18.15, indicated that heat conducts significantly faster through PSGC than through the propellant. This implied that there was a significant phase lag between the heat provided by the heater and the heat put into the propellant, without

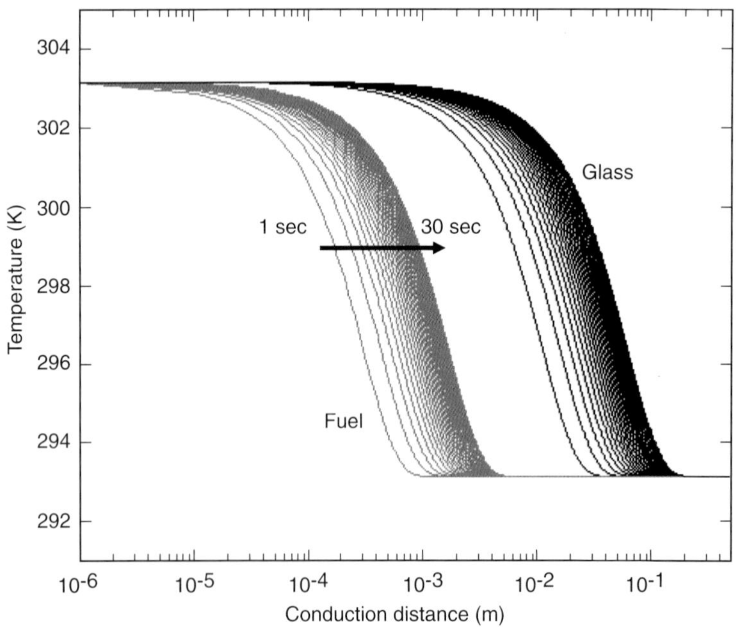

Fig. 18.15. 1-D conductive heat transfer calculations demonstrating the difference in the heat condition through the propellant and the PSGC structure. Given the faster conduction rates in the PSGC, excessive heating into the structure is expected.

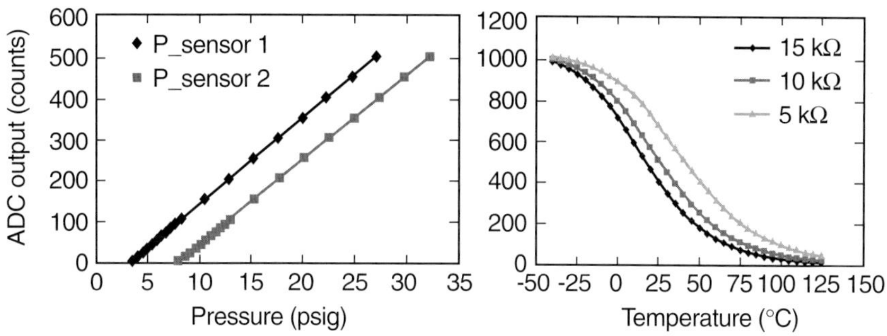

Fig. 18.16. The actual outputs of the pressure sensors (left) and the nominal outputs of the thermistors (right) mounted on the PSGC COSA-PM.

even considering non-adiabatic conditions. Thus, a pressure sensor is needed to directly measure the propellant gas pressure in addition to the temperature sensor that monitors the heater. The characteristics of the pressure and temperature sensors used on the COSA propulsion module are shown in Fig. 18.16. The pressure calibration was determined experimentally, while the thermistor calibration was based on the manufacturer's datasheet. Notice that the pressure sensor has a considerable offset bias while maintaining the same sensor sensitivity (transduction ratio), which is very common for MEMS-type transducers that are not trimmed. No efforts were made to trim the pressure sensor output because all sensors exhibited good linearity above the expected operating regime, ~10 psi. The thermistor (10 kΩ nominal resistance) was chosen based on its ability to provide maximum sensitivity and good linearity at the temperature regime of interest, mainly 0–50°C. Space-qualified COSA units would require calibration and trimming of each individual sensor.

18.4.3 Li-Ion Battery Tests

The laboratory propulsion module used rechargeable lithium polymer batteries to provide experimental flexibility. Although the COSA mission is inherently expendable, rechargeable batteries make sense from the perspective of reducing overall satellite size and weight if recharging opportunities (solar cells) exist. Thus, battery capacity was sized to provide sufficient charge during the dark periods of the LEO mission at 700 km altitude (i.e., orbital periods of ~90 min) with recharging during sun illumination at the rate of 1 C (i.e., full capacity in 1 hr). We tested our batteries by first discharging them using load resistors with nominal powers ranging from 1.024 to 8.192 W, based on the nominal battery voltage of 3.6 V. The results of the load tests are presented in Fig. 18.17 (top), which shows that the LP550 is capable of satisfying a 2 W LEO mission load. Next, charging and discharging of the battery were monitored to indicate any long-term degradation and life-cycle issues. The data are also shown in Fig. 18.17 (middle and bottom). Although these graphs seem to indicate battery degradation after ~50 charge-discharge cycles, this is really not the case. The charging circuit is a constant voltage supply with a series sense resistor that also acts as the load resistor during discharge (nominally at 3.072 W). Notice that this is a more stringent test than the 2 W requirement and represents a charging rate of 1.55 C. Once the battery voltage reaches 4.2 V, the charging current is cut off, and no further charge cycling is used to maximize the charge. The batteries were not charged to capacity, and a corresponding linear loss in the discharge capacity is seen in bottom graph. The battery tests showed that lithium-ion polymer batteries have the charge/discharge characteristics to satisfy the laboratory COSA-PM prototype. A more sophisticated charging circuit will be needed for the COSA space mission. Further characterization in temperature and life-cycle performance of these COTS batteries will also be needed, and such experiments are being conducted for a similar follow-up project in picosatellites.

18.4.4 MEMS Rate Gyro Tests

The MEMS rate gyro and magnetometer form the limited inertial measurement unit onboard the COSA-PM. Our goal was to demonstrate autonomous control of rotation without external telemetry support. During the component testing phase, the rate gyro PCB was mounted on top of a rotating turntable controlled by a stepper motor for characterizing sensor sensitivity, linearity, noise, and full-scale range. Although these parameters were available from the manufacturer's datasheet, we tested the sensor using the same peripheral electronics and the 12-bit data-acquisition system used in the propulsion module for a complete evaluation of the whole gyro system, from sensing input to digitized output. The rate gyro sensitivity, linearity, and full-scale range can be seen in Fig. 18.18, which shows the gyro characteristics for three different averaged rates (5.5, 8.5, and 22 Hz). The output noise from the gyro set at a 40 Hz bandwidth is ±3 counts (±3.6 mV) at 1 standard deviation, which is relatively high but can be filtered out through averaging. Although most of the data followed the manufacturer's specifications, the initial null (zero rate output) and null stability varied from gyro to gyro. These variations are ideally correctable through software. It should be noted that null stability is highly dependent on temperature, and the manufacturer included an onboard temperature sensor to provide for corrections; however, we found that adequate angular position data could still be gathered by integration without temperature corrections, resulting in a drift error of ±1 deg in ~45 sec. This was deemed satisfactory for the terrestrial COSA-PM demonstrations without engaging in time-consuming temperature compensation efforts. Figure 18.19 shows the gyro experiment on the Gen 2, where the rate gyro output was digitized and integrated onboard before transmitting via an RF link to the complementary receiver connected to a laptop.

18.4.5 Foturan Bonding Tests

Because of the stacked layer design of the propulsion module, a relatively high alignment resolution (~100 μm) was required for bonding the wafers together. In the past, PSGC parts had been bonded with commercial epoxy-based adhesives

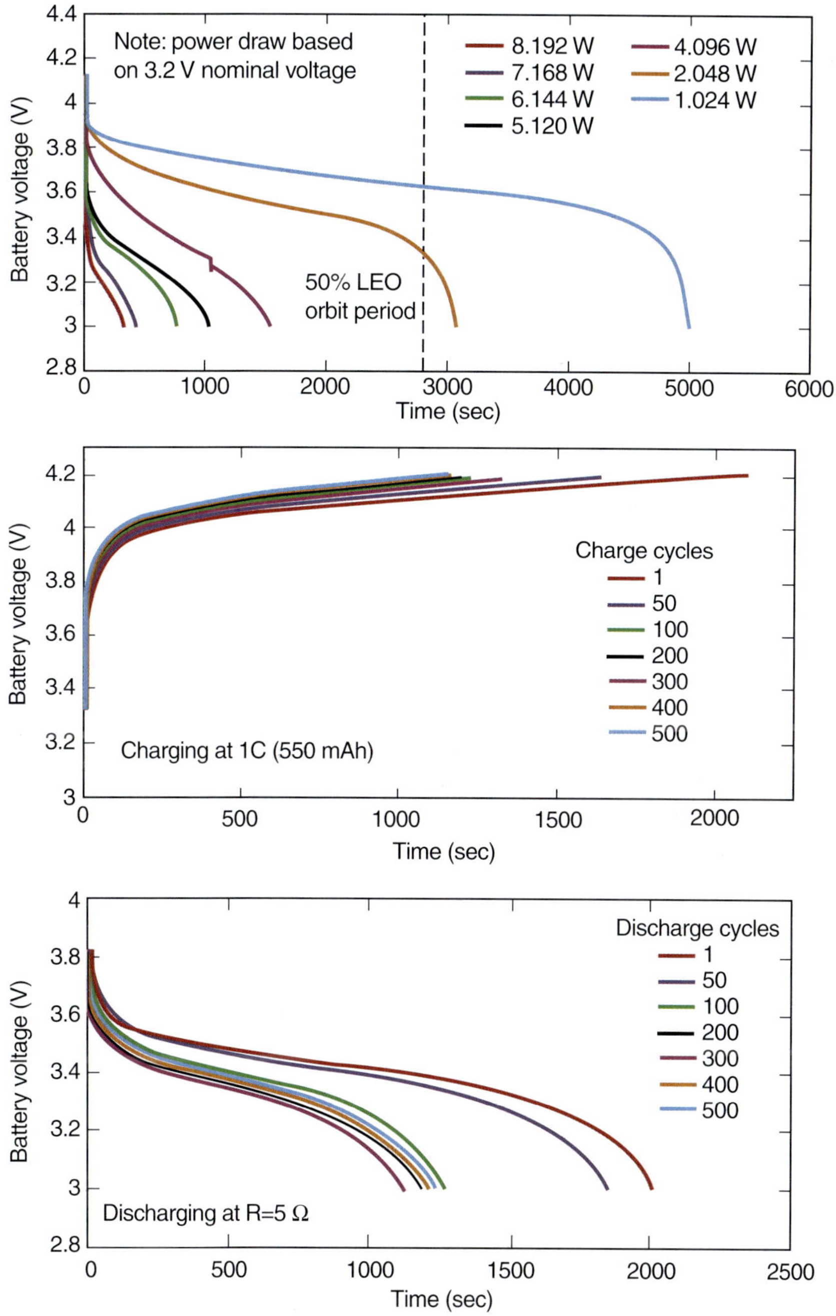

Fig. 18.17. The discharge-curve characteristics at varying loads (top), charging life-cycle characteristics (middle), and, discharge life-cycle characteristics (bottom) of the LP550 550 mAhr lithium-ion polymer battery used on the PSGC COSA-PM.

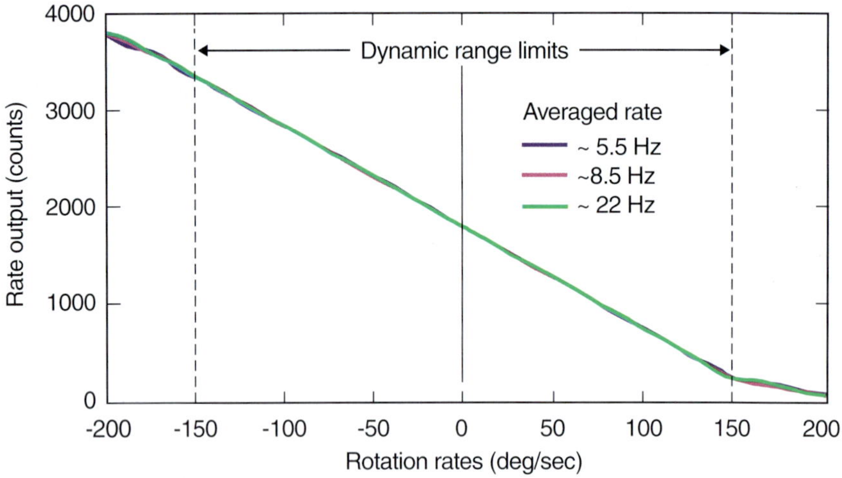

Fig. 18.18. The MEMS rate-gyro output over a ±200 deg/sec rotation rate range at three different averaged rates with the gyro output bandwidth set at 40 Hz. The vertical lines indicate the manufacturer's dynamic range specification.

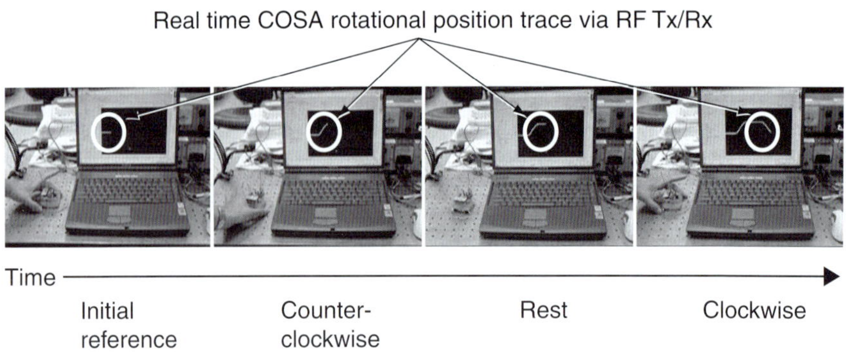

Fig. 18.19. The rate-gyro demonstration, where sensor output is digitized and processed onboard the Gen2 prototype and wirelessly updated to the laptop in real-time.

(Master Bond, EP21TDCHT), which can be replaced by space-qualified versions for flight hardware. For this project, different bonding techniques (pressure, fusion, anodic, and other intermediate adhesive layers) were briefly investigated, with particular emphasis on a proprietary process developed by Schott USA. The Schott process is a low-temperature bonding technique that uses an organic silicate solution that yields optically transparent bonds between glass and a variety of other materials, such as PSGC, LTCCs, and metals.[23] Two slides of PSGC samples bonded to an LTCC substrate are shown in Fig. 18.20. Unfortunately, the Schott process was not usable because of one important parameter—namely, the condensation of water during cold-gas thruster operations. Water can dissolve

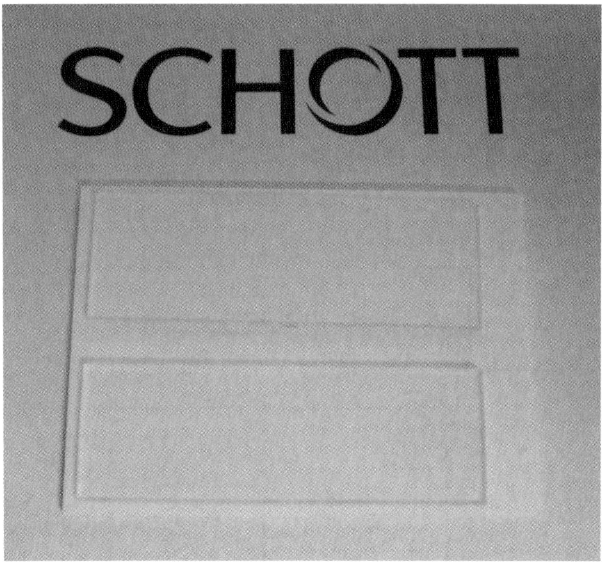

Fig. 18.20. The Schott Foturan test samples showing the PSGC being bonded to LTCC.

the bonding material and eventually cause the layers of the pressurized propellant tank to separate at differential pressures as low as 20 psi. Eventually, precision epoxy dispensing via a motorized adhesive gun mounted on a manual positioning stage was used to bond the PSGC layers of the COSA-PM, and this was satisfactory for laboratory testing. Since the development of the COSA-PM, there have been several Foturan bonding investigations conducted by both The Aerospace Corporation and a commercial company (Invenios Corp.). For terrestrial investigations, we have a solution, but these have to be tested in the space environment. These experiments are ongoing.

18.5 Electronics and Software Development

Figure 18.21 shows the overall system architecture of COSA-PM and the ground station. Every analog signal is digitized, and the digital information is buffered through the use of intermediate processors, which dictate the overall systems architecture of the processing network. The propulsion module incorporates a network of four microcontrollers performing specialized tasks (with one acting as master controller), while the ground station consists of a microcontroller that acts as a data-handling buffer before the command center (simulated by a PC). This decoupled architecture (or systems in a system) provides the ability to test individual components and integrate them digitally in a "plug-and-play" mode at the expense of increased parts count and software coding. This system architecture was found to be extremely easy to debug and expand; it yielded few surprises during hardware and software integration due to the common communication and

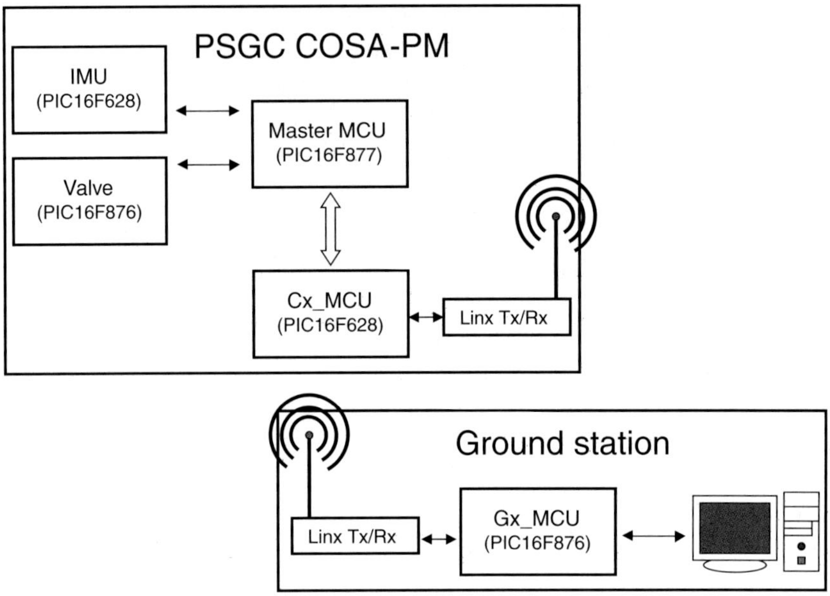

Fig. 18.21. The system architecture of the PSGC COSA-PM and the ground station.

data-checking and data-handling protocol. For example, the network of four on-board microcontrollers can be easily expanded to include peripheral devices such as vision system and other detectors simply by providing more microcontrollers without worrying about unexpected code and hardware overheads. Each microcontroller can be active, sleeping, activated, or bypassed through software.

18.5.1 Multiprocessor Network Architecture

The COSA-PM electronics design is based on an open architecture focused on this particular hardware. Every transducer is handled digitally to allow for direct software access regardless of input or output devices. The four microcontrollers in the master/slave network each serve a specific primary function. The master (PIC16F877-20I/L) communicates with the three slaves via an independently connected half-duplex asynchronous serial communication link at 50 kbps. It oversees data flow between all microcontrollers and directly monitors and controls critical conditions, such as propellant storage state (propellant temperature and pressure through on/off heater control), propellant refuel and bleed valves, thruster failsafe trigger, and transceiver signal-strength monitoring. In addition, it executes closed-loop feedback control codes during autonomous operations. The three slave units are the inertial measurement unit computer (PIC16F628-20I/SO), the transceiver buffer (PIC16F628-20I/SO), and the valve driver (PIC16F876-20I/SO). Each microcontroller also includes a separate half-duplex asynchronous serial channel directly connected to a PC for independent debugging.

The inertial measurement unit digitizes the MEMS rate gyro and magnetometer analog outputs using a 12-bit analog-to-digital converter. For the rate gyro, angular position is monitored by integrating the rate gyro output over time. Integration is done by continuous summing of the output, multiplied by a constant iteration loop time of 10 msec (100 Hz). This is achieved by balancing every branch of the program code. The integrated result is stored in four unsigned-integer data words (16 bits each) and one unsigned-integer data byte for sign tracking. Each data word is only used to represent a three-place decimal number down to 10^{-7} deg. For example, the maximum angle summed is ±(999)(99.9)(999)(999) deg, representing about 278 revolutions in each direction. Hence, the minimum resolution of 10^{-7} deg at 10 msec represents a minimum angular rate of 10^{-4} deg/sec. This resolution is beyond the 12-bit resolution sampled at 100 Hz for the 150 deg/sec full-scale range of the rate gyro and was designed to accommodate future COSAs with higher sampling resolutions and faster integration times.

The transceiver buffer provides transmitter/receiver switching, receiver code error detection, and code balancing tasks for the Linx TR-916-SC-S RF transceiver operating at 916 MHz. By using the same transceiver buffer design on both ends of the RF link (COSA-PM and ground station), few incompatibility issues were experienced that would have required tailoring the communication protocols (i.e., code balancing).

The valve driver is a dedicated device that generates eight simultaneous phase-locked waveforms, one for each thruster valve, with potentially different duty cycles. The minimum "on" time is ~0.1 msec, so a cycle time or period of 10 msec provides a 1% control capability in impulse bit. Longer cycle times (up to 10 seconds) can provide even finer impulse control. The cycle time is constant for all valves, but it can be adjusted after every pulse if required.

18.5.2 LabVIEW-Based Ground Station and Vision System

The COSA-PM demonstration system includes a ground station that provides diagnostics, command uploads and downloads, and man-in-the-loop capabilities. The communication data are passed between the command center (a PC with a 2.4 GHz Pentium IV running LabVIEW 7.1 and IMAQ Vision 7.0 under Windows 2000) and the master microcontroller onboard the propulsion module via an RF transceiver controller and buffer microcontroller on each system. The vision system provides instantaneous position information for the COSA-PM and optionally provides angular orientation data. Digitized but unprocessed data from onboard sensors (i.e., propellant pressure, propellant temperature, rate gyro, magnetometer, and RF signal-strength data) are also downloaded via the RF link to the command center to permit simultaneous onboard and offboard processing. This flexibility allows rapid experimentation of new control codes and commands without constantly reprogramming and debugging the COSA-PM microcontrollers. The command center LabVIEW VI interface is shown in Fig. 18.22. Key features include real-time display and logging of the vision and inertial measurement unit data, direct man-in-the-loop thruster controls, waypoint designation for autonomous modes, fuel-state control and setpoints, and COSA-PM closed-loop feedback rate control. In the waypoint-guided autonomous modes, both onboard inertial measurement unit and offboard vision data can be used, but the processing and thruster control are performed onboard. The autonomous control for z-axis rotation using the onboard magnetometer was not implemented for this project; however, the hardware is fully capable and only requires extra software codes to accomplish this task.

18.5.3 Electronics Integration

Although a single, more capable processor would be able to perform all onboard tasks, the multiprocessor architecture enables individual testing of each subsystem while minimizing the need to account for programming overhead for every subsystem. For example, balancing components with different iteration cycle times—such as the rate gyro integration, valve waveform generation, and RF scheduling—would require significant time-based interleaved programming and interrupts. This can only be solved easily by selecting a processor with significantly increased instruction cycle rates, which translates to increased processor power consumption, code complications, and real estate. The COSA-PM hardware architecture also permits future expansion of subsystems by simply "plugging" in to the master microcontroller using the serial interface and command protocols.

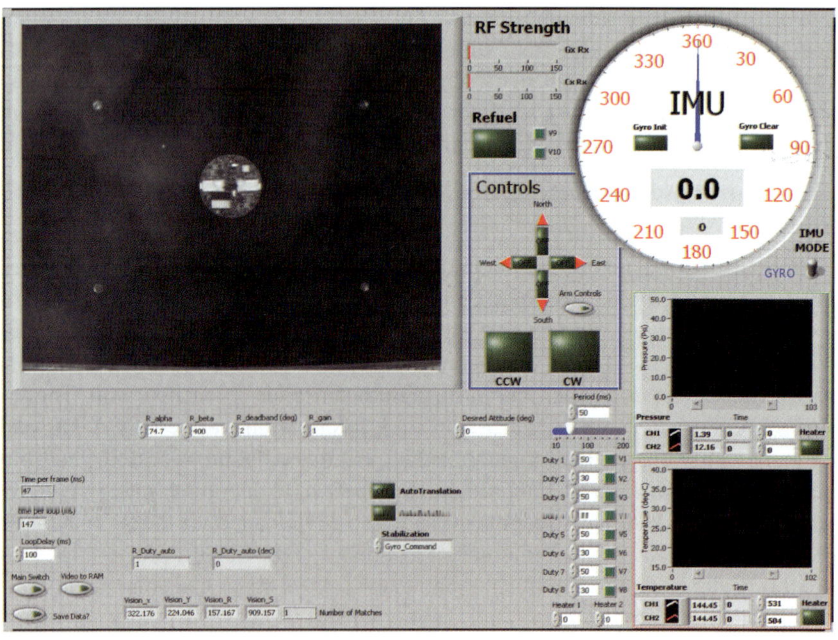

Fig. 18.22. The COSA-PM ground station control panel hosted by LabVIEW. The real-time image, onboard sensor data, and control values are recorded to provide postexperiment data processing and examination capability.

Overall, the electronics integration has proved to be relatively trouble-free and easy to modify during experimentation. The overall schematic of the electronics PCB is shown in Fig. 18.23 (left), including an image of the populated board (right). The diameter of the PCB is the same as the PSGC wafers (100 mm), and the two customized square cutouts were created by a miniature CNC mill to provide access to the pressure sensor ports. The basic electronics subsystem architecture is shown in Fig. 18.24.

18.5.4 Processor Software Architecture and Development

The PICBASIC PRO compiler and EPIC programmer (from microEngineering Labs, Inc.) were used for software development and programming of the microcontrollers. The In-Circuit Serial Programming (ICSP) mode was used to upload the microcontroller program codes while mounted on the application PCB, requiring only two communication wires for synchronous serial data plus power, plus

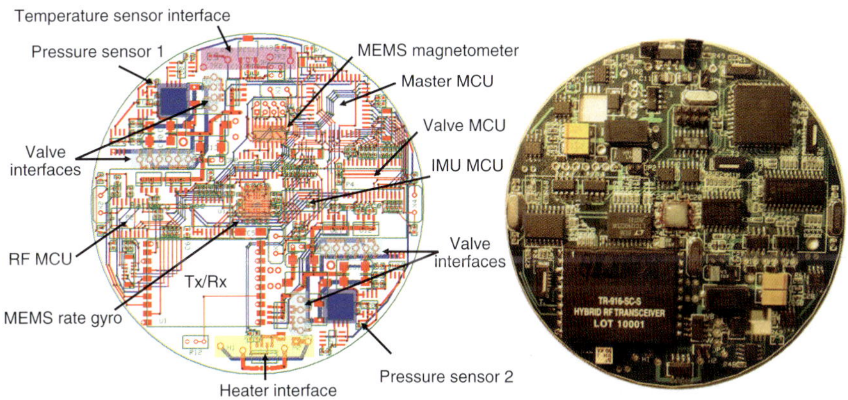

Fig. 18.23. The top (red traces) and bottom (blue traces) layer schematic of the PCB design (left) and the populated PCB with the surface mount components (right).

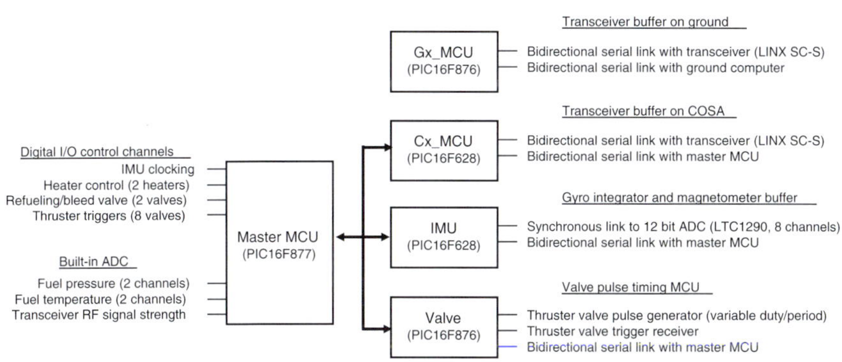

Fig. 18.24. Communication and data paths of the onboard microcontroller networks.

a ground, and a programming mode trigger on the master clear-reset pin. Since power and ground are provided separately on the PCB, each microcontroller only needs three wires allocated to programming functions. A miniature 10-pin flex cable connector (OMRON, XF2J-1024-11 vertical side-lock) was used for the ICSP and serial debugging of the communication between master and slave microcontrollers. The software developed for each subsystem was self-contained and only needed to follow a common protocol for communication with each other. When software was being developed for each microcontroller, a common software communication protocol was first copied and pasted into the program codes, and then task-specific codes required for each subsystem were defined. The communication protocol was based on tagging data with corresponding alphabetic command code, a separator "!", and the binary codes of the data value to be passed. The command codes allowed unformatted data of various lengths to be transmitted by ending the data stream with an end code, "E!". The transmitted code was also based on verbose ASCII codes to ease readability during debugging. For faster transmission-receiving cycles, binary codes could be used for the data values and the optional separator "!" code could be deleted at the expense of code readability.

18.6 COSA-PM "Flight" Test

The COSA-PM flight tests involved both open-loop and closed-loop control experiments on a custom designed air table. Because of the flexibility in the design of the COSA-PM software and hardware, experiments such as man-in-the-loop and offboard/onboard autonomous controls were performed at the same setting by simply toggling between the modes on the command center LabVIEW VI. The goal of the flight test was to demonstrate self-monitoring, autonomous stabilization, and waypoint functions. This last goal was achieved by locking on to a specified yaw angle, rotating to a new attitude uploaded from the command center, and then stabilizing at the final position.

18.6.1 Air-Table Setup

The air table used for the terrestrial "flight" experiments measures 91.4 × 61.0 cm (36 × 24 in.). Its aluminum surface is covered with 1.02 mm (0.040 in.) diameter CNC-drilled holes at a spacing of 25.4 mm (1 in.) in regular square arrays with rows at 12.7 mm (0.5 in.) offset from each other. The air table and the hole patterns are shown in Fig. 18.25, which also shows the relative size of the COSA-PM. The table is designed to provide up to ~60 Pa (0.08 psi) of surface pressure using an air blower controlled through a variable ac transformer; this is required to lift a 0.5 kg, 10 cm diameter disk off the table. The surface of the air table was expected to vary as much as 1 mm in height due to the bowing of the aluminum plate caused by the pressure differential between the top and bottom surfaces. Although this height differential may seem inconsequential, it would impart a 14 cm/sec velocity gain to a floating object under a nominal gravitational field

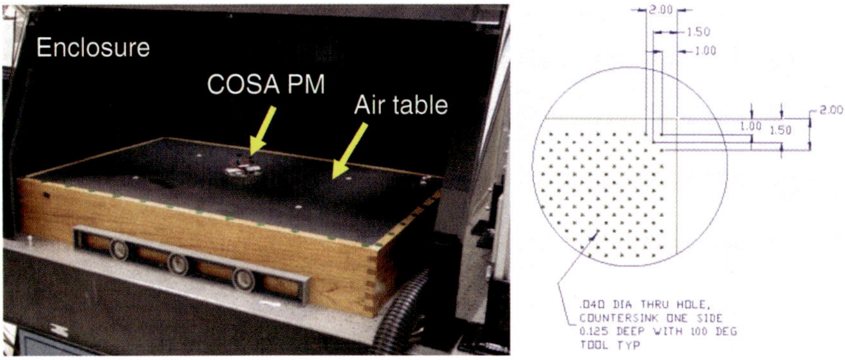

Fig. 18.25. Customized air table design used to support the COSA-PM experiments (dimensions are in inches).

($a = 9.8$ m/sec^2). The surface bowing was used to empirically determine the maximum ~0.02 m/sec^2 in-plane acceleration, which translates to a surprisingly high minimum thrust requirement of ~10 mN for air-table testing. Fortunately, this could be met by the 20 mN thruster design. The surface nonuniformity could be minimized using tie-down bolts at the center and each inset corner of the table top (the bolt heads appear as white dots in Fig. 18.25). With optimal adjustment, the resulting in-plane acceleration was estimated to be ~3–5 mm/sec^2.

18.6.2 Open-Loop Flight with Offboard Control

Open-loop flight control was used to demonstrate thruster authority and to calibrate system response times to within our requirements (>10 Hz). It was also used to demonstrate man-in-the-loop control to provide direct experience for controlling cold-gas thrusters on a picosatellite. Initial tests were performed using offboard N_2 gas through extremely flexible 1 mm inner diameter, 2 mm outer diameter PVC tubing. This permitted validation of the cold-gas thruster software, valve switching, and pressure sensor response. It offered only limited dynamic thruster experiments due to the relatively high restoring torsion moment of the PVC tubing.

Subsequently, onboard stored N_2 gas was used (instead of hexafluoropropane) for dynamic thruster response experiments to simplify refueling at the cost of reduced testing time. A typical subset of thruster data provided by the onboard instrumentation is shown in Fig. 18.26; it indicates the rapid depletion of onboard N_2 gas within 2 seconds after a rotational command is given (using one pair of the rotational thrusters). Notice that the COSA-PM reversed its rotation and accelerated in the opposite direction after the propellant was expended. Furthermore, the module always accelerated naturally in the same direction (clockwise) regardless of the location of the air table. This rotational tendency was originally thought to be caused by the possible warpage of the bottom wafer due to uneven stresses from the stacked PSGC layers. However, the addition of an ultrasmooth surface

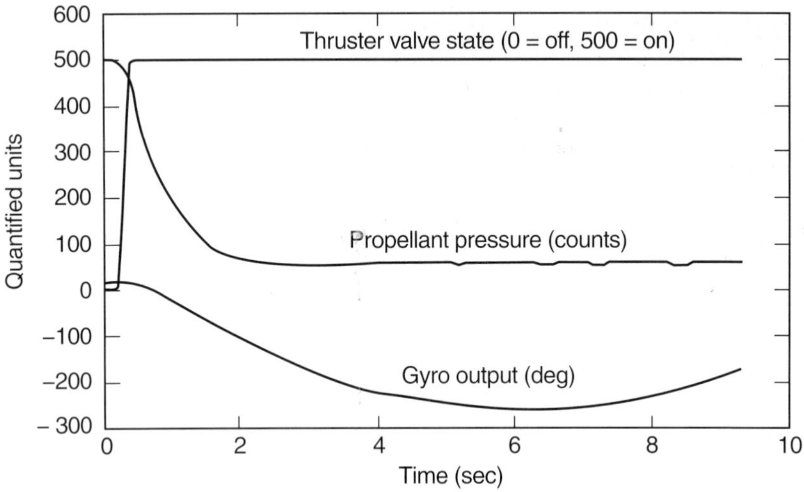

Fig. 18.26. Rate-gyro output vs. propellant pressure in the onboard N_2 experiments.

(i.e., polished silicon wafer) did not help reduce this tendency, and this problem was left unsolved because the thrusters have demonstrated the ability to overcome it. The problem may stem from the way the holes were drilled in the air table.

18.6.3 Closed-Loop Flight with Onboard and Offboard Control

The first task of closed-loop flight was to determine the thruster control coefficients (α_i and β_i). Notice that the simulated control algorithm given below solves for the required thrust, while the hardware modulates the valve duty cycle rather than the thrust. The following is the modified control law at constant propellant pressure (an assumption that will be modified later):

$$\left[(\text{Thrust}_{\max})(\text{Duty})\right] = \text{Thrust}_i = \alpha_i(\delta\Phi_i) - \beta_i \frac{d\Phi_i}{dt}$$
$$\text{Duty} = \alpha_i'(\delta\Phi_i) - \beta_i' \frac{d\Phi_i}{dt}; \textit{ where } \alpha_i' = \frac{\alpha_i}{\text{Thrust}_{\max}} \text{ and } \beta_i' = \frac{\beta_i}{\text{Thrust}_{\max}} \tag{18.2}$$

The key point is that the maximum thrust was not initially known, and this manifested itself in the modified control coefficients. There are basically two ways to determine the coefficients—through direct thrust measurements of the installed thrusters, or through trial-and-error guesswork. The latter method was selected, since it was relatively straightforward when used in conjunction with the fact that α'_i is related to response time and β'_i is related to damping. The first step was to estimate the maximum thrust by turning on a thruster and letting it accelerate. The recorded acceleration was used to estimate α'_i and β'_i, and the control codes were implemented with these coefficients. The corresponding coefficients were then modified as needed in subsequent flights to yield the same step response as the simulated response. This was achieved in very few iterations. For example, yaw coefficients were found after just five test flights, with α'_i = 74.7% duty/rad and β'_i = 400% duty-sec/rad. Four 90 deg step response flights with different damping coefficients are shown in Fig. 18.27. Notice that these coefficients yield units in terms of percent of the valve duty cycle rather than thrust and time units based on milliseconds; therefore, the magnitudes are considerably different.

For the 90 deg slew underdamped case, the rate gyro output and the onboard integrated angular data are shown in Fig. 18.28. The rate gyro also detected the small amplitude oscillations due to thrusters firing for stabilization, which are shown in Fig. 18.29 against the propellant pressure (sensor #2) for the same overdamped case. For space missions, to reduce propellant expenditure, stabilization thrusting must be kept to a minimum, which dictates the controllability of the required impulse bit of the COSA. It should be noted that the valve is switched only when the trigger from the master microcontroller is also enabled. The valve on duty of 50% shown in Fig. 18.29 is simply the reset value of the register and not the triggering of the valves, as indicated by the recovering fuel pressure after triggered events (valve trigger is separately recorded but not shown here). Duty-cycle values other than 50% indicate a nonreset value and are an indication of valve

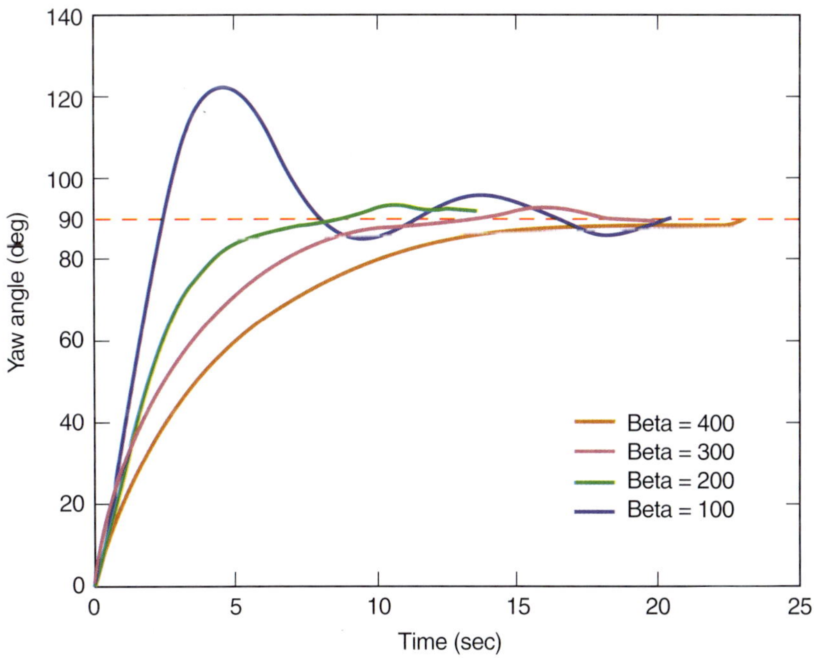

Fig. 18.27. Step response of the thruster system to the 90 deg slew command. The critically damped case (β'_i = 400% duty-sec/rad) is then used to provide the control coefficients for the COSA-PM.

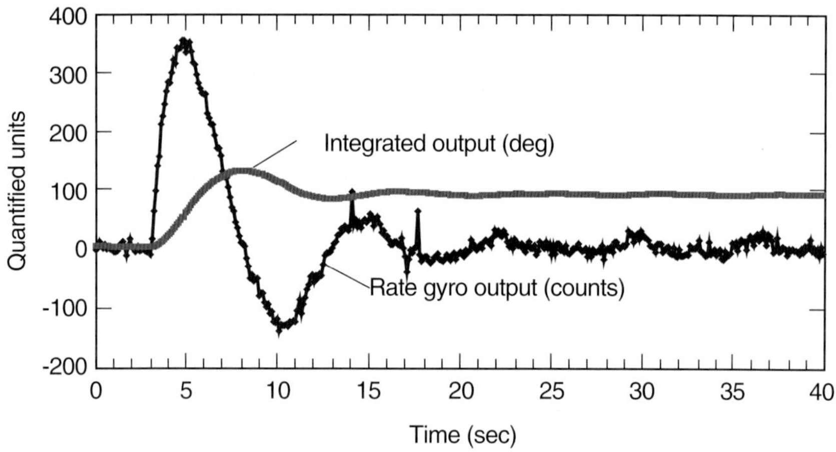

Fig. 18.28. The raw and integrated output of the rate gyro under a 90 deg slew command for the underdamped case to demonstrate the phase-lag of the angular position.

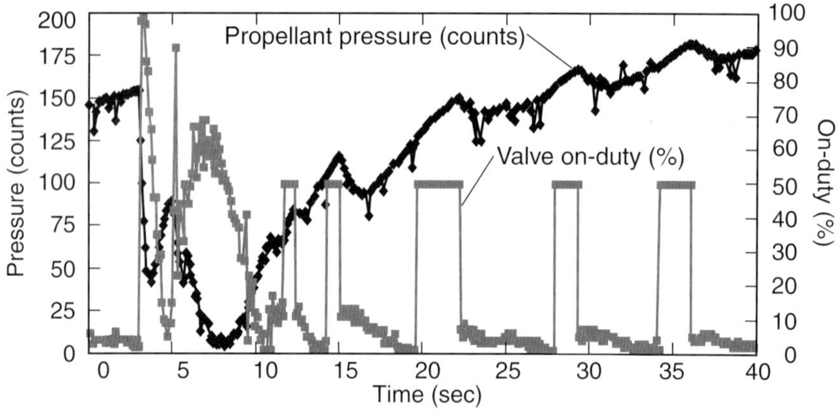

Fig. 18.29. Propellant pressure and on-duty valve command corresponding to the same test case for Fig. 18.28. Notice that the pressure drops rapidly during large angle rotations, and recovers during the successive small angle corrections.

triggering. In the above tests, the heater was not used and is expected to prevent the dramatic pressure loss shown in Fig. 18.29 during the large angle slews.

18.7 Lessons Learned and Future Direction

Overall, our lessons learned were primarily on technical issues. We validated the general applicability and strengths of using multiprocessor hardware and software architecture for rapid development of small systems, cold-gas thruster throttling with high-speed solenoid valves, and the use of PSGC for space systems. The multiprocessor hardware integration and common software communication protocol demonstrated value in significantly reducing development time. The general electronic and software architecture remained the same throughout the project due to the lack of any serious shortcomings. The cold-gas thruster system, in combination with onboard and offboard instruments, demonstrated its ability to provide the required mobility and control of the COSA-PM. Finally, the laser-fabricated PSGC structure verified the ability to directly take CAD/CAM designs and convert them to actual PSGC space system parts with micron-level resolutions and centimeter-level feature sizes.

A few suggestions for future COSA designs lie mainly in the use of RF transceivers, increasing onboard communication bandwidth, and capitalizing on further decreases in electronic component sizes to increase functionality and packing density. Originally, we thought that a half-duplex RF transceiver would allow reduction in electronics-board real estate at the expense of extra software code and microcontroller pins to switch between transmitting and receiving. This was found to be the limiting factor in the bandwidth of the man-in-the-loop (or ground station) control mode due to the combined effects of switching latency and half-duplex communication. Future development should adopt full-duplex communication hardware, such as separate RF transmitter and receiver modules with different frequencies. Although onboard communication between the microcontrollers did not create any control bottlenecks, higher-speed protocols such as synchronous serial communication (i.e., I^2C, SPI, and USB) can be adopted when increased bandwidth is required. Because the PSGC is optically clear, photonic communication buses should also be pursued due to their simplicity, low power requirements, and high immunity to electromagnetic interference.

In terms of reducing the size of the COSA and its electronics, key enablers include smaller hybrid components, followed by System-in-Package (SiP) and System-on-Chip (SoC) solutions, with PSGC as the integrating material. In all of these, MEMS technology can transform the capability of the current COSA-PM design by enabling increased sensing densities and modalities. The ultimate COSA would have MEMS electronics, sensors, and actuators (i.e., pressure sensors, temperature sensors, gyros, accelerometers, various space sensors, valves, and processors) fabricated and embedded directly into the PSGC structure. Furthermore, RF electronics integrated on top of LTCC would enable direct bonding to the PSGC structure. Thus, high-resolution bonding of PSGC with different substrates and direct fabrication of active devices (sensors, actuators, and electronics) on PSGC remains the logical follow-on research toward a truly monolithic and mass-producible miniature satellite.

18.8 Conclusions

Cold-gas propulsion units based on PSGC material and COTS components have been successfully developed at The Aerospace Corporation. Key project goals included demonstrating the mass-producible properties of PSGC, design and testing of a cold-gas thruster module with PSGC, and development of the representative command and control electronics. A pair of these propulsion units will provide a nanosatellite (1 kg) with 30 m/sec ΔV capability for week-long COSA missions at LEO or GEO altitudes. Each propulsion system could also act as a prototype picosatellite capable of autonomous three-axis control with the onboard electronics.

Since its inception, this project has evolved into the development of a highly integrated system that will have full six-axis thrust control (including *z*-translation, pitch, and yaw) and LTCC substrate electronics with wireless photonic and RF communications. Furthermore, we have collected additional data on the material characteristics of PSGC in the space environment. Foturan was flown in LEO as part of the Materials International Space Station Experiment (MISSE). The results of the three-year exposure to the solar, UV, and atomic-oxygen environment have been very promising in allowing us to proceed with more confidence. The primary goal of the COSA-PM project was to develop a technology demonstrator, using PSGC as a structural material, to show the feasibility of implementing digital, direct, development manufacturing schemes for mass-producing small satellites. We believe this revolutionary objective has been achieved.

18.9 Acknowledgments

The authors would like to thank The Aerospace Corporation's Independent Research and Development (IR&D) program, the Air Force Office of Scientific Research (Dr. H. Schlossberg), and DARPA (Dr. W. Coblenz) for funding and support at various stages of this project.

18.10 References

[1] A. Huang, W. W. Hansen, S. W. Janson, and H. Helvajian, "Microengineered Cold Gas Thruster System for a Co-Orbiting Satellite Assistant (COSA)," *MEMS Components and Applications for Industry, Automobiles, Aerospace, and Communication, Proceedings of SPIE,* Vol. 4559, pp. 164–173 (San Francisco, 2001).

[2] S. Janson, A. Huang, W. Hansen, and H. Helvajian, "Development of an Inspector Satellite Propulsion Module Using Photostructurable Glass/Ceramic Materials," *CANEUS 2004: Conference on Micro-Nanotechnologies* (Monterey, CA, 2004), AIAA paper: 2004-6701.

[3] S. Janson, P. Huang, W. Hansen, L. Steffeney, and H. Helvajian, "Development of an Inspector Satellite Using Photostructurable Glass/Ceramic Materials," *Space 2005* (Long Beach, CA, 30 Aug.–1 Sept. 2005), AIAA-2005-6802.

[4] H. Helvajian and S. Janson, "Workshop Reports" in *Proceedings of the International Conference on Integrated Micro/Nanotechnology for Space Applications,* edited by R. Stroud and D. Sutton (Houston, TX, 1995).

[5] H. Helvajian and S. Janson, "Photostructurable Glass/Ceramic Micro Propulsion System and the Co-Orbiting Satellite Assistant (COSA)" Project Final Report, Aerospace Technical Report ATR-2005(5073)-1 (The Aerospace Corporation, 2005).

[6] P. W. McMillan, *Glass Ceramics, 2nd ed.* (Academic Press, London, 1979).

[7] A. Berezhnoi, *Glass-Ceramics and Photo-Sitalls* (English translation), (Plenum Press, NY, 1970).

[8] S. D. Stookey, "Photosensitive Glass," *Industrial and Engineering Chemistry,* **41** (4), 856 (1949).

[9] S. D. Stookey, "Chemical Machining of Photosensitive Glass," *Industrial and Engineering Chemistry*, **45** 115, (1953).

[10] Technical Glass Division Schott Corporation, Schott Technical Products Literature F10 (1999).

[11] F. E. Livingston, W. W. Hansen, A. Huang, and H. Helvajian, "Effect of Laser Parameters on the Exposure and Selective Etch Rate in Photostructurable Glass," *Photon Processing in Microelectronics and Photonics, Proceedings of SPIE*, Vol. 4637, pp. 404–412, (San Jose, June 2002).

[12] F. E. Livingston and H. Helvajian, "Selective Activation of Material Property Changes in Photostructurable Glass Ceramic Materials by Laser Photophysical Excitation," *Journal of Photochemistry and Photobiology A: Chemistry,* **182** (310) (2006).

[13] H. Helvajian, in *Direct-Write Technologies for Rapid Prototyping Applications: Sensors, Electronics and Integrated Power Sources*, edited by A. Pique and D. B. Chrisey (Academic Press, 2002), pp. 415.

[14] P. D. Fuqua, D. P. Taylor, H. Helvajian, W. W. Hansen, and M. H. Abraham, "A UV Direct-Write Approach for Formation of Embedded Structures in Photostructurable Glass-Ceramics," *Materials Research Society Symposium Proceedings,* Vol. 624, p. 79 (2000).

[15] F. E. Livingston and H. Helvajian, "The Symbiosis of Light and Matter: Laser-Engineered Materials for Photo-Functionality," *SPIE Proceedings, Laser Advanced Materials Processing*, Vol. 4830, p. 189 (2003).

[16] A.E. Siegman, *Lasers* (University Science books, Mill Valley, CA, 1986).

[17] S. R. J. Brueck, S. H. Zaidi, X. Chen, and Z. Zhang, "Interferometric Lithography—from Periodic Arrays to Arbitrary Patterns," *Microelectronic Engineering*, 41–42, 145 (1998)

[18] H. F. Dylla, M. J. Kelley, G.R. Neil, L. J. Brillson, D. P. Henkel, and H. Helvajian, "UV FEL Light Source for Industrial Processing," *Lasers as Tools for Manufacturing of Durable Goods and Microelectronics—SPIE Proceedings,* edited by C. Roychoudhuri, L. R. Migliore, R. D. Schaeffer, J. Mazumber, and J. J. Dubowski, Vol. 2703, pp. 15–20 (1996).

[19] A. N. D. Johnson, U.S. Patent No. 6533905 (2003).

[20] The Lee Company, Micro Dispensing Valves, www.theleeco.com (2008).

[21] Redwood MicroSystems Incorporated, NC-1500 Fluistor Microvalve, www.redwood-micro.com (2001).

[22] G. T. A. Kovacs, *Micromachined Transducers Sourcebook* (McGraw-Hill, New York, 1998).

[23] C. Click, R. O'Malley, and L. Gilroy, "Low-Temperature Bonding of Optical Materials," *Frontiers in Optics*, (Optical Society of America, 2004), paper OWA3.

Chapter 19

Toward Space-Based Wireless Sensor Networks

Tanya Vladimirova* and David J. Barnhart*

19.1 Introduction

Since the dawn of the Space Age in 1957, increasing mission requirements have driven up satellite mass from Sputnik's 84 kg to more than 6000 kg for some systems today. Consequently, cost, complexity, program timelines, and management overhead have grown considerably. Reversing this trend, a fast-growing, small-satellite industry rooted in academia has enabled increasingly capable and cost-effective space missions. The success of satellites weighing less than 500 kg is based on developers' embracing sensibly reduced requirements and leveraging commercial technology. Surrey Satellite Technology Ltd. (SSTL), a spin-off company of the University of Surrey in England, has pioneered the small satellite platform. The philosophy of the small satellite community is encompassed by the slogan, "Faster, Better, Smaller, and Cheaper." In 2000, SSTL launched the 6.5 kg SNAP-1 mission, which was the first nanosatellite to demonstrate a complete set of capabilities typically found in larger satellites—at a cost of $1 million[1] (see chapter 10).

During the past decade, novel distributed space mission proposals have emerged,[2] requiring hundreds to thousands of satellites for simultaneous multipoint sensing to accomplish advanced remote sensing and science objectives.[3] Advances in technology have given rise to a class of subkilogram satellites that offer a very capable, low-cost, mass-producible approach to fulfilling these mission concepts.[4] Subkilogram satellites include picosatellites that weigh less than 1 kg and femtosatellites less than one-tenth of a kilogram.

Since 1998, the Surrey Space Centre has been involved in research aimed at applying innovative micro- and nanotechnologies to the small satellite platform. The initial goal was to advance the miniaturization of small satellite subsystems, focusing on data handling, via advanced integration technologies such as system-on-chip (SoC) design. More recently, development of mass-producible fully integrated subkilogram satellite nodes was undertaken, which was aimed at the implementation of low-cost virtual satellite missions. In 2005, the Surrey Space Centre became involved in the ESPACENET project.[5] ESPACENET is targeted at the development of a robust space-based sensor network based on flexible picosatellite nodes using standard commercial-of-the-shelf (COTS) wireless protocols for intersatellite connectivity.

---

*Surrey Space Centre, Department of Electronic Engineering
University of Surrey, Guildford, Surrey, UK

This chapter reports on the research accomplished at the Surrey Space Centre toward realizing the concept of wireless sensor networks in space. It begins by surveying present-day and proposed distributed space mission concepts and technology drivers. This is followed by a comparison of supporting subkilogram satellite technologies; these include traditional picosatellites, microengineered aerospace systems, satellite-on-a-chip, satellite-on-an-MCM (multichip module), and satellite-on-a-PCB (printed circuit board) technologies. After the motivating potential of space-based sensor networks is presented, system-on-chip integration is proposed as a way to realize miniature, capable, and flexible satellite sensor nodes. Finally, enabling technologies for distributed picosatellite missions in low Earth orbit (LEO) are discussed.

19.2 Space-Based Sensor Networks Considerations

Large single-purpose satellites, typically used for Earth monitoring and communication, are highly customized, making them expensive to manufacture and launch. Distributed space systems provide an alternative approach that is gaining popularity. If a low-cost, miniaturized mass-producible satellite were developed in the context of a space-based sensor network, this would unlock a completely new dimension of distributed space missions. Certain types of future missions for multipoint sensing will require autonomous, miniature, intelligent, and massively distributed space systems. Networked constellations of hundreds to thousands of very small subkilogram satellites could be used to enable such new space missions and possibly replace the need for larger satellites. The following section introduces the concept of distributed space systems, identifies suitable missions for space-based wireless sensor networks, and discusses enabling technologies.

19.2.1 Review of Current Distributed Space Missions

There has been a recent explosion of publications discussing distributed missions. In AIAA publications alone, articles featuring distributed satellite systems, satellite formation flying, and/or satellite clusters have increased threefold from 2003–2007. Before 1996, these terms did not exist, with the exception of "satellite cluster," which was used to describe close spacing in geostationary orbit (GEO).

Shaw et al.[2] identify two formal types of distributed satellite systems. The first entails multiple satellites sparsely distributed in a traditional constellation to meet mission requirements. The spacecraft do not typically require precise orientation amongst themselves, but may optionally require propulsive stationkeeping; they are linked via ground relays and systems, with the rare exception of crosslinks or intersatellite links. The second type of distributed satellite system involves a local cluster of satellites intentionally placed close together in the same orbit to train on a common target. These satellites may need to fly in formation, maintaining precise spacing and orientation relative to each other.

A selection of current distributed satellite systems grouped in the four typical mission categories—communications, navigation, remote sensing, and

Table 19.1. Selected Distributed Satellite Systems

Mission Type	System	First Launch	Number of Satellites	Type	Orbit	Satellite Mass (kg)	System Cost ($ millions)
Communication	Iridium	1997	66	Constellation w/ crosslinks	LEO	689	~5,000
	ORBCOMM	1997	29	Constellation w/ ground links	LEO	46	~330
	Globalstar	1998	48	Constellation w/ ground links	LEO	222	unknown
Navigation	GPS	1978	24	Constellation w/ ground links	MEO	989–1,077	>2,000
	GLONASS	1982	24	Constellation w/ ground links	MEO	~1,400	unknown
Remote Sensing	DMC	2002	5	Constellation w/ ground links	LEO	166	40
Science	EOS	1997	17	Constellation w/ ground links	Varied	Varied	unknown
	Cluster/DS	2000	6	Constellation w/ ground links	HEO	1,200/330	~600
	ST-5	2006	3	Constellation w/ ground links	MEO	25	130
	COSMIC	2006	6	Constellation w/ ground links	LEO	69	55
	THEMIS	2007	5	Constellation w/ ground links	HEO	128	200

science—are listed in Table 19.1. With the exception of navigation systems such as GPS and GLONASS, large distributed missions have only just emerged in the past decade, with Iridium being the only commercial constellation with crosslinks.

Distributed science missions have been rare until very recently. The European Space Agency's (ESA) Cluster mission, launched in 2000, is arguably the first satellite "cluster" to gather scientific data on the magnetosphere in three dimensions. Cluster is a maintained constellation of four satellites that form a tetrahedron of various geometries on a periodic basis. The follow-up Double Star (DS) mission added two more satellites. Similarly, the NASA Earth Observation System (EOS) is a coordinated collection of 17 satellites performing various types of remote sensing and science missions. The recent launches of NASA's Space Technology 5 (ST-5) and THEMIS (Mars Odyssey) missions, in addition to Taiwan's FORMOSAT-3/COSMIC mission, indicate a growing interest in distributed science missions; however, mission costs are far beyond the scope of most science budgets.

19.2.2 The Need for Space-Based Sensor Networks

Analysis of present LEO-based distributed-communication missions shows that current realities fall short of previous predictions of a boom in such missions. In fact, Iridium, Globalstar, and ORBCOMM have struggled to survive, although interest in them has been renewed by recent contracts to replenish some of these constellations. Still, large satellites in GEO appear to be the mainstay of high-bandwidth global communications for the near term.

GPS, GLONASS, and the European Union's proposed Galileo system are traditional constellations with ground links. Clusters have not yet been proposed for distributed navigation systems, although crosslinks are being considered for GPS III. Instead, the current focus is on improving accuracy, coverage, and jam resistance. To address these issues, increased RF power output and other techniques will be employed, with a commensurate rise in the satellite mass.

Numerous distributed remote sensing systems have been envisioned,[4] though very few have gone beyond the conceptual or experimental phase due to the lack of an affordable sensor network infrastructure. Some examples of constellation-based Earth observation missions in LEO that require multipoint sensing and that would benefit from the space-based sensor network approach include:

- Volcano, fire, or earthquake warning.
- Treaty monitoring.
- Distress beacon monitoring.
- Space situational awareness, intelligence, or other military uses.
- Imaging with frequent temporal repeats and high spatial resolution.
- Constellation sharing, where contributing members access the services of the entire group.
- Disposable, short-lived, rapid-response sensor networks.

Science and exploration missions have traditionally been dominated by single-spacecraft or interplanetary probe architectures because of limited budgets and resources. Examples of new simultaneous multipoint science missions that could be based on space-based wireless sensor networks include:

- Earth atmospheric monitoring—for example, CO_2 levels at 60–250 km.
- Magnetotail behavioral studies, solar wind variations, and other geospace science.
- Monitoring and warning of large-area space phenomena and weather—including plasma, radiation, and atmospheric density—and aurora.
- Detailed characterization of environments to support interplanetary exploration, such as Mars, asteroids, or other planets.

This list illustrates the science need for space-based sensor networks, but it is by no means exhaustive. These missions can be implemented as short-lived constellations with crosslinks but without any orbital maintenance.

An ideally configured satellite formation would exist for a few days before orbital perturbations disturbed the arrangement. As a result, each spacecraft must have precise real-time location knowledge of all nodes and a propulsion system to maintain the formation. The motivation for formation flying is to synthesize a virtual aperture, antenna, or other sensor to perform tasks that currently cannot be achieved by a monolithic satellite. Most aspects of this concept have been widely studied, but the first implementation has yet to be realized, with the exception of a few initial experiments. Formation flying is a very complex, multifaceted problem.[6] In fact, no complete formation flying missions have been realized beyond a few limited experiments. Several significant proposals, such as the Terrestrial Planet Finder, depend on this technology to achieve sensor aperture sizes that are orders of magnitude larger than current monolithic solutions.

19.2.3 Technology Drivers for Satellite Miniaturization

The concept of space-based sensor networks is tightly coupled with the pursuit of spacecraft miniaturization. Highly integrated, very small subkilogram nodes are needed to create massively distributed satellite networks in space.

Integrated circuits (ICs) are typically fabricated using a complementary metal-on-silicon (CMOS) process in a specialized foundry using lithography processes. Advances in fabrication have reduced the minimum feature size from 350 nm in 1997 to 45 nm in 2007. PCB integration is the standard and most cost-effective technique of integrating traditionally packaged integrated circuits. PCBs are fabricated using a simple lithography process to etch out copper traces with widths on the order of 0.5 mm on a fiberglass board. Multilayer boards are common, with more than a dozen layers possible.

Sophisticated application-specific integrated circuits (ASICs), such as microprocessors, nowadays contain hundreds of millions of CMOS transistors on one silicon die. Advances in very large scale integration have brought to life system-level integration technologies such as SoC, which has allowed PCB-level designs to be implemented as one silicon ASIC. In addition to integrating multiple subsystem functions onto a single chip, SoC solutions feature increased reliability and lower power consumption, due to the reduced number of off-chip and interchip connections.

The field-programmable gate array (FPGA) is a special kind of IC that features reprogrammable connections between key elements, allowing it to be altered after it is fabricated. Initially, FPGAs were used mostly to provide interconnect logic between ASICs, perform signal processing, and prototype SoCs. With the advent of large, fast FPGAs, it is now possible to implement an entire embedded system—such as a central processing unit (CPU), memory, and input/output blocks—on a single FPGA. FPGAs provide flexibility of design, shorter time to market, lower cost, reconfigurability, and other benefits, which make them attractive for use in small satellite systems.

MEMS (microelectromechanical systems) are fabricated by either micromachining or a similar process to IC fabrication, augmented with sacrificial layer etching. Sensors, actuators, and RF components such as accelerometers, gyroscopes, motors, and antennas can be built with MEMS processing. Foundry processing modernization has facilitated the integration of analog, digital, RF, optical, and MEMS devices on one chip, which is commonly referred to as SoC. Advanced packaging techniques, such as MCM, SiP (System-in-Package), and SoP (System-on-Package), have made such heterogeneous multicomponent SoC designs possible.

MCM is a recent technique, where unpackaged ICs are integrated on a variety of substrates with finer interconnects than a PCB. MCM-L uses a substrate similar to traditional PCBs, but does not allow embedding of components in multiple layers. MCM-C, which uses a ceramic substrate, and MCM-D, which uses a thin-film substrate, allow the embedding of components.

SiP is an even more recent proposition, where multiple ICs are integrated in a single package. An SiP can contain several silicon components (dies) and passive components all mounted on the same substrate; for example, a specialized processor can be combined at chip level with DRAM (dynamic random-access memory), flash memory, and resistors/capacitors. This means that a complete functional unit can be built in a chip package so that few external components are needed to make it work.

SoP entails the integration of several materials technologies for multiple functions. With SoP, the package is the system, providing all the system functions in a single module. This is accomplished by codesign and fabrication of digital, optical, RF, and sensor functions in both IC and the package, thus optimizing functions that are accomplished best at IC level and at package level.[7]

19.2.4 Design Challenges of the Space Environment

Compared with their terrestrial counterparts, any device or system intended for space must be designed to withstand a unique spectrum of hazards—mechanical, atmospheric, thermal, energetic, and dynamic.

Erosion, for example, is an issue for LEO, where atomic oxygen can etch certain materials. Space debris is a concern for satellites at any altitude, but collisions are rare events. The main concern for missions involving hundreds to thousands of satellites deployed together is the debris threat they pose to other systems. The only realistic way to solve this problem is to confine these missions to LEO, where the orbital lifetime is very short, essentially making these missions disposable. Currently, no deorbit capability is planned for such small systems, due to the disproportionate size and mass requirements. The vacuum of space introduces several issues such as cold welding and outgassing, but for very small systems, the main concern is limited heat transfer—that is, keeping the satellite warm.

Thermal extremes and cycling are exacerbated in a vacuum, as thermal radiation is the only method available for heat transfer. For some systems proposed in this chapter, bare silicon will be used in space. Silicon wafer thermal properties are well understood, and certain packaging materials can be used to manage the temperature extremes for an SoC.[8] Again, the key challenge is not overheating, but rather capturing and maintaining enough heat to operate correctly.

Radiation and charged particles, whose fluence greatly varies with altitude, is one of the main problems addressed when flying COTS components in space.[9] Long-term exposure to radiation causes a degradation of performance and increased power draw. This will not be a concern for short-lived missions in LEO, where the environment can be as low as 1–1.5 rad (SiO_2) per day. A thin coverglass for the solar cells and a small amount of aluminum spot shielding on the ICs can mitigate the accumulation of total dose.

In addition, high-energy particles can cause single-event effects, which must be tolerated and handled using several strategies. Single-event upsets are the most common phenomena, where a logic bit is toggled. Single-event latch-up is more serious, as parasitic transistors in the circuit, which are normally dormant, can be activated, pulling damaging levels of current and potentially causing burnouts. Gate rupture can also occur in certain circumstances.

Terrestrial sensor networks are composed of relatively fixed nodes. In contrast, orbital velocity in LEO is approximately 7.5 km/sec. Natural perturbations change the orbit over time, altering the arrangement of nodes. This must be fully understood, so key parameters such as communication range can be selected properly. The freefall environment also presents unique challenges. The dominant effect is that objects in orbit "float" and change their orientation based on perturbations from solar pressure, gravity gradients, magnetic fields, and aerodynamic drag. This may not be an issue if the sensor technology does not have pointing requirements.

19.3 Subkilogram Satellite Technology Options

Numerous university and industrial entities from all over the world are actively involved in development of picosatellite missions using the CubeSat platform. A number of picosatellites based on the CubeSat platform have been launched and have operated successfully. The foundation of CubeSat's success is its strict adherence to a common set of spacecraft dimensions along with a standardized space vehicle deployment mechanism.[10]

The use of a standard subkilogram mass-producible satellite design will substantially reduce build and launch costs. One of the elements of the research program at the Surrey Space Centre is to compare subkilogram satellite technologies for cost effectiveness and potential mission suitability. This is essential to focus on the optimal solution for massively distributed missions. In the following section, traditional picosatellites such as CubeSat and emerging microengineered aerosystems are discussed first, followed by discussion of three concepts in development at the Surrey Space Centre: satellite-on-a-chip, satellite-on-an-MCM, and satellite-on-a-PCB. A brief comparative review of these concepts concludes this section.

19.3.1 Traditional Picosatellites

Twenty picosatellites reached orbit from 2000 to 2007. The first six were based on customized satellite buses; however, only two of them were functional after deployment. The next 14 were developed using the CubeSat educational satellite standard, based on traditional satellite design approaches, and deployed with the Poly Picosat Orbital Deployer (P-POD).[11] Nine of these 14 CubeSats achieved some functionality on orbit, with many being a complete success. Unfortunately, 14 additional CubeSat systems were destroyed by a single launch vehicle failure in July 2006. Of all the picosatellite missions, a majority were from educational institutions that only a decade ago would have been considered incapable of developing satellites. The growing popularity of the CubeSat concept has made it now possible to assemble one from COTS components, with the exception of a payload.

To compare this traditional picosatellite technology with others, the Surrey Space Centre developed a simple cost model. As with any cost modeling, the Surrey model made certain assumptions. First, a common reference mission investigating ionospheric plasma depletions or "plasma bubbles" was selected to derive the system requirements. The basic mission architecture would be to deploy a naturally dispersing constellation of subkilogram satellites to act as a space-based sensor network to collect time- and position-tagged *in situ* measurements of plasma density.[12] The candidate payload was the Miniature Electrostatic Analyzer (MESA) developed by the Department of Physics at the United States Air Force Academy. The following list includes other supporting assumptions to the cost modeling:

- Maximum performance parameters would be used.
- System configuration would be equal to or less than 1 kg mass, 10 × 10 × 10 cm.
- Launch costs would be based on historical Dnepr data.
- Nonrecurring engineering would not be included.
- Assembly, integration, and test costs would not be included, unless required to equate systems.
- The proposed integration of components would yield positive design margins for all mission-critical parameters.

For most missions, costs associated with the last two categories can be far greater than those of the actual hardware. As the cost model develops, they will be included.

To determine a cost-versus-performance relationship for comparison, key performance metrics need to be properly identified. Two logical parameters are sunlit average power generation and payload accommodation. Most other system capabilities, such as payload support, onboard computing, and communication range, can be determined from these parameters and the assumptions listed above.

The COTS CubeSat is estimated to have an average sunlit power generation of 2.4 W and an approximate payload capacity of 300 g and 9 × 9.5 × 3 cm. The

Table 19.2. COTS CubeSat Configuration and Costs

Subsystem	Vendor/Model	Mass (g)	Unit Cost @ 1*	Unit Cost @1000**
Payload and interface	USAFA MESA	130	$2,763	$1,517
Structure	Pumpkin skeletonized	155	$1,350	$810
Electrical power	Clyde Space EPS	310	$25,240	$19,252
Data handling	Pumpkin FM430	90	$1,200	$720
Communications	Microhard MHX-2400	75	$750	$345
ADAC	bar magnet/damper	25		
GPS	SSTL SGR-05	20	$20,000	$6,000
Propulsion	none			
Launch	various/P-POD		$40,000	$20,000
Total:		805	$91,303	$48,644

*approximate costs from vendors
**aggregate costs of authors and vendors

MESA payload is composed of an 80 g sensor and 50 g interface PCB. The cost breakdown of the COTS CubeSat is presented in Table 19.2. Single-unit cost data were provided by each vendor in support of this research for simple cost modeling purposes only. Unit costs were also derived for a constellation of 10, 100, 1000, and 10,000 satellites. These costs are purely speculative, jointly developed between the authors and the vendors.

19.3.2 Microengineered Aerospace Systems

Since 1993, Helvajian and Janson [13–16] have pioneered microengineered aerospace systems based on MEMS and microfabrication. One of their first concepts was called "nanosatellite," where satellites are mass-produced by stacking up payloads and subsystems built entirely of silicon wafers. The terms "silicon satellite," "satellite-on-a-wafer," and "spacecraft-on-a-chip" were also introduced at this time. Integrating MEMS with CMOS, which is the most common IC fabrication technology, is highlighted in all these efforts as an essential development for very small satellites.

Helvajian and Janson more recently developed the Co-Orbiting Satellite Assistant (COSA), proposed initially with a 100 g (femtosatellite) mass baseline.[17] (As a side note, no active femtosatellites have flown in space, although much smaller passive satellites flew as part of project West Ford.[18]) COSA is intended for a weeklong satellite inspection mission, with most of the design effort going into a photostructurable glass/ceramic material (Foturan) to mass-produce a multifunctional propulsion and structural subsystem (see chapter 18). COSA has been recently revised, targeting a 1 kg spacecraft configuration with a ΔV capability of 30 m/sec.[19] Startup costs are initially estimated at \$600,000 with a unit cost of \$30,000 for the structural/propulsion subsystem alone.[20] A battery and conventional PCB provide the payload and subsystem functionality.

Since COSA is primarily focused on miniature propulsion system development for spacecraft inspection missions, it would not be fair to compare this technology in a mission scenario where propulsion is not required or desired. Therefore, it is not included in this comparison. However, it should be noted that this promising technology is well suited for a complete propulsion system for picosatellites, which currently does not exist.

In addition to microengineered aerospace systems, the concept of multifunctional structures and architectures was introduced, also backing the idea of inexpensive mass production of satellites.[21] This idea fostered the Operationally Responsive Space (ORS) movement, which proposes that satellites be built and rapidly deployed using streamlined manufacturing processes and modular technologies.[22,23] Similarly, Bruhn et al.[24] have led an effort to reduce satellite mass and volume by "orders of magnitude" using a proprietary architecture termed "multifunctional microsystems." A third party has licensed this technology with the goal of mass producing advanced nanosatellites and picosatellites, but projected costs are above \$1 million.

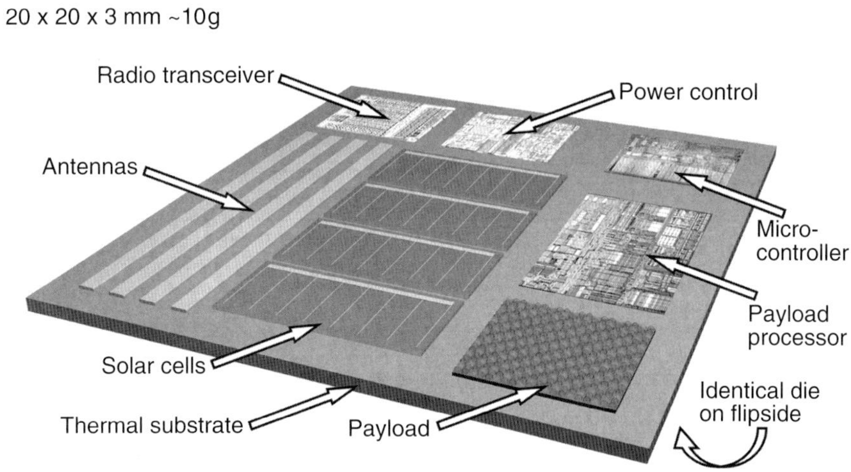

Fig.19.1. SpaceChip concept.

19.3.3 Satellite-on-a-Chip

The satellite-on-a-chip idea has become the ultimate, but seemingly elusive, goal of spacecraft miniaturization since the first known mention of the concept in 1994.[25] In recognition of recent feasibility study results,[26] this effort, called "SpaceChip," is now more focused on developing monolithic sensor nodes, due to the very limited practicality of a satellite-on-a-chip taking a major role in a traditional space mission.[27] However, such a device could support distributed missions in a spectrum of hostile environments on Earth, in space, or in planetary exploration.[28]

In support of these spacecraft miniaturization goals, the Surrey Space Centre adopted a stand-alone SoC approach.[29] The more challenging, but potentially cheaper, monolithic SoC approach was selected over incremental integration approaches such as PCB, MCM, SiP, or the revolutionary approaches discussed in the previous section. PCB-level prototyping efforts supporting satellite-on-a-chip have been recently spun off as a separate project, which is discussed in more detail in the next section.[30]

The initial concept of SpaceChip was crafted using Wertz and Larson's[31] Space Mission Analysis and Design (SMAD) principles and is illustrated in Fig. 19.1.[4] To accomplish the baseline science mission, a large fleet of SpaceChips would be deployed from a host small-satellite relay in LEO. The mission would be performed for as long as the fleet of SpaceChips remained in communication range of the host satellite.

The main advantage of this approach is that it enables a complete turnkey solution, in which the spacecraft design is sent electronically to the foundry, where it is mass-produced. Current studies show that to meet various basic system requirements, two die must be sandwiched on a thermal substrate, which would require only minimal labor. Radiation effects would be mitigated using

Table 19.3. SpaceChip Configuration and Costs

Item	Vendor/Model	Mass (g)	Unit Cost @ 1*	Unit Cost @ 1,000**
SpaceChip	various	10	$2,400	$400
Launch	P-POD		$300	$300
Total:			$2,700	$700

*approximate costs from vendors
**estimated costs by authors and vendors

radiation hardening by design.[32] With the given dimensions in Fig. 19.1, a maximum of 2500 SpaceChips could be deployed from one 2.5 kg P-POD and would have a total mass of 27.5 kg. SpaceChip's costs are summarized in Table 19.3.[4] The SpaceChip concept can be thought of as a modernized or "smart" West Ford experiment.[18]

As one might expect, a satellite at the scale of SpaceChip has some fundamental problems that limit its practicality. Starting with the payload, the sensing technique used in MESA cannot be applied on chip, even with MEMS, due to the geometries required—which is a similar problem for most other payloads, such as optical ones. However, new CMOS-MEMS foundry processes do present the opportunity to integrate a range of other payloads, such as pressure, chemical, thermal, tactile, proximity, flow, force, neural, vacuum, acceleration, gyroscopic, sound, and infrared sensors.[28]

Other challenges are also due to the small scale, such as electrical power generation of only a few milliwatts because of the limited collection area. Consequently, the communication range of an on-chip radio is only 1 km at a data rate of 500 bps with an external antenna, as on-chip antennas at this scale have only a range of a few meters. The final challenge lies in the fact that single-chip GPS solutions are not yet mature enough for integration, and that existing chip-scale propulsion systems cannot be monolithically integrated.

Although SpaceChip may not yet be able to support a meaningful space mission, it is essential to continue the development of this concept until the technology matures and suitable applications arise. Surrey Space Centre has already prototyped and successfully tested a device layout technique for improving the efficiency of on-chip electrical power generation. Additionally, a new strategy of leveraging asynchronous logic and radiation hardened by design techniques has been developed, which facilitates lower-power operations and guarantees operation in harsh radiation and temperature conditions and/or voltage fluctuations.

19.3.4 Satellite-on-a-PCB

The original purpose of satellite-on-a-PCB or "PCBSat" was to serve as a tangible prototype and testbed for the SpaceChip effort. After some development and

Fig. 19.2. PCBSat system configuration and P-POD fit check.

inspiration from the first picosatellites launched in 2000, PCBSat is now a separate effort. Its focus is to design, build, test, and fly a low-cost, mass-producible subkilogram sensor node for a range of massively distributed space missions by leveraging commodity-level system integration infrastructures.

The PCBSat system configuration is 10 × 10 × 2 cm, 260 g, making it compatible with P-POD as shown in Fig. 19.2. A summary of the PCBSat system requirements, configuration, and performance data is shown in Table 19.4, targeting the comparison mission.[12] The average sunlit power generation for PCBSat is 700 mW, and the available payload capacity is 5 × 5 × 0.5 cm, which is sufficient for MESA. The cost model for PCBSat is presented in Table 19.5. Note that the MESA pricing is different from that presented in Table 19.2 (COTS CubeSat), as the electronics are integrated differently. Compared with the COTS CubeSat, PCBSat can meet all mission requirements, but at lower performance levels.

19.3.5 Satellite-on-an-MCM

MCM technology has been used for some time to miniaturize satellite components. Building on that idea, the satellite-on-an-MCM concept, or "MCMSat," which is similar to SpaceChip and PCBSat, would implement an entire satellite using one or more MCM substrates. The MCMSat concept was jointly developed with the United States Air Force Research Laboratory (AFRL).[12] Working under the same assumptions and constraints, the average sunlit power generation also drives the MCMSat configuration, which means that a similar area is needed for solar cells. This then leads to the same 10 × 10 cm configuration to maintain compatibility with the P-POD deployer, but the advantage that MCM encapsulation offers is a reduction in thickness and mass.

MCM-L (chip onboard) was briefly examined, but at the 1 kg scale, it does not offer much of an advantage over traditional PCB because it does not allow embedding of components in multiple layers. Both MCM-C (ceramic) and MCM-D (thin-film) approaches are considered viable; this study focused on MCM-D because of its greater compactness and flexibility.

Table 19.4. PCBSat System Requirements, Configuration, and Performance Data

System	Requirement	Key Components	Performance Data
Top level	All COTS parts and PCB		
Payload	Plasma sensor	MESA	100 mW, 80 g
Orbit	LEO 5 krad (SiO_2) hardness		500 km, 30 deg Rad-tolerant to 20 krad
Configuration and structure	P-POD compatibility	Delrin/aluminum frame payload and bus PCBs	~260 g 10 × 10 × 2 cm
Electrical power	Primary solar power Regulated 3.3V Peak power tracking Battery charge regulation Rechargeable battery Power telemetry	GaAs/Ge 2 × 4 cm cells MAX604 MAX856/982 Panasonic Li-ion MAX4072	$\eta \cong 16\%$ η=92% regulator η=83% PPT/BCR 645 mAhr
Data handling	3.3V RISC CPU 128K flash memory	Atmel Mega128L	3.6864 MHz clock 5 mA current draw 19 kb firmware
Comm	Unlicensed Mesh network 50 km range	MaxStream XTend	900 Mhz, 9600 bps 2 W (500 mW RF) transmit 100 km range
AOCS	Orbit determination Passive attitude control	iTrax-03S GPS Sarantel GeoH-SMP Mass offset/drag tail	Firmware mods req'd 100 mW
Propulsion	None		
Thermal	Passive thermal control 2-channel telemetry	Solar absorbing paint Thermistor	

Two assumptions were made to derive the MCMSat cost using MCM-D. First, all ICs being used in PCBSat are available as bare die. Second, the component type, count, cost, and interconnectivity would be the same as PCBSat. This leads to a thickness of about 1 cm.

The MCMSat design is shown in Fig. 19.3. The primary construction is based on a patterned overlay technology, which allows extremely close component placement (<0.5 mm) and forms a flat, chip-like surface. MCMSat is formed as a sandwich of three MCMs, interconnected through a 2 mm, 80 contact border array between substrates. The RF subsystem is formed using four 1.5 × 80 mm crossed dipoles, and the power subsystem employs eight 20 × 40 mm GaAs solar cells. All components are recessed within the substrates. This arrangement permits the

Table 19.5. PCBSat Configuration and Costs

Item	Vendor/Model	Mass (g)	Unit Cost @ 1*	Unit Cost @ 1,000**
Payload	USAFA MESA	80	\$2,600	\$1,450
Structure	various	41	\$80	\$48
PCBs (2)	various	120	\$1,172	\$811
Comm	MaxStream XTend	18	\$179	\$90
Launch	P-POD		\$8,000	\$6,000
Total:		259	\$12,031	\$8,399

*approximate costs from vendors
**estimated costs by authors and vendors

surface attachment of other components (including payload), but this feature was not exploited in the present design. Small provisions for payloads are made in each layer. Thermal control is implemented passively using thermal coatings to absorb and retain as much solar heat as possible.

In effect, the MCMSat is actually two primitive satellites in one. The top and bottom surfaces are identical autonomous satellites, containing all support except for the battery. The sandwich construction allows for energy collection with up to 50% duty factor per spacecraft, and through appropriate network protocols, the construction permits a higher communication duty factor when in contact with a telemetry station. The top is mounted face up, the bottom face down, positioned at a 90 deg angle relative to each other to simplify interconnection to the center layer. The center layer substrate is dedicated to power storage, supporting two 35 × 45 × 5.5 mm batteries, which are uniformly distributed to each substrate through internal battery charge regulation circuitry.

Based on the assumptions given and the notional design presented, approximate costs are shown in Table 19.6. The sunlit average power generation for MCMSat is identical to PCBSat, which is 700 mW. The available payload capacity is smaller than PCBSat at 1 × 8 × 0.2 cm on two faces, with an optional internal bay of 2 × 8 × 0.6 cm. As with CubeSat and PCBSat, MCMSat can meet all mission requirements, but at lower performance levels than CubeSat.

19.3.6 Comparison of Subkilogram Satellite Technologies

Comparing the preliminary cost models of the four technologies investigated reveals some interesting results. As expected, SpaceChip has the lowest unit cost, and CubeSat has the highest, as shown in Fig. 19.4.

Unit cost does not mean much without a performance metric, such as average sunlit power generation, as shown in Fig. 19.5. In this case, SpaceChip is

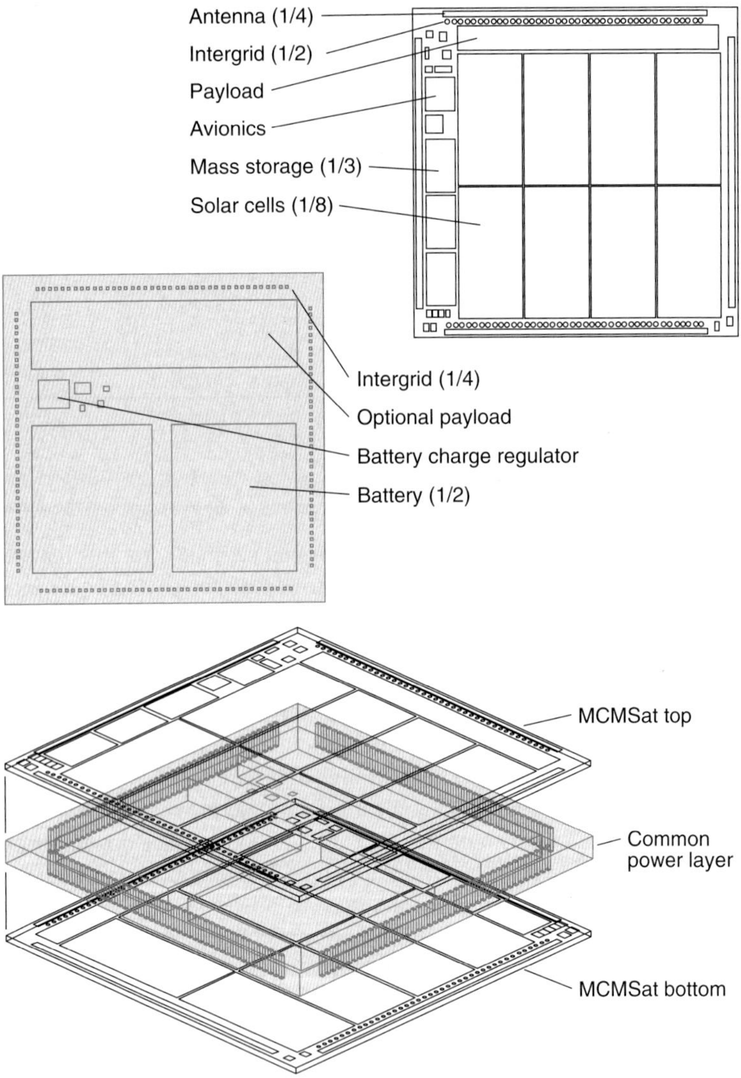

Fig. 19.3. MCMSat notional configuration.

cost prohibitive at $550,800 per watt (off the chart), with a constellation size of 10,000. PCBSat proves to be the most cost effective at any quantity, with MCM-Sat coming in slightly ahead of CubeSat.

Looking at the cost of payload volume in Fig. 19.6, CubeSat turns out to be the most cost-effective, when the maximum available payload volume is considered. SpaceChip is again off the chart at $22 million per cm^3 for a constellation size of 10,000. PCBSat is roughly double the cost, with MCMSat coming in third.

Table 19.6. MCMSat Configuration and Costs

Item	Vendor/Model	Mass (g)	Unit Cost @ 1*	Unit Cost @ 1,000**
Payload	USAFA MESA	80	\$2,600	\$1,450
Structure	various	40	\$15,069	\$9,041
Parts	various	50	\$1,176	\$849
Launch	P-POD		\$4,000	\$3,000
Total:		170	\$22,845	\$14,340

*approximate costs from vendors
**estimated costs by authors and vendors

However, the real measure of value for a small satellite in a distributed space mission is the utility of the system. Demand for such missions is less sensitive to cost than is widely thought, but is highly sensitive to the value of the product delivered. This underlies why ~100 kg microsatellites have proliferated in the small satellite market, even though much smaller satellites would have a lower unit cost. As far as Earth observation is concerned, SSTL has found that a minimum size of ~100 kg is needed to incorporate a useful payload—such as the LandSat ETM (Enhanced Thematic Mapper) data-compatible camera—and to deliver sufficiently timely data to satisfy business needs.

In the case of massively distributed mission discussed here, the large number of *in situ* measurement points requires a low-cost, mass-producible solution, provided that a small enough payload exists. Figure 19.7 compares the results in terms of total costs of the space segment. Recall that launch and hardware expenses are included, with the exception of nonrecurring engineering. The ground segment is not included, but for the comparison mission, a single CubeSat-compatible ground station is all that is required.

19.4 System-on-Chip Approach to Satellite Subsystem Miniaturization

SoC technology is recognized by ESA[33] and NASA[34,35] as an approach to enabling smaller spacecraft. In fact, the NASA X2000 deep space systems program had as one of its the main objectives the development of SoC technology for future flight projects.[36] This section explains the SoC approach in more detail and considers a reconfigurable SoC-based platform for onboard computing.

19.4.1 Introduction to System-on-Chip

SoC design is much more than high-level integration of components, since it requires both "system expertise" and "system know-how" to translate the system functionality to a single-chip implementation. The term "system-on-chip" defines

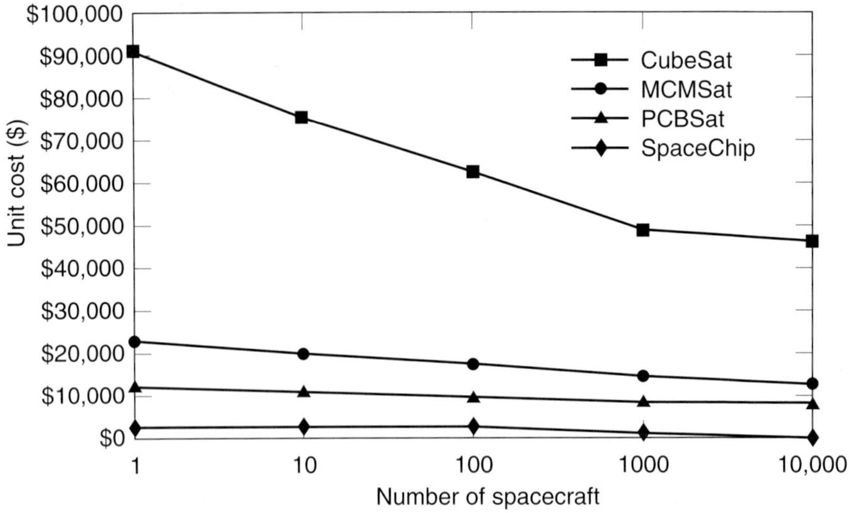

Fig. 19.4. Unit cost vs. constellation size.

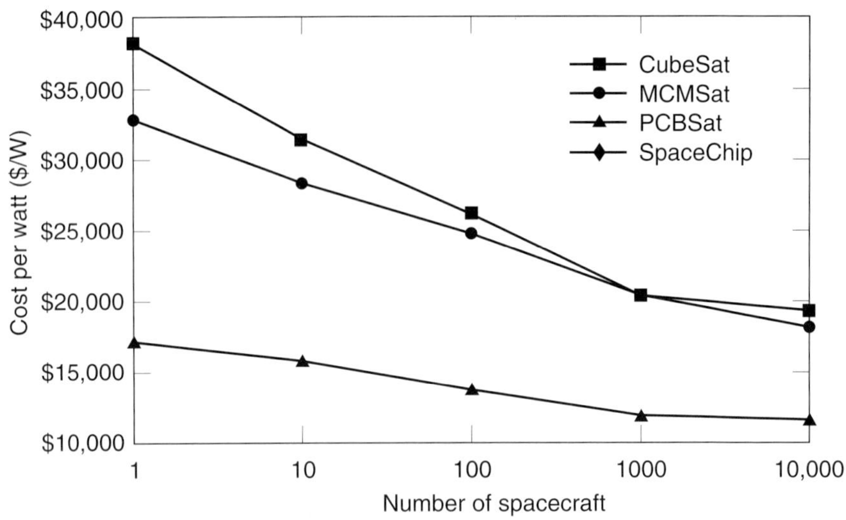

Fig. 19.5. Cost per watt of generated electrical power vs. constellation size.

both a product and a process.[37] As a product, SoC defines specific, targeted applications and contains an entire system—including, for example, processors to control the system, embedded memory to store the system program and data, communication peripherals for interaction with other systems, analog interfaces to the real world, and firmware. As a process, SoC defines system requirements that are modeled, analyzed, and partitioned into hardware and software specifications for design and implementation.

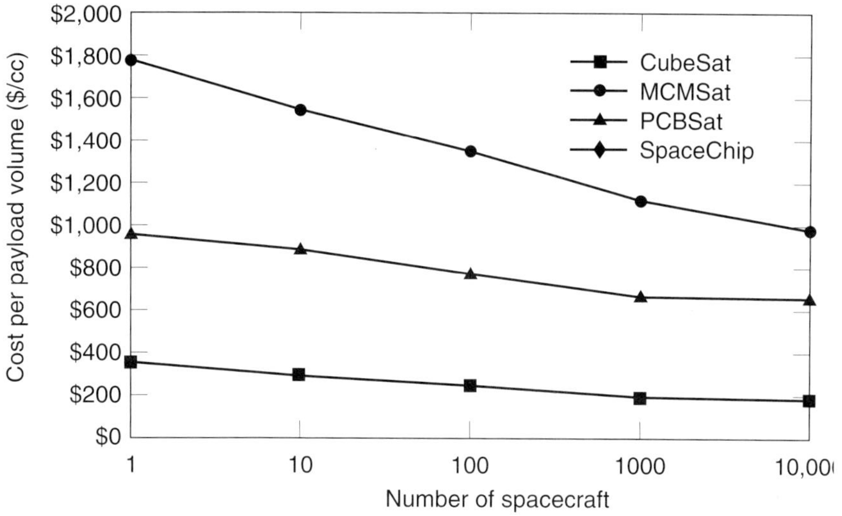

Fig. 19.6. Cost per maximum payload volume (cm³) vs. constellation size.

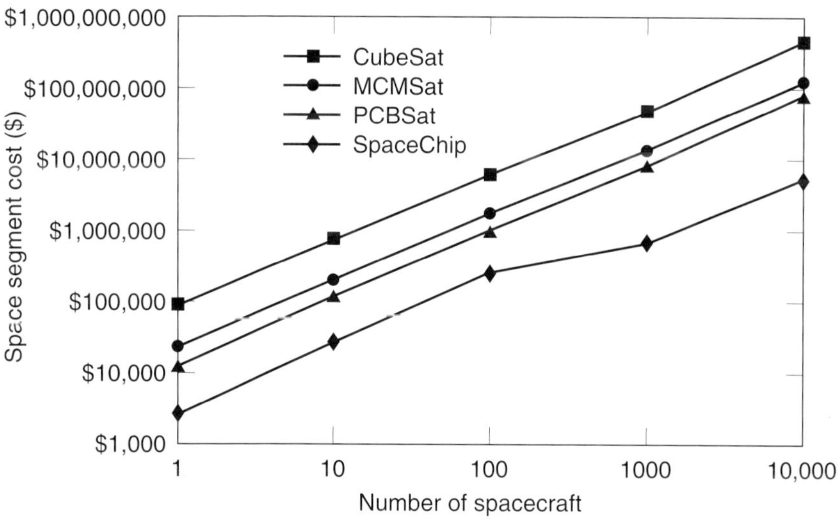

Fig. 19.7. Total space segment mission costs vs. constellation size.

SoC designs are viewed as a selection of modules or intellectual property (IP) cores connected together via common interfaces. The function of a core is normally specified by a hardware description language. Cores can be divided into three classes: soft, firm, and hard.[38] Soft cores are generally offered in a synthesizable Register Transfer Level description, which makes them inherently foundry-process portable. A disadvantage of the soft format is that the core must be reimplemented for each new SoC design; this requires that the design go through another iterative process of synthesis, which includes floor planning (configuring the placement of major design blocks), placement of individual logic cells, routing of signal wires, and back-end verification through simulation. Firm cores are reusable blocks that have been optimized structurally and topologically for performance and area through floor planning and placement. In comparison to soft cores, firm cores are less flexible and more dependent on a specific target technology; however, they feature higher performance and reliability. Hard cores are physically implemented blocks, which are faster and simpler to use, and reuse, in an SoC design. Optimized for area and speed, hard cores offer greater density and higher performance than soft cores. A disadvantage of hard cores is that they are not changeable and are technology dependent: once implemented in a specific target technology, a hard core is "locked" into that silicon process and foundry. Consequently, hard cores are the least flexible and heavily dependent on the technology.

An on-chip bus is needed to interconnect the individual SoC modules. On-chip buses play an extremely important role in an SoC design, providing a common, logical interface between the cores. A well-defined on-chip bus can significantly reduce the interface complexity and the associated development time. Its purpose is to foster design reuse, which improves the portability and reliability of the system and decreases time to market. There are several kinds of on-chip buses for SoC designs: AMBA from ARM Ltd., Avalon from Altera, CoreConnect from IBM, FPI Bus from Infineon Technologies, SuperHighway from SuperH Inc., and WISHBONE from OpenCores. A key task for SoC designers is to ensure that each component in the system obeys the interconnecting bus protocol. It is therefore of paramount importance to exhaustively prove that a proprietary peripheral core in the SoC architecture obeys the bus protocol rules.

With the establishment of SoC technology, core development has become a profitable business. A recently coined term, "fabless," refers to semiconductor companies who do not manufacture chips but who specialize in development, such as ARM Ltd. Almost all FPGA vendors now provide their own soft and hard cores, including various processors. For example, a hard PowerPC 405 processor[39] is embedded in the Xilinx Virtex FPGAs. Xilinx and Altera also offer soft processor cores, such as MicroBlaze and NIOS, respectively.

In 1997, ESA embarked on the development of a high-performance synthesizable soft processor core, named LEON.[40] One of the objectives was to provide an open, portable and nonproprietary processor design capable of meeting the identified requirements for performance, software compatibility, and low system cost. Another objective was to create a synthesizable processor core that could tolerate single-event upsets.

The LEON processor is based on the standard 32-bit SPARC V8 computer architecture. The first version, LEON1-1.0, was released in October 1999. Gaisler Research has since developed two more versions, LEON2 and LEON3. LEON3 has an improved pipeline and multiprocessor support. Starting from LEON2, the processor core comes into two types: standard and fault-tolerant (LEON2-FT). In the fault-tolerant version, data storage elements are protected by triple-modular redundancy, and all internal and external memories are protected by an error detection and correction (EDAC) unit or parity bits.

The first SoC design for onboard computing using the LEON processor core was proposed and prototyped on a Xilinx Virtex FPGA jointly by the Surrey Space Centre and SSTL in 2000.[41] This was followed by several similar SoC activities in Europe, some of which were aimed at an ASIC implementation.[42,43] Since then, the LEON processor core has gained acceptance and popularity in the European space industry for SoC design projects that are either ASIC- or FPGA-based. Notably, the LEON2-FT core has already flown in space—as part of the FPGA-based SoC data processing unit for the miniaturized monitoring camera[44] on the ESA Venus Express mission, launched in 2005. At present, the LEON processor core is the preferred ESA microprocessor, and there are plans to build a multicore processor design based on it.[45]

19.4.2 SoC-Based Platform for Onboard Computing

Computing systems that support a satellite mission are composed of computers onboard and on the ground. In the early days of satellite engineering, onboard computer systems were centralized. Now, the computer architecture is decentralized and is based on a network of embedded microcontrollers. The number of computing units embedded in the various satellite subsystems is growing. For example, the number of CPUs used on SSTL microsatellite missions increased from 3 on UoSAT-1 in 1981 to 25 on TMSAT in 1997, a more than eightfold increase over 16 years. In addition, performance requirements for space-flight electronics are increasing as onboard sensors produce greater amounts of data due to higher-resolution payloads.[46] A unified SoC-based approach to the implementation of the onboard computing could minimize required spacecraft resources and increase performance. This section presents the concept of an SoC-based computer plat form, which could be tailored to the needs of individual onboard subsystems and payloads.[47]

The architecture of the SoC platform is based on the Advanced Microcontroller Bus Architecture v. 2.0 (AMBA),[48] a widely used open standard that supports reusable component design of embedded systems with one or more processors and multiple peripherals. The platform uses a modular "plug and play" design approach using soft cores.

Figure 19.8 shows a block-diagram of the SoC platform architecture, which is centered on the LEON3 processor. The AMBA standard defines a set of buses, which are typically used in a hierarchical fashion. The advanced high-performance bus (AHB) acts as a system bus and is designed for high-performance connection of processors, dedicated hardware, and on-chip memory; it supports multiple bus masters, pipelined operations, and burst transfers. The LEON3 core could also be used for multicore processor SoC designs as shown in Fig. 19.8. Low-speed components are connected to the low-complexity bus interface, named the advanced peripheral bus (APB). The SolidPC software package, which is based on

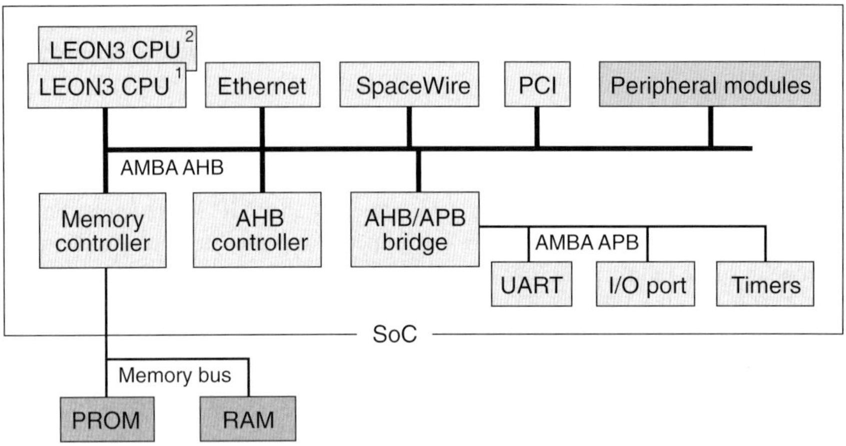

Fig. 19.8. SoC platform for on-board computing.

static functional verification, can be used to check the AMBA bus compliance for Register Transfer Level designs.[49]

In addition to the LEON3 processor core, the SoC structure contains a set of general-purpose and application-specific soft peripheral cores. The general-purpose cores come from the open-source GRLIB library, a collection of parameterizable blocks that are compatible with the AMBA bus and the LEON processor. Application-specific cores can come from any source, provided they're compatible with the AMBA interface.

The GRLIB library includes cores for AMBA bus control, the LEON3 SPARC processor, a 32-bit PC133 synchronous dynamic random-access memory (SDRAM) controller, a 32-bit PCI bridge with direct memory access (DMA), a 10/100 Mb Ethernet media access control (MAC), a programmable read-only memory (PROM) and static random-access memory (SRAM) controller, a controller area network (CAN) bus controller, a universal asynchronous receiver/transmitter (UART), a modular timer unit, an interrupt controller, a 32-bit general-purpose I/O port, and so on. Memory and pad generators are also available for FPGAs and ASICs such as Virage Logic, Xilinx, UMC, Atmel, Actel, etc.

The Surrey Space Centre has developed a library of application-specific AMBA-compatible peripheral soft cores that perform a range of tasks required by the different onboard subsystems and payload instruments. Optimized versions of the cores are also stored in the library to meet different design requirements in terms of area, throughput, and power consumption. Examples include a high-level data-link controller (HDLC) interface for downlink and uplink data transmission; an EDAC; a 32-channel DMA controller;[50] a 32-bit floating-point mathematical coprocessor;[51] a fault-tolerant Advanced Encryption Standard core;[52] and a lossless compression core. The encryption and compression cores are specifically designed to process multispectral satellite images.

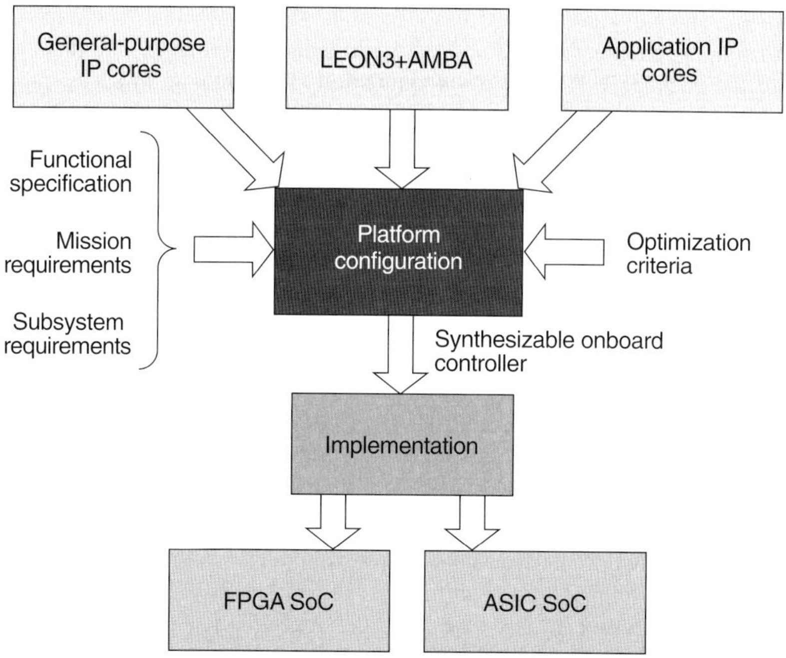

Fig. 19.9. Designing with the SoC platform.

The platform is aimed at generation of onboard controllers that are tailored to the needs of the individual onboard subsystems and payloads. New instances of onboard controllers can easily be generated mixing and matching cores according to the needs of a particular satellite mission. Figure 19.9 illustrates the customization design flow using the SoC platform. The AMBA bus defines a common backbone for the SoC architecture. The general-purpose and application-specific libraries act as repositories of reusable building blocks for platform configuration.

19.4.3 Case Study: Single-Chip Onboard Computer

The onboard computer is the heart of a satellite's onboard data handling system. A project at the Surrey Space Centre aimed at miniaturization of the small satellite data-handling subsystem included reduction of the computer size from a PCB to a single chip.[53] An existing microsatellite computer developed by SSTL (the OBC386) was used as a reference design. A downsized implementation of a single-chip computer, consisting of the LEON2 processor core and two peripheral cores—CAN (controller area network) and EDAC—was prototyped in a Xilinx Virtex XCV800 FPGA chip. A simplified Consultative Committee for Space Data Systems (CCSDS) telemetry and telecommand communication system, which was specifically designed to meet the needs of the SoC onboard computer, was also developed.[54,55] The system represents a streamlined, yet reliable and automated, standalone software implementation of the CCSDS communications protocol. The software package has a modular structure, which facilitates easy expansion of functionality to suit specific mission requirements. The functionality of the package was verified via simulation. The software imposes minimal memory footprint and performance requirements on the onboard computer. The CCSDS software communication system reduces further the hardware overhead of the data-handling subsystem, and in combination with the SoC computer, can provide a cost-effective and flexible communication solution for small satellites.

Figure 19.10 shows a block diagram of the SoC-based onboard computer, which is modeled on requirements derived from SSTL microsatellite missions.[57] The design is prototyped on a Xilinx Virtex II FPGA. Some of the components and interfaces in the SoC are more typical for use in space than for other applications—for example, the CAN, HDLC, and SpaceWire interface controllers, and the EDAC block.

The HDLC protocol is used on SSTL small satellite missions for both uplink and downlink transfers between the spacecraft and the ground station. As the name suggests, the HDLC is a second-layer protocol in the open-systems interconnection model. Data rates up to 10 Mbps are used for the uplink and downlink. The onboard network interface standard used on SSTL satellites is CAN, a serial bus interface developed by Bosch for car applications. Its maximum data transfer rate is 1 Mbps, which is on the low side. SpaceWire is a serial interface for a point-to-point connection developed for spacecraft applications, similar to the FireWire (IEEE 1394) standard on PCs. It is also used as the onboard computer network and supports full-duplex data transmission with higher data rates from 2 to 400 Mbps.

The memory organization of onboard computers is more difficult than that of terrestrial computers. Space radiation can cause bit flips known as single-event upsets; therefore, the memory must be protected. For this protection, an EDAC controller generates parity information while the data is written into the memory, and checks the parity information against the data information while the data is read from the memory. If the EDAC controller detects a correctable error while data is read, it will correct it.

19.4.4 Reconfigurable System-on-a-Chip Payload Controller

While some impressive Earth-bound research applications of reconfigurable computing have been developed, it is still a relatively new field of study for space applications. The first space test of reconfigurable computing is being carried out onboard FedSat.[56] NASA has sponsored an investigation into spaceborne use of FPGAs, and as a result, a reconfigurable data-processing module for space-based remote-sensing applications was developed.[57] The aim was to implement image-compression algorithms in hardware to radically reduce the bandwidth requirements for downloading NASA hyperspectral images.

In this section, a reconfigurable version of the SoC design in section 19.4.3 is presented for the purpose of payload computing. The proposed payload computer is implemented in an SRAM-based FPGA. Many SRAM-based FPGAs,

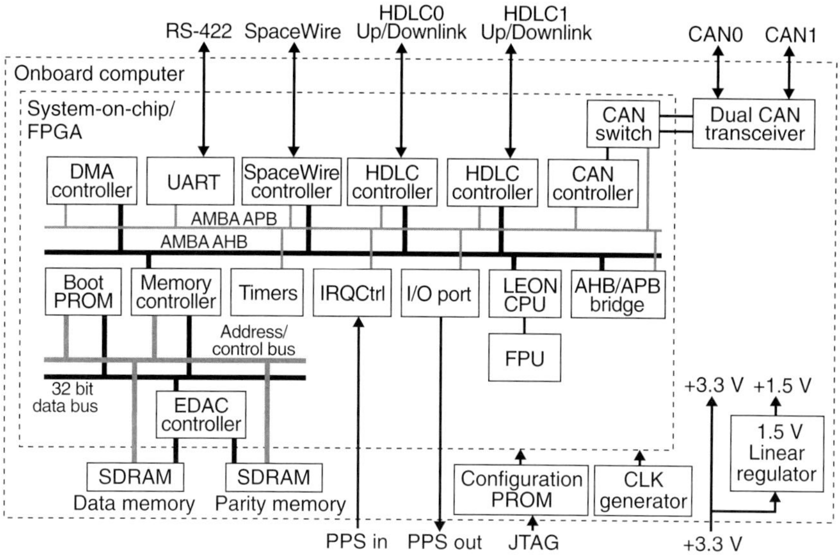

Fig. 19.10. SoC onboard computer block diagram.

such as Xilinx Virtex FPGAs, support partial run-time reconfiguration, which is the ability to update only a part of the logic in an FPGA without affecting the functionality of the unchanged section. Partial run-time reconfiguration could be very beneficial to onboard SoC-based controllers. It could facilitate the plug-in of new components at run time to change or upgrade existing functionality. This capability could also be used to repair malfunctioning blocks.

The reconfigurable SoC design is targeted at the radiation-hardened QPro Virtex II FPGAs, which were released by Xilinx in 2003. The capacity of the largest FPGA of this family is up to 6 million system gates.[58] The total ionizing dose tolerance of the Xilinx QPro Virtex and Virtex II FPGAs are 100 and 200 krad (Si), respectively, which enable them to be used in all but the most extreme orbits. They are immune to latch-up, but are still susceptible to upsets.

A number of strategies for mitigating single-event upsets in SRAM-based FPGAs have been developed.[59] Scrubbing is a procedure of periodically reading back the FPGA's configuration memory, comparing it with a known good copy, and writing back any corrections required. By scrubbing an FPGA, configuration errors can be corrected. Recently, Xilinx provided a design tool[60] that automatically implements triple-modular redundancy in Xilinx FPGA designs, protecting the voting circuits from single-event upsets. However, designs with triple-modular redundancy are at least three times as large as those without, and suffer from speed degradation as well. Power consumption is also tripled along with the logic.

Starting with the Virtex II series, Xilinx Virtex FPGAs have integrated an internal configuration access port (ICAP) into the programmable fabric, which enables the user to write software programs that modify the circuit structure and functionality at run time. The access port is, in fact, a subset of the SelectMAP interface,[61] which is used to configure Xilinx FPGAs.

The block diagram of the reconfigurable SoC is shown in Fig. 19.11. The on-chip peripheral bus is used to connect all the access port modules. The internal configuration access port is connected to the LEON3 processor via a bridge between the on-chip peripheral bus and the advanced high-performance bus. Once the FPGA is initially configured, this access port is used as an interface to reconfigure the FPGA. The driver for the access port is included in the embedded Linux operating system,[62] and the control logic for reading and writing data to it is implemented in the LEON3 processor as a software driver. The Block Select-RAM (BRAM) in the FPGA is used as a configuration cache. Because Virtex II FPGAs support reconfiguration only by frames, the BRAM must be big enough to hold one frame each time. The configuration bitstream of each SoC component can be stored onboard in a flash memory. The bitstream of a new or upgraded SoC component can be uploaded through the satellite uplink from the ground station.

The module-based partial reconfiguration method, developed by Xilinx, is used with the proposed SoC design.[63] The design is divided into modules, which can be developed in parallel and merged into one FPGA design later. The communication between the reconfigurable and fixed modules happens via the AMBA bus through a special bus macro. For the module-based design flow, partial bitstreams can be created using the Xilinx PlanAhead tool,[64] which can then be committed to FPGAs using the SelectMAP interface or the on-chip internal configuration access port module. The on-chip peripheral bus interface to the access port module permits connection of this peripheral to the MicroBlaze soft-core processor or the PowerPC hard-core processor inside the FPGA. For this purpose, Williams et al.[62] developed an internal configuration access port device driver for the uCLinux kernel, running on the MicroBlaze processor. In the proposed SoC, the internal configuration access port will be controlled from the LEON3 processor.

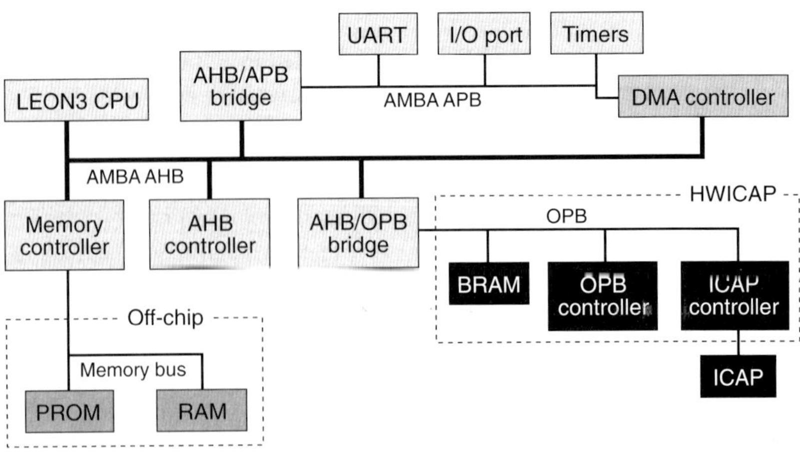

Fig. 19.11. Reconfigurable SoC controller architecture.

Radiation effects such as single-event upsets as well as gate rupture or latch-up induced burnout can be mitigated. This will require that the bitstreams for each peripheral module are pregenerated and stored in the onboard memory, and the design has the ability of self-detecting faulty operation. When a peripheral core is identified as damaged, it will be repaired by partial reconfiguration, using the bitstream of this peripheral from the memory.

The rest of this section presents a simulation experiment that demonstrates partial run-time reconfiguration by adding a direct memory access controller (DMAC) to the SoC onboard controller. The DMAC handles the data transfer between the main memory and the peripherals, bypassing the CPU. The CPU allocates a memory block and assigns it to the DMAC and then writes the transfer mode and the peripheral device address to the DMAC registers. After the DMAC is configured, there are two possible ways of triggering the data transfer process: the CPU sends a start command to the DMAC, or the transfer is triggered via a hardware handshake between the DMAC and the peripheral device.

In the second case, the device must be capable of direct memory access—that is, it must provide the appropriate hardware handshake signals. At a minimum, the handshake requires a request signal, and usually involves an acknowledgement signal as well. If a peripheral device receives data from "outside," it asserts the request signal. The DMAC transfers the received data from the peripheral device controller to the memory and asserts the acknowledge signal. When the transfer is completed a state bit will be set in the DMAC, or the DMAC causes an interrupt.

A fixed SoC module, which consists of the LEON3 processor, internal configuration access port, and BRAM, is first implemented into the Virtex II FPGA. Then, the DMAC is added to the module while it is running. A synthesis tool is used to produce the netlists of the fixed SoC module, the DMAC, and the complete SoC that consists of both the fixed SoC module and the added DMAC. The resultant netlists are floor planned to provide a reference for placing the individual components. The floor plan ensures that the dynamic circuit (the DMAC) is correctly interfaced to the static circuit (the fixed SoC module). Figure 19.12 illustrates the design partitioning between the resultant static and dynamic circuits. Bus macros are inserted to interface signals between the circuit partitions.

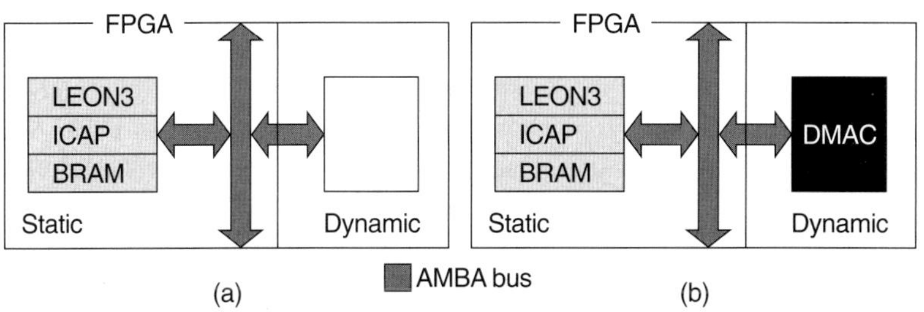

Fig. 19.12. SoC design partitioning for partial reconfiguration before reconfiguration (a) and after reconfiguration (b).

Extensive simulations were carried out, which showed that the DMA controller works according to the specification, which confirms the success of the partial run-time reconfiguration operation.

19.5 Enabling Technologies for Distributed Picosatellite Missions in LEO

This section presents research conducted at the Surrey Space Centre as part of the ESPACENET project,[5] which targets the development of a robust space-based sensor network based on flexible picosatellite nodes. First, a brief overview of the network and its main features is given. Then, initial results on the development of several enabling technologies are reported, including intersatellite links, distributed computing, remote reconfiguration, and multiobjective optimization for signal routing. An approach to building a modular subkilogram satellite via extension of the SoC concept is also discussed.

19.5.1 Picosatellite Network Overview

The network will be realized as a constellation of heterogeneous picosatellites at various locations in the same or different orbital planes. In this work, Surrey adopted a two-level network model based on picosatellite nodes equipped with crosslinks, referred to as intersatellite links.[65] The picosatellites are grouped in clusters: within each cluster, one satellite is configured as a master node, and the others as slave nodes flying in suborbits. All the communications between the slave satellites and the ground station take place via the master satellite. The clustered satellites form an *ad hoc* network, which allows peer-to-peer communication between any two nodes. The master picosatellite nodes are capable of communicating with each other over intersatellite links, forming the second level of the network. Hundreds of clusters can be employed to provide global coverage and continuous terrestrial communication in LEO.

Design of intersatellite links in LEO networks is a challenging topic, as signal transmissions are affected by variable propagation delays, relative velocities, and large Doppler shifts and rates. It is therefore necessary to investigate how changes of orbital parameters influence the use of terrestrial wireless standards for intersatellite communications.

Despite recent advances in wireless technologies, most protocols and algorithms are specifically designed to address challenges in terrestrial networks. In fact, none of the currently used wireless technologies is sufficiently scalable, efficient, or robust to meet the demands of satellite networks. The IEEE 802.11 is designed for terrestrial applications with a maximum transmit power of 1 W for distances of only 300 m. The purpose of this research is to investigate key IEEE 802.11 optimization techniques, with the aim of adapting the standard for intersatellite communication in autonomous distributed spacecraft constellations and formations in LEO.

Formation flying missions are of particular interest for implementation of picosatellite networks because of the close proximity of the nodes in the network. Such a distributed satellite network would require coordination that can be best provided through intersatellite links.

An important feature of such a network is its reconfigurability, which will be manifested at two levels—node level and system level. The piconodes will consist of completely reconfigurable SoC devices to process data from various sensing elements. The SoC controller design platform presented in section 19.4, which encompasses a number of soft cores and driver capability, will be utilized. This will enable SoC customization at run time to best meet the processing requirements of the network. Reconfigurability of the satellite nodes would be required for advanced remote sensing and science missions, such as:

- Beam forming to remotely sense a particular location at optical or radio wavelengths.
- Minimizing power expenditure by dynamically optimizing RF links.
- Measuring ion- or electron-scale space weather events and effects within the magnetosphere.
- Compensating for interference from other sources by frequency hopping.

The picosatellite network must be capable of adapting its topology to achieve efficient distribution of tasks among the resource-constrained piconodes. The network must also be able to concurrently drive both network and hardware resources to effectively and efficiently solve problems in case of node failures and change of mission requirements. The system-level reconfiguration will be guided by a multiobjective evolutionary algorithm to optimize parameter values. Because of the computationally intensive nature of the multiobjective evolutionary algorithms, they will be executed in a distributed fashion.

19.5.2 Intersatellite Links

The intersatellite link research at the Surrey Space Centre has focused on adapting the physical layer and the MAC sublayer of the IEEE 802.11 protocol to space.[66,67] Concerns regarding the short-range and fixed-timing parameters of the IEEE 802.11 standard have been addressed through optimization approaches such as tuning the physical layer parameters and enhancing the data link layer. Investigation of integration techniques of terrestrial smart antenna with both the physical and MAC layers in IEEE 802.11 close-proximity links has been conducted, aimed at increasing the capacity while minimizing interference and collision. Wireless local area network architecture for space networks has been defined to support an increased capacity in terms of throughput and range in dynamic LEO networks. An IEEE 802.11 MAC adaptation algorithm for LEO channel conditions was proposed, and a case study was undertaken to demonstrate the effectiveness of the proposed approaches for satellite clusters flying in formation.[68]

The main outcomes of this research work are:

- Modeling of the impact of intersatellite link variance on communication parameters between directly and indirectly connected satellites.
- Redefinition of the IEEE 802.11 physical and MAC interframe timing parameter for intersatellite link range.
- IEEE 802.11 physical and MAC layer integration with smart antennas for intersatellite link applications.
- Creation of a novel MAC algorithm for intersatellite link applications.
- Creation of a novel wireless communication architecture for LEO satellite formations using the IEEE 802.11 protocol.

These collectively constitute a scalable wireless local area network framework for LEO networks using intersatellite links. The effectiveness of all the proposed schemes has been evaluated through mathematical analysis and simulation.

19.5.3 Distributed Computing

Picosatellites have very limited resources in terms of power, processor speed, and memory. It is not viable to perform computationally intensive tasks such as multiobjective optimization on a single node; however, the processing capability of individual picosatellites could significantly be improved by distributing such tasks to multiple nodes.

There are two different approaches to distributed computing that are applicable to picosatellite networks: address-centric and data-centric. The address-centric approach opens the possibility of using a traditional client-server architecture or a mobile-agent-based architecture. The data-centric approach may be based on publish-subscribe communication schemes.

Distributed computing requires faster development cycles, decreased effort, and greater software reuse, which motivates the creation and use of middleware and middleware-based architectures.[69] Middleware is a software layer that resides between the application programs and the underlying operating systems, including network protocol stacks. Network middleware mediates between an application program and a network and manages the interaction between applications across heterogeneous computing nodes.

Agents are a programming abstraction, used for middleware design in networked systems. By thinking behaviorally (assigning an agent a "role"), communication interactions and autonomous actions become easier to realize. This allows them to work "proactively" and "reactively" to their environment and to any given task. They can be proactive when finding new communications routes in a networked environment and reactive to disconnections, low bandwidths, or high latencies.[70]

There are many potential benefits to using agents onboard spacecraft. These include:

- Increasing the level autonomy through decomposition of high-level goals.

- Increasing the flexibility and adaptability of flight software through dynamic agent uploads.
- Improving the reliability of spacecraft and clusters of spacecraft by incorporating fault detection at both high and low levels, thus reducing the need for large ground support organizations.

In a satellite network, mobile software agents could be used for the following roles:

- Deployment, upgrading, and maintenance. The exploitation of mobile code to support software deployment and maintenance is very useful in a highly distributed environment. Mobile code paradigms would allow configuration of new software and hardware units, such as proprietary cores in an FPGA, or rebuilding an application with new parameters locally at each satellite. Proactive reconfiguration by comparing both hardware characteristics and software features for configuring and installation could be automated. Functionality could be added or removed, depending on the satellite's resources.
- Customization of services. Agents onboard satellites could provide a flexible service for both server and client roles, depending on their current hardware and software status. This allows agents to react in their communications to other satellites.
- Disconnected operations. Without an attitude determination and control system, the picosatellites could be tumbling and moving in differing orbits, making intersatellite communication difficult. Unreliable links with low bandwidth and low reliability require new methods for allowing flexibility in the system. Improving granularity in the client-server paradigm would allow a greater number of operations but would increase local resource use and complexity and reduce flexibility for change.
- Improved fault tolerance. Traditional client-server systems request data and services from remote environments, which can then be executed locally. But if a return result is only partially complete, there are problems in addressing the state and possible recovery. As agents encapsulate all state information in a single component, the information can be traced and recovered. Destructive space environment effects could be mitigated using agents to detect hardware or even software errors.

The JADE-LEAP[71] software environment is used extensively to develop new agent systems and novel applications for resource constrained mobile devices using wireless links in Java. To accommodate the agent computing environment, a Java Virtual Machine for embedded devices is typically required, which can serve as a communication medium between various heterogeneous platforms; however, the Java programming language is not the best choice for real-time critical systems. Instead, Surrey has been working with a Java-specific processor called Java Optimized Processor.[72]

An FPGA-based SoC design is under development, incorporating the Java Optimized Processor with the LEON3 processor[73] as a secondary master processor core. The interface between LEON3 and the Java Optimized Processor is achieved via the AMBA2-advanced high-performance bus. Benefits of using the Java Optimized Processor in an FPGA include moving the Java Virtual Machine software to hardware, reducing the memory footprint, and enabling Java applications, such as agents, for real-time applications.

19.5.4 Remote Reconfiguration

The proposed SoC is also capable of remote partial reconfiguration.[74] This can be realized by sending the partial bitstream from a remote location (either the ground station or other satellites) to the satellite. This bitstream is stored in the onboard memory and is used to reconfigure the corresponding area on the FPGA.

The partial reconfiguration capability is aimed at a picosatellite constellation where satellites are connected via intersatellite links. If a configuration file is too large to be transferred at once, a distributed reconfiguration scheme may be adopted. The file can be split into a number of smaller portions, which can be uploaded to satellites in range of the ground station. After these files are received, they will be sent to the destination picosatellites over intersatellite links. Then, the data will be fused to reconfigure the SoC. This scheme requires the availability of a middleware layer to implement the communication between the servers and the clients.

The SoC configuration file for the Virtex II FPGAs is greater than 10 Mb (10 Mb for a XC2V3000, 20 Mb for a XC2V6000). This can present a problem for the reconfiguration via a ground-to-satellite link. A typical LEO satellite is only visible from its ground station for approximately 10 min during each pass, which occurs four to six times a day. Furthermore, the small satellite uplink is often of low baud rate—from 9.6 to 19.2 kbps, although with some small satellite platforms, the uplink can reach 128 kbps. It may be difficult to upload such a large file with the low uplink rate, even though the partial configuration file is much smaller than the complete one.

One solution is to compress the configuration file before uploading and to decompress it afterward. A configuration file can be compressed up to 25%, even with a routine text compression algorithm. For example, the LEON3 processor, which is the main core in the SoC design, results in a bitstream of 1,311,870 bytes when targeting the XC2V3000 FPGA. After compression using WinRar, the size is reduced to 376,063 bytes, which is 28.67% of the original size. The worst situation occurs when the complete configuration file needs to be uploaded. In this case, it would take less than 3 minutes to upload the compressed file, excluding the control signal transmission overhead.

19.5.5 Signal Routing Using Optimization Algorithms

Picosatellites have a restricted power budget due to their size. For example, the solar arrays of a single or "1U" CubeSat can produce 2.5 W maximum. Satellites performing a virtual satellite mission would be required to carry out computationally intensive processing in addition to internode communication. Hence, the power consumption for the wireless communication should be minimized to meet the total power budget. Surrey has designed a power-aware routing algorithm to reduce the transmission power for signal routing over the network.

Surrey has also been examining an approach to signal routing in a picosatellite network using a multiobjective evolutionary algorithm,[75] which aims at providing a set of Pareto-optimal routes. The satellite network optimization is carried out according to two competing objectives, transmission power and time delay. Because of limited power resources of picosatellites, the optimization of the power consumption is a task of highest priority at the network level. In addition, for virtual satellite missions, some tasks need to meet certain time constraints, including hard real-time constraints. For example, if satellites are flying in formation, the attitude control of a satellite needs to be in a control loop with its neighbor satellites. In this case, delayed signal transmission will disrupt the formation.

Surrey chose an evolutionary algorithm to perform the multiobjective optimization because these have proven to work well with nonlinear objectives.[76] A multiobjective evolutionary algorithm explores the whole search space to find the Pareto-optimal route, which compromises the requirements for power consumption and time delay. A multiobjective evolutionary algorithm starts with a "parent" population of a number of end-to-end routes randomly generated, where each individual is represented by its "chromosome." Each parent individual is randomly mated with another to produce two children. The children are then mutated with a probability p, where each coordinate of the design vector representing each child is modified randomly with p. The transmission power and time delay of the children are then calculated, and a fitness value is assigned to every parent and child. The individuals with best fitness are then passed on to the next generation. The process continues until the maximum number of generations is reached. In the end, an optimized route is obtained.

It is concluded from simulations that multihopping of the sensed data will significantly reduce the transmission power. The time delay for multihopping within a route of n satellites given by the propagation delay and the satellite queuing and processing delay is decided by the amount of buffers, queuing discipline, servers, and the packet arrival rate. On the one hand, more hopping satellites will save more power; on the other hand, the end-to-end time delay will be increased. Using the multiobjective evolutionary algorithm, however, a global optimum can be found.

19.5.6 Extension of the System-on-Chip Concept

Despite the fact that standard wireless COTS protocols are widely used terrestrially, they are not so popular as a communications means onboard spacecraft yet. Suitability of COTS wireless sensor nodes, referred to as mote kits, for communication inside small spacecraft has recently been investigated as part of an ESA sponsored project.[77] Selected mote kits were characterized for use onboard SSTL small satellites via functional, electromagnetic compatibility/interference (EMC/

EMI), vibration, thermal and radiation testing. The study concluded that the use of wireless nodes inside the spacecraft for intrasatellite communication can potentially minimize the use of wires, leading to a significant reduction of the harness complexity and mass as well as the assembly, integration, and verification effort. The criteria for the selection and recommendation of the wireless protocols for intrasatellite communication are robustness of the protocol to fault and interference, support for many nodes, data rate, power requirement, and easy connection to SSTL TCP/IP devices.

ZigBee (IEEE 802.15.4) is particularly appropriate where there is the need for mesh networking of sensor nodes within the spacecraft for monitoring and control. SSTL small satellites use the CAN bus as the onboard network. This bus can transfer up to 1 Mbps, but the existing CAN hardware offers a maximum rate of 32 kbps, in which case the performance of IEEE 802.15.4 is acceptable for intraspacecraft communications. Replacing CAN with a wireless mote system can lead to 80% mass savings, increased data rates, and possible power saving. Requirements for additional communication path onboard with a data rate greater than 1 Mbps will need the use of ultrawideband or IEEE 802.11.

An intermediate solution in the transition to wireless intrasatellite communication could be the realization of a controller area network over Bluetooth (IEEE 802.15.1). Several methods to convert Bluetooth to CAN and vice versa have been proposed for automotive applications. A real-time operating system, such as the TinyOS, could be employed to provide control capabilities in the converter.

These findings suggest that standard wireless protocols are a viable implementation option for intrasatellite communication and data gathering, which can contribute significantly to the spacecraft miniaturization. The SoC design concept presented in section 19.4 deals with the integration of digital components; however, the integration process could be taken a step further with the help of advanced packaging techniques. A module encompassing the SoC onboard computer and a wireless interface can be developed, which could become the core of a unified customizable wireless building block for a picosatellite. A selection of subsystem-specific components—for example, memory, miniaturized MEMS sensors and actuators, converters, RF transceivers, and others—can then be integrated with the core module to realize a complete miniaturized satellite subsystem in the form of an SiP. The radio-enabled SoC onboard computers residing in the individual picosatellite subsystems will form a multinode, onboard, wireless-computer network to support communication, data handling, data gathering, and housekeeping functions as required by the satellite mission.

The proposed approach marries several technologies—SoC, wireless communications, MEMS, SiP—and is therefore very expensive and complex. It will be appropriate to first carry out a feasibility study based on lower-level integration technologies. For example, a modified PCBSat could serve as a customizable SoC-based wireless module for the CubeSat picosatellite platform to prototype the idea.

19.6 Conclusions

Space-based wireless sensor networks can provide a cost-effective approach to the implementation of new remote sensing and science missions that will benefit from miniaturized spacecraft nodes. The Surrey Space Centre evaluated five possible subkilogram satellite technologies for mission suitability and cost effectiveness: CubeSat, microengineered aerospace systems, SpaceChip, PCBSat, and MCMSat. PCBSat proved to be the most cost-effective in terms of average sunlit generated power, while CubeSat offered a clear advantage in available payload volume. However, the most important factor to consider is that the combined mission architecture and system design meet the scientific requirements within the available budget.

An SoC platform that applies reconfigurable technologies to advance onboard data handling capabilities at the subsystem level can easily be adjusted to generate onboard controllers for different onboard subsystems that can be reused in new mission designs. An SoC design targeted at Xilinx QPro Virtex FPGAs and capable of run-time reconfiguration could enable self-repair and remote upgrade of functionality after launch.

The design of space-based sensor networks is a complex and multifaceted area. Initial progress has been made on several enabling technologies, including intersatellite links, distributed computing, remote reconfiguration, and multi-objective optimization for signal routing. Thus, it becomes feasible to build a modular subkilogram satellite with wireless internal interfaces as an extension of the SoC concept.

19.7 Acknowledgments

Funding from ESA, the United Kingdom Engineering and Physical Sciences Research Council (EPSRC), and the United States Air Force Research Laboratory (AFRL) European Office of Aerospace Research and Development (EOARD) in support of the research presented in this chapter is gratefully acknowledged.

19.8 References

[1] C. E. Underwood, G. Richardson, and J. Savignol, "SNAP-1: A Low Cost Modular COTS-Based Nano-Satellite--Design, Construction, Launch and Early Operations Phase," *Proceedings of 15th Annual AIAA/USU Conference on Small Satellites* (Logan, UT, 2001), Paper SSC01-V-1a.

[2] G. B. Shaw, D. W. Miller, and D. E. Hastings, "Generalized Characteristics of Communication, Sensing, and Navigation Satellite Systems," *Journal of Spacecraft and Rockets*, **37** (6), pp. 801–811 (2000).

[3] K. Romer and F. Mattern, "The Design Space of Wireless Sensor Networks," *IEEE Wireless Communications*, **11** (6), pp. 54–61 (2004).

[4] D. Barnhart, T. Vladimirova, and M. Sweeting, "Very Small Satellite Design for Distributed Space Missions," *Journal of Spacecraft and Rockets*, **44** (6), pp. 1294–1306 (2007).

[5] N. Haridas, E. Yang, A. T. Erdogan, T. Arslan, N. Barton, A. J. Walton, J. S. Thompson, A. Stoica, T. Vladimirova, X. Wu, K. D. McDonald-Maier, W. G. J. Howells. "ESPACENET: A Joint Project for Evolvable and Reconfigurable Sensor Networks with Application to Aerospace–Based Monitoring and Diagnostics," *Proceedings of 6th International Conference on Recent Advances in Soft Computing, RASC2006,* edited by K. Sirlantzis (Canterbury, UK, 10–12 July 2006), pp. 410–415.

[6] J. Carpenter, J. Leitner, D. Folta, D., and R. Burns, "Benchmark Problems for Spacecraft Formation Flying Missions," *Proceedings of AIAA Guidance, Navigation, and Control Conference* (Austin, TX, 2003), Paper AIAA-2003-5364.

[7] V. Sundaram, R. Tummala, B. Wiedenman, F. Liu, P. M. Raj, I. R. Abothu, S. Bhattacharya, M. Varadarajan, E. Bongio and W. Sherwood, "Recent Advances in Low CTE and High Density System-on-a-Package (SOP) Substrate with Thin Film Component Integration," *Proceeding of 2006 Electronic Components and Technology Conference* (2006), pp. 1375–1380.

[8] D. Rapp, "Silicon Wafer Properties," (Jet Propulsion Laboratory) http://www.gps.caltech.edu/genesis/Thermal-Coll.html (June 2008).

[9] C. I. Underwood, A. da Silva Curiel, and M. N. Sweeting, "In-Orbit Monitoring of 'Space Weather' and its Effects on Commercial-Off-The-Shelf (COTS) Electronics—A Decade of Research Using Micro-Satellites," *Proceedings of 53rd International Astronautical Congress* (Houston, TX, 2002), IAC-02-IAA.6.3.04.

[10] H. Heidt, J. Puig-Suari, A. S. Moore, S. Nakasuka, and R. J. Twiggs, "CubeSat: A New Generation of Picosatellite for Education and Industry Low-Cost Space Experimentation," *Proceedings of 15th Annual AIAA/USU Conference on Small Satellites* (Logan, UT, 2001), SSC00-V-5.

[11] I. Nason, J. Puig-Suari, and R. Twiggs, "Development of a Family of Picosatellite Deployers Based on the CubeSat Standard," *Proceedings of 2002 IEEE Aerospace Conference* (Big Sky, MT, 2000), Vol. 1, pp. 1-457–1-464.

[12] D. J. Barnhart, T. Vladimirova, M. N. Sweeting, R. L. Balthazor, L. C. Enloe, L. H. Krause, T. J. Lawrence, M. G. Mcharg, J. C. Lyke, J. J. White, and A. M. Baker, "Enabling Space Sensor Networks with PCBSat," *Proceedings of 21st Annual Conference on Small Satellites 2007* (Utah State University, Logan, UT, 13–16 Aug. 2007), SSC07-IV-4.

[13] H. Helvajian and E. Y. Robinson, "Micro- and Nanotechnology for Space Systems," The Aerospace Corporation report, ATR-93(8349)-1 (31 March 1993).

[14] S. W. Janson, H. Helvajian, and E. Y. Robinson, "The Concept of 'Nanosatellite' for Low-Cost Space Systems," *Proceedings of 44th Congress of the International Astronautics Federation* (Graz, Austria, Oct. 1993), IAF-93-U.5.575.

[15] H. Helvajian, *Microengineering Aerospace Systems*, (The Aerospace Press and AIAA, El Segundo, CA, 1999).

[16] S. W. Janson, "Mass-Producible Silicon Spacecraft for 21st Century Missions," *Proceedings of AIAA Space 1999*, (Albuquerque, NM, 1999), pp. 1-10, AIAA-1999-4458.

[17] H. Helvajian, and S. W. Janson, "The Fabrication of a 100 g Co-Orbiting Satellite Assistant (COSA) Using Glass Ceramic Materials and 3-D Laser Processing Techniques," *Proceedings of 9th International Micromachine/Nanotech Symposium* (Tokyo, Japan, 2003), pp. 33–41.

[18] W. E. Morrow and T. F. Rogers, "The West Ford Experiment," *Proceedings of the IEEE*, (1964), **52**(5), pp. 461–468.

[19] S. Janson, A. Huang, W. Hansen, L. Steffeney, and H. Helvajian, "Development of an Inspector Satellite Using Photostructurable Glass/Ceramic Materials," *Proceedings of AIAA Space 2005* (Long Beach, CA, 2005), AIAA-2005-6802.

[20] Private communication with Dr. Henry Helvajian, The Aerospace Corporation.

[21] B. Jackson and K. Epstein, "A Reconfigurable Multifunctional Architecture Approach for Next-Generation Nanosatellite Design," *Proceedings of 2000 IEEE Aerospace Conference* (Big Sky, MT, 2000), Vol. 7, pp. 185–193.

[22] K. Miller, C. Mogensen, and T. Gillespie, "Innovative, Low Cost Microspacecraft," *Proceedings of AIAA Space 2000* (Long Beach, CA, 2000), AIAA-2000-5196.

[23] T. Mosher and B. Stucker, "Responsive Space Requires Responsive Manufacturing-Part II," *Proceedings of AIAA Space 2004* (San Diego, CA, 2004), AIAA-2004-6117.

[24] F. Bruhn and L. Stenmark, "NanoSpace-1: Spacecraft Architecture and Design after Concluding Phase B," *Proceedings of 5th Round Table on Micro/Nano Technologies for Space* (Nordwijk, The Netherlands, 2005), pp. 703–719.

[25] J. Keller, "Startup to Develop Satellite-On-A-Chip," *Military & Aerospace Electronics*, **5**(2), p. 1, (Feb. 1994).

[26] D. J. Barnhart, T. Vladimirova, and M. N. Sweeting, "Satellite-on-a-Chip: A Feasibility Study," *Proceedings of 5th Round Table on Micro/Nano Technologies for Space Workshop* (Nordwijk, 2005), pp. 728–735.

[27] D. J. Barnhart, T. Vladimirova, and M. N. Sweeting, "Satellite-on-a-Chip Development for Future Distributed Space Missions," *Proceedings of CANEUS Micro-Nano Technologies for Aerospace Applications Conference* (Toulouse, France, 2006), CANEUS2006–11045.

[28] D. J. Barnhart, T. Vladimirova, and M. N. Sweeting, "System-on-a-Chip Design of Self-Powered Wireless Sensor Nodes for Hostile Environments," *Proceedings of IEEE Aerospace Conference* (Bozeman, MT, 2007), paper ref. 7.05.01.

[29] T. Vladimirova. "Next Generation Satellites: Picosatellites & Satellites on a Chip," *OSCAR News*, No. 145, 28–37 (Amsat UK, Oct. 2000).

[30] D. J. Barnhart, T. Vladimirova, A. M. Baker, and M. N. Sweeting, "A Low-Cost Femtosatellite to Enable Distributed Space Missions," *Proceedings of 57th International Astronautical Congress* (Valencia, Spain, 2006), IAC-06-B5.6.06.

[31] J. R. Wertz and W. J. Larson, *Space Mission Analysis and Design, 3rd ed.*, (Microcosm Press, El Segundo, CA, 1999).

[32] C. P. Brothers and D. Alexander, "Radiation Hardening Techniques for Commercially Produced Microelectronics for Space Guidance and Control Applications," *Proceedings of 20th Annual American Astronautical Society Guidance and Control Conference* (Breckenridge, CO, 1997), pp. 169-180.

[33] S. Habinc, "Designing Space Applications Using Synthesizable Cores." *Proceedings of 2nd International Conference on Military and Aerospace Applications of Programmable Devices and Technologies, MAPLD 1999* (Laurel, MD, Sept. 1999), D-0.

[34] L. Alkalai, "A Roadmap for Space Microelectronics Technology into the New Millennium," *Proceedings of 35th Space Congress* (Cocoa Beach, FL, Apr. 1998).

[35] L. Alkalai, "Advanced Microelectronics Technologies For Future Small Satellite Systems," *Proceeding of 2nd IAA Symposium on Small Satellites for Earth Observation* (Berlin, Germany, 12–16 April 1999), IAA-B2-0802.

[36] L. Alkalai and E. Kolawa, "Avionics Systems On A Chip for Space Exploration," *Proceedings of International Forum of Space Technology & Applications, STAIF-99* (Albuquerque, NM, Jan 31–Feb 4 1999), p. 721.

[37] J. Pum, "SOC: The Convergence Point for Solutions of the 21st Century," *Keynote, The IEEE Custom Integrated Circuits Conference* (Orlando, FL, 21–24 May 2000), http://www.ieee-cicc.org/2000/conference/program/tp1.html (June 2008).

[38] A. M. Rincon, G. Cherichetti, J. A. Monzel, D. R. Stauffer, and M. T. Trick, "Core Design and System-on-a-Chip Integration," *Design & Test of Computers, IEEE,* **14** (4), pp. 26–35 (Oct.–Dec. 1997).

[39] Xilinx Inc., "Virtex-4 Family Overview," Data sheet, DS112, v 1.2, (8 Dec. 2004), http://direct.xilinx.com/bvdocs/publications/ds112.pdf (June 2008).

[40] J. Gaisler, "A Portable Fault-tolerant Microprocessor Based on the SPARC V8 Architecture," *Proceedings of Data Systems In Aerospace Conference, DASIA 1999* (Portugal, May 1999), pp. 173–178.

[41] H. Tiggeler, T. Vladimirova, D. Zheng, and J. Gaisler, "Experiences Designing a System-on-a-chip for Small Satellite Data Processing and Control," *Proceedings of 3rd International Conference on Military and Aerospace Applications of Programmable Devices and Technologies, MAPLD 2000* (Laurel, MD, Sept. 2000), P-20.

[42] M. Lefebvre and J.-F. Coldefy, "Development and Tests Results of a System-on-a-Chip for a Data Handling System," *Proceeding of Data Systems in Aerospace Conference, DASIA 2003* (Prague, 2003).

[43] R. Weigand, M. Souyri, and M. Lefebvre, "SCOC—A Spacecraft Controller On-a-Chip," *Proceedings of ESTEC Workshop on Spacecraft Data Systems 2003* (Noordwijk, 5–7 May 2003).

[44] B. Fiethe, H. Michalik, C. Dierker, B. Osterloh, and G. Zhou, "System-On-Chip Data Processing Units for Miniaturized Space Imaging Instruments," *Proceedings of Data Systems in Aerospace Conference, DASIA 2006* (Berlin, 2006).

[45] J. Gaisler and E. Catovic, "Multi-Core Processor Based on LEON3-FT IP Core (LEON3-FT-MP)," *Proceeding of Data Systems in Aerospace Conference, DASIA 2006* (Berlin, 2006).

[46] I. Kleyner, R. Katz, and H. Tiggeler, "System-on-Chip Data Processing and Data Handling Spaceflight Electronics," *Military and Aerospace Applications of Programmable Devices and Technologies Conference, MAPLD 1999* (Laurel, MD, 28–30 Sept. 1999), D-2.

[47] D. Zheng, "Reconfigurable System on a Chip Based Platform for On Board Computing," Ph.D. thesis, Surrey Space Centre, University of Surrey, Guildford, June 2005.

[48] ARM Ltd., "AMBA Specification," 1999, Rev 2.0.

[49] SolidPC Datasheet, www.saros.co.uk/amba (June 2008).

[50] M. Meier, T. Vladimirova, T. Plant, A. da Silva Curiel, "DMA Controller for a Credit-Card Size Satellite Onboard Computer," *Proceedings of 7th Military and Aerospace Applications of Programmable Devices and Technologies International Conference, MAPLD 2004* (Washington, DC, Sept. 2004)), P-208.

[51] T. Vladimirova, D. Eamey, S. Keller, and M. N. Sweeting, "Floating-Point Mathematical Co-Processor for a Single Chip On-Board Computer," *Proceedings of 6th Military and Aerospace Applications of Programmable Logic Devices and Technologies International Conference, MAPLD 2003* (Washington, DC, 9–11 Sept. 2003) D5.

[52] R. Banu and T. Vladimirova, "Fault-Tolerant Encryption for Space Applications," to appear in *IEEE Transactions on Aerospace and Electronic Systems,* **44** (4) (2008).

[53] T. Vladimirova and M. N. Sweeting, "System-on-a-Chip Development for Small Satellite On-Board Data Handling," *AIAA Journal of Aerospace Computing, Information, and Communication,* **1**(1), pp. 36–43 (Jan. 2004).

[54] I. Rutter, T. Vladimirova, and H. Tiggeler, "A CCSDS Software System for a Single-Chip On-Board Computer of a Small Satellite," *Proceedings of 15th AIAA/Utah Sate University Conference on Small Satellites* (Logan, UT, Aug. 2001), SSC01-VI-4.

[55] D. Zheng, T. Vladimirova, M. N. Sweeting, "A CCSDS-Based Communication System for a Single Chip On-Board Computer," *Proceedings of 5th Military and Aerospace Applications of Programmable Devices and Technologies International Conference, MAPLD 2002* (Laurel, MD, Sept. 2002), D5.

[56] N.W. Bergmann and P.R. Sutton, "A High-Performance Computing Module for a Low Earth Orbit Satellite using Reconfigurable Logic," *Proceedings of 1st Military and Aerospace Applications of Programmable Devices and Technologies Conference, MAPLD 1998* (Greenbelt, MD, 15–16 Sept. 1998), PCD1.

[57] T. W. Fry and S. Hauck, "Hyperspectral Image Compression on Reconfigurable Platforms," *Proceedings of 10th IEEE Symposium on Field-Programmable Custom Computing Machines, FCCM '02* (Napa, CA, 22–24 Sept. 2002), pp. 251–260.

[58] Xilinx, Inc., "Xilinx: QPRO Virtex-II 1.5V Radiation Hardened QML Platform FPGAs," Data sheet DS124 (v1.1), (San Jose, CA, Jan., 2004), http://direct.xilinx.com/bvdocs/publications/ds124.pdf (June 2008).

[59] M. Stettler, M. Caffrey, P. Graham, and J. Krone, "Radiation Effects and Mitigation Strategies for Modern FPGAs," *Proceedings of 10th Workshop on Electronics for LHC Experiments and Future Experiments* (Boston, 2004).

[60] Xilinx, Inc., "The First Triple Module Redundancy Development Tool for reconfigurable FPGAs," Data sheet, www.xilinx.com/esp/mil_aero/collateral/tmrtool_sellsheet_wr.pdf (June 2008).

[61] B. Blodget, P. James-Roxby, E. Keller, S. McMillan, and P. Sundararajan, "A Self-reconfiguration Platform," *Proceeding of 13th International Conference on Field Programmable Logic and Applications, FPL 2003* (Lisbon, Portugal, 2003), pp. 565-574.

[62] J. A. Williams and N. W. Bergmann, "Embedded Linux as a Platform for Dynamically Self-Reconfiguring Systems-On-Chip," *Proceedings of International Conference on Engineering of Reconfigurable Systems and Algorithms, ERSA 2004* (Las Vegas, NV, 2004).

[63] Xilinx, Inc., "Two Flows for Partial Reconfiguration: Module Based or Difference Based," Application note, www.xilinx.com/bvdocs/appnotes/xapp290.pdf (June 2008).

[64] Xilinx, Inc., "Plan Ahead: The Fastest Route to Better Design," Data sheet, www.xilinx.com/publications/prod_mktg/pn0010825.pdf (June 2008).

[65] T. Vladimirova, X. Wu, A.-H. Jallad, and C. P. Bridges, "Distributed Computing in Reconfigurable Picosatellite Networks," *Proceedings of 2nd NASA/ESA Conference on Adaptive Hardware and Systems, AHS-07* (Edinburgh, 5–8 Aug. 2007).

[66.] K. Sidibeh and T. Vladimirova, "IEEE 802.11 Optimisation Techniques for Inter-Satellite Links in LEO Networks," *Proceedings of 8th International Conference on Advanced Communication Technology, ICACT '06*, "Towards the Era of Ubiquitous Networks and Societies," (Phoenix Park, Korea, 20–22 Feb. 2006), Vol. II, pp. 1177–1182.

[67] K. Sidibeh, T. Vladimirova, and A. Baker, "Adapting the IEEE 802.11 Protocol for Inter-Satellite Links in LEO Constellations," *Proceedings of ESA Wireless for Space Applications Workshop*, ESTEC/ESA (Noordwijk, The Netherlands, 10–13 July 2006).

[68] K. Sidibeh, "Adaptation of the IEEE 802.11 Protocol for Inter-Satellite Links in LEO Satellite Networks," Ph.D. thesis, Surrey Space Centre, University of Surrey, Guildford, UK, 2008.

[69] A.-H. Jallad and T. Vladimirova, "Distributed Computing for Formation Flying Missions," *Proceedings of ACS/IEEE International Conference on Computer Systems and Applications, AICCSA '06* (Dubai/Shariah, UAE, 8–11 March 2006), pp. 616–620.

[70] C. P. Bridges and T. Vladimirova, "Autonomous Software Agents in Wireless Embedded Systems," *Proceedings of 3rd UK Embedded Systems Forum* (Durham University, 2–3 April 2007), p. 167, IET.

[71] JADE, Whitepaper, http://jade.tilab.com/papers/2003/WhitePaperJADEEXP.pdf (June 2008).

[72] M. Schoeberl, "A Java Processor Architecture for Embedded Real-Time Systems," *Journal of Systems Architecture* (21 June 2007).

[73] T. Vladimirova, X. Wu, C. P. Bridges, "Development of a Satellite Sensor Network for Future Space Missions," *Proceedings of 2008 IEEE Aerospace Conference* (Big Sky, MT, 1–8 March 2008), Paper ID: 1263.

[74] T. Vladimirova, P. Davies and M. N. Sweeting, "Reconfigurable Computing for Micro-Satellites," *Proceedings of 10th International Space System Engineering Conference, DAta Systems In Aerospace DASIA '06* (Berlin, 22–25 May 2006), ref. 1B-3.

[75] X. Wu, T. Vladimirova, and K. Sidibeh, "Signal Routing in a Satellite Sensor Network Using Optimisation Algorithms," *Proceedings of 2008 IEEE Aerospace Conference* (Big Sky, MT, 1–8 March 2008), Paper ID: 1264.

[76] K. C. Tan, *Multiobjective Evolutionary Algorithms and Applications* (Springer, 2005).

[77] T. Vladimirova, C. P. Bridges, G. Prassinos, X. Wu, K. Sidibeh, D. J. Barnhart, A.-H. Jallad, J. R. Paul, V. Lappas, A. M. Baker, K. Maynard, and R. Magness, "Characterising Wireless Sensor Motes for Space Applications," *Proceedings of 2nd NASA/ESA Conference on Adaptive Hardware and Systems, AHS '07* (Edinburgh, 5–8 Aug. 2007).

Chapter 20

Picosatellites at The Aerospace Corporation

David Hinkley*

20.1 Introduction

Engineers and scientists at The Aerospace Corporation became interested in the space systems applications of microtechnologies in the early 1990s. Ernest Robinson, Henry Helvajian, and Siegfried Janson saw the potential of microelectromechanical systems (MEMS) for space applications and the possibility of functional, subkilogram active satellites.[1] They advocated the insertion of MEMS and microtechnology into space systems and urged development of a corporate research initiative on MEMS and nanosatellites. The Aerospace Corporation eventually funded a "Microtechnology for Space Systems" research initiative from 1995 to 2000 at a level of $1 million annually. In 1999, as part of the initiative and with additional funding from the NASA Johnson Space Center, 30 different micro- and nanotechnology devices were flown inside a middeck locker on the space shuttle Columbia, Space Transportation System (STS)-93 mission.[2] This was the beginning of our investigations into using MEMS and microtechnology in space.

The Aerospace Corporation began building complete satellites in 1998. Our first picosatellite came about when we were offered a launch tube on the Stanford University picosatellite-deploying spacecraft, OPAL. This opportunity fit in well with the aspiration of a few researchers at The Aerospace Corporation, who for many years had been envisioning the miniaturization of spacecraft. Since then, The Aerospace Corporation picosatellite program has designed, constructed, and flown eleven miniature spacecraft that were delivered to seven different missions (four of them were tethered pairs). The small size, low cost, and high launch frequency of miniature satellites has caused a paradigm shift in the spacecraft business. Now, researchers can take risks and fly more often because the cost of participation is small. With each launch, the quality and capability improves. Since that first mission, expectations have changed from disbelief that they will work at all to mission planning to accomplish a task.

The picosatellites, nanosatellites, and launch tubes (launchers) developed at The Aerospace Corporation are explained in this chapter. Our satellites from 1998 to 2001 were of the so-called OPAL class—namely, 1 × 3 × 4 inches in dimension that fit the Stanford OPAL satellite picosatellite launch tube. Beginning in

*The Aerospace Corporation
El Segundo, California

2001 and continuing until 2006, we developed the MEPSI-class picosatellites, which measured 4 × 4 × 5 inches in dimension and fit a launch tube designed by The Aerospace Corporation exclusively for space shuttle launches—the Space Shuttle Picosatellite Launcher (SSPL). The program has further diversified and is now building CubeSats and nanosatellites. In 2005, The Aerospace Corporation delivered its first CubeSat class satellite, AeroCube-1, followed in 2006 by AeroCube-2, our first solar-powered satellite, and AeroCube-3 in 2008, along with our first nanosatellite (which will not be discussed in this chapter).

20.2 OPAL and MightySat II.1

The Aerospace Corporation's first picosatellite was developed in 1998 for a ride on the Stanford Orbiting Picosatellite Automated Launcher (OPAL) satellite. The OPAL project began in 1995 as a collaboration between Stanford and NASA/JPL (Jet Propulsion Laboratory) to create tiny spacecraft that would be dispersed in space and relay magnetic field measurements back to the OPAL mothership.[3,4] However, NASA/JPL withdrew support, and OPAL remained at Stanford without a ride into space. Robinson created an opportunity to manifest OPAL on the maiden (highly risky) flight of the Minotaur launch vehicle. To accomplish this, he coordinated the support of the U.S. Air Force Space Test Program, the Air Force Research Laboratory (AFRL), the Air Force Office of Scientific Research (AFOSR), and the Defense Advanced Research Projects Agency (DARPA). For Robinson's efforts, Stanford gave The Aerospace Corporation a picosatellite launch tube to populate. Our corporate research initiative and DARPA seed money funded the development.

Our first picosatellites demonstrated MEMS in space and marked a practical step toward realizing constellations.[5] Engineers were envisioning, someday, hundreds or more of these miniature satellites distributed across the sky.[6] For this mission, the constellation effect was created by tethering two identical picosatellites together by 100 ft of Dyneema thread (Fig. 20.1). Each satellite had its own radio and could talk either to the other unit or directly to a ground station on Earth. The MEMS device tested was a DARPA-funded MEMS radio-frequency (RF) switch from Rockwell Science Center (RSC) in Thousand Oaks, California. A small, quarter-inch-square package containing four MEMS RF switches was installed in each picosatellite. The experiment was to open and close the switches repeatedly and measure the contact resistance.

The electronics in each satellite consisted of three circuit boards stacked together using a pair of backplane connectors, one each for MEMS testing, battery power conditioning, and a combined RF transmitter and flight computer. Three primary lithium batteries (model LTC-311) from EaglePicher provided 10 Whr of energy at 3.65 V and had the advantage of a working temperature range from −40°C to +70°C. RSC provided the combined flight computer and RF transmitter board, which was designed around a Rockwell RSST7504 cordless-phone baseband processor. The radio was spread spectrum and transmitted in the 900 MHz

Fig. 20.1. Artist's rendition of the OPAL picosatellites in space.

Fig. 20.2. OPAL (and MightySat II.1) picosatellite (1 of 2) with covers removed.

band at 80 kbps. It had recently been completed for a prior DARPA contract to develop low-power wireless sensor nodes.[7] The Aerospace Corporation designed the power conversion board and the MEMS test board and packaged them, along with the batteries, into an "electronics module." The satellite body was an aluminum structure with two cover panels. The electronics module was installed on Delrin leaf springs that mechanically and thermally isolated it from the satellite body. Also included inside the 1 × 3 × 4 inch satellite were a tether spool, antenna splitter, and power-on plungers. The satellite employed passive thermal control using a combination of polished aluminum and black anodize. Each satellite weighed 250 g. A photograph of the satellite, with covers removed, is shown in Fig. 20.2

Our two largest concerns for mission success were the ability of the United States Space Surveillance Network (SSN) to track the satellites and our communications link budget. The radio in each satellite could only generate 64 mW of RF power, and the circular orbit was 773 km altitude. To work within these constraints, a huge antenna at SRI International in Menlo Park, California, was employed. The 150 ft diameter antenna (Fig. 20.3) had enough gain (49 dB at 900 MHz) to close the link with an estimated +10 dB margin but a very narrow beamwidth (0.65 deg full beamwidth at −3 dB). To point the antenna accurately, the SSN had to detect the satellites and publish two-line elements. To improve the chances of tracking the satellites, the 100 ft long, 0.010 inch diameter Dyneema tether had 0.001 inch diameter gold wire dipoles, tuned to the SSN frequencies,

Fig. 20.3. SRI International 150 ft diameter parabolic antenna.

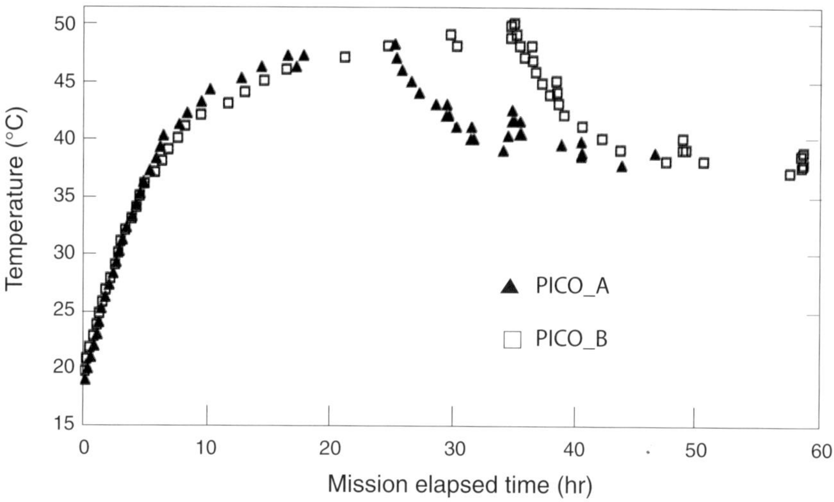

Fig. 20.4. OPAL picosatellite temperature data.

attached. These greatly enhanced the radar cross section of the tethered pair. Furthermore, the SSN station at Mill Stone Hill was engaged to extend an extra effort to find the satellites. At the SRI International antenna, we used a special spiral scan strategy to home-in on the picosatellite beacons.

The Aerospace Corporation developed, packaged, and tested the OPAL picosatellites in less than six months, driven by the integration date to the OPAL host spacecraft. They were delivered in May 1999 and launched on 26 January 2000 into a 100.2 deg inclination, 773 km circular orbit. The OPAL spacecraft ejected all of its picosatellites on 6 February 2000. It took our tracking team two pass attempts (12 hours between passes) to contact the first picosat and three pass attempts (36 hours) to contact the second one for the first time. Data from the MEMS switch actuation and resistance measurements and satellite temperature were downlinked, and new commands uplinked. The temperature profile of the satellites is shown in Fig. 20.4. The rapid rise in heating at the start of the mission was caused by the constant 2 Hz beacon, which drew an average of 0.19 W and was the main power consumer on the satellite. At this rate, the batteries would have lasted less than two days. However, once they were contacted, the satellites were put to sleep and told to awake and begin beaconing only prior to the next good pass. This sleeping mode extended the mission to about 2.5 days.

This mission demonstrated the unique engineering problems with miniature satellites. Our 1 × 3 × 4 inch picosatellites with primary batteries limited how much beaconing could be done to aid in acquisition by our ground station. Also, the low power available for the beacon meant the ground station required significant gain, narrowing its beam and exacerbating the problem of finding them. The SSN was able to track the satellites, but it was our special access to the Mill Stone

Hill radar station that resulted in timely delivery of an ephemeris. In the end, The Aerospace Corporation OPAL picosatellites were successful, and may be the smallest active satellites ever flown.

The Aerospace Corporation delivered its second pair of picosatellites and a launch tube to AFRL in December 1999 to take advantage of an opening on the MightySat II.1 spacecraft.[8,9] The payload was installed in the center of the launch vehicle adapter ring (Fig. 20.5). The plan was to release the picosatellites after MightySat II.1 had completed its primary one-year mission. This would demonstrate storing miniature satellites onboard a host vehicle for use at some point during the host's lifetime. While future picosatellites would have functionality beneficial to the host, such as anomaly inspection, these particular miniature satellites would simply be ejected and depart from the host at a speed determined by the ejection spring inside the launch tube. The tethered 250 g picosatellies would serve as another constellation exercise and would test a newer set of MEMS RF switches.

The MightySat II.1 satellite was launched on 19 July 2000 into a 97.8 deg inclination, 575 km circular orbit. Its picosatellites were released more than one year later on 6 September 2001. The SRI International ground station contacted them on 7 September and downlinked temperature and MEMS-switch resistance data. However, the communications link was worse than the prior mission, and few contacts were made before the mission was declared over on 9 September. Two possible causes were identified for our inability to repeat the OPAL picosatellite mission performance. The first was a system testing error, whereby the satellites, with their flight batteries installed, were thermally cycled too aggressively. This would have reduced the battery capacity, and would explain why there was contact only for 24 hours, compared to almost 60 hours for the OPAL mission. Second, the SSN tracking vectors might not have been very accurate, or the SRI International antenna itself may have not have tracked well. Either reason added urgency to solving the high-gain ground antenna problem—namely, that its beamwidth is so narrow that it affects mission success.

The MightySat II.1 delivery was important for several reasons. First, it demonstrated that miniature satellites can be built very quickly. The picosatellites and the launch tube were built, tested, and delivered in six months. Second, it opened the door for miniature satellites to be installed on a high-value host. MightySat II.1 was a good satellite for this purpose; it was intended for research and development, but it was expensive enough that rigorous mission assurance practices were applied to our payload, thus preparing us for the future. Third, the modified Stanford launch tube (Fig. 20.6) that we built would serve as a foundation for the future launch tubes that we designed and built for the space shuttle. Fourth, because this repeat mission was not as successful as the OPAL mission, it provided insight into our success margin—apparently very small. Fifth, the concept of a daughtership resident on a mothership was realized. A new program would begin on this premise.

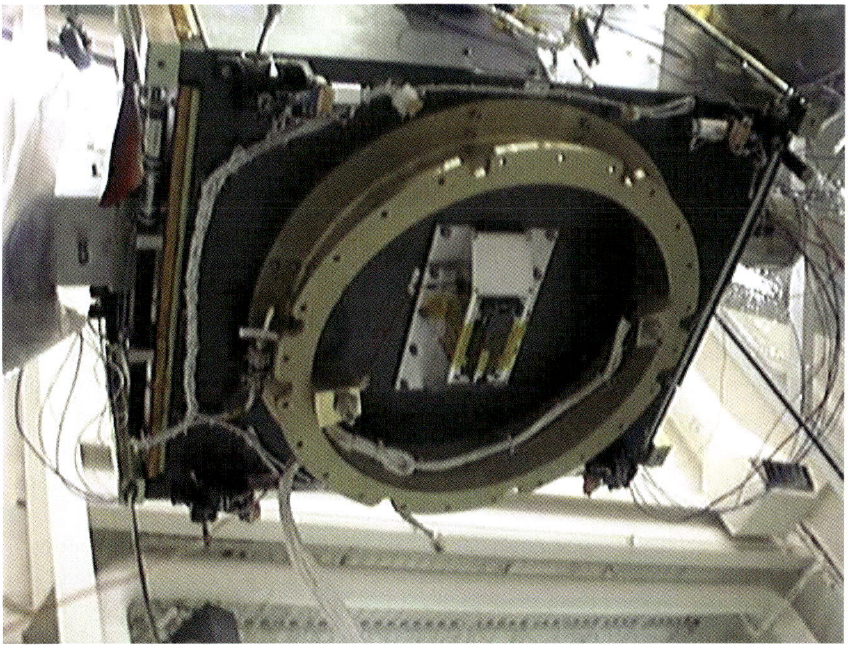

Fig. 20.5. OPAL-class picosatellite launch tube installed in MightySat II.1 spacecraft.

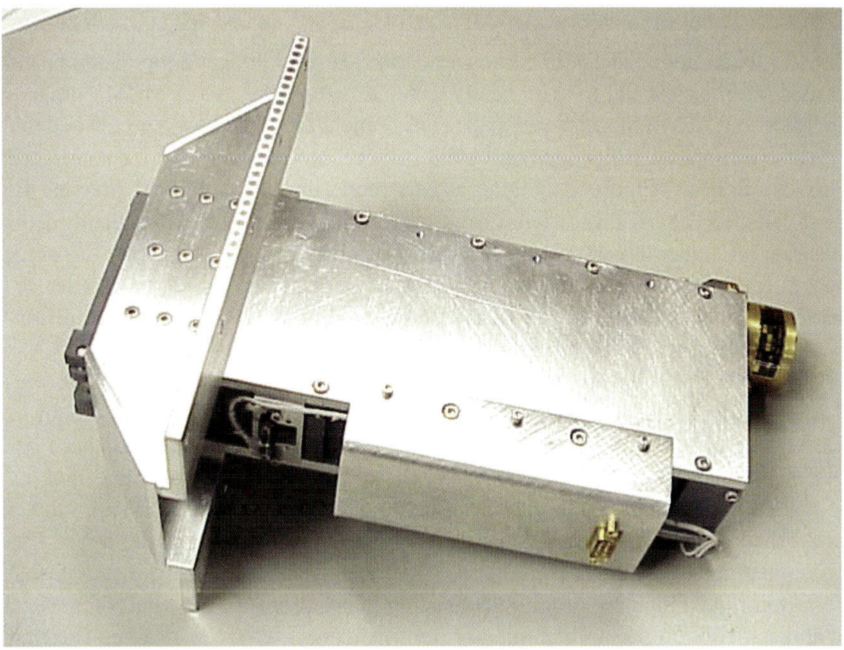

Fig. 20.6. MightySat II.1 picosatellite launch tube assembly.

20.3 MEPSI

The OPAL series of miniature satellites answered the question, "Can this be done?" This next series, developed at The Aerospace Corporation, sought to answer, "What useful task can be done?" Visionaries and planners at the First International Conference on Integrated Micro/Nanotechnology for Space Applications in Houston, Texas, in 1995 discussed this second question at a workshop during the conference. Helvajian's idea of an "untethered flying observer," or UFO, was the focus of this workshop. The UFO would be a miniature daughtership satellite resident on a host satellite. Upon command, it would separate and inspect the host while circumnavigating it for a short time and then prepare for deorbit. With that concept of operations in mind, the workshop participants flushed out a target design: the daughtership would be a 1 kg satellite with primary lithium batteries, avionics, cold gas thrusters, a transmitter, and an inspection camera. The 1 kg mass limitation meant that this UFO spacecraft would rely heavily on micro- and nanotechnology and would therefore not be able to use any traditional spacecraft hardware.

The UFO concept did not move forward until 2000. Riding the momentum of the OPAL picosatellite mission, David Williamson of the Air Force Research Laboratory (AFRL), Information Directorate, along with engineers at The Aerospace Corporation, proposed an updated version called the MEMS-Enabled PicoSatellite Inspector (MEPSI). Williamson was an electrical engineer at AFRL and a DARPA MEMS program agent. He was thus uniquely positioned to lead this proposal, bringing together U.S. Air Force needs and leveraging newly-developed DARPA MEMS technology. The MEPSI concept added several goals on top of the visual inspection mission: (1) to specifically test and mature the use of MEMS devices in spacecraft, (2) to monitor the space environment around a host, and (3) to demonstrate artificial intelligence that would be necessary in the future for automated service missions. Figure 20.7 is an artist's rendering of the MEPSI imaging its host vehicle. The MEPSI differed from other miniature inspector spacecraft demonstrated or under development, such as SSTL's SNAP-1,[10–12] in that the inspector would reside on the host spacecraft and would not be carried up by another vehicle, and would communicate directly to a ground station and not to a nearby satellite or to the host itself (see chapter 10).

The MEPSI program was presented in 2000 to the Space Experiments Review Board (SERB), and was approved. The SERB serves the Department of Defense (DoD) and is charged with vetting and ranking all proposed space experiments that have DoD relevance. The benefit of SERB approval and ranking is that coordination and integration services with the spacelift vehicle are funded and executed by the Space Test Program (STP), currently part of the Space Development and Test Wing of the U.S. Air Force. The SERB granted the request for a spiral development program consisting of six flights of the MEPSI concept, with each flight to be an incremental improvement over the prior one. The first flight retroactively counted the OPAL-class picosatellites already integrated on

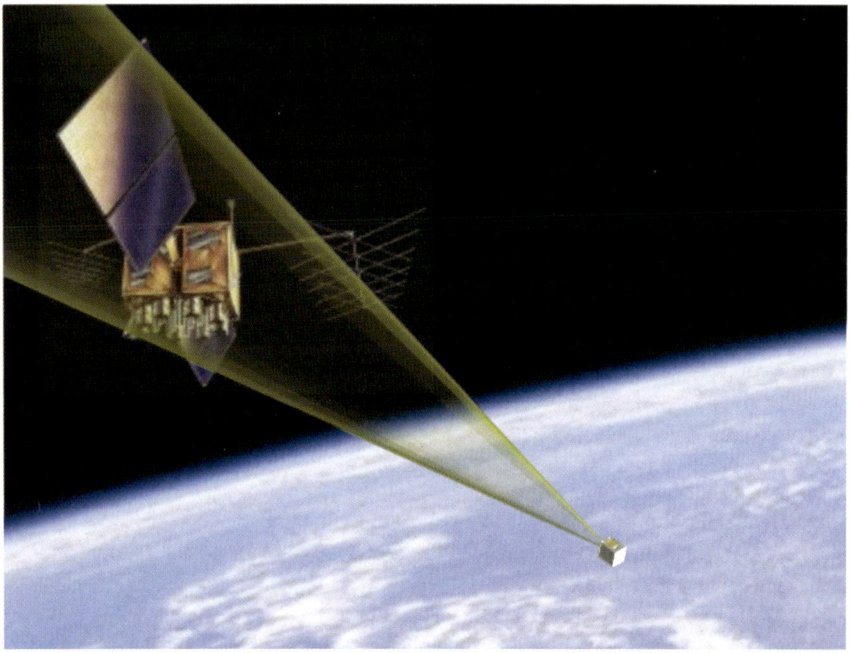

Fig. 20.7. Artist's rendition of MEPSI vehicle imaging its host spacecraft.

MightySat II.1. They would demonstrate long-term storage because they would not be ejected until at least one year after the start of that mission. Flights two through five would use the space shuttle. The sixth and final flight would be a fully functional MEPSI installed on the STPsat-1 spacecraft. The flights were scheduled on six month intervals.

Development costs for the MEPSI series were shared by DARPA and The Aerospace Corporation. The DARPA Microsystems Technology Office (MTO) program managers William C. Tang and Clark C. Nguyen provided generous funding from 2000 through 2003. However, during these years there were delays and frustration because The Aerospace Corporation could not directly accept all of the DARPA funds: only a limited level of funding could come from DARPA, and that amount had to be split with other existing DARPA programs at The Aerospace Corporation. The U.S. Air Force Space and Missile Systems Advanced Planning Directorate took over funding of the program from 2004 to its completion in 2006. That funding was a better fit, and the program developed hardware at a faster pace.

The MEPSI program developed a standard Space Shuttle Picosatellite Launcher (SSPL) in 2001 that was intended to fit all of the MEPSI missions. The benefit of a standard launch tube is that it would ideally only have to be designed once, but it would constrain the growth of the MEPSI satellites as they matured in capability. An educated guess of final MEPSI size determined that a launch

Table 20.1. Capabilities Required to Perform the MEPSI Mission

Requirement	Purpose	MSII.1	STS-113	STS-116
Propulsion	Change location relative to host			X
Communications	Downlink data, uplink commands	X	X	X
Host location detection	Feedback for propulsive maneuvers			
Camera	Sensor			X
Deployment mechanism	Method of attaching to and deploying from host	X	X	X
Attitude control—propulsion	Control attitude via propulsion			X
Attitude control—reaction wheels	Control attitude via reaction wheels			X
Attitude control—inertial meas. unit	Feedback for attitude control		X	X
Bus (C&DH, power management)	Basic functions of autonomous satellite	X	X	X
Ground station and tracking	Earth-bound portion of communications links	X	X	X

tube with a payload volume of 4 × 4 ×10 inches and a maximum payload mass of 3.5 kg would be sufficient. While The Aerospace Corporation was designing the SSPL, we shared design notes with students at the California Polytechnic State University (Cal Poly) at San Luis Obispo, who were concurrently designing the now ubiquitous Poly Picosatellite Orbital Deployer (P-POD) for the CubeSat program. Therefore, both designs have a lot of similarities.

The MEPSI-class satellites had to be very capable spacecraft to carry out the inspection mission. Table 20.1 is a compilation of the predicted necessary functions. Several of the functions—namely, propulsion, closed-loop attitude control, and a radio downlink with sufficient bandwidth to transmit images to a ground station—were hitherto unheard of in a satellite of this size. Furthermore, the MEPSI had to be very reliable because it would co-orbit a high-value satellite, and a collision was unacceptable.

STS-113 in 2002 was the second MEPSI flight, but the first using the space shuttle and the first using the new larger form factor. For that mission, a tethered pair of identical satellites was ejected, and the capabilities featured were the new radio and MEMS attitude determination. STS-116 in 2006 was the third MEPSI flight and the second using the space shuttle. It was released almost exactly four years after the STS-113 mission and featured much of the functionality that a MEPSI mission would require, including propulsion, open-loop attitude control, cameras, and enough bandwidth to downlink images. What was missing was the

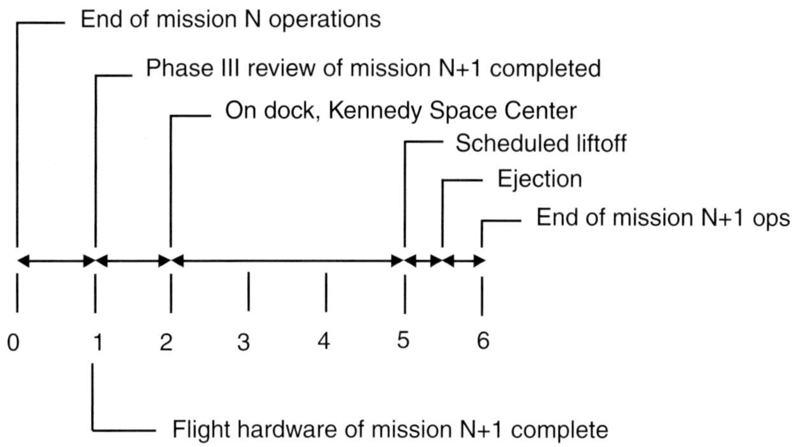

Fig. 20.8. Milestones in a six-month flight delivery schedule under ideal conditions.

closed-loop control and range-finding capability necessary to keep an inspector at the proper distance from its host.

The MEPSI series completed only three of its planned six flights. The primary reason is that the deliveries to the space shuttle program took longer than the six months planned. The shuttle program has four formal reviews named Phase 0, I, II, and III. Sometimes, review phases can be grouped together if the changes are not too drastic between reflights. However, the payload typically has to be delivered three months prior to launch and one month prior to the Phase III review. Figure 20.8 is a timeline graphic of the minimum review milestones that have to be achieved for a six-month delivery. As shown in the figure, when the flight operations from a two-week mission cease, there is only one month available for changes to the next model to be presented at the Phase III review, and only two months to implement them.

Early on in the MEPSI program, the STP office selected STPsat-1 as the host for the final demonstration of MEPSI. STPsat-1 was started with a known finish date, independent of the status of the MEPSI improvement flights on the space shuttle. The MEPSI program maintained both a shuttle development schedule and an STPsat-1 integration schedule. The space shuttle Columbia accident in February 2003 stopped all space shuttle flights until July 2005. The second of the planned four MEPSI development flights using the space shuttle was manifested on STS-116 with a liftoff date in December 2006. Six months later, the flight hardware was due for MEPSI on STPsat-1. This provided no reaction time to fix any flaws or improve the final product based on the STS-116 mission results. In actuality, it probably would have taken at least two more shuttle flights to adequately finish the MEPSI spacecraft. To maintain the integrity of the STPsat-1 launch schedule, a flight SSPL with mass-model picosatellites was installed on the host spacecraft.

Although the MEPSI series was not completed, the program was very important to the miniature satellite program at The Aerospace Corporation. The MEPSI missions provided a set of capability goals that were transferable to other miniature spacecraft. The launch deadlines promoted urgent and efficient hardware development. The 2002 MEPSI launch from the space shuttle, the first of the planned four rides, provided a well publicized image of a launch tube ejecting miniature satellites. While the satellites had functional problems, the iconic image advanced the idea of using miniature satellites for missions and the practical idea of a launch tube as a benign and controlled interface with a launch vehicle or host satellite. The 2006 MEPSI launch from the space shuttle Discovery provided another set of iconic images, this time of the shuttle itself.

20.3.1 MEPSI ON STS-113

The STS-113 mission was the first of an expected four-flight spiral development sequence using the space shuttle as a launch vehicle. DARPA MTO funded The Aerospace Corporation, RSC, and NASA/JPL to create the MEPSI payload for STS-113. Since this was the first shuttle mission, engineers had to develop the infrastructure and NASA procedures necessary to eject picosatellites from the shuttle bay. Furthermore, because this was a human-rated system, NASA safety engineers would scrutinize the launcher design for flaws that might result in a failed ejection that might threaten the health of the shuttle or force an unplanned extravehicular activity. Therefore, launcher design and safety issues consumed the bulk of the engineering time and budget: the satellites were a secondary priority. The kickoff meeting for this mission was March 2001. The hardware, two MEPSI satellites (Fig. 20.9), and a MEPSI launcher (discussed later) were designed, built, tested, and delivered less than two years later in October 2002 and installed on the space shuttle while it was in the vertical position, one month prior to liftoff. Total project cost was approximately $1,200,000 and did not include the cost to the Space Test Program for their integration support.

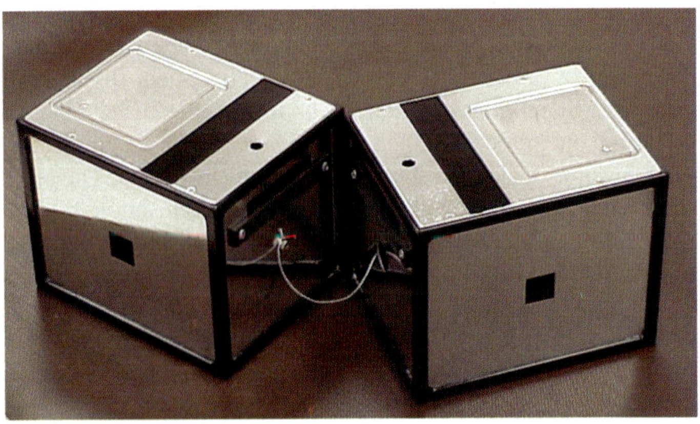

Fig. 20.9. STS-113 MEPSI picosatellites with tail of 50 ft tether joining them.

The concept of operations was to eject two tethered MEPSI picosatellites from the space shuttle. The satellites would power on immediately after ejection and start recording their accelerations and angular rates while the tether was unspooling, and capture the rebound when they reached the end of the tether. Immediately after ejection the picosatellite radios would beacon small data packets at 2 Hz to help the ground station zero in on their position. While in this beaconing mode, the satellites would steadily consume 0.3 W of power. Once they were acquired by the ground station, a contact schedule would be uplinked and the beaconing mode would be shut off to conserve power. The satellites would resume beaconing just prior to the next scheduled pass over the ground station. There were no solar cells on these picosatellites, so acquiring them quickly and turning off the beacon was very important.

Two identical MEPSI spacecraft were used for this mission—both to enable redundancy in case one failed, and to practice initial flight operations for multiple spacecraft in future local cluster configurations. They also served as end masses for a 50 ft long tether that joined them. Identical to that used by the OPAL and MightySat II.1 missions, the tether was laden with gold dipole wires tuned to the SSN radar frequencies to enhance the radar signature and improve the odds of being tracked.

The complement of circuit boards in each of the two identical STS-113 MEPSI picosatellites is listed in Table 20.2. This mission featured a picosatellite inertial rate measurement unit (IRMU) that would measure satellite rotation rates and accelerations. It also had a flight computer, a dc-to-dc converter system and a radio. Each picosatellite was powered by nine 3.65 V, 1 Ahr primary lithium thionyl chloride batteries arranged in series as a single pack. Inside one of the satellites was the tether spool containing the 50 ft of thread, and the free end was tied off to a point on the other satellite. Redundant spring plungers compressed between the two picosatellites kept microswitches open while the picosatellites

Table 20.2. STS-113 MEPSI Picosat Circuit Board List

Subsystem	Board Name	Provider	Description
Computer	Flight computer	RSC	SA1100 microprocessor
Computer	Memory	RSC	1Mx8 flash memory
Radio	Radio (digital)	RSC	Digital half of 915 MHz radio
Radio	Radio (amplifier)	RSC	Analog side of radio
Radio	RF splitter	The Aerospace Corp.	RF splitter
Power	Dc-dc	JPL-The Aerospace Corp.	Primary power converter
IRMU	IRMU pass thru	JPL-The Aerospace Corp.	Spacer board
IRMU	IRMU yaw	JPL-The Aerospace Corp.	IRMU main board
IRMU	IRMU pitch/roll	JPL-The Aerospace Corp.	IRMU orthogonal sensors

were inside the launch tube. Once free of the ejection tube, the spring plungers were free to extend, providing the energy to unspool the tether.

The flight computer and radio developed by RSC were new for this mission. Both of these subystems were designed with the power and resources necessary for future missions. The flight computer was based on a StrongARM 1100 processor running a real time operating system with 1 Mb of dual-ported flash memory. The flight processor code was programmed by engineers at The Aerospace Corporation. The 915 MHz radio was similar to the OPAL picosatellite radios, but had an added power amplifier to increase output power to 0.5 W. Also, per contractual requirement with DARPA MTO, the radio was designed to use MEMS RF switches for the Rx/Tx function. However, at the time of delivery, the MEMS RF switches were not sufficiently reliable, and semiconductor equivalents were used in their place. The radio board used a Conexant RSST7504 chipset with an integral 6502 processor to format the data. It was programmed by engineers at RSC and The Aerospace Corporation. A dual-ported memory simplified data transfer within the satellite: the StrongARM would write mission data to the memory through one port, and the radio would read it out the other port.

The IRMU was designed by engineers at JPL. It consisted of an orthogonal triaxial arrangement of MEMS accelerometers (Analog Devices), MEMS rate sensors (Analog Devices), and a redundant set of micromachined rate sensors (British Aerospace). The mission intended to compare the performance of the British Aerospace gyros to the Analog Devices MEMS gyros. Based on the performance as well as the impact of integrating the two different devices, a decision would be made as to which would be baselined for future MEPSIs. JPL measured the key parameters for both (Table 20.3). The MEMS devices were smaller, used less power, and were easier to procure, but did not have nearly the stability of the British Aerospace device.

Table 20.3. Comparison of MEPSI Picosatellite Angular Rate Sensors

Parameter	British Aerospace	Analog Devices
Model number	RRS01-01	ADXRS150
Rate range	110 deg/sec	150 deg/sec
Sensitivity	18 mV/deg/sec	13 mV/deg/sec
Supply voltage	±5 V	5 V
Current	50 mA	8 mA
Application	Military grade	Automotive
Size	32 × 32 × 17 mm	7 × 7 × 3 mm
Mass	42 g	5 g
Stability*	10 deg/hr	125 deg/hr

*1 sec integration time for each sample; sample taken every 1 msec.

Fig. 20.10. STS-113 MEPSI electronics module assembly.

The MEPSI electronics were packaged to be thermally isolated from the satellite's aluminum case. Since we did not have solar cells, passive thermal control of the electronics was the only option. The circuit boards and batteries were packaged together into an "electronics module" (Fig. 20.10) that stood off from the case by Delrin flexure springs. This technique, called a "Thermos bottle" design, was a carryover from the OPAL-class picosatellites. The satellite exterior was finished with the appropriate ratio of polished and black anodized aluminum to maintain the isolated interior between −20°C and +50°C.

The MEPSI-to-ground-station link margin was approximately 8 dB at a range of 400 miles. The MEPSI satellites each used a pair of patch antennas to achieve an omnidirectional pattern. This was necessary because the satellites had no passive or active attitude control, and the tether would become entangled in a deployed antenna. A simulation of the resulting antenna pattern is shown in Fig. 20.11. For link budget calculations, −10 dB was carried as the satellite antenna gain—i.e., 75% of the spherical pattern equaled this gain or more. The ground station

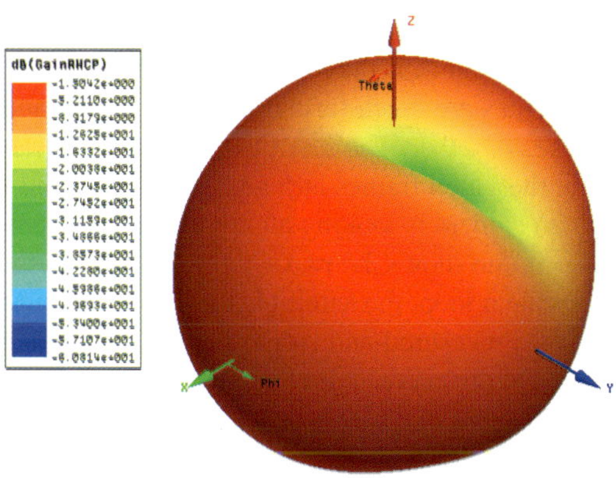

% of total sphere >	dBic
3%	-2
9%	-3
15%	-4
22%	-5
33%	-6
46%	-7
60%	-8
69%	-9
75%	-10
80%	-11
84%	-12

Fig. 20.11. Simulation of MEPSI antenna performance (dBic: decibels isotropic polarization).

Fig. 20.12. Ejecting MEPSI from the space shuttle Endeavour on STS-113.

antenna was the 60 ft diameter parabolic dish at the SRI International facility in Menlo Park, California. This antenna had 1.3 deg beamwidth and 42 dB gain.

The two 800 g MEPSI picosatellites were ejected from the space shuttle Endeavour (see Fig. 20.12) on 2 December 2002 after the shuttle had undocked from the International Space Station, and a day prior to its scheduled return to Earth. Astronauts toggled switches inside the shuttle to fire the nonexplosive device on the launch tube, opening the door and releasing the satellites. Rapid-fire 35 mm pictures of the event captured the satellites departing the shuttle at approximately 3.2 m/sec and separating from each other to deploy the tether at about 23 cm/sec.

A handheld video of the ejection captured the entire tether deployment. The video frames in Fig. 20.13 show the state of the tether deployment at 5, 30, and more than 60 seconds after ejection. After 5 seconds, the satellites are relatively close to the shuttle, and it is estimated that they have unspooled 45 inches of tether. After 30 seconds, the satellites are much farther from the shuttle and have unspooled an estimated 270 inches of tether. The satellites look like they have not moved apart when compared in the video frames at 5 and 30 seconds, but this is an optical illusion since the video camera is looking along the ejection vector. At more than 60 seconds, the video frame shows the two satellites close together and a loop of tether is visible. To create a loop, the satellites must have rebounded at the end of the tether. The satellites measured and stored in onboard memory the rates and accelerations that occurred during the tether deployment for comparison with this imagery.

The shuttle released the picosatellites within two hours of overflying the ground station in Menlo Park, CA. Tracking engineers from The Aerospace Corporation used the shuttle ephemeris and attitude at the time of ejection to create

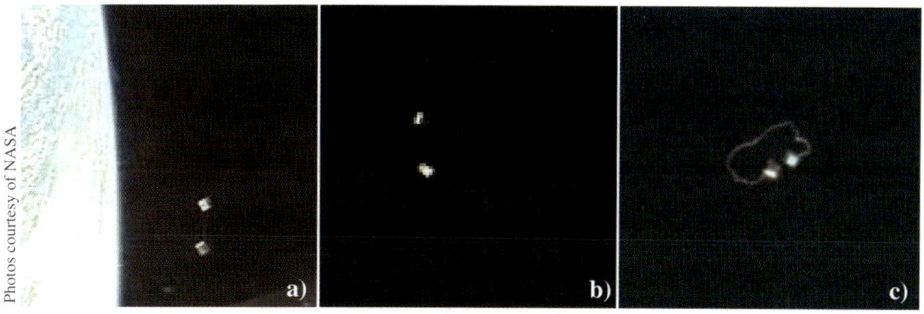

Fig. 20.13. Tether deployment, 5 seconds after release (a), 30 seconds after release (b), and the rebound after more than a minute (c).

a two-line-element set for the released satellites. They also overlaid a spiral scan strategy over the nominal track, looking for the 2 Hz beacons from each satellite. Beacons from one picosatellite were received on the first pass, but not from the second. Several subsequent passes went by with no beacons from either satellite. As time progressed, the error bars on the two-line-element set were growing. Twenty-four hours after ejection, the SSN provided their first two-line-element set based on ground radar observations of the tethered satellites. Using that set and the spiral scan, beacons from the other picosatellite were received. However, a two-way communication link was never closed with either satellite, and thus the inertial rate measurement unit recording of the satellite tether deployment dynamics was not downlinked. Also, the satellites did not receive a command to go to sleep and thus remained in a beaconing mode for two days before running out of primary power, thus ending the mission.

The MEPSI mission on STS-113 was a marginal success. The launch tube developed for shuttle and the NASA procedures to operate it were completely successful. The concept of operations and techniques employed to track miniature satellites also worked: the SSN was able to track the picosatellites, and the tracking engineers and the SRI antenna driver were able to steer the antenna. A satellite radio malfunction prevented complete mission success. The radio problem was traced to an automatic gain issue that was not discovered during the close proximity testing that was routine during development. In short, the satellites were deaf, unable to turn up their gain to hear the ground station commands. A radio test with all of the proper attenuation was not done, mainly due to schedule pressure but also because the radio was very similar to the one used in the OPAL and MightySat II.1 missions. This type of test was done repeatedly for the next MEPSI.

20.3.2 MEPSI ON STS-116

The MEPSI satellites on STS-116 were the next iteration in the satellite inspector series. The Aerospace Corporation was the sole developer of the payload for this mission. The picosatellites on the prior attempt on STS-113 in December 2002 did not perform well, and a comprehensive postmission analysis revealed a number of problems and proposed solutions that were implemented for this next mission.

Fig. 20.14. MEPSI picosatellites with tail of 15 ft tether joining them (covers removed). The "target" is on the left and "inspector" is on the right.

The MEPSI on STS-116 was intended to be a repeat of the STS-113 mission. However, the shuttle Columbia disaster in February 2003 grounded all shuttles for 2.5 years, and with that additional time, more functionality was added and more mission assurance tests were performed. The two MEPSI satellites (Fig. 20.14) were delivered in July 2006 inside of their launch tube, along with an environmental recorder, to the Kennedy Space Center. Analysis of the recorder data showed no vibration or temperatures that exceeded limits, so no further checkout was required. The hardware was directly integrated into the shuttle.

The concept of operations was to eject a tethered pair of 4 × 4 × 5 inch picosatellites from the space shuttle using the launch tube developed for the STS-113 mission. The tether served to increase the radar cross section of the satellite pair via dipole wires embedded along its length, similar to all previous missions. This tether was shorter than ever before—only 15 ft long—so that one picosatellite, called an "inspector" would be in constant proximity to the other picosatellite, called a "target." The inspector had a five-thruster propulsion unit and three-orthogonal reaction wheels so that it could maneuver while two VGA cameras would take pictures of the target. The target had no attitude control, but had a suite of five VGA cameras, on different faces, to take pictures of the inspector. There were no sensors on the inspector or the target for detecting the other picosatellites; those subsystems were not ready at this time, so all camera operations and attitude control changes would have to be commanded within range of a ground station. This idea of one picosatellite inspecting a target picosatellite was an ideal development configuration that would enhance confidence for future missions with much more expensive host satellites.

The complement of circuit boards in each of the two nearly identical STS-116 MEPSI picosatellites is listed in Table 20.4. Each picosatellite had a

Table 20.4. STS-116 MEPSI Picosat Circuit Board List

Subsystem	Board Name	Provider	Description
Computer	Flight computer*	The Aerospace Corp.	PIC + 512k EEPROM
Power	Flight computer*	The Aerospace Corp.	5 V converter
Radio	FGRM*	Freewave	915 MHz, 38.4 kbps
Radio	RFSW	The Aerospace Corp.	MEMS Rx/Tx switch
Radio	RF interface	The Aerospace Corp.	Bus to radio interface
IRMU	IRMU main*	The Aerospace Corp.	IRMU main board
IRMU	IRMU daughter	The Aerospace Corp.	IRMU orthogonal sensors
Attitude control	RWT*†	The Aerospace Corp.	Reaction wheels and thrusters
Camera	Camera main*	The Aerospace Corp.	Camera + 512k EEPROM
Camera	C328	Comedia	Color VGA camera

* has its own processor and local memory
† only on the "inspector" picosatellite

flight computer with an integral 5 V dc-to-dc converter and a new radio. This mission featured an improved picosatellite inertial rate measurement unit (IRMU) that was smaller than the STS-113 version because the British Aerospace rate sensors were eliminated in favor of the smaller but less stable Analog Devices units. The new IRMU used sensors in an orthogonal triaxial configuration to measure satellite rotation rates, accelerations, and magnetic fields. Each picosatellite was powered by two battery packs, each consisting of nine 3.65 V, 1 Ahr primary lithium thionyl chloride batteries arranged in series. This was double the energy of the STS-113 mission. Inside the target picosatellite was a tether spool containing the 15 ft of tether; the free end was tied-off to a point on the other satellite. Redundant spring plungers compressed between the two picosatellites kept microswitches open while the picosatellites were inside the launch tube. Once free of the ejection tube, the spring plungers were free to extend, thus providing energy to unspool the tether and closing the switches that powered-up the satellites.

The STS-116 circuit boards were redesigned around a new architecture to improve mission success. This new architecture featured distributed computing. Before, the picosatellites on STS-113 used a single powerful processor to manage the entire satellite, and a real-time operating system was used to coordinate different processes. This was very complex and required an entirely assembled satellite before the flight code could be tested and debugged. In the new architecture, each main board (labeled in Table 20.4) had its own Microchip processor that was programmed with only the code necessary to run that subsystem. Those boards also had local memory to record data. A flight computer board was the interface with the radio, and it would delegate ground commands to perform a particular task, such as "take pictures," to the proper board. This new architecture protected schedule. Breaking up the code allowed subsystems to be developed in parallel by different programmers and engineers. If a subsystem fell behind, it could be omitted from the architecture without affecting the satellite integrity or requiring code changes on the flight computer. This new architecture also promoted testing. Each subsystem processor, along with the FGRM radio, interfaced with the flight computer via a TTL RS-232 interface. This allowed each subsystem board to interface with a personal computer during development where it was completely tested before it was integrated into the satellite. This kept lead times from stacking up in series.

The radio was the most important new element for this mission. RSC had designed and built the prior picosatellite radios for The Aerospace Corporation. The benefit was that the radio was tailored to its application; the problem was that the previous STS-113 radio did not perform properly. RSC was no longer building radios and was therefore not in a position to fix it. The MEPSI on STS-116 featured a new radio that was engineered to be simple and reliable. Freewave Technologies in Boulder, Colorado, modified its existing miniature 915 MHz terrestrial data radio for use on the satellite. Their commercial radios were already 100% tested for operation between –40°C and +85°C and used a proprietary process to minimize crystal drift over that range. For use in space, the radios were modified to accommodate the anticipated ±25 kHz Doppler shift caused by orbital speeds of 7.7 km/sec and the large latency of approximately 8.9 msec expected for a round trip from a maximum specified range of 800 miles. Their output power was increased to 2 W to accommodate this much longer link range. The radios were tested at the manufacturer prior to delivery to verify that they met all of these requirements. Purchasing a modified existing product had the added benefit that the commercial version could be procured immediately and used as a surrogate during development, while the custom radios were being modified in parallel to meet our unique requirements.

The MEPSI picosatellites featured miniature color cameras with built-in JPEG (Joint Photographic Experts Group) compression. These 10 g, 2.0 × 2.8 cm boards, model C328 from Comedia, Ltd., have a maximum power usage of 200 mW and use a ¼ inch (6.4 mm) 640 × 480 color CMOS imager (OmniVision OV76 series) with a built-in lens (Fig. 20.15). The camera modules send the compressed image over a TTL RS-232 line to a camera main board that stores the images. Credit for discovering the modules and for the first revision of the camera main board belongs to engineering students at Harvey Mudd College in Claremont, California. Their senior design project was to design a camera subsystem for The Aerospace Corporation picosatellite program. The camera main board could accommodate eight camera modules.

The MEPSI inspector featured open-loop attitude control using cold gas thrusters and reaction wheels. Software for a closed-loop attitude control system was not ready for this mission. This was an opportunity to practice integrating and operating propulsion and reaction wheel systems on a picosatellite. When the

Fig. 20.15. Comedia 640 × 480 pixel color camera.

MEPSI program began, the propulsion system was identified as a subsystem that would require significant development and lead time. In 2001, The Aerospace Corporation wrote specifications, and several companies submitted proposals. Vacco Industries in El Monte, California, won the contract, and DARPA then funded the company to develop one flight-ready prototype cold gas propulsion unit for a MEPSI mission. Vacco Industries used a titanium etching and diffusion bonding process to create a compact vaporizing liquid cold-gas propulsion unit with five thrusters fed from a central tank. Vacco embedded the valves and electronics to measure pressure and temperature. Heat from these elements warmed the liquid propellant to help maintain vapor pressure within the propellant tank. The unit, called the Micro-Propulsion System (MiPS), was delivered in 2003. This, however, was not the propulsion system in the MEPSI inspector picosatellite on STS-116. The cost of the MiPS was significant enough that it could not be easily replaced, and it was held in reserve for use in a later mission when the MEPSI avionics would be more mature. In the interim, a substitute was needed.

The Aerospace Corporation flew its internally developed five-thruster cold-gas propulsion system on STS-116.[13] Our propulsion system was cheaper but not as reliable or capable as the MiPS unit, yet it was sufficient for the mission. The idea for this new method of building a propulsion system was discovered when a stereolithography apparatus (SLA) rapid prototyping system was used to create a scale exterior model of the MiPS propulsion system that would help us to design the mounting brackets and wiring harness. The SLA additive process builds a mechanical part from liquid resin, working from a digital file that is "sliced up" into 0.005 inch thick layers. An ultraviolet laser cures each layer and bonds it seamlessly to the previous layer. Voids can be intentionally built into the solid as long as a path exists to drain out the uncured resin at the end of the build. Engineers at The Aerospace Corporation realized that SLA materials might hold a gas

Fig. 20.16. An embedded cold gas thruster in the propulsion manifold.

inside an internal volume without leaking, and that voids built into the solid could include a main tank to store propellant, thin tubes for moving the gas around, and even converging/diverging nozzles (Fig. 20.16). All of the fluidic connections between the integrated elements were leak-tight because they were manufactured from the same material, at the same time, with no distinct joints. Also, the material could be polished so that leaktight O-ring seals could be made with external, commercial grade valves. The SLA process was quick and inexpensive. Parts were designed using 3-D CAD (computer-aided design) software and uploaded to a vendor for fabrication: no paper prints were needed. The turnaround time for parts the size of the MEPSI propulsion unit—approximately 1 × 3.6 × 3.6 inches—was only a few days at a cost of a few hundred dollars.

This process was ideally suited for building a leak-free manifold, but in addition, it was so flexible that other nonpropulsion components could be packaged in the unused portions of the propulsion system volume. The 20 cm^3 main fuel tank was filled with 100 psig of xenon gas, resulting in an estimated maximum ΔV of only 0.4 m/sec. (Had the tank been filled with a saturated liquid as a propellant, it would have produced 20 m/sec total ΔV—enough for a MEPSI mission.[14]) The design used seven LHDA model valves from The Lee Company in Westbrook, Connecticut, with two in series for main tank isolation and one for each of five individual thrusters. A CAD rendering of the final design is shown in Fig. 20.17. The SLA material, DSM Somos Watershed 11120, is transparent when polished, and the internal features are clearly visible, as shown in the figure. Transparency was useful for inspecting the manifold for production defects. Because the finished propulsion system did not fully utilize the 1 × 3.6 × 3.6 inch volume, other subsystems were also embedded. These included three orthogonal miniature reaction wheels and one camera module. Unlike the MiPS unit, there were no sensors to measure the pressure or temperature of the stored gas in the main propellant tank. The final subsystem dry mass was only 188 g, not including the reaction wheels and Comedia camera.

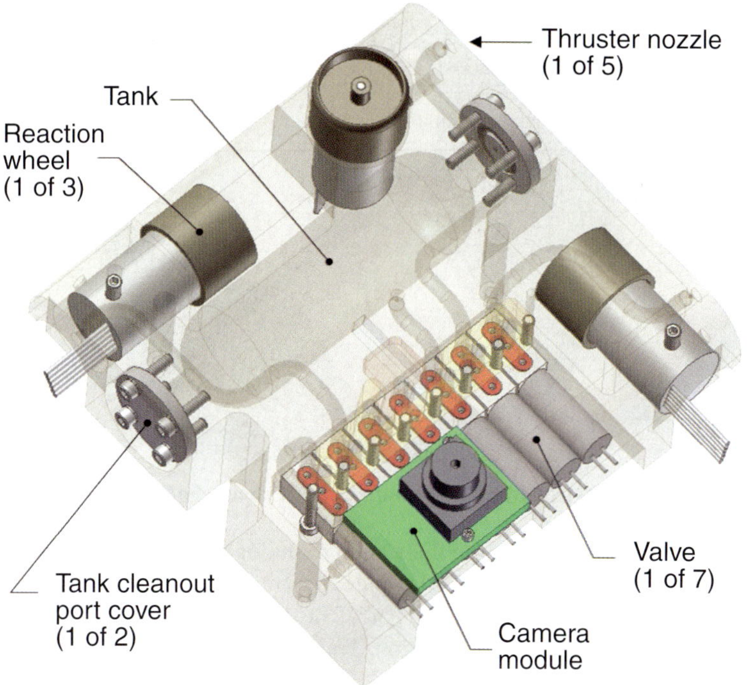

Fig. 20.17. STS-116 MEPSI "inspector" propulsion module assembly.

The three reaction wheels on the MEPSI inspector complemented the attitude control authority of the SLA cold-gas propulsion system. The reaction wheels were MicroMo Electronics, Inc., brushless dc motors (model 1525T009BRE) with a tungsten end mass pressed onto the motor shaft. The end masses were machined into a shape that positioned their center of mass at the forward motor bearing: this minimizes the forces on the motor bearing during liftoff. The motors were used "as is" with no changes to improve heat exchange with the embedded commutation electronics or to reduce outgassing of the lubricants.

The 1.4 kg inspector and 1.1 kg target were ejected from the space shuttle Discovery in December 2006, one day after undocking from the International Space Station and two days before reentry. As with the prior shuttle mission, the astronauts toggled switches to release the picosatellites. Immediately upon release, the satellites began preprogrammed operations that included taking photographs of the shuttle (Fig. 20.18) and recording the tether deployment and subsequent rebounds using the onboard IRMUs. (The vignette in the photograph is due to improper placement of the camera with respect to the aperture in the spacecraft body and the highly reflective edges on the aperture.) They departed the shuttle Discovery at an estimated 2.5 m/sec. Unlike the STS-113 mission, these MEPSI picosatellites did not beacon, so there was no race to shut off the beacon before all of their primary power was exhausted.

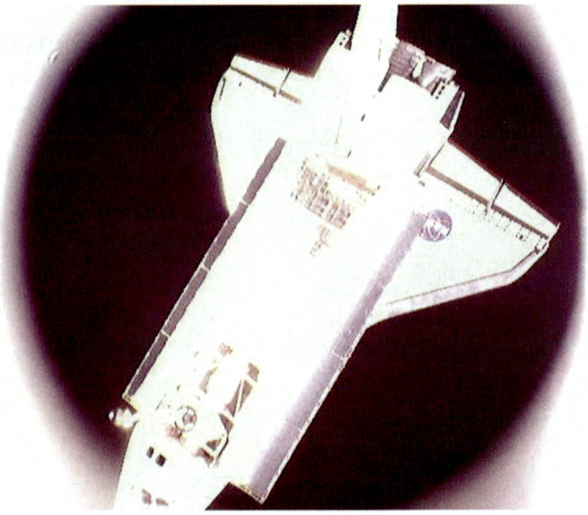

Fig. 20.18. STS-116 MEPSI photograph of space shuttle Discovery.

The tracking of the picosatellites and ground operations were nominal. The shuttle released the picosatellites 20 minutes before overflying the primary ground station at SRI International in Menlo Park. This was the same 60 ft diameter parabolic dish used for the STS-113 mission. Tracking engineers from The Aerospace Corporation and 1Earth Research, LLC, used the shuttle ephemeris and attitude at the time of ejection to create a two-line element set for the released satellites. Later element sets were provided by the SSN. For this mission, a second ground station at The Aerospace Corporation in El Segundo was also used. This 16 ft parabolic antenna is conveniently located, but is much smaller (less gain) than the SRI International antenna and is situated in the Los Angeles area, where the noise floor at 915 MHz is significant.

The MEPSI picosatellites on STS-116 operated nominally, and contacts were routine with both ground stations. From 20 to 22 December, the ground stations had two-way communications with the picosatellites a combined total of 12 times. A total of 1.5 Mb of data was downlinked, consisting of state-of-health data, IRMU experiment data, and imagery (69 pictures). The SRI International antenna had greater link margin and downloaded most of the mission data; however, the El Segundo antenna worked well enough to be baselined for future missions, despite its smaller size and noisy location. The MEPSI state-of-health file consisted of the satellite bus voltage and four temperature readings. The exterior temperature of each picosatellite fluctuated from –25°C to +10°C during an orbit; however, the "Thermos bottle" design that thermally isolated the electronics and batteries from the exterior successfully reduced their temperature variations to between 0°C and +3°C and +7°C and +9°C, respectively, during each orbit.

The MEPSI inspector reaction wheels were tested by operating them open loop and measuring the satellite's response. Engineers turned the reaction wheels on, one by one, and measured the satellite rotation rates at 5 Hz. Fig. 20.19 shows measured angular rotation rates as the *y*, then *x*, and finally *z* reaction wheels were each spun forward and backward at 4000 rpm for approximately 10 seconds. There is a 37 second loss of data in the middle of the test, when the onboard recorder had to be restarted. The ratio of the reaction wheel inertias to the satellite principal inertias was 3200:1 for both the *x* and *y* axes and 2500:1 for the *z* axis. The *z*-axis (the channel with the least noise and the axis with the lowest inertia) had rotation rate changes of about 10 deg/sec, while the *x* and *y* axes had rotation rate changes of approximately 8 deg/sec.

Similarly, a single MEPSI inspector cold-gas thruster was fired, and the satellite's response was measured. The design of the propulsion unit was such that four off-axis thrusters were canted to provide angular rotation about all three satellite axes at once. When the four off-axis thrusters were fired simultaneously, they are equal and opposite to a fifth thruster that was purely axial. By firing the thrusters in other combinations, satellite rotation about a single axis could be achieved. However, in this first test, only one off-axis thruster was fired, which would produce a change in the rotation rate about all three satellite axes. The normalized ratio of thrust about the *x*, *y*, and *z* axes for any of the off-axis thrusters was 3, 1, and 1.7, respectively, after considering the lever arm of the thruster location from the satellite center of gravity and the cosine component of the thrust in each axis. Therefore, the satellite *x* rotation rate should be almost twice that of the *z* rate for a given thrust pulse. In Fig. 20.19, the satellite rotation rate change occurring at approximately 140 seconds into the test was caused by firing one thruster. As shown, the ratio of the *x* and *z* responses is correct. The *y* response is hard to

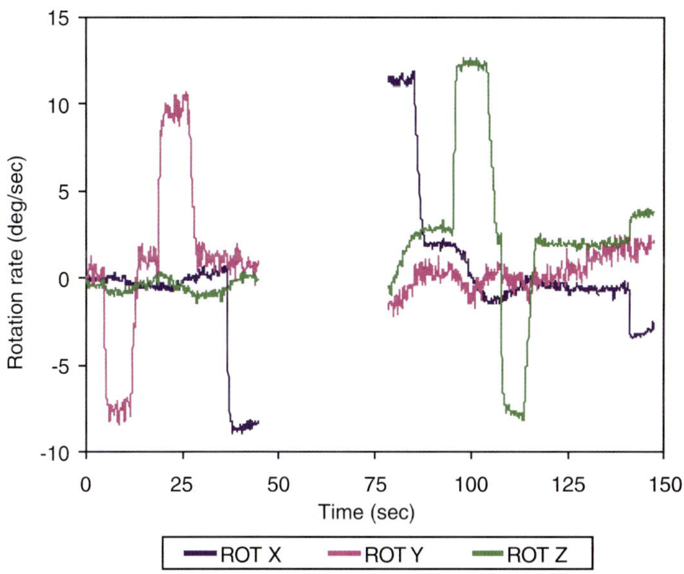

Fig. 20.19. STS-116 MEPSI inspector response to attitude actuation.

read because it is noisy, but it does change at the 140 second mark. From this, engineers concluded that the propulsion system was operating. More tests were planned but were not executed.

Over Christmas, the MEPSI team put the STS-116 picosatellites to "sleep" (i.e., put them in the lowest power mode, in which no communication is possible). They did not hear from them again. The picosatellites had an estimated maximum lifetime of two weeks, based on the energy stored in the primary batteries. They were put to sleep on 23 December until 3 January. From that day on, engineers were unable to talk to either satellite. Because both satellites were not responsive, an investigation into the possible failure mode focused on a systematic error. Using the qualification model hardware, the team discovered that a memory overflow condition caused an infinite loop. The satellites wrote a record into the state-of-health memory every 16 seconds. If that memory was not emptied and reset by ground operations, the flight computer would write data until the buffer eventually overflowed. In that case, it would write to a fictitious address and wait for a response for a successful write that was never going to come. After a time-out period, it would reboot the satellite and again write to that same nonexistent memory location, repeating the process until the batteries were exhausted.

The STS-116 picosatellites were a large step forward in the capabilities required by a MEPSI vehicle. The satellite systems all functioned nominally after launch, and communication with them was more reliable than in any prior mission. Some of the subsystems would need improvement: the rate sensors were noisy, the magnets in the reaction-wheel motors were affecting the satellite's magnetic sensing of Earth's field, and there was no sensor yet for detecting a nearby satellite so that closed-loop attitude control could be accomplished. All of these capabilities would have to work well before a MEPSI vehicle on a host such as STPsat-1 could be expected to work. The scheduled delivery to STPsat-1 was only six months after the STS-116 mission. A MEPSI vehicle was just too far off to be ready for that final flight. The original MEPSI program had four shuttle flights, six months apart. Only two were completed before the designated "final fight" satellite was scheduled for launch.

20.4 Space Shuttle Picosatellite Launchers

Most picosatellites launched today are ejected from a launch tube modeled after those used on the Stanford OPAL spacecraft to eject the first picosatellites in February 2000. That design is simply a tube assembled from four walls with one end closed and the other having a hinged door. Inside the launch tube, a large helical spring or "main spring" is responsible for the energy to kick out the payload when the door opens on orbit. Therefore, when the satellites were loaded into OPAL, as shown in Fig. 20.20, the main spring was being compressed. The satellites have running surfaces designed to slide along four rails in the corners of the launch tube. All other satellite surfaces do not make contact and have a known minimum clearance. Both the satellite and launch tube walls were aluminum. To prevent galling, the OPAL launch tube rails were burnished with Dichronite (tungsten disulfide), and the satellite surfaces that contact the rails were soft anodized. When the launch tube door was closed, a latch would engage to keep it closed. That latch connected via a long rod to a release device on the back of the launch tube, thereby allowing launch tubes to be efficiently stacked inside of the OPAL microsatellite. To keep the picosatellites from rattling around in the launch tube, a 45 deg wedge backed up by a stack of Belleville washers provided a compliant restraining force called a "preload." The rails on the launch tube were also chamfered at 45 deg, as was the door. The chamfers on the door can be seen in Fig. 20.20. In total, the satellites were held firmly and could only move if the accelerations were great enough to defeat the preload applied by the Belleville washer stack. In this design, the actual preload was not measured.

The Aerospace Corporation built an identical copy of the Stanford OPAL launch tube for its picosatellite payload on the MightySat II.1 spacecraft (see Fig. 20.6). The exercise of building and testing the launch tube clarified design details that we would then apply to the design of our space shuttle version. For instance, the launch tube totally encapsulates the picosatellites, protecting the vehicle to which the tube is attached or anything in the vicinity from any debris, outgassing, or radio frequency interference. The MightySat II.1 interface control document for this payload was primarily directed towards the mechanical features, materials, electrical interface, and testing of the launch tube. This was important for future missions, because satellites installed inside the launch tube could have different designs, and yet not require a new interface control document. This also created less work because the launch tube would not change very much between missions, whereas the satellites could change considerably.

In 2001, The Aerospace Corporation created a new launch tube that would accommodate a larger picosatellite and could use the space shuttle—a human-

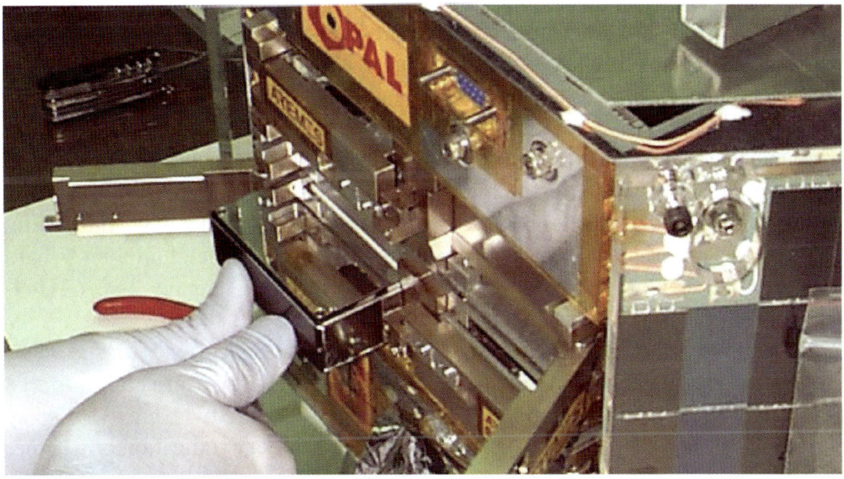

Fig. 20.20. Loading The Aerospace Corporation picosatellites into OPAL in 1999.

Fig. 20.21. Original SSPL model 4410 flown on STS-113.

rated system—as the launch vehicle. The Space Shuttle Picosatellite Launcher (SSPL) model 4410 was specifically designed for the MEPSI series, which had four planned development flights on the space shuttle, starting with the STS-113 mission in December 2002. The SSPL 4410 (Fig. 20.21) is qualified to carry a 2 kg payload that was 4 × 4 × 10 inches in dimension. The launch tube is larger than the Stanford OPAL tube, but had similar features. For example, the satellites slide in and out of the launch tube on rails that touch only the edges of the satellites. A wedge in the back and one on the door as well as a stack of Belleville springs for preload eliminate satellite rattling. The 4410 design is an improvement over its predecessor because it uses a tool instrumented with a load cell to adjust the launch tube preload to the minimum necessary force (plus a safety factor) to keep the satellites from moving during the liftoff vibration. This concept of "zero relative motion" between the satellites and the launch tube rails will eliminate any coating abrasion. The SSPL 4410 also uses a more robust launch tube antifriction coating. Instead of burnishing Dichronite onto aluminum, the aluminum is hard anodized and then coated with Tiolon E-20 from Tiodize Company, Inc., of Huntington Beach, California.

The SSPL 4410 has an electric circuit board that interfaces with the space shuttle payload electrical system. The circuit board has two reflective infrared sensors that detect the payload in the launch tube. These are fed back to a display panel inside the space shuttle. A third feedback provides an indication of whether the release device has been actuated or not. The circuit manages the power to the release device so that three separate astronaut switch throws are needed to actuate the release device and eject the picosatellites. This redundancy is standard for space shuttle payloads. For normal operations, the astronauts would turn on the first switch and check that all of the feedbacks were nominal. The second switch would arm the release device, and the final switch would fire it.

The SSPL 4410 launcher was redesigned for the STS-116 mission because of a failure during requalification testing for a heavier payload. The second MEPSI picosatellites on STS-116 were a tethered pair, and the mass of each satellite exceeded the 1 kg limit imposed by the prior qualification. (To remedy this, the SSPL 4410 was requalified for a 3.5 kg total payload weight. Therefore, if two identical MEPSI picosatellites flew, each could weigh at most 1.75 kg. In actuality, they were not identical, and one weighed 1.1 kg and the other 1.4 kg.) After a vibration test, an ejection attempt failed because the satellites had jammed. A failure analysis revealed that the coating on the rear satellite abraded through its soft anodize coating and deposited aluminum on the launch tube at the highest compression point. This created a high friction point and subsequent jamming of the rear satellite.

The remedy was comprehensive. First, satellites that use the SSPL are now required to have a hard anodize coating where they contact the launch tube rails. Second, the antifriction coating on the launch tube rails was changed to Tiolon S-20, which has better performance at high pressure than the E-20. Third, designers realized that the Belleville washer stack at the rear of the launch tube had exacerbated the failure. It was being applied at the wrong end, and should be applied near the door instead. The Belleville washers were removed altogether, and the preload compliance is now provided by the door bending as a leaf spring. Fourth, the method of setting and measuring the preload force that holds the satellites still during vibration was inadequate. Now, the load cell is incorporated in the launch tube under the release device to measure the preload force. In this location, the load cell value can be checked at any time to verify that it is above the minimum value required for zero satellite motion. Fifth, a second latch was added, called a "safing latch," that secures the launch tube door in case the release device should fire but the satellites do not come out. The safing latch is a swinging arm that rotates out of the way of the door when the launcher is powered on, prior to firing the release device. With the arm out of the way, a successful ejection will release the preload on the door, allowing it to be pushed open by the compressed main spring and eject the satellites. In a failed ejection, the door would not open, despite the preload force on the door having been removed, and the astronauts would power-off the launch tube and the safing latch would swing back over the door, preventing it from opening during reentry (where it was possible that the vibrations would free the stuck satellites).

The modified SSPL 4410 launcher is shown in Fig. 20.22. The second latch is visible, but the permanent load cell is on the back with the release device and

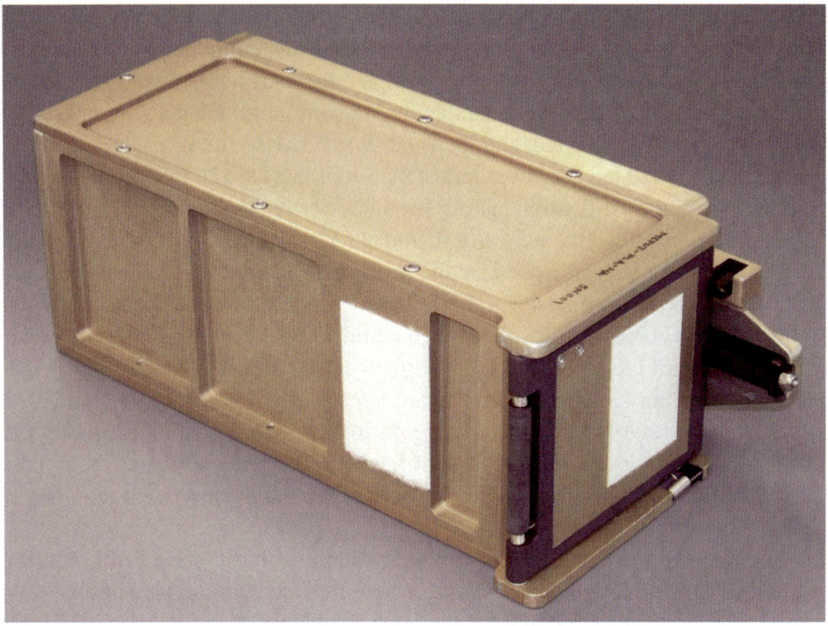

Fig. 20.22. Revised SSPL model 4410 flown on STS-116.

is therefore not visible. These changes to the launch tube design made it a much more robust launcher for picosatellites.

In 2005, The Aerospace Corporation designed and qualified a new, larger SSPL for the Space Test Program (STP). It is identical in almost every way to the 4410 except that it can accommodate a 7 kg payload that measures 5 × 5 × 10 inches and it has heaters to keep the payload above 15°C. It uses the same circuit board to electrically interface with the space shuttle, and it preloads the payload in the same manner. All of the redesigns on the 4410 were incorporated in the new 5510. The two models differ in thermal control of the payload. The 4410 was designed specifically for the MEPSI series, which used primary batteries that operate down to –40°C. Analysis showed that passive thermal control could meet these requirements. The 4410 was heat sunk to the object to which it was bolted and used chromate conversion coatings on its exterior. In contrast, the 5510 is a general purpose satellite launch tube and provides active thermal control. The 5510 is conductively isolated by G-10 standoffs and is isolated from the cold of space by a multilayer-insulation blanket. In Fig. 20.23, the SSPL 5510 is ready for integration on STS-126. It also flew on STS-116 and will fly again on STS-127. The flight article was produced by Oceaneering Space Systems, Inc., in Houston, Texas. Throughout the design and qualification of the 5510, which overlapped with the redesign and requalification of the 4410 after the aforementioned test failure, the STP office in Houston provided invaluable support. The STP office in Houston currently manages use of the SSPL 5510 on the space shuttle.

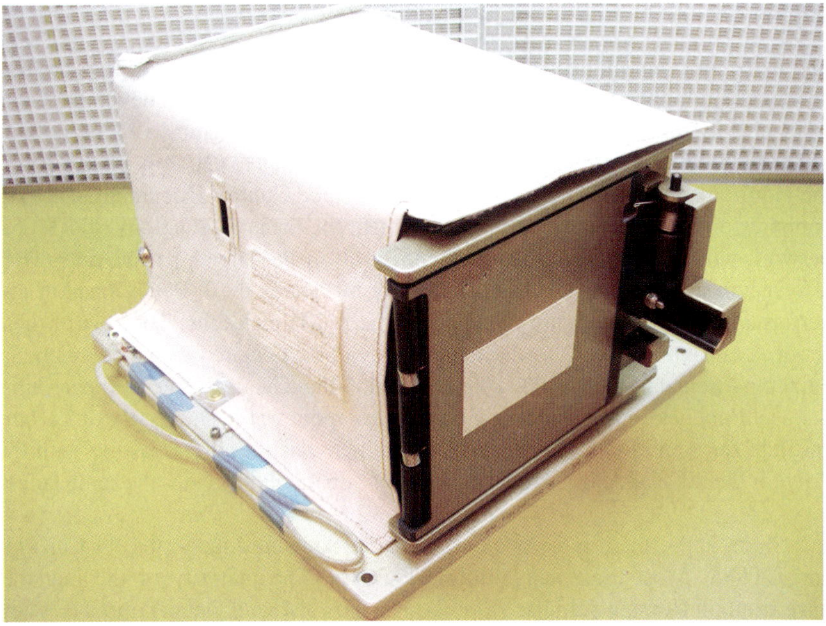

Fig. 20.23. SSPL model 5510 flown on STS-116 and STS-126.

Our work on the SSPL in 2001 was coincident with the development of a launch tube for CubeSats at the California Polytechnic State University, San Luis Obispo (Cal Poly). CubeSats are defined by Prof. Robert Twiggs of Stanford University as 10 × 10 × 10 cm picosatellites that weigh less than 1 kg.[15] Developers of CubeSats anticipated a launch cost of $40,000 each. This was affordable for many universities, and a development explosion began. Cal Poly assisted the CubeSat vision by developing a launch tube called the Poly Picosatellite Orbital Deployer (P-POD). Since its maiden flight in 2003, dozens of copies have been produced, and it is the worldwide standard. The Aerospace Corporation and Cal Poly collaborated during the P-POD and 4410 development, but the designs are still different. The 4410 and 5510 have extra safeguards that are required for space shuttle safety. These include a redundant (safing) latch to hold the launch tube door closed and a load cell for monitoring the preload forces on the payload. Also, the 4410 and 5510 hold the payload firmly in the launch tube for "zero relative motion" to minimize coating wear. This supported the argument for reflights of the same hardware on the space shuttle without disassembly and recoating. In contrast, the P-POD is a one-use device for flights on expendable launch vehicles. The P-POD preloads the CubeSats in the launch tube so that they do not rattle, but the springs are not stiff enough, and the launch tube does not have features such as wedges, to completely arrest their motion. Therefore, the CubeSats can move relative to each other and the launch tube during liftoff vibration. However, it has not been a problem, and all tubes have performed nominally on orbit.

20.5 AeroCubes

The CubeSat program provides relatively easy access to space for picosatellites that conform to the standard. To participate in a CubeSat launch, developers contact Cal Poly for the upcoming manifest schedule. After signing a Memorandum of Understanding, they build and deliver their satellites to Cal Poly after meeting the requirements in the standard CubeSat Interface Control Document. Cal Poly compiles and submits the International Traffic in Arms Regulations (ITAR) paperwork with the U.S. State Department for all of the launch participants. They also secure a launch on a vehicle. Cal Poly fabricates the P-POD, loads it, and performs the acceptance vibration test on the full launch tube. The cost of these services and the payment to the launch provider are shared by all the CubeSat launch participants. The 2003, 2005, and 2006 launches cost $40,000 per CubeSat. Cal Poly has a long-standing Canadian competitor: the University of Toronto Institute for Aerospace Studies (UTIAS), which was initially a partner with Cal Poly for the 2003 and 2005 launches and now separately provides the same launch services.

Since 2003, there have been six CubeSat launches supported by Cal Poly and UTIAS. Most have had multiple P-PODs. The majority of the launches have been on Russian vehicles (Rockot, Kosmos 3M, and Dnepr). India is a new participant and recently launched a CubeSat cluster on the Polar Space Launch Vehicle (PSLV). The United States launched one P-POD in 2006 on a Minotaur vehicle and has two more manifested on another Minotaur launch. Table 20.5 lists the CubeSats that have flown since the first flight in 2003. Participation is international and abundant due to the low cost and easy manifesting procedure. A launch failure in 2006 lost 14 CubeSats.

Table 20.5. CubeSats Completed and Launched*

Country	2003	2005	2006	2007	2008	Total
U.S.	1		~~12~~ 2	6		9
Japan	2	1	~~1~~		2	5
Denmark	2				1	3
Germany		1			1	2
Canada	1				1	2
Norway		1	~~1~~			1
South Korea			~~1~~			0
Colombia				1		1
Holland					1	1
Total	6	3	~~15~~ 2	7	6	24

* strikethroughs indicate CubeSats lost in the 2006 Dnepr crash

In 2004, The Aerospace Corporation began development of its first CubeSat called AeroCube-1 (Fig. 20.24). At that time, the team was in the midst of developing the MEPSI satellites for STS-116. AeroCube-1 had an initial launch date of December 2004 and was proposed as a risk-reduction flight for the next MEPSI. The satellite electronics were the same as for MEPSI on STS-116: a radio, flight computer, IRMU, and several Comedia 640 × 480 pixel color cameras. Also like MEPSI, it used primary lithium batteries that stored approximately 60 Whr of energy—an estimated 10-day lifetime. The 10 × 10 × 10 cm, 995 g CubeSat would test the performance of the new Freewave Technologies radio, quantify the link margin with the ground stations, and mature the mission operation procedures. It would also test the cameras and lenses on orbit. Furthermore, engineers and technicians would benefit from practice assembling a flight article and performing environmental tests. The assembly methods, test procedures, and test results would simplify and reduce risk with the subsequent MEPSI build. The satellite was delivered in April 2005 and was launched 15 months later in July 2006 after a number of primary satellite delays. Unfortunately, the Dnepr launch vehicle crashed.

In 2006, The Aerospace Corporation developed its second CubeSat called AeroCube-2 (Fig. 20.25). The team had just delivered the MEPSI picosatellite pair for STS-116 and was beginning work on a 6.4 kg nanosatellite called the

Fig. 20.24. The AeroCube-1 CubeSat.

PicoSatellite Solar Cell (PSSC) Testbed (which will not be discussed in this chapter). Unlike all of our prior picosatellites (OPAL, MightySat II.1, MEPSI, and AeroCube-1), the PSSC Testbed would have solar cells. AeroCube-2 was proposed as a pathfinder and featured the PSSC Testbed rechargeable power system circuit board and lithium-ion batteries. However, while the PSSC was a 5 × 5 × 10 inch spacecraft with four triple-junction solar cells on each of the long sides, AeroCube-2 was 10 × 10 × 10 cm and had only one triple-junction cell on four of its six sides. Inside were lithium-ion rechargeable batteries with 12 Whr of energy capacity and the usual MEPSI electronics complement: a radio, Comedia 640 × 480 pixel color cameras, an IRMU, and a flight computer. Additionally, AeroCube-2 had a deorbit balloon. The 750 km average altitude would keep the CubeSat in orbit for approximately 30 years. To qualify for a temporary license from the Federal Communications Commission to use the 915 MHz band to communicate with the satellite, our CubeSat needed a deorbit device to reduce the orbit lifetime below 25 years. To this end, a 9 × 6 inch pillow-shaped Kapton balloon was carefully folded and stowed under a releasable cover. (The cover was attached to the balloon so that it would not become debris upon release.) When inflated, the balloon would reduce the orbit lifetime to approximately 20 years. A derivative of the MEPSI cold-gas propulsion system was used to inflate the balloon. It consisted of a leak-free manifold with a single valve filled with DuPont HFC-236fa (Suva 236fa) refrigerant.

Fig. 20.25. The AeroCube-2 CubeSat.

The 959 g AeroCube-2 satellite was delivered in December 2006 and was launched four months later on 17 April 2007 from a Russian Dnepr vehicle. Immediately after release, it powered-on and began taking pictures of the two other CubeSats in its P-POD and of Earth. Figure 20.26 is a photograph of the Cal Poly CP4 spacecraft. Figure 20.27 shows the island of Madagascar. Ground stations at SRI International in Menlo Park and in El Segundo were manned by tracking engineers from The Aerospace Corporation and 1Earth Research, LLC, who directed the antennas using information from the Russian launch provider. The satellite was contacted 12 hours after its release. Both ground stations successfully closed the link and downlinked a combined total of 581 kb of data, including state-of-health information (temperatures, voltages) and 30 camera images in two consecutive passes. After that, the satellite was never contacted again; 12 hours after the first two good passes, AeroCube-2 was drained of power before ground controllers could contact it again and inflate its deorbiting balloon. The satellite solar upconverter board was suspected. That design came from the power-rich PSSC Testbed nanosatellite and was not optimized for the power-starved AeroCube-2. A test of the CubeSat in the solar simulator had shown that the board would work, but the test methods were, in retrospect, optimal. Future designs of our solar-powered spacecraft would have more power margin, and our tests would be more conservative. Despite the mission failure, the satellite image of CP4 is the first on-orbit photograph of a CubeSat and is very popular, and both ground

Fig. 20.26. Cal Poly CP4 CubeSat as seen by AeroCube-2.

Fig. 20.27. Madagascar as seen by AeroCube-2.

stations had a solid communications link. That was particularly important in proving that the El Segundo ground station could work, even though it is located in the heart of Los Angeles, surrounded by all types of RF noise.

In 2008, The Aerospace Corporation developed its third CubeSat, AeroCube-3. Like its two predecessors, it is a 10 × 10 × 10 cm CubeSat and weighs 1.1 kg. (Fig. 20.28). It is our most complicated CubeSat yet. It shares the same radio, cameras, and IRMU as its predecessors but has an improved flight computer and a new power-management system. It will be launched into a 40 deg inclination, 450 km average altitude orbit on a Minotaur launch vehicle with TacSat-3 as the primary payload. The P-POD on the Minotaur is mounted to the upper stage. AeroCube-3 will be the first CubeSat loaded into the P-POD, and will therefore be the last one out. AeroCube-3 has a 200 ft tether inside, with one end tied to the P-POD. When it is ejected, the tether will unspool its entire length but will keep the CubeSat connected to the upper stage. Following P-POD ejection, the upper stage will perform a degassing maneuver that will exhaust all of its remaining propellant. It is unclear how the stage will tumble after this event and even more unclear what will happen to the CubeSat tethered to it. The AeroCube IRMU will record the tumbling. Then, the AeroCube-3 will photograph the upper stage in a MEPSI-like mission: a daughtership inspecting a mothership. In this case, the tether will keep them in proximity in place of a propulsion system, sensors, and closed-loop attitude control. Inside AeroCube-3, a tether reel can be activated by

Fig. 20.28. The AeroCube-3 CubeSat.

ground command to draw in the tether and bring the CubeSat closer to the upper stage, if necessary, for better resolution pictures.

After the tethered part of its mission, AeroCube-3 can cut its tether with the Minotaur upper stage and become a free-flying satellite. A ground command activates a hot wire that will burn through the Dyneema tether. Once it is free, a permanent magnet and hysteresis material on the satellite will orient AeroCube-3, creating North and South faces. (While the satellite is tethered, the magnets will try to do the same thing, but the tether will cause a disturbing force.) A single miniature reaction wheel, taken from the MEPSI series, will spin the CubeSat around an axis normal to the North and South faces. The satellite has a primary sensor suite that is at an angle from the North face so that it will aim toward Earth when the satellite is in the Northern hemisphere. The sensor suite consists of two proprietary sensors and a Comedia 640 × 480 pixel camera. These will sweep the planet at a rate determined by the reaction wheel. In addition to the primary sensor suite, the satellite also has sun and Earth sensors developed at The Aerospace Corporation that will provide additional orientation information. The sun sensor is a tiny device that measures the two angles to locate the sun within a 60 deg cone. The Earth sensor is a single-pixel infrared detector. Finally, AeroCube-3 has a 30 inch diameter deorbit balloon that will theoretically shorten its orbit lifetime from 36 months to 1 month. It uses an inflation system similar to the one on AeroCube-2. AeroCube-3 was delivered in October 2008 and is expected to be launched in

May 2009 (there have already been a number of primary spacecraft delays). Our launch cost for the Minotaur flight is $70,000.

In summary, the AeroCube series of spacecraft have been different from the MEPSI series. The MEPSI missions had a specific purpose, and development was solely targeted toward realizing it. MEPSI was launched from the space shuttle, and as a result, was the subject of a number of safety reviews during the course of a year or two prior to delivery. Changing anything on the satellite between the reviews was discouraged. Therefore, the creativity of the developers was constrained, and they instead focused on the single purpose of the mission at hand. CubeSats are a different story. Changes are allowed throughout development with little or no objection (unless a safety issue is added). Safety issues are generally limited to RF interference, battery chemistry, and pressure vessels. Therefore, if time permits, developers can add more subsystems to their satellites and take full advantage of the spaceflight. The low cost of AeroCube flights also means that the price of failure is low. CubeSats have been a good complement to our space shuttle flights.

20.6 Conclusion

Between 1998 and 2008, The Aerospace Corporation has built 11 picosatellites and nanosatellites. Eight have been tethered pairs, and three were individual CubeSats. We have also designed and qualified two launchers for use on the space shuttle: one to carry a 4 × 4 × 10 inch, 3.5 kg payload, and another to carry a 5 × 5 × 10 inch, 7 kg payload. Each of the 11 successive nano/picosatellites was designed to build upon the lessons of the previous spacecraft. Even though launch delays frequently changed the order of spacecraft launches, thereby invalidating some of the risk-reduction goals, each satellite was sufficiently different, and each advanced the state of the art within the program. In general, the act of building satellite after satellite allowed the team to improve its building methods, resulting in more reliable packaging. Also, with each satellite build, program management skills improved, and the test program matured. All of these skills are directly transferable to much more expensive space programs.

The goal of The Aerospace Corporation picosatellite program is to demonstrate to the satellite community that miniature satellites, launched as secondary payloads, can do a great deal to mitigate risk on much larger programs and thereby save money and time and more quickly advance the exploitation of space. Once convinced, a new industry serving this class of satellite will likely emerge. Another goal is to promote miniature satellites as a vehicle for educating the next generation of satellite builders. Miniature spacecraft are relatively affordable to build and launch and can provide an avenue for young minds to frequently and easily touch outer space.

20.7 Acknowledgments

The people at The Aerospace Corporation who have contributed to building our picosatellites over the last ten years are too numerous to list. The list is similarly long for the many talented contractors who have helped, although the GALCIT model shop at the California Institute of Technology deserves an honorable mention. Many people worked weekends to meet deadlines that are all too strict for secondary payloads like picosatellites. We sincerely thank everyone for participating. I especially thank Ernest Robinson who was my mentor and the reason I became involved in this project. His charisma and charm can make the dullest subjects interesting. Robinson started and ran the picosatellite program at The Aerospace Corporation from 1998 to 2003. I also thank Henry Helvajian and Siegfried Janson, who are visionaries and authorities on miniature satellites. Finally, we thank The Aerospace Corporation management who had the vision to encourage nano/picosatellites to be built and flown. The Aerospace Corporation has all of the necessary disciplines and laboratories for building satellites, and we hope the workforce has enjoyed the diversion from their usual prime contractor oversight duties.

The early picosatellite work at The Aerospace Corporation from 1998 to 2003 was funded equally by DARPA and The Aerospace Corporation Independent Research and Development program. We thank DARPA managers Drs. Albert P. Pisano, William C. Tang, and Clark T.-C. Nguyen for their support. We also thank David Williamson, then at AFRL Rome Laboratories, for his advocacy of MEPSI to DARPA. Since 2004, funding has been provided by the U.S. Air Force Space and Missile Systems Center's Developmental Planning Directorate and other program offices interested in furthering this work. The author thanks John Fujita for advocating the project early on when we were weaning ourselves from DARPA funding, and Jim Gee for providing steady support since then. Without the consistent funding and long-range view provided by the various funding sources at The Aerospace Corporation, this class of satellites would never have achieved critical mass in a timely manner. We also thank the Space Test Program for integrating the MEPSI flights on the space shuttle and the MightySat II.1 spacecraft.

The Aerospace Corporation also acknowledges the pioneering work of Prof. Robert Twiggs of Stanford and Prof. Jordi Puig-Suari of Cal Poly for creating the CubeSat program, ushering in an era of low-cost access to space. We'd also like to express our appreciation for the many universities and companies around the world that are participating in the CubeSat program and making space exciting once again.

20.8 References

[1] S. W. Janson, H. Helvajian, and E. Y. Robinson, "The Concept of 'Nanosatellite' for Revolutionary Low-Cost Space Systems," *Proceedings of the 44th International Astronautics Federation Conference* (Graz, Austria, Oct. 1993); Paper IAF-93-U.5.573.

[2] D. G. Sutton, R. S. Smith, P. Chaffee, S. Janson, L. Kumar, N. Marquez, J. Osborn, B. H. Weiller, and L. Wiedeman, "MEMS Spaceflight Testbed," *AIAA Space Technology Conference & Expo* (Albuquerque, NM, 28–30 Sept. 1999); AIAA Paper 99-4604.

[3] D. S. Clarke, M. T. Hicks, A. M. Fitzgerald, and J. J. Suchman, "Picosat Free Flying Magnetometer Experiment," *Proceedings of the 10th Annual AIAA/USU Small Satellite Conference* (15 June 1996).

[4] J. Cutler and G. Hutchins, "OPAL: Smaller, Simpler and Just Plain Luckier," *14th Annual AIAA/USU Conference on Small Satellites* (Logan UT, 2001).

[5] J. V. Osborn, E. Y. Robinson, D. Hinkley, N. Ho, M. Gorlick, B. Carlson, D. Gilmore, J. Swenson, L. Gurevich, L. Berenberg, C. Chien, J. Yao, V. Panov, B. Schiffer, R. Au, and N. Romanov, "A PicoSAT Network and MEMS Space Platform," *2nd International Conference on Integrated MicroNanotechnology for Space Applications* (Pasadena, CA, 11–15 April 1999).

[6] J. E. Pollard, C. C. Chao, and S. W. Janson, "Populating and Maintaining Cluster Constellations in Low Earth Orbit," *35th AIAA/ASME/SAE/ASEE Joint Propulsion Conference* (Los Angeles, CA, June 1999); AIAA paper 99-2871.

[7] K. Bult, A. Burstein, D. Chang, M. Dong, M. Fielding, E. Kruglick, J. Ho, F. Lin, T. H. Lin, W. Kaiser, H. Marcy, R. Mukai, P. Nelson, F. Newburg, K. Pister, G. Pottie, H. Sanchez, K. Sohrabi, O. Stafsudd, K. Tan, G. Yung, S. Xue, and J. Yao, "Low-Power Systems for Wireless Microsensors," *Proceedings of the International Symposium on Low Power Electronics and Design* (Aug. 1996).

[8] MightySat II.1 Summary, http://www.spaceandtech.com/spacedata/logs/2000/2000-042a_mightysat-2_sumpub.shtml (as of 15 Oct. 2008).

[9] MightySat2, http://space.skyrocket.de/index_frame.htm?http://www.skyrocket.de/space/doc_sdat/mightysat-2.htm (as of 15 Oct. 2008).

[10] H. Choset, "Path Planning and Control for AERCam, a Free-flying Inspection Robot in Space," *ASCE Journal of Aerospace Engineering* (1999); http://www.ri.cmu.edu/pubs/pub_2869.html#abstract.

[11] X-Mir Inspector Mission, http://www.ik1sld.org/inspector.htm (as of 15 Oct. 2008).

[12] C. Underwood, V. Lappas, G. Richardson, and J. Salvignol, "SNAP-1: Design, Construction, Launch, and Early Operations Phase Results of a Modular COTS-Based Nanosatellite," *Smaller Satellites: Bigger Business?* (Kluwer Academic Publishers, 2002).

[13] D. Hinkley, "A Novel Cold Gas Propulsion System for Nanosatellites and Picosatellites," *22nd Annual AIAA/USU Conference on Small Satellites* (Logan, Utah, August 2008); paper SSC08-VII-7.

[14] S. Janson, A. Huang, W. Hansen, and H. Helvajian, "Development of an Inspector Satellite Propulsion Module Using Photostructurable Glass/Ceramic Materials," *CANEUS 2004: Conference on Micro-/Nanotechnologies* (Monterey, CA, 1–5 Nov. 2004); AIAA paper 2004-6701.

[15] H. Heidt, J. Puig-Suari, A. Moore, S. Nakasuka, R. Twiggs, "CubeSat: A new Generation of Picosatellite for Education and Industry Low-Cost Space Experimentation," *14th Annual USU Conference on Small Satellites* (Logan, UT, August 2000).

Chapter 21

Small Satellite Mission Design and Operations

Daniel L. Oltrogge* and Steven L. Hast†

It's a small world, but I wouldn't want to paint it.
—Steven Wright

21.1 Introduction to Small Satellite Mission Design and Operations

Small satellite operators are often confronted with unique and frequently nontrivial mission design, planning, postlaunch operations issues. Many of these exist because most small satellites are launched as secondary payloads, while other issues are strictly a result of satellite size. This chapter will examine some of these issues to provide strategies that small satellite operators can implement to optimize operational capabilities and minimize mission risk. It represents our experience building and flying small satellites on a modest operations budget.

For example, initial acquisition and commencement of tracking and communications is primarily the responsibility of the telemetry, tracking, and communications (TT&C) team. Requirements for optimal TT&C might include:

- Robust and fault-tolerant satellite designs that prioritize satellite operations and communications.
- Proper design, fabrication, and operation of ground-station communications hardware and software.
- Availability of fast and accurate postlaunch state-vector products from the launching agency, third-party satellite-tracking sensors, or onboard inertial navigation units.
- Automatic search and tracking functions on each ground antenna.
- Enlistment of as many ground antennas, external tracking resources, and data providers as possible to provide globally distributed redundant tracking coverage and augmented TT&C access.
- Relocation of small antennas with large beamwidth and low gain to relatively noise-free environments.
- Separate ground-station commanding and receiving antennas.
- Thorough preflight end-to-end testing of all space-to-ground communications segments and interfaces and calibration of antenna pointing.

---

*1Earth Research LLC, Colorado Springs, Colorado

†The Aerospace Corporation, El Segundo, California

From a practical standpoint, however, many of these options are outside the scope of typical small satellite mission budgets. Small satellite builders typically cannot afford geographically diverse antenna locations, dedicated radar tracking, and training exercises. And even in the best of funding circumstances, unreliable preflight knowledge of deployment events and limited third-party radar tracking data can lead to degraded situational awareness.

Initial acquisition and tracking for newly launched small satellites is often a harrowing time. Immediately after satellite deployment, analysts have only estimated prelaunch or (if lucky) postlaunch satellite-based inertial navigation orbit and attitude information (coupled with associated error metrics) with which to conduct satellite operations. The longer the delay between satellite deployment and initial acquisition and commanding, the greater the risk that a mission could fail due to degradation of onboard batteries and limited opportunities to detect and fix critical system anomalies.

If the initial signal acquisition cannot be accomplished based on prelaunch or launcher-provided state vectors because of large errors or degradation of prediction accuracy, the satellite operator must wait until third-party observers encounter favorable viewing geometries to obtain enough observations to estimate orbital states. As more and more observations are collected, the state-vector accuracy will increase. Unfortunately, third-party observers can often "cross tag" objects of similar radar cross section or optical characteristics, especially when those objects are built to a common form factor (e.g., the CubeSat standard).

The problem of acquiring and continually tracking a vehicle is even more pronounced in the case of very small satellites (e.g., picosatellites), which generally lack the solar-array space and battery capacity for a strong transmitter. A relatively small transmitter on the satellite requires a large receiving antenna on the ground to detect the weak signal and close the telemetry link. This, in turn, causes the receiving antenna beamwidth to be quite small. A small beamwidth makes it difficult for the receiving antenna to acquire the satellite in the presence of typically large launch and deployment uncertainties.

21.2 Preflight Planning

To be ready for the satellite operations phase, a variety of astrodynamics analyses, arrangements for organizational coordination, and data management tasks are recommended.

Astrodynamics Analyses

- Analyze launch host position and attitude uncertainty data, and determine how the uncertainties will evolve over time.
- Estimate the attitude and ΔV magnitude of the deployment/separation mechanisms, accounting for any anticipated dependencies on thermal conditions in the space environment.

- Estimate the anticipated drag characteristics for all objects of interest (including other objects that may pose a risk of recontact with your satellite).
- Estimate orbit lifetime and ensure that it is less than the recommended limit of 25 years.
- Generate best-estimate two-line elements based upon fitting of best-available propagation of state vector data in a realistic drag environment.
- Map out the ground-station obstructions (i.e., buildings, trees, and the local terrain) so that predicted and operational visibility and communications timelines are as realistic as possible.
- Perform a recontact analysis, including uncertainties in deployment attitude, ΔV magnitude, and drag.
- Generate multiple antenna scan profiles to cover the nominal trajectory and the most likely anomalous flight conditions.
- Where necessary, generate initial antenna rotation command sequences to avoid tracking-antenna cable wrap.
- Estimate the radar cross section and visual magnitude (at least in a relative sense).

Organizational Coordination

- Generate ground-station contact times and viewing-angle profiles and distribute them to the mission planners, TT&C team members, and external space surveillance agencies. Include any details that may be of use (e.g., solar lighting conditions, visibility to the launch host or other satellites, anticipated rotation rates, etc.).
- Negotiate agreements, partnerships, and contracts to procure tracking services, ground stations operations support, and mission data from external agencies.
- Ensure that the ground station's proposed support plan and FCC registration are approved for the launch rehearsal and flight, and be aware of any limitations on ground station operations that result from FCC registration (e.g., minimum elevation constraints, operations time windows, etc.).
- Test the compatibility of state-vector and covariance data and reference frames with external agencies.

Data Management

- Provide external space surveillance agencies with preflight mission data to help them with their task of observation, identification, and tracking. Include launch windows, trajectories, total number of objects produced during the flight, detailed object descriptions (shape, size, drag, propulsion, mission events, attitude, materials, frequencies and transmissions, and radar cross section and visual magnitude estimations), booster events, satellite deployment attitudes and sequences, state vectors and associated uncertainties, and anticipated relative motions.

- Create an electronic catalog of orbit state vectors from the various pre- and postlaunch orbital data providers for easy retrieval.
- Ensure that the Earth orientation parameters (e.g., leap seconds and UT1/UTC conversions) are current.

Figure 21.1 shows the station pass plan that was generated for the AeroCube-2 mission. Note that the visibility characteristics reflect roughly twice-per-day contact frequency (typical when viewed from ground stations that are more than ~12 deg closer to the equator than the orbit's southernmost or northernmost points—51 deg N, for this mission). The rise, peak, and set characteristics of the pass are provided as a function of local time, including parameters such as ground-antenna viewing elevation angle ("EL"), viewing azimuth angle ("AZ"), relative slant range ("RNG"), ground observer lighting conditions ("OBSLIT"), and satellite lighting conditions ("SC_LIT").

21.2.1 Modeling Free-Flight Satellite Position

Antenna controller hardware is becoming increasingly capable. Some of the latest models incorporate simple "diamond pattern" scan algorithms and two-line element orbit propagation capabilities into the controller unit.

For antennas without such a controller or for orbit analysts who require more accuracy, an orbit propagator must be used. Orbit propagators can be categorized as analytic (e.g., Keplerian motion), semianalytic (e.g., using averaging techniques and analytical rates-of-change for some aspects of the propagation and integrating the rest), and numerical. Of these, numerical integration is the most accurate, as it is the easiest method for incorporating very detailed perturbation models.

DATE	LOCAL	EL	AZ	RNG	OBSLIT	SC_LIT	LOCAL	MaxEL	AZPeak	RNG	OBSLIT	SC_LIT	LOCAL	EL	AZ	RNG	OBSLIT	SC_LIT
20061220	16:46:51	30	285.79	589	N 90.5	SUN	16:47:31	35.55	320.81	518	N 90.6	SUN	16:48:10	30	355	591	N 90.8	SUN
20061221	15:28:27	30	155.1	586	S 77.0	SUN	15:28:54	32.33	132.32	554	S 77.0	SUN	15:29:21	30	109.29	587	S 77.1	SUN
20061221	23:25:33	30	323.71	608	D168.2	UMBRA	23:26:43	72.73	42.03	340	D168.3	UMBRA	23:27:53	30	122.06	609	D168.4	UMBRA
20061222	15:43:33	30	251.3	586	S 79.3	SUN	15:44:36	56.93	318.27	370	S 79.5	SUN	15:45:39	30	24.96	588	S 79.7	SUN
20061223	22:22:25	30	349.94	603	D157.7	UMBRA	22:23:20	43.36	41.29	457	D157.9	UMBRA	22:24:14	30	91.39	604	D158.0	UMBRA
20061224	14:39:50	30	221.26	584	S 69.5	SUN	14:40:58	82.8	134.69	314	S 69.7	SUN	14:42:06	30	50.45	586	S 69.8	SUN
20061224	22:37:41	30	279.03	602	D160.5	UMBRA	22:38:37	44.74	225.84	446	D160.6	UMBRA	22:39:32	30	173.91	602	D160.8	UMBRA
20061226	13:35:42	30	186.76	583	S 62.0	SUN	13:36:37	45.14	132.93	429	S 62.1	SUN	13:37:31	30	80.4	584	S 62.2	SUN
20061226	21:32:58	30	308.5	597	D147.4	UMBRA	21:34:07	81	226.41	322	D147.7	UMBRA	21:35:16	30	139.98	598	D147.9	UMBRA
20061227	13:50:31	30	270.97	583	S 63.3	SUN	13:51:22	41.96	318.92	452	S 63.4	SUN	13:52:13	30	7.95	584	S 63.5	SUN
20061228	20:27:49	30	336.48	593	D133.7	UMBRA	20:28:52	55.33	42.18	380	D133.9	UMBRA	20:29:55	30	107.1	593	D134.1	UMBRA
20061229	12:44:43	30	234.89	582	S 58.1	SUN	12:45:51	76.94	317.78	318	S 58.2	SUN	12:46:58	30	38.86	583	S 58.2	SUN
20061229	20:42:31	30	260.56	591	D136.6	UMBRA	20:43:09	35.13	227.36	524	D136.8	UMBRA	20:43:47	30	194.19	591	D136.9	UMBRA

Fig. 21.1. Station contact pass plan for AeroCube-2 mission rehearsal.

Orbit propagators and station visibility prediction software do not have to be built from scratch. Many capable commercial off-the-shelf packages exist that are used in satellite operations today.

21.2.2 Powered Flight Burn Estimation and Ephemeris Reconstruction

From a commanding and telemetry data perspective, one of the most fruitful and productive times to communicate with a satellite is while it is performing orbit and attitude maneuvers using thrusters or other powered mechanisms. But powered flight introduces further complexities into the ground tracking problem. When the powered flight system is first used, the hardware needs to be properly calibrated under actual flight conditions.

One technique that we have developed to support powered-flight missions is ephemeris reconstruction.[1] It addresses maneuver calibration and ephemeris reconstruction needs, whereby free-flight and powered-flight segments of interest can be reconstructed based on appropriately selected orbit state vectors. The method uses direct integration of numerical partial derivatives governing powered flight, coupled with a robust Marquardt-Levenberg optimization method[2] that interpolates between the Taylor series and gradient methods to minimize postmaneuver errors based upon the initial orbital state. It works very well for powered flight reconstruction, based on finite burn modeling and dynamic thrust-vector directions.

Figure 21.2 shows the full convergence history of such a technique. The Marquardt-Levenberg algorithm trades final position error for final velocity error on several occasions during the search for a six-parameter set that minimizes the weighted final error. The weighted error decreases with each iteration, and the final velocity error eventually reduces to 1.5 m/sec, whereas the final position

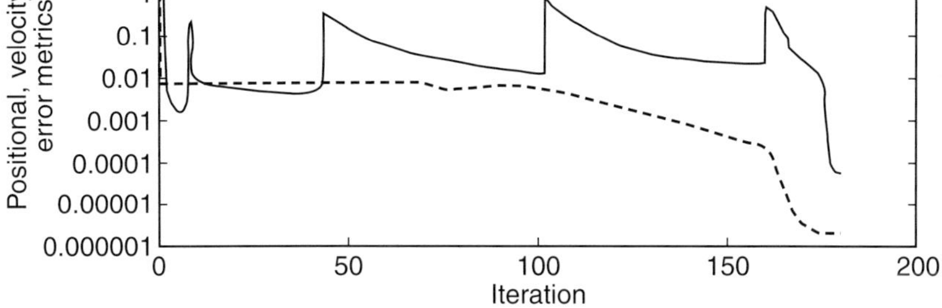

Fig. 21.2. Typical convergence characteristics for modeling of an operational satellite maneuver.

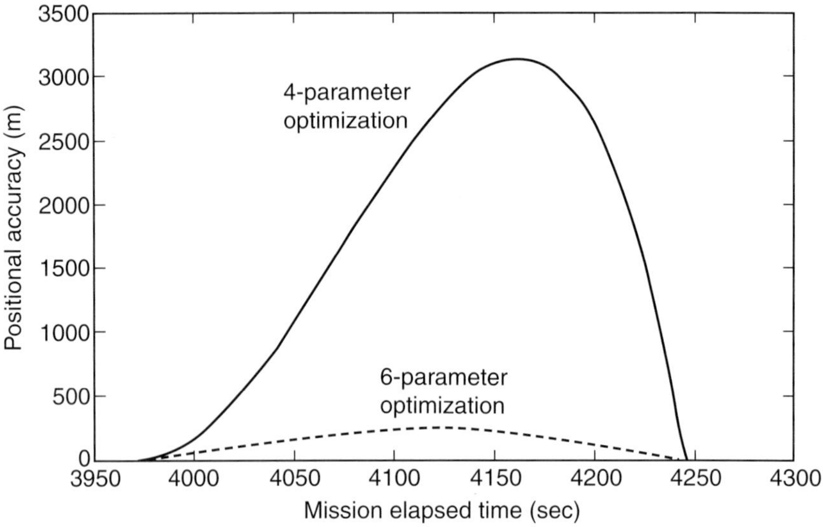

Fig. 21.3. Resulting ephemeris reconstruction accuracy for modeling of an operational satellite maneuver.

error remains under 1 m. Figure 21.3 shows the positional accuracy obtained using the reconstruction techniques.

Once the powered flight models have been optimally determined for each powered flight segment, vehicle acceleration can be determined as a function of time. This acceleration can be combined with the other orbit accelerations (e.g., two-body, drag, solar radiation pressure, etc.) to permit the reconstruction of the satellite's trajectory during all flight segments. These summed accelerations also permit the propagation and reconstruction of both state-transition matrices and error-covariance matrices (discussed in the next section).

21.2.3 Modeling and Propagating Satellite Position Uncertainty

Operators should incorporate knowledge of measurement and state uncertainties into their mission analyses. This reduces risk, enhances understanding, and decreases mission cost.[3,4] This position uncertainty can be represented by a normal distribution around the nominal predicted position. Predicting how this uncertainty shape will evolve over time is critical in ground-station antenna tracking, communications, recontact, and other analyses.

Modeling the uncertainty volume shape is the subject of some debate. A simple, reasonably accurate way to model the distribution is to assume that it is Gaussian in three dimensions about the predicted location. Although the Gaussian function continues on to infinity, the error volume can typically be represented by a 3-sigma cutoff, which implies the object is about 99% likely to be inside this boundary (Fig. 21.4).

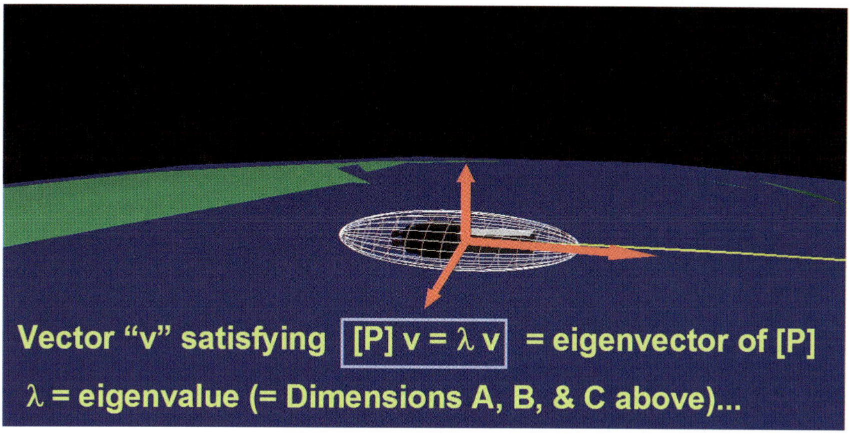

Fig. 21.4. Error covariance volume in three dimensions.

The arbitrarily chosen 3-sigma boundary forms a triaxial ellipsoid that moves with and is centered on the predicted position. This ellipsoid can be computed directly from the object position covariance, if such information is available. An eigenvalue/eigenvector decomposition can determine the lengths (eigenvalues) and orientation (eigenvectors) of the ellipsoid's three principal axes.

If the covariance is not available, it is usually safe to assume for nonpowered flight that most of the error lies in the velocity-vector direction and the least amount in the radial direction. A rule-of-thumb ratio between the principal axis lengths for nonpowered tracked objects with a well established observational history (e.g., a month or more) is 10:2:1 for the in-track: out-of-plane: radial directions. Longer histories of observational data can also be empirically processed to produce reasonably good covariance estimates.

Other uncertainty distribution representations have been proposed to incorporate the nonlinear dynamics of the orbital system.[5] These yield a crescent-shaped volume or a narrow torus centered on the predicted position. Although physically more appealing, these are more difficult to model than the traditional Gaussian distribution. And the linearized Gaussian-distribution covariance matrix (with its corresponding three-dimensional ellipsoidal representation) is the most readily available uncertainty data from launch and tracking agencies. This linearized matrix may also be approximated if a time-ordered series of tracking states are available.[6] As a result, we assume in the remainder of the chapter that 3-sigma Gaussian-distribution, linearized ellipsoids represent uncertainties in position.

A covariance matrix contains all the information needed to describe the ellipsoidal error volume of an object at a given time. A covariance matrix is a symmetric matrix of linearized error relationships. The diagonal elements contain the mean-square errors in the state variable knowledge, and the off-diagonal elements indicate the cross-correlation between the state variables. If the covariance reference frame corresponds to the principal directions of the ellipsoid, then the off-diagonal terms of the covariance will be zero.

Be aware that engineers can choose to represent the covariance matrix at any sigma level they choose. The most common are 3-sigma and 1-sigma covariances. The "gotcha" is that the engineers often forget to label the adopted error level!

A covariance matrix can be propagated to a new epoch by one of several methods. The first is the continuous time-matrix differential equation, which allows the direct integration of the covariance matrix via

$$\dot{P}(t) = F(t)P(t) + P(t)F^{T}(t) + G(t)Q(t)G^{T}(t) \tag{21.1}$$

where P(t) is the current covariance at time t, $F(t)=\partial\dot{x}/\partial x$, and $G(t)Q(t)G^{T}(t)$ is the contribution of process noise to the covariance rate of change.

A second covariance propagation method, using a state-transition matrix approach, allows the direct mapping of a covariance from one frame or state to another with partial derivatives using Eq. (21.2):

$$P(t) = \Phi(t, t_0) \cdot P_0 \tag{21.2}$$

The state-transition matrix $\Phi(t,t_0)$ contains the mapping from the state (t_0) to state (t_1). The value $\Phi(t,t_0)$ can be computed using analytic formulas, such as W.H. Goodyear's formulation,[7] and integration of numerical state-transition matrix formulas, such as $\{d\Phi/dt = F(t)\ \Phi(t)\}$.

Covariance matrixes can be mapped from one frame (for example, the Earth-centered inertial frame) to another (e.g., the ground station topocentric frame) via computed partial derivatives as shown in Eq. (21.3). This is a crucial step to the creation of antenna scan patterns, as discussed later in section 21.10.1.

$$P_B = \left[\frac{\partial B}{\partial A}\right] P_A \left[\frac{\partial B}{\partial A}\right]^{T} \tag{21.3}$$

As a final caution, note that covariance matrices are often supplied for specific events, such as at the end of a powered-flight segment for which the time is also unknown. This means an event covariance that includes uncertainties in both position and velocity vectors should have a 7 × 7 dimension (time being the seventh parameter). This 7 × 7 covariance must be converted into a 6 × 6 covariance to get the position uncertainty at a given time, which is required in Eqs. (21.1) and (21.2).[8]

To see the uncertainty volume evolution in action, we obtained error covariances for a recent CubeSat flight on the Dnepr launch vehicle and implemented them in the OPUS software. In Fig. 21.5, it can be seen that the initial orientation of the covariance matrix has the greatest error in the cross-plane direction, but this evolves within half an orbit to be predominantly in-track. This is typically

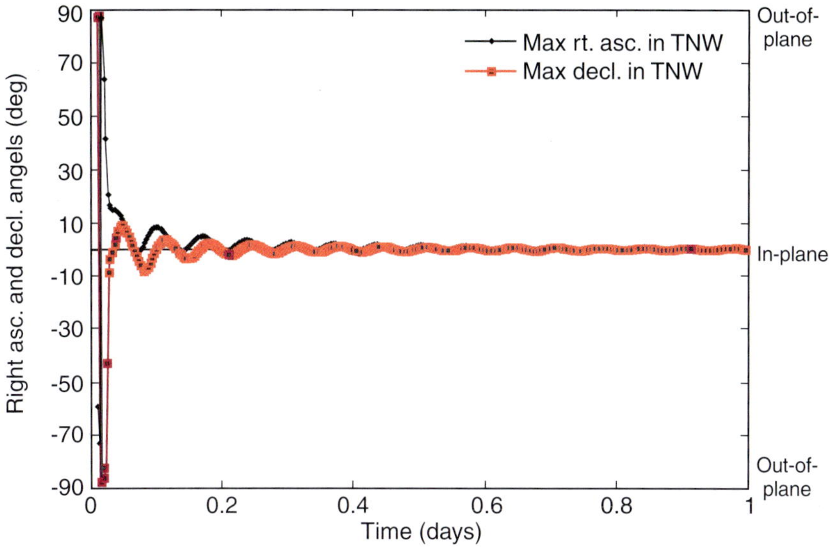

Fig. 21.5. Error volume alignment transitions to in-track errors within 1 hr.

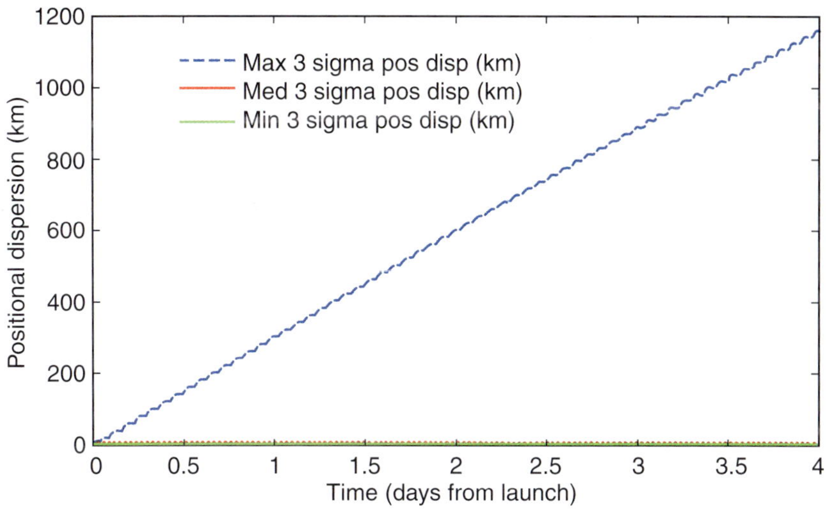

Fig. 21.6. Error dispersion 3-sigma eigenvalues (medium and minor curves overlay the *x* axis).

because booster velocity errors at injection eventually lead to in-track timing errors that dominate the initial injection position errors. In Fig. 21.6, a 30 km/day error growth rate is observed in the in-track direction while the radial and cross-track directions oscillate about a roughly constant value. These sample uncertainty volumes will be reexamined in the next section.

21.2.4 Working with Covariance Reference Frames

As Fig. 21.4 illustrated, a covariance matrix is associated with an ellipsoid in three-dimensional space. To work with these uncertainty volumes, they must be referenced to the same frame. Often, we must first rotate covariances into the same frame before error source contributions can be summed. In Fig. 21.7, an Excel spreadsheet handles these covariance-matrix reference-frame conversions to permit the addition of the launch host's state covariance (specified in a local-vertical/local-horizontal reference frame) with ΔV uncertainties for a canted deployment tube spring. Uncertainties in the velocity imparted by deployment springs or deployment thrusters are a key contributor to future tracking difficulties.

21.3 Orbit Selection and Predicted Orbit Lifetime

Achieving the intended satellite mission is the single most important consideration of a small satellite owner fortunate enough to be able to select the mission orbit. Key factors to consider are Earth coverage, communications links, and orbit lighting (which affects sensor performance, thermal control, and power systems). These factors are governed by orbit semimajor axis, eccentricity, inclination, right ascension of ascending node, and argument of perigee. By optimizing these orbital parameters, you can select the orbit to maximize mission performance. It is

	A	B	C	D	E	F
1	Sample Post-Deployment Covariance Construction Worksheet					
2						
3	1-Sigma TRTN Covariance (Without Deployment Uncertainty):					
4						
5	50000.00	0.00	0.00	0.00000000	-10.00000000	0.00000000
6	0.00	350000.00	0.00	-30.00000000	0.00000000	0.00000000
7	0.00	0.00	89000.00	[illegible]	[illegible]	[illegible]
8	0.00000000	-30.00000000	0.00000000	0.50000000	0.00000000	0.00000000
9	-10.00000000	0.00000000	0.00000000	0.00000000	0.12000000	0.00000000
10	0.00000000	0.00000000	0.00000000	0.00000000	0.00000000	0.20000000
11						
12	LVLH Deployment Ejection Angles:					
13						
14	Roll (deg):	300				
15	Pitch (deg):	90				
16	Yaw (deg):	0				
17						
18	Rotation Matrix from XYZ-Body to RTN:					
19						
20	R		1.00000000	0.00000000	0.00000000	X-body
21	T	=	0.00000000	-0.86602540	0.50000000	Y-body
22	N		0.00000000	-0.50000000	-0.86602540	Z-body
23						
24	Deployment Ejection Velocity and 1-sigma Uncertainty and Correlation:					
25						
26	DVX-body (mps):	0		DVR (fps):	5.02439E-16	
27	DVY-body (msp):	0		DVT (fps):	-8.202099738	
28	DVZ-body (mps):	-5		DVN (fps):	14.20645347	
29				DV-Right Asc (deg):	270	
30				DV-Declination (deg):	60	
31						
32	VX-1Sig-Uncert (mps)	0		DVR-1Sig-Uncert (fps	0.70710678	
33	VY-1Sig-Uncert (mps	0		DVT-1Sig-Uncert (fps	0.38328811	
34	VZ-1Sig-Uncert (mps)	0.1		DVN-1Sig-Uncert (fps	0.52983896	
35						
36				DVR-DVT-Corr:	0.00000000	
37				DVT-DVN-Corr:	-0.22950992	
38				DVN-DVR-Corr:	0.00000000	
39						
40	Post-Deployment 1-Sigma TRTN Covariance (Including Deplyment Uncertainty):					
41						
42	50000.00	0.00	0.00	0.00000000	-10.00000000	0.00000000
43	0.00	350000.00	0.00	-30.00000000	0.00000000	0.00000000
44	0.00	0.00	89000.00	[illegible]	[illegible]	[illegible]
45	0.00000000	-30.00000000	0.00000000	0.50000000	0.00000000	0.00000000
46	-10.00000000	0.00000000	0.00000000	0.00000000	0.14690978	-0.04660910
47	0.00000000	0.00000000	0.00000000	0.00000000	-0.04660910	0.28072933

Fig. 21.7. Error covariance before and after deployment.

assumed that you have a basic understanding of selecting an orbit to meet mission goals and satisfy mission constraints. You may want to consider two additional factors: international postmission orbit lifetime guidelines, and the potential for radio frequency interference (RFI).

21.3.1 Basic Coverage Angles, Relationships, and Limits

As an orbit selection aid, some basic coverage angles, relationships, and limits are shown in Figs. 21.8 through 21.13. From the parameters defined in Fig. 21.8, Eq. (21.4) can be derived, which can be evaluated for a given planetary radius *R* and altitude *h* (Fig. 21.9) to show the relationship between the amounts of ground coverage a satellite sensor would provide as a function of the elevation angle σ at which a person on the ground would look above the local horizon to see the satellite. In Fig. 21.10, the relationship between a satellite sensor's boresight off-nadir angle η and the same ground-based viewer elevation angle σ is shown.

$$\cos(\theta) = \cos^2\sigma\left(\frac{R}{R+h}\right) + \sin(\sigma)\sqrt{1-\left(\frac{R}{R+h}\right)^2\cos^2\sigma} \tag{21.4}$$

For small satellites hosting an optical sensor, it's useful to determine the relationship between optical resolution (called ground sample distance,[8] or GSD), altitude, and the ground viewing elevation angle (σ). This relationship is characterized in detail in textbooks;[9] Eq. (21.5) is a simplified version that assumes symmetric pixel dimensions, where "DetectorPitch" is the average individual pixel dimension of the CCD, "EffFocalLength" is the effective focal length of the optics, and σ, ρ, and *h* are the elevation angle, slant range, and altitude, as defined in Fig. 21.8.

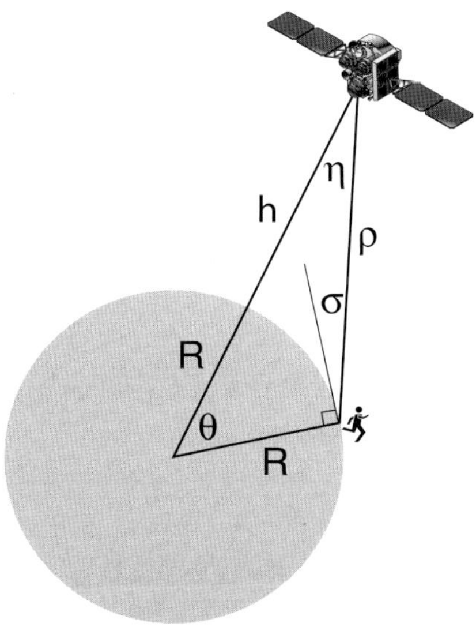

Fig. 21.8. Satellite coverage geometry.

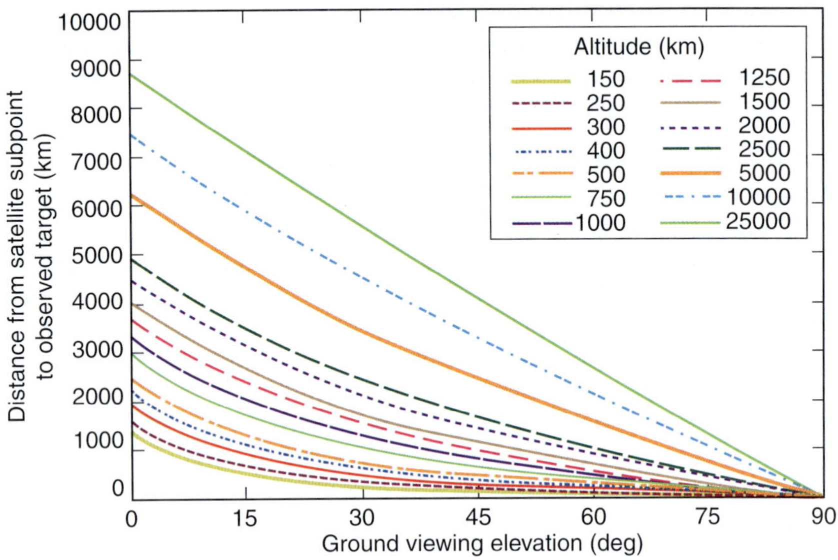

Fig. 21.9. Relationship between ground observer's elevation angle and distance from satellite subpoint (planet's surface location directly below the observing satellite) to that observer.

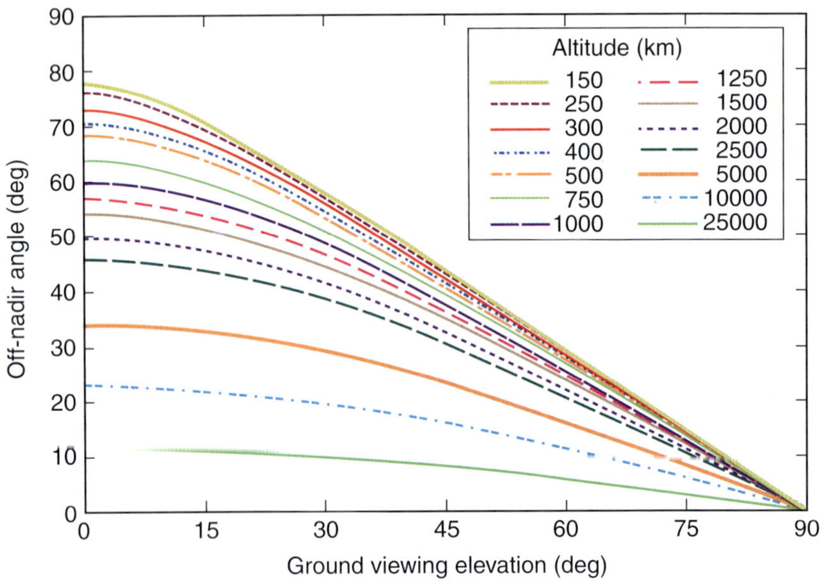

Fig. 21.10. Relationship between off-nadir (sensor) viewing angle and ground elevation angle.

At an elevation angle of 0 deg, the optical sensor is grazing the surface of Earth in a tangential fashion, and the sensor's "away resolution" grows to infinity, as the sensor is unable to resolve anything in the ground local horizon. At this viewing geometry, the sensor can basically detect only objects that are "in profile" or are flying in front of the atmospheric and space backgrounds. Conversely, a viewing elevation of 90 deg indicates that the observing platform is directly above the ground target, at which the optical sensor will be able to provide its maximum viewing resolution.

$$\mathrm{GSD}_{\mathrm{Mean}} = \left(\frac{\mathrm{DetectorPitch} \cdot \rho}{\mathrm{EffFocalLength} \cdot \sqrt{\sin \sigma}} \right) = \left(\frac{\mathrm{GSD}_{\mathrm{Nadir}} \cdot \rho}{h \sqrt{\sin \sigma}} \right) \tag{21.5}$$

Using Eq. (21.5), the relationship between ground viewing elevation and degradation in observing sensor resolution may be characterized (see Fig. 21.11).

Alternately, if you have multiple small satellites that are designed to communicate with each other via crosslinks, the distance between crosslinked satellite subpoints can be determined, which provides the desired crosslink grazing altitude, as shown in Fig. 21.12. The results are shown in Fig. 21.13; a grazing altitude threshold of 1.5 km is commonly adopted to avoid local terrain variations and RFI from ground transmitters.

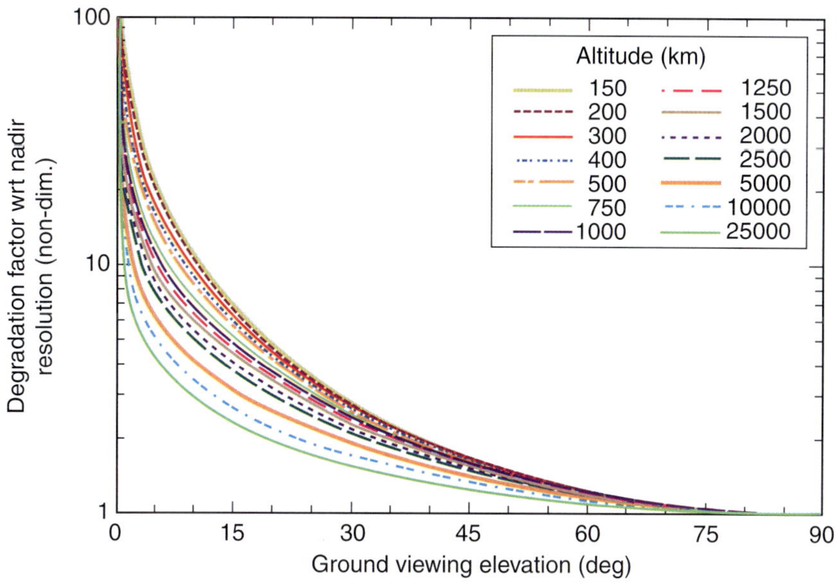

Fig. 21.11. Degradation in optical sensor resolution.

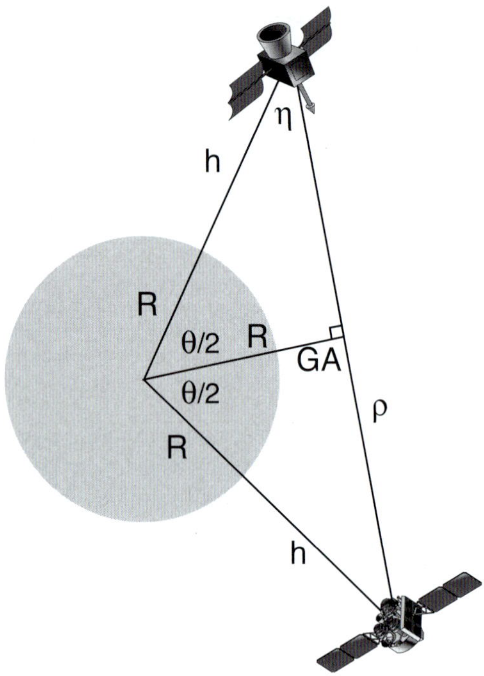

Fig. 21.12. Satellite crosslink geometry.

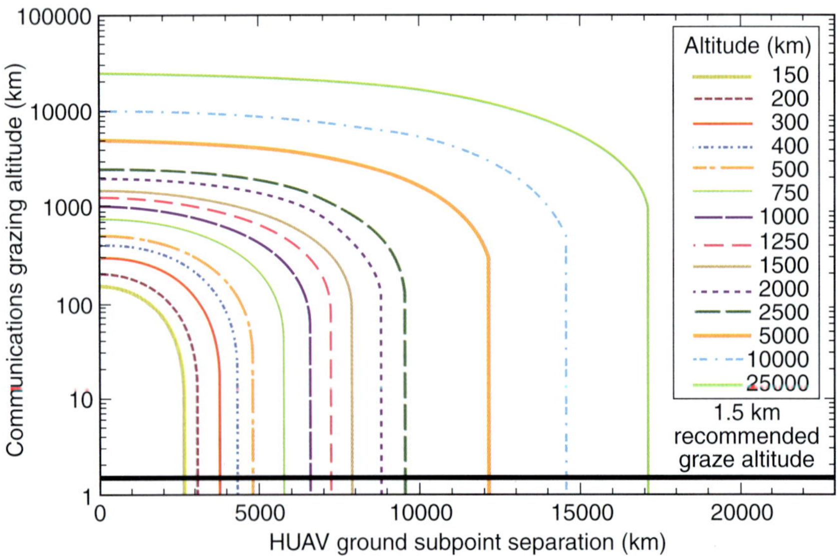

Fig. 21.13. Crosslink geometrical relationships.

21.3.2 Postmission Orbit Lifetime

The international community is concerned about the proliferation of debris in space—in particular, small satellites, because of the rapid and inexpensive methods available to build and deploy swarms of them. The small satellite community must address these concerns by developing hardware, mission designs, and concepts of operation that minimize orbital debris population growth and reduce its impact to both current and future satellite operators.

As discussed in chapter 22 of this book, national and international standards recommend that satellite designers and operators take steps to mitigate or at least minimize production of orbital debris. This can be accomplished by preventing deployment sequences from creating additional pieces of debris, taking steps to actively or passively deorbit the satellite upon completion of its mission, and selecting orbits that will experience reasonably rapid decay and reentry.

The mission designer must address each of these three strategies. Although reducing long-term orbital debris and minimizing postmission time on orbit makes logical environmental sense, in practice it presents the small satellite owner with several difficulties. The first is that it can be very difficult to find a launch platform for small satellites without paying excessive launch costs. The most economical method is to launch small satellites as secondary payloads, but this means that the orbit provided by the launcher is designed specifically for the primary satellite. Secondary payloads often have little or no voice in the orbit selection process. In this case, the use of inflatable drag enhancement devices or electrodynamic deorbit tethers may be the only viable alternative.

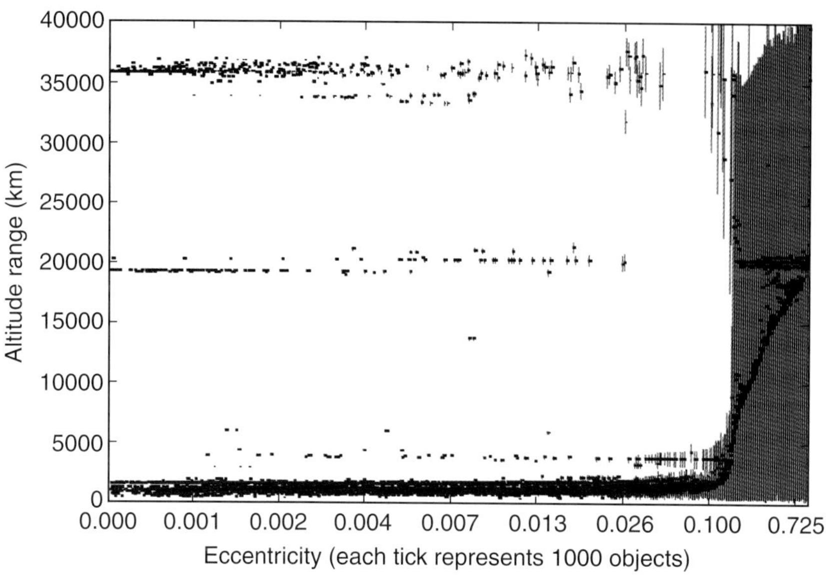

Fig. 21.14. Mapping resident space objects as a function of orbit eccentricity. Vertical lines represent altitude range (perigee to apogee).

If you can select your own mission orbit, then the Inter-Agency Space Debris Coordination Committee (IADC) recommendations[12,13] should be met to ensure that your small satellite does not occupy the LEO-crossing orbit regime for more than 25 years after the end of its life. The crowded LEO regime is easily seen as the dark banded region in Fig. 21.14; the "error-bar-type" lines depict the range of altitudes the satellite experiences during its orbit.

Unfortunately, the accurate assessment of orbital lifetime is not a simple task, as shown in Fig. 21.15. Higher-order, long-term orbit perturbations and solar/geomagnetic variations can drastically affect the lifetime of selected orbits. Guidance and methods to help the mission planner through these difficulties and assess orbital lifetime for all LEO-crossing orbit classes can be found in the draft "ISO Orbit Lifetime" standard[14] and accompanying technical documents.[15,16]

In a recent study,[16] orbit lifetimes were computed for a large portion of the LEO-crossing spacecraft, accounting for nonlinearities of solar and geomagnetic indices, solar-cycle phasing, drag coefficient variations, orbit-to-sun geometry variations, and the recently observed thermospheric cooling effect.[16] The study was done to support both the spacecraft designers and the ISO standards development effort. Figures 21.16 and 21.17[16] categorize the resulting apogee, perigee, and ballistic coefficients that could yield a 25-year orbit lifetime. The 25-year region's large size is due to the wide range in ballistic coefficients, which were simulated, in addition to the large variations in solar and geomagnetic activity and cycle timing uncertainties.

A fair prediction of orbit lifetime, averaged over all historical solar activity cycles and launch-time phasing within the solar cycles, can be obtained using the following equations.[16] Note, however, that it is always preferable to predict orbit lifetime using the latest available orbit and solar activity predictions and numerical analysis tools.

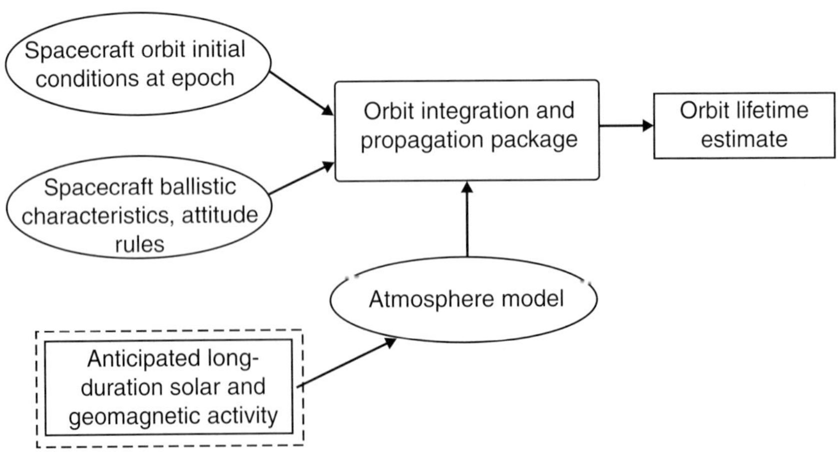

Fig. 21.15. Orbit lifetime assessment process.

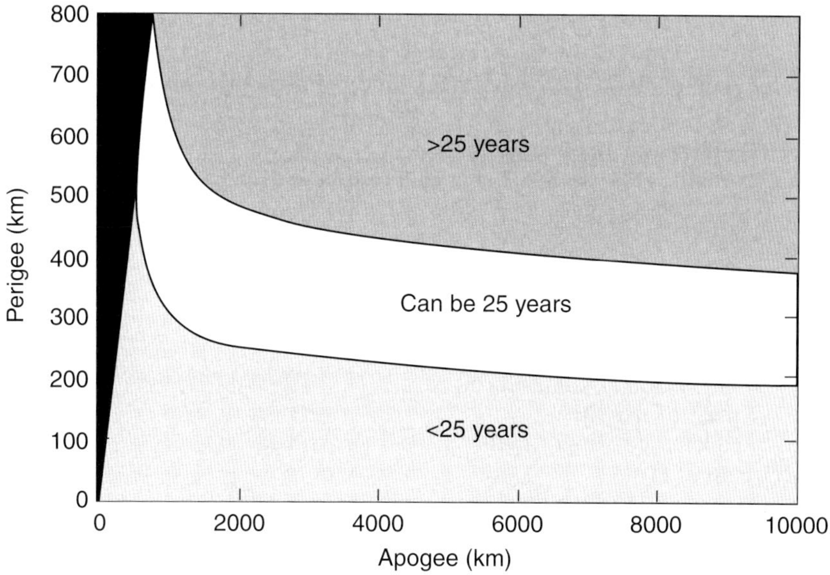

Fig. 21.16. Perigee versus apogee boundaries for 25-year orbit lifetime conditions ($25 < CDA/m < 500\ cm^2/kg$).

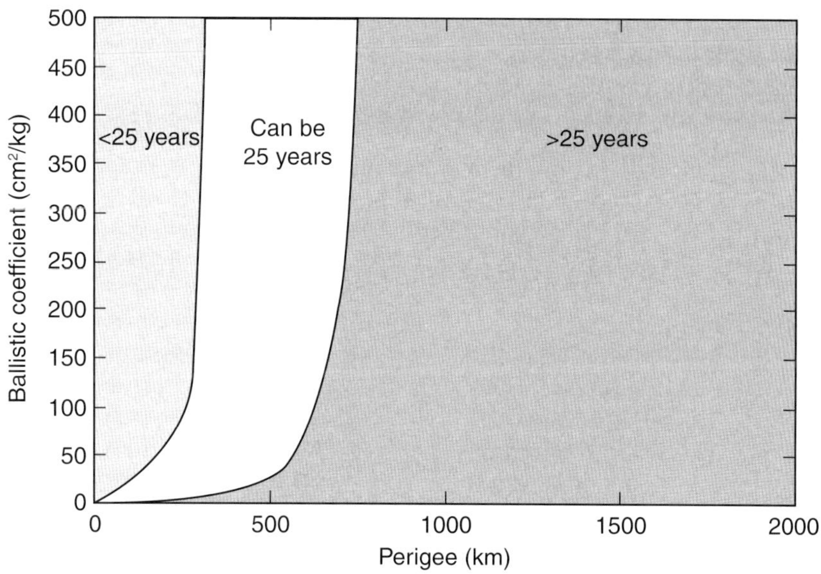

Fig. 21.17. Ballistic coefficient versus initial perigee altitude for all 25-year orbit lifetime cases (apogee <10,000 km).

Orbit Lifetime Estimation Equation Set:

Orbit Life in Years (H_p, β, inc & Ecc) = (21.6)

Life(circ alt of H_p) * Factor(H_{avg}, inc) * Factor(β) * Factor(Ecc)

where

a = orbit semimajor axis
A = satellite cross-sectional area with respect to the relative wind
β = ballistic coefficient of satellite = $C_D * A / m$
C_D = satellite drag coefficient
Ecc = orbit eccentricity
H_a = apogee altitude = a (1 + e) – R_e
H_{avg} = average of apogee and perigee altitudes
H_p = perigee altitude = a (1 – e) – R_e
inc = inclination of satellite orbit
m = mass of satellite
R_e = equatorial radius of Earth

Life(circ alt of H_p) =

- if: H_p < 550 km: Life(circ alt of H_p) = $F_1 * (H_p \text{ in km})^{F_2}$;
- else: Life(circ alt of H_p) = $F_1 * (H_p \text{ in km})^{F_2} * [1.0 + F_4 * (H_p - F_3)^{F_5}]$

Factor(H_{avg}, inc) =

- if: H_{avg} < 155 km: Factor (H_{avg}, inc) = 1.0;
- else if: H_{avg} > 550 km: Factor (H_{avg}, inc) = 1.3 – 0.00333333 * abs(90 – inc, deg);
- otherwise: Factor (H_{avg}, inc) = [-0.9307 * $\cos^2$(inc) – 0.2286 * cos(inc) + 2.1011] exp [K * H_{avg}],

where

K = (2.E-07 * (90 – inc, deg)^2 + 4.E-06 * abs(90 – inc, deg) – 0.0009) if 30 < inc < 150 deg
or K = –1.33333E-06 * (inc, deg) if inc < 30 deg
or K = –1.33333E-06 * (180 – inc, deg) if inc > 150 deg

Factor(β) = $[F_6 + F_7 * (\beta \text{ in cm}^2/\text{kg})^{F_8}]$

Factor(Ecc) = $\{1.0 + [F_9 * (H_p \text{ in km})^2/(F_{10} + (H_p \text{ in km})^3] * (F_{11}^{(F_{12} * Ecc)} - 1.0)\}$

Coeff:	Initially, use:	If resultant lifetime < 1 year:
F_1	6.350724982689956e-020;	6.350724982689956e-020;
F_2	6.295460674885141e+000;	5.946396392270378e+000;
F_3	4.820503248857688e+002;	3.185325139041694e+004;
F_4	3.919897205867653e-006;	3.137120106945147e+004;
F_5	2.367023890310601e+000;	3.137356808942190e+004;
F_6	-6.612928883060383e+000;	6.049861647189439e+001;
F_7	2.285691478658406e+004;	1.203161061917634e+005;
F_8	-8.492321907172038e-001;	-9.158828054010361e-001;
F_9	5.303219229102377e+005;	1.522209799457603e+005;
F_{10}	-8.474659230907487e+005;	2.718074721276698e+006;
F_{11}	1.000349462601627e+000;	1.000780241340044e+000;
F_{12}	8.790821659324631e+002;	2.065283320597687e+003;

21.3.3 Radio Frequency Interference

Because small satellites are typically deployed into LEO, the potential for RFI with respect to other background operational satellites and their respective ground stations is relatively low compared with the geosynchronous (GEO) regime. Instead, the primary RFI threat for a small satellite would probably come from the other satellites that were deployed with it.

RFI can occur both at the satellite (caused by two ground stations simultaneously attempting to communicate with their collocated satellites) or at the ground (caused by two collocated satellites attempting to communicate with nearby ground stations). Interference will occur if the desired transmitting antenna does not have sufficient link margin to overcome the sum of all other frequency-collocated transmission and noise sources.[17] The signal from a transmitting or interfering source received at a given antenna is computed by using standard link-margin equations, after accounting for the satellite transmission power, antenna gain, polarization losses, modulation losses, atmospheric losses, rain losses, and free-space loss, as well as the receiver's same link-budget items. These items depend on the system and usually are readily available for an operator's satellites and antennas.

Of the link-budget items, the antenna pattern often leads to the most variability in RFI analysis. As Figs. 21.18 and 21.19 show, the APS8 antenna model from the International Telecommunications Union (ITU) conservatively estimates antenna gain in terms of off-boresight interference. That is to say, if an antenna boresight (the main lobe or peak in Fig. 21.18) is directed at a desired satellite, the APS8 model allocates too much gain to an interfering satellite's slightly off-boresight signal. This result may be good (conservative) because an RFI prediction table that uses the APS8 model should capture all RFI events and more. But as seen in Fig. 21.19, the APS8 model may not be conservative when the signal is 40 deg or more off the boresight. As an interfering satellite approaches the position of the transmitting satellite, asymmetry in the actual measured antenna pattern (Fig. 21.19) and its many sidelobes make RFI a recurring cycle during an overall RFI event.

Figure 21.20 shows a sample plot of effective signal-to-noise ratio as a characterization of RFI and angular separation between two satellites as viewed by the receiving antenna. Note the general but not absolute correlation between angular separation and RFI. When the angular separation is small, RFI (lower values of signal-to-noise ratio) may occur. A communications link may require, for example, that the ratio of the carrier signal to the sum of the noise and interference be greater than 10 dB. In our sample case in Fig. 21.20, there are five time intervals when RFI would be most likely; during these times, satellite operators should avoid joint communications with their satellites. Users may also wish to assess the potential for RF power impingement on mission hardware[18] caused by terrestrial high-power transmitters, which may have sufficient power to damage delicate space hardware.

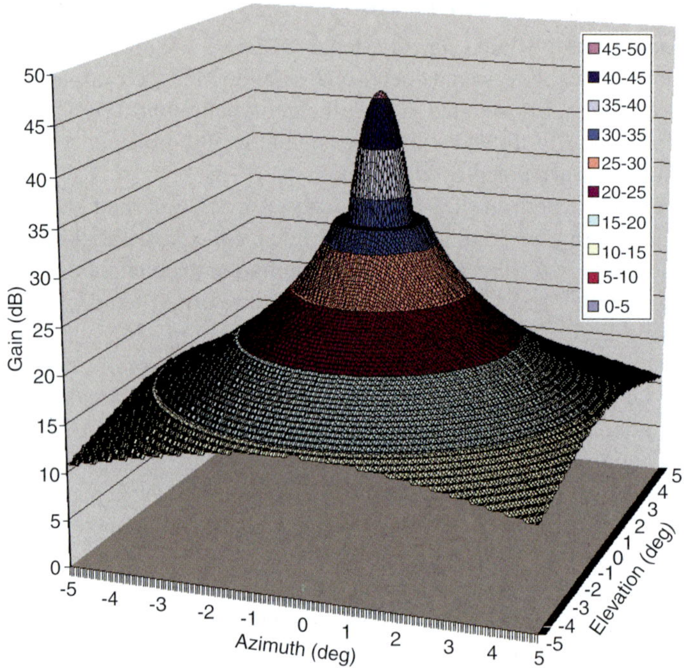

Fig. 21.18. APS8 gain pattern (ITU).

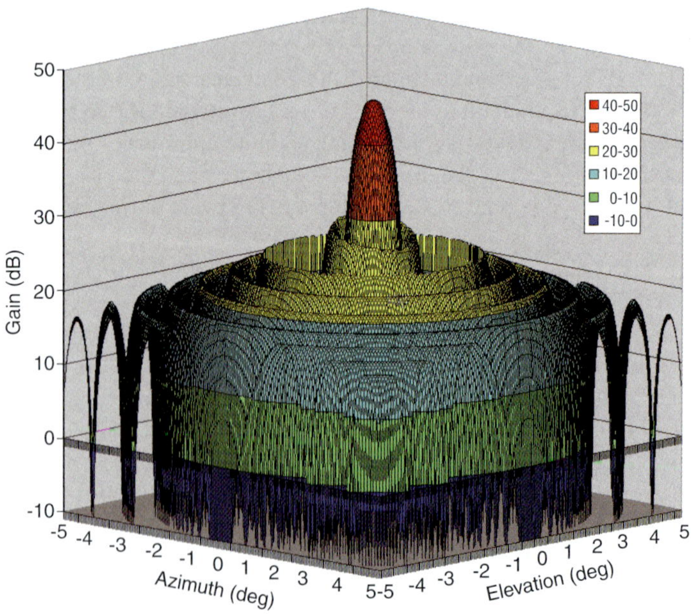

Fig. 21.19. Real antenna gain pattern (note radial asymmetry) in three dimensions.

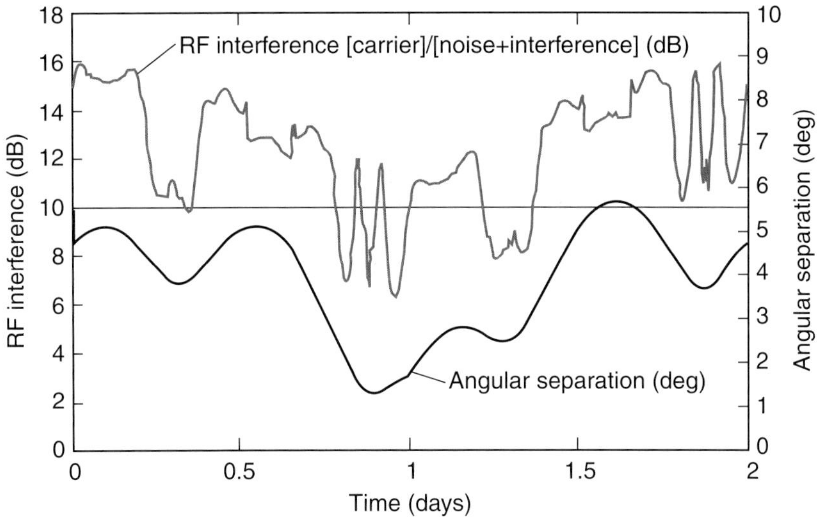

Fig. 21.20. Ratio of RF interference as defined by [carrier]/[noise + interference] vs. time for two GEO satellites.

21.4 Ground Station Selection, Placement, and Anticipated Link Margin

Small satellite operators are often limited by funding, geography, or other operational resources when exploring options for ground antennas and ground stations. Options seem to fall into three categories: new or existing ground stations co-located with existing satellite owners' facilities; external existing ground stations exhibiting availability, high performance, and reasonable pricing structure; or new ground stations placed at selected remote sites optimized to provide best satellite coverage for the specific mission.

A funding hierarchy and decision tree exists in satellite mission design for ground station placement:

- Mission goals, booster capabilities, primary payload constraints, and coverage requirements can dictate that the orbit inclination must be less than a certain limit. If so, then the ground station latitude must be very near to or less than that of the orbit inclination limit.
- Conversely, if the mission designer can specify the desired inclination or if the launcher-dictated inclination is already very near to or greater than the latitude of any ground site under consideration, then ground stations can be placed to optimize cost, ground coverage of the satellite, and physical access and convenience.

Typically, placing an antenna at northerly or southerly latitudes very near to the inclination of the orbit (or the complement of the inclination, for retrograde orbits) will provide optimal coverage of the mission. For example, for an orbit inclined at 50 deg, placing the ground station between 48 deg and 52 deg (either north or south latitude) would provide very good coverage; such placement provides consecutive-revolution coverage as Earth rotates beneath the satellite's inertially fixed orbit. This effect can be seen in Fig. 21.21, where the ground tracks transition to an easterly heading at the northernmost portion of the orbit.

The figure also shows that the specific orbit achieved by the launch vehicle may set up resonance areas of viewing with respect to certain ground locations. For example, placing a ground station in Vancouver would provide excellent coverage during the first four mission days. Unfortunately, the final planned set of orbital parameters may not be available in time to select an optimal location for a transportable ground station or to select among a set of potential stations. If foreign antenna placement is desired, insufficient time may remain between orbit definition and launch to obtain the necessary permits (e.g., country clearances, FCC approvals). Figure 21.22 shows the initial orbit revolution viewing geometry for the AeroCube-1 flight for a number of candidate ground station locations. The effects of subsequent viewing resonance can be seen in Fig. 21.23 as the pass peak elevation evolves with time; due to resonance between the satellite's in-plane motion and Earth's rotation, each ground station's set of passes will vary in a predictable fashion.

Yagi-Uda, helix, and parabolic dish antennas (Fig. 21.24) are the principal options for the small satellite user. Yagi antennas are relatively inexpensive and easy to maintain, have a very wide beamwidth (e.g., 30 deg), and can communicate in the presence of large satellite positional uncertainties; however, their gain is substantially lower than larger parabolic dish antennas (e.g., 15 dB vs. 40 dB).

The gain of a parabolic dish increases with dish diameter, while beamwidth decreases. Some missions require the use of a large-area antenna to receive extremely weak satellite signals. The 150 ft (46 m) diameter "Big Dish" antenna

Fig. 21.21. First four days of flight for the AeroCube mission.

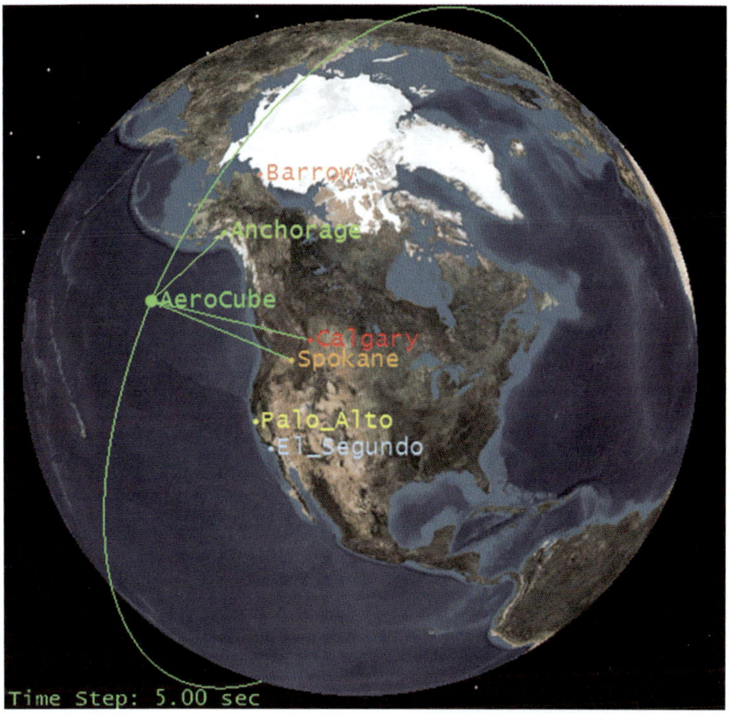

Fig. 21.22. Relative ground station viewing geometry.

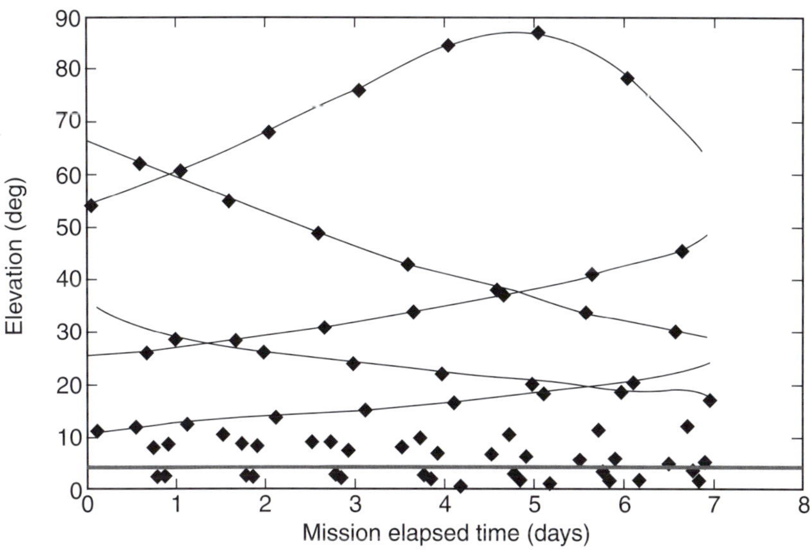

Fig. 21.23. Sample antenna peak elevation angles vs. mission elapsed time, with fared-in resonance curves depicting time variation of families of peak elevation.

Fig. 21.24. Example of low-gain, large-beamwidth helical and Yagi antennas (a). The "Big Dish" 46 m SRI antenna in Menlo Park, California (b). Canadian Defence Research Establishment Ottawa (DREO) 9.1 m dish antenna (c).

(Fig. 21.24c), located in the foothills behind Stanford University and operated by SRI International, was used to transmit and receive telemetry from the first picosats, described in section 21.6.1. It is sometimes useful to have such a dish with a very narrow beamwidth. The antenna can be pointed at the peak azimuth and elevation of a station pass to listen for a small satellite's signal. This has shown to eliminate the large in-track orbit uncertainty, dramatically reducing reliance upon third-party sensor and orbit data tracking providers.

21.4.1 Anticipated Communications Link Margin

Satellite communications subsystem design, ground station placement and performance, and anticipated antenna pointing accuracy are assessed to predict a satellite's anticipated link margin. These joint dependencies can make link-margin assessment a slowly converging, iterative process. All aspects must perform sufficiently well to establish successful communications. For an FM telemetry system, the link margin *M* (dB) of power received above noise is given by

$$M = \mathrm{EIRP} - L_S - L_O + G_R - L_R - \mathrm{Sen} \tag{21.7}$$

where

$$\begin{aligned} \mathrm{EIRP} &= \text{effective isotropic radiated power (dBm)} \\ &= P_T + G_T - L_T \end{aligned} \tag{21.8}$$

P_T = transmitter power (dBm)
G_T = transmitter antenna gain (dBi)
L_T = transmitter circuit loss (dB)
L_S = free-space path loss (dB)

$$= 20\log_{10}\left(\frac{4\cdot\pi\cdot\rho\cdot f}{c}\right) \tag{21.9}$$

ρ = slant range (distance in meters)
f = frequency (Hz)
c = speed of light (m/sec)
L_O = other losses (dB) (e.g., atmospheric and polarization losses)
G_R = receiver gain (dBi)
L_R = receiver circuit loss (dB)
Sen = receiver sensitivity (dBm)

Note that off-boresight pointing reduces receiver gain, as illustrated in Figs. 21.18 and 21.19.

As an aid to the reader, the parameter ρ (slant range) has been characterized as a function of orbital altitude and viewing elevation in Fig. 21.25. Assuming a transmission frequency of 1.0 GHz, Fig. 21.26 depicts free-space path loss as a function of elevation and altitude. For other transmission frequencies, the free-space path loss shown in Fig. 21.26 can be adjusted to the correct frequency via

$$\begin{aligned} &L_S \text{ at user frequency (dB)} \\ &\quad = L_S \text{ at 1 GHz (dB, from plot)} - 20\log_{10}(\text{user freq in GHz}). \end{aligned} \tag{21.10}$$

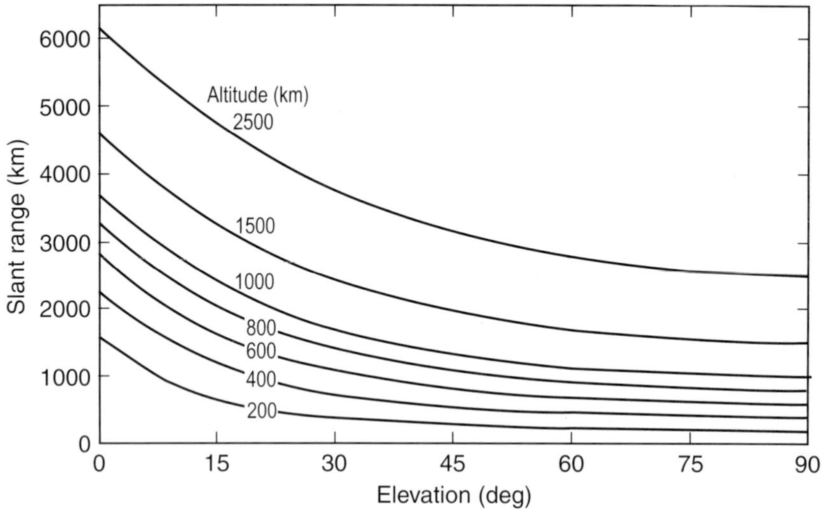

Fig. 21.25. Slant range (ρ) as a function of viewing elevation, by altitude.

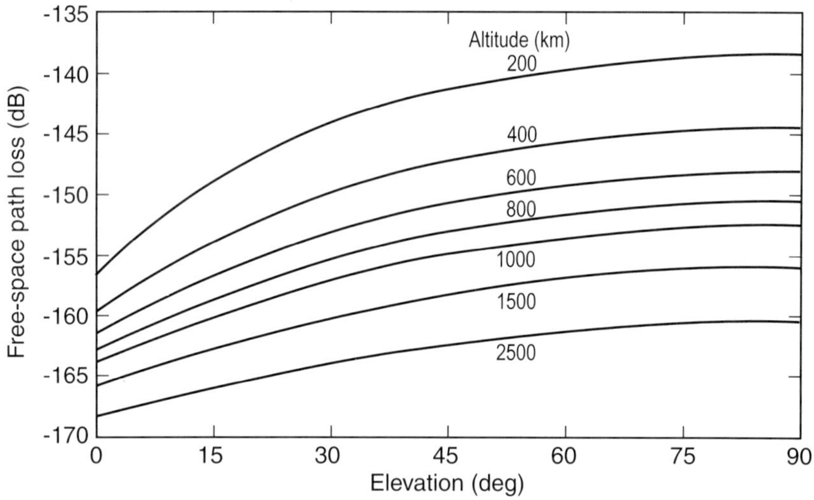

Fig. 21.26. Free-space path loss (L_s) at 1 GHz as a function of viewing elevation, by altitude.

21.5 Radar, Laser, and Optical Tracking

Many government, civil, and educational agencies operate radar, laser, and optical tracking facilities that may be able to support small satellite missions. These include the Space Surveillance Network (SSN), government research laboratories (e.g., the Air Force Research Laboratory), and universities. These sites can provide position data, attitude data (in some instances), and health and status of your small satellite.

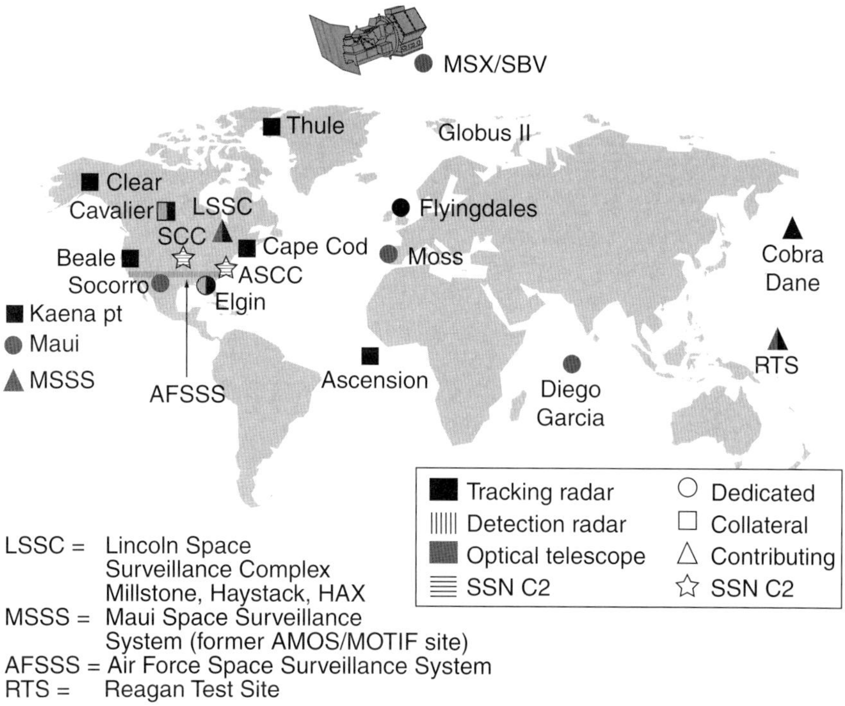

Fig. 21.27. The U.S. Space Surveillance Network (SSN).

The SSN[19] is a network of sensors located at approximately 20 sites worldwide and jointly operated by U.S. Army, Navy, and Air Force personnel as well as UK personnel (Fig. 21.27). The SSN uses three primary types of sensors to monitor Earth's population of artificial satellites: conventional radars, phased-array radars, and an optical system known as the Ground-based Electro-Optical Deep Space Surveillance system (GEODSS). UHF radars can most readily be utilized for SSN tracking and provide the most coverage.

Details on how to request space surveillance data can be found at http://www.space-track.org. In some cases, you may be able to aid the SSN in tracking your spacecraft (thereby yielding more accurate orbital data) by providing as much prelaunch relative-motion data as possible: e.g., event timing sequences, numbers and sizes of objects deployed, and other relevant information. This may also help the SSN resolve and properly tag the deployed objects.

Those using or operating a laser tracking site need to be aware of the relevant regulations. Department of Defense Instruction 3100.11, "Illumination of Objects in Space by Lasers, March 2000," directs all DOD laser owners and operators to coordinate laser activities directed above the horizon or in space with CINCSPACE. Responsibility for this coordination is currently assigned to the United States Strategic Command (USSTRATCOM). Civil and commercial laser owners and operators are also encouraged to coordinate with USSTRATCOM, which will support them as resources allow.

To fulfill these responsibilities, the Laser Clearing House was established to support the safe and responsible conduct of laser activities above the horizon to protect satellites and crewed spaceflight, and to be the primary interface between USSTRATCOM and satellite owners/operators for such concerns.

The Laser Clearing House registers each laser and assesses whether it poses a potential hazard to on-orbit objects. If the mode of operations does not present a hazard, the laser receives a waiver and does not require further USSTRATCOM coordination prior to use. For potentially hazardous lasers, the Laser Clearing House provides centralized predictive avoidance support for the laser owner and operator. If predictive avoidance tools determine that a susceptible satellite may be illuminated during the planned laser activity, then a control is mandated with the laser owner and operator to prevent laser emission at that time or place.

21.6 Satellite Design to Maximize Operations Success

When possible, the satellite operations team should be consulted early in the satellite design process. Several key components of the design process will subsequently help to ensure successful mission operations.

21.6.1 Radar Observability Enhancement

Large tracking radars bounce an RF signal off a satellite and then detect and process the returned signal to determine the satellite's position. Such radars are accurate at determining the relative range and range rate, but less accurate at determining angular position.

To facilitate tracking, a small satellite must be made as observable as possible to third-party tracking assets—which, from a radar-tracking standpoint, is accomplished by enhancing its radar cross section (external agencies have facilities[20] to help designers measure a satellite's current radar cross section across a wide range of frequencies for all viewing orientations).

An illustration of radar cross section enhancement can be seen in the pair of tethered picosatellites built by The Aerospace Corporation and flown in January 2000.[21] Analyses[22] of the Reagan Test Site ALTAIR radar's anticipated performance for this mission showed that it could not track at or below –25 dBsm (decibels relative to one square meter) below 50 deg elevation angle for the picosatellite's anticipated 720 km altitude circular orbit, as shown in Fig. 21.28. In fact, even in the best case (a direct overhead view at 90 deg elevation), the radar could not detect objects below approximately –28 dBsm.

A single picosatellite-sized box-shaped object (as viewed along orientation angles ϕ = azimuth and θ = declension—Fig. 21.29) may have a radar cross section that never exceeds –26.5 dBsm and spends much of the time below –28 dBsm (Figs. 21.30 and 21.31). Thus, without enhancements or other observable features, it may be difficult to detect, discriminate, and track. Options for dimensional

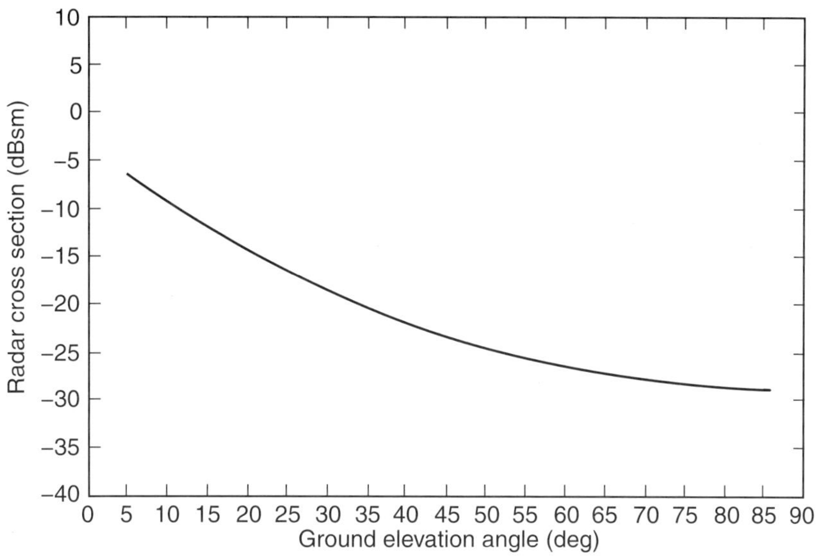

Fig. 21.28. Sample minimum detectable radar cross section vs. elevation angle for 720 km circular orbit, single 120 msec pulse.

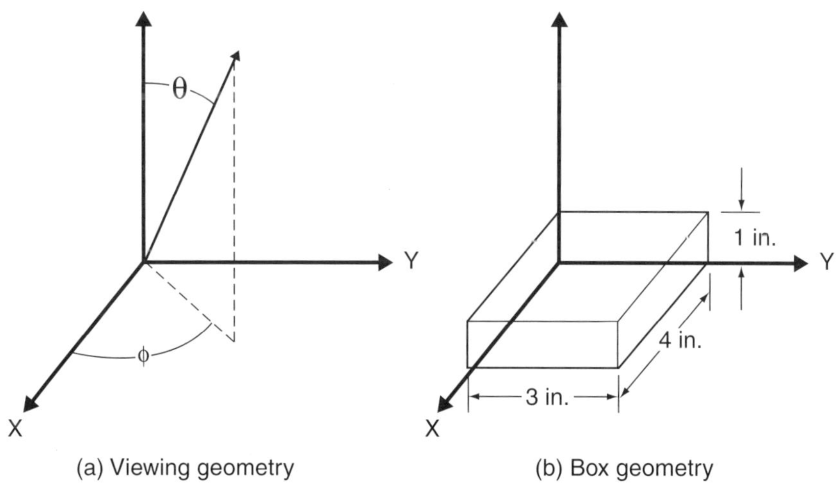

Fig. 21.29. Geometry of a picosat.

modifications to improve radar observation typically conflict with picosatellite mission goals, deployment constraints, and power and propulsion limitations.

Observability enhancements can help mitigate this issue. These can include the addition of spring-loaded or "measuring tape" metal dipoles, wire dipoles that can be woven into tether thread if the satellite uses a tether, or concave shapes incorporated into the design.

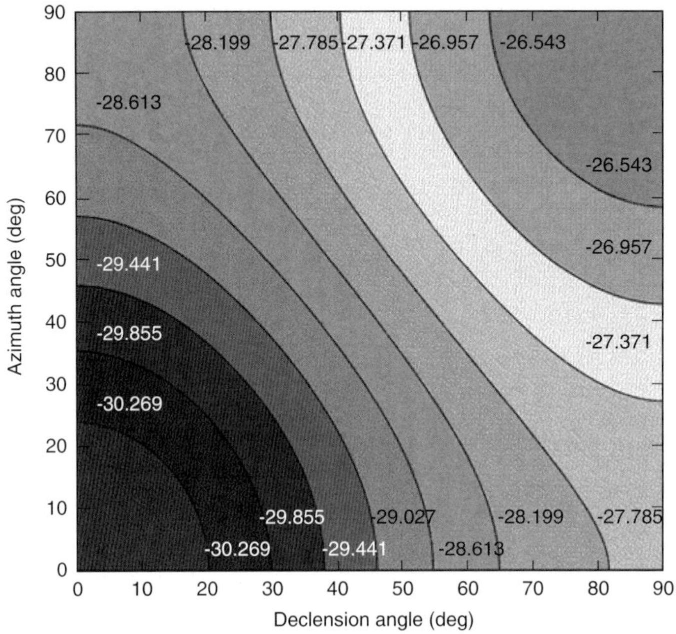

Fig. 21.30. Picosat radar cross section (dBsm), orthogonal polarization.

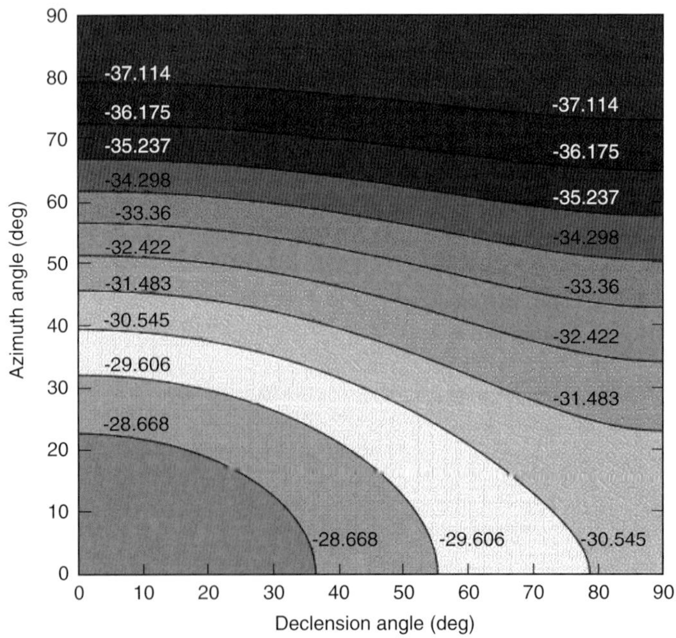

Fig. 21.31. Picosat radar cross section (dBsm), parallel polarization.

The radar cross section of the Aerospace picosatellites was enhanced via the introduction of tether dipoles, which offered the advantage of being light, compact, and easy to make. The small size of the picosatellites (4 × 3 × 1 in.) prevented other enhancement approaches. The specific goal was to enhance radar visibility at 217 MHz (NAVSPASUR search fence) and 440 MHz (UHF L-band radars). This was accomplished by weaving gold wire dipoles into the tether between the pair of picosatellites, sized at half wavelengths (65.74 cm for NAVSPASUR and 32.41 cm for L band).

If wire dipoles are used, it is recommended that a number of redundant dipoles be incorporated, sized at 0%, ±5%, and ±10% of the ideal half-wavelength dimension, to compensate for curled or coincident tether dipoles, inherently narrow dipole bandwidths, Doppler shift, dipole polarization dependency, and random dipole orientation angles.

Figure 21.32 shows the geometry of a single dipole of length l impinged by a radar's electric vector E^i of incident angle θ. When the radar signal's polarization plane contains the half-wave dipole (the parallel polarization), the calculated radar cross section at 440 MHz varies as a function of impinging angle, as shown in Fig. 21.33. Note that the dipole has virtually no cross section for the other polarization or when viewed end-on.

Figure 21.34 shows the frequency response of the dipole viewed from the side for several viewing angles. The resonances at multiple half-wavelengths are narrow. It should be noted that the dipole does not always have its peak radar cross section precisely at broadside when not at a resonance frequency.

The average of dipole radar cross section over all viewing angles versus frequency is shown in Fig. 21.35 for the parallel polarization. If one assumes there are multiple dipoles in random orientations and with random radar polarization, the curves in Fig. 21.35 represent radar cross section versus the number of dipoles. Maximum radar cross section, average radar cross section, and the 3-sigma upper and lower bounds on the average radar cross section are depicted.

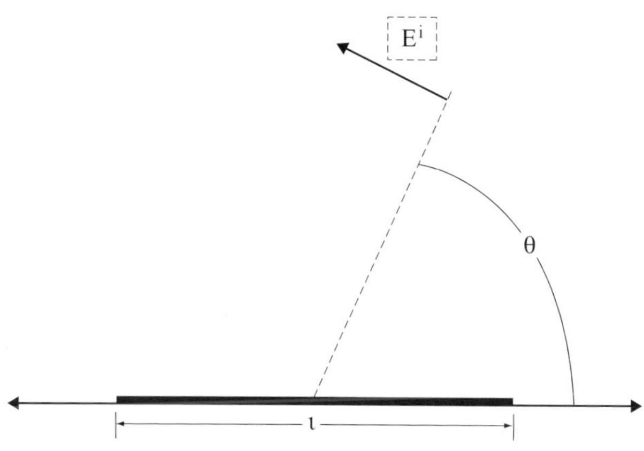

Fig. 21.32. Dipole geometry.

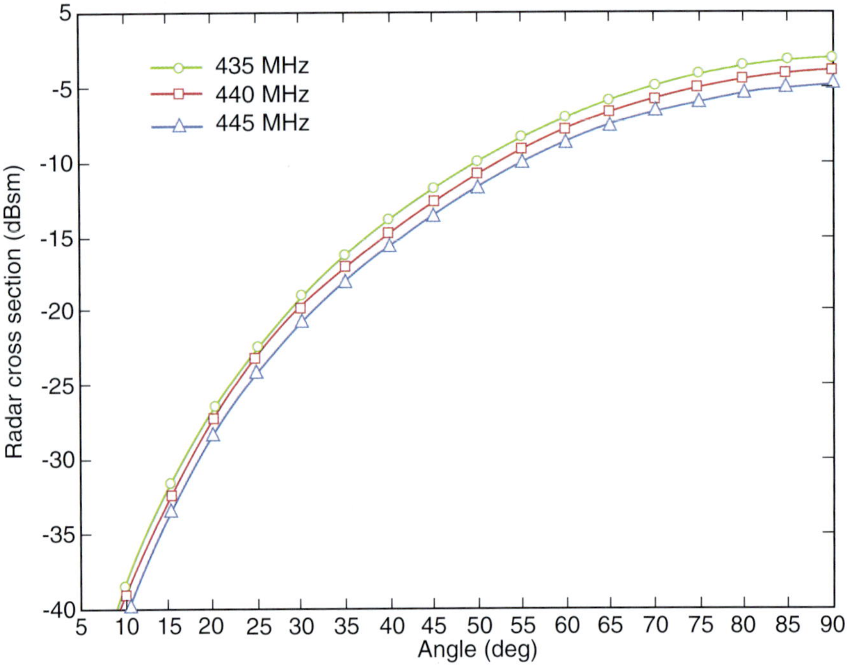

Fig. 21.33. 440 MHz dipole radar cross section vs. incidence angle θ.

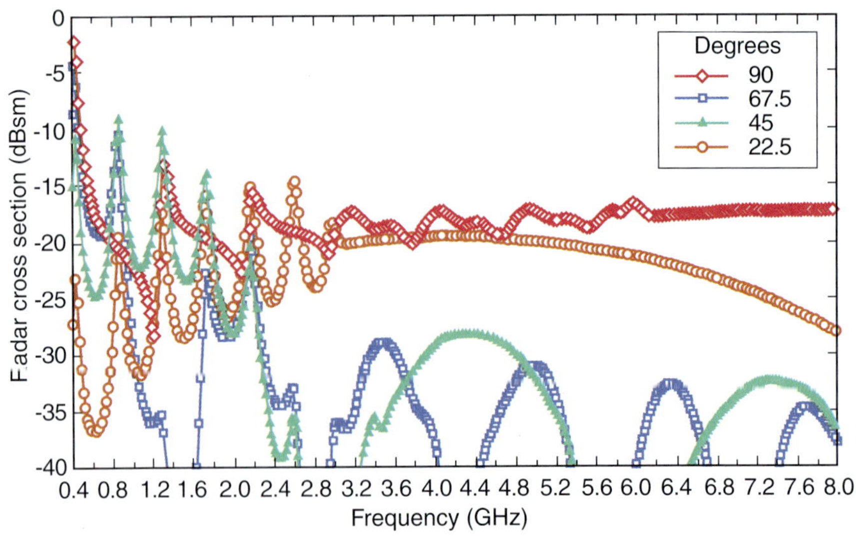

Fig. 21.34. 440 MHz dipole radar cross section vs. frequency.

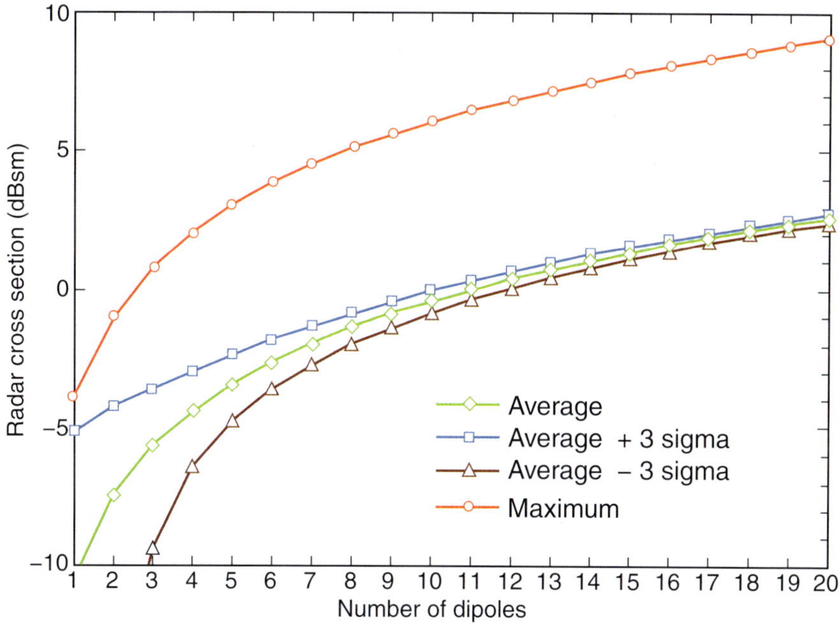

Fig. 21.35. Number of dipoles vs. radar cross section.

21.6.2 Optical Observability Enhancement

For optical trackers, the parameter corresponding to radar cross section is visual magnitude. Since small satellites generally do not have onboard laser emitters or other light sources, the best way to enhance visual magnitude is by increasing the satellite reflectivity. Optical sensors can then make use of either ambient light (i.e., sun or albedo) or the ground sensor's own active light source (i.e., a laser).

Satellites with highly reflective outer coatings will help optical tracking. For example, the well known Iridium satellite "flares" occur as the sun glints off highly reflective solar panels during the satellites' standard attitude orientation profiles.

Reflectivity augmentation can also be accomplished through the use of reflective tape such as 3M Scotchlite high-reflectivity sheeting[23,24] (though this tape is not necessarily space-qualified and may require testing or waivers prior to application in the flight environment). For a previous nanosatellite flight, 3000× luminosity "closed-cell" sheeting was rejected in favor of "exposed lens" retroreflective sheeting (with only 900× luminosity) because little was known about depressurization of closed-cell sheeting in space. Even for exposed-cell sheeting, outgassing can be a concern for satellites with high-quality optics. Potential users of the reflective sheeting might wish to consult existing outgassing test results (e.g., ASTM E 595, the "Standard Test Method for Total Mass Loss and Collected Volatile Condensable Materials from Outgassing in a Vacuum Environment").

Laser retroreflectors may also be used to augment reflectivity for active light sources. For example, Project Starshine flew three small, optically reflective spherical student satellites covered by ~1000 small, front-surface aluminum polished mirrors coated with a scratch-resistant, antioxidizing layer of silicon dioxide; faint sunlight flashes were visible to the naked eye against the star background and during certain recurring morning and evening twilight periods. Alternately, laser retroreflectors can be arrayed in characteristic patterns that may help with spin rate and attitude determination.

On one mission, concern that a lens mounted in the laser corner reflector device might break away during severe launch vibrations led the satellite designers to choose reflective sheeting instead.

21.6.3 Telemetry Tracking Observability Enhancement

Selecting communications and telemetry characteristics that set your satellite apart from other satellites deployed by the same host vehicle also aids the initial signal acquisition and tracking process. If certain telemetry signal characteristics (e.g., a directional antenna coupled with a certain deployment spin rate or attitude) are anticipated prior to launch, these can also be used when examining received communications to help identify your vehicle.

Another useful approach is to employ multiple communications frequencies and permit amateur radio operators to attempt to contact the vehicle to obtain health data. This is an inexpensive and potentially effective way to greatly expand the geographic coverage of your "virtual ground station" network. In this case, the satellite owner should take steps (e.g., authentication and firewall schemes) to prevent unauthorized radio operators from commandeering the satellite.

21.6.4 Satellite Identification Enhancement

As deployment scenarios become more complex and increasing numbers of satellites are deployed on a single launch (e.g., the 2007 Dnepr launch deployed 15 independent space objects), satellite designers should enlist all possible techniques to help third-party sensor and tracking agencies properly identify and discriminate among the space objects. This is particularly critical when the deployed satellites have an identical form factor. Third-party agencies work hard to ensure proper cataloging and tracking of each satellite, but complex missions with many picosatellites (e.g., 10 or 20) may prove to be unmanageable without further identification aids. These include unique signal characteristics (frequency, beaconing, on/off timing sequences, etc.), laser corner cube mounting configurations, and radar characteristics. Distinguishing radar cross section, visual magnitude, and telemetry characteristics are necessary but not sufficient for successful discrimination; you must make third-party radar, optical, and telemetry tracking sites aware of the unique features of your satellite. That is, any identifying characteristics unique to your satellite need to be properly analyzed, characterized, and conveyed to third-party trackers so that the initial search and track sequences can be as effective as possible. This should also include full analysis and disclosure of anticipated relative motion of all mission-deployed space objects. To be most helpful, you can even cast the relative motion analysis results into the same graphical formats that the radar operators are likely to see and use during the flight.

21.7 Operations Testing and Training

Two of the essential ingredients for a successful mission are testing and training. This means conducting as much personnel training and end-to-end systems testing as possible, with participation by all third-party agencies that will be involved during the mission. Limited funding often restricts these aspects in small satellite missions, and this is compensated by the hard work of a small group of dedicated and talented support personnel. As a result, training, testing, and rehearsals are often the first mission preparations to be eliminated. But, experience has taught us that prelaunch testing and training are vital to mission success.

At the ground antenna, steps should be taken to run end-to-end tests if possible (for example, from a nearby hilltop as shown in Fig. 21.36). Ideally, natural or manufactured far-field platforms can be used (e.g., a distant ridge near the antenna site) to conduct more realistic end-to-end antenna pointing, calibration, and communications tests. It's a good idea to incorporate attenuation into the cabling system to approximate on-orbit free-space loss and make the test more realistic

Fig. 21.36. Initial immediate vicinity communications test with test article on top of car.

(just don't forget to take the attenuators back out!). Use of a handheld GPS receiver at both the remote site and at the antenna boresight permits computation of the actual elevation, azimuth, and relative range reference where the antenna is pointing. The antenna can then be steered slowly away from the actual pointing direction to examine the communications system performance in off-nominal conditions. Be aware that antenna sidelobes, if not properly considered, can give the tester a false sense of antenna performance, pointing direction, and accuracy.

Visualization software and telemetry displays permit the operator to more easily monitor information in real time. Mission analysts use these tools to prepare for a mission and may ultimately use the same tools to support the mission. The fast-paced operations environment often dictates that software tools be rapidly reconfigured to adapt to a changing situation. An example of this kind of display is the visualization screen captured during the support of The Aerospace Corporation's picosat mission (Fig. 21.37).

21.8 Obtaining State Vectors and Generating Two-Line Elements

Prior to booster launch and small satellite deployment, the availability of state vectors, trajectories, and uncertainty metrics varies widely depending upon the launch provider. For expendable booster launches, this is often limited to prelaunch positional and uncertainty estimates, which are often of marginal accuracy for precise in-track pointing of narrow beam antennas. For on-orbit launcher hosts (such as the space shuttle), high-accuracy vectors may be obtainable based upon ongoing sensor and telemetry tracking.

The small satellite operator should always request all available state vectors and deployment orientation information from the launching or deployment

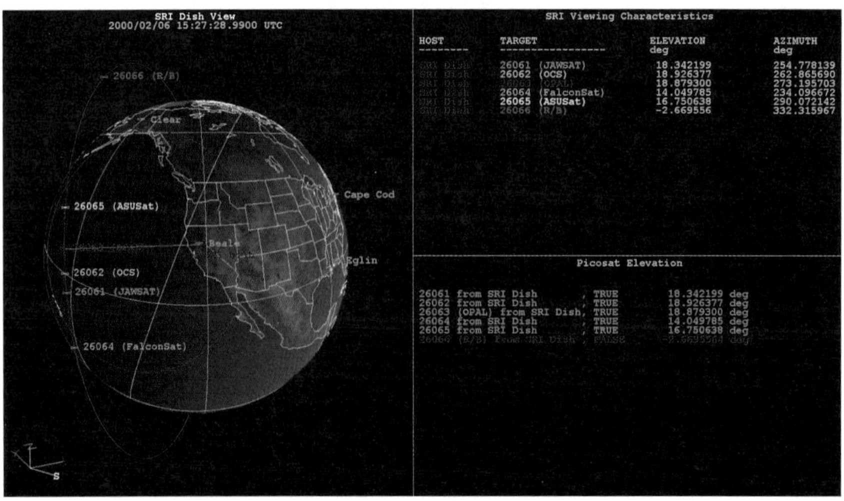

Fig. 21.37. Real-time visualization of picosatellite orbits and communications links.

agency irrespective of their identity. This approach gives the antenna tracking engineer options, should the primary tracking data source fail.

Depending on the size of your operations budget, you could perform your own orbit determination and TLE generation using GPS measurements from your satellite, ground-based measurements made from laser retro-reflectors, or other tracking methods. This would give you more control of your tracking data and might offer more precise estimates of your mission orbit.

21.8.1 Obtaining Two-Line Element Sets

USSTRATCOM tracks all Earth-orbiting spacecraft, rocket bodies, and debris, maintaining a catalog of orbital elements for each object. These sets of elements are formatted into two lines of text containing mean (average) orbital parameters analogous to classical orbit elements. They are known as two-line elements, or simply TLEs.

Once a small satellite is detected and tracked by several space surveillance radars, USSTRATCOM begins to estimate these orbit parameters and adds them to the catalog (which can be obtained from various sources). Initially, the orbit fit is imprecise, and occasionally, small satellites may be misidentified or confused with other satellites—especially those released from the same launch vehicle. Visibility predictions using early TLEs may not be accurate enough for precise antenna pointing. Generally, after several days of tracking, USSTRATCOM will collect enough radar data to create more accurate orbital parameters, and published TLEs can be used for more precise tracking.

Tracking and communications are most critical immediately following the small satellite launch, when the TLEs from USSTRATCOM are least accurate. If your tracking software requires TLEs, you may be forced to create your own using predicted orbit elements or state vectors provided by the launching agency.

21.8.2 Creating Two-Line Element Sets

Care must be taken not to convert an osculating (instantaneous) set of elements into classical orbit parameters and simply insert those into the TLE format. USSTRATCOM generates its mean element sets by removing periodic variations in a particular way, and you must use a similar method to generate a set that is consistent with the propagation model. Hoots and Roehrich[25] describe the propagation models used by USSTRATCOM and present FORTRAN source code for the programs. Independent users have translated this source into other computer languages that can be found with a web search. The current propagation model is known as SGP4 (Simplified General Perturbations Model #4), which comprises two subparts, one for near Earth and the other for deep space.

The process of creating a TLE from an osculating set will require one or more position and velocity vectors spanning the period of interest. These can be created using any of the many available orbit propagation programs, or the launching agency may provide an ephemeris based on prelaunch simulations or postlaunch vehicle telemetry. In either case, a more accurate force model (i.e., inclusion of higher-order gravity perturbations, third-body effects, and solar radiation pressure, or use of a better atmosphere model) will result in a better orbit prediction and, consequently, better pointing accuracy. These elements are known as osculating because they are the instantaneous state of the satellite. Mean elements, on the other hand, can have a vague definition. The definition of "mean" can only be precisely given in the context of the general perturbation theory from which they are derived. For our application, that is SGP4.

The simplest process for creating a TLE is a fixed-point iteration that is performed on a single-state vector. An initial set of mean element values are assumed, and the osculating elements resulting from the USSTRATCOM SGP4 propagation are compared against the epoch vector. Differences are computed and the process is repeated, differentially correcting the mean elements until the resulting osculating elements are sufficiently close to the given epoch state. This is known as an epoch point conversion (EPC) and is described by the formula

$$\underline{X}_m^{k+1} = \underline{X}_m^k + \left[\underline{X}_o - \underline{f}\left(\underline{X}_m^k\right)\right] \tag{21.11}$$

where

$\underline{X}_m$ = six vector of mean elements at epoch

$\underline{X}_o$ = six vector of given osculating elements at epoch

$\underline{X}_m^k$ = current estimate of the epoch state vector

$\underline{f}\left(\underline{X}_m\right)$ = six vector of osculating elements obtained by evaluating the SGP4 theory using the current estimate of the mean elements

k = iteration counter

This will converge to a consistent set of mean elements that will match the given osculating elements. Unfortunately, this does not incorporate the information contained in the remaining state vectors in the ephemeris. The accuracy of the propagated state using an epoch point conversion TLE will degrade rather quickly over a long prediction span.

A better method for incorporating the entire ephemeris is known as precise conversion of elements (PCE), a term coined by Paul Cefola of Draper Laboratory to denote the process of least-squares correction when applied to the orbit initialization process. It is possible in some cases to estimate the ballistic coefficient when the ephemeris file contains the effects of atmospheric drag. The theory of batch orbit determination can be found in standard astrodynamics texts. The basic iteration is expressed by the formula

$$\underline{X}_m^{k+1} = \underline{X}_m^k + \left[H^T W H\right]^{-1} H^T W \left[\underline{Z} - \underline{f}\left(\underline{X}_m^k\right)\right] \tag{21.12}$$

where

$$H = \frac{\partial \underline{f}\left(\underline{X}_m^k\right)}{\partial \underline{X}_m^k} = \text{Jacobian matrix}$$

$\underline{f}\left(\underline{X}_m^k\right)$ = six vector of osculating elements obtained by evaluating the SGP4 theory with the current estimate of the mean elements

$\underline{Z}$ = vector of observations (state vectors in ephemeris file)

W = diagonal weighting matrix of the observations

Note if $H_k \equiv I$ and $W \equiv I$, this reduces to an epoch point conversion for a fixed vector at epoch.

While epoch point conversion is certainly less desirable than a precise conversion of elements, it does have the advantage of being simple to implement. It requires only the initial osculating state vector rather than the full ephemeris file of predicted state vectors for precise conversion. We have found that precise conversion of elements produces accurate predictions over several days using only the simple gravitational force model contained in SGP4. The TLE parameters are adjusted in the differential correction process to "smooth out" errors in the fit.

21.9 Deployment Sequencing and Recontact Avoidance

By properly planning and executing a sound launch deployment sequence, numerous satellite operations difficulties can be avoided. If possible, strive for the following in your deployment sequence planning:

- Deploy your satellite individually, rather than as part of a group, with suitably long delays between subsequent deployments.
- Deploy the satellites during thrusting portions of launch vehicle flight, to place each satellite into a slightly different orbit; for example, Fig. 21.38 depicts the simulated deployment sequence for the Dnepr launch of 2006, which occurred during powered flight, yielding the ordered and predictable subsequent relative motion shown in Fig. 21.39.
- Ensure that all co-deployed satellites receive some form of relative motion with respect to your satellite, using springs, microthrusters, or different deployment directions.
- Ensure that deployment mechanisms using multiple springs cannot work to counteract themselves and render a zero-net deployment rate.
- If time permits, let the launch host achieve and maintain orbit for a suitable time to permit external organizations to track and refine the orbit prior to deployment.

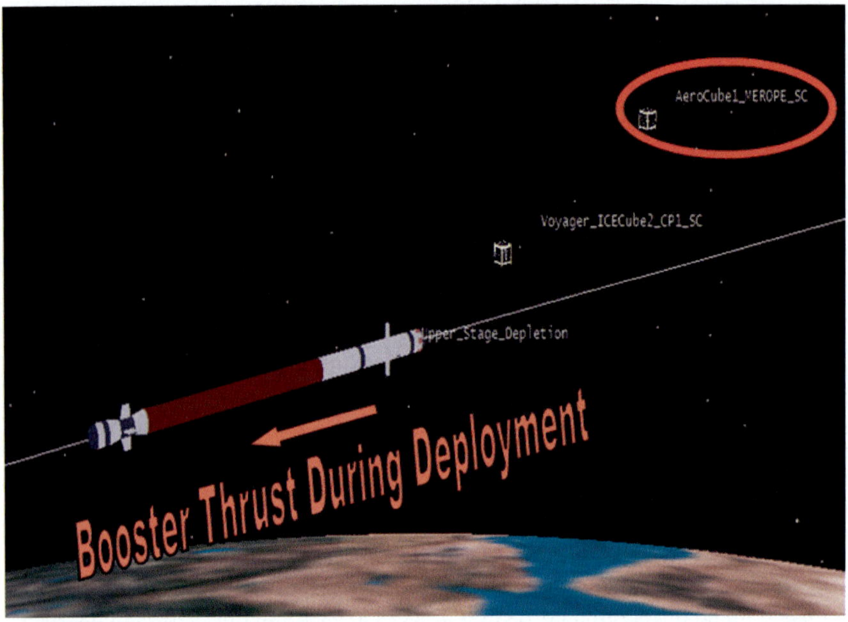

Fig. 21.38. Dnepr deployment of AeroCube.

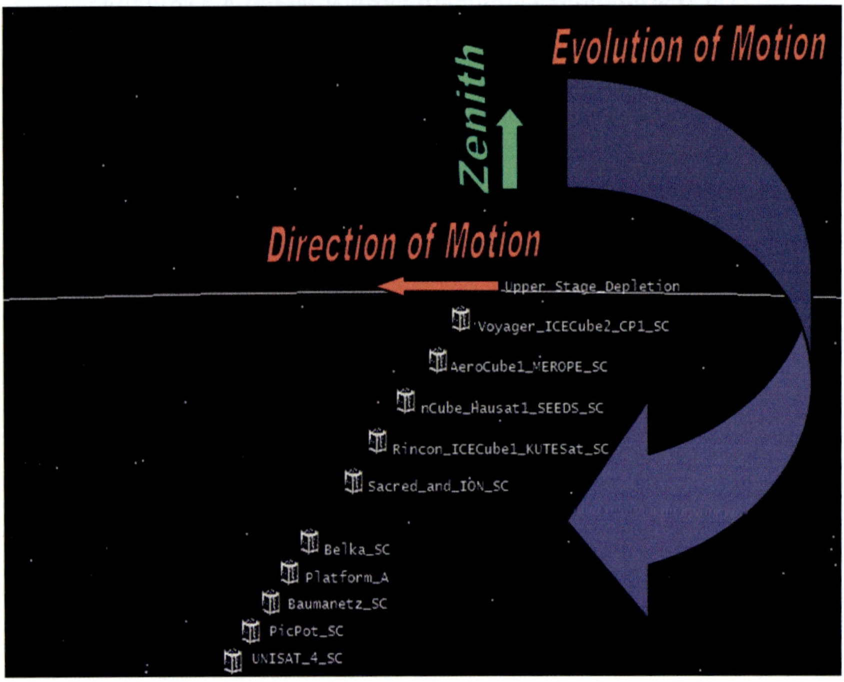

Fig. 21.39. Evolution of satellite relative motion roughly one-half orbit postdeployment.

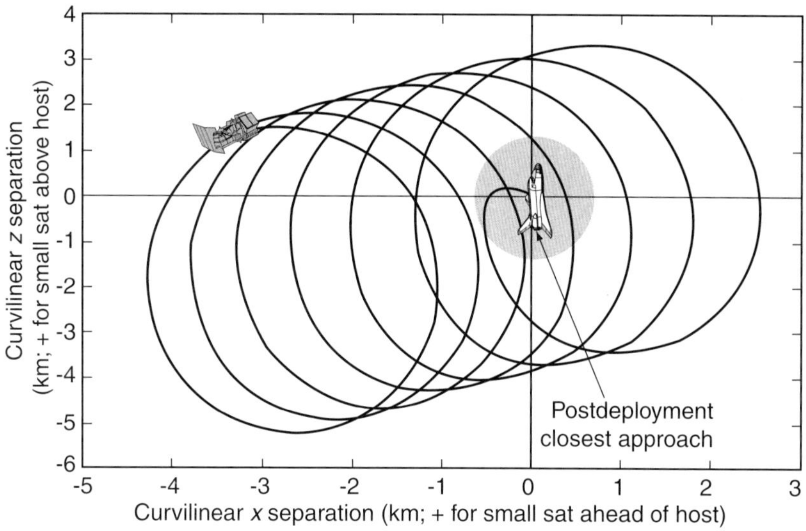

Fig. 21.40. Deployment spring ΔV-imparted relative range.

- Select the deployment times to be just prior to or during a sequence of high-elevation telemetry antenna passes.
- Avoid complex host-vehicle propulsive and reorientation maneuvers prior to your satellite's deployment because each maneuver can introduce large uncertainties in subsequent deployment event positions and times.
- Analyze the predicted relative motion of all deployed objects and supply this information to any external tracking agencies to aid in object identification.

Operations planners should also seek to minimize the risk of postdeployment recontact with other deployed satellites and associated debris. As shown in Fig. 21.39, deployed satellites are in a constant state of motion with respect to each other. Relative motion plots (Fig. 21.40) can be used to assess potential recontact risks.

Relative motion is initiated by deployment springs, thrusting, differential drag, and solar radiation pressure, or other orbit perturbations differences. For example, Fig. 21.41 illustrates the in-plane relative motion caused only by changes in drag coefficient. The goal of such analyses is to design the deployment sequence and select deployment spring velocities and deployment attitudes to minimize the risk of recontact.

When recontact is predicted by such analyses, the analyst can characterize this risk as a function of deployment magnitude, direction, and timing. A sample deployment attitude characterization is shown in Fig. 21.42. The plot shows the closest approach between the launch host and the deployed satellite during the three days following deployment as a function of deployment attitude; deployment velocity is held constant for all attitudes. To minimize recontact risk, the

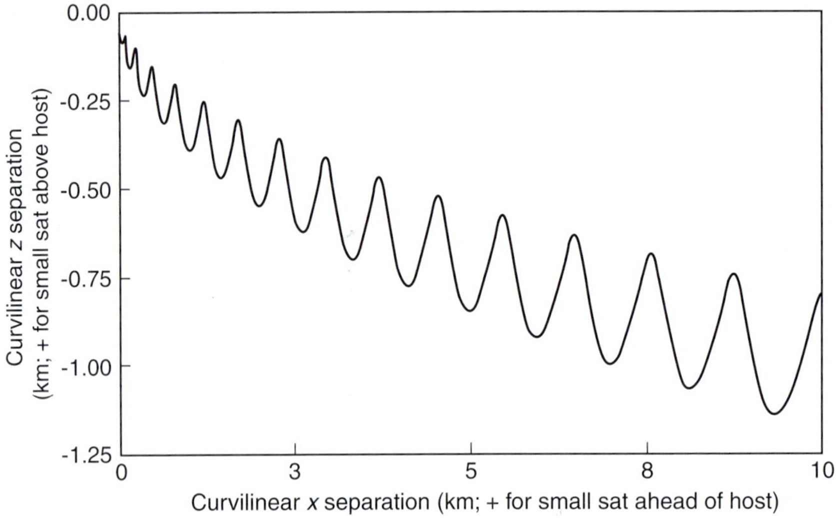

Fig. 21.41. Differential drag-induced relative motion.

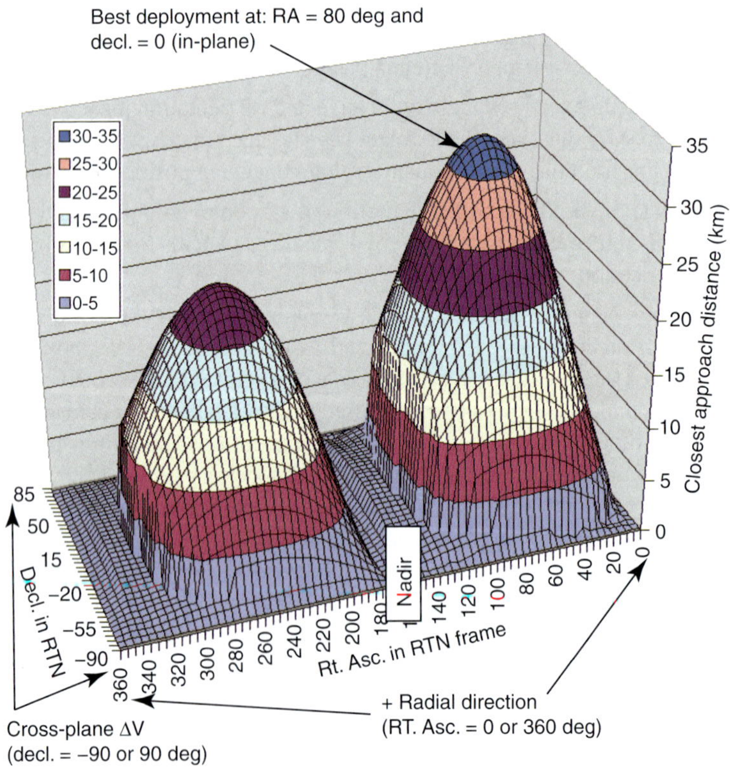

Fig. 21.42. Closest postdeployment approach vs. deployment direction.

deployment attitude—depicted as right ascension "Rt. Asc." and declination "Decl" in a radial transverse normal frame—that maximizes the subsequent closest approach distance (i.e., the top of the highest peak) should be selected.

The following observations can be deduced from Fig. 21.42:

- Deploying the satellite in a purely cross-plane (i.e., out of the orbit plane) direction is risky and will result in a very near miss, if not recontact and collision.
- Deploying in a purely radial or nadir direction is risky and will result in a very near miss, if not recontact and collision.
- Deploying in-plane, with a combination of along-track and radial directions, will maximize relative miss distance, thereby minimizing the chance of postdeployment recontact during the following three days on orbit.

21.10 Ground Telemetry Antenna Operations and Pointing

Even if a small satellite is successfully acquired upon initial deployment, a communications-based method of tracking is prudent as a backup because of continued postlaunch error growth. To maintain contact regardless of SSN tracking availability, a method must be implemented that quickly detects and tracks the small satellite's orbit after deployment. One should apply the best obtainable orbit and uncertainty data, coupled with the best analysis algorithms and software, to create an automated spiral search pattern for tracking with the ground telemetry/downlink dish.[26,27]

21.10.1 Ground Antenna Track Scan Generation

In the launch and early orbit phase of flight, viewing dispersions (Figs. 21.43 and 21.44) can be accommodated with software designed to improve acquisition and ensure continued satellite tracking, using simple conical, elliptical, spiral, and vermicular search scanning techniques (Figs. 21.45 and 21.46). For example, these techniques were applied in the complex picosatellite deployment in 2000 in which an initial satellite (JAWSAT) deployed a daughter satellite (OPAL) that then deployed a number of granddaughters (picosats, including the two Aerospace picosats).

Covariance matrix techniques and the mapping of resultant ellipsoidal position uncertainties into platform viewing-angle variations permits such software to account for uncertainties from the launch process, separation velocity magnitude and orientation dispersions, attitude uncertainties, prediction errors, and any other known uncertainty to determine a bounding volume that conservatively encloses the satellite's actual position.

Returning to the Dnepr flight covariance example contained in the previous section, an antenna scan can be constructed (as characterized by Figs. 21.47, 21.48, and 21.49) that ensures that the antenna's high-gain pointing region (e.g., Fig. 21.19) suitably covers any position dispersion that the vehicle is likely to

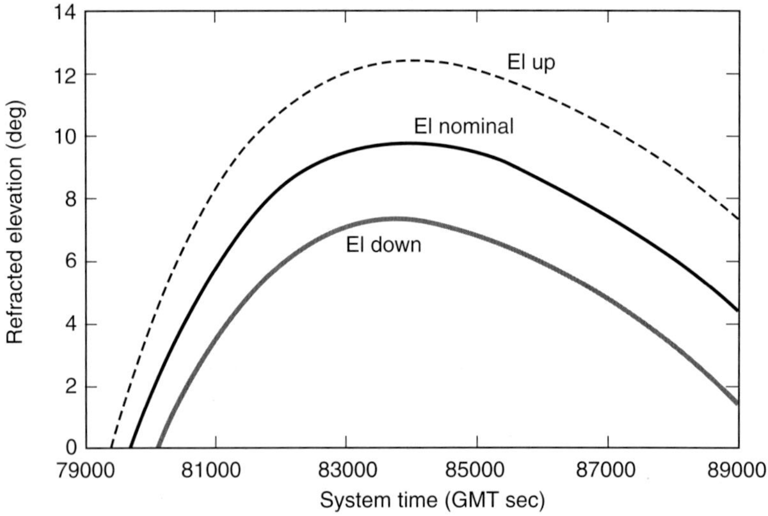

Fig. 21.43. Elevation uncertainty at ground station as a function of time.

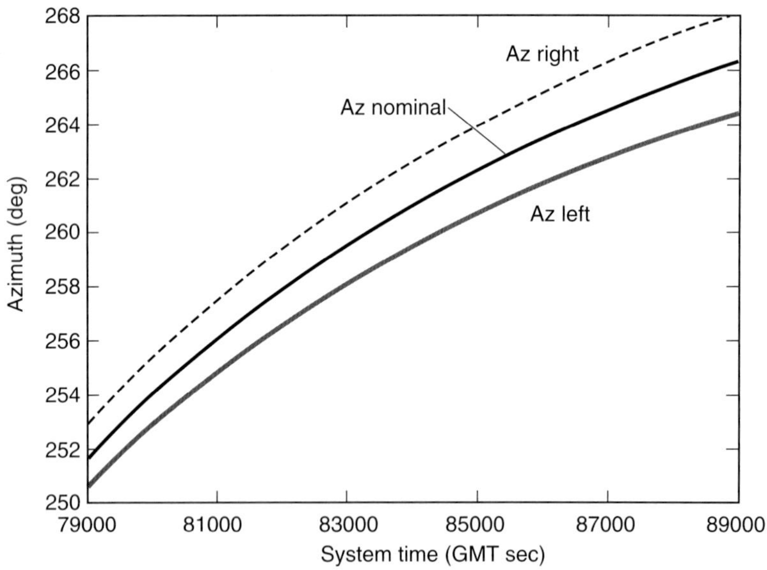

Fig. 21.44. Azimuth uncertainty at ground station as a function of time.

experience. Figures 21.47 and 21.48 represent the shift from the nominal azimuth and elevation tracking profiles, respectively.

The analysis software can dynamically adjust the scan frequency and sigma level of the uncertainty region that the scan covers, as shown in Fig. 21.48. Note that the azimuth and elevation profiles have been created using a repetition period of 30 sec, with the sigma values ranging from 3-sigma (at rise and set) down to

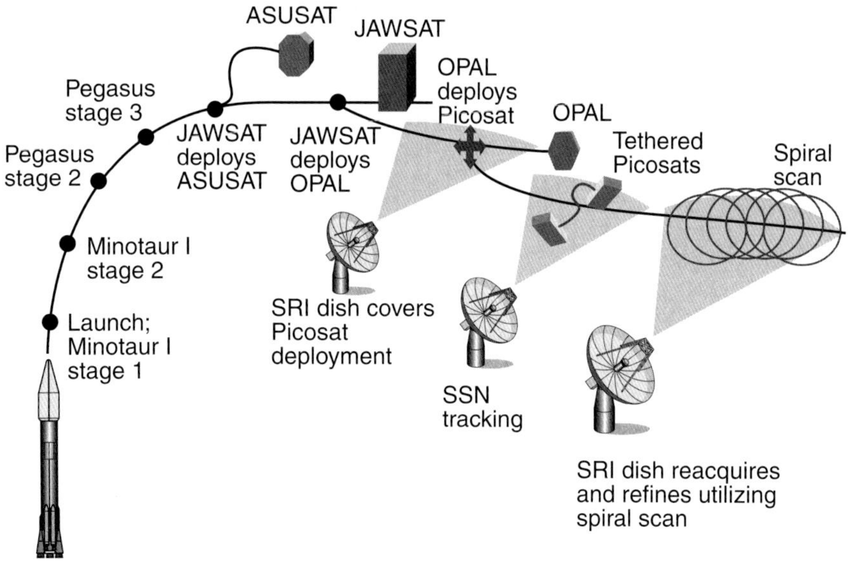

Fig. 21.45. Complex picosat deployment sequence with spiral scanning technique.

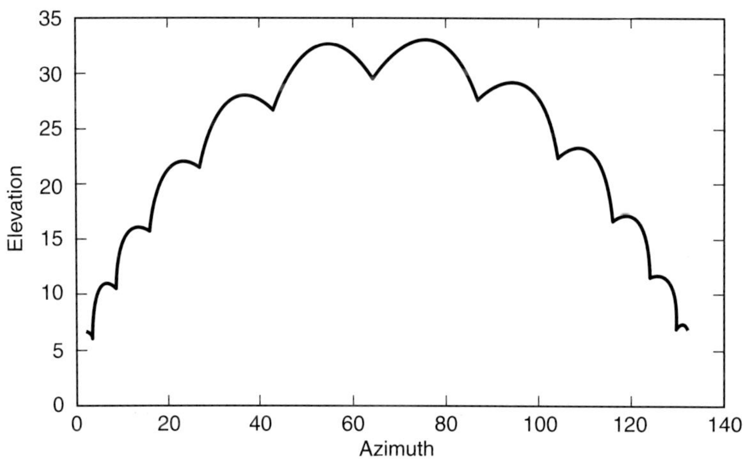

Fig. 21.46. Sample application of a suitable scanning technique.

1.5-sigma at the peak of the pass. Reducing the sigma value during the midpass portion of the contact helps ensure that telemetry antenna elevation and azimuth tracking rate hardware limits are not exceeded, as shown in Fig. 21.49.

There are several practical considerations in using this antenna scan technique: first, the antenna must dwell long enough to permit the satellite to be detected in its high-gain region during any frequency search scan; second, azimuth and

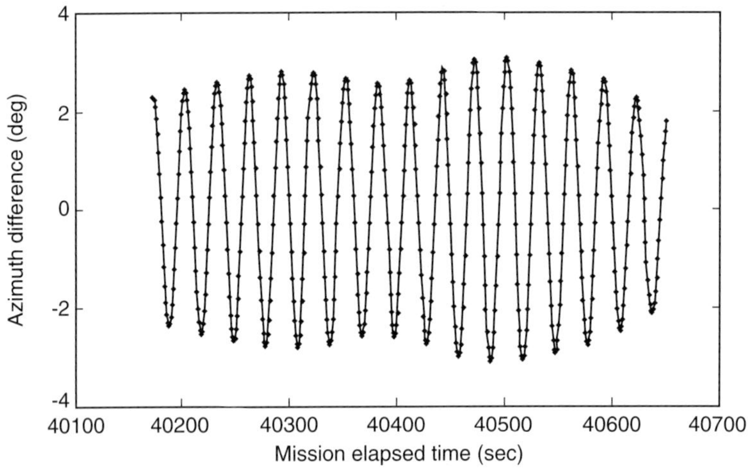

Fig. 21.47. Azimuth shift vs. mission elapsed time for elliptical scan.

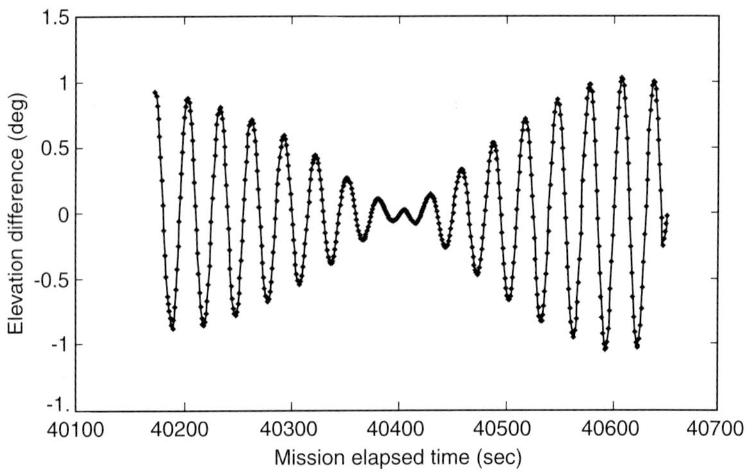

Fig. 21.48. Elevation shift vs. mission elapsed time for elliptical scan.

elevation angles, rates, and accelerations must not exceed antenna hardware and software constraints; third, the antenna servos and tracking mechanisms tend to experience more pointing error as angular tracking rates increase (Fig. 21.50).

On a more humorous note, the first practical application of this technique was accomplished at the SRI Big Dish (Fig. 21.24b), which has a control house that is integral to the antenna structure—that is, when the antenna moves during the track, the personnel controlling the antenna move as well. Suffice it to say that there were a number of people whose faces were a ghastly shade of green upon completion of a high-frequency scan exercise.

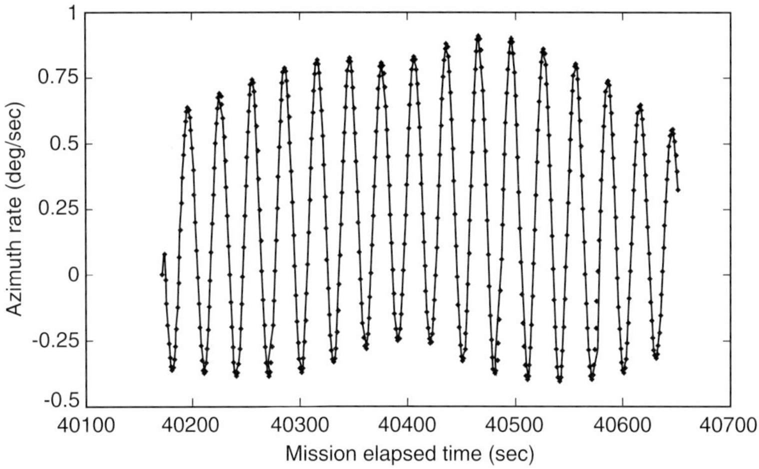

Fig. 21.49. Azimuth rate vs. mission elapsed time for elliptical scan.

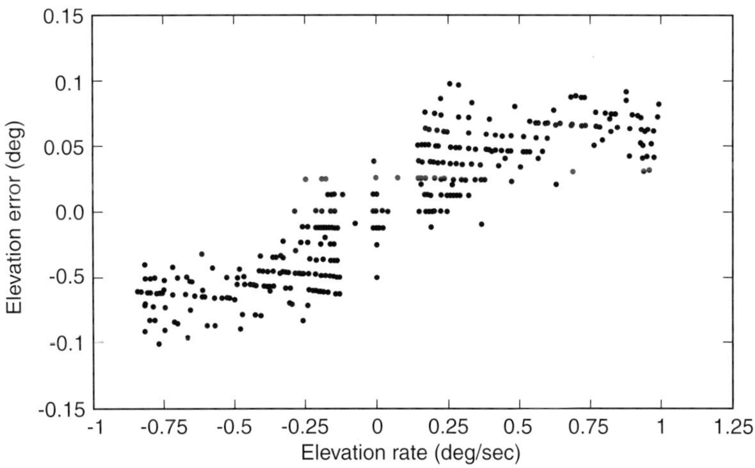

Fig. 21.50. Correlation between antenna elevation pointing error and elevation rate.

21.11 Postlaunch Drag and Maneuver Performance Assessment

21.11.1 Drag Characteristics Estimation

It can take a while for small satellite drag characteristics to be observable, due to their frequent lack of solar-array appendages; this yields a relatively low ballistic coefficient. Once the satellite is launched and after a number of reliable orbit state vectors have been obtained spanning approximately a 1-week time period (depending upon orbit altitude), the satellite's actual ballistic coefficient characteristics can be empirically determined.

Table 21.1. Ballistic Coefficient Estimation via Iteration

Iteration	Ballistic Coefficient (cm^2/kg)	Mean Argument of Latitude (initial epoch+482287.150944 sec)
#0: Actual result/goal	—	172.71 deg
#1	310.52	177.05 deg
#2	210.00	171.36 deg
#3	226.25	172.70 deg
#4	226.38	172.71 deg
#5	226.47	(solution/convergence)

As a simple example, we look at two CubeSat vectors spaced 5.6 days apart. Based upon initial mass and dimension and adopting the MSIS E2000 atmosphere model, a rough value between 228 and 310.5 cm^2/kg was anticipated. Propagation of the first orbital state up to the epoch of the second orbital state allows direct comparison of actual versus predicted orbital elements. Constructing the partial derivatives of the orbital element differences with respect to ballistic coefficient permits a ballistic coefficient iteration to match a singly-averaged mean argument of latitude value with the actual value. Ballistic coefficient iteration (Table 21.1) yielded 226 cm^2/kg for this sample case.

21.11.2 Deployment Separation ΔV Estimation

There are several ways to estimate the velocity change that occurs during deployment. Depending upon the deployment mechanism, methods may include video capture, iteration between bounding orbital states, or ephemeris reconstruction (discussed earlier in this chapter). We now examine application of the first two methods to the 2005 shuttle deployment of the tethered MEPSI satellites. For this mission, the nominal spring-induced ΔV was estimated at 2.5 m/sec ±10%.

Our first attempt was to determine the ΔV from an examination of deployment videos. If such video footage is available, they provide a reasonably good method for assessing separation velocity. Pictures from a camera mounted at the front of the shuttle cargo bay are shown in Figs. 21.51 and 21.52. As seen from the snapshot in Fig. 21.52, a frame does exist showing the time the first MEPSI vehicle reached the tip of the orbiter tail at a time of 2.14 sec after passage of the beginning of the tail. Using the orbiter dimensions, approximate shuttle bay geometry, and MEPSI deployment configuration, an approximate ΔV of 2.34 m/sec was determined.

An alternate method for estimating the deployment ΔV was to couple the previously estimated ballistic coefficient with the highly accurate NASA deployment vector, and then propagate the postdeployment state to the next available orbital state. Again, comparing mean argument of latitude at a known point, the deployment ΔV was estimated as shown in Table 21.2. Adopting this ΔV value, a further refinement of ballistic coefficient yielded 232 cm^2/kg.

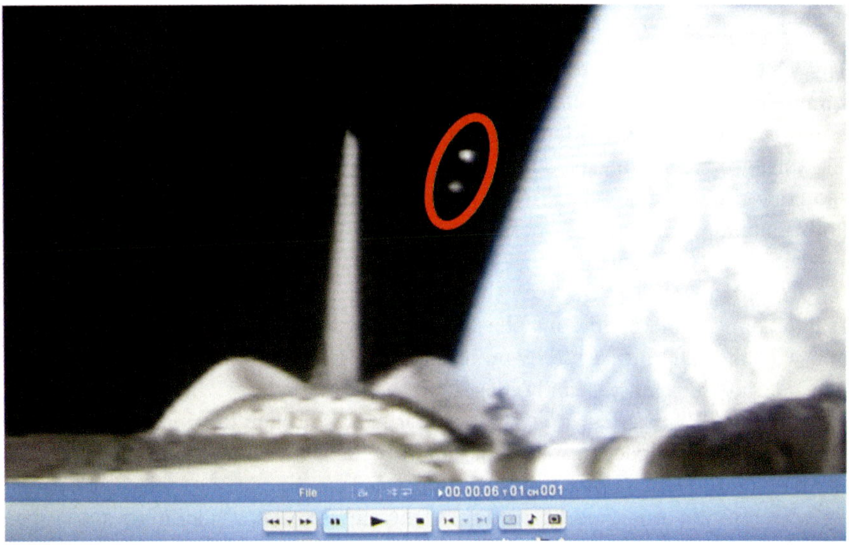

Fig. 21.51. End-on perspective of MEPSI deployment.

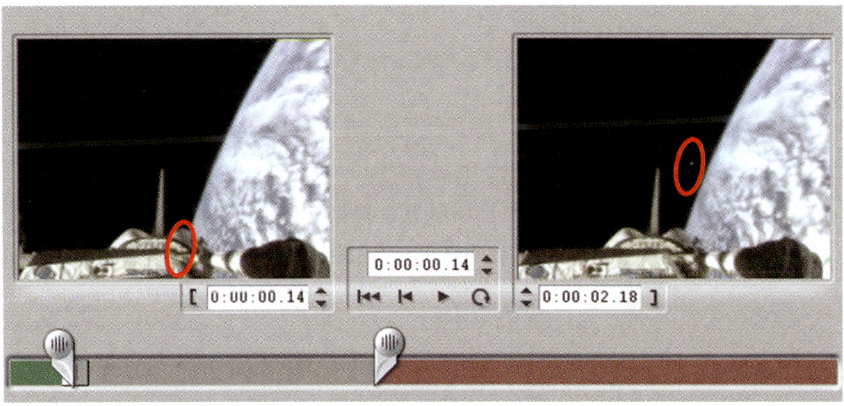

Fig. 21.52. End-on perspective of MEPSI deployment in high-quality video editor.

Table 21.2. Deployment ΔV Estimation Iteration

Iteration	ΔV (m/sec)	Mean Argument of Latitude (deploy epoch+164622.048056 sec)
#0: Actual result/goal	—	6.61529 deg
#1	2.5	6.96352 deg
#2	2.0	4.83326 deg
#3	2.418267	6.61526 deg
#4	2.418276	6.61529 deg

21.12 Summary

This chapter has repeatedly touched upon the common themes of preparation, communications, and adaptability.

Preparation begins with the design of the satellite and continues through orbit analysis, ground station selection, software tool development, equipment testing, and compiling as much data on the launch and satellite deployment as possible. It also includes training the operations team and rehearsing with all the different agencies involved in the mission.

Communications is the key to information exchange throughout the entire design, launch, and operations phases of the mission. This includes communication with the satellite design team, the launching agency, other ground stations, outside tracking agencies, and other small satellite operators.

Adaptability ensures you will be able to work through the unexpected events that often occur in space missions, particularly those involving small satellites and small budgets. Develop operations software that can be easily updated and reconfigured. Provide a method of displaying the status of your satellite in real time, including the time and visibility of upcoming station passes. Consider possible launch vehicle and satellite anomalies, and develop contingency plans.

We recommend the following actions to reduce risk and maximize chances for a successful mission.

Satellite

- Create a robust and fault-tolerant communications system design.
- Include a radio beacon with unique characteristics.
- Ensure sufficient power to permit unconstrained operation.
- Include radar cross section and optical signature enhancements to enable identification.
- Incorporate onboard inertial navigation units.
- Conduct thorough preflight testing of the communications system ("test like you fly, fly like you test").

Ground Stations

- Incorporate automatic search and tracking capabilities.
- Develop manual search techniques based on ground station performance while tracking is underway.
- Assemble a globally distributed, geographically diverse set of tracking stations.
- Locate ground stations in environments that are relatively free from RF noise.
- Conduct thorough preflight end-to-end testing of ground system hardware and software.

- Enforce configuration control/lockdown of resources upon successful completion of testing.
- Enlist tracking and communication from third-party agencies.
- Communicate to other agencies what makes your satellite unique to aid in identification and discrimination.

Deployment

- Design sequences and operations concepts to ensure successful acquisition and tracking.
- Analyze relative motion of all deployed objects to aid in discrimination.
- Communicate deployment sequence and anticipated relative motion to all third-party tracking agencies.
- If possible, deploy your satellite individually.
- When deployed in a group, ensure all objects are ejected with unique separation velocities or directions.
- Avoid host-vehicle maneuvers and reorientation prior to deployment, since this introduces uncertainties in deployment time and state.
- Deploy during or just before a sequence of high-elevation ground station passes.
- Schedule deployment after external agencies are expected to have accurately determined the host-vehicle orbit.
- Analyze and test deployment mechanisms to ensure sufficient separation velocity.

Operations

- Assemble a team large enough to handle round-the-clock operations and station passes at odd times of the day.
- Enlist several teams for multiple ground stations or extended missions.
- Train as a team and rehearse with other participating agencies.
- Emphasize station passes that occur shortly after deployment, since these are the most likely to yield successful communication with your satellite.
- Document all that occurs during each station pass in a log for later review and analysis.
- Include an evaluation of what you did right and what you did wrong in the pass.
- Consider when to "call it quits" if you have been unsuccessful in repeated attempts to communicate with your satellite.
- Celebrate your success and acknowledge the dedication and hard work of your team!

21.13 References

[1] S. L. Hast, D. L. Oltrogge, and M. J. Hart, "Compact Finite Burn Targeting and Reconstruction Algorithm," *Journal of Guidance, Control, and Dynamics*, **19** (4) (July–Aug. 1996).

[2] *Numerical Recipes in C++: The Art of Scientific Computing, 2nd Edition*, pp. 688-693 (Cambridge University Press, 2002).

[3] D. L. Oltrogge and R. G. Gist, "Collision Avoidance and Radio Frequency Interference," in *Space Systems Modeling and Simulation,* L. B. Rainey, ed. (The Aerospace Press/ AIAA, 2004), ch. 15.

[4] D. L. Oltrogge, "Covariance-Based Mission Operations," Stanford University and Naval PostGraduate Engineering Schools (April 2000).

[5] S. Alfano, "Addressing Nonlinear Relative Motion For Spacecraft Collision Probability," AIAA Paper No. 2006-6760, *15th AAS/AIAA Astrodynamics Specialist Conference* (Keystone, CO, 21–24 Aug. 2006).

[6] G. E. Peterson, R. G. Gist, and D. L. Oltrogge, "Covariance Generation for Space Objects using Public Data," Paper AAS 01-113, *AAS/AIAA Space Flight Mechanics Meeting* (Santa Barbara, CA, 11–14 Feb. 2001).

[7] W. H. Goodyear, "A Generalized Method for the Computation of Cartesian Coordinates and Partial Derivatives of the Two-Body Problem," NASA Report CR-522 (1965).

[8] R. G. Gist and D. L. Oltrogge, "Collision Vision: Covariance Modeling and Intersection Detection for Spacecraft Situational Awareness," *AAS/AIAA Astrodynamics Conference* (Girdwood, AK, 16–19 Aug. 1999).

[9] R. G. Driggers, *Encyclopedia of Optical Engineering*, pp. 2371–2389 (CRC Press, 2003).

[12] *Proceedings of the 15th Inter-Agency Space Debris Coordination Committee Meeting* (NASA JSC, Houston, TX, 1997).

[13] "IADC Space Debris Mitigation Guidelines," IADC-0201, (15 Oct. 2005); http://www.iadc-online.org/docs_pub/IADC-101502.Mit.Guidelines.pdf, 25 June 2007.

[14] D. L. Oltrogge, H. Klinkrad, D. Finkleman, J.-C. Benech, A. da Silva, G. Fujta, C. Pardini, and Y. Stryzhak, ISO Draft International Standard 27852, "Space Systems: Determining Orbit Lifetime" (working document, 31 March 2009).

[15] D. Finkleman and D. L, Oltrogge, "Progress in International Space and Astrodynamics Standards," *20th AAS/AIAA Space Flight Mechanics Conference* (Tampa, FL, 24–26 Jan. 2006), Paper AAS 06-234.

[16] D. L. Oltrogge, and C. C. Chao, "Standardized Approaches for Estimating Orbit Lifetime after End-of-Life," *2007 AAS/AIAA Astrodynamics Specialist Conference* (Mackinac Island, MI, 19–23 Aug. 2007), Paper AAS 07-261.

[17] D. L. Oltrogge and R. G. Gist, "Collision Avoidance and Radio Frequency Interference," in *Space Systems Modeling and Simulation,* L. B. Rainey, ed. (The Aerospace Press/ AIAA, 2004), ch. 15.

[18] D. L. Oltrogge, "Launch and On-Orbit RF Power Impingement Characterization," *16th Space Control Conference* (MIT Lincoln Laboratory, 14–16 April 1998).

[19] LtCol G. Shepherd, "Space Surveillance Network," (Air Force Space Command, Sept. 2006); http://www.cdi.org/PDFs/SSA2Shepherd[1].pdf.

[20] B. M. Kent and D. Turner, "RCS Capabilities and Assessment for Space Tracking" (Air Force Research Laboratory, 8 Oct. 2004).

[21] D. L. Oltrogge and J. Jamison, "Initial Postlaunch Station Acquisition of Nanosatellites," *Second International Conference on Integrated Micro/Nanotechnology for Space Applications* (Pasadena, CA, 11–15 April 1999).

[22] "Manual of Mission Planning for KREMS Radars" (MIT/Lincoln Labs, July 1992).

[23] 3M Scotchlite High Gain Reflective Sheeting 7610 Data Sheet (3M Corporation, 2005).

[24] 3M Scotchlite Very High Gain Sheeting 3000X Data Sheet (3M Corporation, 2005).

[25] F. R. Hoots, and R. L. Roehrich, "Models for Propagation of USSTRATCOM Element Sets," *Spacetrack Report No. 3.* (ADCOM, Dec. 1980).

[26] D. L. Oltrogge, "Spiral Scan Acquisition/Tracking Method for Picosats," *Aerospace Forum on Space Debris, Collision Avoidance, and Reentry Hazards* (The Aerospace Corporation, 1–3 Nov. 2000).

[27] D. L. Oltrogge and D. Dichmann, "Improved Covariance-Based Search & Acquisition Strategies for Launch & Early Orbit," *3rd NRO/AIAA Workshop on Space Launch Integration* (Chantilly, VA, 23 July 2001).

Chapter 22

Small Satellites: Space Debris and Reentry Hazards

W. Ailor*, V. Chobotov*, G. Peterson*, R. Patera*

22.1 Background

The term "space debris" refers to nonfunctioning, human-made items in Earth orbit. Typical items include dead satellites, rocket stages, lens covers, fragments from exploded stages, debris from collisions, small flecks of paint, and the occasional wrench lost by astronauts.

Space debris is of increasing concern for operators of space systems—even very small objects can create appreciable damage, primarily because of the very high relative velocities at impact. For example, Fig. 22.1 shows a photo of a pit on the windshield of the space shuttle formed by impact of a fleck of white paint at a relative velocity estimated between 3 and 6 km/sec. There have been 795 window craters during more than 24 shuttle missions, with windshields replaced every two to three missions because of damage caused by bits of space debris and micrometeoroids. The International Space Station has shielding to prevent damage from 1 cm or smaller space debris and micrometeoroids, and critical systems on many satellites are also shielded to prevent damage by small particles.

While there have been only three known collisions of large, tracked objects (as of late 2008), only one of these involved damage to an operating satellite: the French Cerise satellite, built by Surrey Satellite Technology Ltd., had its gravity-gradient boom clipped by a rocket fragment[1] (Cerise did recover some of its function). As the shuttle windshield example shows, even very small objects pose a threat in that they can seriously damage an instrument or a critical system if they impact at critical areas of a spacecraft. Given these concerns, and the reality that the population of orbiting objects—both operating satellites and debris—is growing, there is increasing focus on ensuring that satellites are deployed and operated in ways that minimize the possibility that they will be direct or indirect (through collisions with other objects) sources of debris.

From the space debris perspective, small satellites have advantages and disadvantages. One advantage is they have a smaller cross section than larger satellites and, hence, a reduced threat of collision in general. A disadvantage is that the lower costs associated with small satellites makes it possible to put more of them into space, increasing the population of orbiting objects. And very small satellites may be untrackable—other operating satellites may not be able to see them coming and avoid a collision.

*The Aerospace Corporation
El Segundo, California

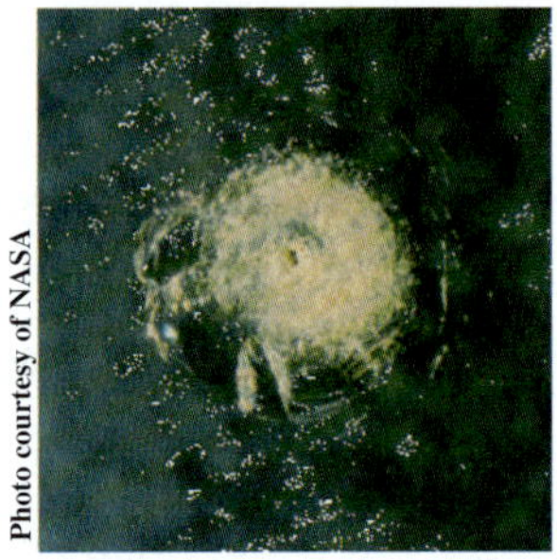

Fig. 22.1. A 4 mm crater in the windshield of the space shuttle orbiter caused by a 0.2 mm fleck of white paint; the relative velocity at impact was about 3–6 km/sec.

22.1.1 Space Debris Population

There are two categories of space debris: tracked and untracked. Tracked debris are operational satellites, dead satellites, fragments of satellites, and released hardware that are generally larger than about 10 cm. This debris is tracked by ground-based sensors, with orbit ephemeris information included in the Resident Space Object catalog maintained by the U.S. Air Force. Because each object's orbit is known, the hazard it poses to other objects can be computed, and operating satellites can, with sufficient notice, take action to avoid a possible collision.

Figure 22.2 shows the growth in the number of tracked objects in Earth orbit since the space age began in 1957, and Fig. 22.3 shows the breakdown of the current catalog by object type. Note that active satellites constitute only 7% of the catalog, with the remaining 93% being various kinds of debris. The peaks in the debris curve relate to debris created by the noted events: the explosion of the Ariane V16 in 1986, the explosion of a Pegasus stage in 1996, and, most recently, the explosions in early 2007 of a Russian Briz-M rocket booster and the Chinese antisatellite test that destroyed the Feng Yung (FY1C) weather satellite. Each of the 2007 events created more than 1500 tracked fragments. Such debris-producing events increase operational costs by the need to maneuver away from debris but incur no penalties other than adverse publicity.

Note that in Fig. 22.2, several periods show declines in the overall population of debris. These dips reflect increases in the atmospheric density (resulting from heightened solar activity) that caused debris in low Earth orbit (LEO) to decay and reenter at somewhat higher rates.

Table 22.1 shows the estimated numbers of orbital debris objects in various size ranges as of 2003.[2] The estimated total mass is based on the Resident Space Object catalog. Today, of approximately 13,000 tracked objects larger than 10 cm, more than 3000 were created by the explosion of the Russian rocket stage and the Chinese antisatellite test in early 2007.

Table 22.2 shows the breakdown of objects by the launching nation or entity, and Table 22.3 lists objects launched to the geosynchronous (GEO) ring. The term "payload" in Table 22.2 refers to the satellite placed in orbit to perform a specific

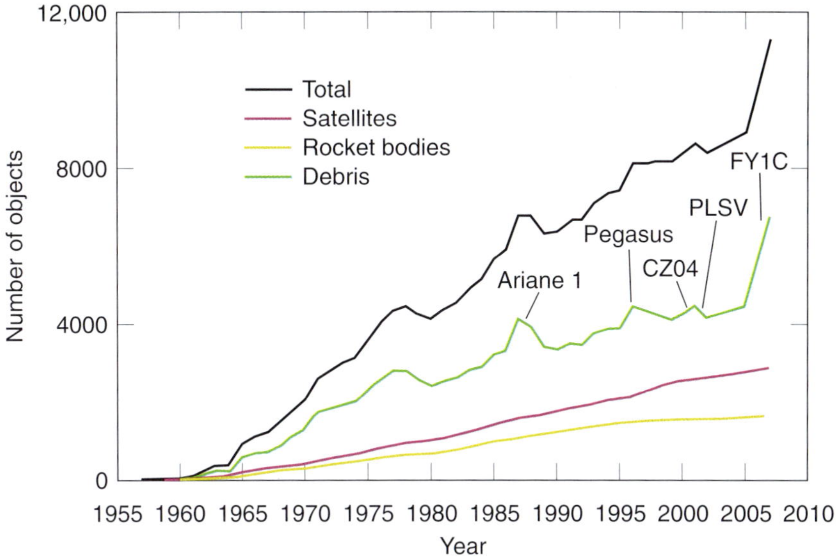

Fig. 22.2. Number of tracked objects in Earth orbit.

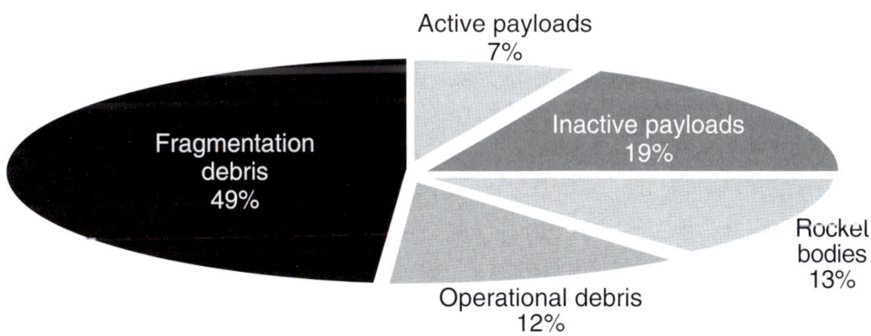

Fig. 22.3. Breakdown of tracked objects by object type (note that the term "payload" refers to a satellite designed to perform a specific mission).

Table 22.1. Estimated Number of Orbiting Objects in Size Ranges

Size (cm)	Number of Objects	Percent Total Number	Percent Total Mass	Mass (kg)
0.1–1.0	35,000,000	99.67	0.035	
1.0–10.0	110,000	0.31	0.035	
>10.0	8000	0.02	99.93	
Total:	35,118,000	100.0	100.0	1,400,000

Table 22.2. Breakdown of Satellite Ownership as of January 2008

Source	Payloads on Orbit	Reentered	Total	Debris on Orbit	Reentered	Total	All
ArabSatCom	8	1	9	0	0	0	9
AsiaSat	4	0	4	0	0	0	4
Algeria	1	0	1	0	0	0	1
Argentina	10	2	12	0	0	0	12
Australia	11	2	13	2	0	2	15
Brazil	11	0	11	0	0	0	11
Canada	25	1	26	2	2	4	30
China/Brazil	3	0	3	60	5	65	68
Chile	1	0	1	0	0	0	1
CIS/Russia	1413	2440	3853	2952	10603	13555	17408
Columbia	1	0	1	0	0	0	1
Czechoslovakia	5	1	6	0	0	0	6
Denmark	3	0	3	0	0	0	3
Egypt	3	0	3	0	0	0	3
ESA	42	6	48	36	14	50	98
ESRO	0	7	7	0	3	3	10
EUMETSAT	2	0	2	2	0	2	4
EUTELSAT	28	0	28	0	0	0	28
France/Germany	2	0	2	0	0	0	2
France	44	8	52	324	655	979	1031
Germany	27	13	40	1	1	2	42
Globalstar	60	0	60	0	1	1	61
Greece	2	0	2	0	0	0	2
INMARSAT	11	0	11	0	0	0	11
India	34	9	43	107	266	373	416
Indonesia	10	1	11	0	0	0	11
Iran	1	0	1	0	0	0	1
Iridium	0	1	1	0	12	12	13
Israel	8	3	11	1	4	5	16
India	1	0	1	0	0	0	1
Intl. Space Station	4	0	4	4	46	50	54

Table 22.2 (cont.). Breakdown of Satellite Ownership as of January 2008

Source	Payloads on Orbit	Reentered	Total	Debris on Orbit	Reentered	Total	All
Italy	14	9	23	2	1	3	26
Intelsat	64	1	65	0	0	0	65
Japan	111	18	129	73	168	241	370
Luxembourg	15	1	16	0	0	0	16
Malaysia	4	0	4	0	0	0	4
Mexico	7	0	7	0	0	0	7
NATO	8	0	8	0	0	0	8
Netherlands	5	1	6	0	0	0	6
ICO	1	0	1	0	0	0	1
Nigeria	2	0	2	0	0	0	2
Norway	3	0	3	0	0	0	3
Globalstar	35	0	35	0	0	0	35
Pakistan	1	1	2	0	0	0	2
Portugal	1	0	1	0	0	0	1
China	64	52	116	2694	374	3068	3184
Taiwan	8	0	8	0	0	0	8
Philippines	2	0	2	0	0	0	2
South Africa	1	0	1	0	0	0	1
Saudi Arabia	12	0	12	0	0	0	12
Sea Launch	1	0	1	20	3	23	24
South Korea	10	0	10	1	0	1	11
Spain	9	1	10	0	0	0	10
Singapore/Taiwan	1	0	1	0	0	0	1
Sweden	11	0	11	0	0	0	11
Thailand	6	0	6	0	0	0	6
Turkey	5	0	5	0	0	0	5
United Arab Em.	3	0	3	0	0	0	3
United Kingdom	25	9	34	1	4	5	39
United States	1040	754	1794	3194	4254	7448	9242
USA/Brazil	1	0	1	0	0	0	1
Total	3245	3342	6587	9476	16416	25892	32479

Table 22.3. Geosynchronous Catalog as of June 2007

Owner	Spacecraft	Rocket Bodies	Debris
ArabSatCom	8	0	0
AsiaSat	4	0	0
Argentina	3	0	0
Australia	7	0	0
Brazil	6	0	0
Canada	16	0	0
CIS/Russia	144	142	4
Egypt	2	0	0
ESA	14	6	0
EUMETSAT	2	0	1
EUTELSAT	28	0	0
France/Germany	2	0	0
France	11	0	0
Germany	4	0	0
Greece	2	0	0
INMARSAT	11	0	0
India	21	0	0
Indonesia	9	0	0
Israel	2	0	0
Italy	4	0	0
Intelsat	61	0	0
Japan	46	3	0
South Korea	4	0	0
Luxemburg	15	0	0
Malaysia	3	0	0
Mexico	6	0	0
NATO	8	0	0
Netherlands	5	0	0
Nigeria	1	0	0
Norway	3	0	0
China	28	4	0
Philippines	2	0	0
Spain	6	0	0
Singapore/Taiwan	1	0	0
Sweden	4	0	0
Thailand	5	0	0
Turkey	4	0	0
United Arab Emirates	2	0	0
United Kingdom	9	0	0
United States	228	24	6
USA/Brazil	1	0	0
Total:	742	179	11

mission; "debris" refers to any nonfunctioning items released in orbit as part of the launch and deployment of the satellite. Many of the spacecraft were actually placed into orbit using launch services provided by the United States, Russia, China, or other launching nation, as well as by commercial launch services.

Figure 22.4 shows the number of satellites passing through various altitude ranges. Note that the population of satellites and debris objects is the highest at altitudes less than 2000 km. The peaks noted at approximately 20,000 km and 36,000 km are the GPS navigation satellites and GEO satellites, respectively.

Figure 22.5 expands the distribution of objects in LEO, and Fig. 22.6 shows the distribution as a function of orbit inclination. Note the peaks in density in the neighborhoods of the Iridium and Globalstar communication satellite constellations and Russian Parus and Strela navigation satellite constellations. The "FY1C" peak is associated with debris from the Chinese antisatellite test in 2007.

22.1.2 Small Satellites

As of April 2007, 832 satellites with mass less than 150 kg were in orbit. These included dead satellites and 66 operational satellites (listed in Table 22.4). More small satellites have been placed in orbit, but these are either no longer active or have reentered the atmosphere. Figure 22.7 shows the mass distribution of active satellites in 2007. Note that the majority have masses below about 30 kg, and three have masses of approximately 1 kg. Figure 22.8 shows that many of these small satellites are in approximately 800 km circular orbits—important for satellite orbit lifetime considerations, as will be discussed in section 22.4. Figure 22.9 shows the inclinations of the small satellite orbits.

22.1.3 Projections for the Future

Predictions show that even with immediate actions to limit the growth of space debris, the number of orbiting objects will continue to grow—with an increasing component being fragments from collisions among the orbiting population. Results of NASA analyses,[3] which assume that spacecraft are not removed from their orbits after end of life and that explosions and related debris generation follow recent trends, are shown in Fig. 22.10. The analyses predict that "collisions, especially intact-intact and intact-fragment catastrophic collisions, will become the most important debris generation mechanism in the future."

Based on these and similar results from other researchers, efforts to prevent collisions among larger objects will likely increase in the future; speculation is that a collision avoidance service might become available. Such a service would provide warnings of upcoming close approaches, and operators would use available data to plan avoidance maneuvers.[4,5]

In addition, there is an increasing emphasis on vehicle and mission designs and on-orbit operations that will limit debris creation. End-of-life disposal actions to move spacecraft from areas of high operational utility to areas where the long-term risk of collisions is minimized will also become more common and may become mandatory.

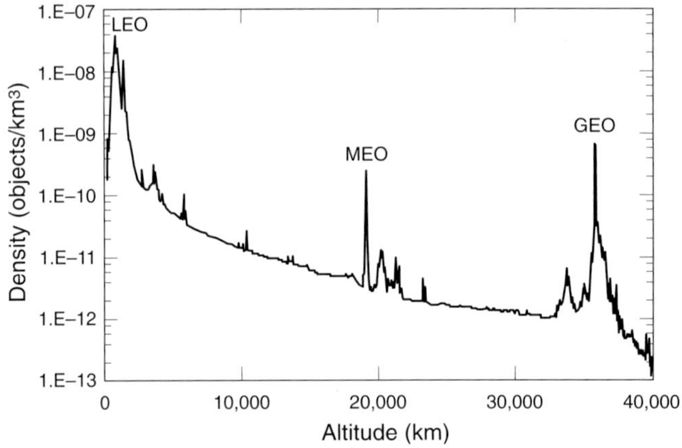

Fig. 22.4. Altitude distribution of objects in the near-Earth region.

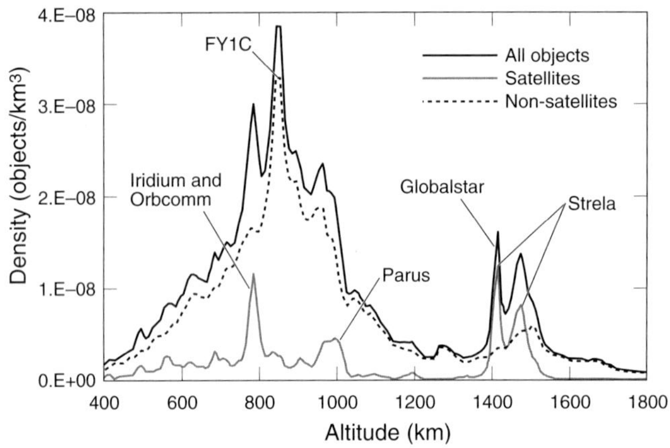

Fig. 22.5. Altitude distribution of objects in low Earth orbit.

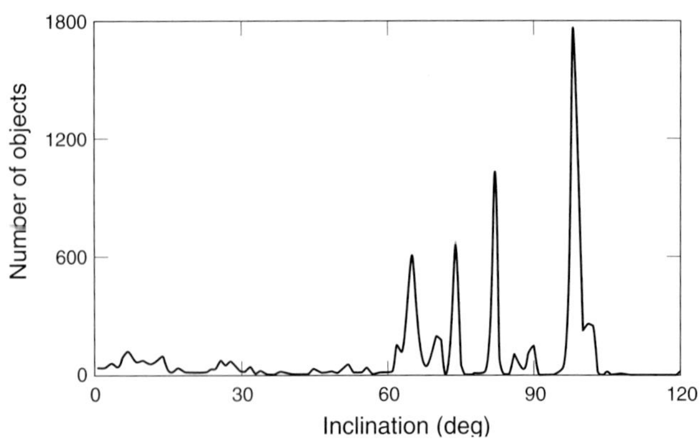

Fig. 22.6. Distribution of objects in low Earth orbit as a function of inclination.

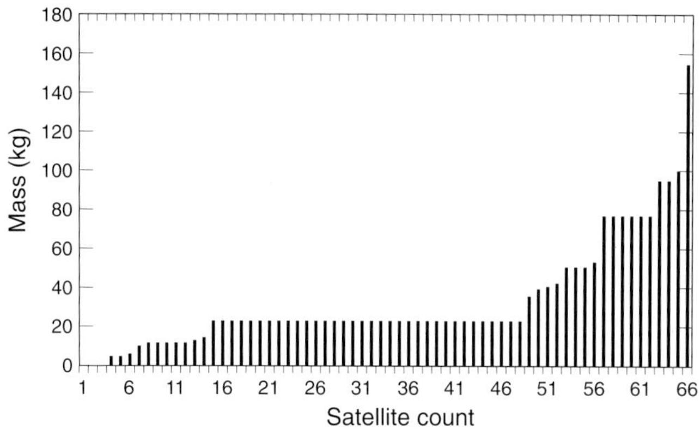

Fig. 22.7. Population of operational satellites with mass of 150 kg and less in orbit as of April 2007.

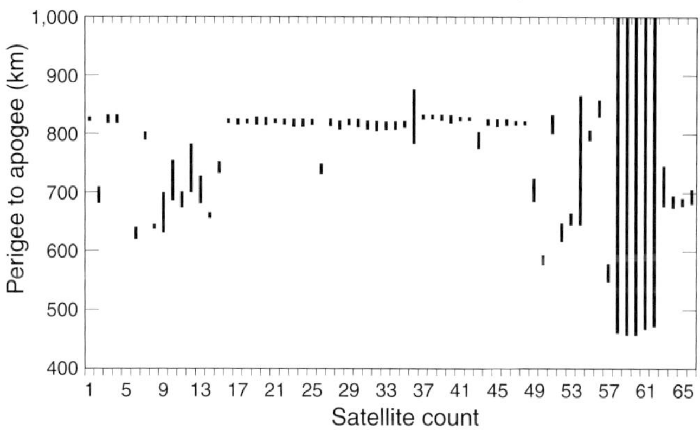

Fig. 22.8. Orbits where operating satellites with mass of 150 kg and less reside. The vertical bar shows the altitude range traversed by each satellite.

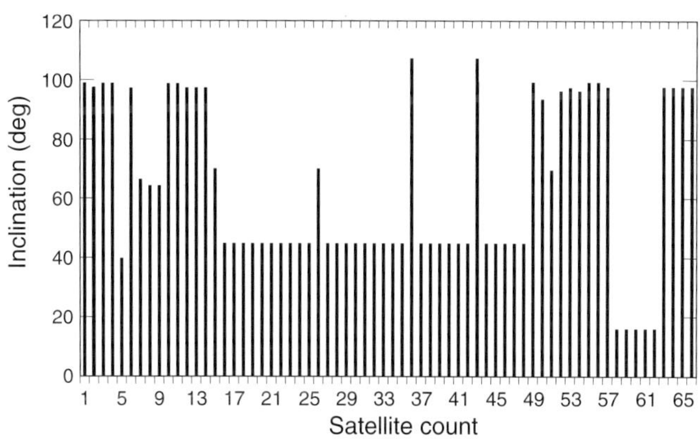

Fig. 22.9. Inclinations of orbits where operating satellites of 150 kg and less reside.

Table 22.4. Microsatellites to 155 kg; source: UCS Satellite Database (9 April 2007)

Satellite	Owner	Orbit	Perigee (km)	Apogee (km)	Incl. (deg)	Mass (kg)
CubeSat XI-IV	Japan	LEO, sun-sync	822	828	98.7	1
CubeSat XI-V	Japan	LEO, sun-sync	682	708	98.1	1
Cute-1	Japan	LEO, sun-sync	819	831	98.7	1
QuakeSat	USA	LEO, sun-sync	821	832	98.7	5
GeneSat-1	USA	LEO	410	410	40	5
PehuenSat 1	Argentina	LEO, sun-sync	622	640	97.94	6
PCSat	USA	LEO/I	790	801	67	10
LatinSat A	Argentina	LEO/I	638	644	64.5	12
LatinSat B	Argentina	LEO/I	631	701	64.5	12
LatinSat-C	Argentina	LEO, sun-sync	687	753	98.3	12
LatinSat-D	Argentina	LEO, sun-sync	673	699	98.3	12
Unisat-3	Italy	LEO, sun-sync	700	779	98.2	12
AcrimSat	USA	LEO, sun-sync	682	727	98.2	13
Nanosat-1	Spain	LEO, sun-sync	658	666	98.1	15
ORBCOMM FM-1	USA	LEO/I	733	750	69.9	22
ORBCOMM FM-10	USA	LEO/I	818	824	45	22
ORBCOMM FM-11	USA	LEO/I	817	824	45	22
ORBCOMM FM-12	USA	LEO/I	818	824	45	22
ORBCOMM FM-13	USA	LEO/I	816	827	45	22
ORBCOMM FM-14	USA	LEO/I	816	827	45	22
ORBCOMM FM-15	USA	LEO/I	819	824	45	22
ORBCOMM FM-16	USA	LEO/I	817	825	45	22
ORBCOMM FM-17	USA	LEO/I	813	824	45	22
ORBCOMM FM-18	USA	LEO/I	813	824	45	22
ORBCOMM FM-19	USA	LEO/I	814	824	45	22
ORBCOMM FM-2	USA	LEO/I	733	748	69.9	22
ORBCOMM FM-20	USA	LEO/I	814	825	45	22
ORBCOMM FM-21	USA	LEO/I	810	822	45	22
ORBCOMM FM-22	USA	LEO/I	814	823	45	22
ORBCOMM FM-23	USA	LEO/I	813	824	45	22
ORBCOMM FM-24	USA	LEO/I	810	822	45	22
ORBCOMM FM-25	USA	LEO/I	806	820	45	22
ORBCOMM FM-26	USA	LEO/I	807	820	45	22

Table 22.4 (cont.). Microsatellites to 155 kg; source: UCS Satellite Database

Satellite	Owner	Orbit	Perigee (km)	Apogee (km)	Incl. (deg)	Mass (kg)
ORBCOMM FM-27	USA	LEO/I	807	820	45	22
ORBCOMM FM-28	USA	LEO/I	811	821	45	22
ORBCOMM FM-3	USA	LEO/I	782	874	108	22
ORBCOMM FM-30	USA	LEO/I	825	833	45	22
ORBCOMM FM-31	USA	LEO/I	825	833	45	22
ORBCOMM FM-32	USA	LEO/I	823	832	45	22
ORBCOMM FM-34	USA	LEO/I	820	831	45	22
ORBCOMM FM-35	USA	LEO/I	824	829	45	22
ORBCOMM FM-36	USA	LEO/I	824	829	45	22
ORBCOMM FM-4	USA	LEO/I	777	804	108	22
ORBCOMM FM-5	USA	LEO/I	814	824	45	22
ORBCOMM FM-6	USA	LEO/I	813	824	45	22
ORBCOMM FM-7	USA	LEO/I	815	823	45	22
ORBCOMM FM-8	USA	LEO/I	816	822	45	22
ORBCOMM FM-9	USA	LEO/I	816	822	45	22
Saudisat-2	S. Arabia	LEO, sun-sync	686	723	98.3	35
CHIPSat	USA	LEO/P	577	594	94	40
FORTÉ	USA	LEO/I	800	833	69.9	41
HAMSat	India	LEO, sun-sync	618	646	97.2	42
Clementine	France	LEO, syn-sync	647	664	98	50
Ørsted (Sunsat)	Denmark	LEO, sun-sync	646	865	96.4	50
Posat-1	Portugal	LEO, sun-sync	790	806	98.6	50
MOST	Canada	LEO, sun-sync	831	855	98.7	53
Bird 2	Germany	LEO, sun-sync	550	580	97.7	77
THEMIS 1	Int'l	Elliptical	461	87,304	15.7	77
THEMIS 2	Int'l	Elliptical	458	87,298	15.7	77
THEMIS 3	Int'l	Elliptical	459	87,054	15.68	77
THEMIS 4	Int'l	Elliptical	467	87,260	15.7	77
THEMIS 5	Int'l	Elliptical	474	87,526	15.7	77
Alsat-1	Algeria	LEO, sun-sync	680	744	98.2	95
NigeriaSat-1	Nigeria	LEO, sun-sync	676	694	98.2	95
Bilsat-1	Turkey	LEO, sun-sync	677	692	98.2	100
BeijinGalaxy-1	China	LEO, sun-sync	682	705	98.18	155

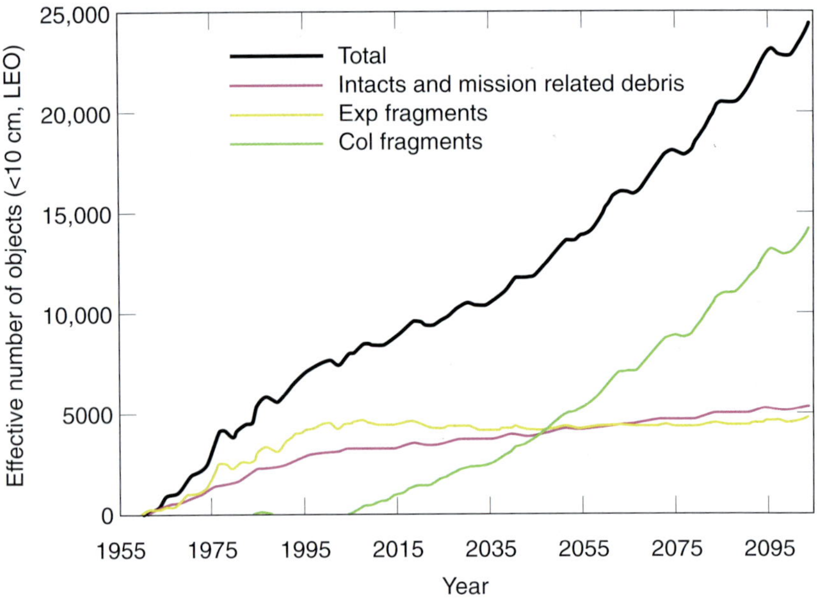

Fig. 22.10. Projected number of objects in low Earth orbit[3] ("Exp fragments" are fragments generated by explosions; "Col fragments" are generated by collisions).

22.2 Debris Minimization Guidelines and Requirements

Some years ago, studies indicated that space debris was an emerging problem,[6–9] a recognition that led to the formation of the Inter-Agency Space Debris Coordination Committee (IADC) in 1993. The IADC's goal was to bring technical experts together to assess the problem, coordinate supporting research, and recommend approaches that would limit the growth of space debris. Specifically, the IADC was to encourage an international dialog to "exchange information on space debris research activities between member space agencies, to facilitate opportunities for cooperation in space debris research, to review the progress of ongoing cooperative activities, and to identify debris mitigation options." Initial founders were NASA, the European Space Agency (ESA), the Japan Aerospace Exploration Agency, and the Russian Federal Space Agency. Membership has since expanded to include virtually all space-faring nations.

Since its founding, the IADC has developed a number of recommendations, which in many cases have been adopted by space and licensing agencies. As an example, the U.S. Federal Communications Commission (FCC)[10] has issued a ruling[11] for all spacecraft types and Earth orbits that requires FCC licenses to file end-of-life disposal plans that essentially comply with IADC recommendations. In addition, IADC recommendations are reflected in reference documents issued at agency, national, or European level, either as standards, standard practices, policy documents, or handbooks.[12–19] In some cases (e.g., NASA-STD-8719.14 and DODI 3100.12), these documents reflect requirements, not voluntary guidelines. In addition, ISO and other standards-setting bodies are developing standards that capture best practices related to debris minimization. Given the projected growth of space debris, it is expected that governments and corporations acquiring satellites will require satellite manufacturers and launch providers to provide hardware and mission designs that follow these standards.

The basic ground rules[20] developed by the IADC for minimizing the growth of space debris are:

- Preventing on-orbit breakups.
- Removing decommissioned spacecraft and spent stages from the useful densely populated orbit regions.
- Limiting the objects released during normal operations.
- Preventing on-orbit collisions.

22.2.1 Preventing Breakups

Preventing on-orbit breakups includes preventing explosions of stages and spacecraft—major sources of debris—both during and after end of life. For example, several Delta II second stages have exploded in orbit long after completing their missions, primarily because they contained residual propellant, pressurized gas, or other energy sources.

Currently, in accordance with space debris minimization guidelines, stages and spacecraft are "safed"—tanks vented, batteries discharged, energy sources depleted—at end of life to prevent future explosions. These ground rules recognize that these types of events, which may occur many years after a spacecraft has been deactivated, can create and have created large numbers of debris objects that will remain a hazard for years, centuries, or beyond. The 1996 Pegasus rocket-stage explosion released more than 700 trackable objects,[21] and the 2007 Russian Briz-M rocket-stage explosion released more than 1500.[22] Figure 22.11 notes that these types of events can substantially increase the risk of collision in nearby orbits for decades.

22.2.2 Removal of Hardware at End of Mission

Regulatory agencies in several countries have developed regulations that require satellites to be reentered into Earth's atmosphere or moved away from designated "protected regions" and safed at end of life. In Fig. 22.12, region A is the protected region for objects in LEO, region B for satellites in GEO. The protected region for GEO actually consists of three subregions: the first represents the region where a GEO satellite will be confined under stationkeeping; the second is where a GEO satellite may be located when its inclination is not controlled, but remains under a mission-specified value. The third represents the inclination excursion region for a nonoperational GEO satellite and is the outer boundary of the GEO protected

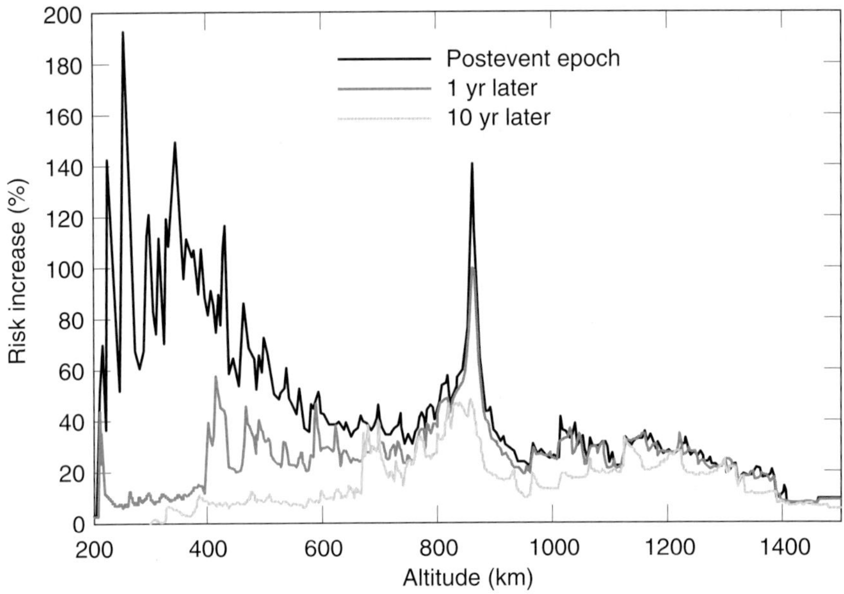

Fig. 22.11. The increase in the risk of collision as a function of altitude as a result of the 2007 Chinese antisatellite test and explosion of the Russian rocket stage.[23]

zone. The protected regions are areas of heavy utility for space missions and have been designated as regions to be kept as clean as possible, from the space debris perspective.

Several options are available for disposal of space hardware at the end of its mission, and the choice of the disposal option for a particular space vehicle depends on the altitude range of the mission orbit, the lifetime of the mission orbit, the health and status of systems critical to executing maneuvers required for disposal, and the amount of propellant remaining.

Hardware in LEO or hardware whose perigee is within the LEO protected region may be removed from interfering with the protected region by reentry into the atmosphere or transfer to an orbit whose perigee stays higher than 2000 km and whose apogee remains lower than 19,700 km (500 km below the semisynchronous or medium Earth orbit, MEO, altitude). The apogee constraint is to prevent interference with GPS and other satellites in semisynchronous orbits with periods of approximately 12 hours. Similarly, objects may be disposed by moving them to orbits with perigees above 20,700 km (500 km higher than the semisynchronous altitude) and apogees lower than 35,300 km (500 km lower than the geosynchronous altitude). Satellites in geosynchronous orbit or in orbits passing through the GEO protected region may be disposed to orbits whose perigees are greater than 36,100 km (approximately 300 km above the GEO protected region). The recommended separation constraints are in accordance with the *U.S. Government*

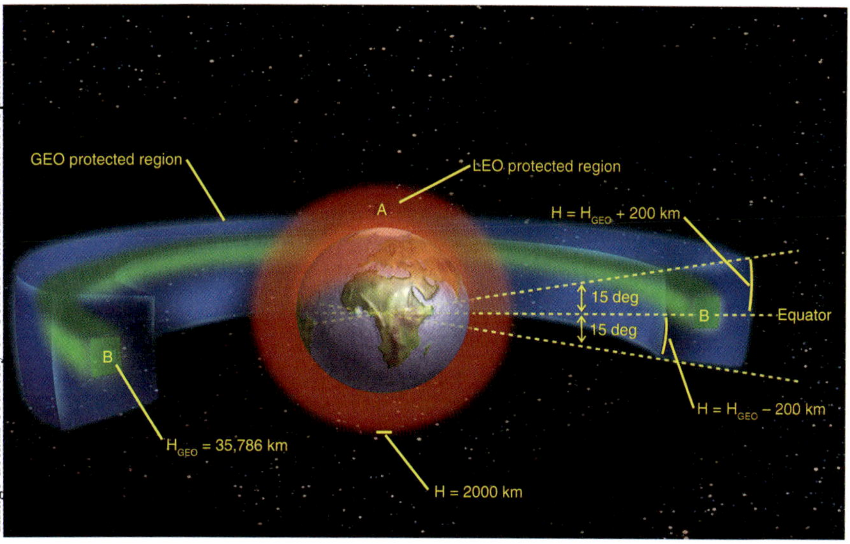

Image details courtesy of Centre National d'Etudes Spatiales

Fig. 22.12. Three-dimensional view of protected zones.

Orbital Debris Mitigation Standard Practices[24]; there are some small variations in the specific constraint altitudes adopted by other spacefaring nations.[25] There are additional orbit eccentricity constraints that should be considered for disposal of MEO and GEO satellites.[26,27]

If reentry disposal is chosen for LEO satellites, there are two options:

- If the expectation for human casualties caused by debris surviving a random reentry is low (see discussion of reentry hazard in section 22.3), the satellite may be placed in an orbit whose lifetime is limited, and the satellite can reenter randomly sometime in the future.
- If the casualty expectation exceeds 1 in 10,000 (i.e., 1 injury or death in 10,000 reentries), the spacecraft should be reentered into an ocean area via a controlled deorbit. The latter option requires the spacecraft to execute a specific deorbit maneuver and to have an operating propulsion system.

One method of deorbiting a satellite whose perigee is in the LEO protected region is to move it to a disposal orbit—lowering the perigee such that atmospheric drag causes the orbit to slowly decay to the point where the object reenters the denser atmosphere and is destroyed. While the overall lifetime of the disposal orbit can be estimated, variations in atmospheric density due to solar and other effects make precise estimates of the reentry time impossible. As a result, the spacecraft can reenter anywhere within a latitude band defined by its orbit inclination. For example, if the spacecraft is in an orbit inclined at 28.5 deg, it can reenter anywhere in the area between 28.5 deg N and 28.5 deg S latitudes.

Disposal by controlled deorbit provides certainty (subject to reliability of the propulsion system during the required maneuvers) in the impact location. Controlled deorbits are most desirable when

- the mass of the satellite is so large that the number of surviving debris objects is thought to pose an unacceptable risk to life and property,
- the satellite contains hazardous materials, or
- the satellite contains sensitive components, and recovery of these is not desirable.

Disposal via controlled deorbit requires a relatively healthy spacecraft and more propellant than the orbit decay method. The reentries of NASA's Compton Gamma Ray Observatory and the Russian Mir space station[28] are examples of controlled deorbits.

22.2.3 Limiting Objects Released During Normal Operations

The third space debris mitigation ground rule seeks to prevent or minimize the material ejected during standard mission operations: injection, deployment, operations, disposal, and in some cases, postdisposal. In past years, lens covers, deployment devices, and other hardware were regularly expelled during satellite activation and normal operations. This debris could remain in orbit for years or generations, depending on their release altitude.

Currently, deployment and other hardware are being designed to minimize the debris released. For example, lens covers are tethered so that they do not float free. Similarly, bolt catchers now contain fragments from separation devices. Other sources of operational debris include ejecta from solid rocket motors (burning of aluminized propellants produces many very small particles during the burn plus a few larger objects chuffed out at the end) and paint flakes and bits of materials released due to weathering in space.

22.2.4 Preventing On-Orbit Collisions

Any exposed surface in space is subject to impact from micrometeoroids that are passing through the near-Earth environment as well as orbital debris and larger objects, including operating satellites. Small satellites are no exception. This final debris mitigation objective seeks to avoid collisions with other objects and to limit the damage and debris created by unavoidable collisions with micrometeoroids and small bits of orbiting debris.

22.2.4.1 Collisions with Small Particles

Micrometeoroids and small, untracked particles and fragments from human activities in space can damage a spacecraft and possibly terminate its mission if they puncture a critical component. Table 22.5 illustrates the effect of impacts of small objects at 10 km/sec, a reasonable average velocity for collisions between spacecraft and orbital debris.[31] Of course, the specific damage to a particular spacecraft depends on where the small object strikes it and on the material used in that area.

Table 22.5. Effects of Small Debris Impacts at 10 km/sec

Diameter of Impactor (cm)	Effect
<0.01	Surface erosion
<0.1	Potentially serious damage to spacecraft
0.3	Equivalent to being struck by a bowling ball traveling at 100 km/hr
1.0	Equivalent to being struck by a 180 kg safe traveling at 100 km/hr

For example, a small piece of debris striking a heat pipe or critical electrical connection could cause a loss of cooling fluid or loss of power, whereas a strike in a noncritical region might not be noticed.

Figure 22.13 shows the flux for micrometeoroids and space debris estimated by several environmental models.[30] The figure includes data measured by the Long Duration Exposure Facility (LDEF), an actual count of debris impacts on an orbiting object over a 69-month period. ORDEM2000 and MASTER are orbital debris models developed by NASA and ESA, respectively. MEM is Marshall Space Flight Center's meteoroid model that will shortly be replacing the older NASA SSP (Space Station Program) model used in shuttle and International Space Station analyses.

The risk to small satellites can be estimated by using the velocity and density of an object to compute a critical diameter, above which debris objects will penetrate a specific single element of a satellite. The flux of particles whose size exceeds that critical diameter is then used to compute the probability of penetration.

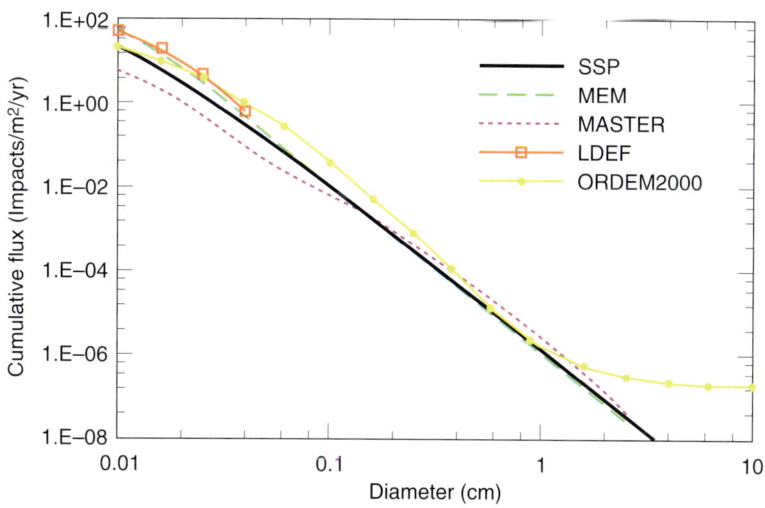

Fig. 22.13. Measured and predicted flux of micrometeoroids and space debris.

The power relation of the dependency between velocity, density, and the critical mass varies with the type of panel that is being impacted and is found through empirical curve fits to laboratory hypervelocity test results. From an examination of the literature,[31–33] this dependency can be expressed as:

$$M_c \propto \rho^{1-3\alpha} V_n^{-\beta} \tag{22.1}$$

where M_c is the critical mass, ρ is the density of the particle, V_n is the component of the relative velocity that is normal to the surface, α ranges from 1/3 to ½, and β ranges from 1/3 to 2/3. The particular values of these parameters will depend upon the type of wall (single, double, Whipple shield, etc.) and the material composition of the wall. For particles smaller than about 1 cm in diameter, the micrometeoroid and orbital debris fluxes are similar; for particles larger than this, the orbital debris flux becomes dominant over the meteoroids. However, the density and velocity distribution of the natural and human-made environments can differ by a considerable amount, and this affects just how dangerous the two populations are.

The orbital debris threat is assumed to consist primarily of metallic objects with a mean density close to aluminum (2.8 g/cm^3), while the micrometeoroid particles are composed mostly of ice with density ranging from 0.5 to 2 g/cm^3. These densities may be adequate to describe the bulk of micrometeoroid and orbital debris particles, but it is known that not all orbital debris is composed of aluminum (and the shape of the particles can substantially deviate from spherical), and some meteoroid particles can be composed mostly of iron with densities approaching 8 g/cm^3. Similarly, the velocity for debris ranges from 6–16 km/sec, while meteoroids can have velocities in excess of 70 km/sec. As a consequence of the greater kinetic energy of impact, smaller meteoroids can have as big a negative influence on spacecraft as larger orbital debris.

It is important to note that a large fraction of the orbital debris and essentially all of the micrometeoroid flux occur at velocities above the routine capabilities of impact test facilities (~7–8 km/sec) used for establishing the ballistic limit equations critical to these analyses. In addition, the majority of hypervelocity impact testing is conducted using types of particles consistent with orbital debris (i.e., aluminum) and with spherical shapes. Little has been done to examine icy material or the types of odd shapes expected from actual orbital debris and micrometeoroids. As a result, there is substantial uncertainty in the ballistic limit equations due to the fact that meteoroid (very fast icy particle) impacts are being covered by equations derived using experimental results from slower, hard material impacts. Therefore, the ballistic limit equations quoted in the literature should be viewed as approximations.

22.2.4.2 Collisions with Large Objects

Any object greater than 10 cm (large enough to be tracked) that strikes any operating satellite is likely to cause catastrophic damage to that satellite; if the large object itself is an operational vehicle, then it too could be lost. Risk assessment in this case is concerned with computing and possibly reducing the probability of collision, P_c. Once operations are under way, comparisons between the small satellite and the ephemerides of the tracked objects can be used to look for upcoming close approach events.

For mission planning purposes, if the debris environment is approximated as a gas, the Kinetic Gas Theory provides a means of approximating the probability of the small satellite suffering an impact during its mission. In essence, the small satellite is flying through a cloud of debris, and the density of objects within the cloud affects the probability of collision over the long term. If there is control over the final orbit, this parameter can be used to place a small satellite in an orbit where the probability of collision is low (see Fig. 22.4).

The fundamental equation governing the average risk of a collision from the Kinetic Gas Theory can be derived from the Poisson distribution:[34,35]

$$P_{KGT} = 1 - \exp(-\rho V_{ave} A_{KGT} \Delta t) \tag{22.2}$$

where P_{KGT} is the average probability of collision over the mission duration, ρ is the spatial object density (number of objects/km^3; this will be a function of not only the altitude but the inclination), V_{ave} is the average relative velocity of encounter (km/sec), A_{KGT} is the effective cross-sectional area of the passing objects (km^2), and Δt = mission duration (seconds). The time span of interest must be long enough to make an averaging of the spatial density relevant. Typically, this requires durations of about a year or more. For durations shorter than that, the Kinetic Gas Theory can produce an average probability, but it should be recognized as an average probability. Deterministic propagations using the actual objects' orbits should be utilized to get a truer measure of the risk for short time periods.

The average velocity of encounter in a randomly distributed cloud of orbiting particles is going to be greater than the velocity of the small satellite by about a factor of the square root of 2 (assuming the average relative approach angle is 45 deg). The average encounter velocity can therefore be approximated by:

$$V_{ave} \approx \sqrt{\frac{2\mu}{R_e + alt}} \text{ (km/sec)} \tag{22.3}$$

where μ is the Earth gravitational parameter ($\mu = GM = 398601.8$ km^3/sec^2 where G is the gravitational constant and M is the mass of the Earth), R_e is the Earth radius, and *alt* is the altitude at the middle of an arbitrary shell.

Since collisions involve finite objects and not infinitesimal cloud particles and the relative attitude of passing objects is unknown, then the areas of the two objects should not simply be added. Instead, the desired area for purposes of risk is the area of a circle defined by the sum of the radii of the two objects.

The average radius of catalog objects was found by examining radar cross sections of orbiting objects.[36] To account for the fact that the radar only "sees" the reflective surfaces of the spacecraft and not the circle that is desired for collision risk purposes, an adjusting factor must be multiplied by the radar cross-sectional area. This factor is produced by assuming the object can be described as a tumbling "box" and then averaging over all possible lines of sight for the observing radar: (r_p is the primary object radius, r_s is the average size for the secondary objects, r_{RCS} is the average size as found from the radar data)[37]:

$$A_{KGT} = \pi\left(r_p + r_s\right)^2 = \pi\left(r_p + \sqrt{\frac{\pi}{2}} r_{RCS}\right)^2 \tag{22.4}$$

The average radius for the secondary objects, r_s, in LEO was found to be 0.92 m. Since this average is based upon all objects (including small debris), it is not representative of operational spacecraft. Therefore, for the primary, a more realistic operational size of 2 m for LEO was chosen (i.e., 4 m wingtip-to-wingtip). The resulting Kinetic Gas Theory area was thus 30 m^2.

Figure 22.14 shows the collision risk results for the chosen LEO satellite for a 10-year mission. Note the similarity between this plot and Fig. 22.11: There is not an exact one-to-one correspondence because there is a slight dependency upon the altitude in the average relative velocity of encounter. Other than in specific regions where LEO constellations already reside, most of the risk derives from debris. The average risk for objects large enough to be tracked in LEO from the Kinetic Gas Theory is 7×10^{-4} over a 10-year mission, although at certain altitudes, the risk can be several times larger (or smaller). For objects of different size regimes, the Kinetic Gas Theory approach can be applied, but the spatial object density at that size regime should be used (Fig. 22.13).

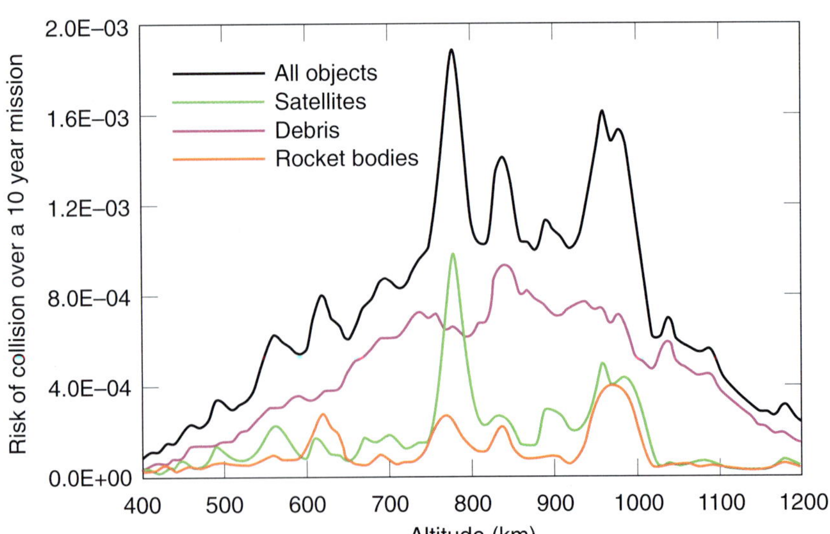

Fig. 22.14. Kinetic Gas Theory risk for objects in LEO.

22.2.4.3 Small Satellites as a Collision Risk

Collision of an operating satellite with even a very small satellite would cause significant damage to both, and warnings of impending collision may pose a challenge if the small satellite is difficult to track. Designers of very small satellites should consider including tracking enhancing features into their vehicles (see chapter 21).

Collision avoidance requires sufficient information, including ongoing tracking data on the two objects that might be involved, to enable prediction of an upcoming close approach. Such predictions must leave enough time for an operating satellite to plan and execute an avoidance maneuver (of course, not all operating satellites have maneuver capability).

The likelihood of a collision is computed by means of tracking data and a covariance matrix providing uncertainties in the tracking estimates.[38] Tracking data is maintained in the Resident Space Object catalog, which is also available on the Internet.[39] Generally, collision avoidance considerations require high-accuracy data on both objects, and high-accuracy data is not generally available—but in some cases, can be derived by the operators of a satellite or from updated tracking data.

Given good data on the current position of an orbiting object, the position can be propagated into the future and compared with predicted positions of other tracked objects. The predicted positions may be influenced by uncertainties in the drag due to atmospheric effects, or by sun-moon gravitational effects, or by errors in the initial tracking and orbit determination process. As the object's position is propagated into the future, these errors must also be propagated, and any predictions of close approaches or possible conjunctions must take these factors into account. The propagated position of an object at its closest approach is its mean position. Since there are errors, the true position at the closest approach is normally distributed about the mean position. As a result, the position-error probability density takes the form of a position-error ellipsoid, with each axis of the ellipsoid having an associated standard deviation. Its associated position-error covariance matrix is diagonal, with elements being the position-error variances. The position-error ellipsoid is also referred to as the covariance ellipsoid, but this may not be equivalent to the mathematical covariance derived from orbital determination techniques.

Probabilistic techniques have been developed that can take the position and error covariance information for two approaching objects into account and predict a probability of collision.[38] Satellite operators can use this information to assess the risk and decide whether a maneuver is desirable. Figure 22.15 illustrates the close approach of two objects represented by their covariance ellipsoids. Note that the illustration of the satellites, placed at the centroid of each error ellipsoid, are not to scale. A typical 1 σ error in position of an object may be 12 km along the velocity vector, 3 km along the radius from Earth, and 8 km cross-track at the measurement point, but may grow substantially as time from the last measurement increases.[40] Since atmospheric drag effects increase for satellites in LEO, the position errors may grow more quickly than for satellites at higher altitudes.

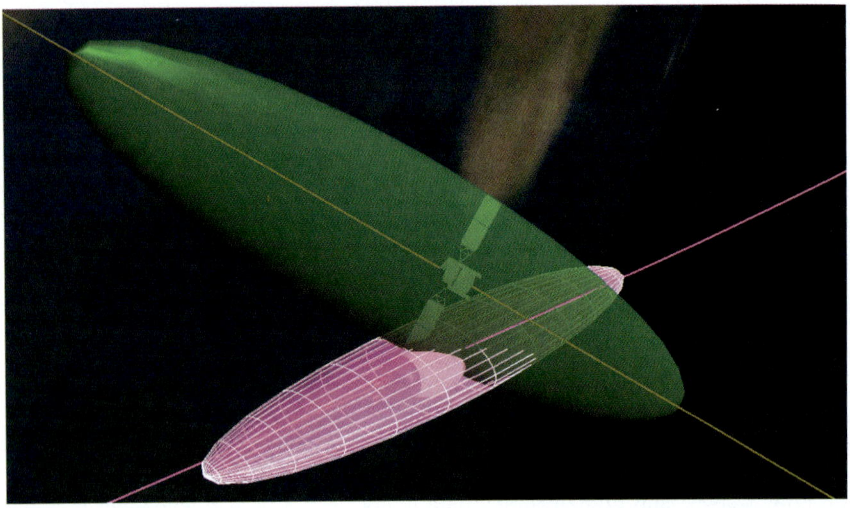

Fig. 22.15. Intersection of covariance ellipsoids for two orbiting objects.

A second approach for collision avoidance for operators who experience repeated close approaches with the same object is to utilize passive techniques to avoid such situations. For example, two satellites in GEO might experience periodic close approaches, and operators of the satellites can slightly modify the orbits of the satellites such that the close approaches do not occur. Techniques that can be used to ensure that satellites flying in close formation will avoid collisions have also been developed.[41]

Another approach for minimizing the possibility of collisions is to place the satellite in an orbit regime where the population is low. As Figs. 22.4 and 22.5 show, the density of orbiting objects, and hence the collision probability, varies with orbit altitude, and selecting mission orbits in the valleys of this profile help minimize the probability of collision over the long term.

22.2.4.4 Increasing Trackability

Many experts expect "space traffic control" services, including collision avoidance, to become available in the near future. Even if collision avoidance services are available, tracking becomes an issue as the size of the satellite decreases to a meter or less, with the general limit of tracking ability set at objects whose smallest dimension is approximately 10 cm. Solar panels and long appendages can increase the trackability (see chapter 20). The Aerospace Corporation picosats, each about the size of a deck of cards, were connected by a 100 ft tether with embedded gold wire dipoles to increase trackability. The 2 kg RAFT satellite included a transponder that "beeped" when observed by radar (see chapter 8). This allowed radar location of a satellite that is physically too small to be detected. Small satellite tracking is discussed in greater detail in another chapter of this book (see chapter 21).

22.3 Reentry Breakup Overview

Before discussing the expected casualties associated with reentry, it is instructive to provide an overview of the reentry breakup process to illustrate that fragments from reentering vehicles can and do survive reentry. While small satellite reentries would be similar in nature, clearly the quantity and size of surviving debris would be substantially less. It is expected that a mass threshold exists, below which the ground hazard associated with the reentering object would always be less than the casualty expectation discussed in this section. This threshold has not been determined.

Figure 22.16 shows a Delta II second stage used to place a large spacecraft in orbit. A similar unit was separated from its payload and remained in orbit for nine months. That stage reentered the atmosphere over Canada and followed the trajectory shown in Fig. 22.17. During descent, it experienced the severe heating and loads typical of atmospheric reentries and broke apart, shedding lightweight fragments like the one that brushed Ms. Lottie Williams on the shoulder (see Fig. 22.18), as well as the larger tanks that landed further downrange. The largest tank had a mass of approximately 250 kg and landed about 50 m from a farmer's house in Texas. The debris was spread over a ground footprint more than 400 km long. This is typical for large reentering objects, and current estimates are that from 10% to 40% of a spacecraft's dry mass at reentry will survive to hit the ground, depending on the types of materials used. Additional examples of surviving debris are included in Refling and Stern.[42]

The amount of surviving material depends on the materials used in the construction of the reentering body and on the aerodynamic and mass properties of each major component. For example, experience has shown that aluminum structures and components typically do not survive reentry unless they are protected from reentry heating by other components. On the other hand, titanium propellant and pressurized gas tanks and items fabricated from steel and beryllium have been found on the ground after reentry.

Should debris from a space object survive reentry and cause injury or property damage, United Nations resolutions state that, "A launching State shall be absolutely liable to pay compensation for damage caused by its space object on the surface of the Earth or to aircraft in flight,"[43] and, "Upon request of the launching authority, objects launched into outer space or their component parts found beyond the territorial limits of the launching authority shall be returned to or held at the disposal of representatives of the launching authority, which shall, upon request, furnish identifying data prior to their return."[44]

As a result of increasing recognition that debris does survive and that it does pose a potential hazard, recent guidelines and regulations[45,46] state that space hardware must be deorbited in a controlled fashion if the casualty expectation (number of predicted injuries or deaths) for an uncontrolled reentry exceeds a defined threshold—in the United States, this threshold is 1×10^{-4} casualties per reentry. These requirements place increased emphasis on the ability to develop accurate

Courtesy of NASA

Fig. 22.16. Delta II second stage, similar to that which reentered on 22 January 1997.

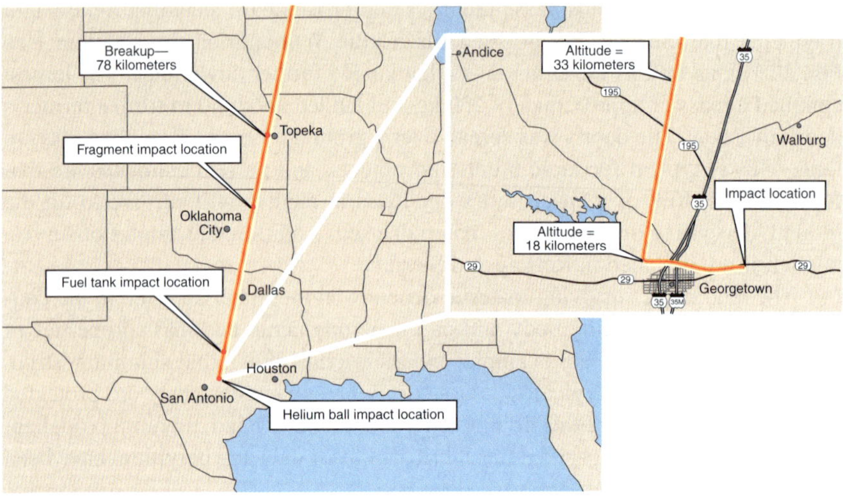

Fig. 22.17. Flight path for Delta II second-stage reentry.

Courtesy of Tulsa World

Courtesy of NASA

Fig. 22.18. Debris from 22 January 1997 reentry of Delta II second stage. *From left*: lightweight fragment, propellant tank, thrust chamber.

Table 22.5. Reentry Survivability Models

Model Name	Methodology	Originator
ORSAT[48]	3-DOF, initial breakup at 78 km typically assumed	NASA
SCARAB[49]	6-DOF, predicts breakup	ESA
AHaB[50]	3-DOF, predicts breakup	The Aerospace Corp.
DAS[51]	3-DOF, initial breakup at 78 km typically assumed	NASA

estimates of what will and will not survive reentry and to estimate the final hazard associated with each surviving fragment.

Several models of varying complexity have been developed to estimate the size and mass of surviving debris, and these are listed in Table 22.5 (ORSAT and DAS are available from NASA; SCARAB is available from ESA). Models vary from those that assume a reentering object breaks into simple components at a specified altitude to models that simulate the full 6 degree-of-freedom motion of an object and the heating and loads to individual components to the greatest extent possible. In all cases, engineers use best judgment as a substitute for relevant *in situ* data to calibrate these models. The 78 km breakup altitude assumed by several models is based on data from observed spacecraft deorbits.[47]

22.3.1 Estimation of Reentry Casualty Expectation

The casualty expectation for a random reentry can be computed by first finding the casualty expectation per unit casualty area based on the inclination of the reentering object's orbit and then multiplying by the total casualty area of the surviving debris that impacts the Earth's surface.[52] The other critical parameters for computation of this reentry hazard are the population density that might be affected by the reentry and the number, mass, and size of surviving debris fragments.

Figure 22.19 shows the ground area that could have been affected by the reentry of the Mir space station if it had reentered randomly (Mir was deorbited into the South Pacific in 2001). The area is defined by the inclination of Mir's orbit, approximately 51 deg; that is, debris from a reentry could have fallen anywhere between latitudes of 51 deg N and 51 deg S, and people living within this band could have been affected. Thus, an object's orbit inclination defines the population that may be affected by its reentry, and as a first estimate, the population within these bands is assumed to be evenly distributed. It should be noted that more refined casualty expectations can also be generated. For example, rather than evenly distributing the population, orbits can be identified that pass over heavily populated areas, and the actual population density can be used. Generally, a more specific or detailed analysis is performed when a specific impact area is known.

The casualty area, in which an impacting object has a possibility of striking a standing human, is defined by the projected area of the object increased by a

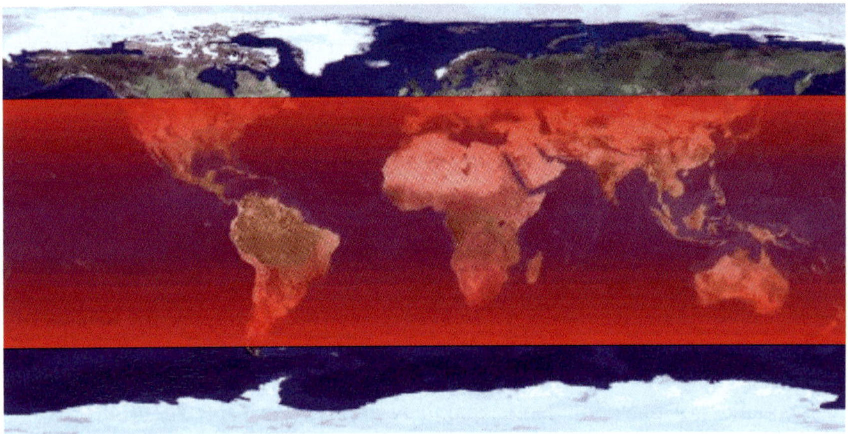

Fig. 22.19. Ground area that could have been affected by random reentry of Mir.

"human border" representative of the radius of a standing human. NASA specifies a human-border of 0.34 m be used.[12]

Figure 22.20, which uses population data (in this case, from 1995[53]) and incorporates the debris impact probability density using a method derived in *Space Modeling and Simulation*,[52] shows the casualty expectation per unit area for a given orbit inclination. For our Mir example, an orbit inclination of 51 deg yields ~1.3×10^{-5} casualties per square meter if the reentry were to occur in 1995. If the total casualty area, A_C, for Mir was ~450 m^2 (only a guess), the casualty expectation, E_c, for a 1995 impact would have been 5.8×10^{-3}, as shown in Eq. (22.5).

To estimate the casualty expectation for the actual reentry year, 2001, one can either use an actual population database that year or extrapolate from a database for a prior year by assuming a growth rate in the population (see "Total Midyear Population for the World: 1950–2050,"[54] for information on past and projected growth rates). To illustrate the second approach, a uniform population growth rate of 1% per year is used. Assuming uniform population and population growth rate over all latitudes, one simply increases casualty expectation according to Eq. (22.6). Thus, for 2001, the casualty expectation is 6.21×10^{-3} as given in Eq. (22.7).

$$E_{c(1985)} = (E_c/A)A_c = 1.3 \times 10^{-5}(450) = 5.85 \times 10^{-3} \tag{22.5}$$

$$E_{c(YR)} = E_{c(1995)()}\,(1.01)^{YR-1995} \tag{22.6}$$

$$E_{c(2001)} = (E_{c(1995)}(1.01)^{2001-1995} = 5.85 \times 10^{-3}(1.01)^6 = 6.21 \times 10^{-3} \tag{22.7}$$

Figure 22.21 illustrates one method for how the casualty area of an impacting object is computed, and the total casualty area is the sum of the casualty areas

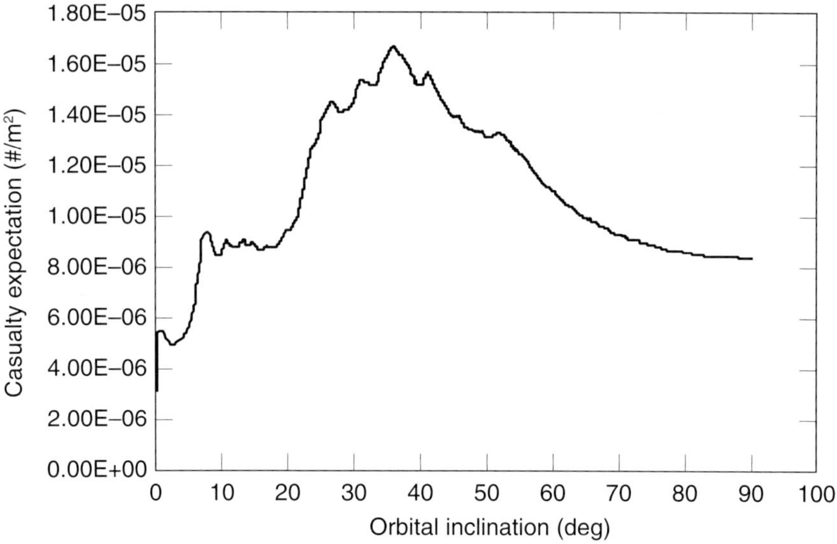

Fig. 22.20. Casualty expectation for the year 1995 as a function of orbit inclination.

for each surviving fragment. The total casualty expectation is the product of this value times the casualty expectation per square meter, as discussed earlier.

Since some pieces of debris are not very massive and may have a low impact velocity, they may not have sufficient kinetic energy to pose a hazard. Therefore, debris pieces should be screened based on impact energy before they are included in the casualty expectation calculation. One can exclude objects having impact kinetic energies less than a threshold of 15 J.[24]

The impact velocity (V_I) and kinetic energy (E_I) can be expressed in terms of object mass, ballistic coefficient, β, and atmospheric density at the impact

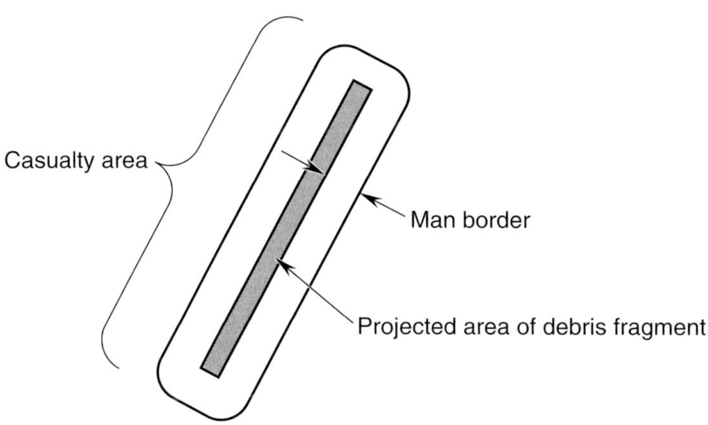

Fig. 22.21. Casualty area for debris fragment.

altitude, ρ_0. The equations relating these terms are given in Eqs. (22.8–22.10), where m is the mass of the debris piece, g is the gravitational acceleration, C_D is the drag coefficient, A is the drag reference area (defined according to C_D), and

$$V_I = \sqrt{\frac{2g\beta}{\rho_o}} \tag{22.8}$$

$$E_I = \frac{mg\beta}{\rho_o} \tag{22.9}$$

where,

$$\beta = \frac{m}{C_D A} \tag{22.10}$$

As an example, a 1 kg object with a ballistic coefficient of 10 kg/m^2 (note that this ballistic coefficient is the inverse of that used for orbit lifetime assessments—see section 22.4) impacts sea level at 13 m/sec with kinetic energy of 80 J. Since the kinetic energy of this fragment exceeds the threshold value noted earlier, it would be included in the calculation of casualty expectation.

22.4 Orbit Lifetime Estimation and Reduction

As noted, an accepted disposal technique for space hardware at end of mission is to move it to an orbit that will lead to a reentry into Earth's atmosphere within a specified time frame. The objective is to ensure that the hazard posed to other space objects ends within a reasonable time; in the United States, 25 years is the requirement.[24]

Operational orbits of many small satellites are dependent upon the launching mission profile. That is, they hitch a ride on a preexisting mission, and therefore may not be able to control their injection altitude. They may carry little to no fuel and therefore cannot change their orbit to any substantial degree; as such, they are completely dependent upon natural decay when meeting the 25-year requirement. It is therefore useful to know what the maximum altitude that a small satellite can be injected into and still meet the orbit lifetime requirement. There are two such general cases to consider: small satellites put into near-circular LEO, and small satellites riding on geosynchronous transfer orbits (GTO).

22.4.1 Modeling Issues for Decay Analysis

The maximum satellite injection altitude that meets the 25-year decay requirement is a function of several factors: the atmospheric model used for the prediction, the solar activity index, and the ballistic coefficient of the satellite. The two relevant parameters in commonly used atmospheric models (e.g., MSIS-90E,[55] Jacchia-

Walker-Bruce[56]) are F10.7 and Ap. The F10.7 is a measure of the solar flux at the 10.7 cm wavelength that impinges upon Earth. This wavelength has been shown to be the main exciter of atmospheric response. The Ap is a measure of the response of Earth's magnetic field to solar activity and is of lesser significance than F10.7 in its impact on the atmosphere.

Predictions of these parameters are required in performing orbit lifetime analysis for mission planning purposes, and predictions for solar activity in 5, 50, and 95 percentiles are available to the public.[57] These are obtained by smoothing historical data from previous cycles to generate a minimum, mean, and maximum prediction for F10.7 and Ap for the next solar cycle plus enough time to reach either the following solar minimum or maximum. Figures 22.22 and 22.23 show recent historical behavior of the F10.7 and Ap values along with the current 5%, 50%, and 95% predictions. It appears from Fig. 22.22 that the solar flux can vary by a factor 2 within the predictive range.

A difficulty for decay analysts lies in the level of predictions to use and in how long the predictions are valid. The 95% levels of activity will produce the greatest amount of decay, but that estimate may not be an accurate predictor of the actual orbit lifetime.

A second difficulty is that the F10.7 and Ap predictions themselves are made for one full solar cycle (11 years) from the present plus an additional amount of time to get to the next solar minimum/maximum. Therefore, for decay analysis purposes, the predictions do not cover the necessary 25 years—current predictions go out for 15 years until 2022, for example. However, given that the predictions are based upon historical values that have been smoothed, statistical analysis implies that the F10.7 and Ap cycles can be simply repeated out to the required

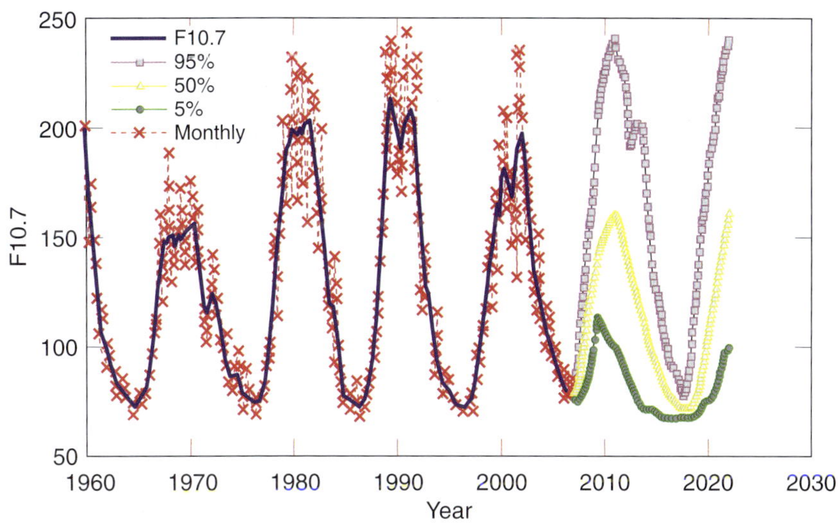

Fig. 22.22. Solar flux F10.7 historical behavior and predictions. Units are in 10^{-22} W-s/m^2.

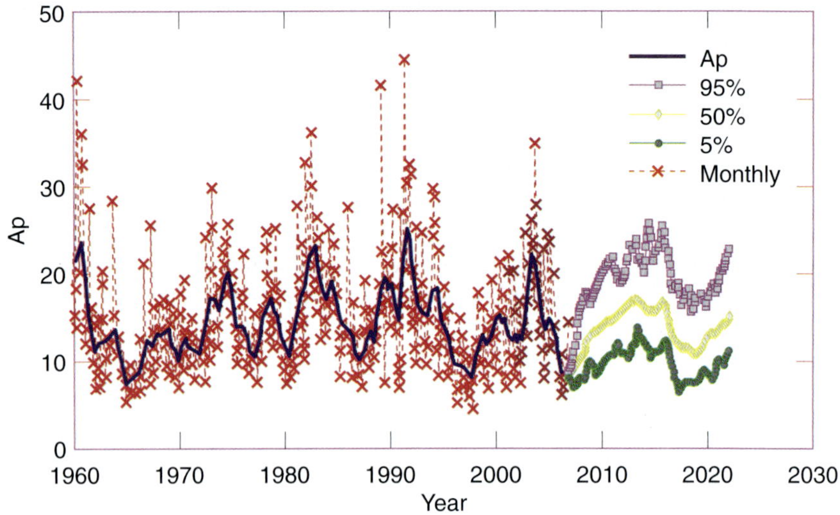

Fig. 22.23. Dimensionless geomagnetic index Ap historical behavior and predictions.

25 years without significant loss of accuracy. It must be noted that this repetition cannot be done ad infinitum, and analysts examining very long-term decay issues must be aware of this limitation; but, for the duration required for the 25-year analysis, it is sufficient for producing mean (50%), optimistic (95%), and pessimistic (5%) decay profiles.

For these purposes, the satellite's ballistic coefficient is now defined as

$$\beta_o = C_D \frac{A}{m} \tag{22.11}$$

where C_D is the free-molecular coefficient of drag, A is the drag reference area (m^2), and m is the mass (kg). The value of C_D is typically taken to be 2.2 with A defined as the cross-sectional area. Note that the on-orbit ballistic coefficient is the inverse of the ballistic coefficient used in atmospheric flight.

It should be noted that the decay analysis begins once the satellite's mission is over; therefore, the average cross-sectional area for a tumbling object should be used for orbit lifetime analyses rather than a minimum or maximum value.

22.4.2 Near-Circular Orbits

Once F10.7, Ap, and β_o have been determined for a decay analysis, the maximum allowable injection or operational altitude for a satellite that will use orbit decay for disposal can be found through simple propagation schemes. Figure 22.24 shows a sample altitude profile for a satellite with a β_o of 0.02 m^2/kg and a Jacchia-Walker-Bruce atmospheric model using a 95% level of solar activity. The higher-frequency oscillations are due to the nonspherical nature of Earth's

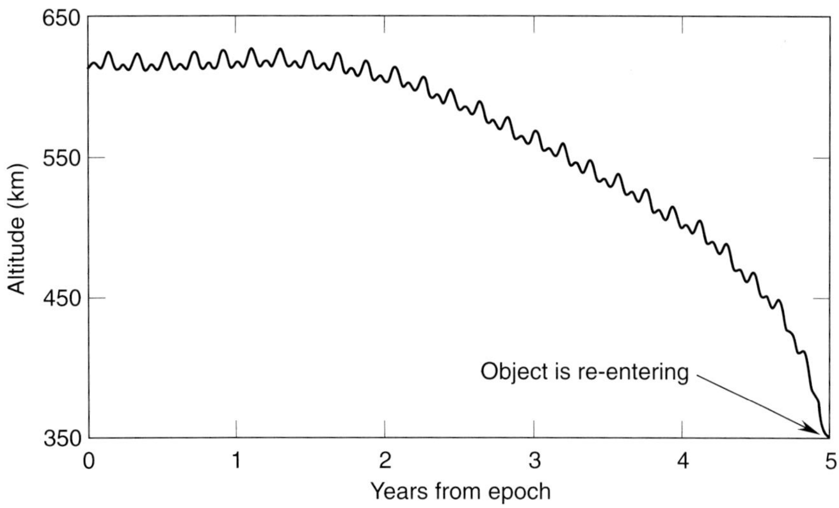

Fig. 22.24. Sample altitude profile for a LEO satellite with a ballistic coefficient of 0.02 m²/kg, 95% atmosphere, Jacchia-Walker-Bruce atmosphere with an epoch of 1 January 2008 (near minimum of solar cycle).

gravitational field. The maximum allowable altitude is determined by making an initial guess, determining the resulting lifetime, and iterating until a decay of 25 years is predicted. The epoch of Fig. 22.24 corresponds to January 2008, which is near the minimum of the solar cycle. As expected, near solar minimum, the drag initially is very slight and increases as the solar activity increases, causing the precipitous drop in altitude. Studies have shown that the 25-year maximum allowable altitude is not dependent upon the epoch (i.e., the time during the solar cycle) of disposal (for 25 years, the satellite goes through more than two complete solar activity cycles regardless of actual disposal epoch), nor is there a strong dependence upon the inclination of the orbit or other orbit parameters.

Figure 22.25 shows the lifetime (up to 25 years) of the sample satellite defined above for various initial altitudes for the three solar activity levels for a β_o of 0.015 m²/kg. For a satellite with this ballistic coefficient to decay in less than 25 years under a 95% atmosphere, the deployment altitude must be at or below 700 km.

The one parameter (other than altitude) under the control of the small satellite designer that does influence the decay time is the ballistic coefficient. Figure 22.26 shows the maximum allowable altitude as a function of this parameter for the various solar activity levels. As expected, the higher the solar activity, the higher the small satellite can be deployed. The differences between the activity levels are large (>50 km), so the activity level must be chosen carefully. As is evident, drag enhancement is an option available to decrease orbit lifetime. For example, if the ballistic coefficient is doubled from 0.02 to 0.04 m²/kg, then the maximum allowable altitude will increase by ~50 km.

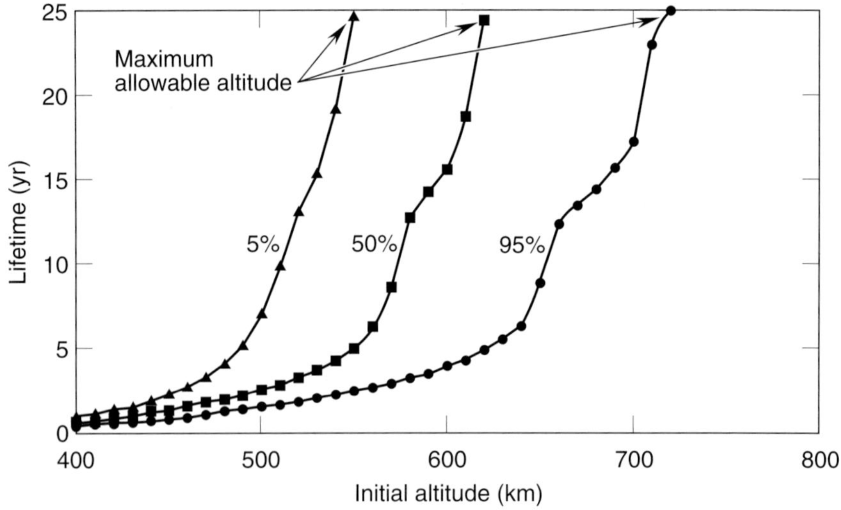

Fig. 22.25. Decay time for sample LEO satellite for a ballistic coefficient of 0.015 m²/kg.

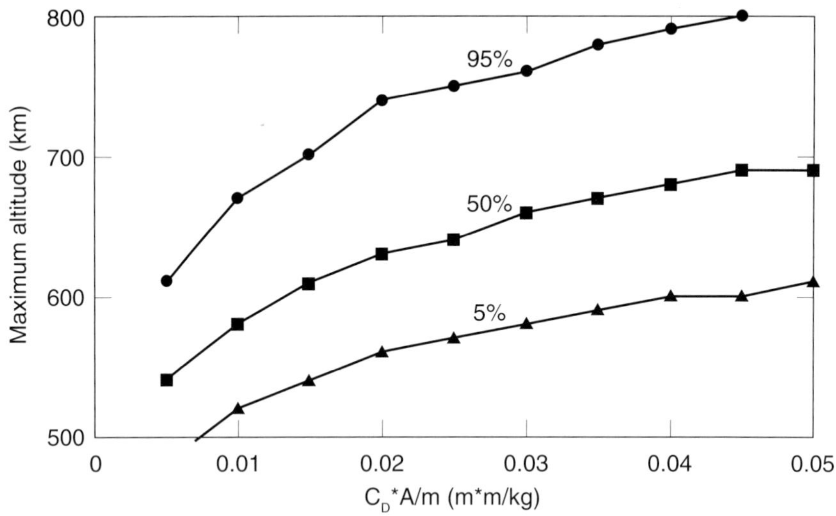

Fig. 22.26. Maximum allowable altitude as a function of ballistic coefficient.

22.4.3 Highly Eccentric Orbits

Another possibility for small satellite deployment is to piggyback onto missions to GEO or to be launched directly into highly eccentric orbits (HEO). GTO and HEO trajectories are problematic for small satellites because they spend most of their time away from perigee and hence are less influenced by drag. But since they cross the LEO region, they are subject to the same 25-year decay requirement.

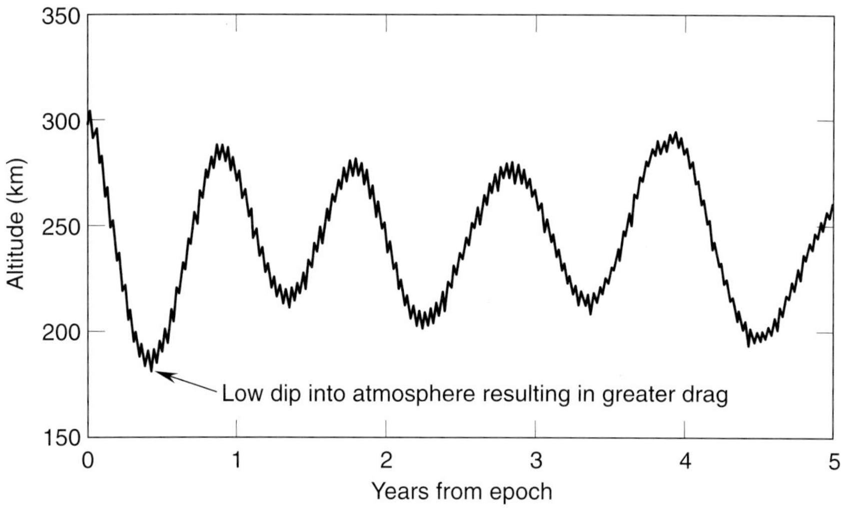

Fig. 22.27. Perigee altitude time history of a sample GTO orbit.

HEO is also more complex to model because solar and lunar perturbations cause large fluctuations in the eccentricity (and therefore perigee altitude). These fluctuations are not uniform, as Fig. 22.27 shows for GTO. Note that for this case, the first dip to a low-perigee altitude is lower (~180 km) than subsequent dips, while the highest dip occurs at an altitude of ~220 km. Thus, the small satellite will experience much greater drag during its first year on orbit than on the following years.

As is evident, determining the maximum allowable perigee altitude for this type of deployment is potentially a very sensitive analysis. Just when, and if, a deep dip (or lack of a deep dip) in the perigee altitude behavior occurs has a significant effect on the resulting lifetime. This is exhibited in Fig. 22.28 for a GTO satellite with a ballistic coefficient of 0.2 m^2/kg and a 95% Jacchia-Walker-Bruce atmosphere. For initial perigee altitudes below ~290 km, the satellite's lifetime will rise smoothly as a function of increasing initial altitude; above 290 km, the satellite's lifetime continues to rise as altitude increases until the 25-year maximum is reached at 330 km. However, at 290 km, the satellite's perigee altitude stays high for a much longer period of time than at other altitudes, resulting in a violation of the 25-year requirement.

Unfortunately, this sensitivity is not merely a function of initial altitude. Analysis has shown that the dependence upon the initial inclination and time of deployment within the annual solar and monthly lunar cycles can result in maximum allowable perigee altitudes that vary by as much as 70 km. Figure 22.29 shows that the dependency upon the initial argument of perigee and right ascension of ascending node (RAAN) is even more pronounced. For example, for an argument of perigee and RAAN of 180 deg, the maximum allowable altitude

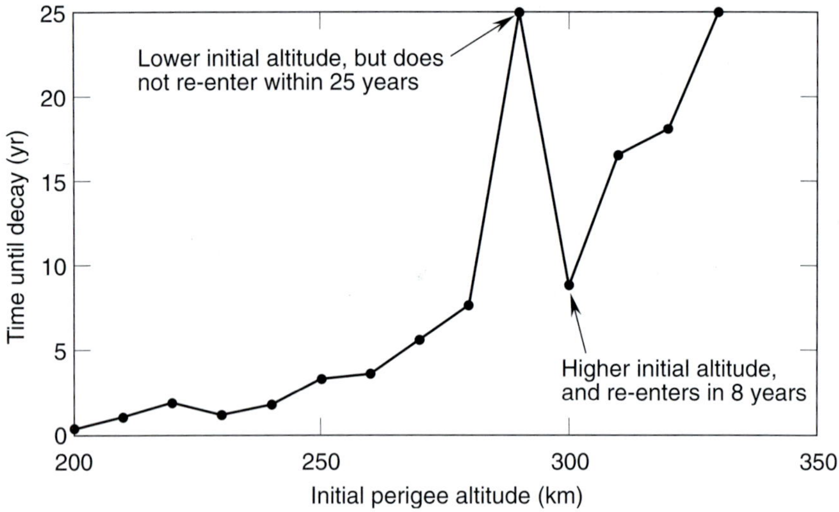

Fig. 22.28. Decay time for sample geosynchronous transfer orbit as a function of initial perigee altitude.

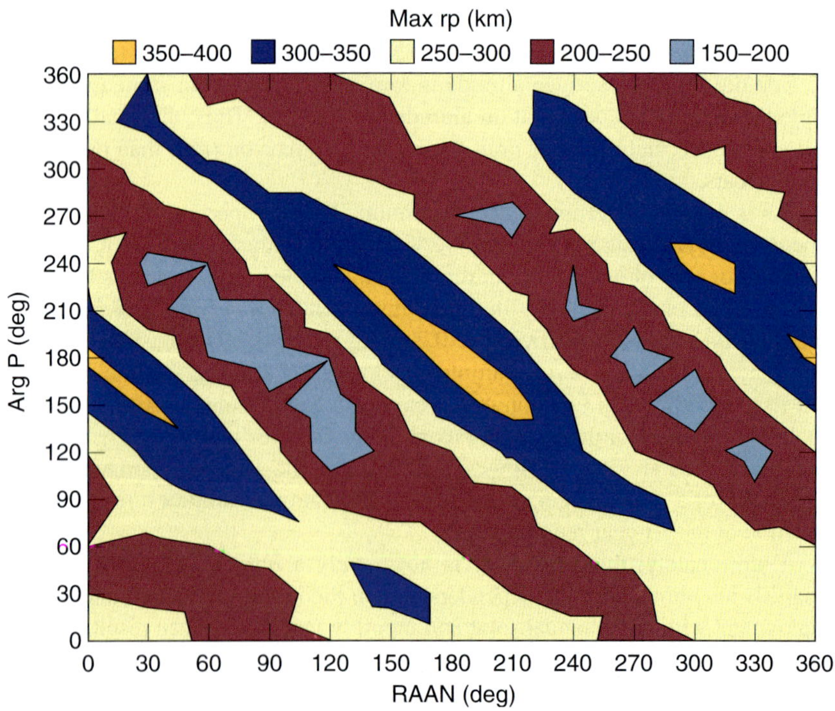

Fig. 22.29. Maximum allowable initial perigee altitude as a function of argument of perigee (Arg P) and right ascension of ascending node (RAAN).

would be between 350 and 400 km; simply shifting the RAAN to 90 deg would drop the maximum allowable altitude to between 150 and 200 km.

The basic conclusion for GTO (and by extension, other HEO trajectories) is that these orbits are not particularly well suited for small satellites relying on orbit decay to meet disposal guidelines.

There are two additional considerations that must be addressed before a small satellite can be attached to a GTO mission: first, the low initial perigee altitudes that would be required to attain the 25-year requirement are not typical of current GTO profiles; second, since many small satellites are passive passengers on missions whose own constraints will take priority, the choice of particular GTO will be out of the control of the small satellite operator. It is unlikely that the small satellite perigee altitude constraint would be allowed to alter the main mission, either in terms of orbit design or launch date.

If the small satellite can be directly launched into HEO without having to piggyback on a preexisting mission, then the possibility of these orbits becomes more likely. Care must still be taken to examine the individual mission profile to ascertain the likelihood of reentry within 25 years. This mission profile will be sensitive to all of the orbit elements, the initial epoch of disposal, and the ballistic coefficient of the satellite. Failure to meet the 25-year requirement would have the technical consequence of increasing the likelihood of collision with other on-orbit objects, the legal consequence of liability for such damage, and the political consequence of a lower likelihood of licensing approval for future missions.

22.5 Issues for Small Satellites

Small satellites (<150 kg) present several problems and opportunities from the space debris perspective. Problems include:

- Tracking difficulty increases as size decreases. Satellites whose smallest dimension is less than 10 cm may be very difficult to track, and as a result, Resident Space Object catalogs may not have data of sufficient accuracy to allow collision avoidance by other operating satellites. These objects may remain untracked for their lifetime in orbit. As noted earlier, objects in LEO can impact at speeds up to 14 km/sec, and at these speeds, even an object the size of a softball can destroy a satellite and generate clouds of debris. Untracked objects are generally smaller than 10 cm in size. This means that operating satellites are "blind" to their existence and are at risk of collision.
- A single launch vehicle can carry and deploy several individual satellites. Each additional satellite slightly increases the probability of collision for other satellites and debris sharing that region of space.
- Small satellites are relatively inexpensive, but debris mitigation would require a system to move them to a disposal orbit and possibly deorbit

them at end of mission. The addition of these capabilities may increase the cost and complexity of building and operating these vehicles.

- Some small satellites are opportunists, when it comes to launches; that is, their owners look for a launch with available margin and try to hitch a ride. In these cases, the small satellites are deployed at the pleasure of the host vehicle, and might be deployed in orbits with long lifetimes and remain there long past their operational lifetime, posing hazards for many years unless disposed of properly.
- The cost of a ride to space might significantly increase if a small satellite has propellant aboard. Owners of a host satellite will require that the small satellite go through an extensive process to ensure that it does not threaten the successful deployment or mission of the host. Availability of inexpensive rides to space might be threatened if a small satellite is required to have a propulsion system for disposal.
- Analysis may well show that small satellites have an advantage from the reentry safety perspective. Reentry hazards *decrease* as the size and mass *decrease*, and below a certain size and mass, reentry breakup hazards may be below the mandated 1 in 10,000 casualty expectation. Thus, disposal by uncontrolled reentry would automatically be acceptable for satellites below a specified threshold. However, until eliminated by development of specific guidelines, reentry breakup and hazard analyses might still be required to ensure that orbit decay is an acceptable disposal option for small satellites.

Opportunities include:

- Small satellites may offer relatively inexpensive opportunities for testing space debris minimization and removal techniques, and small satellite technologies may also enable new reentry testing opportunities. For example, tethers and other proposed end-of-mission lifetime reduction techniques might be tested using small satellites. Such systems might offer the advantage of increasing the trackability of the small satellites as their orbits decay.

22.6 Deorbit Data

There is currently a lack of quality-measured data from spacecraft undergoing uncontrolled reentry. The availability of such data would improve calibration and validation of reentry survivability models. One possible means of collecting such data is the Reentry Breakup Recorder (REBR) being developed by The Aerospace Corporation and NASA Ames Research Center; it uses memory, controller, and sensor boards developed for the Aerospace PicoSats. REBR electronics, which also includes GPS, batteries, and an Iridium modem, are shielded by a NASA-developed thermal protection system that enables the vehicle to survive reentry.

REBR, illustrated in Fig. 22.30, is approximately 0.33 m in diameter and weighs less than 2 kg. It is designed to attach to a host vehicle and remain there until the host reenters the atmosphere. During reentry, REBR records data from sensors internal and external to the device and ultimately detaches from the disintegrating host vehicle. REBR decelerates, reaches a subsonic velocity at approximately 60,000 ft, and during the next 5 min before impact, transfers recorded data via the Iridium system. The device is not designed to survive impact and does not need to be recovered (for reference, the casualty expectation for each REBR reentry is estimated to be approximately 1×10^{-6}, well below the 1×10^{-4} threshold described in section 22.3).

One major advantage of the REBR approach is the use of the commercial Iridium system for communications. Iridium is global, so a reentry can occur anywhere on Earth and data will be returned. Deorbit of the host vehicle into a predefined and instrumented range is not required—a major cost savings. This eliminates the need to recover REBR and also eliminates the need for deorbiting the host vehicle into a specified location for a ground recovery team—both expensive prospects.

Should recovery be desired, a more robust version of REBR might be designed, but some control over the landing area would probably be required. Another concept would be to build a "flyable" REBR—a device the size of a model airplane that would fly a space-shuttle-like trajectory to a specified location. Such a vehicle might also enable the return of sample materials and experiments from orbit. It might also be used to prototype and qualify new heat-shield materials, sensor systems, structural health monitoring systems, and other concepts ultimately destined for crewed vehicles and interplanetary missions.

Courtesy of NASA

Fig. 22.30. Exploded view of the Reentry Breakup Recorder.

22.7 Summary

The population of objects orbiting Earth is growing, and with that growth comes an increasing risk of collision, RF interference, or other interaction that will affect missions of operating satellites. As a result, satellite designers, manufacturers, and operators will see increasing efforts to limit the growth of space debris, including mandatory requirements and international standards.

Current guidelines and requirements include preventing explosions during missions and after end of life, removing hardware from protected orbital regions at end of mission, limiting release of debris in orbit, and preventing on-orbit collisions. Small satellites must adhere to these same requirements. In some cases, debris mitigation requirements may demand that special features be included in small satellites. For example, preventing on-orbit collisions may result in adding features to very small satellites that increase their trackability.

One technique for removing a satellite from the LEO protected orbital region is to return it to Earth's atmosphere for destruction by reentry heating. Unfortunately, not all of a spacecraft's mass will be destroyed, and some will survive and pose a potential hazard to people and property. Objects that have casualty expectations exceeding 1 in 10,000 must be deorbited into ocean areas. Small satellites may have an advantage in this regard—casualty expectations may be below this threshold in most cases.

If the reentry hazard is low, orbit decay may be used to remove hardware from orbit. In the United States, the goal is to complete the decay and reentry within 25 years after the end of a spacecraft's life. Small satellites may serve as excellent test beds for innovative techniques to increase the decay rate of a satellite's orbit. Possibilities include using tethers and balloons for increasing atmospheric drag.

Satellites in GEO must be moved to a higher orbit that will not intersect the GEO protected region for at least 100 years. Small satellites placed in GEO must meet this requirement.

A real advantage of small satellites is their low mass. Low mass translates into lower launch costs, opening the possibility for specialized satellites (and reentry vehicles) to be developed by universities and research institutions and placed into orbit as secondary payloads. As a result, small satellites and reentry vehicles using miniature sensor and communications systems may be attractive testbeds for new techniques and technologies.

22.8 Acknowledgments

Significant contributions were made to this chapter by Dr. Michael A. Weaver of The Aerospace Corporation.

22.9 References

[1] F. Alby, E. Lansard, and T. Michal, "Collision of Cerise with Space Debris," *Second European Conference on Space Debris, Organized by ESA,* ESA-SP 393, p. 589 (ESOC, Darmstadt, Germany, 17–19 March 1997).

[2] P. Anz-Meador, "International Guidelines for the Preservation of Space as a Unique Resource," (Embry-Riddle Aeronautical University Dept. of Physics, Nov. 2003).

[3] J.-C. Liou, "A Statistical Analysis of the Future Debris Environment," *Acta Astronautica*, **62**, 264–271 (2008).

[4] W. Ailor, "Space Traffic Control: A View of the Future," *Space Policy*, **18** (2), pp. 99–105 (June 2002).

[5] W. Ailor and G. Peterson, "Collision Avoidance as a Debris Mitigation Measure," IAC-04-IAA.5.12.3.01, *55th International Astronautical Congress* (Oct. 2004).

[6] *Space Debris*, Report of the ESA Space Debris Working Group, ESA SP-1109 (Nov. 1988).

[7] *Position Paper on Orbital Debris*, International Academy of Astronautics (1993; updated 2000).

[8] *Report on Orbital Debris*, Report of the Interagency Group (Space) to U.S. National Security Council (Feb. 1989).

[9] *Orbital Debris—A Space Environmental Problem*, Office of Technology Assessment, U.S. Congress (Sept. 1990).

[10] The Federal Communications Commission (FCC), an independent U.S. Government agency, regulates interstate and international communications by radio, television, wire, satellite and cable. The FCC's jurisdiction covers the 50 states, the District of Columbia, and U.S. possessions (see www.fcc.gov).

[11] "Second Report and Order in the Matter of Mitigation of Orbital Debris," U.S. Federal Communications Commission (21 June 2004).

[12] "NASA Procedural Requirements for Limiting Orbital Debris," NASA NPR 8715.6 (17 August 2007) and "Process for Limiting Orbital Debris," NASA-STD-8719.14 (8 Aug. 2007).

[13] "CNES Standards Collection; Methods and Procedure; Space Debris—Safety Requirements," CNES, RNC-CNES-R-13-501, Issue 2 (9 March 2000).

[14] "NASA Procedural Requirements for Limiting Orbital Debris," NASA NPR 8715.6 (17 Aug. 2007) and "Process for Limiting Orbital Debris," NASA-STD-8719.14 (8 Aug. 2007).

[15] "U.S. Government Orbital Debris Mitigation Standard Practices" (Jan. 2000).

[16] "NASDA Space Debris Mitigation Standard," NASDA-STD-18 edition (original issue: 28 March 1996).

[17] "Russian Aviation & Space Agency (RASA) Branch Standard—General Requirements for Mitigation of Space Debris Population," Date of enforcement: 1 July 2000.

[18] *ESA Space Debris Mitigation Handbook, Second Edition, Issue 1.0* (European Space Agency, 3 March 2003).

[19] "Space Support," Department of Defense Instruction Number 3100.12 (14 Sept. 2000).

[20] "IADC Space Debris Mitigation Guidelines," IADC-02-01 (Sept. 2007) http://www.iadc-online.org/index.cgi?item=docs_pub.

[21] T. J. Settecerri, E. G. Stansbery and M. J. Matney, "Haystack Measurements of the Orbital Debris Environment," *Advances in Space Research*, **23** (1), pp. 13–22 (1999).

[22] "Rocket Explosion," *Space Weather* (27 Feb. 2007); http://spaceweather.com.

[23] Courtesy Heiner Klinkrad, ESA.

[24] "U.S. Governmnet Orbital Debris Mitigation Standard Practices" http://orbitaldebris.jsc.nasa.gov/library/USG_OD_Standard_Practices.pdf.

[25] H. Klinkrad, *Space Debris: Models and Risk Analysis* (Springer-Praxis Books in Astronautical Engineering, 2006), pp. 193–197.

[26] C. C. Chao and R. A. Glick, "Long-Term Evolution of Navigation Satellites in Orbit: GPS/GLONASS/GALILEO," COSPAR02-A-02858. PEDAS1-B1.4-0051-02, *World Space Congress* (Houston, TX, 18 Oct. 2002).

[27] "Space Systems, Disposal of Satellites at Geosynchronous Altitudes," ISO CDV 26872 (Dec. 2008)

[28] "The End is Mir" (NASA, 10 March 2001); http://science.nasa.gov/headlines/y2001/ast10mar_1.htm.

[29] P. Anz-Meador, "International Guidelines for the Preservation of Space as a Unique Resource" (Embry-Riddle Aeronautical University Dept. of Physics, Nov. 2003).

[30] "BUMPER II Micrometeoroid and Orbital Debris Risk Assessment Independent Technical Inspection Report" (RP-05-66) (NASA NESC, April 2005).

[31] E. L. Christiansen, "Design and Performance Equations for Advanced Meteoroid and Debris Shields," *Int. J. Impact Engr.* **14,** pp. 145-156 (1993).

[32] B. J. Anderson and W. J. Cooke, Jr., "Calculation of Area Loss Due to Meteoroid Penetration," *Proc. of First Leonid Meteoroid Storm and Satellite Threat Conference* (AIAA/Aerospace Corporation, Manhattan Beach, CA, April 1998).

[33] S. Scott, M. Fisher, H. Evans, W. Schonberg, L. Compton, J. Williamsen, G. Peterson, C. Poe, and S. Pandya, "Independent Technical Inspection of the Bumper II Micrometeoroid and Orbital Debris Risk Assessment Final Report," *RP-05-66* (NASA Engineering and Safety Center [NESC], April 2005).

[34] Kaman Sciences Corporation, "On-Orbit Collision Hazard Analysis in Low Earth Orbit Using the Poisson Probability Distribution" (Federal Aviation Administration, 26 Aug. 1992).

[35] D. J. Kessler and B. G. Cour-Palais, "Collision Frequency of Artificial Satellites: Creation of a Debris Belt" (American Geophysical Union, 1978).

[36] R. Lambour, N. Rajan, T. Morgan, I. Kupiec and E. Stansbery, "Assessment of Orbital Debris Size Estimation from Radar Cross-Section Measurements," *Advances in Space Research*, **34** (5), pp. 1013–1020 (2004).

[37] G. E. Peterson, "Determining Object Sizes from Radar Cross Section for Collision Avoidance," *AAS/AIAA Astrodynamics Specialist Conference* (Lake Tahoe, CA, Aug. 2001).

[38] R. P. Patera, "General Method for Calculating Satellite Collision Probability," *Journal of Guidance, Control, and Dynamics*, **24** (4), pp. 716–722 (Aug. 2001).

[39] "Space Track: The Source for Space Surveillance Data," http://www.space-track.org/perl/login.pl.

[40] W. H. Ailor and G. E. Peterson, "Collision Avoidance as a Debris Mitigation Measure," IAC-04-IAA.5.12.3.01, *55th International Astronautical Congress* (2004).

[41] C. C. Chao, *Applied Orbit Perturbation and Maintenance* (The Aerospace Press and AIAA, El Segundo, CA, 2005).

[42] O. Refling and R. Stern, "Review of Orbital Reentry Risk Predictions," Aerospace Report No., ATR-92(2835)-1 (The Aerospace Corporation, July 1992).

[43] *Convention on International Liability for Damage Caused by Space Objects* (United Nations Resolution 2777 (XXVI), annex)—adopted on 29 Nov. 1971, opened for signature on 29 March 1972, entered into force on 1 Sept. 1972. See "United Nations Treaties and Principles on Outer Space," http://www.unoosa.org/pdf/publications/STSPACE11E.pdf.

[44] *The Agreement on the Rescue of Astronauts, the Return of Astronauts and the Return of Objects Launched into Outer Space* (United Nations Resolution 2345 (XXII), Annex)—adopted on 19 December 1967, opened for signature on 22 April 1968, entered into force on 3 Dec. 1968. See http://www.unoosa.org/pdf/publications/STSPACE11E.pdf.

[45] U.S. National Space Policy of 2006 (paragraph 11); http://www.ostp.gov/html/US%20National%20Space%20Policy.pdf.

[46] U.S. Government Orbital Debris Mitigation Standard Practices (February 2001); http://orbitaldebris.jsc.nasa.gov/library/USG_OD_Standard_Practices.pdf.

[47] R.P Patera and W.H. Ailor, "The Realities of Reentry Disposal," Paper 98-174, *8th AAS/AIAA Space Flight Mechanics Meeting* (Monterey, CA, 9–11 Feb. 1998).

[48] W. C. Rochelle, B. S. Kirk, B. C. Ting, L. N. Smith, R. N. Smith, E. A. Reid, N. L. Johnson, and C. B. Madden, "Modeling of Space Debris Reentry Survivability and Comparison of Analytical Methods," IAA-99-IAA.6.7.03, *50th International Astronautical Congress* (Amsterdam, Oct. 1999).

[49] T. Lips, B. Fritsche, G. Koppenwallner, and H. Klinkrad, "Spacecraft Destruction during Re-entry—Latest Results and Developments of the SCARAB Software System," *34th Scientific Assembly of COSPAR,* PEDAS1-B1.4-0030-02 (Houston, TX, Oct. 2002).

[50] Atmospheric Heating and Breakup (AHaB) Tool, The Aerospace Corporation.

[51] J. N. Opiela, E. Hillary, D. O. Whitlock, B. Hadjisavvas, and M. Hennigan, "Debris Assessment Software Version 2.0 User's Guide" (Orbital Debris Program Office, NASA Johnson Space Center, Nov. 2007).

[52] *Space Modeling and Simulation: Roles and Applications Throughout the System Life Cycle*, L. Rainey, ed., pp. 722–725 (The Aerospace Press and AIAA, El Segundo, CA, 2004).

[53] W. Tobler, U. Deichmann, J. Gottsegen, and K. Maloy, "Global Demography Project," Technical Report TR-95-6 (Department of Geographic Information and Analysis, University of California at Santa Barbara, April 19950.

[54] "Total Midyear Population for the World: 1950–2050," U.S. Census Bureau, http://www.census.gov/ipc/www/idb/worldpop.html.

[55] A. E. Hedin, "Neutral Atmosphere Empirical Model from the Surface to the Lower Thermosphere: MSISE," *J. Geophys, Res.*, **96**, 1159–1172 (1991).

[56] Jacchia, L.G. 1971. *Revised Static Models for the Thermosphere and Exosphere with Empirical Temperature Profiles.* SAO Special Report No. 332 (Smithsonian Institution Astrophysical Observatory, Cambridge, MA).

[57] "Predicted Sunspot Numbers and Radio Flux," NOAA Space Weather Prediction Center, http://www.swpc.noaa.gov/ftpdir/weekly/Predict.txt.

Chapter 23

The Future of Small Satellites

Siegfried W. Janson*

23.1 Introduction

Small satellites started the Space Age, and applications for small and ultrasmall spacecraft—microsatellites, nanosatellites, picosatellites, femtosatellites, and attosatellites—will continue to expand to include a surprising array of missions, including Earth observation, lunar and near-Earth exploration, interplanetary probes, and communications. Applications for these spacecraft are, however, limited by technical factors, such as physical scaling, orbital mechanics, economics, and technology readiness. In this chapter, examination of technical factors is followed by discussion of some of the possible future applications of these small spacecraft in the near, mid, long, and very long term. To quote a famous lecture by physicist Richard Feynman, "There's plenty of room at the bottom."[1]

23.2 Satellite Scaling Issues

Spacecraft power, inertial stability, image sensor resolution, and optimal radio communications frequencies are functions of physical geometry and size. For many properties, such as available power, bigger is better; some properties, such as required propellant mass fraction for a given mission, are independent of size; for the rest, smaller *is* actually better. Scaling issues ultimately drive feasibility and optimal spacecraft configuration for a given set of mission requirements using a fixed set of available technologies.

23.2.1 Power

Maximum and average power levels limit the types of missions that small satellites can perform. The vast majority of satellites use solar cells for primary power. Solar-cell output is proportional to the projected area of the solar cells with respect to the sun, times the energy conversion efficiency of the cells, times the solar constant of 1350 W/m^2. For commercially available, 28% efficient, triple-junction solar cells oriented directly toward the sun, each square centimeter can produce 38 mW$_e$, where W$_e$ is electrical power in watts. Creating a stable bus voltage requires a voltage converter with about 80% efficiency, which results in an effective maximum spacecraft power of approximately 30 mW$_e$ per square centimeter of solar cell. In low Earth orbit (LEO), spacecraft are typically in sunlight only about

*The Aerospace Corporation
El Segundo, California

60% of the time. Factoring in the eclipse fraction and an additional 20% energy loss for the battery charging system results in a maximum orbit average power level of about 16.5 mW_e per square centimeter of triple-junction solar cell. This power density drops further if the solar cells are not pointing directly at the sun, or if less-efficient cells are used.

Figure 23.1 shows the orbit average spacecraft electric power level in eclipsing LEO orbits as a function of side length for cubic, tumbling spacecraft with body-mounted cells, and for a sun-pointing cubic spacecraft with four side panels deployed toward the sun. The solar capture efficiency η_{sc} is the product of solar-cell efficiency times the fractional area coverage of solar panels; it does not include cosine losses due to illumination from off-normal angles. A cubic spacecraft with 28% efficient cells and 50% surface coverage would have an η_{sc} of 0.14, while deployed solar panels with the same cells and a 90% surface coverage would have an η_{sc} of 0.25. Figure 23.1 uses a single solar-capture efficiency of 0.25 for the deployed solar panels and assumes sunlight hitting the panels at normal incidence. Body-mounted cells will, in general, have lower surface coverage fractions, and hence lower η_{sc}, than deployed panels because optics, antennas, thrusters, and similar elements take up surface area on the spacecraft body.

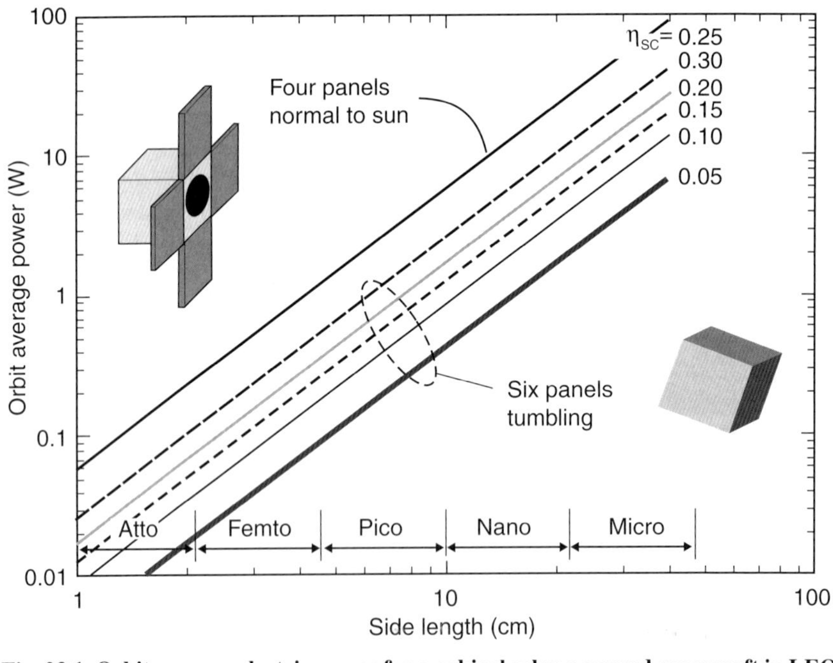

Fig. 23.1. Orbit average electric power for a cubical solar-powered spacecraft in LEO as a function of side length. The solar collection efficiency η_{sc} is the solar-array energy conversion efficiency times the surface coverage fraction for each side. This illustration assumes a 40 min eclipse in a 100 min orbit with an array-output-to-bus-power conversion efficiency of 70%. The mass classifications along the bottom are based on an average spacecraft density of 1 g/cm^3.

Figure 23.1 therefore has multiple capture efficiencies for body-mounted solar cells and assumes an orbit average electric power level based on random orientations with respect to the sun. Figure 23.1 also assumes a 40 min eclipse in a 100 min orbit with an orbit average conversion efficiency of 70%. This is representative of 80% efficient voltage converters, with one converter operating during sunlight and effectively two converters in series during eclipse (one to charge the battery during sunlight and another to reconvert the battery voltage during eclipse).

If we assume a typical small satellite average density of 1 g/cm^3 and a cubic body, specific size ranges can be associated with different mass classifications. These classifications are listed on the bottom axis of Fig. 23.1. Table 23.1 summarizes basic electric power properties for an ideal cubic spacecraft with body-mounted triple-junction cells covering 50–90% of the available surface area, and for a sun-oriented solar array consisting of four panels with 90% surface coverage. For body-mounted cells, microsatellites typically have more than 10 W, nanosatellites have 1–10 W, picosatellites have roughly 1 W, femtosatellites have less than 0.5 W, and attosatellites have less than 0.1 W of average electric power. Using four deployable side panels boosts these levels by a factor of 2.6, enough to enable a 1 W femtosatellite or a 0.25 W attosatellite.

Satellite average power-capture level scales as l^2, where l is a characteristic dimension of the spacecraft. Satellite mass, for constant density, scales as l^3, so power-to-mass ratios scale as l^{-1}. This assumes that any required rechargeable batteries and power-conversion systems are part of the spacecraft mass. As illustrated in Table 23.1, small satellites have higher power-to-mass ratios than larger satellites if spacecraft geometries, solar cell technologies, and average densities are maintained. Power-to-mass ratios play a key role in determining spacecraft acceleration using electric thrusters and in determining the economy of geosynchronous Earth orbit (GEO) communications satellites. In the former case, thrust is proportional to input power for a given electric thruster and specific impulse, and acceleration is simply thrust divided by mass. In the latter case, spacecraft fabrication and delivery costs to GEO are proportional to mass, but total transponder bandwidth (which generates profit) is proportional to available power.

Table 23.1. General Power Levels for Cubic Small Satellites in LEO Using Solar Cells

Satellite Class	Power Using Body-Mounted Cells (W_e)	Power Using Array (W_e)	Four-Panel Array Power/Mass Ratio (W_e/kg)
Microsatellite	5.6–46	26–120	2.6–1.2
Nanosatellite	1.2–10	5.7–26	5.7–2.6
Picosatellite	0.26–2.1	1.2–5.7	12–5.7
Femtosatellite	0.06–0.47	0.27–1.2	27–12
Attosatellite	<0.1	<0.27	>27

Power-to-mass ratio is therefore proportional to the rate of return on investment. Commercial GEO communication satellites have the highest power-to-mass ratios of any satellites, with current values of about 6 W_e/kg. The XM-1 satellite, for example, transmits XM radio to mobile users with an initial on-orbit mass of 2950 kg and an initial power level of 18 kW_e; the power-to-mass ratio is 6.1 W_e/kg.[2] Note that a collection of picosatellites or smaller spacecraft, if they could provide the same function collectively, could have a higher aggregate power-to-mass ratio with a potentially higher return on investment.

Solar-array specific power levels in the >100 W category have traditionally been below 40 W_e/kg, but a recent flight using a thin-film array had a specific power of approximately 150 W_e/kg at a power level of 380 W_e per panel.[3] This technology can provide up to a kilowatt for a microsatellite, thus enabling microsatellite power-to-mass ratios of 10–20 W_e/kg.

For interplanetary missions, solar-cell power output is roughly proportional to the inverse square of the distance to the sun, R_s. Spacecraft within Earth's orbit enjoy higher levels of power output, while those outside have reduced power output. A decade ago, solar power at distances beyond Jupiter's orbit was considered impractical due to significant reductions in cell energy-conversion efficiency under low-intensity, low-temperature conditions. Recent designs and improved cell selection techniques have overcome this issue, so the remaining problem is just the $1/R_s^2$ power loss.[4] Figure 23.2 shows the impact of this loss as a function of R_s in astronomical units (AU, the average distance of Earth to the sun, or

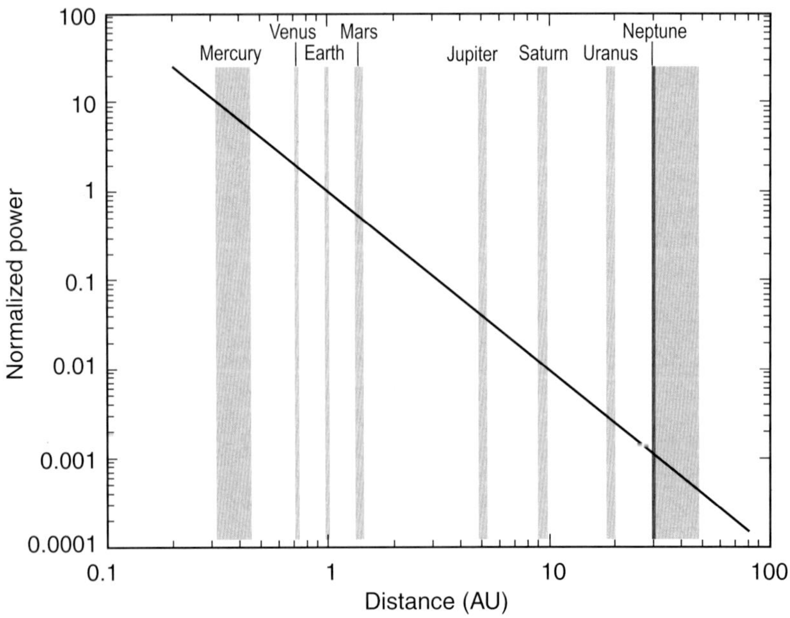

Fig. 23.2. Solar power density normalized to 1 at 1 AU.

1.496 × 10^8 km), normalized to a solar power level of 1 at R_s = 1 AU. At Jupiter, solar cells will produce only 4% of the power they produce in LEO, while at Neptune, it drops to 0.25%.

Table 23.2 shows the power levels for solar-powered small spacecraft at Mars, Jupiter, and Saturn using sun-pointing deployed panels as shown in Fig. 23.1. These calculations assume a close orbit with 60% sunlight, 40% eclipse fractions as in LEO, plus a 70% conversion efficiency between array power and bus power. Solar-powered spacecraft beyond Mars require significantly more array area than provided by four square side panels. A nominal 100 W_e/kg solar array provides effective power densities of 44, 4, and 1.2 W_e/kg at Mars, Jupiter, and Saturn, respectively. This would enable approximately 17 and 5 W_e orbit average powers at Jupiter and Saturn for a 10 kg array suitable for a microspacecraft.

Solar power is not the only option for powering small spacecraft. Nuclear power in the form of radioisotope thermoelectric generators (RTG) was used in small satellites and is still the primary power option for interplanetary probes heading to Jupiter and beyond. Table 23.3 lists U.S. microsatellites that have used RTGs. Since output power decays slowly over time once the RTG is fueled, electrical output powers are given at beginning of mission (BOM). Although Soviet and U.S. reactors have also flown in orbit, their masses exceed the 100 kg limit of

Table 23.2. General Power Levels for Cubic Interplanetary Spacecraft Using Sun-Pointed Square Solar Panels

Spacecraft Class	Power at Mars using a Four-Panel Array (W_e)	Power at Jupiter using a Four-Panel Array (W_e)	Power at Saturn using a Four-Panel Array (W_e)
Microspacecraft	9.4–43	1–4.8	0.26–1.2
Nanospacecraft	2.1–9.4	0.23–1.0	0.058–0.26
Picospacecraft	0.43–2.1	0.048–0.23	0.012–0.058
Femtospacecraft	0.1–0.43	0.011–0.048	0.0027–0.012
Attospacecraft	<0.1	<0.011	<0.0027

Table 23.3. U.S. Microsatellites That Have Used RTG Power Sources[5]

Spacecraft	Launch Date	Spacecraft Mass (kg)	RTG Mass (kg)	BOM Power (W_e)
Transit 4A	29 June 1961	79	2.1	2.7
Transit 4B	15 Nov. 1961	79	2.1	2.7
Transit 5BN-1	28 Sept. 1963	70	12.3	26.8
Transit 5BN-2	6 Dec. 1963	75	12.3	26.8
Triad	2 Sept. 1972	94	13.6	35.6

microsatellites and will not be considered. RTGs for U.S. space applications use plutonium-238 in either metallic or ceramic form, surrounded by thermocouples and a radiator that convert heat from nuclear decay into electricity with an efficiency of approximately 7%.[6] Plutonium-238 has a half-life of 87.7 years and a power density of 0.56 W/g; it produces primarily alpha particles, thus making it attractive for long-term space missions.[7] Figure 23.3 shows power and mass for RTGs flown by the United States. These generators have flown in Earth orbit, landed on the moon and Mars, orbited Jupiter and Saturn, or are on their way to the outer solar system. The dashed line in Fig. 23.3 is a simple power fit to the data:

$$P = 0.93M_g^{1.4} \tag{23.1}$$

where P is electrical output power in watts and M_g is generator mass in kilograms. Based on this power scaling, a 20 kg RTG has a higher specific power than a 20 kg, 100 W_e/kg solar array at a heliocentric distance in excess of 5.7 AU, slightly beyond Jupiter's orbit. A 20 kg RTG suitable for a microsatellite would produce 62 W of electric power, a 2 kg RTG suitable for a nanosatellite would produce 2.5 W, and a 0.2 kg RTG for a picosatellite would produce 0.1 W. These power sources can be made small enough to power an attosatellite, or even part

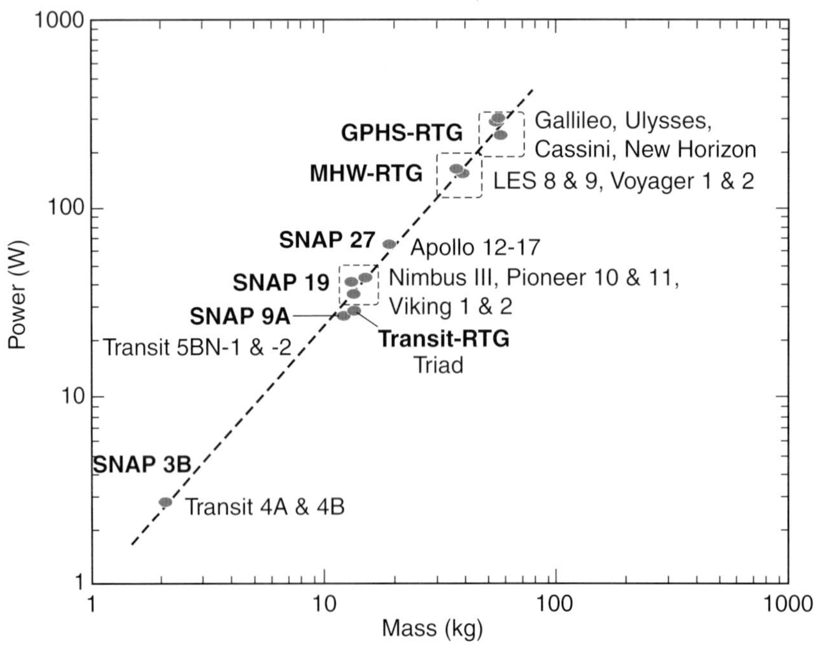

Fig. 23.3. Power and mass for U.S. RTGs flown in space.

of an attosatellite. The Medtronic Implantable Demand Isotopic Pulse Generator, manufactured as a power source for heart pacemakers starting in 1968, used plutonium-238, had a mass of 30 g, and had an initial power output of 250 μW.[8] The power output is more than an order of magnitude lower than that predicted by Eq. (23.1), but this source was developed for a very low-power application with no need to maximize power-to-mass ratio. Nuclear power sources are no longer used in Earth orbit (or in pacemakers) because of safety, environmental, and political concerns, but they are ideal for interplanetary missions beyond Jupiter where solar arrays become too feeble, or for missions with long periods of "night."

A third power option for small satellites is the use of primary batteries. In this case, energy storage is proportional to battery mass, and mission lifetimes tend to be short. Commercial lithium thionyl chloride "C" batteries have energy densities of approximately 260 $W_e$hr/kg at room temperature and 0.3 A discharge rate (4 Ahr capacity at 3.4 V with a mass of 51 g).[9] Available energy density in these batteries is a function of discharge rate; at 1 A, the energy density drops to 190 $W_e$hr/kg, while at a 10 mA discharge rate, it increases to 410 $W_e$hr/kg. Figure 23.4 shows battery lifetime as a function of spacecraft mass, assuming 25% of the spacecraft mass is battery mass, using an average energy density of 260 $W_e$hr/kg. Atto- and femtosatellites will have to operate at average power levels of a few hundred milliwatts or less to provide lifetimes in excess of a few days. Primary battery power becomes useful for multiday lifetimes at approximately 1 W_e power

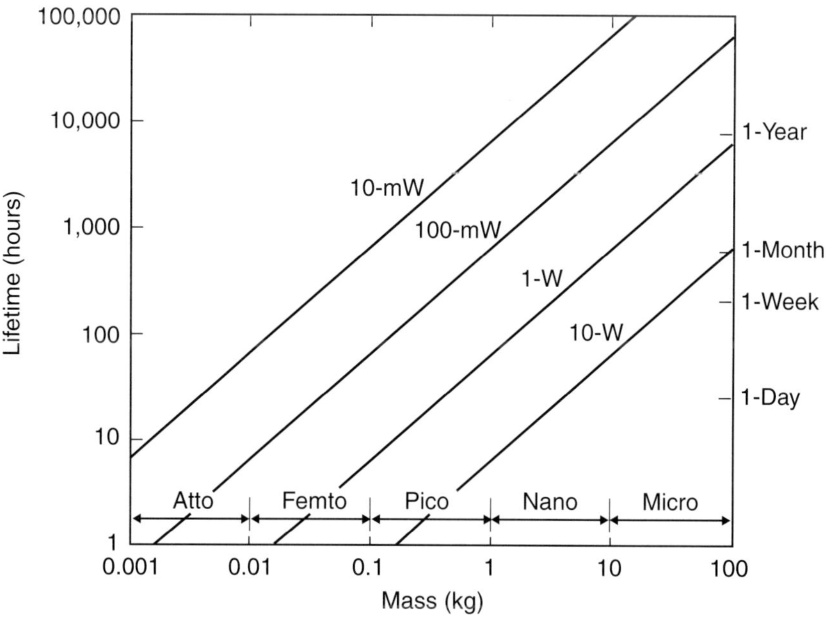

Fig. 23.4. Battery lifetime as a function of spacecraft mass at different average power levels for primary batteries with a battery mass fraction of 25% and an energy density of 260 W_e hr/kg.

levels starting with picosatellites. At a 1 W average power level, a 1 kg picosatellite can operate for about 65 hours, a 10 kg nanosatellite for almost 1 month, and a 100 kg microsatellite for almost 9 months. The MEPSI missions (see chapter 20) were powered solely by primary batteries, with approximately one-week lifetimes at average power levels of 0.5 W_e.

23.2.2 Attitude Control

Angular accelerations for a rigid body such as a spacecraft are proportional to torque T and inversely proportional to moment of inertia. Moment of inertia scales as mass M times the square of the radius of gyration R_g. For a fixed spacecraft geometry and density, these moments scale as l^5, since the R_g is proportional to the size of the vehicle l. The basic result is that small satellites are much easier to rotate than large satellites.

Magnetic attitude control using open magnetic coils can provide rapid orientation changes for LEO nanosatellites and smaller spacecraft. The torque T produced by a circular coil of radius R_c with axial unit vector $\hat{e}_c$ is given by

$$T = \pi N I R_c^2 \hat{e}_c \times B \qquad (23.2)$$

where N is the number of turns in the coil, I is the current in the coil, and B is the local magnetic field vector. Equation (23.2) can be recast in terms of coil mass M_c and power P_c by incorporating coil material resistivity K_c, coil material density ρ_c, and coil radius R_c:

$$T = \frac{1}{2}(M_c / \rho_c)^{1/2}(P_c / K_c)^{1/2} R_c \hat{e}_c \times B \qquad (23.3)$$

If coil mass is set at a particular fraction of the spacecraft mass M (e.g., 10%), and coil power P_c is set at a particular fraction of spacecraft orbit average power P (e.g., 10%), and coil radius R_c is set at a particular fraction of spacecraft side length l (e.g., 90%), magnetic torque T becomes proportional to $l^{3.5}$. Figure 23.5 shows the angular acceleration of a cubic spacecraft with body-mounted cells on all sides as a function of side length l using a copper wire coil for several values of solar collection efficiency η_{sc}. Coil power was set at 10% of orbit average power, coil mass was set 10% of spacecraft mass, coil diameter was set at 90% of spacecraft side length l, and the local magnetic field was set at 0.3 G. Note that the angular acceleration scales as $l^{-1.5}$, torque scales as $l^{3.5}$, while moment of inertia scales as l^5. Small spacecraft have higher angular acceleration rates than larger spacecraft. The right vertical axis indicates the time required to change attitude by 30 deg. Under these coil conditions, CubeSats can perform a 30 deg slew maneuver in about 30 sec, while microsatellites require minutes. At LEO speeds of about 7.6 km/sec, a minute lost to repointing of the spacecraft represents about 460 km of lost ground track. Femtosatellites will require only about 15 sec to perform this attitude-changing maneuver, while attosatellites will require less than 10 sec.

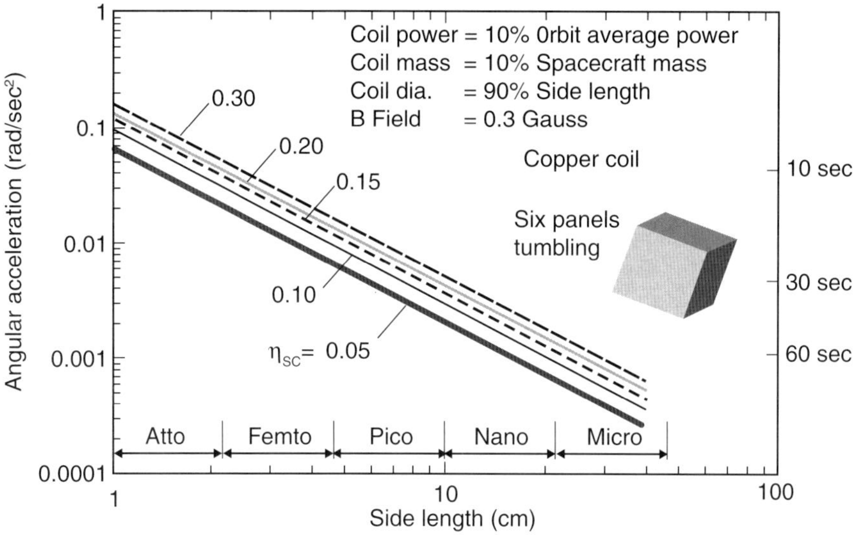

Fig. 23.5. Maximum angular acceleration for a cubical spacecraft as a function of side length. Assumes a copper wound coil with power of 10% of the levels shown in Fig. 23.1.

Microthrusters are another option for attitude control on small satellites. Torque produced by thrusters scales as $l_t * F_t$, where l_t is the distance between the thruster and the spacecraft center of mass, and F_t is the force produced by the thruster. Since l_t is proportional to the size of the vehicle l, and spacecraft moments of inertia scale as l^5, angular accelerations scale as l^{-4} for constant thrust level. Propellant flow rate *dm/dt* through a thruster is given by

$$dm / dt = F_t / I_{sp} \tag{23.4}$$

where I_{sp}, specific impulse, is determined by the type of thruster and its operating conditions. Adjusting the thrust level to scale as l^3, for constant I_{sp}, will provide a mass flowrate proportional to spacecraft mass; this provides a constant maximum thrusting time for a fixed initial propellant mass fraction (initial propellant mass divided by initial spacecraft mass). With this thrust level scaling, angular acceleration then scales as l^{-1}. Therefore, propulsive attitude control requires less propellant, in terms of propellant mass fraction per maneuver, as spacecraft get smaller.

23.2.3 Image Resolution

Image resolution in terms of minimum resolvable angle α under perfect conditions is a function of optic diameter D and wavelength λ:

$$\sin\alpha = 1.22\lambda / D \tag{23.5}$$

For most visible (0.4–0.7 μm wavelengths) and infrared (0.7–30 μm wavelengths) optical systems, where wavelengths are measured in microns and apertures are measured in centimeters, $\lambda/D << 1$, sin α is small, and α is directly proportional to λ/D. Figure 23.6 shows this diffraction-limited ground resolution in meters as a function of wavelength for 1, 2, 5, 10, 20, and 50 cm diameter primary optics at a range of 500 km. For visible wavelengths, ground resolutions less than 21 m are theoretically possible using a 2 cm diameter optic, 4.3 m is possible using a 10 cm diameter optic, 2.1 m is possible using a 20 cm diameter optic, and 1 m could be possible using a 43 cm diameter optic. Figure 23.6 indicates that at low (300–500 km) altitudes, microsatellites with approximately 50 cm dimensions could support ground resolution down to 1 m, nanosatellites with approximately 20 cm dimensions could support ground resolution to 2.5 m, and picosatellites with approximately 8 cm dimensions could support ground resolution to 5 m.

In practice, spacecraft orbital motion, detector sensitivity (which determines minimum exposure time), focal length limitations, detector pixel size, and atmospheric aberration will degrade ground resolution from the theoretical values given in Fig. 23.6. Table 23.4 gives aperture diameter, panchromatic ground resolution, image system mass, spacecraft mass, and orbit altitude for three small Earth imaging spacecraft: TopSat, OrbView-3, and RazakSat. The panchromatic ground resolution includes both visible and near-infrared radiation spanning from 0.4 μm to at least 0.9 μm in wavelength. TopSat could provide 1.9 m resolution at

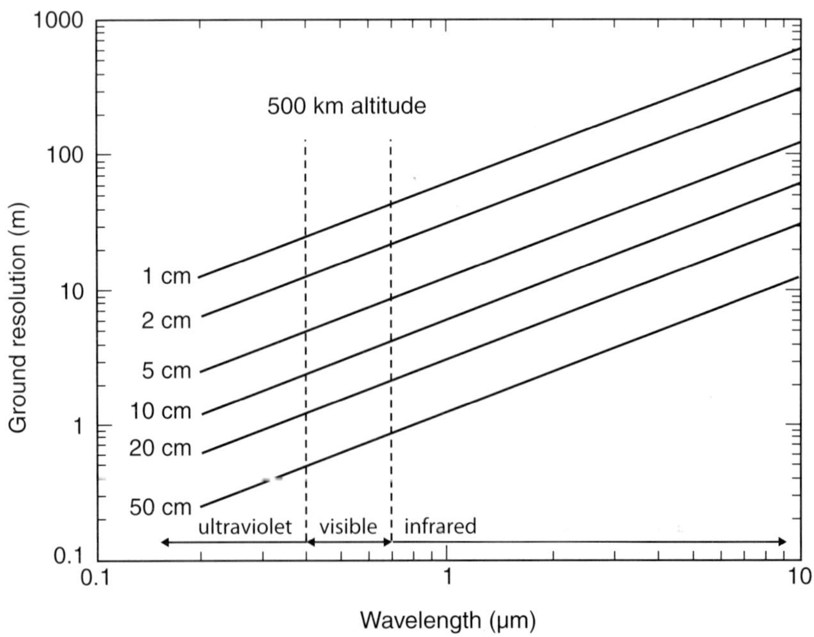

Fig. 23.6. Diffraction-limited ground resolution as a function of wavelength for 1, 2, 5, 10, 20 and 50 cm diameter primary optics at a range of 500 km.

Table 23.4. Representative Small Satellite Earth Imagers[10–14]

Satellite	Aperture Diameter (cm)	Ground Resolution (m)	Image System Mass (kg)	Spacecraft Mass (kg)	Orbit Altitude (km)
TopSat	20	2.5	36	120	600
RazakSat	30	2.5	50	180	685
OrbView-3	45	1	66	304	470

450 km altitude, demonstrating that a 2 m ground resolution microsatellite should be possible in the near term. Creating a 1 m resolution microsatellite still requires a 3× reduction in mass from the OrbView-3 spacecraft; this may be possible in 5–10 years using advanced materials, improved optical designs, and improved electronics. Similar time scales should apply to approaching theoretical resolution limits for nanosatellites (2.5 m) and picosatellites (5 m).

Focal length is another issue with small satellites, because the required length typically exceeds spacecraft dimensions by a factor of 2 to 4. Focusing a 5 m diameter spot at a range of 500 km onto a 5 μm diameter detector element, for example, requires a focal length of 0.5 m. Folded optics such as Cassegrain telescopes are typically used to enable physical lengths of 30–50% of the focal length. A relatively new development is the annular folded optic, which can provide a telescope thickness of one-eighth the focal length; a 35 mm focal length lens with a 5 mm physical thickness has been demonstrated.[15] This type of optic is ideal for payload-limited volumes within small satellites.

23.2.4 Communications

The wavelengths used for radio frequency (RF) satellite communications range from a few centimeters to a few meters. The range determines the minimum size of antennas and how much gain is possible for a given area and/or length. Antenna gains are typically given as decibels above isotropic, or dBi. Figure 23.7 shows the length of a 3 dBi quarter-wave whip, the diameter of a 6.5 dBi patch antenna, the length of a 12 dBi helical antenna, and the diameter of a 20 dBi parabolic dish antenna as a function of frequency. The patch calculations assumed a vacuum dielectric with 1 mm spacing between the ground plane and active element; this constant spacing accounts for the slight nonlinear behavior of this curve at the higher frequencies. Figure 23.7 indicates that microsatellites can support moderate- to high-gain antennas (gain above 6.5 dBi) for frequencies above 400 MHz. However, nanosatellites, picosatellites, femtosatellites, and attosatellites will require frequencies above 850 MHz, 1.8 GHz, 4 GHz, and 8 GHz, respectively.

Figure 23.7 indicates what antenna gain is possible for a given frequency and scale length. Transmit antenna gain times RF power yields a key communications link parameter: the effective isotropic radiated power (EIRP). Usable bandwidths, and hence data rates, for line-of-sight links are proportional to EIRP and the gain

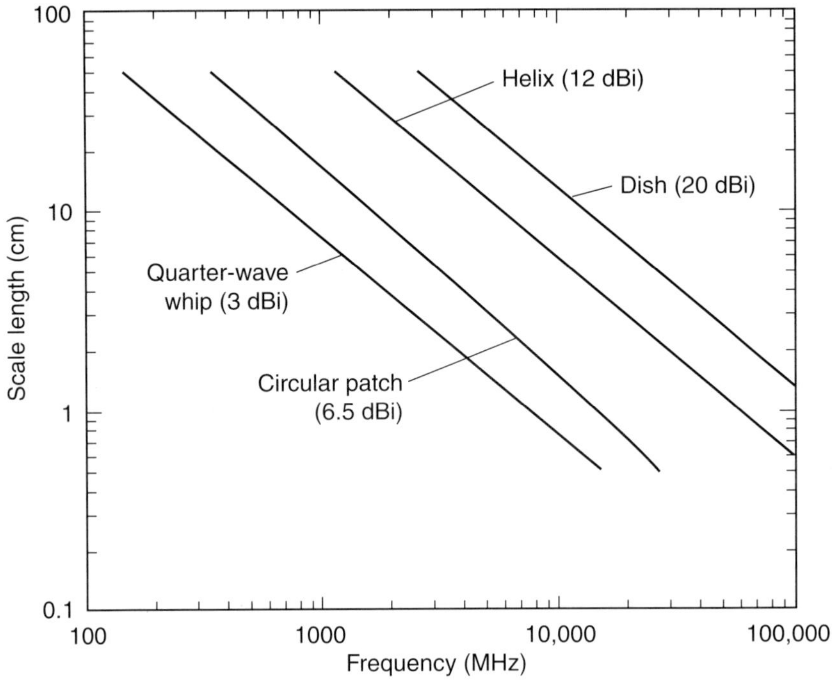

Fig. 23.7. Scale lengths for monopoles and circular patch antennas as a function of resonant frequency. The patch antenna uses a 1 mm thick vacuum insulator.

of the receiving antenna, and inversely proportional to the square of the distance between the transmitter and receiver. Since ground stations can radiate kilowatts of power using large antennas, while satellites have limited power and antenna sizes, the downlink is usually the limiting factor in satellite communications. Figure 23.8 shows calculated data rates in kbps for 2 GHz transmissions from an orbiting satellite as a function of range and satellite EIRP in watts. The calculations assume typical conditions and a modest ground station: a 2 m diameter receiving antenna on the ground, a 440 K receiver system temperature, a 50 K sky temperature, a 12 dB signal-to-noise ratio, and binary phase shift keying (BPSK) modulation. Figure 23.8 is accurate to first order for frequencies ranging from 1 to 15 GHz as long as the receive antenna diameter is fixed.

Figure 23.8 shows that data rates for small satellites in LEO can be surprisingly high. Table 23.5 gives available data rates in Mbps for 2 and 4 m diameter ground station antennas using a 10 dBi transmit antenna on the spacecraft (20 cm diameter at 2 GHz, 4 cm diameter at 10 GHz); 2500 km range, body-mounted solar cells; and a spacecraft RF output power level 25% of the orbit average power from the first column in Table 23.1 (i.e., 50% of orbit average power dedicated to transmitting data times 50% efficiency for the RF power amplifier). Microsatellites can accommodate data rates from 2.8 to 100 Mbps, nanosatellites from

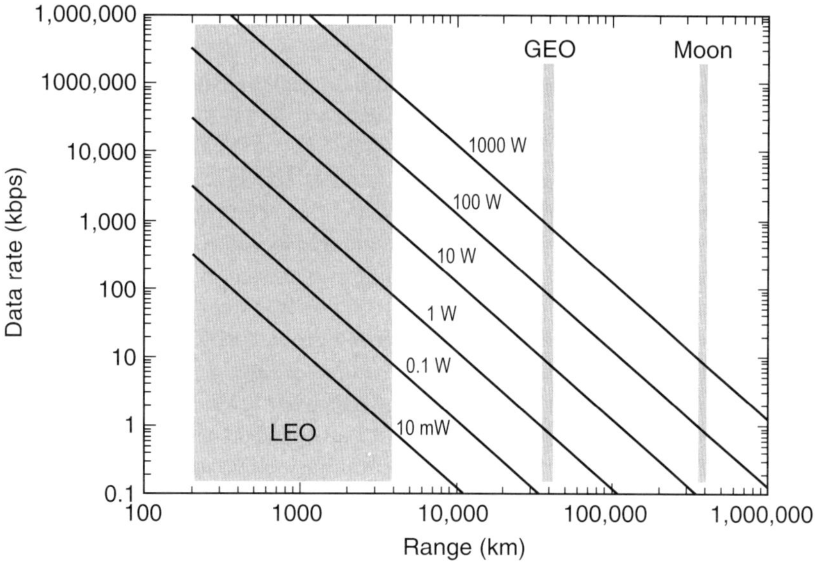

Fig. 23.8. Data rate as a function of range for different spacecraft isotropic radiated powers. Frequency is 2 GHz.

0.6 to 20 Mbps, picosatellites from 0.1 to 4 Mbps, femtosatellites from 0.03 to 1 Mbps, and attosatellites below 0.2 Mbps. The surprising result is that picosatellites can support 1 Mbps data rates and that femtosatellites can provide data rates on the order of 100 kbps to a modest ground station and rates up to 1 Mbps to a 4 m diameter ground station. Microsatellites through picosatellites can accommodate the required antenna diameter at 2 GHz, while femtosatellites and attosatellites will have to operate between 5 and 10 GHz.

Note that the previous calculations used a 12 dB signal-to-noise ratio. By using additional onboard computing capability, forward-error-correction codes

Table 23.5. Data Rates at 2500 km Range for Small Satellites in LEO Using a 10 dBi Transmit Antenna with RF Output Power at 25% of Spacecraft Orbit Average Power

Satellite Class	RF Output Power (W)	Data Rate for 2 m Diameter Receiver (Mbps)	Data Rate for 4 m Diameter Receiver (Mbps)
Microsatellite	1.4–12	2.8–24	11–97
Nanosatellite	0.30–2.5	0.61–5.1	2.4–20
Picosatellite	0.065–0.53	0.13–1.1	0.53–4.3
Femtosatellite	0.015–0.12	0.031–0.24	0.12–0.97
Attosatellite	<0.025	<0.051	<0.20

can be implemented in the data stream. Forward error correction adds additional bits to a packet of data that enable detection of errors in the data packet after it is received and reconstruction of the original data packet if the number of errors is low. These codes significantly lower the required signal-to-noise ratios at the receiver, thus further increasing effective data rates by up to a factor of 10.[16]

While link budgets are good for LEO satellites, the spacecraft are visible to ground stations for only 10–20 min per pass, and the passes may occur only four to six times per day for ground stations at low to mid latitudes. GEO satellites, on the other hand, are constantly in view of a single ground station. Figure 23.8 shows that at GEO, an EIRP of 100 W provides an 80 kbps data rate to the 2 m diameter ground station. At 2 GHz, 100 W EIRP can be provided by a 30 cm diameter transmit antenna with 5 W of RF power. A microsatellite or large nanosatellite can support this size antenna and provide this RF power level. With forward error correction, the effective data rate increases to at least 500 kbps; a 1-megapixel full-color image can be sent every 5 sec, assuming 10× image compression. A large picosatellite can support a 20 cm diameter antenna and 1 W of RF power; this would provide about 50 kbps with forward error correction. Images of 1 megapixel could be transmitted at a rate of one per minute to a modest ground station.

Femtosatellites can also provide low, but potentially useful data rates from GEO. The main problem is that an approximate 10 cm diameter antenna, suitable for a large femtosatellite, provides less than 4 dBi at 2 GHz. Increasing the frequency to 10 GHz boosts the gain to almost 18 dBi, thus providing a 6 kbps data rate with an RF output power of 0.2 W. This data rate can be useful for transmitting space science measurements or for sending a megapixel image once every 7 min. For a GEO attosatellite, though technically challenging, the maximum data rate drops to approximately 300 bps.

The next step outward is cislunar space. The moon is almost 10 times further away than GEO, so data rates from lunar orbit or the lunar surface will be about 100 times lower. Lunar data rates can equal GEO data rates if the ground station antenna for lunar reception is 10 times larger than the antenna for GEO reception. An additional complication is that the moon generates more RF noise than deep space at 1–15 GHz frequencies, so slightly larger antenna diameters are required. A 24 m diameter ground station antenna can provide data rates ranging from 500 kbps for microspacecraft through 300 bps for attospacecraft at lunar range.

Moving to interplanetary space can increase communications range by up five orders of magnitude compared with GEO satellites. Larger ground-station antennas, exceptionally quiet receivers in those ground stations, higher transmit antenna gain on spacecraft, and aggressive forward error correction codes are required to enable acceptable interplanetary communications. Figure 23.9 shows calculated data rates as a function of range for 1 W of RF power into a parabolic transmit antenna with a diameter of 0.1, 0.2, 0.5, 1, and 2 m. Distance to the planets and Pluto from Earth are shown. Data rate calculations were performed for NASA's Deep Space Network using the 70 m diameter ground-station antennas at

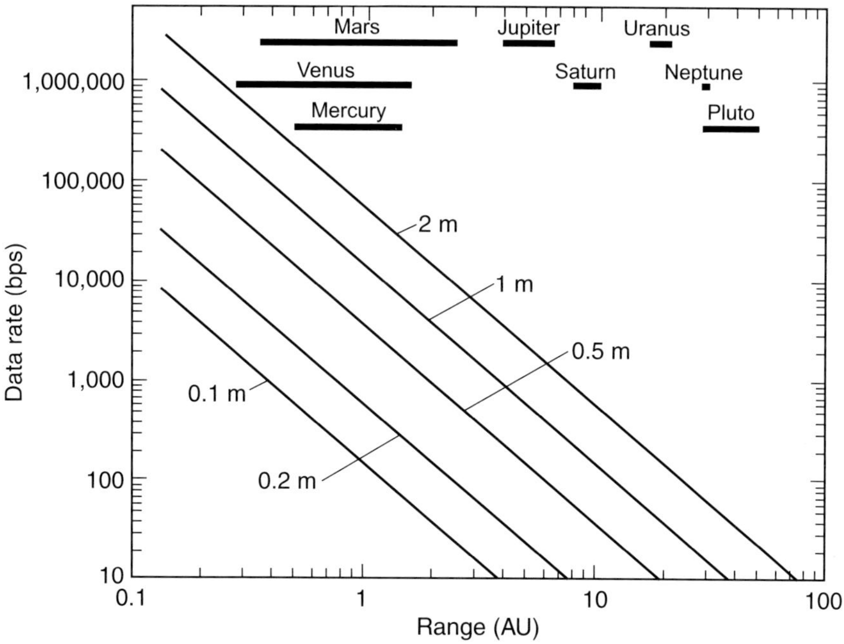

Fig. 23.9. Data rate as a function of range and spacecraft transmit antenna diameter for 1 W of RF power, using a 70 m diameter ground station receive antenna at 8.4 GHz.

X-band frequencies (~8.4 GHz). Higher receive bandwidth (about 4 times greater) is available using the Ka-band downlink (32 GHz), while a lower receive bandwidth (about 4 times lower) would result from using the more numerous 34 m diameter dishes at X band in the Deep Space Network.

Table 23.6 lists general transmit power, antenna size, and spacecraft class recommendations for different mission ranges with data rates of at least 1 kbps. Solar-powered nanospacecraft are suitable for near-Earth exploration, solar

Table 23.6. General Recommendations for Transmitter RF Power, Transmit Antenna Diameter, and Smallest Spacecraft Class for Transmitting to Earth at Data Rates of at Least 1 kbps

Mission Location	Output Power (W)	Antenna Diameter (m)	Spacecraft Class
Near Earth (within 1 AU)	1	0.3	Nanospacecraft
Mercury through Mars	3	0.5	Microspacecraft
Mars through Jupiter	4	0.7	Microspacecraft
Jupiter through Saturn	6	1.2	Microspacecraft
Saturn through Neptune	15	2.2	Minispacecraft
Neptune to 50 AU	20	3	Minispacecraft

microspacecraft are suitable for the inner solar system through Jupiter, solar or nuclear-powered microspacecraft are suitable for Jupiter, and nuclear-powered minispacecraft are suitable for Saturn and beyond. The New Horizons spacecraft that is heading toward Pluto, for example, uses 12 W at 8.4 GHz into a 2.1 m diameter antenna to provide an approximately 600 bps data rate at a distance of 36 AU.[17] The 478 kg spacecraft contains 69 kg of propellant and has an RTG that will produce approximately 200 W during the encounter with Pluto.[18]

23.2.5 Ultrasmall Satellites

Picosatellites have flown in LEO, but active femtosatellites and attosatellites are in the future. Based on calculations from the previous sections, simple attosatellites in LEO could operate at average power levels of 10–50 mW and be able to downlink data at rates of several hundred to a few thousand bps. Explorer I and Vanguard 1 had similar power and communications capabilities. The main difference between a 1950s Explorer I and a 21st century version is the tremendous advancement in miniaturization and microelectronics. Low-power, ultrasmall microprocessors and microcontrollers are now available with processing performance on the order of 1000 MIPS/W (MIPS = million instructions per second). A few milliwatts can provide approximately 3 MIPS. The Microchip PIC10F222, shown in Fig. 23.10, requires less than 350 μW when operating at 4 MHz (1 MIPS) and fits within a 2 × 3 × 0.9 mm volume. A late 1950s-era equivalent computer using hundreds of thousands of "new" transistors would not have fit within the Explorer-1 volume. The remaining challenge for ultrasmall satellite designers is to develop centimeter-scale payloads, attitude determination, and attitude control systems with mass of 1–10 g.

Attitude determination and control will enable a wide variety of mission applications for small satellites, especially pico- through attosatellites. A number of sun sensors, Earth horizon sensors, and star cameras are commercially available

Fig. 23.10. The Microchip PIC10F222 microcontroller on a U.S. dime.

Fig. 23.11. Sun and Earth sensor board designed by The Aerospace Corporation for use on a 10 cm CubeSat.

for microsatellites and nanosatellites, but sensors for picosatellites and smaller spacecraft must be custom built. Figure 23.11 shows an example of a two-axis sun sensor, based on a position-sensitive detector, and a single thermopile detector operating as an Earth sensor on a single 2.5 × 3.3 cm circuit board.[19] This board, designed, fabricated, and tested at The Aerospace Corporation, is currently flying on the PicoSatellite Solar Cell (PSSC) Testbed and is also scheduled to fly in AeroCube-3 (a CubeSat). This board could be further reduced in size to fit on a 2.0 × 2.5 cm board to readily fit in a femtosatellite.

Even smaller sensor systems can be fabricated by moving key circuitry off of circuit boards and into sensor dice. The Aerospace Corporation has used the MOSIS (Metal-Oxide Semiconductor Implementation Service) custom CMOS (complementary metal-oxide semiconductor) prototyping service to fabricate 2.2 mm^2 active pixel detector chips for sun sensors and Earth sensors that could fit on attosatellites and larger spacecraft.[20–22] This approach enables the sensor, signal conditioning electronics, and even serial communications circuits to be fabricated on a single silicon die to get the smallest possible sensor. The sun sensors used custom-designed p-n junctions as photodetectors, and the Earth sensors used series-connected polysilicon/metal junctions as thermocouple temperature sensors. Figure 23.12 shows three versions of the sun sensor chips. The left and middle versions are single-axis detectors designed for use with a horizontal slit. The right version was an initial attempt at a two-dimensional array using 100 detectors. These chips are 2.2 mm^2 and 0.5 mm thick; they have a mass of only 6 mg. The required optics and packaging bring the volume up to about 0.1 cm^3 with a mass of approximately 0.3 g.

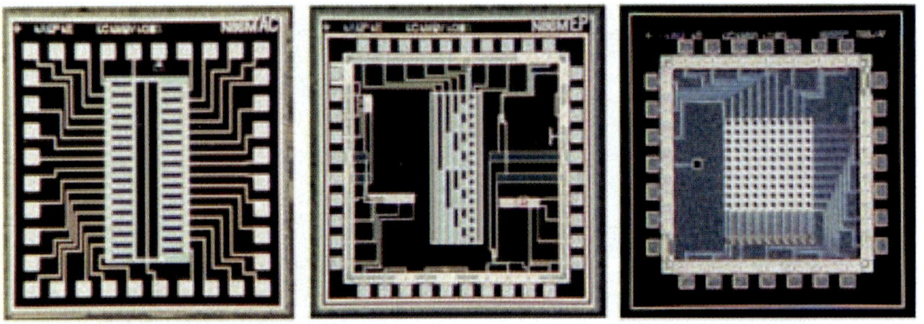

Fig. 23.12. Three CMOS chip-scale sun sensors, each 2.2 mm², designed at The Aerospace Corporation and manufactured using the MOSIS.

Current-generation miniature star cameras can potentially fit in a CubeSat, but they would require a significant fraction of the internal volume. Smaller star cameras are needed to enable high-accuracy (tens of arcseconds) pointing of picosatellites and smaller spacecraft. Initial tests and calculations at Aerospace indicate that a star camera with approximately 2.5 cm diameter optic should be able to see stars down to 4th magnitude in a 0.1 sec exposure using commercial CMOS image sensors.[19] This should easily fit on a picosatellite and may fit on a femtosatellite. A star camera for attosatellites is currently not feasible.

23.3 Orbital Mechanics of Multisatellite Systems

Small satellites enable the cost-effective use of tens, hundreds, and potentially thousands of individual spacecraft to perform a mission. The ORBCOMM constellation (see chapter 14) is one example of a communications system that was enabled by the development of relatively inexpensive microsatellites. Another example mentioned in chapter 2 is the ejection of Romb spacecraft from larger host vehicles to periodically calibrate ground-based radar systems. Multiple spacecraft can be used in constellations or local clusters, or repeatedly dispensed as short-term disposable spacecraft. Orbital mechanics plays a key role in defining what's possible.

23.3.1 Constellations

John Walker in 1984 developed the Walker Delta Pattern, a notation for labeling a class of circular orbit patterns that has since become a standard.[23] The Walker Delta Pattern notation is $i: T/P/F$ where i is the inclination in degrees, T is the total number of satellites, and P is the number of evenly spaced planes. Each plane is separated from the next one by $360/P$ deg, the change in right ascension of the ascending node (RAAN). Note that the ratio T/P is equal to the number of satellites per plane. The Walker number F, an integer between 0 and $P-1$, determines the phase relationship between adjacent planes; the change in true anomaly for equivalent satellites in neighboring planes is equal to $F*360/T$ deg.

Figure 23.13 shows an example 90 deg: 12/3/0 Walker constellation for a 1000 km altitude.[24] The three orbit planes and visible satellites within each orbit plane are shown. Note that there are four satellites per plane, spaced 90 deg apart, and that the true anomalies of each satellite are shifted by 0 deg when moving from one plane to an adjacent plane; the closest satellites in each plane are equidistant to the orbit intersection above the North Pole.

Walker Delta Pattern constellations are typically optimized to provide maximum continuous Earth coverage. At any given time, individual spacecraft are well distributed with respect to longitude and latitude within the positive and negative inclination limits—for example, within ±60 deg latitude.

When just one or two spacecraft are used per plane, Walker constellations can produce simple satellite topologies that appear to defy orbital mechanics. Figure 23.14a shows a time sequence of a 60 deg: 50/50/0 Walker constellation that produces a "bouncing halo" effect. The frames are 15 min apart. All spacecraft have identical true anomalies that change with time, resulting in a horizontal ring that oscillates in the north-south direction. A complete north-south oscillation takes one orbit period; 105.1 min at 1000 km altitude. Since each spacecraft must follow a Keplerian orbit, the ring rotates once about the north-south axis per bounce.

Figure 23.14b shows a 60 deg: 50/50/49 constellation that produces a figure-eight global structure that revolves around Earth once every 105 min; Fig. 23.14c shows a 60 deg: 50/50/48 constellation that produces a "baseball seam" structure that revolves around Earth once every 210 min. All three constellations, and many more with different *F* numbers, are closely related; corresponding spacecraft in each constellation have the same orbit inclination, altitude, and RAAN. Changing from one constellation to another requires rephasing individual

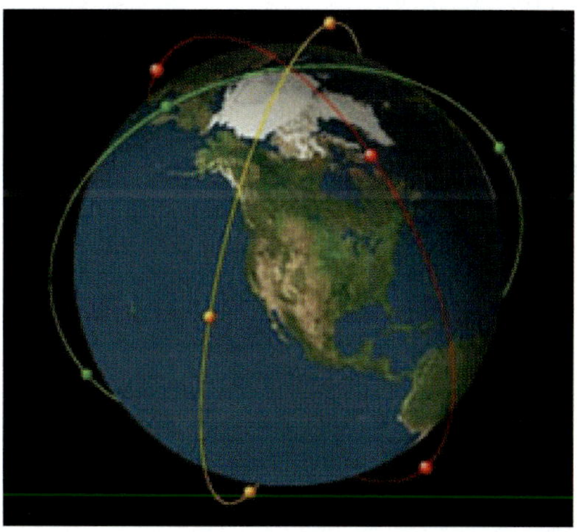

Fig. 23.13. An example 12/3/0 Walker constellation with an inclination of 90 deg and an altitude of 1000 km.

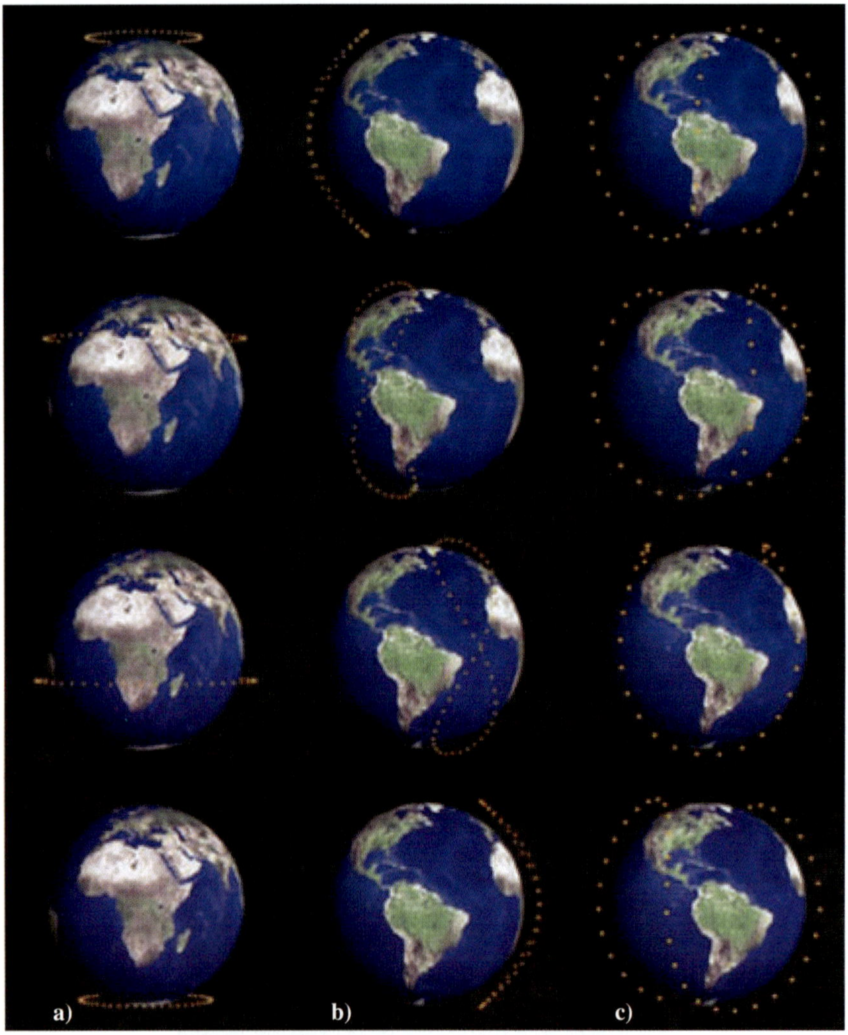

Fig. 23.14. Time sequence images of a 60 deg: 50/50/0 (a); a 60 deg: 50/50/49 (b); and a 60 deg: 50/50/48 (c) Walker constellation at 1000 km altitude.

satellites in their orbits. Rephasing over a two-week period using electric thrusters has modest propulsion requirements of about 10 m/sec.[25] Tens or hundreds of "morphings" between constellation configurations should be possible.

Figure 23.14 shows three examples of planetary-scale structures that can be created through orbital mechanics. The bouncing halo and baseball structures provide a simple ring networking architecture for linking individual spacecraft. This enables straightforward data forwarding between spacecraft for downloading to a single ground station each time the ring crosses the appropriate latitude.

23.3.2 Co-Orbital Motion

A number of small satellite applications involve maintaining relatively close contact with other spacecraft. Relative motion in the reference frame that follows and rotates with the target spacecraft as it orbits Earth can be calculated using the linearized equations of relative orbital motion for circular orbits with no external accelerations:[26]

$$s = s_0 + 2\frac{\dot{r}_0}{\omega}(1 - \cos\omega t) + \left(4\frac{\dot{s}_0}{\omega} - 6r_0\right)\sin\omega t + \left(6\omega r_0 - 3\dot{s}_0\right)t \tag{23.6}$$

$$4r_0 - 2\frac{s_0}{\omega} + \left(2\frac{s_0}{\omega} - 3r_0\right)\cos\omega t + \frac{r_0}{\omega}\sin\omega t \tag{23.7}$$

$$z = \frac{\dot{z}_0}{\omega}\sin\omega t + z_0\cos\omega t \tag{23.8}$$

$$\dot{s} = 2\dot{r}_0\sin\omega t + \left(4\dot{s}_0 - 6\omega r_0\right)\cos\omega t + 6\omega r_0 - 3\dot{s}_0 \tag{23.9}$$

$$\dot{r} = \left(3\omega r_0 - 2\dot{s}_0\right)\sin\omega t + \dot{r}_0\cos\omega t \tag{23.10}$$

and

$$\dot{z} = \dot{z}_0\cos\omega t - z_0\omega\sin\omega t \tag{23.11}$$

where s is the in-track (flight) direction, r is the radial or outward direction, z is the orbit-normal direction, t is time, ω is orbit period, and the zero subscript denotes a starting condition. These equations apply to relative orbital motion for a rotating reference frame in a circular orbit with period ω. They can be used for following the path of a screwdriver or other object lost by an astronaut (as seen by an astronaut), for proximity operations near another spacecraft, and for docking. These are simplified equations that treat Earth as a perfect point source of mass and are valid for a few orbital periods of motion. Although the equations look complicated, they are just algebraic functions of time given an initial position and velocity vector.

The simplest local cluster can be formed by placing spacecraft at several different in-track locations (different s_0) in a single orbit. Since all spacecraft are essentially in the same orbit, this type of formation is very stable, and the local cluster size can range from meters to more than 10,000 km. These string-of-pearl clusters allow multiple spacecraft to provide nearly simultaneous measurements of a particular location. NASA's "A-Train" has the Aura, Parasol, Calipso, CloudSat, and Aqua spacecraft flying in this formation at 705 km altitude with separation between the first and last satellite of 15 min (6760 km).[27] CloudSat and Calipso are near the middle and fly only 12.5 ± 2.5 sec (94 km) apart. This approach creates an upgradable and highly powerful one-dimensional distributed satellite for Earth observation; the Glory spacecraft will be added in 2009. NASA's ST-5 mission used three microsatellites with intersatellite distances ranging from 5000 to 50 km.[28] The primary payload was a three-axis magnetometer, and the use of three spatially separated spacecraft enabled uncoupling of temporal and spatial structures in magnetic field measurements.

Cooperative spacecraft can use the string-of-pearls formation to probe the ionosphere and atmosphere using radio occultation as shown in Fig. 23.15. When a radio wave traverses the ionosphere, a phase shift is generated by the presence of free electrons in the plasma. The total phase shift can be related to the line-integrated electron density along the transmission path.[29] When a radio wave traverses the atmosphere, significantly larger phase shifts and propagation delays result, which can be used to determine atmospheric density, pressure, and temperature.[30] This technique is normally used with GPS receivers on satellites as accurate phase and delay monitors for signals that skim the limb of Earth. Use of two different GPS frequencies (1.575 and 1.228 GHz), which will have slightly different phase shifts, greatly improves the accuracy of line-integrated measurements. Up to 500 atmospheric profiles can be generated per day using a single GPS receiver in orbit.

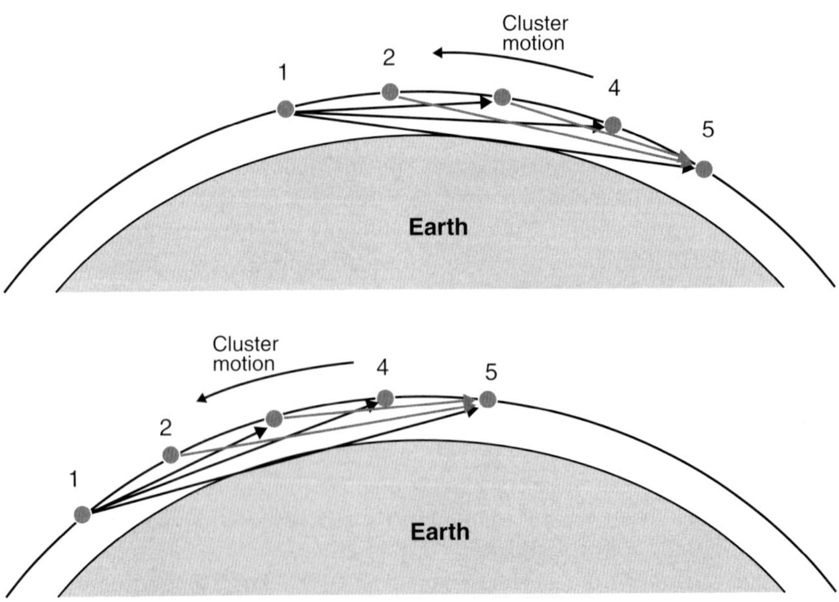

Fig. 23.15. Schematic of a string-of-pearls local cluster in the in-track and radial motion plane. Cooperative spacecraft using multiple transmitters and receivers can measure propagation path phase differences and attenuation to determine spatially resolved atmospheric and ionospheric properties using tomographic reconstruction.

With the local cluster shown in Fig. 23.15, measurements can be taken continuously. In this case, satellites 1 and 2 are transmitters, while satellites 3, 4, and 5 are receivers. Multiple measurements taken between different pairs of spacecraft and at different times as the local cluster moves along its orbit enable tomographic reconstruction of spatially resolved atmospheric and ionospheric properties. Similar techniques using absorption measurements between different spacecraft at optical through radio frequencies can be used to provide spatially resolved maps of specific molecular species such as water vapor, carbon dioxide, sulfur dioxide, and methane.

Two-dimensional clusters can be generated by adding orbit-normal motion to a string-of-pearls cluster. Equation (23.8) shows that in a co-orbital reference frame, a displacement in the orbit normal direction will result in a harmonic oscillation about the orbit plane. Equations (23.6) and (23.7) show that this orbit-normal motion is decoupled from in-track and radial motions. Figure 23.16 shows the time evolution of two different two-dimensional local clusters formed in the in-track and orbit-normal plane at an altitude of 600 km. The orbit period is 96.69 min (5801 sec). The left sequence shows the evolution of a cluster with 15 satellites, where the satellites are started at fixed positions with no orbit-normal velocities; the right sequence shows a random distribution with initial velocities chosen to maintain a somewhat filled rectangular area. Rigid geometries cannot be maintained, but a specific geometry can be created twice per orbit (e.g., the 0 sec and 2400 sec frames in Fig. 23.16). Orbiting billboards composed of tens to hundreds of highly reflecting spacecraft spelling out a name or forming a logo while passing over certain parts of the Earth are possible.

Addition of radial motion to orbit-normal motion enables the creation of essentially rigid geometries under special conditions.[31,32] At any position on the in-track vector, two specific radial velocities (one positive, one negative) exist that generate a closed co-orbit centered about the origin. At 700 km, for example, this velocity is ±0.0263 m/sec at $s = \pm 50$ m. Proximity operations within a few hundred meters of another object optimally require velocity increments less than a meter per second per maneuver; limited operations can be readily accommodated using cold-gas thrusters.

Figure 23.17 shows three co-orbital paths starting at $s_0 = 50$ m with initial radial velocity of 0.0263 m/sec. The initial orbit with no orbit normal velocity (0 deg path) is elliptical with the maximum in-track range twice as great as the maximum radial range. As the orbit-normal velocity is increased, the plane tilts and the orbit becomes more circular until the tilt equals 60 deg. This relative tilt provides constant range from the origin or host vehicle and is ideal for satellite inspection. Since the co-orbital period is equal to the orbit period and is independent of range for ranges of a few hundred kilometers or less in LEO (range << orbit radius), multiple spacecraft can be placed in this common plane, and the configuration will rotate as if it were a solid body. This is ideal for generating sparse-aperture antennas composed of formation-flying spacecraft.

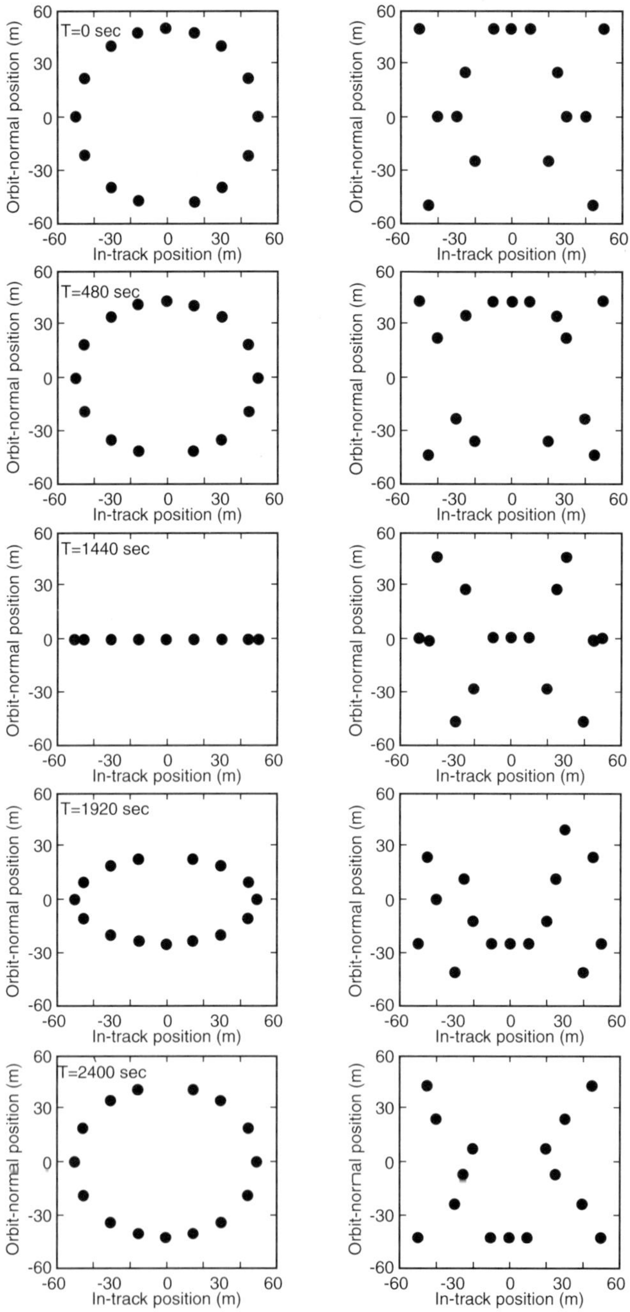

Fig. 23.16. Temporal evolution of two local clusters formed in the in-track/orbit normal plane over almost half an orbit period. In the sequence on the left, the satellites start at fixed positions. The sequence on the right is a random distribution.

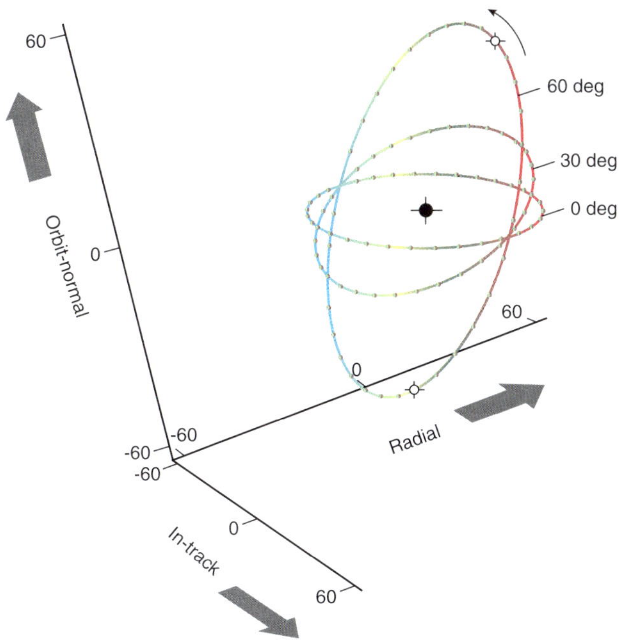

Fig. 23.17. Examples of co-orbital motion that enable spacecraft to "orbit" about another vehicle. The tick marks are 180 sec apart.

Since the origin (or host vehicle) and the spacecraft in tilted co-orbits actually have different orbit eccentricities and inclinations, orbital perturbations such as apsidal rotation and nodal regression will slowly destabilize the local cluster. Perfectly rigid formations are not possible, but nearly rigid formations can be maintained. For spacecraft co-orbiting a host vehicle at 1 km radius and 700 km altitude, the velocity increment requirement is 67 m/sec/year with a standard position deviation of 14 m if the relative position determination error is 10 cm and the relative velocity determination error is 1 mm/sec.[31] With a further reduction in position determination error to 3 cm and velocity error to 0.3 mm/sec, the velocity increment requirements drop to 42 m/sec/year, and the positional standard deviation drops to 7 m. Positional deviations of approximately 10 m within a kilometer-radius cluster would be barely perceptible to the human eye. Efficient formation flying still requires accurate, real-time, relative position determination systems that may not fit on the smallest satellites.

23.4 Spacecraft Stabilization

The spacecraft attitude stabilization method ultimately determines the pointing capabilities and pointing accuracy of payload sensors and antennas and the ability to change orbit. This directly affects mission capabilities. A spacecraft with a pointing accuracy of only 1 deg, for example, can't effectively use a high-resolution imager with a field-of-view less than 1 deg. Table 23.7 lists relevant stabilization methods for small satellites and their limitations. No stabilization was needed on Vanguard, many picosatellites, and a number of 1960s-era nanosatellites because they used near-omidirectional antennas and had no payload pointing requirements. Passive magnetic stabilization uses a permanent magnet that aligns the spacecraft along the local magnetic field vector. Spacecraft with passive magnetic stabilization will point north (or south) near the equator, south (or north) near the North Pole, and south (or north) near the South Pole. A spacecraft in a nearly polar orbit, therefore, rotates twice per orbit period, thus providing a slow rotation for thermal control, averaging of solar-cell power, and nadir pointing near the poles. This technique has been used in quite a few nanosatellites and CubeSats.

Gravity-gradient stabilization requires that spacecraft be composed of two masses physically separated by at least several meters. In LEO, spacecraft masses line up along the nadir-zenith direction to provide nadir pointing within a few degrees. Unfortunately, gravity-gradient stabilization, like passive magnetic stabilization, only works well in LEO because the stabilizing forces are inversely proportional to orbit radius cubed. The Surrey Satellite Technology Ltd. microsatellites successfully used this technique for three decades.

Spin stabilization was used on Explorer-1 and a large number of GEO communications satellites during the 1980s. Spin stabilization generally requires an additional actuator to adjust the spin axis over time. Active magnetic control systems require low Earth orbits, magnetic field sensors, and a sophisticated control scheme to calculate coil currents based on the local magnetic field direction and strength. This is further complicated by time-varying magnetic field strengths

Table 23.7. Basic Stabilization Schemes for Small Spacecraft

Name	Designation	Accuracy	Difficulty	Requirements
No Stabilization	NS	None, random pointing	None	Good thermal design
Passive Magnetic	PM	A few degrees for one axis only	Very easy	Permanent magnet
Gravity Gradient	GG	A few degrees for nadir pointing	Moderate	Deployment mechanism
Spin Stabilization	SS	< 0.1 deg	Easy	Spin-up mechanism
Active Magnetic	AM	< 1.0 deg	Hard	Magnetic and optical sensors
Three-Axis Momentum Wheel	3A	< 0.01 deg	Moderate	Attitude sensors
Active Thruster	AT	< 0.1 deg	Moderate	Attitude sensors

and directions created by spacecraft orbital motion, and the inability to directly rotate about the instantaneous magnetic field line. Fortunately, active magnetic reorientation maneuvers can be accomplished within minutes for picosatellites and small nanosatellites, thus minimizing the local magnetic field variation during the maneuver.

Three-axis stabilization requires at least three momentum wheels or a control moment gyro, but the control scheme is much simpler. The disadvantage is that thrusters or magnetic coils are still required because external torques on spacecraft can eventually cause a momentum wheel to reach its maximum rotational velocity. Finally, active control using thrusters is a viable option for small satellites, especially those that require propulsion systems for orbit adjustment, maneuvering, or transfers. This technique was chosen for the COSA mission (see chapter 18).

23.5 Small Satellite Missions

Missions and potential mission benefits drive the development of all spacecraft technologies. This section looks at different mission applications for small satellites, ranging from single satellite missions, through distributed satellites, to "smart" propellant for larger spacecraft that could enable more cost-effective human exploration of the solar system.

23.5.1 Historical Trends

In the last 50 years, more than 860 microsatellites, 680 nanosatellites, and 38 picosatellites have been launched. The majority of the microsatellites were used for low-data-rate communications—for example, the U.S. Initial Defense Communication Satellites, the Soviet Strela series, and the ORBCOMM system.[33,34] The vast majority of nanosatellites, the Soviet Romb series, were passive radar calibration spheres ejected by larger spacecraft.[35] Active picosatellites emerged in the year 2000, primarily as low-cost experimental test beds. To date, no active femtosatellites or attosatellites have flown.

One of the key factors in the nanosatellite rebirth during the 1990s and the launch of active picosatellites in 2000 was the ongoing development of microelectronics, solar-cell technologies, and rechargeable batteries. Of these, microelectronics had the greatest impact.

23.5.2 Future Trends

In his book, *Advanced Space System Concepts and Technologies:2010–2030+*, Ivan Bekey presents a list of high-leverage principles to pursue in space concepts.[36] The main ones are:

- Replace structures with information.
- Adopt distributed space systems. Use coherent cooperation among many spacecraft to implement coherent sparse apertures.
- Transport energy and information, rather than mass, through space.

- Replace chemical combustion in propulsive devices with electromagnetic and electrostatic forces and plasmas.
- Service, repair, and upgrade large and complex spacecraft.
- Exploit nanotechnology, MEMS, and NEMS (nanoelectromechanical systems).

To this list, one might add:

- Use the GPS or similar constellation to provide timing, position, and velocity information.
- Recycle consumables (such as propellant) whenever possible.
- Design for cost-effective mass-production of common systems or vehicles.

The functional density of spacecraft and their systems will inevitably rise as Moore's law and microfabrication techniques continue their evolution. This will enable small spacecraft to perform many, but not all, missions previously performed by the next-higher class; for example, nanosatellites will perform microsatellite missions, and picosatellites will perform nanosatellite missions.

23.5.3 Near-Term Applications

Table 23.8 lists past, present, and near-term (within 5 years) missions that can be accomplished using individual microsatellites, nanosatellites, or picosatellites. Most of these are based on previously flown missions. Those missions requiring three-axis stabilization will be limited to microsatellites and nanosatellites, while most of the others could be accommodated using picosatellites.

Microsatellites will continue to increase in functionality. One possibility for the near term is an Earth observation microsatellite with a 2.5 m ground resolution. This represents a modest (17% mass reduction) improvement over the TopSat spacecraft (see Table 23.2). In addition, 5 m ground resolution should be possible using a nanosatellite with 10 cm diameter optics, and 10 m ground resolution using a picosatellite with 5 cm optics. One recent concept uses modified catadioptric mirror lenses for 35 mm cameras to provide 7.5 m resolution from an altitude of 540 km in a triple-cube nanosatellite.[54] The missing attitude determination component for serious Earth observation using ultrasmall satellites, a star camera that fits on a picosatellite, could be developed in the near term.

Small satellites are ideally suited for LEO communications constellations. The Strela-1M series, ORBCOMM, AprizeSat, and SaudiComSat constellations (see chapter 2) demonstrated the utility of microsatellite-based store-and-forward messaging. In the near term, we should expect further expansion of the AprizeSat and SaudiComSat constellations to approach more continuous Earth coverage. Within five years, these microsatellites could be redesigned to fall within the nanosatellite mass range.

With nadir pointing, downlink data rates can be 4 to 10 times higher than with omnidirectional transmit antennas, or the same downlink data rate can be used

Table 23.8. Single Satellite Missions

Mission	Stabilization	Orbit	Examples
Radar calibration	NS	LEO	ODERACS[37]
Atmospheric density measurements	NS	LEO	ODERACS[38]
On-orbit radiation testing of components	NS	LEO, GTO, HEO	ERS-27[39]
Space biology experiments	NS	LEO, GTO, HEO	GeneSat[40]
On-orbit radiation monitoring	NS	LEO, GTO, HEO	ERS-27[39]
General relativity experiments	NS, SS	LEO	LAGEOS[41]
Local space plasma measurements	NS, PM, SS, GG	LEO, GTO, HEO	Munin[42]
Low data rate communications relay	NS, PM, SS, GG	LEO	ORBCOMM[43]
Low data rate store and forward communications	PM	LEO	UoSAT-2[44]
Medium data rate communications	GG	LEO	MUBLCOM[45]
Tether experiments	GG	LEO	MAST[46]
Seismic monitoring via VLF wave reception	GG	LEO	QuakeSat[47]
Solar studies	SS, AM, 3A	LEO, GTO, HEO	SOLRAD-3[48]
Solar cell testing	SS, AM, 3A	LEO, GTO, HEO	ERS-27[39]
Ionospheric mapping using GPS attenuation	AM, 3A	LEO	PICOSAT[49]
Medium to high resolution, targeted Earth imaging	AM, 3A	LEO	TopSat[50]
Stellar magnitude monitoring	AM, 3A	LEO, GTO, HEO	MOST[51]
Lunar and asteroid rovers		Interplanetary	MINERVA[52]
Gravitational field mapping	AT	Lunar orbit	RSTAR[53]
Cislunar or interplanetary probes	AT	Interplanetary	

with a simpler low-gain fixed antenna on the ground. In addition, uplinks with approximately 10 kHz bandwidth using 1–10 W at UHF, or VHF frequencies with omnidirectional ground antennas, can be received by the spacecraft. This enables store-and-forward communications from low-power terrestrial transmitters on ocean buoys, migratory animals, unattended battlefield or border monitors, high-altitude balloons, and others. The NCUBE mission, for example, was to monitor ships using the Automatic Identification System mandated by the International Maritime Organization.[55] Position, velocity, heading, and identification packets are periodically transmitted by vessels at 2 or 12.5 W at 9600 bps at 161.975 MHz. A LEO satellite can receive these signals and forward them to a ground station for tracking, thus providing "over-the-horizon" monitoring capability for the ground station. The simple receiver and modem fit on an approximately 7 cm^2 circuit board within the CubeSat. This system was adapted to track reindeer by designing collars for these animals that contained a GPS receiver and data transmitter. More picosatellites with this capability could be flown within the next five years.

Active femtosatellites could also be flown in the near term. The basic technology exists, the average on-orbit power levels are low but acceptable, and communications downlink rates of tens of kbps are possible using modest ground stations. Ten or more femtosatellites can fit within the volume of a CubeSat, thus offering ultralow-cost (~$5,000 each, as opposed to $40,000 for a single CubeSat) access to space for universities, high schools, and individuals. The combination of low cost and easy space access should result in a rapid increase in femtosatellite launches in the future. Currently, picosatellite launch rates are higher than nanosatellite launch rates, and five years from now, femtosatellite launch rates could exceed picosatellite launch rates.

One of the main drivers for cislunar micro-, nano-, or even picospacecraft during the next five years will be the Google Lunar X Prize. The first nongovernment team to land a robot on the moon, have it travel 500 m over the lunar surface, and send back images and data will receive $20 million.[56] This is somewhat challenging from a propulsion standpoint. Table 23.9 gives minimum ΔV requirements for a number of maneuvers within cislunar space. Getting a satellite from a 400 km circular low Earth orbit to a 400 km altitude circular low lunar orbit requires almost the same ΔV as putting a satellite into GEO. An additional 1840 m/sec is required to go from lunar orbit to a soft landing on the lunar surface. The net result is that the cost of landing a payload on the moon is on the order of $100,000/kg. This is a strong incentive to minimize payload mass. Section 23.2.4 showed that a picospacecraft with a 20 cm diameter antenna and 1 W of RF power could transmit a compressed, 1 megapixel image to a 24 m diameter Earth station from the moon in about a minute, so a nano-class lunar lander is possible.

Interplanetary microspacecraft are also possible for exploring near-Earth objects. The "Smallsat Intercept Missions to Objects Near Earth" (SIMONE) is one concept based on a 120 kg spacecraft with approximately 1 kW ultralight

Table 23.9. Minimum ΔV Requirements for Some Missions within Cislunar Space

Mission	ΔV (m/sec)
Launch a satellite into a 400 km circular LEO	9000
Move satellite from a 400 km circular LEO to 1000 km altitude circular orbit	318
Move satellite from a 400 km circular equatorial LEO to geostationary orbit	3854
Put satellite into an Earth escape trajectory starting from a 400 km altitude LEO	3175
Put satellite into an Earth escape trajectory starting from a geotransfer orbit	778
Put a satellite into a 400 km circular low lunar orbit starting from a 400 km altitude LEO	3900
Land on surface of moon starting from a 400 km low lunar orbit	1840

solar arrays and ion propulsion for Earth escape, interplanetary maneuvering, and asteroid rendezvous.[57] Five spacecraft could fit on an Ariane-5 ASAP (Ariane Structure for Auxiliary Payloads) to provide low-cost launching of multiple interplanetary spacecraft.

An interplanetary picospacecraft, MINERVA, was deployed 12 November 2005 by the Japanese Hayabusa spacecraft near the small asteroid Itokawa.[52] MINERVA was intended to operate as a hopping rover in the approximately 10 μg surface gravity of Itokawa. It had a mass of 591 g, a diameter of 12 cm, and a height of 10 cm; it carried three CCD cameras, six photodiodes, and thermometers as sensors. Although MINERVA was improperly deployed with a velocity in excess of the asteroid's escape velocity, it proved that picospacecraft could function as interplanetary spacecraft assistants. Future missions to asteroids or airless moons should carry multiple picospacecraft or femtospacecraft to provide additional surface measurements to the host spacecraft.

23.5.4 Midterm Applications

In the midterm (5–10 years), further evolution of micro- through femtosatellite capabilities should occur. As attitude determination and control systems improve, especially for nanosatellites and picosatellites, satellite inspectors similar to Mini AERCam should become operational (see chapter 16). The MEPSI and COSA efforts at The Aerospace Corporation have developed cold-gas propulsion and attitude control systems suitable for kilogram-class inspector vehicles with mission lifetimes of about two weeks[58,59] (see chapters 18 and 20). A NASA/JPL study concluded that a 5 kg inspector for crewed or interplanetary spacecraft could be built.[60] With an improved, higher I_{sp} propulsion system, the satellite inspector or assistant can provide support throughout the host vehicle's entire lifetime.

Satellite augmentation using co-orbiting subsatellites is one application that should grow rapidly in the midterm. If satellites are designed and fabricated with a local broadband connection, similar to terrestrial WiFi, co-orbiting subsatellites can communicate and interact with a host satellite. RF power levels required to transmit 10 Mbps at 2.4 GHz across 100 m is only 1 mW for line-of-sight links with omnidirectional antennas. This can be useful for augmenting onboard processing power and memory. The technology freeze date for large satellites is typically about five years before launch, so after five years on orbit, the onboard computer and memory systems are already 10 years out of date. A co-orbiting small satellite with a rapid design, fabrication, and launch cycle can provide a 10 to 32 times improvement in processing speed and memory storage if Moore's Law continues to hold. Additional functions that can be added to the host spacecraft over time are updated environment sensors, new uplink capabilities, and new crosslinks. Large spacecraft can effectively become distributed satellites with an entourage of assistants.

Active attosatellites may be fielded in the midterm as small satellite technologies improve further. More than 100 could be ejected by a single picosatellite—which would represent a significant space debris threat if they are ejected within most of LEO. At low altitudes where orbital lifetime is limited to months, hundreds of attosatellites with magnetometers or other miniaturized space weather sensors could be used to provide time-stamped, simultaneous, distributed measurements at very low cost. Downlink rates to a 4 m diameter ground station dish can be about a 100 kbps, thus providing about 7 Mb of data per 10 min pass.

Evolution of interplanetary microspacecraft to smaller sizes or higher capability should also occur during the next five to ten years. Nanospacecraft are capable of exploring individual near-Earth objects, while microspacecraft are capable of exploring the thousands of known objects in the main asteroid belt (from Table 23.6). More advanced missions, such as monitoring Martian space weather and atmosphere, using multiple microspacecraft to obtain distributed, simultaneous measurements have already been proposed.[61]

23.5.5 Long-Term Applications

A long-term (10–20 years) application of small-satellite technology will be distributed, "fractionated," or "virtual" spacecraft, where mission functionality is distributed across a large number of satellites in a local cluster. Overall capability is distributed, much like the Internet, to enable graceful degradation and multifunctionality through physical or computational rearrangement. Examples of distributed spacecraft include multiple spacecraft distributed along a single orbit; local clusters that form sparse-aperture antennas several kilometers in diameter; and entire constellations that reconfigure their topology by rephasing individual spacecraft within their assigned orbits.

The sparse-aperture antenna is one form of distributed spacecraft composed of hundreds or thousands of subsatellites that co-orbit a mother ship. In this system, the individual subsatellites contain a low- to moderate-gain subantenna, a receiver, and a transmitter to broadcast the received bandwidth to the mother ship at the center of the cluster. This is a space-based version of multiple phased antennas, similar to the 27-antenna VLA (Very Large Array) used in radioastronomy.[62] The mothership accepts all signals and performs coherent signal addition based on the instantaneous position of each subsatellite and the required aggregate antenna beam pattern. The advantage of this approach is that the effective antenna diameter is equal to the diameter of the subsatellite distribution—hundreds of meters to hundreds of kilometers. A 10 km diameter, solid aluminum antenna dish a mere micron thick would have a staggering mass of 210,000 kg, while a sparse aperture array could have a mass of a few hundred to a few thousand kilograms. One disadvantage of a sparse-aperture array is that the individual elements should not be arranged with any periodicity in order to avoid grating lobes. The logarithmic spiral distribution shown in Fig. 23.18 eliminates this problem.

The major disadvantage of a non-periodic sparse-aperture array is that random side lobes appear with an average strength of $1/N$ of the main beam, where N is the number of subantennas. Figure 23.19 shows calculated gain patterns at 300 MHz for a 1 km radius logarithmic spiral sparse-aperture antenna with 128 and 512 subsatellites. Each subsatellite carries a simple dipole antenna at 300 MHz (0.5 m length) with a maximum gain of 3 dBi. Note the main beam angular half width of only 0.05 deg. The 512-subsatellite pattern has a side-lobe/main-lobe ratio that is 4 times better than the 128 subsatellite pattern. To obtain

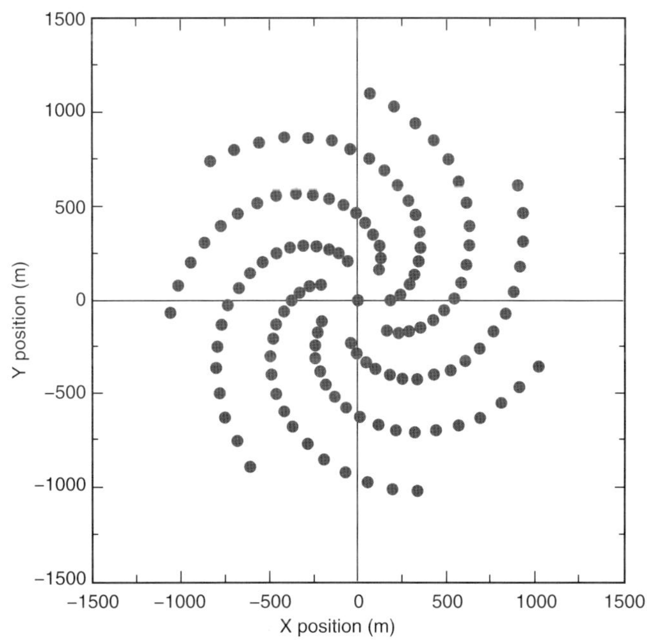

Fig. 23.18. Logarithmic spiral subsatellite distribution to eliminate grating lobes in an orbiting sparse-aperture array.

side lobes that are 30 dB down from the main lobe, at least 1000 subsatellites are required. To make the total mass of the antenna reasonable, these subsatellites must be nanosatellites or smaller. Finally, to make the system cost effective, the subsatellites should be mass producible (see chapter 24). Orbiting sparse-aperture antennas will be enabled by mass-produced nanosatellites, picosatellites, or femtosatellites. Possible applications include jam-proof communications at UHF (300–1000 MHz) using narrow-angle beams; readout of unattended ground sensors from GEO; high-throughput, high-speed wireless Internet connections using a GEO satellite; and high-resolution surface mapping radiometry. These and other applications are covered in Bekey.[36]

Another mode of operation for distributed spacecraft is to physically connect them. Thousands of mass-produced picospacecraft or nanospacecraft can link together on orbit to produce a large spacecraft with dimensions of tens of meters or larger. A terrestrial example of this is a self-configuring modular robot.[63,64] Spacecraft can grow over time as new modules are added, change physical geometry, disassemble at end of mission, and form one or more new spacecraft. Obvious applications are phased-array antennas and upgradable communication satellites to suit evolving mission needs (see chapter 24).

Microspacecraft should also be used for outer solar system exploration in the coming decades. While minispacecraft and larger vehicles are currently required to explore Saturn and beyond, further advancements in structures, thermoelectric conversion, and ground-station technologies could enable microspacecraft or at least 200 kg versions of the New Horizons spacecraft. Beyond Neptune lies the Kuiper belt, a torus-shaped region that extends from 30 to 50 AU (4.5–7.5 billion

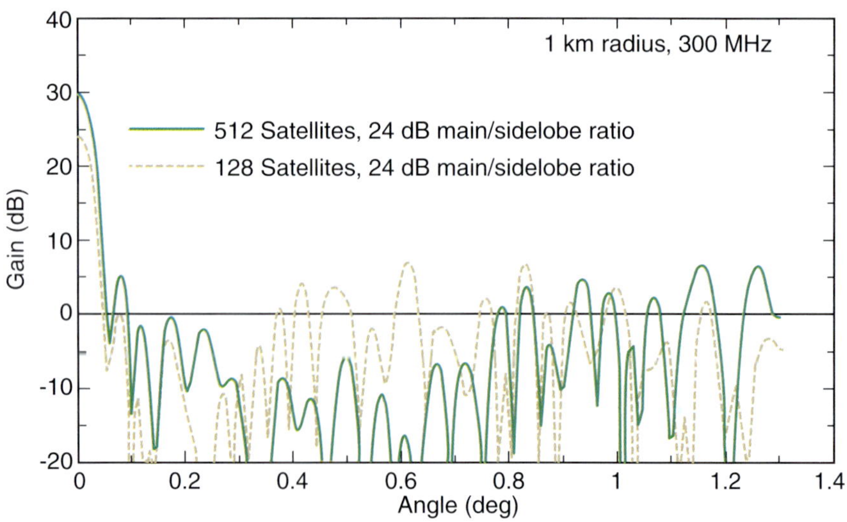

Fig. 23.19. Calculated antenna gain patterns at 300 MHz for a 1 km radius logarithmic spiral sparse-aperture antenna with 128 and 512 subantennas.

km) from the sun, which contains more than a thousand known objects, most of which have been discovered since 1992.[65] Table 23.10 lists the names, diameter, and basic orbit information for these trans-Neptunian objects (including Pluto's moon Charon) that are at least 1000 km in diameter. Eris is larger than Pluto, and all of these bodies are larger than Ceres, the largest dwarf planet in the main asteroid belt. Many trans-Neptunian dwarf planets also have moons; Pluto, Eris, and Makemake each have one, while Haumea has two.[66]

More than a thousand interesting targets exist in the outer solar system for nuclear minispacecraft or microspacecraft. Since a Hohmann transfer suitable for placing a spacecraft in orbit about Pluto takes at least 32 years, and longer for objects beyond 31.5 AU, missions to these trans-Neptunian objects will use faster flyby trajectories with 10–20 year transfer times for the foreseeable future. The disadvantage of a flyby trajectory is that the period of close-in observation is limited, and at least half of the object is not in view during closest approach where imaging resolution is highest. To extend the period and the physical extent of close-in observations, a number of primary battery-powered picospacecraft and nanospacecraft could be ejected before target encounter to provide close-up views and measurements of regions that are not in view of the main spacecraft during closest approach. These spacecraft assistants can store gigabytes of data and send them to the main spacecraft for relay back to Earth. Spacecraft lifetimes at a 1 W power level using primary lithium batteries range from a few days to a month. An additional option for the pico/nanospacecraft is to impact the minor planet to obtain close-up images, much like the 1960s-era Ranger probes of the moon, while transmitting real-time data back to the main spacecraft.

Table 23.10. Semimajor Heliocentric Axis, Orbit Eccentricity, Diameter, and Heliocentric Distance from Sun for Trans-Neptunian Objects of at least 1000 km Diameter (Heliocentric Distances as of June 2008)[67]

Object	Semimajor Axis (AU)	Orbit Eccentricity	Diameter	Distance from Sun (AU)
Eris 2003 UB313	67.90	0.436	2400 km	96.75
Pluto	39.45	0.250	2300 km	31.52
Haumea 2003 EL61	43.13	0.195	~2000 × 1500 × 1000 km	51.09
Makemake 2005 FY9	45.43	0.161	~1600 km	52.07
Charon	39.45	0.250	1270 km	31.52
Sedna 2003 VB12	494.0	0.845	~1500 km	88.15
Orcus 2004 DW	39.19	0.226	~1600 km	47.83
Quaoar 2002 LM60	43.58	0.039	~1200 km	43.22
Ixion 2001 KX76	39.69	0.244	~1000 km	41.75

23.5.6 Very Long-Term Applications

An interesting but speculative very long-term (20–50 years) application of ultra-small satellites is to use them as propellant on translunar and interplanetary spacecraft. Once again, the basic concept is to substitute intelligence for mass, since "intelligence doesn't weigh much." A significant propulsion challenge is to land humans or equipment on the moon and return them to Earth. The velocity increment is the sum of the last two mission ΔVs in Table 23.9 times two (going and returning), or about 11,500 m/sec. This is a very large velocity increment for an in-space mission.

Sending a spacecraft to GEO from an initial 400 km circular LEO has a minimum propellant mass fraction of 83% using 220 sec I_{sp} thrusters and 59% using 440 sec I_{sp} thrusters. Note that the final mass includes empty propellant tanks, thrusters, and similar components, so the final mass can be larger than the delivered spacecraft mass. This becomes critical when propellant mass fractions exceed 80%, since the "dry" mass, consisting of thrusters, valves, plumbing, and propellant tanks, can be about 10% of the propellant mass. This doesn't leave much for the actual spacecraft or payload mass.

For the 400 km LEO to moon landing and return scenario with a ΔV of 11,500 m/sec, the propellant mass fraction is 93% using the best chemical thruster I_{sp} of 440 sec. The only way to accomplish this mission using chemical thrusters is to use multiple stages and throw away unneeded mass such as empty propellant tanks whenever possible. The Apollo moon landings of the late 1960s and early 1970s, for example, started with 118,000 kg in a LEO parking orbit and put 47,000 kg in low lunar orbit using a 400 sec $I_{sp,}$ thruster. The remaining maneuvers used space-storable propellants with a reduced $I_{sp,}$ of 315 sec. The mission ended with a 5800 kg capsule entering Earth's atmosphere at near escape velocity; less than 5% of the original mass in LEO was returned. A sustainable, long-term lunar exploration strategy for the 21st century should use propulsion with higher effective I_{sp} to reduce the amount of propellant mass required, and to make expensive spacecraft available for continual reuse.

Dramatic reductions in required propellant mass occur as I_{sp} is increased. Unfortunately, electric thrusters are required to attain higher I_{sp}, and the input energy requirements, for constant thrust, scale as I_{sp}. This is why electric thrusters with high $I_{sp,}$ generally have low thrust levels. A 1000 N (224 lb-force) thruster, for example, would require 30 MW of electric power at 5000 sec I_{sp} and 60 MW at 10,000 sec I_{sp}. As $I_{sp,}$ increases, the propellant mass savings is eliminated at some point by an increasing mass of the power supply; an optimal $I_{sp,}$ exists for a given mission and power system technology. If propellant mass could be returned to the spacecraft after ejection and recycled, the rocket equation would no longer apply, need for very high I_{sp} would be eliminated, and higher average thrust levels could be maintained with significantly reduced power. How can propellant mass be recycled once it has left the spacecraft?

Conventional rocket propulsion is analogous to trying to propel yourself, while standing on a frictionless skateboard or roller skates, by throwing baseballs horizontally. Your final velocity will be proportional to how many baseballs you threw and how fast you threw them. A more efficient method would be to throw baseballs at a wall, catch the reflected balls that come back, and reuse them many times. Catching a reflected ball not only returns mass for reuse, but also provides an additional impulse from the ball's momentum. In space, orbital mechanics can provide the virtual wall that reflects mass back to the host spacecraft. We can't get molecules or atoms to follow precise trajectories and return, but we can eject small spacecraft at high velocities that are programmed to return. Each spacecraft would have navigation determination, attitude determination and control, and thruster systems to fine-tune trajectories; they could be considered as "smart propellant." We already pay thousands of dollars to put a pound of propellant into orbit; why not replace that "dumb" propellant with more expensive, but much more useful "smart" propellant? This approach could make human and robotic exploration of the moon and Mars in the next decades more economical by drastically reducing the amount of propellant needed; it will require mass production of tens of thousands of nano-, pico-, or possibly even femtospacecraft.

The conceptually simplest propellant recycling maneuver is called apoapsis reflection. It is based on the conservation of energy (kinetic plus gravitational potential energy) and orbital angular momentum for satellites. Orbital angular momentum is equal to the instantaneous azimuthal spacecraft velocity times the distance to the center of mass of the gravitationally attracting primary body (e.g., Earth, moon, sun, etc.). Note that the kinetic energy, and hence orbital velocity, of a satellite in a closed or open orbit decreases as it moves away from the primary body. For an object in a parabolic escape trajectory, the azimuthal and radial velocities approach zero at infinite distance at infinite time. A small orbit perturbation near infinity can cause a satellite to retrace its parabolic orbit in a retrograde fashion; a propellant mass ejected along a parabolic or nearly parabolic escape trajectory can return to the original spot at which it was ejected from an orbiting spacecraft. In practice, propellant could be ejected at a velocity slightly lower than escape velocity, and the perturbation would be supplied by onboard thrusters or solar sail at or near apoapsis (furthest distance from the gravitating body). Escape velocity for a circular orbit is $2^{1/2}\,V_{circ}$, where V_{circ} is the circular orbit velocity. The "smart" propellant has a minimum velocity at apoapsis, so this is the preferred point for reversing direction in a closed (elliptical—not parabolic or hyperbolic) orbit. Use of a closed orbit allows the propellant to return in a finite time. By properly adjusting the orbit and hence orbit period for the smart propellant, both the smart propellant and spacecraft can return to the same point in space at the same time for rendezvous and recycling.

The apoapsis reflection maneuver is illustrated in Fig. 23.20. Consider a 1000 kg spacecraft in a posigrade circular orbit about Earth at an altitude of 400 km as illustrated in Fig. 23.20a. Spacecraft orbital velocity is a constant

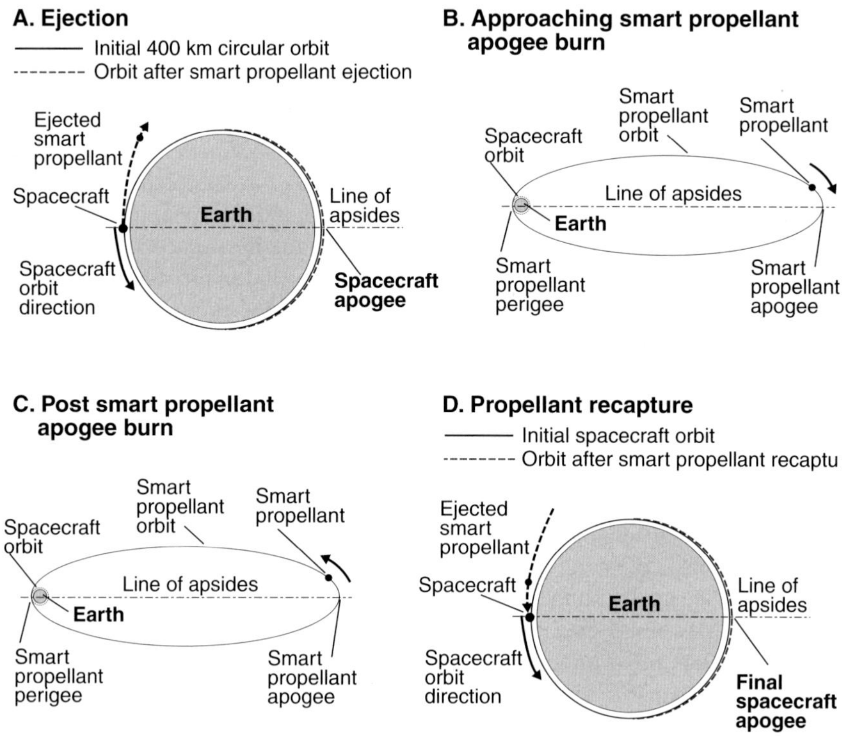

Fig. 23.20. Four stages of the apogee reflection maneuver.

7.669 km/sec, and escape velocity for this orbit is 10.613 km/sec. If the spacecraft ejects a 1 kg smart propellant mass at 18.406 km/sec in the retrograde direction, the spacecraft velocity will increase to 7.687 km/sec. For these high ejection velocities, an electromagnetic accelerator would be required. This new velocity puts the spacecraft in an elliptical orbit with a perigee of 400 km and an apogee (apoapsis for an Earth-centered orbit) of 465.5 km. The ejected smart propellant mass has a retrograde velocity of 10.719 km/sec, putting it on a retrograde orbit with perigee of 400 km and apogee of 280,400 km. This is illustrated in Fig. 23.20b. After 77.7 hr, the smart propellant is at apogee with an altitude of 280,400 km and a significantly reduced velocity of 0.2534 km/sec. The smart propellant now fires a thruster in the antiflight direction to produce a velocity increment of 0.5067 km/sec. This reverses the flight path and puts the smart propellant in a posigrade orbit (see Fig. 23.20c). After 77.7 hr, the smart propellant returns to perigee with an instantaneous velocity of 10.613 km/sec. The 1000 kg spacecraft has performed 100 orbits during this interval, and if the propellant orbit period was adjusted properly, the host spacecraft will be present and both will collide with a relative velocity of 3.032 km/sec (Fig. 23.20d). Using an electromagnetic accelerator as a power generator should allow deceleration of the incoming smart propellant mass while extracting the excess kinetic energy. We want to recapture the smart propellant intact so that we can reuse it over and over again. This imparts an additional impulse to the spacecraft, further boosting its apogee to 476.7 km (~11.2 km increase). The spacecraft has received a velocity increase of 21 m/sec with a miniscule reduction in mass.

The apoapsis reflection maneuver is extremely efficient in the utilization of propellant mass. Since smart propellant maneuvering occurs near apoapsis where velocities are low, thrusting periods of many hours are possible, and high I_{sp} electric thrusters can be used to advantage. If the smart propellant module uses a 1000 sec I_{sp} electric thruster for the 0.507 km/sec apogee reflection maneuver, the mass loss based on the rocket equation is only 50.4 g. This yields an effective I_{sp} for the spacecraft plus smart propellant system of 43,000 sec with an energy expenditure equal to that of an 1880 sec I_{sp} system. The limitations of the rocket equation and the power equation for electric thrusters have been reduced by more than an order of magnitude. In practical terms, much greater mass can be delivered throughout cislunar and interplanetary space using the smart propellant concept. Key enablers are inexpensive, mass-produced smart propellant spacecraft and an electromagnetic accelerator/decelerator capable of generating the required exit velocities. For this particular maneuver, exit velocities up to 14 km/sec are required. Other maneuvers require significantly reduced exit velocities. Ross Jones's dream of electromagnetically launched kilogram-mass spacecraft for space exploration (see chapter 2) has been revived with a new twist.

Smart propellant maneuvers exist for satellite rephasing (moving satellites around within their orbit), injection of spacecraft into low lunar orbit from an Earth escape trajectory, and injection into lunar orbit from the surface of the moon. This is an "out of the box" concept with a large number of technological hurdles to overcome, but the basic physics shows enticingly large payoffs. Small satellites started the Space Age and will probably drive human expansion into the solar system when we approach the 100th anniversary of Sputnik 50 years from now.

23.6 Conclusions

About 860 active microsatellites have been launched, but most of the 680 nanosatellites and 38 picosatellites were passive spacecraft used for radar calibration and atmospheric density monitoring. Today, there are more capable microsatellites, mission-capable active nanosatellites, and experimental active picosatellites such as CubeSats. Mission applications for micro-, nano-, and picosatellites now range from space environment monitoring to medium-resolution Earth observation, store-and-forward communications, and surface exploration of asteroids. More technology development is needed to provide attitude determination and microthruster systems required for more demanding missions such as high-resolution Earth imaging, satellite inspection, and distributed satellites. Thousands of nano- or picosatellites will be required for some of these missions. Active femtosatellites will be flown in the next few years, and their numbers could skyrocket due to their relatively low cost. Even active attosatellites may appear in the coming decades for modest space missions. In the long run, we should see picospacecraft and nanospacecraft as interplanetary spacecraft assistants, and even mass-produced nano-, pico-, and femtosatellites used as recyclable "smart" propellant in larger space vehicles.

23.7 Acknowledgments

I thank The Aerospace Institute for sponsoring this chapter, and I gratefully acknowledge The Aerospace Corporation's Independent Research and Development program and its former Corporate Research Initiative in MEMS and Microtechnology for funding research in silicon nanosatellites, micropropulsion, formation flying, CMOS-based spacecraft sensors, and parts of The Aerospace Corporation's PicoSat program. I also want to thank Henry Helvajian, Ernest Y. Robinson, the late Seymour Feuerstein, and David Hinkley for their efforts in advancing nano/picosatellite technologies. Finally, I'd like to thank Jennifer—the love of my life and my inspiration to keep pushing the envelope.

23.8 References

[1] Richard P. Feynman, "There's Plenty of Room at the Bottom," reprinted in *J. Microelectromechanical Systems*, **1** (1), pp. 60–66, March 1992.

[2] D. H. Martin, P. R. Anderson, and L. Bartamian, *Communications Satellites, Fifth Edition*, 2007, pp. 363–364 (The Aerospace Press and AIAA, El Segundo, CA, 2007).

[3] J. W. Zuckermandel, S. Enger, N. Gupta, and J. Summers, "Modular, Thin Film Solar Arrays for Operationally Responsive Spacecraft," *Proc. of the 2007 IEEE Aerospace Conference* (2007).

[4] P. Stella, R. Mueller, G. Davis, and S. Distefano, "The Environmental Performance at Low Intensity, Low Temperature (LILT) of High Efficiency Triple Junction Solar Cells,", AIAA paper 2004-5579, *2nd Int. Energy Conversion Eng. Conference* (Providence, RI, 16–19 Aug. 2004).

[5.] Gary L. Bennet, "Space Nuclear Power: Opening the Final frontier," AIAA paper 2006-4191, *4th Int. Energy Conversion Eng. Conference and Exhibit* (San Diego, CA, 26–29 June 2006).

[6] Roger D. Launius, "Powering Space Exploration: U. S. Space nuclear power, Public Perceptions, and Outer Planetary Probes," AIAA paper 2008-5638, *6th Int. Energy Conversion Eng. Conference and Exhibit* (Cleveland, OH, 26–30 July 2008).

[7] J. A. Angelo and D. Buden, *Space Nuclear Power*, pp. 136–139 (Orbit Book Company, Malabar, FL, 1985).

[8] A. R. Kahn, J. D. Hixson, J. E. Puffer, and E. E. Bakken, "Three-Years' Clinical Experience with Radioisotope Powered Cardiac Pacemakers," *IEEE Trans. On Biomedical English,* **BME-20** (5) 326–331 (Sept. 1973).

[9] LSH-14 Primary Lithium Battery Datasheet, Saft Specialty Battery Group, Bagnolet, France, 2007; http://www.saftbatteries.com/doc/Documents/primary/Cube655/LSH14_0607.dab6764e-f7e1-432f-8f3a-a6ee6a21eb70.pdf (10 Dec. 2008).

[10] S. Cawley, "TopSat: Low Cost High-Resolution Imagery from Space," *Acta Astronautica*, **56**, 147–152 (2005).

[11] Norhan Mat Yusoff, "RazakSAT–Technology Advent in High Resolution Imaging System from Small Satellite," *Proc. of the 26th Asian Conference on Remote Sensing* (Hanoi, Vietnam, 7–11 Nov. 2005).

[12] E-E, Kim, Y-H Choi, E. D. Kim, S-K Jeong, M-S Kang, H. M Nasir, M. R. Rosdi, and A. H. A. Hai, "Integration and Testing of a High-Resolution Camera for Small Satellites," *Proc. of the 2nd International Conference on Resent Advances in Space Technologies* (9–11 June 2005), pp. 551–554.

[13] Orbital Sciences Corporation, OrbView-3 Fact Sheet, Orbital Sciences Corporation, Dulles, VA (2006); http://www.orbital.com/NewsInfo/Publications/OV3_Fact.pdf (10 Dec. 2008).

[14] K. Kohm, "Modulation Transfer Function Measurement Method and Results for the OrbView-3 High Resolution Imaging Satellite," *Proc. of the International Society for Photogrammetry and Remote Sensing* (Istanbul, Turkey, 12–23 July 2004).

[15] E. J. Tremblay, R. A. Stack, R. L. Morrison, and J. E. Ford, "Ultrathin Cameras Using Annular Folded Optics," *Applied Optics*, **46** (4), 463–471 (Feb. 2007).

[16] C. Wang, D. Sklar, and D. Johnson, "Forward Error Correction Coding," *Crosslink*, **3** (1), 26–29, http://www.aero.org/publications/crosslink/winter2002/index.html (The Aerospace Press, El Segundo, CA, Winter 2001/2002).

[17] C. C. DeBoy, C. B. Haskins, T. A. Brown, R. C. Shulze, M. A. Bernacik, J. R. Jensen, W. Millard, D. Duven, and S. Hill, "The RF Telecommunications System for the New Horizons Mission to Pluto," *Proc. of the 2004 IEEE Aerospace Conf.* (6–13 March 2004).

[18] D. W. Kusnierkiewicz, G. Fountain, Y. Guo, and C. B. Hersman, "The New horizons Mission to the Pluto System and the Kuiper Belt," *Proc. of the 2008 IEEE Aerospace Conf.* (1–8 March 2008).

[19] Siegfried W. Janson, "Micro/Nanotechnology for Picosatellites," paper SSC08-VII-6, *Proc. of the 22nd Annual USU Conference on Small Satellites* (Logan, Utah, 2008).

[20] The MOSIS Service, "Welcome to MOSIS" Home Page, Marina del Rey, CA, http://www.mosis.com (11 Dec. 2008).

[21] S. W. Janson, "Nanotechnology—Tools for the Satellite World," Keynote Address, *6th ISU Annual International Symposium; "Smaller Satellites, Bigger Business?"* (Strasbourg, France, 21–23 May 2001, Kleuwer Acedemic, Norwell, MA, 2002), pp. 21–30.

[22] S. W. Janson, "Micro/Nanotechnology for Micro/Nano/Picosatellites," AIAA paper 2003-6269 (2003).

[23] J. G. Walker, "Satellite Constellations," *J. British Interplanetary Society*, **37**, 559–572 (1984).

[24] D. Y. Stodden and G. D. Galasso, "Space System Visualization and Analysis Using the Satellite Orbit Analysis Program (SOAP)," *1995 IEEE Aerospace Applications Conference Proceedings*, Vol. 1, pp. 369–87 (1995).

[25] S. W. Janson, "Electric Propulsion for Low Earth Orbit Constellation Morphing," AIAA paper 2002-3669, *38th AIAA/ASME/SAE/ASEE Joint Propulsion Conference & Exhibit* (Indianapolis, IN, 7–10 July 2002).

[26] V. A. Chobotov, *Orbital Mechanics*, AIAA Education Series, (AIAA, Washington, DC, 1991), p. 178–179.

[27] S. Q. Kidder, J. A. Kankiewicz, and T. H. Vonder Haar, "The A-Train: How Formation Flying is Transforming Remote Sesning," *Joint AMS/EUMETSAT Satellite Metrology Conf.* (Amsterdam, 24–28 Sept. 2007).

[28] G. Le, T. E. Moore, and J. A. Slavin, "Space Technology 5—Enabling Future Constellation Missions Using Micro satellites for Space Weather," paper SSC07-IV-6, *Proc. of the 21st Annual USU Conference on Small Satellites* (Logan UT, 2007).

[29] P. A. Bernhardt, R. P. McCoy, K. F. Dymond, J. M. Picone, R. R. Meier, F. Kamalabadi, D. M. Cotton, S. Charkrabarti, T. A. Cook, J. S. Vickers, A. W. Stephan, L. Kersley, S. E. Pryse, I. K. Walker, C. N. Mitchell, P. R. Straus, H. Na, C. Biswas, G. S. Bust, G. R. Kronschnabl, and T. D. Raymund, "Two-Dimensional Mapping of the Plasma Density in the Upper Atmosphere With Computerized Ionospheric Tomography (CIT)," *Physics of Plasmas*, **5** (5), 2010–2021 (May 1998).

[30] T. P. Yunck, "An Overview of Atmospheric Radio Occultation," *Journal of Global Positioning Systems*, **1** (1), 58–60 (2002).

[31] J. E. Pollard, C. C Chao, and S. W. Janson, "Populating and Maintaining Cluster Constellations in Low Earth Orbit," AIAA paper 99-2871, *35th AIAA/ASME/SAE/ASEE Joint Propulsion Conference* (Los Angeles, CA, June 1999).

[32] C. C Chao, J. E. Pollard, and S. W. Janson, "Dynamics and Control of Cluster Orbits for Distributed Space Missions," paper AAS 99-126, *American Astronautical Society/ AIAA Space Flight Mechanics Meeting* (Breckenridge, CO, 7–10 Feb. 1999).

[33] D. H. Martin, P. R. Anderson, and L. Bartamian, *Communication Satellites, Fifth Edition* (The Aerospace Press, El Segundo, Cailifornia, 2007), pp. 187–189.

[34] B. Harvey, *Russia in Space: The Failed Frontier?* (Springer, New York, 2001), p. 126.

[35] N. L. Johnson, D. O. Whitlock, P. Anz-Meador, M. E. Cizek, and S. A. Portman, "History of On-Orbit Satellite Fragmentations, 13th Edition," NASA–Johnson Space Center report JSC-62530 (May 2004).

[36] I. Bekey, *Advanced Space System Concepts and Technologies: 2010-2030+* (The Aerospace Press, El Segundo, CA, 2003).

[37] NASA Orbital Debris Program Office, The Orbital Debris Radar Calibrations Spheres, (2007); http://www.orbitaldebris.jsc.nasa.gov/measure/oderacs.html (11 Dec. 2008).

[38] G. H. Cress, A. E. Potter, T. J. Setticeri, G. P. Sherrill, E. G. Stansbery, D. L. Talent, "Radar and Optical Ground Measurement, Final Report," *Orbital Debris Calibration Spheres*, NASA Report JSC-27241 (NASA Johnson Space Center, Houston 1996).

[39] M. C. Chapman, J. B. Gardner, and R. J. Martin, "Octahedral Research Satellite Mark III OV5-1 (ERS-27)," Air Force Cambridge Research Laboratories report AFCRL-68-0265, (Bedford, Mass. April 1968).

[40] NASA, GeneSat-1 Technology Demonstration Mission, (2007); http://genesat.arc.nasa.gov. (11 Dec. 2008).

41. T. Otsubo, R. A. Sherwood, P. Gibbs, and R. Wood, "Spin Motion and Orientation of LAGEOS-2 from Photometric Observation," *IEEE Trans. On Geoscience and Remote Sensing*, **42** (1), 202–208 (January 2004).

42. S. Barabash, I. Kiryushkin, O. Norberg, M. Ovchinnikov, and V. Penkov, "The Nanosatellite Munin: A Simple Tool for Auroral Research," *Advances in Space Research*, **31** (2), 313–318 (2003).

[43] M. Deckett, "Orbcomm—A Description and Status of the LEO Satellite Mobile Data Communication System," AIAA paper 1994-1135-CP, *AIAA Int. Communications Satellite Systems Conf.* (San Diego, CA, 28 Feb.–3 Mar. 1994).

[44] M. N. Allery, H. E. Price, J. W. Ward, and R. A. Da Silva Curiel, "Low Earth Orbit Microsatellites for Data Communications Using Small Terminals," *Proc. of the 10th Int. Conf. on Digital Satellite Communications*, Vol. 2 (May 1995), pp. 457-465.

[45] J. Bravman, B. Coffman, S. Whelan, S. Koutsoutis, and D Hampel, "Multiple Path Beyond-Line-of-Sight Communications (MUBLCOM)," *Proc. of the IEEE Military Communications Conf.* Vol. 1 (18–21 Oct. 1998), pp. 35–39.

[46] R. Hoyt, and Slostad, "The Multi-Application Survivable Tether (MAST) Experiment," AIAA paper 2003-5219 (July 2003).

[47] S. Flagg, T. Bleier, C. Dunson, J. Doering, L. DeMartini, P. Clarke, L. Franklin, J. Seelbach, J. Flagg, M. Klenk, V. Safradin, J. Cutler, A. Lorenz, and E. Tapio, "Using Nanosats as a Proof of Concept for Space Science Missions: QuakeSat as an Operational Example," paper SSC04-IX-4, *Proc. of the 18th Annual USU Conference on Small Satellites* (Logan, UT, 2004).

[48] W. R. Corliss, "Scientific Satellites," NASA report SP-133, (1967), pp. 786–787.

[49] P. R. Strauss, "Ionospheric Remote Sensing with the Ionospheric Occultation Experiment (IOX): First Results," *Eos. Trans. AGU,* **82** (47), Fall Meet. Supp., (2001).

[50] W. A. Levett, Laycock, J. and Baxter, E. "The TOPSAT Demonstrator, Opening the Way for High Quality and Timely Imagery from an Affordable Constellation," paper IAC-06-B5. 4. 08, *57th Int. Astronautical Conf.* (Valencia, Spain, 2–6 Oct. 2006).

[51] S. C. O. Grocott, R. E. Zee, and J. Matthews, "The MOST Microsatellite Mission: One Year in Orbit," paper SSC04-IX-1, *Proc. of the 18th Annual USU Conference on Small Satellites* (Logan, UT, 2004).

[52] T. Yoshimitsu, T. Kubota, and I. Nakatani, "Operation of MINERVA Rover in Hayabusa Asteroid Mission," paper IAC-06-A3. 5. 01, *57th Int. Astronautical Conf.* (Valencia, Spain, 2–6 Oct. 2006).

[53] T. Iwata, and N. Kawano, "Development of Selene Small Sub-Satellites for Lunar Gravity Observation," paper IAC-05-B5. 2. 01, *56th Int. Astronautical Conf.* (Fukuoka, Japan, 17–21 Oct. 2005).

[54] A. Kalman, A. Reif, D. Berkenstock, J. Mann, and J. Cutler, "MISC: A Novel Approach to Low- Cost Imaging Satellites," paper SSC08-X-3, *Proc. of the 22nd Annual USU Conference on Small Satellites* (Logan, UT, 2008).

[55] E. Eide and J. Ilstad, "NCUBE-1 The First Norwegian CubeSat Student Satellite," *16th ESA Symposium on European Rocket and Balloon Programmes and Related Research* (Sankt Gallen, Switzerland, 2–5 June 2003), pp. 85-88.

[56] X Prize Foundation, "About the Google Lunar X Prize" (2007); http://www.google lunarxprize.org/lunar/about-the-prize (11 Dec. 2008).

[57] N. Wells, R. Walker, S Green, and A. Ball, "SIMONE: Interplanetary Microsatellites for NEO Rendezvous Missions," *Acta Astronautica*, **59**, 700–709 (2006).

[58] S. Janson, A. Huang, W. Hansen, and H. Helvajian, "Development of an Inspector Satellite Propulsion Module Using Photostructurable Glass/Ceramic Materials," AIAA paper 2004-6701, *CANEUS 2004: Conference on Micro- Nanotechnologies* (Monterey, CA, 1–5 Nov. 2004).

[59] D. Hinkley, "A Novel Cold Gas Propulsion System for Nanosatellites and Picosatellites," paper SSC08-VII-7, *Proc. of the 22nd Annual USU Conference on Small Satellites* (Logan, UT, 2008).

[60] H. Goldberg, J. Mueller, and L. Alkalai, "Micro-Inspector Spacecraft: An Overview," paper SSC-06- II-6, *Proc. of the 20th Annual USU Conference on Small satellites* (Logan, UT, 2006).

[61] R. Shotwell, A. Gray, P. Illsley, M. Johnson, R. Sherwood, M. Vozoff, and J. Ziemer, "Mars Aeronomy Explorer (MAX): Study Employing Distributed Micro-Spacecraft," paper 2. 0301, *Proc. of the IEEE Aerospace Conf.* (Big Sky, MT, 5–12 March 2005).

[62] National Radio Astronomy Observatory, Very Large Array (2007); http://www.vla.nrao.edu (11 Dec. 2008).

[63] M. Yim, Y. Zhang, and D. Duff, "Modular Robots," IEEE Spectrum, **39** (2), 30–34 (Feb. 2002).

[64] W-M. Shen, M. Krivokon, H. Chiu, J. Everist, M. Rubenstein, and J. Venkatesh, "Multimode Locomotion via SuperBot Reconfigurable Robots," *Autonomous Robots*, **20** (2), 165–177 (March 2006).

[65] D. Jewitt, "The 1000-km Scale KBOs," (2008); http://www.ifa.hawaii.edu/~jewitt/kb.html (Dec. 2008).

[66] M. E. Brown, M. A. van Dam, A. H. Bouchez, D. Le Mignant, R. D. Campbell, J. C. Y. Chin, A. Conrad, S. K. Hartman, E. M. Johansson, R. E. Lafron, D. L. Rabinowitz, P. J. Stomski Jr., D. M. Summers, C. A. Trujillo, and P. L. Wizinowich, "Satellites of the Largest Kuiper Belt Objects," *The Astrophysical Journal*, **639**, L43–L46 (1 March 2006).

[67] NASA Near Earth Object Program "Orbit Diagrams" (May 2006); http://neo.jpl.nasa.gov/orbits (11 Dec. 2008).

Chapter 24

The Generation After Next: Satellites as an Assembly of Mass-Producible Functionalized Modules

Henry Helvajian*

Forerun thy peers, thy time, and let
Thy feet, millenniums hence, be set
In midst of knowledge, dream'd not yet.[1]

24.1 Introduction

This chapter addresses the practicality of manufacturing satellites as an assembly of mass-producible functional units through approaches currently labeled as mass manufacturing, mass customization, and design for manufacturability. Some concepts presented are natural extensions of the personal computer industry and are actively being pursued by small satellite manufacturers; others are more in the forefront.

The purpose is to investigate manufacturing methodologies to develop a means of mass production and mass customization of satellites. Solutions are sought for the satellite mass class of 1–10 kg (nanosatellite) and by extension mass classes that approach 100 kg (microsatellites) or higher. This treatise does not present the business case for mass production, which will be critical to the viability of the venture. This lack of economic analysis is not the Achilles heel, however, because it is the specifics of the particular mission that justify the business rationale. To meet the requirements for conducting a viable business, the mass manufacturing methodologies must include processes that allow for designs that keep costs low and provide high reliability.

24.1.1 The Power and Limitations of Modularity

In a series of publications in the late 1990s, Baldwin and Clark presented the advantages of modularity in design, using the U.S. computer industry as an example.[2,3] In modular design, elements are separated into modules by a formal architecture that maps function to module. Some of the modules are "hidden," in that design changes to these modules do not affect other modules. Other types of modules must exist to make the hidden modules interoperable; these are the "visible" modules that embody design rules that must be adhered to by the hidden module designers. Modular designs have advantages over interdependent designs because the "hidden" modules can be upgraded continually without incurring the cost of a major

*The Aerospace Corporation
El Segundo, California

system design change. Modularity also engenders commonality, thereby delivering additional cost savings. Finally, modularity also increases the overall system reliability or robustness with respect to maintaining operations over long durations.

To varying degrees, modularity has been applied to many industries, among them consumer electronics, automotive, and aerospace. It is not widespread, partly because of the apparent inherent trade-off between modularity and overall system technical performance.[4] In standard mechanical systems, an uncoupled or modular system favors the interface with its commensurate losses (e.g., power/signal losses, inefficient packaging), while coupled or interdependent systems favor efficiencies in mass and force transfer to enhance technical performance.[5] In general, it has been argued that modularity favors "business performance," while more integral architectures favor "technical performance."[6] In satellite systems, where low mass and low power are desired and technical performance is the primary aim, it is difficult to see how modularity and its demonstrated advantages can ever be wholly adopted. And yet, the limitations of modular architecture have been mitigated somewhat, as first demonstrated in very-large-scale integrated microelectronics and then by the development of microelectromechanical systems (MEMS). In both technologies, modularity in design and manufacturing is used in the context of functional basic devices (e.g., amplifiers, adders, processors, mechanical resonators). However, the inefficiencies that commonly exist at the device interfaces are reduced by the ability to co-fabricate the devices onto a common substrate. The microelectronics industry demonstrated that efficient interfaces can be built to pass current and voltage, and the MEMS industry demonstrated that mechanical force and torque can equally be transferred efficiently across an interface, at least at microdimensions.

Another limitation is the large capital investment needed to convert a non-modular design and manufacturing process into one that implements modularity. Notwithstanding the advantages of modularity, the cost of this conversion has been deemed too prohibitive in some industries. This would also be true for the satellite manufacturing industry. Perhaps a means to gain a foothold into the manufacturing of satellites as an assembly of modular units is through the nanosatellite concept and in the development of modules that can perform multiple functions. In this approach, the nanosatellite can serve as a modular building block for the manufacture of larger satellite systems. Furthermore, if the nanosatellite served as a multifunctional module, the desire to offer on-orbit mission capabilities for the lowest possible mass would be further realized.

24.1.2 Nanosatellites, Microsatellites, and Functional Modules

Nanosatellites are small space systems having a defined total mass between 1–10 kg. To date, most have been experimental vehicles used for validating specific technologies in space or for educating future space professionals—e.g., the CubeSat[6] (see chapter 5). No existing operational missions use nanosatellites, but there have been experimental missions that could result in a fully operational mission. For example, the first QuakeSat was an experimental nanosatellite based on a triple-length CubeSat. Launched in 2003, it was designed to detect extremely low-frequency earthquake precursor signals from space[7] (see chapter 5). Given this example and what has been made clear in a number of articles and ongoing developments, the size and volume of a nanosatellite should not be perceived as limiting[8] (see chapter 23). Obviously, nanosatellites have limited propellant volume and power; but this volume limitation becomes less of a handicap with regard to packing computational processing capability. The miniaturization trend in microelectronics, along with the development of MEMS and advancements in batteries, enables significant processing capability to be embedded within a 10 × 10 × 10 cm volume. One only needs to view a current cellphone, a personal digital assistant, a camera, and the integration of these systems onto a common platform (e.g., Apple iPhone[9]) to appreciate the possibilities when a systems-level integration is applied. Consequently, there are missions that a nanosatellite can accomplish.

Perhaps those satellite pundits who do not yet perceive the nanosatellite as a viable and practical space vehicle could defer judgment, just for now, and agree to view the nanosatellite as simply a functional module with some level of control and autonomy. Current microsatellite manufacturers have used the concept of functional modules or "drawers" to assemble larger spacecraft (>10 kg).[10,11] The concept has both practical and economic appeal because each module could potentially be manufactured and tested at different locales. The modules are then integrated onto a common scaffolding at final assembly. The concept is a derivative of typical procedure in the instrumentation and electronics packaging industry, where an assortment of multichip modules are integrated into a plug-and-play module, and the integrated system delivers a higher-level function using a common backplane tray (i.e., bus) as the "scaffolding."

Modularity is sure to be accelerated by the recent development and standardization of SpaceWire,[12] which enables high-speed serial communication between connected modules. Still, the common backplane analogy fails to consider the true functional aspects of a satellite. In a computer, the common backplane is only an interface for supply voltages and data lines; by contrast, a space system would consist of an assembly of modules that would require common backplanes for voltage and data lines, radio frequency (RF), fluidics, and thermal control. The problem is further exacerbated by the need for a low-current and a high-current backplane. The solution implemented by most microsatellite manufacturers is to separate the functions (i.e., the backplanes) into individual modules. Each is self-contained and uses only power and data lines from a common satellite bus. In essence, the subsystems are modular and individually tested. There are practical outcomes of this satellite manufacturing approach. These are the decrease in the development time and an increase in overall system reliability; both of these benchmarks have been realized.[13]

Even though a modular subsystem approach is viable, the question arises—using Lego blocks as an analogy—whether there is a need to develop a multifunctional block or module that has elements of multiple subsystems, thereby enabling the assembly of small and large satellites alike. Of course, unlike the Lego block—which is a passive module—the envisioned satellite block would also generate and store power locally for both internal use and external distribution. The modular subsystem may even have propulsion for limited locomotion or for physical attachment to nearby units. It has a primary function but also several secondary functions. Two good examples are the CONRO project and the SuperBot from the University of Southern California's Polymorphic Robotics Laboratory.[14] The SuperBot consists of an assembly of modules, each self-contained and self-sufficient (power, intelligence, control, locomotion) but designed to work together. The advantage of developing such blocks for space applications is that because they are nearly identical, they can be mass produced, and because they can be joined to form larger structures, they enable the assembly of space systems in very complex shapes.

24.1.3 Overview of the Satellite Manufacturing Industry

When a satellite is a manageable size, it becomes feasible to establish a manufacturing process flow whereby the satellite is physically moved through manufacturing stations. This process is an assembly line, though it need not be automated. Assembly line production has increasingly been used for the satellites intended for low Earth orbit (LEO) and medium Earth orbit (MEO; see chapter 14). Modern satellites also contain an increasing number of commercial-off-the-shelf (COTS) components that are put into assemblies, which are typically manufactured by subcontractors. Quality control has shifted from testing all components and systems to testing sample units, though components and systems are still tested when possible. The implementation of these process improvements has reduced satellite delivery times to less than 18 months—a major improvement over the three-year delivery times typically quoted just five years ago. Modern satellites are also expected to do more. There is a drive toward more onboard data processing, increased use of reconfigurable antennas that involve both electronics and robotics, and higher communications bandwidth—for example, very high Ka-band (18–40 GHz) and V-band frequencies (50–75 GHz) or laser communications.

These technological advances drive the need for more satellite power. The objective has been to provide more power without increasing the overall mass. Harnessing power in space—at least as solar energy—is ultimately related to capture area, while power storage is related to volume. Large satellites that serve complex and power-demanding missions have continued to grow in mass.[15] The same fate need not fall on the nanosatellites and microsatellites. Perhaps the satellite shape could be optimized to maximize solar-cell area. Cubic and rhomboid shapes do not capture energy efficiently, but are derivative of the shape of the internally modularized units. Satellites that are designed to get by with only a fraction of the available solar power do have the advantage of not requiring an attitude pointing system. However, the complexity associated with an attitude pointing systems need not be considered a limitation. Small satellite technology has evolved, and sun-pointing attitude control systems can now be included.

In comparison to many terrestrial markets, the space industry is a niche market even though it generates $100 billion in revenue worldwide.[16] Under construction are 100 satellites having an estimated value of $11 billion; another 175–200 new satellites will be required by 2010 to replace the current satellites or to augment service.[17] As transponder requirements grow (estimated requirement is 11,129 transponders by 2010 or more than a 78% increase from 2002), either more geosynchronous (GEO) satellites will be needed or even larger space systems will have to be built. Most of these transponders are designated to support HDTV, but other applications—such as store-and-forward communications or vehicle and container tracking—are being shifted to smaller LEO satellites, which have advantages over GEO satellites. For example, because LEO satellites are 25–40 dB closer, Earth stations can use omnidirectional handheld antennas. However, the primary limitation of LEO satellites is their intermittent service. This limitation can be surmounted by the deployment of large constellations. Several recent commercial attempts to provide nearly continuous coverage via constellations have been less than successful (e.g., Teledesic, Celestri, Ellipso, Globalstar, Odyssey, Iridium), and these failures have rippled through the industry.

Given that existing satellites are lasting longer than their predecessors and the overall demand forecast of nearly 200 satellites over the years 2001–2010 is nearly half the build capacity for the U.S. manufacturers alone, why would the industry consider a change in the manner by which satellites are manufactured?[18] The answer lies in the fact that the models used to make these forecasts do not adequately describe the present state. *Information has become a resource, and the value of a deliverable increases with the information content.* Satellites working in concert with localized ground sensors can provide far more information worldwide and with more immediacy than is currently possible using any other scheme. Concentrating on celestial rather than terrestrial applications, small satellites in large numbers can also provide a quicker overall view of our solar system and of Earth's space environment than is possible with the current one-vehicle-one-mission approach. The economist Paul Hawken[19] considers information synonymous with traditional mass-based resources and in a concise way has defined the *value-of-a-deliverable* as the ratio of the information-content it can hold to its physical mass.

The GPS constellation has proved that information content sells. For example, analyzing the total GPS market in North America, the aviation segment grew around 10%, the military and timing markets grew just under 25%, while the land (including marine) market grew 35%.[20] In the revenue scale, the land market makes up 62% of the total revenues.[20] Those less-than-successful commercial satellite constellation providers knew the value of information content; they failed to implement their vision because their approach to manufacturing was not sufficiently cost-effective.

Bringing revolutionary innovation to the satellite industry is inherently difficult because the financial barrier to entry is high. Using the space imaging market as an example, the entry cost estimates range from $97 million to $497 million.[21]

This includes the satellite, ground segment, launch, and insurance costs. Furthermore, because the industry is heavily regulated by government, it becomes somewhat immune to market forces. The issue should not seem hopeless; indeed, such government involvement may be the saving grace, as governmental sponsorship may be required to significantly alter the existing manufacturing approach.

Government sponsors should be interested in mass manufacturing of satellites for two reasons. First, by implementing mass manufacturing concepts, nonrecurring costs should be reduced. Second, once the infrastructure for mass manufacturing has been established, the cost to design and manufacture a new satellite should also be reduced. These reasons are mere representations of the economies of scale. Two more factors should be of interest to a government sponsor. First, with mass manufacturing, proven quality control methodologies can increase product reliability without incurring additional cost. Examples are easy to find (history of laptop computers, for instance) where a reduction in unit cost also yielded an increase in unit reliability. The key is to reach unit commodity numbers where economies of scale can be a factor and established quality control methodologies can be applied. The satellite industry, as it is constituted today, cannot hope to achieve such numbers *if the commodity unit is defined as the complete satellite*. However, if the commodity unit is defined as the *module or block* by which complete satellite systems can be assembled, then economies of scale can be achieved. A second factor that should interest government sponsors is that with the modular approach to satellite manufacturing, customization, adaptability, upgrading, and replacements are easier to implement with the natural consequence that there is "graceful" degradation of a system as opposed to abrupt and catastrophic loss.

The problem is not only with the satellite manufacturing industry but also with the launch industry. The cost of launch to orbit, after 50 years, still hovers around $15,000 per kilogram to polar orbit, and this cost is only realized by using a modest, not small, launcher (e.g., the U.S. Delta II-7920, ~3300 kg to polar orbit).[22]

In summary: Because of the large cost of manufacturing, launch, and operations, high reliability has become a paramount concern to enable a reasonable return on investment. Brooding over this issue eventually leads to complete risk aversion. This change in perspective has an ironic component in regard to innovation: As a venture becomes more efficient, it also tends toward more rigidity and can evolve toward a "frozen state" in which substantive innovations are not easily made.[23] Studies show that frozen-state industries are more likely to allow innovative approaches in good economic times but quickly shift toward efficiency mode when the economic outlook is weak or the return on the investment becomes precarious. To mitigate this dilemma and allow innovation back into the aerospace industry, several problems must be tackled simultaneously, and governments have to be involved. These include

- The cost per kilogram to orbit must be reduced by a factor of 2 or 3.
- Smaller launchers must be developed to allow lower total cost to orbit.

- Satellites must be designed as an assembly of mass-producible units.
- Very large space systems must be assembled in space using robotics.
- Modern design methodologies, such as design for manufacturability and concurrent engineering, must be promulgated throughout a space systems design and manufacturing phase.
- Risk must be managed rather than eliminated.
- The *total overall system cost* must be minimized, rather than the cost of each independent segment (i.e., satellite manufacturing, launch, support, end-of-life operations).

This chapter delineates the required basic criteria and the type of manufacturing technology that must be utilized to economically manufacture a multifunctional module or block.

24.2 Mass Manufacture, Mass Customization, and Build-to-Order

David Anderson, author of *Design for Manufacturability & Concurrent Engineering*, defines design for manufacturability as "the process of proactively developing products by optimizing all the manufacturing functions" and concurrent engineering as "a proactive practice of designing products to be built on standard processes or concurrently developing new processes while developing the product."[24]

Both design for manufacturability and concurrent engineering are procedures that need to be implemented prior to and during the design and manufacturing cycle. Furthermore, these procedures become more significant with the sobering fact that 80% of the total system cost (i.e., design, manufacture, operations, service, and repair) of a typical product *is committed by the end of the product design phase*, before a single deliverable unit has yet to be manufactured. Even more germane is Anderson's conclusion that nearly 60% of a product's cumulative lifetime cost is already committed at the concept/architecture phase, *before design*.[24] Therefore, there should be a strong impetus to get a design right before proceeding to the manufacturing phase. By the time manufacturing has commenced, 95% of the total cost has been committed. Any design changes at this stage is associated with a high-penalty cost.

One approach to ensuring the right design and system architecture is to bring to the table at the outset all the available knowledge, ideas, issues, and concerns, as well as resource allocation. In traditional approaches, the number of people involved is initially very small and continues to grow as the product development proceeds through design, prototype, and release. Various experts are brought in at various stages to deal with the design modifications, customer issues, and, after release, problem solving and servicing. An alternative but counterintuitive approach is to have a very large team at the outset that represents all segments of the product development phases, including the customer, vendors, and marketing. This large team is involved in the architectural design phase, and as the design progresses, is reduced until a core group remains that is responsible for the product release. The first approach initially appears to be more economical but *is actually found to be more* expensive when viewed from the perspective of the total system cost because changes made as a consequence of expert opinion cost more when applied at the later development stages. The second approach initially appears to be costly but actually provides a more reliable product with an overall reduction in system cost.

24.2.1 Methods of Manufacturing: an Overview

The methods of manufacturing changed in the 20th century to reduce total system cost. Mass production (i.e., the assembly line), popularized by Henry Ford and the Model T automobile, became the focus. Mass production is capital intensive, with a high proportion of machinery to workers. One characteristic, however, is product uniformity. A key element is that resource inventory must be maintained to sustain the production-line flow at demand levels: Ford wrote that for efficient and cost-effective operations, inventory should be carefully monitored to be just enough for immediate needs.[25]

In the 1950s, Toyota Corporation explored and then instituted a variant of this method known as "just-in-time" or "kanban" manufacturing. It is an inventory flow process that uses a series of signals that identify key transition points in the production process. These signals ensure that resources are controlled to maintain a balance of no overstock or understock. A subsequent refinement is now called Lean Manufacturing;[26] it is associated with the moniker Six-Sigma and includes a series of practices and techniques that were developed by Bill Smith of Motorola[27] in the 1980s, but were embraced and applied with success by Jack Welch at General Electric. One key element in Six-Sigma is the continuous effort to reduce process variation by rigorous measurement and the systematic removal of manufacturing defects. The industries where these manufacturing methods have been successful are those that deal with commodities, where the unit production quantities are on the order of thousands to millions. It has not fared as well in industries with a high level of human interaction (i.e., retail), where the production quantities are lower and where some level of customization is considered a feature.

Various other manufacturing methods have also been developed to contain cost and increase quality. One common term is agile manufacturing, which refers to an organization that has established flexible processes that enable it to respond quickly to customer needs and market changes. For example, a machine or tool would be adapted to produce a range of parts or conduct a series of manufacturing processes based on a common platform. Another example is the utilization of several machines that can be altered to perform the same operation and thereby absorb rapid increases in demand.

Rapid Manufacturing is a derivative whereby the sequential delivery of either energy or material to specific places enables the fabrication of a part. The technique is usually ascribed to material-additive fabrication (e.g., inkjet printing of "parts") but can also be utilized, in certain cases, to a material-subtractive process.

The key element that differentiates Rapid Manufacturing from other approaches is that it is controlled by a computer and commonly utilizes a mathematical model of the relevant piece part, which is created by another computer. The technique falls under the general heading of CAD/CAM (computer-assisted design and computer-assisted manufacturing). Rapid Manufacturing is inherently a serial process, but if essential elements can be done in parallel (i.e., batch production) it provides a large advantage in speed and cost while offering customization.

One consequence of Rapid Manufacturing is the ease of product customization. Under certain circumstances, flexible manufacturing systems can offer an approach to mass customization, defined as "the technologies and systems to deliver goods and services that meet individual customer's needs with near mass production efficiency."[28] Mass customization should not be confused with personalization. Whereas personalization, with its desire to fulfill the customer specifications down to the letter, is an intensive customer-provider process that can work outside the solution space, mass customization works within a fixed and predefined solution space of stable processes to offer a range of customizations. Mass customization has been successfully applied in numerous industries. Examples are in apparel (e.g. Land's End clothing) and in consumer electronics (Dell Computers).

Satellite manufacturing has traditionally adhered to the approach known as build-to-order. Products are only built after a confirmed order is received. It's the oldest style of production and is considered the most appropriate for highly customized or low-volume products. The advantage of build-to-order is the ability to manufacture a product to exact customer specifications and therefore achieve personalization. The disadvantage is that fluctuations in demand can lead to underutilization of manufacturing capacity.

Most governments view space systems manufacturing as a strategic asset, and they have a strong motivation to maintain and support manufacturing capabilities in times of reduced demand. At this point, market forces no longer apply, and manufacturing methods that should be altered are given extended life. The practice leads to an overall added cost to systems procurement, but the additional cost only appears on the government's balance sheet. Measuring quality and reliability may also be more costly. Build-to-order manufacturing personalizes a product; therefore, because there may not be sufficient unit volumes, statistical methods to measure quality may not be possible. Consequently, the manufacturer has to initiate quality control by costly testing of individual components.

24.2.2 Insights Regarding New Manufacturing Approaches for Space Systems Development

To what extent can these management philosophies, manufacturing systems, practices, and techniques be applied to the satellite manufacturing industry, which serves both governments (with emphasis on reliability over innovation) and civilians (with emphasis on innovation over reliability)? Several insights may provide some clues:

- Product uniformity increases reliability.
- High reliability and cost-effective production requires a concerted effort to reduce process variation in manufacturing outputs.

A new approach to the manufacturing of space systems should strive for product uniformity. But how is this to be done in a statistically significant manner when production volumes are small? Furthermore, one wonders how much of satellite manufacturing today still involves component modifications or local "fixes" throughout the manufacturing phases to realize a customer specification. These local fixes or changes normally require innovative solutions, but these solutions may not have the maturity to be labeled as a stable process for reliability verification.

- Manufacturers and government sponsors are more apt to pay for innovative solutions, but are less likely to support process development efforts that transform the innovative solution into a stable manufacturing process.
- If increasing reliability and reducing costs become ends in themselves, they can become counterproductive to innovation.[29]

It appears that manufacturing can survive on a series of innovative solutions that then make up a patchwork frame of operations instead of defining and developing a viable solution space of stable operations. However, manufacturing via a series of innovative solutions can run counter to processes that are designed to increase reliability and reduce cost. Therefore, innovation must be separated from reliability and cost reduction. Reliability and cost-reduction must be applied only after an innovative solution is converted by investigative characterization into a stable process.

- One must first be able to define and measure a variable in a manufacturing process before it can be controlled.
- Human interaction reduces reliability, whereas computer control increases it.
- Innovation is necessary for product development and enhancement, but it must be documented, and the solution must be represented by the commensurate development of a proven process.

An all-robotic manufacturing line controlled by a network of computers with established processes would progressively lead to better product reliability. But innovation, which is a distinctly human process that enables revolutionary changes, cannot haphazardly be inserted into the manufacturing process. Innovations must either be inserted at the beginning (in the design/architecture development phase) or if it is a concurrent action, only after a reliable, stable process based on that innovation has been realized and configured for computer control.

- Product uniformity enables mass production and a stable market.

- Defining a *solution space* of stable processes with a limited number of permutations of functionality enables mass customization.
- Mass customization engenders customer loyalty because customers do not pay the full premium for customization.

Given that a major aspect of the satellite industry is driven by customer specification, it is more likely that customization is the norm rather than the instance. The trend is to standardize a satellite bus and limit customization to payloads that can be mounted on the bus. This approach will not realize the production efficiency values commonly associated with mass production because the unit production volumes of the standardized systems are still small. An alternative approach might be to explore a solution whereby the standardized unit is a multifunctional module (MFM), and satellites are developed as an assembly of modules. The actual number of modules would depend on the satellite size and mission. If this conceptual framework were to be made a reality, then defining the manufacturing solution space would entail deriving the stable processes that enable the reliable development of the MFM. Because the MFM would be the common elemental unit in all satellites, a large number of them would have to be manufactured. Mass manufacturing efficiencies would result from the large production volumes. For example, a distributed sparse-aperture array would need hundreds to thousands of units, and a large satellite may need 10,000 units. Therefore, a yearly production quantity close to 100,000 is conceivable. Mass customization would be determined by how many MFMs could be conjoined and the ensuing functional permutations that could be derived. Given these large production quantities, the quality level per unit could be enhanced.

- To increase the quality level of a manufacturing line, piece-part count in a product must be minimized.
- The design philosophy must encompass the total system.

Product quality can be determined by the part quality raised to the exponent of the number of parts in the product. For example, if an MFM is composed of an assembly of the same components, then the product quality value can mathematically be represented as $Q_s = (Q_c)^n$, where Q_s represents the quality of the total system, Q_c represents the quality of the component and n is the number of components. Assume the quality value for a component, Q_c, is 0.993 (i.e., 99.3% of units pass) and a 100 such components form a system, then the overall system quality factor is ~0.50 (i.e., 50% of system units pass). The obvious conclusion is that to establish a manufacturing process with a desired quality value, the quality values of the fabrication techniques must inherently be much higher.

The goal is then to reduce the number of processing steps that can come about by reducing the materials list, integrating subsystems onto a common platform, or designing subsystems in a manner that permits multiple functions to be derived from the same unit. The first notion realizes an advantage in space applications, where a reduction of the materials list should also reduce problems resulting from differences in thermal expansion. The second notion has the basic advantage of a lower number of interfaces. In the extreme level of integration, the entirety of the device components and functionalities are somehow integrated on a common substrate. In the microelectronics lexicon, it is called wafer-scale integration. Therefore, a substrate material that offers some multiplicity in the types of functions it can support (e.g., structural, electronic, and photonic) would lend itself to the development of a more reliable system because more integration would be possible. Therefore, the choice of the base substrate material and the availability and precision of the processing tools for that material class should be a key element in designing the MFM.

- A serial fabrication or manufacturing process is not as economical as a batch process.
- However, a serial manufacturing process does add flexibility (in essence, customization can be added at a unit level).
- One approach to gaining an agile manufacturing capability economically is to integrate a batch fabrication process into the serial manufacturing process. In this case, customization is applied during the serial process, but it is during the batch process where the effect of the customization is realized.

The microelectronics fabrication industry has proved the value and power of combining lithography and batch chemical etching. During the lithography step, only the pattern *information* is transferred. Material removal or modification is actually accomplished by chemical etching, ion implantation, thermal processing, and so on, which operates in batch mode. The combination permits some flexibility in the manufacturing (i.e., different patterns, masks). The big advantage is the cost savings in that multiple devices can be manufactured at once with very high reliability.

- The Internet presents a continuous web of connectivity.
- Software "travels" and can engage multiple stations simultaneously worldwide.
- Manufacturing facilities and controllers need not be at the same location.

Computers and the Internet have profoundly changed how business will be done.[30] Instead of a manufacturing center where all the tools are collocated, distributed minifactories or processing centers can exist in widely different locations and can conduct near-synchronous operations to provide a virtual manufacturing system. An example is the Modular Integrated Robotized System developed by Pirelli Tires Corporation.[31] A more subtle variation is the MEMS Exchange,[32] a service consisting of hundreds of foundry processes that are physically distributed throughout the United States. In cases where multiple processes must be done to a piece part, the part is shipped from location to location to accommodate the desired processes. It is not inconceivable that MFMs could be manufactured in a similar manner, either using the distributed factory or distributed foundry concepts. The advantage to having distributed operations is the reduced startup costs and the possible utilization of excess capacity in manufacturing foundries. The disadvantage is that the manufacturing solution space representing stable processes must also include reliable processes that seal or encapsulate units for transfer or shipping.

24.3 The MFM and Microelectronics Fabrication, Glass Ceramics, 3-D Patterning

MFM is envisioned as having functions that are implemented by the addition or integration of electronics and photonics (e.g., intelligence, communications), a primary-function device that is aided by a sensor or actuator, and a secondary-function device that can be placed into action in the event of failure of a neighboring MFM. The envisioned MFM also includes ancillary subsystems such as propulsion, energy capture and storage, and attitude control. The energy needed to conduct its primary and secondary functions would be generated by photoconversion and stored in chemical batteries. The MFM has nearly all the attributes of a current satellite, but the extent of its functional capabilities is limited. Units are not designed to operate alone but in concert with others. They must therefore be mass-producible. When they are physically connected, information and power can flow among them as necessary. In more advanced versions, fuel or propellant can also be routed to specific locations. In the simplest example, the MFM is a block that serves as scaffolding with drawers to hold instrument trays. The concept of trays and drawers is already used in the microsatellite industry, and the concept of small blocks, each with space system functionality, is suggestive of CubeSats and picosats. Even though these approaches can result in the building of reliable space systems in modular fashion, it is not what we mean when we speak of a mass producible MFM.

As envisioned for the *generation-after-next* spacecraft, the MFM is intended to be an active element working in concert with a large number of MFMs. In various sizes and configurations, the ensembles provide different space system capabilities. The metamorphic robot technology described above can provide a viable precedent toward this distributed space systems vision. Within a SuperBot, for example, each module is "self-sufficient and is autonomous with regard to the use of its resources. For example the use of its sensors and actuators."[14] Consequently, a group of these units becomes reconfigurable, and there have been demonstrations of various walking, sliding, and rolling "gaits" implemented by autonomous SuperBots. A recent interesting technological development is the application of a biologically inspired method to establish control and induce self-organization among a module swarm.[33] In this development, a digital algorithm mimics a simple hormone response found in most biological systems, and this is used to control the tasking and execution of program steps among a robot module swarm. The consequence is a significant simplification in the control system and a reduction in the communication bandwidth necessary to conduct an operation.

To enable the development of a reconfigurable space system, it may be useful to explore the manufacturing aspects of an MFM that is first and foremost designed for manufacturability and, second, exists as an integrated system in which the structural material can be used for additional functions beyond that of structural support. The manufacturing and fabrication tools for the MFM must have precision and tolerance over a wide dimensional range because the unit will have macroscopic (1–10 cm), mesoscopic (0.1–1 cm), and microscopic (0.001–0.1 cm) features.

What characteristics should be included in a late-generation MFM? It must have a series of standard attributes and a few series of functional attributes. Some examples of these attributes are given below.

Standard Attributes

- Can be passive or active but can be bypassed.
- Has the innate ability to physically connect with other units.
- Has local intelligence to govern health.
- Can be governed by a master controller (i.e., subordinated to master/slave relationship).
- Can serve as an information router.
- Can serve as a power router.
- Can serve as fluidic router.
- Has sensors to judge local awareness.
- Can share resources with nearby neighbors.
- Can generate some power locally (sun/solar cells).
- When nearing the end of its life, it transfers local information to neighbors, reduces its connections, and decouples from neighbors to enable replacement.

Functional Attributes

- Has particular main task (i.e., primary "payload") that can be performed either as a solitary unit or via concerted effort with others.
- Majority of its resources are focused on main task.

At the conceptual level, the manufacturing approach for the envisioned MFM must at least draw insights from modern manufacturing. Recapitulating, it is clear that product uniformity increases reliability, which dictates a form of mass production. However, it is customization that engenders customer loyalty, and because customization is more akin to the requirements of the satellite industry, mass customization may in fact be the viable manufacturing model. The overall product quality can be increased by either reducing the number of piece parts or fabrication processes. Alternatively, product quality can also be increased through systems integration, or, in the extreme, fabricating integrated components and devices on a common substrate. For a given stable manufacturing process, automation increases reliability, given that the fabrication techniques that are applied have high quality factors. Finally, the microelectronics industry shows that lithography, patterning, and chemical etching can surpass traditional machining approaches both in terms of fabrication quality and cost. Therefore, in laying out the conceptual framework of an MFM manufacturing line, microelectronics manufacturing is a good place to start.

24.3.1 Microelectronics, Metals, and Plastic Fabrication Technologies

Microelectronics fabrication is based primarily on silicon and silicon-dioxide materials. Even though the technology is mature for fabricating electronics and reasonably mature in fabricating actuators, it is not clear that silicon is ideal as a substrate structural material for the envisioned MFMs. First, the MFMs will mostly be used in Earth orbit—and given that the estimated mass of an MFM is between 1–10 kg, large temperature fluctuations is a concern, at least for LEO missions. There are, of course, approaches for encapsulating temperature-sensitive components, and these are currently employed in CubeSats and other small space systems. Still, a conceptual analysis should consider the major effects of the space environment on the chosen structural substrate material. Bulk silicon has high thermal conductivity (149 W/mK),[34] which could exacerbate the rate of cooling and heating in LEO missions. Furthermore, many of the useful physical properties in bulk silicon that have been applied for making sensors (e.g., Hall effect, Seebeck effect, and piezoresistance) will have substantial temperature sensitivity. These temperature-dependence effects are typically compensated via engineering the design (e.g., Wheatstone bridge circuits). This is clearly possible if necessary. But is it necessary?

From the perspective of fabrication of structures and systems integration, there are other features of bulk silicon that make it less attractive as a substrate for the envisioned MFM.

- Lithographic patterning in bulk silicon is based on applying a patterned protective layer (e.g., photoresist, oxide, nitride). It would be preferable if the material itself were photosensitive (i.e., photo patternable), and, therefore, there would be no need for applying a protective layer photoresist application step.
- The principal dimension of the MFM is going to be related to the availability of bulk crystalline silicon boule sizes (currently ~ 300 mm diameter). Consequently, the MFM shape is governed to some extent by the maximum diameter of a boule.
- Because bulk silicon is crystalline by nature, the ability to alter macroscopic shapes smoothly by molding is constrained.

- Bulk silicon processing is based on 2-D mask lithography; therefore, the structures that can nominally be fashioned by chemical etching (e.g., cavities, tunnels, beams) appear as extruded shapes of the lithography mask. It would be preferable to have a material where full 3-D machining or excavating and structuring is possible.
- To fabricate deep undercut structures in bulk silicon, chemical etching anisotropy among the crystalline planes is used.[35,36] For example a <100> cut wafer and a pattern mask can undercut a structure leaving the <111> plane side walls at 54.7 deg relative to the wafer surface.[35] These inward sloping walls produce a lot of unused real estate on the surface. It would be more useful if undercut structures could be fabricated without needing wafers with specific crystalline cuts or needing to carefully align masks to the crystallographic axis.
- To fabricate deep wells with near vertical walls by chemical etching in bulk silicon, <110> wafers have to be used. An alternative to chemical etching is to use a deep reactive ion etching (DRIE) tool (1–5 μm/min rates). It would be preferable to rely on standard chemical etching since it is the least costly approach.
- There is an advantage to the integration of systems if the substrate material for the MFM is amenable to implementation of a photonic bus. Silicon becomes optically transparent in the infrared (1–2 μm wavelengths), and therefore, the insertion of a visible to near-infrared photonic bus becomes less practical.

Notwithstanding these issues, bulk silicon does have advantages as a structural material for manufacturing MFMs. These are the maturity of the technology with regard to quality and reliability and the level of detail in material and process characterization that has been conducted worldwide. Furthermore, the transitioning of a microelectronics foundry to an MFM manufacturing line can be done economically if the approach uses outmoded or prior-generation silicon processing tools.

Aside from silicon, other candidate structural materials are metals, plastics, and glass-ceramics, and in each of these material classes, established tooling and manufacturing technology exists. Of these three classes, the metals are the least viable from the perspective of integrating systems on the "wafer-scale." Outside of the desirable mechanical properties (high strength, toughness, noncatastrophic failure, high degree of plasticity) and its ability to withstand space radiation, metals integrated with other materials (glasses, semiconductors, plastics) to produce multifunctional properties is still an area of active basic research. Therefore, an MFM design using metal as the base structural material will likely yield a surface or scaffolding upon which other units are physically attached, which is very close to the current methods used in satellite production. However, for the MFM, there will be a need for high-precision 3-D processing. In this regard, the metals have an advantage over silicon materials. It is now possible to fabricate precise 3-D structures in metal via a number of material additive processes (e.g., EFAB,[37] selective laser sintering,[38] microstereolithography[39]) and subtractive processes (EDM[40]). Another approach is via LIGA (from the German for lithography, electroforming, and molding). Using LIGA and injection molding technology, high-precision metal components can be manufactured, but this manufacturing process is not amenable to mass production because the fabrication and assembly processes outside of the LIGA step are strictly serial and expensive. Furthermore, LIGA imparts the added cost of the x-ray lithography.

As a material class, plastics and composites have made significant inroads in the past few decades, often replacing metal. High-resolution injection molding and casting, extrusion, and laser direct-write polymerization have allowed cost-effective fabrication of precision plastic components in 3-D. If cost were the only driver, fabrication with plastics would be hard to beat. Plastics also offer the significant advantage of chemical engineering to modify the material properties and make them suitable for the application. Dense polymers have been developed for armaments (e.g., the Glock polymer-frame gun), and conducting polymers have also been developed[41] along with light-guiding polymers for photonic applications.[43,44]

However, for space applications, the current formulations of plastics have drawbacks when considered as a structural material. That is not to say that a space-qualified plastic cannot be developed; but the general properties of plastics are that over long periods, they are *not* dimensionally stable with the temperature fluctuations encountered in LEO missions. The coefficient of thermal expansion for common non-reinforced plastics is ~25–50 times that of silicon, ~8–20 times that of metal, and ~6–16 times that of glass. For application as a structural material, dimensional stability is an important factor, as all the subsystems will be connected to the base material. Second, in comparison with metals, glass, and silicon, the plastics have a lower temperature of operation. Consequently, plastics would not be the material of choice in high-temperature areas (e.g., combustion chambers and propulsion nozzles). Third, plastics have the potential for outgassing in vacuum, which raises the issue of contamination of nearby optical structures (although this problem can be mitigated by design and by reinforcing the plastic with additives). Fourth, plastics have low yield-strength values: ~2% that of silicon and some metals (e.g., tungsten), and ~13% that of quartz.[44] This issue would be a factor when MFMs are physically interconnected and the compounding forces generated by neighboring attached units are exerted. Finally, plastics having carbon-carbon bonds tend to show degradation over time, especially in LEO (i.e., in the vicinity of the 7 eV atomic oxygen) and in sun illumination (i.e., UV photochemistry). Historically, the major barrier to space application of plastics has been the poor properties as a radiation barrier for electronics packaging.[45] However, with additives it is possible to tailor the properties of plastics over a wide range. For example by adding fluorocarbon fibers or glass fibers and whiskers, dimensional stability and material-creep issues can be minimized, and the material can be strengthened and toughened.

The polymer material class is nominally processed either by conventional machining, molding, casting, and extrusion. However, in the past 20 years, laser stereolithography, a technique by which polymeric fluids harden under light excitation, has matured and now offers true 3-D fabrication of polymer structures with near micron resolution.[46] Laser stereolithography, albeit with its serial fabrication process and its limitation in manufacturing throughput, still offers advantages from the manufacturing perspective; the key is to develop a polymeric material that has ultimate properties useful in space applications.

24.3.2 Glass Ceramics: a New "Old" Material for Space

Glass ceramics are a unique class of materials that combine the special properties of sintered ceramics with the characteristics of glass.[47] They are manufactured in the amorphous homogeneous glass state and can be transformed to a composite material via heat treatment. The treatment provides the controlled nucleation and crystallization of the constituents. Since their invention nearly 60 years ago, glass ceramics have been used in a wide range of scientific, industrial, and commercial applications. They are particularly well suited for aerospace engineering, biotechnology, and photonics due to their attractive chemical and physical properties (e.g., no porosity, optical transparency, high-temperature stability, limited shrinkage, corrosion resistance, and biocompatibility).[48] A key advantage is that the constituents that form the glass can be varied, altering the chemical and physical properties of the material. The manufacturing technology is sufficiently mature to allow the variation of select physical properties that can be derived from the controlled growth of crystalline matter within an amorphous glass. These include modifications to the material strength, density, thermal conductivity, maximum temperature of operation, electrical insulation and RF transmission, coloring, transparency in the optical wavelengths, and susceptibility to chemical etching. Consequently, the material has found wide applications. Furthermore, many new exciting applications of glass-ceramic systems are yet possible with the likelihood of incorporating "intelligence" in the form of micro- and opto-electronics.

Glass ceramics can be classified as a phase-change material, and it has been the focus of worldwide research—for example, as insulating materials in the development of optical switches[49] and metallic films in the development of shape-memory alloys.[44] Glass ceramics have more in common with the so-called ceramic-matrix composites than with monolithic ceramics (e.g., silicon nitride, silicon carbide). Ceramic-matrix composites include continuous or discontinuous reinforcing materials that reduce the inherent brittleness typical of monolithic ceramics. In the specific class of glass ceramics, the strength reinforcement is derived from phase transformations that occur during thermal treatment. However, these reinforcements affect the physical property globally. In a global transformative process, where everything changes everywhere and to the same degree, the ability to integrate various devices or structures on the same substrate is limited. This is because each device or structure will derive its functionality from the local physical property of the substrate, and there can be little variation of the properties in a globally transformed system. Consequently, the usefulness of glass ceramic as a substrate becomes limiting for the envisioned MFM, where one may want to simultaneously integrate various functionalities (e.g., enhanced IR optical transmittance, reduced RF transmittance, increased electrical conductivity) on the same substrate.

24.3.3 Photosensitive Glass Ceramics: Glass Ceramics that can be Microstructured for Increased Functionality

There is a subclass of the traditional glass-ceramic family known as the *photo-structurable* or *photosensitive* glass ceramics (PSGC), in which the material phase transformation may not necessarily be global but can be locally controlled.[50,51] Consequently, the issues that limit glass ceramics with regard to integration of multiple functions are somewhat mitigated. The PSGCs differ from glass ceramics by the addition of photosensitive compounds to the base matrix. The physical phase transformation still takes effect during the thermal treatment step, but *only in the exposed regions*. The material operates much like photoresist except that the patterned material can be controllably transformed into different material phases, each with unique mechanical, thermal, and chemical properties. In one common PSGC composition, the possible phases are a low-temperature (~600°C) semiceramic phase that is soluble in dilute hydrofluoric acid, a high-temperature (~800°C) compatible semiceramic phase that is not soluble in hydrofluoric acid, and an amorphous glass state that is optically transparent and only slightly soluble in hydrofluoric acid.

The three primary processing steps are depicted in Fig. 24.1. Step 1 is the exposure process—fundamentally, a photo-ionization event that generates a trapped electron in the glass. The exposure step has far more versatility when conducted with a pulsed laser in a direct-write mode, as opposed to traditional lithography using a continuous light source. With a pulsed laser, the incident laser light can be focused to a particular spot, and the amount of irradiance within the focal volume can be maintained just above that required for state transformation.[52] This allows embedded patterns to be made, thereby inducing embedded property changes.[53] Step 2 entails a thermal treatment. It begins with the chemical reduction of a constituent metallic ion species by the photo-induced trapped electron. The result is neutral metal atoms. This chemical reduction process that constitutes the "fixing" of the exposure pattern is the fundamental difference between the photostructurable and regular glass ceramics. Subsequently and during the thermal treatment stage, the enhanced mobility of the neutral metal atom induces the formation of metal clusters and then metal nanoparticles. When the particle size has reached ~8 nm for some common PSGC compositions, it can serve as a nucleating site for the growth of a crystalline phase. Step 3 in the figure depicts the third processing stage, and the processing particulars of this stage strongly depend on the thermal treatment protocol chosen in step 2. The equation for the chemical reduction is also depicted in step 3 for one common PSGC composition. If the maximum temperature during the thermal treatment stage reaches higher than 750°C, the resulting

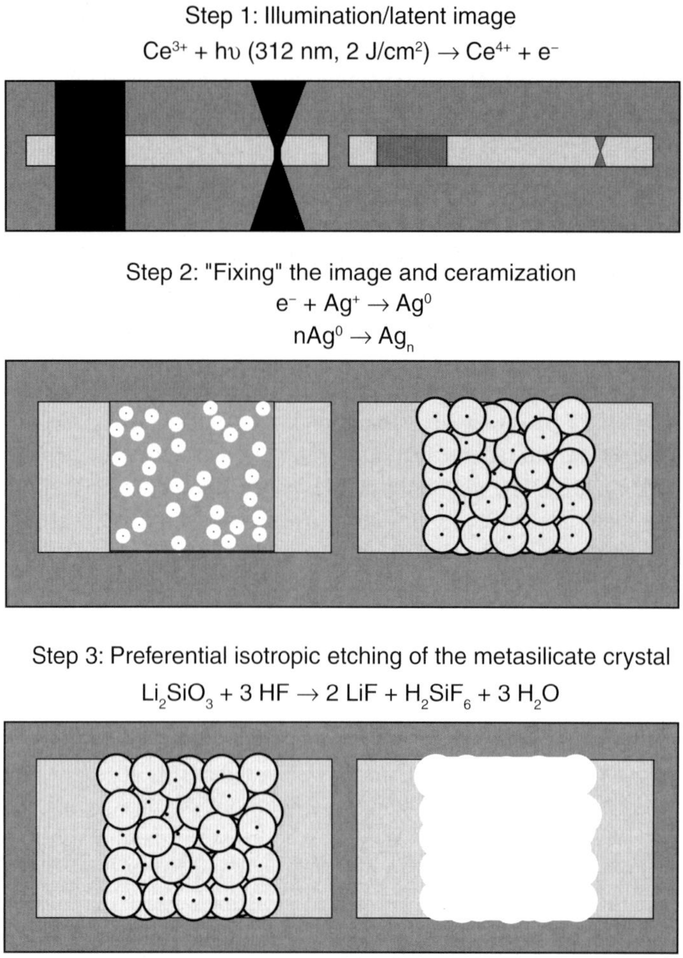

Fig. 24.1. Primary processing steps for PSGC materials. Step 1 describes the first stage and the exposure process. Step 2 entails a thermal treatment. In step 3 the processing particulars strongly depend on the thermal treatment protocol chosen in step 2.

patterned material is transformed to a high-temperature glass-ceramic phase. This material is described as an opal—not optically transparent and not chemically soluble in hydrofluoric acid. However, if the maximum temperature in step 2 is less than 750°C or more commonly at ~600°C, then a low-temperature glass-ceramic phase grows in the exposed regions. This material is not optically transparent but is chemically soluble in hydrofluoric acid, with an etch contrast that approaches 50:1 with respect to an unexposed area. However, if the maximum temperature reached in step 2 is ~500°C, then the exposed regions contain trapped metallic nanoparticles that show optical absorption in the UV (< ~300 nm), and under certain exposure conditions it is possible to imbue the patterned area with color.[50]

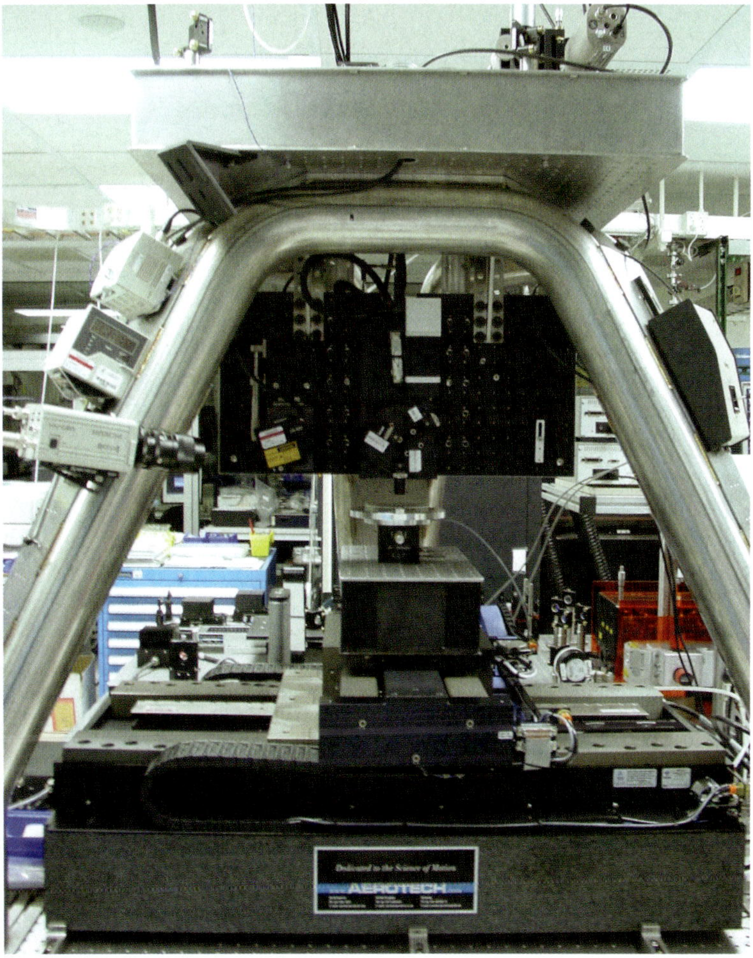

Fig. 24.2. Maskless laser direct write patterning tool. The fabrication process is fully automated from design to manufacturing.

24.3.4 Three-Dimensional Volumetric Exposure Patterning

The physical property changes as described can be patterned by a laser direct-write tool in a process known as laser volumetric exposure processing (LVEP). Figure 24.2 shows a maskless *x-y-z* laser direct-write patterning tool that can pattern in 3-D over a 0.5 × 0.5 m area with velocities approaching 0.5 m/sec and overall accuracy of ±0.75 μm (repeatability ±0.1 μm). Figure 24.3 shows a true 3-D microfabricated structure that has been exposed, baked, and etched.

A series of recent experiments has determined that some of the physical property changes depend on the irradiance of the photo-exposure tool.[54] With a laser direct-write patterning tool, the delivered photon flux can be controllably

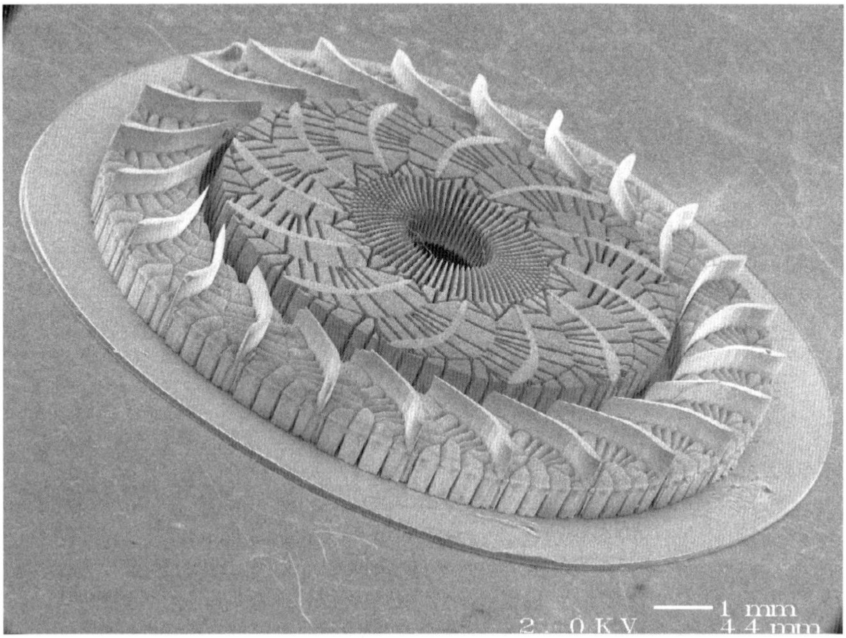

Fig. 24.3. Scanning electron microscope (SEM) image of a true 3-D microfabricated double turbine structure that has been exposed, baked, and released. The turbine is a centimeter across with tapered vanes that get as thin as 30 μm. The design is made to look as if it were fabricated by traditional "cut-glass" processes.

altered commensurate with the 3-D tool path pattern.[55] Consequently, the physical property changes can have gradation, which can be used to advantage in making devices or structures on the same material substrate. The physical and chemical properties that currently can be varied over a limited range by controlling the optical excitation include optical transmission in the IR,[56] material strength/compliance,[54] optical index,[57] and chemical solubility in hydrofluoric acid.[58]

Figure 24.4 shows an optical microscope photograph of target patterns showing the gradation of optical transmission that can be patterned. The data below the images show the measured optical transmission near IR telecom wavelengths as a function of the incident irradiance during the exposure step. This dependence allows for metering the optical transmission in the material locally. Figure 24.5 also shows an optical photograph of a variegated height structure that has been patterned and chemically etched by just metering the laser irradiance to control locally the chemical etch rate (i.e., solubility in hydrofluoric acid). No masks were used in the exposure or in the chemical etch processing stages. The measured data in Fig. 24.6 show the change in chemical etch contrast (as defined by the ratio of chemical etch rates of an exposed region to that of the unexposed region) as a function of laser irradiance. These data show an etch contrast of 30:1. However, more recent results show that a 50:1 etch contrast is possible. The

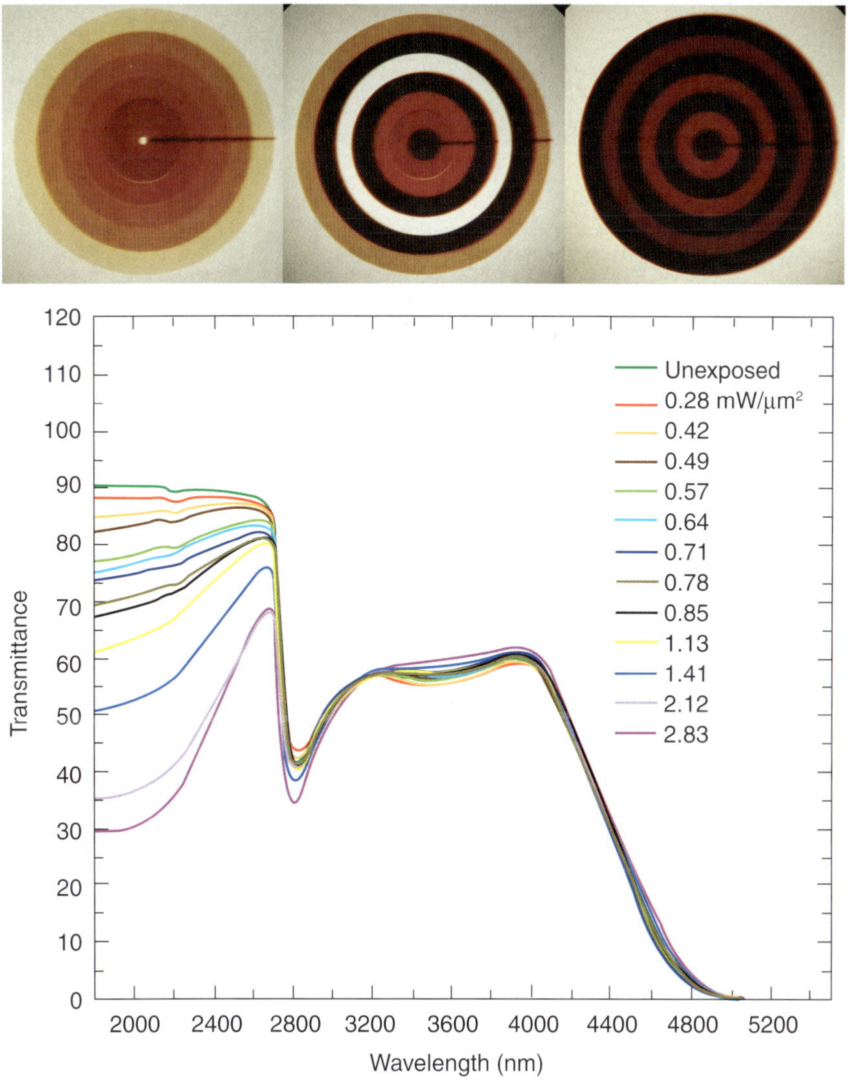

Fig. 24.4. Optical photograph of three optical microscope images of target patterns with data showing the measured optical transmission near the IR telecom wavelengths for the structure shown in upper left.

data also show a threshold irradiance that is defined by an etch contrast of 1:1. Clearly, below this value no discernible exposure results in any enhancement of chemical etching in comparison to the unexposed material. Using this threshold irradiance value to advantage, it is possible to pattern embedded fluidic channels using a laser direct-write tool. Figure 24.7 shows two scanning electron microscope pictures of embedded channels that have been patterned, phase-changed

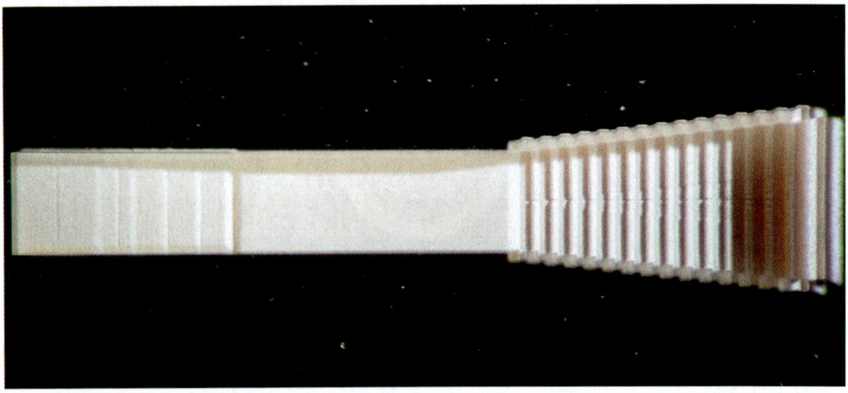

Fig. 24.5. Optical photograph of a variegated height structure patterned and chemically etched by metering the laser irradiance. The structure is a step-down waveguide at THz frequencies. Not clearly observable but on the right are a series of slotted and appropriately spaced curved baffles.

by baking, and subsequently etched (note: the process requires vias to transport the chemical etchant to the embedded structure). The exposure process was conducted from above the surface through the material; a focused laser beam was used to provide a high flux only within a small, localized volume around the beam waist. The laser irradiance was set during the patterning step to exceed the exposure threshold only within the embedded volume. Figure 24.7 (left) shows two large-opening reservoirs that are connected by a tunnel (shown with a human hair threaded through). The tunnel is roughly 100 × 100 μm wide and approximately 500 μm long and is etched within a 1 mm thick PSGC wafer. In Fig. 24.7 right, 10 stacked embedded channels are shown also within a 1 mm thick PSGC wafer. These stacked channels were not patterned at a wafer edge, as the photograph can falsely imply; they were patterned away from an edge and through the material in a series of stacked exposures. Connected to the stacked exposures was a pattern called a chemical etch line that itself etched through the material and subsequently provided a chemical transport path to the stacked channels (i.e., the vias for chemical transport). By proper choice of the laser wavelength and the type of pulsed laser, it is possible to confidently pattern structures very near the surface or very deep within a solid (<30 μm deep for 193 nm wavelength, <500 μm deep for 266 nm wavelength, <5 mm deep for 355 nm wavelength, many cm deep for 800 nm wavelength).

An important aspect for space systems is the ability to pattern mechanical compliance where necessary. This is less of an issue for on-orbit operations but becomes paramount during the launch phase. It is expected that the MFM will contain sensors and devices that could be susceptible to vibration and shock. It is possible to vary, over a limited range, the mechanical stiffness of a PSGC substrate. This property arises as a result of the *in situ* growth of the crystals, which, if not etched away can make the overall structure stiffer. Because the density and

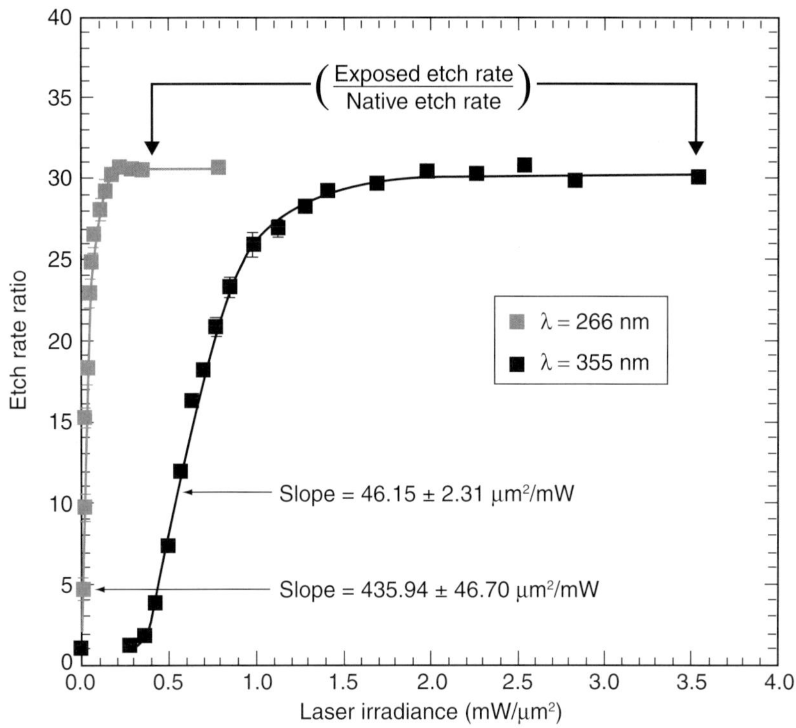

Fig. 24.6. Change in chemical etch contrast as a function of laser irradiance at two laser exposure wavelengths.

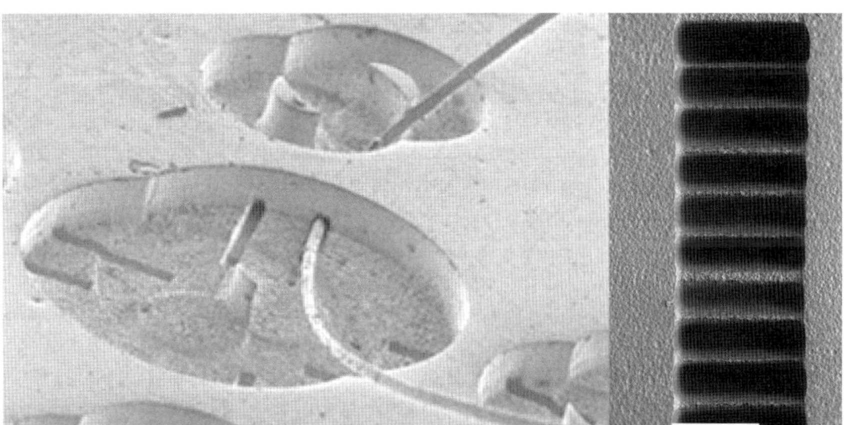

Fig. 24.7. Two scanning electron microscope pictures of embedded channels patterned then phase changed and subsequently etched out. The left image shows two large opening reservoirs connected by a tunnel (shown with a human hair threaded through to denote the path). The right image shows 10 stacked embedded channels also within a 1 mm thick PSGC wafer.

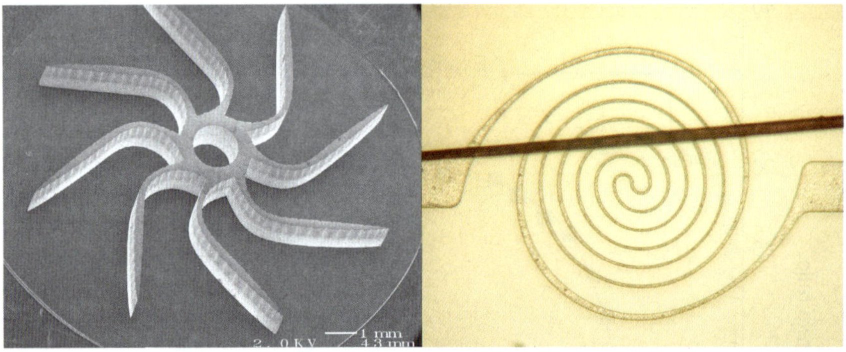

Fig. 24.8. Two optical microscope photographs of mechanically compliant structures fabricated from PSGC material. The image on the left shows a turbine blade that has been patterned and released from a wafer. The turbine blade vanes are roughly 155 μm wide and nearly 1 mm tall. The image on the right shows a common hotplate shape, spring fabricated and released, overlaid with a human hair for size comparison. The measured width at the uniform sections is approximately 10 μm.

size of the grown crystals can be controlled to some extent (i.e., via exposure and thermal treatment), mechanical properties can also be controlled. Figure 24.8 shows two images of mechanically compliant structures fabricated from PSGC. Figure 24.8 (left) shows a scanning electron microscope photo of a turbine blade that has been patterned and released from a wafer. The turbine blade vanes are roughly 155 μm wide (thickest point) and nearly 1 mm tall. Using an air jet, the turbine was spun to 180,000 rpm without failure (the experiment was stopped at this rpm). The turbine survives the initial start and acceleration because the material composition along the length of the blade varies, stiffer at some sections (i.e., more crystalline matter) and more ductile at others (i.e., less crystalline matter). Figure 24.8 (right) shows an optical microscope photograph of a common coil-shaped spring fabricated and released that is overlaid with a human hair for scale. The measured width at the uniform sections is around 10 μm. Other flat piston springs have also been fabricated for providing local compliance to an adjoining mechanically sensitive structure.

With the ability to pattern—either by traditional lithography or with higher versatility by laser direct writing—micro- to macro-scale modification of material property changes in 2-D and 3-D becomes possible. The three-step process of controlled optical exposure, thermal treatment, and chemical etching enables the processing of material at more than square-meter scales with near micron resolution. Furthermore, with the current understanding of the photophysical process in PSGCs, there is the hope that additional material properties can similarly be controlled. This would be possible if the base composition is slightly altered during PSGC manufacturing to engender a family of materials that have various functionalities that can be patterned through photo exposure.

24.3.5 PSGC Material Attributes, Space Exposure, and Comparison with Other Fabrication Approaches

There have been at least three commercially prepared PSGCs in the past 30 years, all based on a similar composition. They are: Fotoform, manufactured by the Corning Glass Corporation in Corning, New York, USA; PEG3, synthesized by the Hoya Corporation of Tokyo, Japan; and Foturan, manufactured by the Schott Corporation of Mainz, Germany. Foturan is the only PSGC that is commercially available today. However, extensive data supports the conclusion that other compositional varieties are possible.[59,47] A brief summary of pertinent material characteristics and processing advantages is given below.

- PSGC can be molded/shaped before patterning (e.g., antenna and optical mirror dish shapes can easily be molded and then patterned).
- PSGC can be manufactured in any size or shape and then patterned.
- A PSGC substrate has low thermal conductivity; therefore, it is less affected by on-orbit temperature fluctuations in comparison to aluminum and silicon (at 20°C: Foturan ~ 2W/mK; aluminum 250 W/mK; bulk silicon ~149 W/mK).
- PSGC has a moderate coefficient of thermal expansion (at 20–300 K: Foturan 8.6×10^{-6} to 10.5×10^{-6} depending on material phase; aluminum 25×10^{-6}; bulk silicon 2.9×10^{-6}).
- PSGC is lighter than most metals (Foturan 2.4 g/cm^3; Al-2017 2.8 g/cm^3).
- PSGC can be made tough (Knoop hardness: Foturan 530; nickel 560).
- The strength of PSGC can be tailored (Foturan modulus of rupture: 60–150 MPa).
- PSGC has a stiffness-to-weight ratio, E/ρ, (Foturan $E/\rho = 35$) that is better than that of steel (ASTM-A36 $E/\rho = 25$), aluminum (Al-1100-H14 $E/\rho = 26$), and is 42% that of silicon ($E/\rho = 83$).
- A PSGC substrate is an electrical insulator with a resistance of 8.1×10^{12} to 5.6×10^{16} (Ω-cm at 25°C) depending on material phase.
- PSGC has higher emissivity (Foturan ~0.96) than bulk silicon (~0.64), glass (~0.95), and anodized aluminum (0.77).
- PSGC has better RF transparency than metals.
- The visible and IR transparency of PSGC can be locally controlled (~0% to 100%).
- PSGC does not outgas in vacuum (as opposed to sol-gel–based glasses).
- Compositional variation of the base material can act as a radiation shield for gamma and x rays (contains cerium).[60–62]
- Zero porosity in ceramic state allows containment of high-pressure gas.

- Four material processing procedures exist for metallization, allowing for patterned high- and low-current lines, embedded lines, and a thin "film" covering on the outside to mitigate space charging effects.

The effects of LEO space environment on Foturan has been examined as part of a Materials International Space Station Experiment (MISSE). Both exposed and unexposed samples together with photostructured samples were flown in a 3-year mission that tested the effects of impinging atomic oxygen (ram direction) and UV light (sun direction). The returned samples show no major changes. Material tests are ongoing.

To assess the feasibility of mass production or mass customization, the LVEP technique for PSGC materials must be compared with other materials processing techniques. Table 24.1 shows a comparison of LVEP with conventional machining and three microfabrication techniques (bulk silicon micromachining, surface silicon micromachining, and LIGA). A few points are worth noting.

First, all the techniques are amenable to automation, which is a necessary requirement for achieving mass throughput efficiencies while maintaining component reliability. For example, the LVEP technique, along with the associated tooling, can be configured for fully electronic CAD/CAM compliance. The laser direct-write tooling shown in Fig. 24.2 operates under full CAD/CAM control.

LVEP can process material with nearly the same feature size, device thickness, and relative tolerance as LIGA and bulk silicon micromachining. Surface micromachining can fabricate a smaller feature size with a commensurate reduction in the range of device thicknesses. As for conventional machining (e.g., milling), it would be too costly and impractical to achieve the small feature sizes that are possible by all the other techniques; however, the relative tolerance of conventional machining supersedes all other techniques because it is possible to process very large linear dimensions with ultrafine diamond milling.

With regard to fabricating structures beyond the base material, LIGA and LVEP share an advantage over bulk and surface micromachining in that they can fabricate precision components in other materials via electroplating or serving as an injection mold tool. Conventional machining accommodates the largest range of materials because it is a direct subtractive process using abrasive force. LVEP shares with conventional machining and surface micromachining the attribute of scalability to processing large areas.

Whereas microelectronics can easily be integrated with surface and bulk silicon micromachining techniques, for LVEP and the current class of available PSGCs, only thin-film electronics integration is possible. However, with some technology development, PSGCs can perhaps be made compatible with low-temperature cofired ceramic (LTCC). LTCC is the electronics and RF substrate that has superseded printed circuit-board technology; it offers the advantage of 3-D electronics packaging on a ceramic substrate. With technology development, there is the hope that the coefficient of thermal expansion of PSGC (e.g., Foturan ~8.6×10^{-6} at 300°C) can be reduced to match that of LTCC, currently designed to match

Table 24.1. Comparison of LVEP with Conventional Machining and Three Microfabrication Techniques Evolved from Microelectronics and MEMS Technologies

Processing Attributes	LIGA	Bulk Silicon Micromachining	Surface Micromachining	Conventional Machining	Volumetric Exposure Processing of PSGC
Amenable to automation	Yes	Yes	Yes	Yes	Yes
Feature size	~3.5 µm	~3.5 µm	~1 µm	~10–25 µm	~3.5 µm
Achievable device thickness	>1 mm	>1 mm	< 100 µm	>>1 mm	>1 mm
Lateral dimension	>2 mm	>2 mm	2 mm	>10 m	>100 mm
Relative tolerance	$\sim 10^{-2}$	$\sim 10^{-2}$	$\sim 10^{-1}$	$\sim 10^{-3}$	$\sim 10^{-2}$
Materials	Electroplated metals, injection-molded plastics	Limited material suite	Limited material suite	Large material suite	Glass, ceramics, electro-plated metals, injection molded plastics
Scalability	Limited	Limited	Yes	Yes	Yes
Microelectronics integration	No	Yes for SOI bulk process	No	No	Yes for thin-film deposition
Photonics integration	Yes	Limited	Limited	Limited	Yes
Device geometry	2-D high aspect ratio	2-D high aspect ration	2-D	3-D high aspect ratio	3-D high aspect ratio
Processing methodology	Parallel at wafer level	Parallel at water level	Parallel at water level	Serial processing	Serial and parallel merged
Material property control via patterning	No	No	No	No	Yes

silicon (~3.4×10^{-6} at 300°C). If this were feasible, the result would allow direct wafer-scale anodic bonding of LTCC embedded electronics onto PSGC. There is a viable but less elegant solution that mitigates the bonding problem: It is the use of an interlayer that has intermediate thermal properties and can support the mismatch in the thermal expansion. Regardless of how electronics are integrated into PSGCs, the LTCC approach offers a significant advantage because of its maturity and wide use. The LVEP technique has an advantage when it comes to integrating photonics. Leaving the integration of the photonics source aside for the moment, the control and guiding of light within PSGC has already been accomplished.[63] Using a variant of the LVEP technique, one can create 3-D optical waveguides within the PSGC glass. Furthermore, optical reflecting surfaces integral to the PSGC wafer have been fabricated.[64] Figure 24.4 shows that the optical transmission in the IR can be controllably varied. These results (along with other data not presented[65]) suggest that it should be possible to pattern optical filters integral to the device and thereby enable intrasatellite communication via wavelength-division multiplexing. The integration of a photonic source is problematic for the PSGC materials and the LVEP technique. Automated pick-and-place solutions that use a variant of LVEP are being explored,[66] but the most obvious solution is to integrate the photonic source as a surface device in LTCC. Subsequently, the emitted light would then be coupled and guided within the PSGC.

Table 24.1 also shows that the LVEP technique, like conventional machining, can fashion structures in 3-D with high aspect ratio (~50:1). It also shares with conventional machining the ability to fabricate components without the need of designing, manufacturing, and testing a lithography mask set. In conventional machining, maskless operation leads to a slow serially driven process that can be time consuming and therefore costly—especially when fabricating small feature sizes. Unlike conventional machining, in LVEP, the serial patterning segment can be done quickly (at m/sec velocities with galvanometer motion) because material is not being removed; only the pattern information is being written. The material transformation and material removal steps are done in a parallel or batch process. Serially driven processes have the advantage of not requiring a mask, which means that pattern customization can easily be implemented. For example, in a serial process, it is possible to implement multiple processes concurrently, and the fabrication of structures with variable height in true 3-D can easily be accomplished.

On the other hand, batch or parallel processes have the advantage of process and component uniformity over large areas and consequently low cost when numerous units are fabricated at once. In the LVEP technique, a serial and a batch process are merged with the consequence that advantages from both survive. First, by replacing the mask and lithography procedure with a serially driven direct-write variable-exposure procedure, a sequence of processing steps is immediately removed from the overall set. Figure 24.9 shows a schematic comparing a mask lithography process with that of LVEP, where the laser irradiance is dynamically

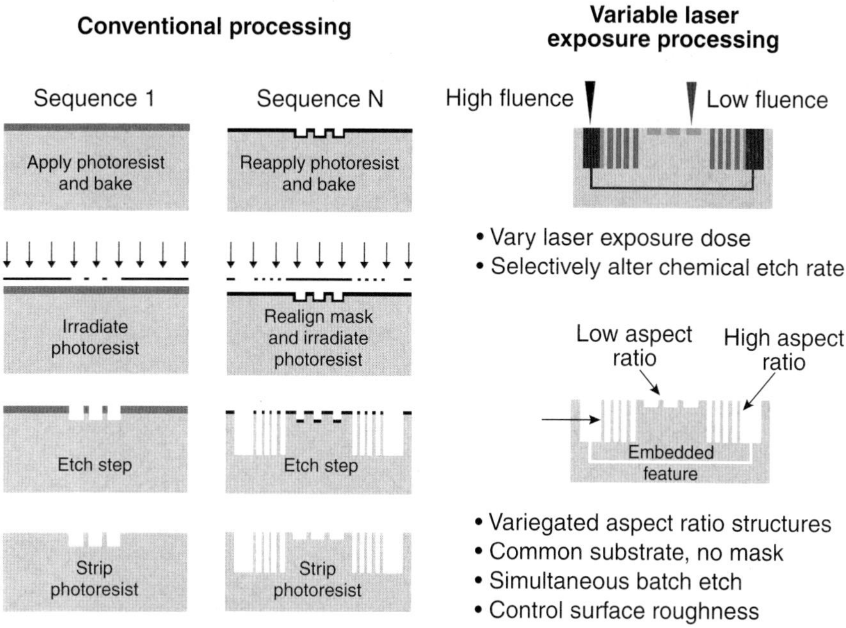

Fig. 24.9. Schematic comparing a mask lithography process with that of LVEP where the laser irradiance is dynamically varied during the patterning to match the desired property changes.

varied during the patterning to match the desired property changes. The example shown is the fabrication of a shallow well next to a deep well, which requires two exposure masking steps. There are five basic processing steps using the conventional lithography mode (i.e., apply the resist, bake, perform the lithography, etch, and strip the resist) that have to be repeated for each specific etch depth (i.e., mask). Using the LVEP technique, there are only three (i.e., apply all patterns with variable exposure, bake, conduct a single timed etch to fabricate all features). Seven processing steps are saved in this simple example that uses just two masks. With increasing number of masks, N, the number of sequential steps in the conventional mode increases as $5N$, while that for the LVEP remains at three.

A reduction in processing steps is usually commensurate with increasing fabrication quality. Figure 24.10 shows an optical microscope image of a microfabricated structure that would require four masks (i.e., 20 processing steps) using conventional lithography. A cross section of the fabricated part is shown in the adjacent panel in the figure. The structure is a four-level structure with the deepest feature being 600 μm. Plotted on top of the cross-section profile that depicts the design (i.e., calculated etch depth profile) is the measured profile. The variation between the measured and design is less than 10% in this prototype and can be further improved with additional processing controls.

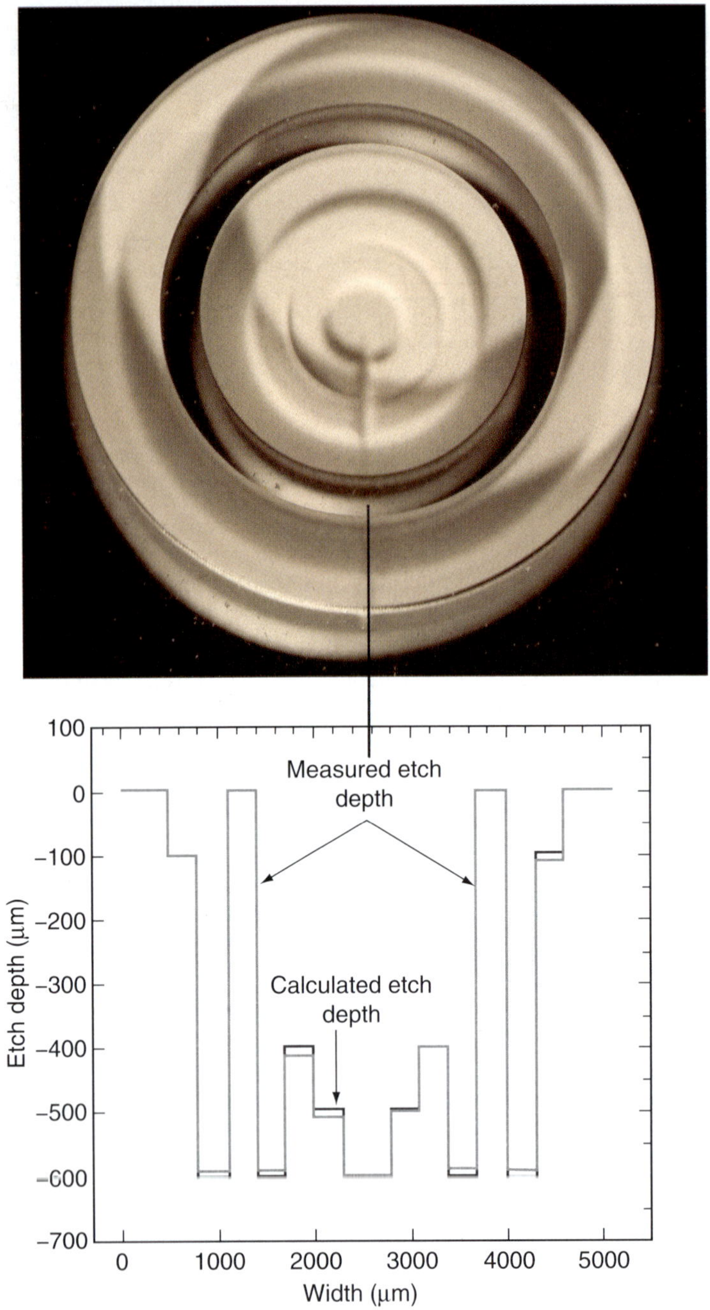

Fig. 24.10. Optical microscope image of a microfabricated structure that would require four masks (i.e., 20 processing steps) to fabricate using conventional lithography (top). A cross section of the part with measured and calculated design depths (bottom).

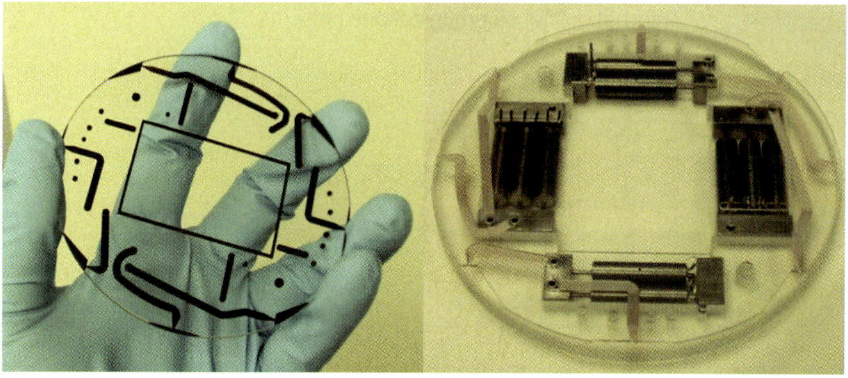

Fig. 24.11. Optical photograph of 10 cm diameter PSGC wafer patterned with a complex sequence of both 2-D and 3-D microstructures. The wafer is an integrated system representing a subsystem of a glass/ceramic micropropulsion prototype unit for a 1 kg nanosatellite. The left photograph shows the exposed and baked wafer before etching. The pattern is of the fluidic lines and the eight thruster nozzles. The right photograph shows two processed wafers integrated with the microvalves.

Finally, as shown in Table 24.1, LVEP is the only technique that offers the ability to vary the substrate material property during the exposure patterning step. This is primarily a consequence of the PSGC material itself. However, these material property changes can only be articulated if there is precision in the delivery of the photons with respect to irradiance (photons/time-area), flux (photons/time; time is per laser pulse), and fluence (energy/area) at the specified volume element (i.e., voxel). A control scheme for pulsed laser direct-write processing has been developed that can deliver scripted photon irradiance, flux, and fluence to each predefined voxel during patterning.[67] The delivery of this exposure prescript can be done at patterning speeds approaching 0.5 m/sec. Figure 24.2 shows such a tool located at The Aerospace Corporation. In Fig. 24.11 is an optical photograph of a 10 cm diameter PSGC wafer that has been patterned with a complex sequence of both 2-D and 3-D microstructures. The wafer is an integrated system and represents a subsystem of a glass-ceramic micropropulsion prototype unit for a 1 kg nanosatellite. Details of the propulsion prototype unit and its function are described in chapter 18.

24.4 Thoughts on Development of a Mass-Customized MFM Foundry and Nanosatellites in PSGC

Commercially produced PSGCs and their variants can be implemented in a diverse array of new applications in which the material acts as the host substrate or support structure, and the desired "instrument" is fabricated within it or on top of it. Therefore, instruments or devices can be fabricated by direct patterning and local alteration of the material. Alternatively, the PSGC can be a substrate such as that in a multichip module (MCM) and allow the direct attachment of microelectronics,

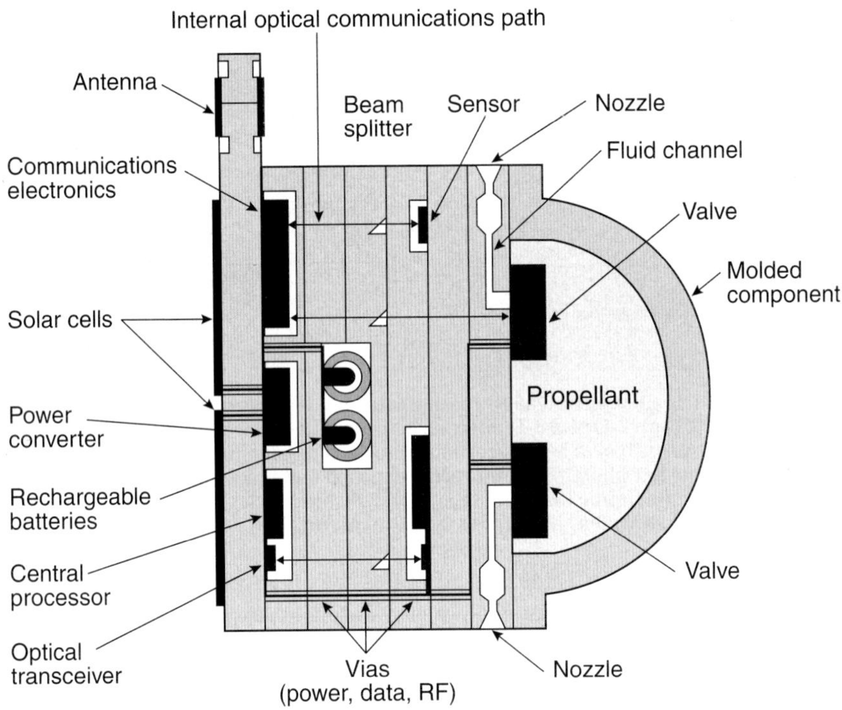

Fig. 24.12. Schematic of a possible primarily glass ceramic satellite that implements many of the capabilities described in the prior section.

photonics, fluidic MEMS, micro-optoelectronic systems (MOEMS), and high-frequency RF communication systems. In this latter application, the PSGC is used as a multipurpose substrate that goes beyond the capabilities of current substrate materials (e.g., silicon, plastics, and even LTCC). The consequence is the potential to fabricate complex systems such as MFMs. An assembly of MFMs would subsequently be a complete small mass-producible satellite that is constructed almost entirely from glass ceramic, with added advantage that a photonics-based intrasystem communication bus can also be implemented, obviating the need for some electronic vias.

Figure 24.12 shows a schematic of a possible nanosatellite that is produced primarily of glass ceramic. The figure is intended to illustrate many of the capabilities described in the prior section. It shows a molded component shaped to form the propellant tank. A molded component can also be shaped with negative curvature to become an antenna or, with better surface figure control and smoothness, an optical focusing mirror. A nozzle is shown being fed by an embedded fluidic channel that goes to the propellant tank. This channel can also be routed through a number of stacked "wafers," as in the case of the glass-ceramic schematic prototype described in chapter 18 of this book. Three kinds of vias are also depicted in Fig. 24.12. First, there is a high-power bus—a low-gauge metallized via and data lines that could be manufactured by various metallization schemes described above. Second, there is an RF bus, which could either be an embedded hollow and metalized waveguide or an index-guided RF waveguide. Third, there is an internal optical communication bus that is depicted by a series of line-of-sight systems comprising an optical transceiver chip and a sensor. The line-of-sight optical paths also show beam splitters integrally fabricated to direct and distribute the optical signal. What is not depicted in Fig. 24.12 but relevant to an internal optical bus is the integral fabrication of cutoff filters and the fact that the optical paths need not always run in the line of sight. The inclusion of optical filters allows for wavelength division multiplexing. Using pulsed laser compaction techniques, single-mode optical waveguides can be integrally patterned in 3-D to connect any two points. Also not depicted are integrally fabricated microlenses that focus solar power onto a solar cell or a CCD chip in the case of a camera.

A first-order attempt to design a manufacturing line for MFM production would utilize many of the concepts already in use in mass production. Some of these concepts are present in the software used in foundry design.[68] Furthermore, the first-order design will also have to borrow procedures and approaches currently in use in select small-satellite manufacturing centers where mass customization concepts are being implemented. Because an MFM can essentially become a satellite subsystem, the mission definition and design phase—the basis of most satellite development methodologies—has less relevance. An interesting approach being developed by the Air Force Research Laboratory (AFRL) is the satellite data model.[69] In this model, satellite devices and applications are developed separately; when there is a need for satellite components appropriate to a particular application or mission, they are selected and added to the "system." One goal is to reduce nonrecurring costs in the design, engineering, and manufacturing phases.[70] From the manufacturing side, reducing nonrecurring cost essentially means minimizing assembly steps and reducing manufacturing techniques. With regard to assembly, automation should be the aim. However, this does not necessarily argue for robotic pick-and-place machines; ultrasonic consolidation techniques[71] with self-interlocking microstructures should also be explored.

Modeling and simulation tools should be used throughout the design phase, in conjunction with cost-estimating software. Included in the simulation "toolbox" is software for systems integration and its impact on the manufacturing process development. However, it is imperative that the modeling and simulation software have a factual basis that is taken from both ground and space experiments. The current technology demonstrations being conducted with small satellites (e.g., CubeSats, picosatellites, nanosatellites, microsatellites) are key to ensuring the validity of these future simulation tools. The design of the MFM and the subsequent family of derivable satellites would be the goal, addressing mission requirements at the satellite system level. The mission requirements then flow down to the MFM design, as a customization procedure. Furthermore, if the design software (e.g., solid modeling) were to be coupled with the space systems orbital dynamics tools, it could help couple the mission needs to the requirements that are to be levied on the MFM. The modeling and simulation software industry is following a trend in which the software goes beyond CAD/CAM and animation and includes engineering and physics (e.g., thermal modeling, mechanics) and could include space environment specifics (e.g., available power, radiation level).

The design of the MFMs themselves must implement a modular approach where each segment is a separable submodule that can be tested individually using an adopted universal interconnection scheme. To make this design approach feasible, intelligence by way of processors is locally embedded, and computation is distributed to enable plug-and-play assembly and disassembly.[72] If the onboard computation is distributed, intrinsic in the design is a level of fault tolerance that can be implemented because the MFM can be "operated or tested" via multiple control routing schemes (techniques toward realizing this approach via wireless communications can be found in chapter 19). In subsequent design generations, if the range over which control, via distributed computation protocols, is increasingly expanded, then it could serve as the basis for connecting multiple MFMs to form larger and more complex systems that are designed to work in concert. Ultimately, this process is realized in a functional satellite. Finally, within this distributed control "fabric," neural network analysis software could be used to monitor system trends and highlight functional areas that are out of the nominal range.[73] A satellite nervous system is then essentially born.

The MFM manufacturing line will be similar to a microelectronics foundry—in essence, a series of automated cluster tools, pick-and-place machines, and diagnostics instruments in a cleanroom. For a PSGC-based MFM manufacturing line, an abbreviated list of the major processing stations are given below with the understanding that these stations may include multiple processes and that quality-control testing is conducted at each one.

- Shaping and molding layers/wafers for more mission-specific applications.
- Patterning in 2-D and 3-D (RF conduits, optical conduits, fluidic conduits, MEMS sensors and actuators, mechanical devices).
- Batch chemical baking.
- Batch chemical etching.
- Patterning for thin-film electronics.
- Electronics deposition.
- Segment qualification.
- Segment assembly with LTCC electronics.
- Segment packaging/bonding.
- Plug-and-play segment testing.
- MFM as an assembly of plug-and-play segments.

- MFM testing.
- MFM vibrational testing.
- MFM thermal vacuum testing.

24.5 Steps Forward

To realize the mass customization and manufacture of an MFM, several initial steps must be undertaken, starting with a more extensive engineering analysis along with the development of a cost-modeling tool that marshals a rational business case. Government space-policy experts, along with the national laboratories and the major satellite manufacturers, are in a position to evaluate the concept from the perspective of both engineering viability and profitability. Expert opinion should also be drawn from producers of both microelectronics and consumer electronics. This insight will help clarify what is in fact possible with regard to modularity, packaging, automated assembly, and distributed manufacturing. Following the results of this interdisciplinary team analysis, national governments should sponsor small space research centers to explore and demonstrate aspects of the mass-production concepts.

A key element in the proposed concept is the use of glass and ceramic materials as the primary structural material replacing metal; PSGC is the material of choice because it enables lithographic patterning. With lithography comes the consequent batch fabrication technologies that have enabled the microelectronics industry to increase capability without losing reliability. Still, a fundamental problem with PSGC must initially be addressed: Even though the materials have been available for nearly 60 years, issues of nonuniform photosensitivity remain. Given what the material properties of bulk silicon were at the infancy of the microelectronics industry compared with what is now available, it's likely that the glass-ceramics industry will be able to solve this problem. However, far better than refining the composition and uniformity of the existing PSGC materials would be to develop a material specification and requirements list for a class of PSGC that would truly be suitable for the development of an MFM (e.g., rad-hard variants of current PSGC). An argument can then be made to establish a PSGC materials manufacturing line that develops materials with specific and tailored requirements. The estimated required quantities of material will be measured on the order of 10–100 tons per year. In materials manufacturing, there is an inherent cost in terms of the energy expended to mine, refine, and produce a select material. This energy factor (MJ/kg) roughly defines the associated environmental cost for a given material and can be used to compare various MFM structural candidates. Figure 24.13 shows this comparison for a select number of materials.[74] The PSGCs are manufactured as a glass as opposed to a glass-ceramic composite. Silicon carbide is included as a representative monolithic crystalline material. The data show that glass (~25 MJ/kg) followed by the glass-ceramic composites (~100 MJ/kg) require the lowest amount of energy to produce.

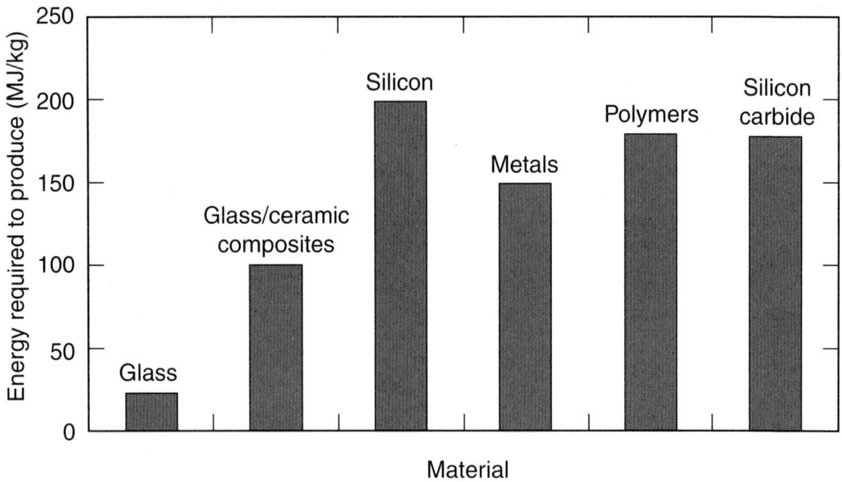

Fig. 24.13. Environmental energy cost of material production

The LVEP approach with PSGC holds advantages that can aid in the design and development of integrated MFMs. Though research-grade fabrication tools do exist, these are far from true manufacturing-quality tools. However, there are no real technological barriers to developing a manufacturing tool that uses lasers and processing techniques as described above. For example, the manufacturing setup shown in Fig. 24.2 incorporates lasers (Spectra-Physics OEM Models J40-BL6-266Q and J40-BL6-355Q) used in other product manufacturing lines, so their reliability has been established. The motion systems are taken directly from the microelectronics manufacturing industry, and the CAD/CAM software tools used are taken from the conventional machining industry. Following the initial exposure step, the processes (thermal treatment and chemical etching) are directly acquired from the microelectronics industry. Regardless of where these technologies originate, a recommendation would be to develop a manufacturing-class LVEP tool that can demonstrate the necessary speed and piece-part fabrication throughput.

The development of a pilot manufacturing plant is not considered a first step. However, following the interdisciplinary team overview, it would be prudent to begin appraising processing capabilities worldwide where subsystems of the MFM can be designed and manufactured for assembly or bonding at a central facility.

An approach toward developing a distributed and agile pilot plant is to enable all the physical machines to "communicate" via a common software interface for example, the Common Object Request Broker Architecture (CORBA).[75]

The MFM should be developed in stages. Picosatellites, nanosatellites, and CubeSats can serve a role in this development process. Low-cost space testing can be conducted using these small space systems, and as each subsequent MFM generation grows in complexity, so will these small spacecraft grow in capability. In time, there will be a transition from each small spacecraft initially serving as a bus for the MFM package to the spacecraft evolving into an entity that is merely an assembly of MFMs. One positive consequence of this approach is that these new generations of small spacecraft will be mass producible, and large satellites can begin carrying the mass-producible small satellites either as assistants or components of complex distributed missions.

Current producers of small satellites and various research groups worldwide are developing technologies intended to make future space systems more reliable and easier to integrate. Many of these technologies are presented in the pages of this book. Some of them could find direct application to the MFM concept—for example, communication protocols that enhance plug-and-play operation, or a generic and reconfigurable system-on-a-chip to handle onboard and local computation. Other developing technologies that would have to be adapted to the MFM concept include, for example, miniature propulsion and attitude sensors. Finally, some technologies must be developed that would be specific to the MFM and the potential assembly of larger structures in space. Examples include:

- Development of a universal microconnection scheme that can couple fluidics, power, and photonics from one MFM to another.
- Development of proximity sensors that would enable self-assembly in space.
- High-precision propulsion for self assembly only.
- An integral unit that generates power for local and distributed use.
- Use of celestial x-ray sources for attitude sensing. This would allow each MFM with a modified star tracker to have an autonomous sense of location.[76]

24.6 Conclusions

Changing the way we use space can entail more than simply replacing a single large satellite with a counterpart that is just more integrated. On the contrary, satellites as an assembly of mass-producible multifunctional units will have a far more pervasive impact on the aerospace industry by increasing reliability, reducing cost, and enabling new missions that are desired but not possible today. This chapter briefly outlines the idea of mass production of satellites. It describes one approach toward realizing the goal of economies of scale—defining the production unit not as the satellite but as an element of the satellite, the "multifunctional module." The MFM provides the basis for designing and building satellites in a novel manner and permits the adoption of manufacturing procedures and practices that can increase overall reliability. This proposed concept is a far-term vision.

This chapter goes beyond the proposition and suggests the properties that an MFM should have and how it should be developed to acquire the benefits that the microengineeering and microelectronics industries have enjoyed. Presented is the use of glass-ceramics materials, specifically the family of photostructurable glass ceramic, as a candidate structural material and as a substrate for integrating the fluidic, RF, electrical, and possibly photonic systems. Glass and ceramics have an advantage over silicon because they can be manufactured in large quantities and can be macroscopically shaped using cost-effective molding and casting technology. In addition, these materials can further be processed with near micron precision using lithography and batch chemical etching techniques.

It should be possible to leverage the vast resources and technologies of both the glass-ceramics and microelectronics industries and apply them to the development and mass manufacture of MFMs, thereby enabling development of a satellite as an assembly of mass-processed modules. One outcome of this approach is that future spacecraft could have any nominal size and shape.

24.7 Acknowledgments

The author acknowledges The Aerospace Corporation and The Aerospace Press for the opportunity to write this chapter. The author also acknowledges Dr. Howard Schlossberg, Program Manager in the Air Force Office of Scientific Research (AFOSR), who has provided continuous sponsorship of the fundamental investigations and the characterization of the PSGC using laser exposure techniques. The Aerospace Corporation's Independent Research and Development Program is also acknowledged for providing funding to demonstrate the capabilities of the laser material processing technique. Furthermore, the Department of Energy's Jefferson Laboratory is thanked for sponsoring the development of the LVEP tool. These investigations would not have been possible without the help and discussions with Siegfried Janson, Frank Livingston, William Hansen, Ernest Robinson, and Lee Steffeney of The Aerospace Corporation. Finally, the author acknowledges the late Seymour Feuerstein of The Aerospace Corporation, whose belief in the power of miniaturization helped to promote the microengineering of aerospace systems.

24.8 References

[1] A. Tennyson, "The Two Voices" (1842).

2 C. Y. Baldwin and K. B. Clark, "Managing in an Age of Modularity," *Harvard Business Review*, **75** (5), 84–93 (Sept.–Oct. 1997).

3 C. Y. Baldwin and K. B. Clark, *Design Rules: The Power of Modularity* (MIT Press, 2000).

[4] D. E. Whitney, Working paper, ESD-WP-2003-01.03-ESD, Massachusetts Institute of Technology, 2004 (unpublished).

[5] K. Hölttä, E. S. Suh, and O. deWeck, "Trade-off between Modularity and Performance for Engineered Systems and Products," *ICED 2005: The 15th International Conference on Engineering Design* (Melbourne, Australia, 15–18 Aug. 2005).

[6] D. E. Whitney, Working paper, ESD-WP-2003-01.03-ESD, Massachusetts Institute of Technology, 2004 (unpublished).

[7] S. Flagg, T. Bleier, C. Dunson, J. Doering, L. DeMartini, P. Clarke, L. Franklin, J. Seelbach, J. Flagg, M. Klenk, and V. Safradin, "Using Nanosats (e.g., QuakeSat) as a Proof of Concept for Space Science Missions," *18th Annual/USU Conference on Small Satellites* (AIAA, Logan, UT, 2004), Vol. SSC04-IX-4.

[8] S. W. Janson, in *Micro- and Nanotechnology for Space Systems: An Initial Evaluation*, edited by H. Helvajian and E. Y. Robinson (The Aerospace Press, El Segundo, 1993); Aerospace Report ATR-93(8349)-1, Second printing AIAA press 1997, pp. 138; S. W. Janson, *30th AIAA/ASME/SAE/ASEE Joint Propulsion Conference* (AIAA, Washington DC, Indianapolis, IN, 1994), Vol. AIAA-94-2998, pp. 14; S. W. Janson, in *Microengineering Technology for Space Systems*, edited by H. Helvajian (The Aerospace Press, El Segundo, 1995); Aerospace Report ATR-95(8168)-2, Second printing AIAA 1997, pp. 309; S. Janson, H. Helvajian, and E. Y. Robinson, *44th Congress of the International Astronautics Federation* (IAF, Graz, Austria, 1993), Vol. IAF-93-U.5.573, p. 5.

[9] iFixit, *iPhone Disassembled* (IFixit, Atascadero, CA, 2007).

[10] M. Sweeting, "25 Years of Space at Surrey–Pioneering Modern Microsatellites" (IAA, Melbourne, Australia, 1998).

[11] Rapid Spacecraft Development Office, http://rsdo.gsfc.nasa.gov (23 July 2008).

[12] B. M. Cook, C. Paul, and H. Walker, "SpaceWire Plug and Play, An Early Implementation and Lessons Learned," *2007 Conference and Exhibit* (AIAA, Rohnert Park, CA, 2007), Vol. AIAA 2007-2930, p. 6.

[13] M. Sweating, A. D.-S. Curiel, and W. Sun, "The 'Personal Computer' Revolution in Space," *20th Annual Small Satellite Conference* (AIAA, Ogden, UT, 2006).

[14] A. Castano, W-M. Shen, P. Will, "CONRO: Towards Deployable Robots with Inter-Robots Metamorphic Capabilities," *Autonomous Robots*, **8**, 309–324 (2000).

[15] S. O. Kennedy Jr, T. Mosher, and Q. Young, "The Rise and Fall of the Capital Asset—An Investigation into the Aerospace Industry Dynamics and Emerging Small Satellite Missions," *20th Annual Small Satellite Conference* (AIAA, Logan, UT, 2006), Vol. SSC06-I-5.

[16] ISBC, *State of the Space Industry* (ISBC, Bethesda, MD, 2005).

[17] C. Blalock, S. Busby, W. Chapman, M. Evans, and E. Helton, "Space Industry Study Industrial College of the Armed Forces National Defense University," Report No. A682524 (Ft. Lesley J. McNair, Washington, DC, 2002).

[18] J. H. Saleh, "Will We See a Regional or Global Duopoly in the Satellite Manufacturing Industry? The European Perspective," *Space Policy,* **21** (4), 277–285 (Nov. 2005).

[19] P. Hawken, *The Next Economy* (Holt, Rinehart, and Winston, 1983).

[20] Futron Corporation, Industrial College of the Armed Forces (28 March 2002).

[21] Y. A. Dehqanzada and A. M. Florini, *Secrets for Sale: How Commercial Satellite Imagery Will Change the World* (Washington, DC: Carnegie Endowment for International Peace, 2000).

[22] Committee on Earth Studies Space Studies Board Commission on Physical Sciences Mathematics Applications, *The Role of Small Satellites in NASA and NOAA Earth Observation Programs* (National Academy Press, Washington, DC, 2000).

[23] W. J. Abernathy and J. M. Utterback, "Patterns of Industrial Innovation," *Technology Review*, **80**, 40–47 (June–July 1978).

[24] D. M. Anderson, *Design for Manufacturability & Concurrent Engineering* (CIM Press, 2004).

[25] H. Ford, *My Life and Work* (Doubleday, Page & Company, New York, 1923).

[26] J. P. Womack, D. T. Jones, and D. Roos, *The Machine That Changed The World. The Story of Lean Production* (Simon & Schuster Ltd., 1990).

[27] P. S. Pande, R. P. Neuman, and R. R. Cavanagh, *The Six Sigma Way: How GE, Motorola, and Other Top Companies are Honing their Performance* (McGraw-Hill Professional, 2000).

[28] M. M. Tseng and J. Jiao, *Mass Customization* (New York: Wiley, 2001).

[29] B. Hindo, "At 3M, A Struggle Between Efficiency and Creativity," *BusinessWeek* (14 Sept. 2007).

[30] T. L. Friedman, *The World is Flat: A Brief History of the 21st Century* (Farrar, Straus and Giroux, New York, 2006).

[31] S. Brusoni and A. Prencipe, "Modularity as an Entry Strategy: The Invasion of New Niches in the LAN Equipment Industry," CESPRI Working Papers 171, (CESPRI, Centre for Research on Innovation and Internationalisation, Università Bocconi, Milano, Italy, revised July 2005).

[32] MEMS and Nanotechnology Exchange, www.mems-exchange.org (23 July 2008).

[33] W.-M. Shen, P. Will, A. Galstyan, and C.-M. Chuong, "Hormone-Inspired Self-Organization and Distributed Control of Robotic Swarms," *Autonomous Robots,* **17** (1) pp. 93–105 (2004).

[34] J. Fraden, *AIP handbook of Modern Sensors, Physics, Design and Applications* (American Institute of Physics, New York, 1993).

[35] K. E. Petersen, "Silicon as a Mechanical Material," *Proceedings IEEE*, **70** (5) p. 421 (1982).

[36] J. B. Angell, S. C. Terry and P. W. Barth, "Silicon Micromechanical Devices," *Scientific America*, **248** (4), p. 44 (April 1983).

[37] A. Cohen, G. Zhang, F. Tseng, F. Mansfield, U. Frodis, and P. Will, EFAB: "Rapid, Low-Cost Desk Micromachining of High Aspect Ratio of True MEMS," *12th IEEE International Conference on Micro Electro Mechanical Systems* (1999), pp. 244–251.

[38] Y. Tang, H. T. Loh, Y. S. Wong, J. Y. H. Fuh, L. Lu, and X. Wang, "Direct Laser Sintering of a Copper-Based Alloy for Creating Three-Dimensional Metal Parts," *Journal of Materials Processing Technology*, **140** (1–3), pp. 368–372 (2003).

[39] J. W. Lee, H. Lee, and D.-W. Cho, "Development of Micro-Stereolithography Technology Using Metal Powder," *Microelectronic Engineering* **83**, 1253–1256 (2006).

[40] K. P. Rajurkar, "Technology and Research in Electrodischarge and Electrochemical Machining," *American Society of Mechanical Engineers*, **43**, 309 (1990).

[41] *Handbook of Conducting Polymers, 2nd edition,* edited by T. A. Skotheim, R. L. Elsenbaumer, and J. R. Reynolds (CRC Press, New York, 1998).

[42] K.-H. Haas and H. Wolter, *Synthesis, Properties, and Applications of Inorganic-Organic Copolymers (Ormocers)* (Elsevier, 1999).

[43] R. Buestrich, F. Kahlenberg, M. Popall, P. Dannberg, R. Muller-Fiedler, and O. Rosch, "Ormocers for Optical Interconnection Technology," *Journal of Sol-Gel Science and Technology*, **20** (2), 181 (2001).

[44] M. Madou, *Fundamentals of Microfabrication* (CRC Press, New York, 1997).

[45] J. J. Simonne, "A Systems Approach to Microsystems Development," in *Microengineering Aerospace Systems*, edited by H. Helvajian (AIAA and The Aerospace Press, El Segundo, CA, 1999), pp. 227–258.

[46] D. Kochan, "Solid Freeform Manufacturing: Possibilities and Restrictions," *Computers in Industry*, **20** (2), 133–140 (1992).

[47] P. W. McMillan, *Glass Ceramics*, 2nd ed. (Academic Press, London, UK, 1979).

[48] W. Holland and G. Beall, *Glass-Ceramic Technology* (American Ceramic Society, Westerville, OH, 2002).

[49] S. H. Messaddeq, V. K. Tikhomirov, Y. Messaddeq, D. Lezal, and M. S. Li, "Light-Induced Relief Gratings and a Mechanism of Metastable Light-Induced Expansion in Chalcogenide Glasses," *Physical Review B*, **63**, 224203 (2001).

[50] S. D. Stookey, "Photosensitive Glass: A New Photographic Medium," *Industrial and Engineering Chemistry*, **41**, 856 (1949).

[51] S. D. Stookey, "Chemical Machining of Photosensitive Glass," *Industrial and Engineering Chemistry*, **45**, 115 (1953).

[52] H. Helvajian, "3D Microengineering via Laser Direct-Write Processing Approaches," in *Direct-Write Technologies for Rapid Prototyping Applications: Sensors, Electronics, and Integrated Power Sources*, edited by A. Pique and D. B. Chrisey (Academic Press, NY, 2002), ch. 14.

[53] P. D. Fuqua, D. P. Taylor, H. Helvajian, W. W. Hansen, and M. H. Abraham, in *Materials Development for Direct-Write Technologies*, edited by D. B. Chrisey, H. Helvajian, and D. P. Taylor (Materials Research Society, Pittsburgh, PA, 2000), Vol. 624, pp. 79.

[54] F. E. Livingston and H. Helvajian, "Selective Activation of Material Property Changes in Photostructurable Glass Ceramic Materials by Laser Photophysical Excitation," *Journal of Photochemistry and Photobiology A: Chemistry,* **182** (3), 310–318 (2006).

[55] F. E. Livingston, L. F. Steffeney, and H. Helvajian, "Genotype-Inspired Laser Material Processing: a New Experimental Approach and Potential Applications to Protean Materials," *Applied Physics A: Materials Science & Processing*, vol. 93, pp. 75–83 (2008).

[56] F. E. Livingston and H. Helvajian, "Photophysical Processes that Lead to Ablation-Free Fabrication in Glass-ceramic Materials," in *3D Laser Microfabrication: Principals and Applications*, edited by S. J. H. Misawa (Wiley-VCH, Weinheim, 2006), Ch. 11.

[57] V. R. Bhardwaj, E. Simova, P. B. Corkum, D. M. Rayner, C. Hnatovsky, R. S. Taylor, B. Schreder, M. Kluge, and J. Zimmer, "Femtosecond Laser-Induced Refractive Index Modification in Multicomponent Glass," *Journal of Applied Physics*, **97**, 083102 (2005).

[58] F. Livingston, W. Hansen, A. Huang, and H. Helvajian, "Effect of Laser Parameters on the Exposure and Selective Etch Rate in Photostructurable Glass," *Proceedings of the SPIE*, **4637**, 404 (2002).

[59] A. Berezhnoi, *Glass-ceramics and Photo-Sitalls* (Plenum Press, NY, 1970).

[60] P. W. Levy, "The Kinetics of Gamma-ray Induced Coloring of Glass," *Journal of the American Ceramic Society*, **43** (8), 389–395 (1960).

[61] G. S. Monk, "Coloration of Optical Glass by High-Energy Radiation," *Nucleonics*, **10**, 52–55 (1952).

[62] S. M. Brkhovskikh, Y. N. Viktorova, Y. L. Grinshteyn, and L. M. Landa, (National Technical Information Service, U.S. Dept. Commerce, 1973).

[63] K. Sugioka, Y. Cheng, and K. Midorikawa, in *Photon-Based Nanoscience and Nanobiotechnology*, edited by J. J. Dubowski and S. Tanev (Springer, Dordrecht, The Netherlands, 2006), Vol. 239.

[64] Y. Cheng, K. Sugioka, K. Midorikawa, "Microfluidic Laser Embedded in Glass by Three-Dimensional Femtosecond Laser Microprocessing," *Optics Letters*, **29** (17) 2007–2009 (2004).

[65] F. E. Livingston and H. Helvajian, "Photophysical Processes that Activate Selective Changes in Photostructurable Glass Ceramic Material Properties," in *Photon-based Nanoscience and Nanobiotechnology*, edited by J. J. Dubowski, and S. Tanev (Springer, Dordrecht, The Netherlands, 2006), pp. 225–266.

[66] G. K. I. Zergioti, G. Koundourakis, N. A. Vainos, and C. Fotakis, "Laser-Induced Forward Transfer: An Approach to Single-Step Microfabrication," *Direct-Write Technologies for Rapid Prototyping Applications: Sensors, Electronics, and Integrated Power Sources*, edited by A. Pique and D. B. Chrisey (Academic Press, New York, 2002), Ch. 16.

[67] F. E. Livingston, L. F. Steffeney, and H. Helvajian, "Tailoring Light Pulse Amplitudes for Optimal Laser Processing and Material Modification," *Applied Surface Science*, **253** (19), 8015–8021 (2007).

[68] G. Qiao, R. Lu, and C. McLean, "Flexible Manufacturing System for Mass Customization Manufacturing," *International Journal of Mass Customisation*, **1** (2/3), 374–393 (2006).

[69] K. Sundberg, S. Cannon, T. Hospodarsky, D. Fronterhouse, and J. Lyke, "A Satellite Data Model for the AFRL Responsive Space Initiative," *20th Annual AIAA/USU Conference on Small Satellites* (AIAA, Ogden, UT, 2006), paper number SSC06-II-9.

[70] S. Eagleson, K. Sarda, S. Mauthe, T. Tuli, and R. E. Zee, "Adaptable, Multi-Mission Design of CanX Nanosatellites," *20th Annual AIAA/USU Conference on Small Satellites* (AIAA, Ogden, UT, 2006), paper SSC06-VII-3.

[71] J. Clements, J. George, and B. E. Strucker, "Rapid Manufacturing of Satellite Structures and Heat Pipes Using Ultrasonic Consolidation," *20th Annual AIAA/USU Conference on Small Satellites* (AIAA, Ogden, UT, 2006), paper SSC06-V-4.

[72] M. Wilkinson, "The Changing Paradigms of Satellite Reconnaissance, Creating Opportunities in the Small Satellite Industry," *20th Annual AIAA/USU Conference on Small Satellites* (AIAA, Ogden, UT, 2006), paper number SSC06-III-7.

[73] M. Hammond and R. Jobman, "Satellite-as-a-Sensor Neural Network Abnormality Classification Optimization," *20th Annual AIAA/USU Conference on Small Satellites* (AIAA, Ogden UT, 2006), paper SSC06-III-2.

[74] U. Cambridge, http://www-materials.eng.cam.ac.uk/mpsite/interactive_charts/energy-cost/IEChart.html (July 2008).

[75] C. M. Pancerella and R. A. Whiteside, "Using CORBA to Integrate Manufacturing Cells to a Virtual Enterprise," *Proceedings of SPIE: Plug and Play Software for Agile Manufacturing* (SPIE, Bellingham, WA), Vol. 2913, 1996, pp. 148–173.

[76] D. Beckett, "Overview of the XNAV Program, X-Ray Navigation Using Celestial Sources," *20th Annual AIAA/USU Conference on Small Satellites* (AIAA, Ogden, UT, 2006), paper SSC06-VI-6.

Index

B

C

D

E

G

H

I

J

K

L

M

N

S

U

V

W

X

Z